Toxicological Survey of African Medicinal Plants

Toxicological Survey of African Medicinal Plants

Edited by

Victor Kuete
Faculty of Science
University of Dschang, Cameroon

AMSTERDAM • BOSTON • HEIDELBERG • LONDON • NEW YORK • OXFORD
ELSEVIER   PARIS • SAN DIEGO • SAN FRANCISCO • SINGAPORE • SYDNEY • TOKYO

Elsevier
32 Jamestown Road, London NW1 7BY
225 Wyman Street, Waltham, MA 02451, USA

First edition 2014

Notices
Knowledge and best practice in this field are constantly changing. As new research and experience
broaden our understanding, changes in research methods, professional practices, or medical treatment
may become necessary.

Practitioners and researchers must always rely on their own experience and knowledge in evaluating and
using any information, methods, compounds, or experiments described herein. In using such information
or methods they should be mindful of their own safety and the safety of others, including parties for
whom they have a professional responsibility.

To the fullest extent of the law, neither the Publisher nor the authors, contributors, or editors, assume
any liability for any injury and/or damage to persons or property as a matter of products liability,
negligence or otherwise, or from any use or operation of any methods, products, instructions, or ideas
contained in the material herein.

British Library Cataloguing-in-Publication Data
A catalogue record for this book is available from the British Library

Library of Congress Cataloging-in-Publication Data
A catalog record for this book is available from the Library of Congress

ISBN: 978-0-12-800018-2

For information on all Elsevier publications
visit our website at http://store.elsevier.com

This book has been manufactured using Print On Demand technology. Each copy is produced to order
and is limited to black ink. The online version of this book will show color figures where appropriate.

Working together
to grow libraries in
developing countries

www.elsevier.com • www.bookaid.org

Contents

Chapter 1

Preface

Poisonous medicinal plants can affect the entire spectrum of organ systems in humans, with some plants having several toxic principles that affect different systems. Africa has a rich and varied flora that includes a wide variety of plants with the potential to cause poisoning of animals and humans. The last three decades have experienced the boom of published scientific data on the pharmacological activities of medicinal plants. Furthermore, evidence of the toxicological properties is being intensively provided with considerable efforts by African scientists. Although there is a large amount of information in the veterinary field, human poisoning appears to be less well documented especially when medicinal plants are concerned. Until now, there is no global standard book highlighting the toxic potential of African medicinal plants. In the previous standard book on *African Medicinal Plant* published by Elsevier 2013, we documented the pharmacological potencies of the medicinal plants of Africa, as well as their chemical constituents. The interrelationship of pharmacology and toxicology is important as therapeutic efficacy occurs at a lower dose, where overdosing can induce poisoning or severe side effects. However, toxic plants may contain active molecules that display useful pharmacological effects. With the current emphasis on research and development of phytomedicines throughout the continent, it is imperative to be aware of and have some information on the harmful potential and toxic components of African plants. Prior to any pharmacological screenings, we strongly encourage scientists to search available literature for known toxic properties of plants of interest. In this book, we produced and compiled scientific data on all toxicological as well as beneficial aspects of African plants. In the first part of the book, from Chapters 1–4, we provide baseline information for a comprehensive approach of the topics discussed on African plants. Hence, we discussed at the continental level, the toxicological societies (Chapter 1), the ethical issues (Chapter 2), the review of the guidelines and methods (Chapter 3), as well as discordant results reported (Chapter 4). In the second part, from Chapters 5–18, we discussed the acute and subacute (Chapter 5) as well as the subchronic and chronic (Chapter 6) toxicities of plants used for human therapy in the continent, and we finally brought out information about the toxic plants (Chapter 7), with emphasis of the effect at cellular level (Chapter 8). We also reported the effect of African medicinal plant on human genome as well as their ability to lead to deleterious effects on development known as teratogenesis (Chapter 9), their abilities to lead to cancer, known as carcinogenesis, and their toxic effects on the genetic material and the inheritance of these effects, known as mutagenesis (Chapter 10). In the second part, we also discussed the organ toxicity as well as protective effects of African medicinal plants, considered effects at the level of organ function such as the liver (Chapter 11), kidney (Chapter 12), heart (Chapter 13), central nervous system (Chapter 14), reproductive system (Chapter 15), skin and eyes (Chapter 16), spleen and lung (Chapter 17). In Chapter 18, we attempted to highlight African medicinal plants with good pharmacological potential and low side effects. In the third part of the book, we included knowledge on the potential toxic and protective constituents of African plant (Chapters 19−21) in order to predict their possible side effects, even if their toxicity profiles were not investigated. Finally, in the fourth part, we analyzed the physical (Chapter 22) and biochemical (Chapter 23) parameters involved in therapy with African medicinal plants. To place African research globally and for a possible academic use, emphasis was put on the general knowledge in toxicology, principle of the methods used in the screenings as well as their limitation, and tools for data interpretations. The topics of this book are of interest for scientists of several fields including Pharmaceutical Science, Pharmacognosy, Complementary and Alternative Medicine, Ethnomedicine, Pharmacology, Medical and Public Health Sciences, Phytochemistry, and Biochemistry. The highlight of this book is an exhaustive compilation of scientific data related to the toxicological and safety survey of African plants by up to top scholars from several countries. Finally, I would like to thank Molly McLaughlin, the Editorial Project Manager at Elsevier (225 Wyman Street, Waltham, Massachusetts 02451), Vignesh Tamil, the technical assistant, and Stalin Viswanathan, the production manager, for their help and fruitful collaboration.

Victor Kuete

List of Contributors

Adejuwon Adewale Adeneye Department of Pharmacology, Faculty of Basic Medical Sciences, Lagos State University College of Medicine, Ikeja, Lagos, Nigeria

Akanji Musbau Adewunmi Phytomedicine, Toxicology and Reproductive Biochemistry Research Laboratory, Department of Biochemistry, University of Ilorin, Ilorin, Nigeria

Roland E. Akhigbe Department of Physiology, Ladoke Akintola University of Technology, Ogbomoso, Oyo state, Nigeria

Afolabi Clement Akinmoladun Phytomedicine, Drug Metabolism and Toxicology Unit, Department of Biochemistry, School of Sciences, The Federal University of Technology, Akure, Nigeria

Jules C.N. Assob Department of Biomedical Sciences, Faculty of Health Sciences, University of Buea, Cameroon

Doriane E. Djeussi Department of Biochemistry, Faculty of Science, University of Dschang, Dschang, Cameroon

Jean P. Dzoyem Department of Biochemistry, Faculty of science, University of Dschang, Cameroon; Phytomedicine Programme, Department of Paraclinical Sciences, Faculty of Veterinary Science, University of Pretoria, South Africa

Martins Ekor Department of Pharmacology, School of Medical Sciences, University of Cape Coast, Cape Coast, Ghana

Wafaa El Sayed Abd El-Aal Professor of Pathology, National Research Center, Cairo, Egypt

Esameldin E. Elgorashi Toxicology and Ethnoveterinary Medicine, ARC-Onderstepoort Veterinary Research Institute, Onderstepoort, South Africa

Jacobus N. Eloff Phytomedicine Programme, Department of Paraclinical Sciences, Faculty of Veterinary Science, University of Pretoria, Onderstepoort, South Africa

Aimé G. Fankam Department of Biochemistry, Faculty of Science, University of Dschang, Dschang, Cameroon

Ebenezer Olatunde Farombi Drug Metabolism and Toxicology Unit, Department of Biochemistry, College of Medicine, University of Ibadan, Ibadan, Nigeria

Louis L. Gadaga Drug and Toxicology Information Service (DaTIS), School of Pharmacy and Department of Clinical Pharmacology, College of Health Sciences, University of Zimbabwe, Avondale, Harare, Zimbabwe

Ngueguim K. Glawdys Department of Biochemistry, Faculty of Science, University of Dschang, Dschang, Cameroon

Rebecca Hamm Department of Pharmaceutical Biology, Institute of Pharmacy and Biochemistry, University of Mainz, Mainz, Germany

Alfred Ekpo Itor Department of Biochemistry, Faculty of Science, University of Dschang, Dschang, Cameroon

Victor Kuete Department of Biochemistry, Faculty of Science, University of Dschang, Cameroon

Namrita Lall Department of Plant Sciences, University of Pretoria, Pretoria, South Africa

Faustin Pascal Tsagué Manfo Department of Biochemistry and Molecular Biology, Faculty of Science, University of Buea, Cameroon

Armelle T. Mbaveng Department of Biochemistry, Faculty of Science, University of Dschang, Dschang, Cameroon

Lyndy J. McGaw Phytomedicine Programme, Department of Paraclinical Sciences, Faculty of Veterinary Science, University of Pretoria, Onderstepoort, South Africa

Edouard Akono Nantia Department of Biochemistry, Faculty of Science, University of Bamenda, Cameroon

Neville Mvo Ngum Department of Biomedical Sciences, Faculty of Health Sciences, University of Buea, Cameroon

Jaurès A.K. Noumedem Department of Biochemistry, Faculty of Science, University of Dschang, Dschang, Cameroon

Dickson S. Nsagha Department of Biomedical Sciences, Faculty of Health Sciences, University of Buea, Cameroon

Emeka C. Okereke Department of Pharmacology and Toxicology

Theophine Chinwuba Okoye Department of Pharmacology and Toxicology

Ajiboye Taofeek Olakunle Antioxidants, Free Radicals, Functional Foods and Toxicology Research Laboratory, Department of Biological Sciences, Al-Hikmah University, Ilorin, Nigeria

Mary Tolulope Olaleye Phytomedicine, Drug Metabolism and Toxicology Unit, Department of Biochemistry, School of Sciences, The Federal University of Technology, Akure, Nigeria

Collins A. Onyeto Department of Pharmacology and Toxicology

Armel J. Seukep Department of Biochemistry, Faculty of Science, University of Dschang, Dschang, Cameroon

Germain S. Taïwe Department of Zoology and Animal Physiology, University of Buea, Cameroon

Dexter Tagwireyi Drug and Toxicology Information Service (DaTIS), School of Pharmacy and Department of Clinical Pharmacology, College of Health Sciences, University of Zimbabwe, Avondale, Harare, Zimbabwe

Jean-de-Dieu Tamokou Faculty of Science, Department of Biochemistry, University of Dschang, Dschang, Cameroon

Simplice B. Tankeo Department of Biochemistry, Faculty of Science, University of Dschang, Dschang, Cameroon

Gerald Ngo Teke Department of Biomedical Sciences, Faculty of Health Sciences, University of Bamenda, Bambili, Cameroon

Francesco K. Touani Department of Biochemistry, Faculty of Science, University of Dschang, Dschang, Cameroon

Yakubu Musa Toyin Phytomedicine, Toxicology and Reproductive Biochemistry Research Laboratory, Department of Biochemistry, University of Ilorin, Ilorin, Nigeria

Danielle Twilley Department of Plant Sciences, University of Pretoria, Pretoria, South Africa

Phillip F. Uzor Department of Pharmaceutical and Medicinal Chemistry, Faculty of Pharmaceutical Sciences, University of Nigeria, Nsukka, Enugu State, Nigeria

Igor K. Voukeng Department of Biochemistry, Faculty of Science, University of Dschang, Dschang, Cameroon

Qiaoli Zhao Department of Pharmaceutical Biology, Institute of Pharmacy and Biochemistry, University of Mainz, Mainz, Germany

1 Toxicological Societies in Africa: Roles and Impact in Policy Making and Living Conditions

Jules C.N. Assob[1], Dickson S. Nsagha[1], Neville Mvo Ngum[1] and Victor Kuete[2]

[1]Department of Biomedical Sciences, Faculty of Health Sciences, University of Buea, Cameroon, [2]Department of Biochemistry, Faculty of Science, University of Dschang, Cameroon

1.1 Introduction

Africa and other developing countries are experiencing a rise in both informal mining especially artisanal gold mining and other fluctuating informal industrial activities that are often situated within communities and surrounded by the mainstream population. This, together with the increasing population migration and rapid urbanization that encompasses a wide range of anthropogenic activities, contribute to environmental degradation and pollution. The high prevalence of infectious diseases such as lung diseases, tuberculosis, human immunodeficiency virus/acquired immunodeficiency syndrome coupled with the malaria endemicity make the populations of Africa and other developing countries more susceptible to the toxic effects of pollutants [1].

Human exposure to persistent toxic substances (PTS) including both toxic metals exposure and persistent organic pollutants (POPs) exposure can be from either natural sources, current industrial activities, past industrial activities, or anthropogenic activities. PTS have the ability to exert negative health effects that are often subtle, long term, sometimes transgenerational, and difficult to measure, even in large epidemiological studies. Furthermore, the continuous consumption of unsafe food following pesticide and chemical exposure remains a major global public health challenge, especially in Africa, where there is heavy application of a variety of chemicals yet so little available toxicological data on the chemicals [2].

Although necessary measures have been taken by developed countries to curb both the exposure and the effects to toxic substances, developing countries in general and African countries in particular still lag behind [2]. Generally, the various governments in Africa have enacted policies and institutions with regard to the use and control of

Toxicological Survey of African Medicinal Plants. DOI: http://dx.doi.org/10.1016/B978-0-12-800018-2.00001-7

chemicals with potential hazards to humans. However, most of these governmental policies and institutions are weak and should be accompanied in their endeavors by nonprofit nongovernmental organizations (NGOs) that can gather experts in academia, research institutions, government, and multinational societies to reinforce strategies necessary to curb human and environmental toxicity. To this effect, Africa has witnessed the formation of many organizations at the continental, regional, and national levels such as the African Society for Toxicological Sciences (ASTS), Society of Environmental Toxicology and Chemistry (SETAC)-Africa, West African Society of Toxicology (WASOT), Cameroon Society for Toxicological Sciences (CSTS), *Association Tunisienne de Toxicologie* (ATT), and the Egyptian Society of Toxicology (EST). This chapter aims to highlight the role of these NGOs and other major projects in creating awareness of the existing and potential toxicological hazards for populations and the ecosystems on the African continent and how their available data on specialized knowledge can be used to liaise with governments, industries, and other NGOs. The missions, goals, histories, and achievements of the various organizations are presented in this chapter and the mentoring role of worldwide organizations such as the International Union of Toxicology (IUTOX) and Society of Toxicology (SOT) in accompanying African experts in developing new skills through trainings and traveling grants have been acknowledged.

1.2 Toxicological Societies in Africa

1.2.1 African Society for Toxicological Sciences

The ASTS is a not-for-profit organization of scientists in academia, industry, and governmental organizations from around the world. Members of ASTS include scientists and policy makers who are interested in health and environmental issues that affect the continent of Africa. ASTS seeks to promote the acquisition and utilization of knowledge in toxicological sciences relevant to the continent of Africa [3].

1.2.1.1 Mission

The organization's mission is to create awareness of existing and potential toxicological issues that constitute hazards for the populations and ecosystems on the African continent. The Society strives to promote human health and a safe environment through research and education. Members are available to share their specialized knowledge with government, industry, and NGOs.

1.2.1.2 Goals

The goals of the organization are (i) to promote the development of relevant curricula in toxicological sciences among institutions of higher learning in Africa; (ii) to create awareness of African toxicological issues in scientific and medical circles worldwide; (iii) to provide expertise in medical, veterinary, and environmental toxicology to African governments, organizations, institutions of higher learning, and the international community on issues of toxicology relevant to Africa; (iv) to encourage and foster self-sustaining partnerships between the pharmaceutical and chemical industries and government, private, and NGO sectors in Africa that promote the safety and health of Africans and their resources; (v) to act as a liaison with other toxicological and scientific organizations worldwide on behalf of the African continent; and (vi) to create public awareness of activities that might lead to toxicological problems or health hazards on the continent of Africa [3].

1.2.1.3   History and Achievements of ASTS

Twenty African scientists attending the SOT meeting in Seattle, WA, in 1998 found it necessary to form ASTS, which was later founded in March 1999 in New Orleans, LA. It organized its first workshop in Accra, Ghana, in January 2001 followed by a satellite meeting of ASTS members and advisors in Salt Lake City, UT, to discuss achievements and the future direction of the association. The Third International ASTS Conference was held in Abuja Nigeria on April 22–27, 2003; the main sponsor was the National Institute for Environmental Health Sciences (NIEHS). In 2004, an ASTS mini-symposium was held in Baltimore, MD, on the theme: "Sustainable development and the management of industrial chemicals and pharmaceuticals in Africa." The Director General, Dora N. Akunyili, PhD of the National Agency for Food and Drug Administration and Control, Nigeria, was the keynote speaker [3].

In 2005, a conference was held at the Hilton Riverside Hotel, New Orleans, LA, on the theme: "Emerging health and environmental adverse effects of pharmaceuticals and industrial chemicals and the need for good regulatory policy in Africa." The keynote speaker was John O'Donoghue, DVM, PhD, DABT, former Director of Health and Environmental Laboratories of Eastman Kodak Co. Representatives for SETAC and SOT participated, and travel awards to three African scientists to attend the SOT annual meeting were awarded. The awardees were Prof. Isaac U. Asuzu of the Department of Veterinary Physiology & Pharmacology, University of Nigeria Nsukka, Enugu State, Nigeria, who presented a paper titled "Alternative treatment of snakebites from plant-extracts used in folklore medicinal practices in Nigeria"; Dr. Orish E. Orisakwe of the Department of Pharmacology, College of Health Sciences, Nnamdi Azikiwe University, Nnewi, Nigeria, who presented a poster titled "Heavy metal exposure from herbal supplements in Nigeria"; and Dr. Balarabe Magaji Jahun of the Veterinary Teaching Hospital, Faculty of Veterinary Medicine, Ahmadu Bello University Zaria Nigeria, who presented a poster titled "Efficacy of piroxicam in alleviating tetracycline-induced muscle damage."

In 2006, a conference took place at Marriot Hotel & Marina, San Diego, CA, on "The millennium development goals and the role of toxicologists and environmental health scientists." The keynote speaker was Dr. Gerald Fisher, senior vice president, Wyeth, who presented on "Moxidectin safety and efficacy studies for the elimination of river blindness in Africa, a WHO/Wyeth collaboration." Two travel grants were awarded to attend the SOT meeting in San Diego, CA: to Eric Banseka

Tungla of the Department of Biochemistry, University of Yaounde I, Cameroon, who presented a poster entitled "Toxicity risk assessment of *Carica papaya* leaves aqueous extract in albino rats (*Rattus norvegicus*)," and to Mamadou Fall of the Laboratory of Toxicology and Hydrology, Faculty of Medicine, University of Dakar, Dakar, Senegal, who presented a poster titled "Mutagenicity of benzyl chloride in the AMES test depends on exposure conditions."

An International Conference on "Risk Assessment and Quality Assurance Training" and a workshop was later held on October 21–28, 2006 in Limbe, Cameroon, chaired by Dr. Ken Olden and was attended by 47 participants from Cameroon, Nigeria, Sudan, South Africa, the United States, the United Kingdom, Italy, and Belgium. This conference led to the formation of the CSTS. Several organizations including IUTOX, NIEHS, British Society of Toxicology, and Wyeth, sponsored and participated in the workshop.

In 2007, a conference was held in Charlotte, NC, on the theme: "Working together to promote hazard identification and influence environmental policies and regulations in Africa," which spearheaded the formation of the Toxicologists of African Origin Special Interest Group with SOT and promoted collaboration with the National Library of Medicine on the World Library Project.

In 2008, an informational session at the SOT conference titled "Toxicological and public health challenges in Africa." Three travel grants were awarded to Lyndy McGaw from South Africa, Gerard Ngueta from Belgium, and Evans Afriyie-Gyawu from Texas A&M University, to attend the SOT meeting in Seattle, WA. In its 10 years of existence, ASTS has sponsored eight scientists to attend SOT meetings in the United States at a cost of more than $20,000. This cost is in addition to those invested in organizing conferences in Africa and the United States [3].

1.2.2   Society of Environmental Toxicology and Chemistry

SETAC Africa is a geographic unit of the SETAC established to promote and undertake activities of SETAC in the Africa region [4]. The SETAC world council approved full geographic unit status for Africa in May 2012 at the sixth SETAC world congress. The decision came after a decade of SETAC activity in Africa, including the formation of an SETAC Africa branch within SETAC Europe and a series of biannual meetings, most recently in Buea, Cameroon, in May 2011, and prior to that in Kampala, Uganda (2009), Arusha, Tanzania (2007), and in South Africa [4].

1.2.2.1   Goals of SETAC Africa

SETAC is designed (i) to promote research, education, training, and development of the environmental sciences, specifically environmental toxicology and chemistry, hazard assessment, and risk analysis; (ii) to encourage interactions among environmental scientists and disseminate information on environmental toxicology and chemistry and its application to the disciplines of hazard and risk assessment; (iii) to sponsor scientific and educational programs and provide a forum for communication among professionals in government, business, academia, and other segments of the environmental science community involved in the use, protection, and management of the environment, and the protection and welfare of the general public; and (iv) to promote the development and adjustment of principles and practices for sustainable environments, considering appropriate ecological, economic and social aspects adapted to African problems and conditions. SETAC Africa's membership approaches 100.

1.2.2.2   Organized Workshops and Trainings

So far, SETAC has organized a series of workshops which include: Guidance on passive sampling methods to improve management of contaminated sediments; Pesticide risk assessment for pollinators; Ecological assessment of selenium in the aquatic environment; Science-based guidance and framework for the evaluation and identification of PBTs and POPs; Evaluation of persistence and long-range transport of organic chemicals in the environment; A multistakeholder framework for ecological risk management; Use of sediment quality guidelines and related tools for the assessment of contaminated sediments; and Pure water toxicity testing: biological, chemical, and ecological considerations with a review of methods and applications, and recommendations for future areas of research [4].

1.2.3   West African Society of Toxicology

The WASOT is a professional society of intellectuals involved in the study of the adverse effects of chemical, physical, or biological agents on people, animals, and the environment and the communication of the nature of such effects in collaboration with relevant stakeholders and government agencies [5].

1.2.3.1   Trainings and Workshops

WASOT has organized many workshops on a variety of topics including: Practical methods for *in vitro* toxicology workshop in 2013 in Maryland [6], as well a workshop on "Introduction to toxicology—the science of poisons 2012—2013" accessible [7]. WASOT intends to organize in 2014 a workshop on global understanding of chemicals in health, diseases, and economics.

1.2.4   Cameroon Society for Toxicological Sciences

The CSTS is an autonomous and NGO formed in October 2006 in Limbe following the successful workshop organized by ASTS. Since its creation, it has successfully hosted two conferences in 2007 (Buea) and 2009 (Dschang) [8].

1.2.4.1   Objectives

The goals of the organization are (i) to sensitize the government and the populace on the importance of toxicology as a discipline and the role it plays in our daily lives; (ii) to disseminate the importance of toxicology to primary, secondary, and high schools through lectures quiz and prizes; (iii) to eventually create Departments of

Toxicology in Cameroonian universities and possibly the Central Africa subregion; and (iv) to build future reputable toxicologists in Central and West Africa.

CSTS has devised a roadmap and some programs:

- First, CSTS plans to begin offering short talks to selected schools in the country each year to deliver lectures on toxicology.
- Second, CSTS wishes to institute a Toxicological Symposium to take place prior to the CSTS conference of 2013 in Yaounde.
- Third, in collaboration with some Cameroonian universities, CSTS plans to carry out certificated short courses in toxicology.
- Fourth, CSTS plans to host a Risk Assessment Summer School (RASS) in the future, for toxicologists within the Central and West Africa subregions: Cameroon, Chad, Gabon, Congo, Equatorial Guinea, Central Africa Republic, and Nigeria.

In May 2011, CSTS/SETAC/NEF jointly organized a conference on the theme: "Searching for African solutions to human and environmental toxicological challenges" at the University of Buea. The conference was attended by about 150 participants, 50 of whom were international participants. In addition to representation from academia (which was unsurprisingly dominant), there were participants from important industries, regulatory bodies and NGOs such as SONARA, the oil refinery industry in Cameroon, and CSIR (Council for Scientific and Industrial Research) South Africa, Environmental Council of Zambia, Africa Education Foundation (NEF), Judson Foundation UK, Health Lions Nigeria, Ministry of Scientific Research and Innovation Cameroon, and Ministry of Environment and Nature Protection Cameroon.

1.2.5 The "Association Tunisienne de Toxicologie"

The Tunisian Association of Toxicology (ATT) is a young association created by university researchers and scientists in 2008. It has the ambition and the desire to advance knowledge in the fields of environmental science and particularly in the field of toxicology with its multiple facets (analytical, molecular, cellular, clinical, nutritional, etc.) [9].

The ATT responds to a need by providing Tunisian researchers and experts in the field a platform that allows them to express themselves, to share their work, and exchange ideas and experiences with colleagues in Tunisia, Maghreb, or others from various countries. The ATT will participate, with the agreement of the authorities, in establishing a scientific network of environmental monitoring and awareness [9]. The ATT is able to engage and encourage research activities in the fields of environmental science, particularly those that emphasize the role of prevention.

1.2.5.1 Objectives

The Association is strongly committed to encouraging and disseminating research in the fields of toxicology, especially those that affect the environment, health, and food. It also aims to promote and publicize all scientific advances on prevention, and it will work to introduce and apply them to specific problems of Tunisia. The Association will be an interface between the fields of clinical sciences, biological sciences, and environmental sciences in order to achieve comprehensive preventive solutions. The Association will serve the interests of young researchers in the fields of molecular toxicology, giving them the opportunity to showcase their work at scientific conferences and regular seminars. The Association plans to launch a journal in the fields of Toxicology and Environmental Sciences. The ATT is already a member of European Toxicology Societies, and it intends to partner with several other international associations.

1.2.5.2 Activities, Workshops, and Trainings

In 2009, activities performed included several workshops focusing on "Culture and primary cancer cells and opportunities for their use in cases of poisoning"; "Study of apoptosis of cells in culture and its importance in the elimination of cancer cells"; "Test comets and its importance in the identification of DNA damage"; and "Biological and toxicological study of marine organisms (Jellyfish and Stripes): Chracteristics and extraction of venom." Conferences included "Adult and embryonic stem cells: differentiation into hepatocytes and prospects biotherapy of liver disease" and "Expression of a transgene in a eukaryotic cell." It was during this year that the first scientific day of the ATT was held on "Toxic environmental health and economic benefits." In 2010, two scientific days of ATT and the Fifth International Symposium were held in Monastir on "Process Toxic, their relationship with the environment and health" and a Workshop: "Functional genomics of gliomas: from cell culture to bedside" moderated by Prof. Laurent Pelletier. The same year, a conference was held on "Report TRAIL and chemotherapy: the impact of TRAIL decoy receptors" moderated by Prof. Olivier Micheau. In 2012, a third scientific meeting of the ATT was held on "Toxicology—environment—health." On November 9, 2013 under the "Organic Caravan," The Association organized a workshop on the theme: "Opening the world of research in biology." On March 17−19, 2013, she organized the fourth Scientific Meeting of ATT and first Conference of the Federation Maghreb Toxicology on the theme: "Toxic process of urban and environmental origins pathologies induced" [9].

1.2.6 The Egyptian Society of Toxicology

The EST was founded on February 14, 1983. It is a multidisciplinary society with the participation of biological sciences, pharmacy, medical and veterinary medicine, and agriculture specialists. Its purpose is to coordinate and promote activities in the field of toxicology, particularly in education, research, publication, and meetings [10].

The Society is in intimate contact with specialized institutions and intends to inform the scientific public on opinions as well as authorities on national concerns in its field. The Society has an important goal of supporting international contacts and cooperation and joining specialized unions in its field [10].

1.2.7   The Africa Education Initiative (NEF)

The mission of the Foundation is to further the advancement of science and engineering in Africa. It hopes to achieve its mission by providing educational materials and essential tools necessary to advance educational institutions [11].

1.2.7.1   Goals

Through its education assistance program, the Foundation strives to promote academic research in institutions. The Foundation supports scholars pursuing a vast spectrum of professions. Funding also assists students at critical stages of their education. Advancement of science and engineering is not limited to the continent of Africa; other schools around the world may also request assistance. Assistance will be given based on need and available resources. The Foundation assists professors and students with contacting institutions or colleagues in other countries. Students in need of internships are assisted with making the necessary contacts. In addition, professors seeking information about institutions abroad offering research opportunities also are aided [11].

1.2.7.2   NEF Toxicology Internship Program

The NEF organizes an annual 3-month prestigious all-expense-paid Toxicology Internship Program (TIP) at the National Veterinary Research Institute in Vom, Nigeria, aimed at helping interns gain hands-on scientific research and technical experience by participating in structured projects guided by a mentor. The 6-year-old TIP has successfully trained over 36 graduate students in biomedical research and undergraduate medical or veterinary students from across West (Ghana and Nigeria), Central (Cameroon), and North (Egypt) African regions [11].

After the course, NEF continues to network with ex-interns and facilitate their attending scientific meetings/congresses to present their internship toxicology research projects. International scientific forums, such as SOT, IUTOX, RASS, just to mention a few, have witnessed NEF ex-interns present TIP research projects; publication of TIP papers is not left out.

Applications are usually open in February of each year, and students all over Africa are encouraged to apply [11].

1.2.8   Environmental Protection Agency—Program in Africa

1.2.8.1   Mission

Environmental Protection Agency (EPA)'s environmental program in Africa is focused on addressing Africa's growing urban and industrial pollution issues, such as air quality, water quality, electronics waste, and indoor air from cookstoves. EPA expertise is assisting Africa with pollution-related issues before they become overwhelming and before their impacts irreparably damage the African and global environment. EPA programs in Africa are designed to protect human health, particularly vulnerable populations such as children and the poor, and to strengthen good environmental governance [12]. The key activities of EPA in Africa include the following.

1.2.8.1.1 Good Environmental Governance

Many countries in Africa are in the early stages of developing their environmental governance structures. EPA is focused on strengthening environmental laws and regulations, building capacity for enforcement and compliance, and promoting public participation in environmental decision making. EPA is supporting the East Africa Network on Environmental Compliance and Enforcement and is in partnership with the International Network on Environmental Compliance and Enforcement, the Danish aid agency, and the Kenyan Environmental Management Authority [12].

1.2.8.1.2 Water and Sanitation

Poor sanitation, unsafe water, and unhygienic environments are leading causes of illness and death among children in Africa. USEPA plans to improve public health through increasing the capacity of urban providers in Africa to deliver safe drinking water in a sustainable way through piped water supply systems, through the development and implementation of water safety plans (WSPs) [12]. WSPs are a comprehensive, "catchment to consumer" approach, which uses a health-based risk assessment methodology for identifying the greatest vulnerabilities for contamination within a drinking water supply system, thereby allowing the drinking water providers the ability to effectively operate and manage their systems and target their investments to gain the greatest health impact possible. WSPs also can be seen as an adaptation of asset management that is used globally, primarily in developed countries, throughout the water sector, both for drinking water and sanitation [12].

1.2.8.1.3 Ambient Air Quality

Growing dirty transportation fleets, expanded and uncontrolled industrial production, and resource extraction are increasingly significant problems in Africa. EPA is involved in improving vehicle fuels and promoting emissions control technologies to improve air quality in urban areas. As a founding Partner of the Partnership for Clean Fuels and Vehicles (PCFV), EPA was involved in providing support to Sub-Saharan Africa (SSA) in the phasing out of leaded gasoline, which occurred throughout the region in January 2006. Building on this success, EPA is working through PCFV to support SSA countries to lower sulfur in fuels. EPA is working with United Nations Environment Program (UNEP) on eliminating lead in gasoline in North Africa, where several countries still use lead in gasoline. At the same time, EPA is providing assistance to the PCFV on promoting clean vehicles technologies, particularly the use of catalytic converters [12].

1.2.8.1.4 Clean Cookstoves and Indoor Air Quality

In Africa, EPA is working to reduce the negative health impacts of indoor air pollution for the more than 75% of Africans who burn wood, charcoal, dung, crop residue, and coal for their home cooking and (in some places) heating. EPA is currently managing scale-up projects in Ethiopia (Addis Ababa with Project Gaia) and Kenya (in Kisumu in Western Kenya with the NGOs Solar Cookers International and Practical Action). Prior to these grants, EPA managed pilot projects in Mauritania, Nigeria, and Uganda. The goal of the pilot was to demonstrate effective approaches for increasing the use of clean, reliable, affordable, efficient, and safe home cooking and heating practices that reduce women's and children's exposure to indoor air pollution [12]. In addition to the above results, the pilot in Uganda with "Ugastove" was the first cookstove organization to be certified under the Gold Standard to receive voluntary carbon credits.

1.2.8.1.5 Climate Change
The Global Methane Initiative is an international initiative that advances cost-effective, near-term methane abatement and recovery and use as a clean energy source. The goal of the Initiative is to reduce global methane emissions in order to enhance economic growth, strengthen energy security, improve air quality, improve industrial safety, and reduce emissions of greenhouse gases. The Initiative now focuses on five sources of methane emissions: agriculture (animal waste management), coal mines, landfills, oil and gas systems, and wastewater treatment facilities.

1.2.8.1.6 Toxic Substances
1.2.8.1.6.1 E-Waste Africa is increasingly becoming a destination for used electronics, with little capacity to safely manage what legally or illegally enters the countries, as well as recycling the electronics that are being used within their borders once they reach the end of their useful life. Open-air burning and acid baths used to recover valuable materials from electronic components expose both the workers and the communities to high levels of contaminants such as lead, mercury, cadmium, and arsenic. These exposures, which threaten vulnerable populations such as children and the poor, can lead to irreversible health effects, including cancers, miscarriages, neurological damage, and diminished IQs.

1.2.8.1.6.2 Mercury Use in Artisanal Gold Mining EPA has been working with Senegal and other West African countries to reduce the use and release of mercury in the artisanal gold mining sector under the UNEP Global Mercury Partnerships. The approximately 10,000 artisanal gold miners in Senegal use mercury to amalgamate fine gold particles; this mercury is then burned off, resulting in significant exposures via inhalation to miners and their families.

The project provides information and tools to encourage the development of regional and national action plans to reduce mercury exposures from small-scale gold mining, to improve the health of miners and their families, protect the environment in West Africa and globally, and improve the economic conditions in this sector.

An initial workshop was held for Francophone West African countries which brought together miners, NGOs, government officials, local authorities, industry, gold buyers and sellers, medical personnel, and donors to develop a multistakeholder "strategic plan" for mercury reduction in the artisanal and small-scale gold mining and processing sector. A similar workshop will be held for Anglophone West Africa [12].

1.3   Interactions Between African Societies of Toxicological Sciences with European and American Societies

Many foreign societies and organizations have played key roles in mentoring and assisting African organizations in achieving their goals [9]. The interaction and cooperation between some foreign societies have been highlighted.

1.3.1   *International Union of Toxicology*

IUTOX is the voice of toxicology on the global stage, founded on July 6, 1980. The organization seeks to increase the knowledge base of toxicological issues facing humankind and to extend this knowledge to developing societies and nations. Founded in 1980, IUTOX now has 51 affiliated societies representing six continents and over 20,000 toxicologists from industry, academia, and government as members; offering a diverse and challenging perspective on every issue and development relating to the profession [13]. National toxicology groups are encouraged to become an IUTOX member society; to submit session proposals, abstracts, and volunteer to speak at IUTOX meetings; to help translate toxicology materials; and to contribute financially to IUTOX Programs [13].

1.3.1.1   *Mission*

The IUTOX main mission is to improve human health through the science and practice of toxicology worldwide.

1.3.1.2   *Vision*

Its vision is to foster international scientific cooperation for the global acquisition and utilization of knowledge in toxicology for improvement of the health of humans and their environment.

1.3.2   *General Objectives of IUTOX*

The main objectives of IUTOX are (i) to serve as the scientific voice of toxicology in the world; (ii) to provide leadership as a worldwide scientific organization that objectively addresses global issues involving the toxicological sciences; (iii) to broaden the geographical base of toxicology as a discipline and a profession to all countries of the world; and (iii) to pursue capacity building in toxicology, particularly in developing countries to utilize its global perspective and network to contribute to the enhancement of toxicology education and the career development of young toxicologists.

1.3.3   *The Society of Toxicology*

The SOT is a professional and scholarly organization of scientists from academic institutions, government, and industry representing the great variety of scientists who practice toxicology in the United States and abroad. SOT is committed to creating a safer and healthier world by advancing the science of toxicology. The Society promotes the acquisition and utilization of knowledge in toxicology, aids in the protection of public health, and facilitates disciplines. The Society has a strong commitment to education in toxicology and to the recruitment of students and new members into the profession [14].

To increase the diversity and inclusiveness, the SOT promotes and facilitates the formation of subgroups of members with a common ethnicity, country of origin, and gender. Thus far, six Special Interest Groups have been established to develop, propose, and conduct programs and educational activities focused on career development opportunities for toxicologists and recognition of the accomplishments of toxicologists within each group [15].

1.4   The Role of African Societies of Toxicological Sciences in Government Policy Making

Most countries in the African region have either developed or are in the process of developing policies and regulations in the management of chemicals, including PTS. It is possible that the low level of awareness among the stakeholders and the poor dissemination of available information of the adverse effects of PTS on humans and the environment are responsible for the slow pace in developing regulations and policies on PTS. Even then some of the existing national policies need to be reviewed in response to new challenges and international obligations within existing conventions such as the Stockholm Convention on POPs [16].

It is regrettable that whereas most of the national legislations are either too general or too fragmentary and nonspecific to PTS, some countries do not have any laws regarding hazardous chemicals. It is important that national legislations are enacted and/or harmonized to deal with hazardous chemicals in general and PTS in particular. It is also evident that most African countries have established or are developing institutions to manage the environment but lack management strategies regarding hazardous chemicals coupled with the lack of adequate capacity and resources. A major constraint toward sustainable chemical management is the lack of and/or weak enforcement of regulations. For the region to contribute effectively in the global effort to reduce PTS, there is a need to establish and/or strengthen existing institutions and legal frameworks through capacity building and putting in place necessary mechanisms for compliance monitoring and enforcement.

The monitoring of PTS and other chemical levels in the environment varies from country to country depending on the level of development and financial resources available. The few established organizations and research institutions lack adequately trained scientists and proper equipment to monitor and assess PTS in various media. Data that might have been generated by research is rarely published and disseminated to relevant authorities that might use such data to establish control measures or perform enforcement. It also must be noted that most generated data, if not all, are from individual studies and are not ongoing [16].

Despite these limitations, the increasing awareness about PTS is stimulating cooperation among the various research institutions and other stakeholders including NGOs. This may be a good indication of proper future PTS management in SSA in particular and Africa in general. International agencies are encouraged to join hands with countries of the SSA region in addressing the potential effects of PTS [16].

1.5 Information Resources in Toxicology

1.5.1 Toxic Remnants of War

Toxic Remnants of War (TRW) is a research-orientated web site that aims to consider and quantify the detrimental impact of war, military operations, and munitions on the environment and human health. The site is intended to act as a resource for policy makers, humanitarian organizations, and members of the public concerned with mitigating the effects of war on victims and communities [17]. The site also presents many informative reports and discussion papers on toxic arms. Some few topics are presented below.

- TRW—Discussion paper—Toxic harm: humanitarian and environmental concerns from military origin contamination—by Dr. Mohamed Ghalaieny.

This paper is the outcome of research done by the TRW Project into the scope of the problem from military-origin contamination. The paper overviews current health and environmental problems resulting from conflict and military activities before presenting limitations on the study of such problems and discussing existing legal and practical measures for environmental protection. The work also presents a methodology developed to study military-origin contamination in the appendix; this includes examples of problematic substances [17].

- TRW Legal Workshop—Defining toxic remnants of war: sources, properties and examples—by Dr. Mohamed Ghalaieny.

This presentation on defining examples of TRW, their sources, and properties was part of the TRW workshop exploring a legal framework for TRW held at the Free University of Berlin in June 2012.

- US National Research Council—Toxicity of Military Smokes and Obscurants, Volumes 1, 2, and 3 [17].

A variety of smokes and obscurants have been developed and used to screen armed forces from view, signal-friendly forces, and mark positions. Smokes are produced by burning or vaporizing particular products. Following concerns over exposure during training, the Office of the Army Surgeon General requested that the US National Research Council review data on the toxicity of smokes and obscurants and recommend exposure guidance levels. Volume 1 evaluates data on the toxicity of four obscurant smokes: fog oil, diesel fuel, red phosphorus, and hexachloroethane. Volume 2 evaluates data on the toxicity of four obscuring smokes: white phosphorus, brass, titanium dioxide, and graphite. Volume 3 evaluates data on the toxicity of seven colored smokes used for signaling, marking, and, in some cases, simulating exposure to chemical warfare [17].

1.6 Conclusions and Recommendations

In recent years, there have been significant increases in population growth all over the world, particularly in Africa. This has been accompanied by rapid urbanization, an increase in industrial activities, and greater exploitation of natural resources. These revolutions have provoked varieties of pollutants such as food additives, pesticides, industrial wastes, heavy metals, noxious gases, plant and animal toxins, and TRWs resulting in huge waste discharges into the environment. Most of these untreated wastes end up in soils, rivers, lagoons, or the atmosphere and cause undesirable human and environmental health hazards. The monitoring of PTS and other chemical levels in the environment varies from country to country in Africa, depending on the level of development and financial resources available. Many regions and countries are now witnessing the formation of many NGOs and societies to address these toxicological issues. These NGOs can play a huge role in guiding Africa through the process of environmental preservation by providing a platform for communication and exchange of scientific knowledge in the field of environmental toxicology and chemistry. Given the missions and goals of most of the national and international organizations presented here, it is possible that the low level of awareness among the stakeholders and the poor dissemination of available information of PTS on humans and the environment might be a springboard in creating a link between science and policy in Africa, hence boosting the development of regulations and policies on PTS [16]. Through these policies, Africans will be able to revise their priorities, curricula, and their way of thinking when dealing with drugs, pesticides and all toxicants, solid wastes, hazardous waste, and pollutants so as to be able to shoulder their responsibilities in the near future.

References

[1] Medical Research Council South Africa website, <http://www.mrc.ac.za/healthdevelop/healthdevelop.htm>; [accessed 17.12.13].
[2] UNEP Chemicals. Regionally based assessment of persistent toxic substance. Global report; Switzerland. 2003. 211 pages.
[3] Africa Society for Toxicological Sciences' website, <http://www.asts.org>; [accessed 17.12.13].
[4] Society of Environmental Toxicology and Chemistry' website, <www.setac.org>; [accessed on December 2013].
[5] West Africa Society for Toxicological Sciences' website, <www.wasot.net>; [accessed on December 2013].

[6]   Institute for *In Vitro* Testing Sciences' website, <http://www.iivs.org/education/training/ practical-methods-for-in-vitro-toxicology-workshop-2013/>;   [accessed   on   December 2013].

[7]   <http://toxicologytraining.com/training-courses-2012-2013/>; [accessed on December 2013].

[8]   Cameroon   Society   for   Toxicological   Sciences'   website,   <www. cameroonSocietyforToxicologicalSciences.org>; [accessed on December 2013].

[9]   Site Web de l'Association Tunisienne De Toxicologie, <www.attox.org>; [accessed on December 2013].

[10] The Egyptian Society of Toxicology's website, <www.est.org>; [accessed on December 2013].

[11] The Africa Education Initiative's website, <http://www.nef3.org>; [accessed on December 2013].

[12] Environmental Protection Agency Program in Africa's website, <www.epa.gov/oita/ regions/Africa/programs>; [accessed on December 2013].

[13] International Union of Toxicology's website, <www.iutox.org>; [accessed on December 2013].

[14] Society of Toxicological Sciences, <www.sot.org>; [accessed on December 2013].

[15] Vision   and   Overview,   <http://www.toxicology.org/gp/aboutsot.asp>;   [accessed 28.12.13].

[16] UNEP Chemicals. Regionally based assessment of persistent toxic substances. Sub-Sahara Africa report; Switzerland. 2002. 132 pages.

[17] Toxic   Remnants   of   War   website,   <www.toxicremnantsofwar.info>;   [accessed 28.12.13].

2 Ethical Issues for Animal Use in Toxicological Research in Africa

Wafaa El Sayed Abd El-Aal

Professor of Pathology, National Research Center, Cairo, Egypt

2.1 Introduction

Africa is the world's second-largest and second-most-populous continent. The African Regional Health Report of the World Health Organization (WHO) 2014 stresses that Africa can move forward on recent progress only by strengthening its fragile health systems [1]. One of the most positive signs for Africa has been the recent increase in scientific research being conducted by local African scientists. From 1996 to 2012, the number of research papers published in scientific journals with at least one African author more than quadrupled (from about 12,500 to over 52,000). During the same time, the share of the world's articles with African authors almost doubled from 1.2% to around 2.3% [2]. Plants have been used for medicinal purposes in Africa for many centuries [3]. The practice of traditional medicine (TM) is as old as the human race itself [4]. TM still remains the main recourse for a large majority of people for treating health problems [5]. In Sub-Saharan Africa, the traditional healers still play a major role in the provision of healthcare. This has been attributed in part to the unavailability of healthcare facilities and affordability [6]. It is crucially important that regulatory studies are conducted prior to exposure of human beings to herbal products [7]. The regulation of herbal medicine practice is still a major challenge in Africa [8]. However, the most encouraging fact is that there are several publications with regard to the products obtained from the extracts of these plants [9]. Toxicological evaluations of all medicinal plants are important in order to ascertain their safety. The primary aim of toxicological assessment of any herbal medicine is to identify adverse effects and to determine limits of exposure levels at which such effects occur [10]. As a result of this increase in toxicological research in Africa, many ethical issues involving the use of animals have been brought to the attention of different universities and institutions conducting research. Also, research on many diseases that affect Africa has been conducted in animal models in various African countries. Indeed, research and academic activities conducted in Africa that use different types of animals are too many to enumerate, which highlight the importance of collective and continuous efforts to enhance mechanisms of protecting the welfare of

Toxicological Survey of African Medicinal Plants. DOI: http://dx.doi.org/10.1016/B978-0-12-800018-2.00002-9

animals used in research or teaching. Although there may not be empirical data on the existence or adequacy of national and/or institutional policies and guidelines on the use of animals in research in Africa, most African countries are not yet at the same level as developed countries [11]. Initiatives to build research ethics capacity in developing countries must attempt to avoid imposing foreign frameworks and engage with ethical issues in research that are locally relevant [12]. The use of animals in science is a global practice, and one that continues to generate considerable political and public concern. Efforts to reconcile public concern with demands for freedom of scientific inquiry have resulted in legislation on animal procedures, in some, but not all, countries [13]. With research projects that use animals on the increase worldwide and in Africa in particular, animal research ethics should continue to be reviewed to improve the welfare of animals used in research [11]. The development of knowledge is necessary for the improvement of the health and well-being of humans as well as other animals requires *in vivo* experimentation with a wide variety of animal species. Many advances have been made to develop and disseminate information and guidelines for the care and use of laboratory animals in many parts of the world aiming to improve the health, welfare, and psychological well-being of the research animals. These advances are based on the 3 Rs and also improve the science by increasing the accuracy and reproducibility, and by ensuring quality control of the validity of animal-based results. However, this trend is lacking in most developing countries. In many African countries, ethics review capacity needs to be strengthened. Inadequate training, limited infrastructural facilities, and lack of funding remain a big challenge. As the number of clinical trials conducted in Africa is growing, so is the requirement to empower the countries of Africa to conduct their own safety reviews of those trials [14]. In this chapter, we review the importance of animal research in toxicological studies, the relevant guidelines, and ethical regulations of animal research and the animal research ethics committees (RECs) in developed countries, then we describe the situation of research ethics and challenges in Africa.

2.2 Definitions

The following terms are defined according to the 2013 Australian code [15].

- **Alternative**: encompasses replacement alternatives, reduction alternatives, and refinement alternatives as a whole.
- **Animal**: any live non-human vertebrate (that is, fish, amphibians, reptiles, birds and mammals encompassing domestic animals, purpose-bred animals, livestock, wildlife) and cephalopods.
- **Animal carer**: any person involved in the care of animals that are used for scientific purposes, including during their acquisition, transport, breeding, housing and husbandry.
- **Animal ethics committee (AEC)**: a committee constituted in accordance with the terms of reference and membership laid down in the Code.
- **Animal welfare**: an animal's quality of life, which encompasses the diverse ways an animal may perceive and respond to their circumstances, ranging from a positive state of well being to a negative state of distress.

- **Application**: a request for approval from an animal ethics committee to carry out a project or activity. An application may be for commencement of a project or activity, or an amendment to an approved project or activity.
- **Biological product**: any product derived from animals, including blood products, vaccines, antisera, semen, antibodies and cell lines.
- **Competent**: the consistent application of knowledge and skill to the standard of performance required regarding the care and use of animals. It embodies the ability to transfer and apply knowledge and skill to new situations and environments.
- **Compliance**: acting in accordance with the Code.
- **Conflict of interest**: a situation in which a person's individual interests or responsibilities have the potential to influence the carrying out of his or her institutional role or professional obligations, or where an institution's interests or responsibilities have the potential to influence the carrying out of its obligations.
- **Consensus**: the outcome of a decision-making process whereby the legitimate concerns of members of the animal ethics committee are addressed and, as a result, all members accept the final decision, even though it may not be an individual's preferred option.
- **Death as an endpoint**: when the death of an animal is the deliberate measure used for evaluating biological or chemical processes, responses or effects—that is, the investigator will not intervene to kill the animal humanely before death occurs in the course of a scientific activity. 'Death as an endpoint' does not include the death of an animal by natural causes or accidents, or the humane killing of an animal as planned in a project or because of the condition of the animal.
- **Distress**: an animal is in a negative mental state and has been unable to adapt to stressors so as to sustain a state of well being. Distress may manifest as abnormal physiological or behavioural responses, a deterioration in physical and psychological health, or a failure to achieve successful biological function. Distress can be acute or chronic and may result in pathological conditions or death.
- **Ethics**: a framework in which actions can be considered as good or bad, right or wrong. Ethics is applied in the evaluation of what should or should not be done when animals are proposed for use, or are used, for scientific purposes.
- **Facility**: any place where animals are kept, held or housed, including yards, paddocks, tanks, ponds, buildings, cages, pens and containers.
- **Governing body**: the body or person responsible for the administration and governance of the institution (e.g. university council or senate, board of an organization, school board) or, where appropriate, its delegated officer.
- **Harm**: a negative impact on the well being of an animal.
- **Humane killing**: the act of inducing death using a method appropriate to the species that results in a rapid loss of consciousness without recovery and minimum pain and/or distress to the animal.
- **Investigator**: any person who uses animals for scientific purposes. Includes researchers, teachers, undergraduate and postgraduate students involved in research projects, and people involved in product testing, environmental testing, production of biological products and wildlife surveys.
- **Monitoring**: measures undertaken to assess, or to ensure the assessment of, the well being of animals in accordance with the Code. Monitoring occurs at different levels (including those of investigators, animal carers and animal ethics committees).
- **Pain**: an unpleasant sensory and emotional experience associated with actual or potential tissue damage. It may elicit protective actions, result in learned avoidance and distress, and modify species-specific traits of behavior, including social behavior.

- **Procedure**: a technique employed when caring for or using animals for scientific purposes. One or more procedures may be used in an activity.
- **Project**: an activity or group of activities that form a discrete piece of work that aims to achieve a scientific purpose.
- **Reduction alternatives**: methods for obtaining comparable levels of information from the use of fewer animals in scientific procedures or for obtaining more information from the same number of animals.
- **Refinement alternatives**: methods that alleviate or minimize potential pain and distress, and enhance animal well being.
- **Replacement alternatives**: methods that permit a given purpose of an activity or project to be achieved without the use of animals.
- **Research**: as defined in the Australian code for the responsible conduct of research.
- **Reuse**: the use of an individual animal more than once for a procedure, activity or project.
- **Standard operating procedure (SOP)**: detailed description of a standardized procedure or process.
- **Well being**: an animal is in a positive mental state and is able to achieve successful biological function, to have positive experiences, to express innate behaviors, and to respond to and cope with potentially adverse conditions. Animal well being may be assessed by physiological and behavioral measures of an animal's physical and psychological health and of the animal's capacity to cope with stressors, and species-specific behaviors in response to social and environmental conditions.

2.3   Ethical Aspects of Medicinal Plants in Africa

Plants have been used for medicinal purposes in Africa for many centuries [3]. It is crucially important that regulatory studies are conducted prior to exposure of human beings to herbal products. Until toxicological, pharmacodynamic, and pharmacokinetic data are available, traditional health practitioners and clinicians must exercise caution in prescribing concurrently to their patients [7]. Governments should actively promote the rational use of herbal medicines that have been scientifically validated. To do so, they need a national policy for approving those that are safe and effective for specified clinical indications. The adoption of such policy will help to overcome some of the legal barriers against the use of herbal medicines which in some countries may still be inadequately standardized [16]. The documentation of medicinal uses of African plants is becoming increasingly urgent because of the rapid loss of the natural habitats of some of these plants because of anthropogenic activities [10]. The Beijing Declaration, which was the key outcome of the first WHO Congress on TM held on that organization's sixtieth anniversary and the thirtieth anniversary of the Declaration of Alma-Ata, for the first time offered international acknowledgment of the role of TM in healthcare. The Declaration clearly reaffirmed the need to regard these practices as something that "should be respected, preserved, promoted, and communicated widely and appropriately based on the circumstances in each country" and the responsibility of individual governments to "ensure appropriate, safe, and effective use of TM" through suitable policies, regulations, and standards [17]. Legislation concerning procedures for the registration of herbal medicine can play a very important role in ensuring that medicinal plant preparations are of acceptable quality, safety, and efficacy. Research on herbal medicines, which is necessary to ensure their improved utilization by the public, would benefit from strong governmental endorsement. Research on animals must be carried out with respect for their welfare and consideration must be given to using *in vitro* laboratory methods that may reduce experimentation on intact animals [16]. A successful effort to preserve both plants and knowledge of how to use them medicinally is required in order to protect TM [18]. Given that the vast majority of the African population uses TM as the primary, if not the only source of healthcare, some countries in the region have been particularly keen to assess its results. Several countries are currently drafting a legislative and legal framework based on national policies and regulations [17].

2.4   The Importance of Animals in Toxicological Studies

The historical importance of animal models cannot be ignored. The use of animal models in research has contributed to the massive amount of medical knowledge on human and animal diseases [19].

Drug testing using animals became important in the twentieth century. In 1937, a pharmaceutical company in the United States created a preparation of sulfanilamide, using diethylene glycol (DEG) as a solvent, and called the preparation "Elixir Sulfanilamide." DEG was poisonous to humans, but the company's chief pharmacist and chemist was not aware of this. He simply added raspberry flavoring to the sulfa drug, which he had dissolved in DEG, and the company marketed the product. The preparation led to mass poisoning, causing the deaths of more than 100 people. No animal testing was done. The public outcry caused by this incident and other similar disasters led to the passing of the 1938 Federal Food, Drug, and Cosmetic Act requiring safety testing of drugs on animals before they could be marketed [20].

Another tragic drug fiasco occurred in the late 1950s and early 1960s with thalidomide. It was found to act as an effective tranquilizer and painkiller and was proclaimed a "wonder drug" for insomnia, coughs, colds, and headaches. It was found to have an inhibitory effect on morning sickness, and hence, thousands of pregnant women took the drug to relieve their symptoms. Consequently, more than 10,000 children in 46 countries were born with malformations or missing limbs. The drug was withdrawn in 1961 and 1968 after a long campaign [20].

Animal and other toxicity studies are conducted according to generally accepted principles, referred to collectively as Good Laboratory Practice, which should be consulted in order to design appropriate studies [16].

According to Society of Toxicology (SOT), 1999 [21]:

- Research involving laboratory animals is necessary to ensure and enhance human and animal health and protection of the environment.
- In the absence of human data, research with experimental animals is the most reliable means of detecting important toxic properties of chemical substances and for estimating risks to human and environmental health.
- Research animals must be used in a responsible manner.

• Scientifically-valid research designed to reduce, refine or replace the need for laboratory animals is encouraged.

A full complement of toxicity tests for a successful pharmaceutical compound that proceeds to the market, involving single dosing, repeat subchronic and chronic dosing, reproductive testing, genotoxicity, and carcinogenicity testing, can involve between 1500 and 3000 animals. The preclinical drug development process can be performed in two disciplines: General Pharmacology and Toxicology [22].

Toxicity studies are highly variable in design, and where they involve the use of animals, the implications for animal welfare must be considered. The value of animal use for predicting human outcomes also has been questioned in the regulatory toxicology field, which relies on a codified set of highly standardized animal experiments for assessing various types of toxicity [22].

The whole animal usually is presumed to be closely correlated to human toxicity as the system incorporates pharmacokinetic (absorption, distribution, and metabolism) disposition of the test substance when administered by a route similar to its intended use. It also takes into consideration other physiological events in an organism that influence toxicity [23].

The adverse effects on animals that may arise specifically in toxicity tests, as opposed to other forms of animal research, are due mainly to dosing procedures and the toxic effects of the treatments [24]. Toxicity testing has a range of welfare implications for test animals, some of which can be severe. These effects are minimized by the "buildup" approach in which severe reactions can be detected at an early stage (acute toxicity followed by chronic toxicity). However, it is an intrinsic part of most toxicity tests to cause some form of harm to animals.

Although there is increased acceptance and utilization of medicinal plants worldwide, many are used indiscriminately without recourse to any safety test. Thus, there is a great need for toxicity tests to determine the safe dose for oral consumption. Toxicological evaluations of all medicinal plants are important in order to ascertain their safety [10]. According to Ifeoma and Oluwakanyinsola [23], before conducting a safety study of any herb or its product in animals, some major factors need to be considered:

• *Preparation of test substance*: Herbal products can be prepared into different dosage forms such as capsules, tablets, ointments, creams, and pastes.
• *Animal welfare considerations*: Guidance on the use of clinical signs as humane end points for experimental animals used in safety evaluation.
• *Animals*: Different rodent and non-rodent species are used in animal toxicity tests. In chronic studies, justification often is required for choice of species or strain of animals used.

2.5 Guidelines of Animal Research Ethics

2.5.1 History: Development of Guidelines and Legal Frameworks

Animals have been used repeatedly throughout the history of biomedical research. Early Greek physician scientists, such as Aristotle (384−322 BC) and Erasistratus

(304−258 BC), performed experiments on living animals. Likewise, Galen (129−199/ 217 AD), a Greek physician who practiced in Rome and was a leader in the history of medicine, conducted animal experiments to advance the understanding of anatomy, physiology, pathology, and pharmacology. Ibn Zuhr (Avenzoar), an Arab physician in twelfth-century Moorish Spain, introduced animal testing as an experimental method for testing surgical procedures before applying them to human patients [20]. In 1831, Marshall Hall [25] outlined five principles to govern animal experimentation:

1. An experiment should never be performed if the necessary information could be obtained by observations.
2. No experiment should be performed without a clearly defined and obtainable objective.
3. Scientists should be well informed about the work of their predecessors and peers in order to avoid unnecessary repetition of an experiment.
4. Justifiable experiments should be carried out with the least possible infliction of suffering (often through the use of lower, less sentient animals).
5. Every experiment should be performed under circumstances that would provide the clearest possible results, thereby diminishing the need for repetition of experiments.

During the nineteenth century, however, there was a gradual buildup of demands for legislation to protect animals used in research [19]. There have been laws on the use of animals for scientific purposes in the United Kingdom and Ireland since 1876, including the Cruelty to Animals Act 1876 by the British Parliament, which related specifically to scientific experiments [26]. In 1966, the United States adopted regulations for animals used in research with the passage of the Laboratory Animal Welfare Act [22]. Within the last decade or so, other countries in Africa, Asia, and South America have enacted national legislation [27].

Legislation was established to regulate the way in which animals were treated in 1959, when William Russell and Rex Burch published their book, *The Principles of Humane Experimental Technique*, which emphasized reduction, refinement, and replacement of animal use, principles which have since been referred to as the "3 Rs." These principles encouraged researchers to work to reduce the number of animals used in experiments to the minimum considered necessary, refine or limit the pain and distress to which animals are exposed, and replace the use of animals with nonanimal alternatives when possible [22].

2.5.2   3 Rs (Reduction, Replacement, and Refinement)

A major issue in toxicity testing is "animal welfare." The use of animals in research gave rise to the adoption of the critical 3 Rs to consider before conducting animal-based toxicity testing of herbals. There is a need to reduce the number of animals, refine the tests methods used in order to minimize pain and suffering of experimental animals, and replace animal tests with validated alternatives employing human cells where possible. Additionally, the number of rats used for LD_{50} tests can be significantly reduced by the adoption of *in vitro* cell-based assays and chemicals shown to be harmful to cultured cells are excluded from any further

LD_{50} tests and animal tests. It is no longer news that cellular models of toxicity are more rapid and can easily be adapted to high-throughput screening [23].

The "Rs" are defined as replacement of animals with alternative techniques, reduction of the number of animals required for an experiment, and refinement of the experimental techniques in order to use fewer animals. The use of alternative techniques is increasing. However, until these techniques can duplicate all the interacting complex physiological factors of a living animal, animal models or humans will still be a necessary part of our biomedical research [19].

According to the University of South Africa (UNISA) 2012 [28] in dealing with experiments with animals, strategies to refine the methodology and operative procedures should be applied as a fundamental obligation with a view to reducing the number of animals used and, where possible, replacing the use of animals with an alternative approach or experimental system.

"Reduction" refers to the use of an appropriate sample size. This is primarily aimed at using the fewest number of animals necessary to obtain statistically valid results. However, use of too few animals is also considered a waste of animal life.

According to the Directives 2010 [29], animal tissue and organs are used for the development of *in vitro* methods. To promote the principle of reduction, Member States should, where appropriate, facilitate the establishment of programs for sharing the organs and tissue of animals that are killed.

The numbers of animals involved in experiments can be minimized:

- By design:
  - Of the program.
  - Of each experiment.
- By minimizing variability (by inducing minimal stress, for example).
- By collaboration to reduce numbers of experiments—and by full publication.
- By maximizing use of tissue.
- By reusing animals [27].

"Refinement" refers to the requirement that you use the most refined methods and techniques available. For example, you may no longer immobilize wildlife using paralytics; there are currently much better methods to capture wildlife. The duration of activities must be no longer than required to meet the aim(s) of the project and must be compatible with supporting and safeguarding animal well-being. Animals must not be held for prolonged periods as part of an approved project before their use, without AEC approval [15].

"Replacement" has a few meanings depending on the type of research you are doing. From a biomedical research perspective you are expected to, whenever possible, replace the use of live vertebrates with species lower on the phylogenetic scale or to replace the use of living creatures with computer models, cell culture techniques, or cadavers. Another component of replacement or the use of alternatives is tied in with "refinement." Whenever possible, you are expected to replace painful procedures that may cause animal suffering with nonpainful procedures.

Replacement can be considered at two levels:

1. Complete replacement, where the method does not require any animal-derived material, for example, the Ames test which uses bacteria for screening for mutagens.

2. Incomplete replacement, in which the method requires biological material obtained from living or killed animals or uses embryonic stages or invertebrates. An example would be using orientated brain slices instead of the whole animal for studies on the visual system [27].

There are also general replacement possibilities to consider.

- Computer models, physicochemical characteristics can be used to predict mechanism of action or toxicity. If sufficient *in vivo* data is available, simulation models can be used to explore possible interactions between components of a body system, effects of substances, distributions of wildlife populations, and so on.

Human volunteers, human tissue, or tissue fractions, though not relevant for wildlife studies and much animal work these possibilities may be particularly important to consider for nonhuman primate investigations [27].

2.5.3   International Guidelines for Use and Care of Animals in Scientific Procedures

International Guidelines are intended to assist institutions in caring for and using animals.

The principles set out in these Guidelines are for the guidance of investigators, institutions, AECs to plan and conduct animal experiments in accord with the highest scientific, humane, and ethical principles.

Some of the famous International Guidelines are as follows:

- The Canadian Council on Animal Care's Guide to the Care and Use of Experimental Animals, Volume 1 (1993) [30].
- The National Research Council's Guide for the Care and Use of Laboratory Animals (1996) [31].
- New Zealand's Good Practice Guide for the Use of Animals in Research, Testing and Teaching (2002) [32].
- The Australian Code of Practice for the Care and Use of Animals for Scientific Procedures (2004) [33].
- The US Public Health Service Policy on Humane Care and Use of Research Animals (2002) [34].
- National Advisory Committee for Laboratory Animal Research (NACLAR) Guidelines on the Care and Use of Animals for Scientific Purposes (2011) [35].

2.5.4   Guidelines for Managing and Supervising Breeding and Holding Facilities

A very important factor ensuring high standards of animal care is a sufficient number of well-trained, knowledgeable, and committed staff [36]. The person responsible for the overall management of a facility used for breeding and holding animals (the facility manager) must be competent, with appropriate animal care or veterinary qualifications or experience. The person providing oversight of the program of veterinary care, including the care, husbandry, and health of animals and biosecurity in a facility, must be competent and must hold appropriate veterinary qualifications [15]. All individuals responsible for the care and use of animals must be appropriately trained on the natural biology and proper laboratory handling of the species under study.

The welfare of the animals used in procedures is highly dependent on the quality and professional competence of the personnel supervising procedures, as well as of those performing procedures or supervising those taking care of the animals on a daily basis. Member States should ensure through authorization or by other means that staff are adequately educated, trained, and competent [26].

People involved in research with animals furthermore have the obligation to respect the interests of laboratory animals and to recognize that they are sensitive to pain, that they may become anxious, and that they may experience fear if they remember such experiences [28].

Researchers must be committed to the welfare of the animals used and must respect the contribution those animals make to research. Researchers must ensure that procedures which will cause hunger, thirst, injury, disease, discomfort, fear, distress, deprivation, or pain to the animals involved in the studies are limited to the absolute minimum. The elimination or reduction of the total of those conditions experienced by an animal will be achieved by the application of the "3 Rs" principles [28].

Institutions must encourage and promote formal training of all Staff in animal science or technology [36].

2.5.5 Hazards for Animal Care Personnel

Potential hazards include experimental hazards such as biologic agents (e.g., infectious agents or toxins), chemical agents (e.g., carcinogens and mutagens), radiation (e.g., radionuclides, X-rays, lasers), and physical hazards (e.g., needles and syringes). The risks associated with unusual experimental conditions such as those encountered in field studies or wildlife research also should be addressed. Other potential hazards—such as animal bites, exposure to allergens, chemical cleaning agents, wet floors, cage washers and other equipment, lifting, ladder use, and zoonoses—that are inherent in or intrinsic to animal use should be identified and evaluated [37].

2.5.6 Training Guidelines

These Training Guidelines are to aid institutions in implementing an educational and training program that will meet the expectations of the guiding. In particular, these Training Guidelines are intended to assist institutional officials and Institutional Animal Care and Use Committees (IACUCs) or any others assigned the responsibility for coordinating Training Programs (Training Coordinators) in determining the scope and depth of such programs. These Training Guidelines encompass all staff caring for and working with live laboratory animals for scientific purposes. It is important to ensure that these personnel are educated, trained, and qualified to use animals in a manner that would be humane and ethical.

Training and education of all staff concerned is mandatory, with oversight by the IACUC and financial assistance from the Institution [36].

The following categories of personnel must be involved:

a. Animal facility staff
   - Laboratory animal caretakers
   - Laboratory animal technicians
   - Laboratory animal managers
   - Laboratory animal veterinarians
b. Research staff
   - Principal investigators
   - Research fellows
   - Postdoctoral and postgraduate students
   - Research technicians
c. IACUC members
d. Service personnel

2.6   Animal Welfare Issues

2.6.1   What Is Animal Well-Being or Welfare?

According to the American Veterinary Medical Association (AVMA), "all aspects of animal well-being, including proper housing, management, nutrition, disease prevention and treatment, responsible care, humane handling, and, when necessary, humane euthanasia" [38].

New scientific knowledge is available in respect of factors influencing animal welfare as well as the capacity of animals to sense and express pain, suffering, distress, and lasting harm. It is therefore necessary to improve the welfare of animals used in scientific procedures by raising the minimum standards for their protection in line with the latest scientific developments [29]. For instance, rough handling of animals negatively affects them psychologically and physically. Whenever possible, research procedures should be scheduled in such a way that the animals will have time to recover and rest before the next procedure is due to be done [11]. Animal welfare issues cannot be viewed in isolation from culture, values, and economic conditions—all of which affect how animals are perceived and treated [39].

2.6.2   Strategies to Safeguard Animal Well-being, Australian Code, 2013 [15]

Investigators and teachers have personal responsibility for all matters related to the welfare of the animals they use and must act in accordance with all requirements of the Guidelines. This responsibility begins when an animal is allocated to a project and ends with its fate at the completion of the project.

- Practices and procedures used for the care and management of animals must be appropriate for the situation, the species and strain of animal, and the activities to be undertaken, and must be based on current best practice.
- Identifying known and potential causes of adverse impact on animal well-being, taking into consideration both intended and unforeseen consequences.
- Ensuring that all relevant people are aware of and accept their responsibilities regarding the well-being of the animals.

2.6.3 Some Possible Ways of Protecting the Welfare of Animals Used in Research or Teaching

Nyika 2009 [11] suggested the following procedures for protection of animal welfare:

- Promotion of the implementation of 3 Rs through development of alternatives to animals.
- Dissemination of information about the existing alternatives to animals for the benefit of researchers and teachers.
- Training of researchers on ways of protecting the welfare of animals.
- Animal Ethics Committees to review and monitor research on animals.
- Development of guidelines and legal frameworks.

2.6.4 Protection of Animals During Transport and Related Operations

Correct transport of animals is an important feature of animal welfare. Many laboratory animals may never leave the animal unit where they were born, and even if they do, the time spent in transport will be a very small fragment of their total life. Nevertheless, such transportation may prove to be the most traumatic experience of their whole life, particularly at times of loading and unloading [39].

Methods and arrangements for the transport of animals must support and safeguard the well-being of the animals before, during, and after their transport, and take into account the health, temperament, age, sex, and previous experiences of the animals; the number of animals traveling together and their social relationships; the period without food or water; the duration and mode of transport; environmental conditions (particularly extremes of temperature); and the care given during the journey [15].

Animal Housing, according to *The Guide for the Care and Use of Laboratory Animals*, 2011 [37].

- No overcrowding is permitted
  - Review size requirements for proper caging.
  - Delayed or incomplete weaning is the most common cause of overcrowded conditions.
- Identification
  - Cage cards must be visible and information completed.
  - Animals must be identified.
- No housing of animals outside of the designated animal facility space is permitted.

2.6.5 Acclimatization and Conditioning

Before a project commences, the animals should be conditioned to the handling, experimental conditions, and people who will conduct the procedures [15].

2.6.6 Food and Water

Animals must receive and be able to access appropriate, uncontaminated, nutritionally adequate food of a quantity and composition that maintain normal growth of immature animals and normal weight of adult animals, and meet the requirements of pregnancy, lactation, or other conditions. Clean, fresh drinking water must be available at all times, as suitable for the species [15].

According to Dolan [19], scientists using water restriction should keep the following records for each animal:

- Daily water consumption, food consumption, and body weight.
- Frequency of surgical intervention (if appropriate).
- Frequency of infections/treatments (if appropriate).
- Duration of study and future plans for the animal (if appropriate).
- A record of all treatments given.

2.6.7 Avoiding or Minimizing Discomfort, Anxiety, and Pain

Despite advances in the development of alternatives to using animals in research, scientists still often cite a need to use live animals in experiments. Once that need has been confirmed by funding agencies and/or the local IACUC or other ethics committee, but before the research plans are finalized, the potential for animal pain and distress must be assessed, and plans laid to minimize animal suffering [40].

2.6.8 Anesthesia and Analgesia

The giving of an anesthetic or analgesic or other substance to sedate or dull the perception of pain of a protected animal for scientific purposes is itself a regulated procedure. Likewise, decerebration, or any other procedure to render a protected animal insentient, if done for scientific purposes, is a regulated procedure [19]. Procedures that involve serious injuries that may cause severe pain shall not be carried out without anesthesia [29].

The selection of appropriate analgesics and anesthetics should reflect professional veterinary judgment as to which best meets clinical and humane requirements, as well as the needs of the research protocol. The selection depends on many factors, such as the species, age, and strain or stock of the animal, the type and degree of pain, the likely effects of particular agents on specific organ systems, the nature and length of the surgical or pain-inducing procedure, and the safety of the agent, particularly if a physiologic deficit is induced by a surgical or other experimental procedure [41].

2.6.9 Euthanasia

Euthanasia is the act of humanely killing animals by methods that induce rapid unconsciousness and death without pain or distress. Euthanasia may be planned and necessary at the end of a protocol or a means to relieve pain or distress that cannot be alleviated by analgesics, sedatives, or other treatments. Criteria for euthanasia include protocol-specific end points (such as degree of a physical or behavioral deficit or tumor size) that will enable a prompt decision by the veterinarian and the investigator to ensure that the end point is humane and, whenever possible, the scientific objective of the protocol is achieved [37].

The methods selected should avoid, as far as possible, death as an end point due to the severe suffering experienced during the period before death. Where possible, it should be substituted by more humane end points using clinical signs that determine the impending death, thereby allowing the animal to be killed without any further suffering (Directives, 2010). All methods of euthanasia should be reviewed and approved by the veterinarian and IACUCs [37].

In euthanasia, three questions are to be answered according to the AVMA, 2013 [42]:

- What methods do you use to ensure death?
- Physical methods of euthanasia require specialized training and justification.
- Use of a guillotine for euthanasia requires that it be properly maintained.

2.7 Institutional Animal Care and Use Committees

AECs or IACUCs aim to review, approve, and oversee animal experiments and to balance the interests of researchers, animals, institutions, and the general public [43]. The IACUC is responsible for assessment and oversight of the institution's program components and facilities. It should have sufficient authority and resources (e.g., staff, training, computers, and related equipment) to fulfill this responsibility [37]. An ideal AEC should be composed of members with diverse educational and professional background and should always include expertise in animal-related fields such as veterinary medicine, animal science, and agriculture [11].

Because of the important role these committees play, there has been significant research interest in the way IACUCs function in practice, many studies have explored the subjective experiences of IACUC members and documented the dynamic process of how full committees deliberate protocols under review [44].

According to the Guide, 2011 [37], Committee membership includes the following:

- A Doctor of Veterinary Medicine either certified or with training and experience in laboratory animal science and medicine or in the use of the species at the institution.
- At least one practicing scientist experienced in research involving animals.
- At least one member from a nonscientific background, drawn from inside or outside the institution.

- At least one public member to represent general community interests in the proper care and use of animals.

2.7.1   Responsibilities of AECs, Australian Code, 2013 [15]

The primary responsibility of an AEC is to ensure, on behalf of the institution for which it acts, that all activities relating to the care and use of animals are conducted in compliance with the Code.

The AEC must:

- Review applications for projects and approve only those projects that are ethically acceptable.
- Conduct follow-up review of approved projects and activities and allow the continuation of approval for only those projects and activities that are ethically acceptable and conform to the requirements of the Code.
- Monitor the care and use of animals, including housing conditions, practices, and procedures involved in the care of animals in facilities.
- Take appropriate actions regarding unexpected adverse events.
- Take appropriate actions regarding non-compliance.
- Approve guidelines for the care and use of animals on behalf of the institution.
- Provide advice and recommendations to the institution.
- Report on its operations to the institution.

2.7.2   Institutional Animal Care and Use Committees Monitoring System, According to the Australian Code, 2013 [15]

AEC procedures should cover the delegation of authority to suitably qualified people to monitor animal care and use, including projects and activities.

Procedures should include how reports of such monitoring are to be provided to the AEC.

- The AEC monitors the care and use of animals by inspecting animals, animal housing, and the conduct of procedures, and/or reviewing records and reports.
- The AEC must monitor all activities relating to the care and use of animals (including the acquisition, transport, breeding, housing, and husbandry of animals) on a regular and ongoing basis to assess compliance with the Code and decisions of the AEC.
- The AEC must ensure that identified problems and issues receive appropriate follow-up and, if necessary, refer suspected breaches of the Code to the institution.
- The AEC should monitor activities that are likely to cause pain or distress at an early phase during the conduct of the activity.

2.7.3   Institutional Animal Care and Use Committees Records, According to the NACLAR, 2004 [36]

The following IACUC records should be maintained:

- Minutes of IACUC meetings, including records of attendance, activities of the Committee, and Committee deliberations.

- Records of proposals involving animals, including proposals for significant.
- Changes in ongoing activities involving animals, and whether IACUC approval was given or withheld.
- Records of all IACUC reports and recommendations (including minority views).

All records and reports shall be maintained for at least 3 years. Records that relate directly to a project, including proposals for significant changes in ongoing activities, reviewed and approved by the IACUC shall be maintained for the duration of the project and for an additional 3 years after completion of the project.

2.7.4 Responsibilities of the Institutions Regarding Animal Research, According to the NACLAR, 2004 [36]

All institutions that conduct animal research have a great responsibility for protection of animal welfare; these responsibilities can be addressed in the following:

- Establish one or more IACUCs.
- Ensure, through the IACUC, that the care and use of animals for Scientific Purposes comply with the Guiding Principles and relevant legislation.
- Provide the IACUC with facilities, powers and resources to fulfill its terms of reference and responsibilities. Resources include the purchase of educational materials, access to training courses for IACUC members and access to administrative assistance.
- Refer to the IACUC for comment on all matters which may affect animal welfare including the building and modification of Housing and Research Facilities.
- Review annually the operation of the IACUC. This review includes an assessment of the annual report from the IACUC and a meeting with IACUC Chairman. This annual review of the IACUC is to help ensure that the IACUC is adjusting its operations in light of their experiences and circumstances, and according to continuing developments in care and use of animals for scientific purposes.
- Respond effectively to recommendations from the IACUC to ensure that the facilities for the housing, care and use of animals are appropriate for the maintenance of the health and well-being of the animals and that the disposal of the animals are appropriate.
- Respond promptly and effectively to recommendations from the IACUC to ensure that all care of the animals and use of animals for Scientific Purposes remains in accord with the Guiding Principles and relevant legislation.
- Provide all relevant Staff with details of the Institution's policy on the care and use of animals, and the relevant legal requirements.
- Establish grievance procedures for IACUC members and Investigators who are dissatisfied with IACUC procedures or decisions.

2.7.5 IACUCs in Developed Countries

In many countries, the detailed operation of these committees is controlled by agencies that fund research. The United States has an extensive system of IACUCs, created by the Animal Welfare Act and its regulations. The Act covers the use of warm-blooded animals in research but excludes rats, mice, and birds. The IACUCs operate according to the more detailed policies and guidance published by the National Institutes of Health. Australia uses a similar system of AECs, created under state legislation, but operating in accordance with the code of practice produced by the National Health and Medical Research Council.

Within Europe, there are two, almost identical, legal instruments. They are the Council of Europe Convention for the protection of vertebrate animals used for experimental and other scientific purposes. Member States of the Council of Europe can decide whether or not to ratify the Convention by implementing it in their national legislation. By contrast, Member States of the EU are legally obliged to implement the goals set out in the Directive. All have transposed the Directive in their national or regional legislation, although the European Commission has referred several countries to the European Court of Justice to ensure that their legislation is fully in accordance with the Directive [29].

2.7.6 Ethical Review

During the last decade, new developments and initiatives have been introduced to improve the quality and translational value of animal research. Systematic reviews of animal studies should be conducted routinely. Funding agencies should subsidize systematic reviews, not simply for transparency, but also to avoid waste of financial resources and unnecessary duplication of animal studies [14].

Scientific research involving animal use should always take place within a framework that allows for ongoing critical evaluation of the ethical, scientific, and welfare issues. Scientists are increasingly convinced of the need of systematic reviews of animal studies. However, systematic reviews of animal studies are still relatively rare [45].

2.7.7 Principles of Ethical Review, According to Fry, 2012 [27]

The basic ethical questions may be summarized as:

- Should animals be used at all for scientific investigation?
- If yes, should this apply to all animals?
- If not to all animals, on what basis should particular animals be excluded?
- Should all scientific work be treated the same?

Whenever scientific proposals to use animals are reviewed, it is vital that applicants provide adequate information and argument on which to base the ethical review. Ethical review processes must involve a wide enough range of expertise and perspective to facilitate comprehensive and detailed review of the factors that are relevant in the ethical evaluations [46]. While the responsibility for scientific merit review normally lies outside the IACUC, the committee members should evaluate scientific elements of the protocol as they relate to the welfare and use of the animals. For example, hypothesis testing, sample size, group numbers, and adequacy of controls can relate directly to the prevention of unnecessary animal use or duplication of experiments [37].

Approval for each research project involving animals must be based on considerations on whether the project is justified and whether the potential benefit outweighs the potential harmful effects on the welfare of the animals used. Researchers must submit written proposals to the Animal ERC for all projects involving animals. These proposals must address the expected value of knowledge to be gained, the justification of the project, and an ethical analysis regarding the animal welfare aspects under consideration of the "3 Rs" principles [28].

There is a need to consider at least five questions [26]:

1. What are the goals of research?
2. What is the probability of success?
3. Which animals are to be used?
4. What effect will there be on the animals used in the experiment?
5. Are there any alternatives?

The principal investigator, the IACUC, and the animal care staff must be aware that they are working in an environment in which there are ongoing changes in scientific knowledge and public values, which in turn will require regular reevaluation of protocols. Strong, ongoing communication between the IACUC, the veterinarian, the animal care staff, and the investigator is essential to managing these changes smoothly [47].

Improving the quality and translation of animal research requires cooperation from the wider scientific community, journals, researchers, regulators, funding bodies, peer reviewers, and patients. An international register for animal studies should be established and funded. Journals should publish "negative and neutral" results and promote data sharing [14].

2.8   Health Research Ethics in Africa

Health research is definitely important for the enhancement of public health, as it is instrumental to salutary and acceptable advances in science and medicine for diseases that highly burden a population. In Africa, a continent inundated with diseases such as HIV, malaria, tuberculosis, and a myriad of other neglected tropical diseases, health research is undoubtedly of rather critical importance [48].

During the past 5 years, there has been a dramatic increase in the number and type of initiatives to build capacity in research ethics in developing countries. Increasingly, the WHO is supporting improved ethical standards and review processes for research involving human beings. Several departments have undertaken training programs for researchers and RECs and have supported other capacity-building initiatives at local, regional, and international levels. Increasingly, WHO is supporting improved ethical standards and review processes for research involving human beings [49].

The Mapping African Research Ethics Review Capacity (MARC) project is developing an interactive map of health research ethics review capacity and drug regulatory capacity in Africa. This ongoing project invites self-uploading of information on African RECs and Drug Regulatory Authorities. This information is then integrated into an existing country-based research system mapping structure to facilitate efficiency, sustainability, and linkage of ethics "maps" to health research system capacity. This integration allows for ethics capacity analysis in relation to general research system development, encourages comparisons between countries inside and outside Africa, and facilitates sustainability and knowledge sharing throughout the project [50].

South Africa has two programs funded by the Fogarty International Center: both are making highly valued contributions to capacity building in international research ethics in Southern Africa with the formation of a network of Chairs of South African Human Health RECs [51]. Egyptian Network of Research Ethics Committee (ENREC) was established in 2008 to raise the harmonization between Egyptian RECs in reviewing of research proposals and to augment sharing of information and intellectual resources, policies, and review strategies. The main outcomes of the eighth meeting of ENREC were comparing the achievements and performance of RECs in different medical faculties and institutes in Egypt. The ninth meeting of ENREC was held in collaboration with the Ministry of Health, Ministry of Scientific Research, and WHO, under the theme of "Strengthening the Role of Ethics Review Committees." A successful trial of standardization of protocol reviewing checklist and informed consent form was a distinguished product of this meeting. There are 55 RECs in different universities and research centers in Egypt; 33 RECs are members of ENREC [52].

Wassenaar [53] estimated 100 (and increasing) RECs in Africa, 45 countries with current Federal Wide Assurance approvals, and about 64% of African countries have RECs. In the First African Conference for Administrators of Research Ethics Committees (AARECs) in September 2011 in Kasane, Botswana, participants representing 40 research ethics committee administrators (RECAs) from 21 African countries highlighted the need to have a training body in Africa that would draft a curriculum and offer continued professional development and accreditation of RECAs (AAREC Report, 2011). Countries mapped—MARC has collected information on 155 RECs in 34 countries and this information has been uploaded to the Research Ethics Web [50].

Kass et al. [54] stated that additional information on how African RECs function, including their staffing, operating procedures, strengths, and challenges, would be useful for African and international researchers working within Africa and for growing efforts to enhance ethics capacity on this vast continent. The Welcome Trust's Ethics of Biomedical Research in Developing Countries grant scheme shares this commitment to building research capacity and aims to increase the number of experts with the experience and training needed to address ethical questions raised by such research. Its new research funding initiative supports project grants, research studentships, and seminars among other capacity-building initiatives [49]. In many African countries, ethics review capacity needs to be strengthened. Inadequate training, limited infrastructural facilities, and lack of funding remain a big challenge. As the number of clinical trials conducted in Africa is growing, so is the requirement to empower the countries of Africa to conduct their own safety reviews of those trials. Currently, a number of organizations such as EDCTP, WHO, UNESCO, AMANET, COHRED, and others are working hand in hand to help those countries to create and strengthen their institutional and national bioethics committees [55].

2.9   Challenges of Research Ethics in Africa

Africa presents many challenges when it comes to the application of the ethical principles of health research and this can be summarized as:

*   Insufficient funds for public health.
*   Limited local infrastructures as research centers and other facilities.
*   Inadequate training opportunities.
*   Absence of national legislation and guidelines on Research Ethics.
*   Absence of laws that control the use of animal models in scientific research.
*   Insufficient control of quality and distribution of medicines.
*   Illiteracy or language problems in relation to medical and healthcare communication.
*   Limited number of qualified investigators and trained researches who are familiar with the ethical guidelines of health research.

According to Massele [56], concerns are being expressed about the validity of applying international ethical principles that may be questionable in an African culture. This sometimes creates friction, especially in international collaborative research. A western collaborator may have totally different ethical approaches, which may be unacceptable in African communities.

2.10   Laws of Animal Research Ethics in Africa

Currently, there is scarce empirical data from Africa on the existence or adequacy of national or institutional policies and guidelines on the use of animals in research. However, the little evidence available indicates that most African countries lack relevant legislation and guidelines [11]. Policies, legislative frameworks, and guidelines on the use of animals are all poorly developed in Africa [57].

Most African countries have laws aimed at protecting the welfare of animals, but for some countries, the laws seem to cover the welfare of animals in general without specifically addressing animals used in research or teaching. Indeed, various African countries have been revising their laws in order to enhance the protection of research animals [11].

2.10.1   Situation in Some African Countries

In South Africa, research is guided by the South Africa Medical Research Council Act, No. 58 (1991), which developed internal guidelines on the use of animals in research in 2004. In Kenya, all vertebrates are protected under Cap 360 (the Prevention of Cruelty to Animals Act) (1963, revised 1983). However, enforcement remains a challenge. Egypt is an active participant and was one of the first signatories of the World Organization for Animal Health (OIE) mandate governing animal welfare [58]. The Faculty of Science, Cairo University, Egypt, established IACUC in February 2013. In Egypt, some institutions have RECs that review or monitor protocols using animals, for example, the Medical Research Ethics Committee of the NRC of Egypt, since 2003 [52].

Various research and training institutions in Tanzania have established some guidelines on animal use, including establishing AECs. However, most institutions have not established oversight committees. In institutions where there may be guidelines and policies, there are no responsible committees or units to directly oversee if and how these guidelines and policies are enforced; thus, implementation becomes difficult or impossible [59].

The absence of legal and ethical frameworks and committees to review protocols that involve animals in research and education leaves major gaps in the protection of the animals involved [26]. The internationally accepted 3 Rs framework (Reduction, Refinement, and Replacement) forms a useful guide for assessing, monitoring, and promoting animal welfare in institutions in many African countries [52,57]. It is possible that some African countries may have developed guidelines that are not available in the public domain and are therefore not known by the authors and generally unknown to the larger scientific community [5].

A proposed Universal Declaration on Animal Welfare [60] seeks an agreement among people and nations to recognize that animals are sentient beings and can suffer, to respect their welfare needs, and to end animal cruelty. Introduced in 2003 at an international meeting in Costa Rica and currently endorsed by five countries, this declaration is awaiting adoption by the UN General Assembly and subsequent ratification. African animal welfare organizations from 28 countries have supported this initiative through a proposed African Declaration of Animal Welfare, and it is pending adoption by the African Union [57].

2.11 Conclusions

As the majority of the African population still relies on herbal medicines to meet its health needs, legislation concerning procedures for the registration of medicinal plants in Africa is required and clinical trials are important to validate the safety and efficacy of these plants.

Development of national ethical guidelines and laws that control the use of animals in scientific research in African countries is required. It is essential to establish Animals Ethics Committees in all universities and research centers in African countries to review experimental protocols or to provide oversight during the use of animals in toxicological research.

Efforts should always be made to apply the principles of the 3 Rs in Animal Research, the number of animals involved must be minimized, the well-being of the animals must be supported, and harm, including pain and distress, in those animals must be avoided or minimized.

There is a great need to establish training programs for researchers and policy makers in African countries on scientific and ethical considerations regarding the use of animals in research.

Acknowledgments

Many thanks are due to the members of Pathology Department in the National Research Center, Cairo, Egypt, for their help.

References

[1] WHO Bulletin. The African regional health report: the health of the people. Bulletin 2014;92(January): <http://www.who.int/bulletin/africanhealth/en/>.

[2] Schemm Y. Africa doubles research output over past decade, moves towards a knowledge-based economy, <http://www.researchtrends.com/issue-35-december 2013/africa-doubles-research-output/>; 2013.

[3] Adewunmi CO, Ojewole JAO. Safety of traditional medicine, complementary and alternative medicine in Africa. Afr J Trad Complement Altern Med 2004;1 (1):1−3.

[4] Vickers A, Zollman C, Lee R. Herbal medicine. Western J Med 2001;175:125−8.

[5] Togola A, Diallo D, Dembélé S, Barsett H, Paulsen BS. Ethnopharmacological survey of different uses of seven medicinal plants from Mali, (West Africa) in the regions Doila, Kolokani and Siby. J Ethnobiol Ethnomed 2005;1:7.

[6] WHO. WHO traditional medicine strategy 2002−2005. Geneva: World Health Organization; 2002.

[7] Miller KL, Liebowitz RS, Newby LK. Complementary and alternative medicine in cardiovascular disease: a review of biologically based approaches. Am Heart J 2004;147:401−11.

[8] Madiba SE. Are biomedicine health practitioners ready to collaborate with traditional health practitioners in HIV and AIDS care in Tutume sub district of Botswana. Afr J Trad Complement Altern Med 2010;7:219−24.

[9] Kigen GK, Ronoh HK, Kipkore WK, Rotich JK. Current trends of traditional herbal medicine practice in Kenya: a review. Afr J Pharmacol Ther 2013;2(1):32−7.

[10] Asare GA, Bugyei K, Sittie A, Yahaya ES, Gyan B, Adjei S, et al. Genotoxicity, cytotoxicity and toxicological evaluation of whole plant extracts of the medicinal plant *Phyllanthus niruri* (Phyllanthaceae). Gen Mol Res 2012;11(1):100−11.

[11] Nyika A. Animal research ethics in Africa: an overview. Acta Trop 2009;112(Suppl. 1): S48− 52.

[12] Tshikala T, Mupenda B, Dimany P, Malonga A, Ilunga V, Rennie S. Commentary: engaging with research ethics in central Francophone Africa: reflections on a workshop about ancillary care. Philos Ethics Humanit Med 2012;7:10.

[13] Taylor K, Gordon N, Langley G, Higgins W. Estimates for worldwide laboratory animal use in 2005. Altern Lab Anim 2008;36(3):327−42.

[14] Hooijmans CR, Ritskes-Hoitinga M. Progress in using systematic reviews of animal studies to improve translational research. PLoS Med 2013;10(7):e1001482.

[15] National Health and Medical Research Council. Australian code for the care and use of animals for scientific purposes. 8th ed. Canberra: National Health and Medical Research Council; 2013.

[16] WHO. Research guidelines for evaluating the safety and efficacy of herbal medicines. Manila: World Health Organization Regional Office for the Western Pacific; 1993.

[17] UNESCO. Report of the International Bioethics Committee (IBC) on traditional medicine system and their ethical implications. Paris: UNESCO; 2013.

[18] WHO. Traditional medicine, WHO fact sheet, N°134. Geneva: World Health Organization; 2008.

[19] Dolan K. Laboratory animal law: legal control of the use of animals in research. 2nd ed. Blackwell Publishing Ltd.; 2007.

[20] Hajar R. Animal testing and medicine. Heart Views 2011;12(1):42.

[21] Society of Toxicology (SOT). The importance of animals in research—public policy statement. Society of Toxicology (SOT); 1999, <http://www.toxicology.org/ai/air/AIR_Final.pdf>.

[22] Ferdowsian HR, Beck N. Ethical and scientific considerations regarding animal testing and research. PLoS ONE 2011;6(9):e24059.

[23] Ifeoma O, Oluwakanyinsola S. Chapter 4: Screening of herbal medicines for potential toxicities. New insights into toxicity and drug testing, <http://dx.doi.org/10.5772/54493>; 2013.

[24] Stephens ML, Conlee K, Alvino G, Rowan AN. Possibilities for refinement and reduction: future improvements within regulatory testing. Inst Lab Anim Res J 2002;43:S74−9.

[25] Marshall H. Dictionary of national biography. London: Smith, Elder & Co; 1831. p. 1885−900.

[26] Nuffield Council on Bioethics. The ethics of research involving animals. London; 2005.

[27] Fry DJ. Ethical review of animal experiments—a global perspective report commissioned by the world society for the protection of animals, for consideration by the international whaling commission. World Society for the Protection Animals 2012;64:1−39.

[28] UNISA. Policy on research ethics: University of South Africa, <www.unisa.ac.za/...dept/.../Ethics%20and%20research_Policy_2012.pdf>; 2012.

[29] Directive 2010/63/EU of the European Parliament and of the Council on the protection of animals used for scientific purposes. Off J Eur Union L276:33−79.

[30] The Canadian council on animal care's guide to the care and use of experimental animals, Volume 1, <http://www.ccac.ca/Documents/Standards/Guidelines/Experimental_Animals_Vol1.pdf>; 1993.

[31] The National research council's guide for the care and use of laboratory animals, <http://www.nap.edu/openbook.php?record_id=5140>; 1996.

[32] New Zealand's good practice guide for the use of animals in research, testing and teaching, <http://www.waikato.ac.nz/research/ro/ethics/Good%20practice%20guide-for-animals-use%202008.pdf>; 2002.

[33] The Australian code of practice for the care and use of animals for scientific procedures, <http://www.nhmrc.gov.au/_files_nhmrc/publications/attachments/ea16.pdf>; 2004.

[34] The U.S. public health service policy on humane care and use of research animals, <http://grants.nih.gov/grants/olaw/Guide-for-the-care-and-use-of-laboratory-animals.pdf>; 2002.

[35] National Advisory Committee for Laboratory Animal Research (NACLAR). Guidelines on the care and use of animals for scientific purposes, <http://www3.ntu.edu.sg/Research2/Grants%20Handbook/NACLAR-guide%20Lines.pdf>; 2011.

[36] National Advisory Committee for Laboratory Animal Research (NACLAR). Guidelines on the care and use of animals for scientific purposes, <www3.ntu.edu.sg/Research2/.../NACLAR-guide%20Lines.pdf>; 2004.

[37] The guide for the care and use of laboratory animals. 8th ed. Washington, DC: The National Academies Press; 2011.

[38] ANON. Animal Welfare Committee looks at animal rights. J Am Vet Med Assoc 1990;196(1):17.

[39] Fakoya FA. Who is concerned about animal care and use in developing countries? Proceedings of the 8th world congress on alternatives and animal, <www.altex.ch/resources/003008_Contents221.pdf>; 2011.

[40] Carbone L. Pain in laboratory animals: the ethical and regulatory imperatives. PLoS ONE 2011;6(9):e21578.

[41] Kona-Boun JJ, Silim A, Troncy E. Immunologic aspects of veterinary anesthesia and analgesia. JAVMA 2005;226:355−63.

[42] American Veterinary Medical Association (AVMA). Guidelines on euthanasia. American Veterinary Medical Association, <https://www.avma.org/KB/Policies/Documents/euthanasia.p>; 2013.

[43] Hansen LA, Goodman JR, Chandna A. Analysis of animal research ethics committee membership at American institutions. Animals 2012;2:68−75.

[44] Schuppli CA, Fraser D. Factors influencing the effectiveness of research ethics committees. J Med Ethics 2007;33:294−301.

[45] Korevaar DA, Hooft L, ter Riet G. Systematic reviews and meta-analyses of preclinical studies: publication bias in laboratory animal experiments. Lab Anim 2011;45:225−30.

[46] FELASA Working Group. Principles and practice in ethical review of animal experiments across Europe: a report prepared by the FELASA working group on ethical evaluation of animal experiments, <http://www.felasa.org/recommendations.htm>; 2005.

[47] Morrison AR, Evans HL, Ator NA, Nakamura RK. National Institute of Mental Health. Methods and welfare considerations in behavioral research with animals: report of a National Institutes of Health Workshop. Washington, DC: NIH Publication; 2002.

[48] Ouwe-Missi-Oukem-Boyera O, Munungb NS, Ntoumic F, Nyikae A, Tangwaf GB. Focus: developing world bioethics. Bioethica Forum 2013;6(1):18.

[49] Eckstein S. Efforts to build capacity in research ethics: an overview, <http://www.sci-dev.net/global/policy-brief/efforts-to-build-capacity-in-researchethics-an-ov.html>; 2014.

[50] MARC. Mapping African Research Ethics Review Capacity (MARC), <http://www.researchethicsweb.org/>; 2011.

[51] Benatar S. Research Ethics committees in Africa: building capacity. PLoS Med 2007;4(3):e135.

[52] Abdel- Aal W, Ghaffar EA, El Shabrawy O. Commentary: review of the Medical Research Ethics Committee (MREC), National Research Center of Egypt, 2003−2011. Curr Med Res Opin 2013;29(10):1411−7.

[53] Wassenaar. Challenges: regional perspectives: Africa Fogarty International Center, National Institutes of Health grant number 2 R25 TW01599-08, <http://www.up.ac.za/sareti.htm>; 2010.

[54] Kass NE, Hyder AA, Ajuwon A, Ndebele P, Ndossi G, Sikateyo B, et al. The structure and function of research ethics committees in Africa: a case study. PLoS Med 2007;4:e3.

[55] Van den Brink. Gabon establishes national ethics committee. Putting the well-being of African study participants first. Announcement, <http://www.edctp.org/Announcement.403 + M5f6ccfda9e6.0.html>; 2009.

[56] Massele A. The African perspective on global research ethics in the training program on: "Critical research ethics issues in the Era of HIV in Tanzania", <www.dmsfogarty.org/docs/massele_african_perspective.pdf>; 2005.

[57] Kimwele C, Matheka D, Ferdowsian H. A Kenyan perspective on the use of animals in science education and scientific research in Africa and prospects for improvement. Pan Afr Med J 2011;9:45.

[58] Aidaros H. Global perspectives—the middle east: Egypt. Rev Sci Tech Off Int Epiz 2005;24(2):589–96.

[59] Seth M, Saguti F. Animal research ethics in Africa: is Tanzania making progress? Dev World Bioeth 2013;13:158–62.

[60] World society for the Protection of Animals. The universal declaration for animal welfare, <http://www.animalsmatter.org/en/campaign>; 2010.

3 Critical Review of the Guidelines and Methods in Toxicological Research in Africa

Louis L. Gadaga and Dexter Tagwireyi

Drug and Toxicology Information Service (DaTIS), School of Pharmacy, College of Health Sciences, University of Zimbabwe, Avondale, Harare, Zimbabwe

3.1 Introduction

Since ancient times, humanity has depended on the diversity of plant resources to treat diseases [1,2]. In many developing countries, a large proportion of the populace relies heavily on traditional practitioners and herbal plants to meet primary healthcare needs [3]. Recently among developed countries, there has been a cultural renaissance toward more natural methods of healing [3,4], and many people in Western settings are turning to alternative and complimentary medicines [5]. Similarly, in many African countries, due to the high cost of medicines, and concurrent shortage of drugs, many people are reverting to the use of traditional herbal concoctions and decoctions for their ailments [6].

The increased use of traditional medicines or phytomedicines has led to increasing emphasis on research and development focused toward these preparations [5]. Assurance of safety, efficacy, and quality of medicinal plants and herbal products has thus now become a key issue in many countries, with the World Health Organization (WHO) publishing a series of monographs on selected medicinal plants [3,5].

Africa, and especially Southern Africa, boasts an exceptionally rich plant diversity with an estimated 30,000 species of flowering plants representing almost one-tenth of the world's higher plants. There are 10 endemic families, while 80% of the species and 29% of the genera are endemic [7−9]. In South Africa, research on traditional plants has led to the development of several marketed phytomedicinal products. This is evidenced by an increasing body of scientific literature on the use and properties of medicinally important African plants [5,10−13].

Although there has been a huge increase in ethnopharmacological publications on African medicinal plants, their healing values, and bioactivity [12,14], most of the traditional medicinal plants have never been the subject of exhaustive

Toxicological Survey of African Medicinal Plants. DOI: http://dx.doi.org/10.1016/B978-0-12-800018-2.00003-0

toxicological tests such as are required for modern pharmaceutical compounds [15−17]. Based on their past traditional use for long periods of time, these medicinal plants are often assumed to be safe [16,18,19]. However, phytochemical research has shown that a significant proportion of the purported "safe plants" have *in vitro* mutagenic [20−22] or toxic and carcinogenic properties [19,23].

It is imperative, therefore, that initiatives be put in place in all countries using traditional medicines, including African countries, to provide capabilities for safety, efficacy, and quality evaluation of these herbal medicines. The regulatory status of herbal medicines varies quite widely across the world. African countries have a large number of traditionally used herbal medicines, and a lot of ethnomedical information about these plants, but the majority of them have hardly any legislative criteria to establish these traditionally used herbal therapies as part of national drug legislation [24].

Although many African countries have realized the important role of herbal medicine in their primary healthcare systems, very few countries have made an effort to carry out research based on WHO guidelines on the safety and efficacy of herbal medicines.

While theoretically all African countries have some capacity to conduct rudimental research on medicinal plants, only a few which include South Africa, Egypt, Ghana, Nigeria, Mali, Ethiopia, Tanzania, Kenya, Uganda, and Zimbabwe have sophisticated research facilities that use animal models to ascertain the safety of medicinal products. These countries have facilities that are at different stages of advancement, for example, the Medical Research Council (MRC) of South Africa laboratory can do full-scale quality, safety, and efficacy evaluation following internationally accepted ethical and quality standards [25], whereas the laboratory at the University of Zimbabwe is more suited to do biosafety assays as demonstrated in some publications [26,27].

3.2   Need for Toxicity Testing

While there are reports of severe cases of toxicity resulting from the use of herbs, the potential toxicity of herbs and herbal products has not been acknowledged [6,28]. The toxicity and adverse effects of African medicinal plant use may arise due to a number of factors or issues including inherent poisonous phytochemicals, adulteration of the medicines, contamination with various chemicals and heavy metals, herb−drug interactions, and poor quality control of herbal products [29]. Therefore, quality control and evaluation for the efficacy and safety of herbal products is essential.

The best way of predicting the clinical and adverse effects on human health is to test potentially toxic substances directly on human subjects. However, this approach is often difficult and in many situations is unethical [30]. Some pharmacological and toxicological data can be derived directly from humans, through case reports of accidental exposures to industrial chemicals, cases of food-related poisoning, epidemiological studies, and clinical investigations. However, often the nature and extent of available human toxicological data are too incomplete to serve as the basis for an adequate assessment of potential health hazards [31]. Consequently, in preliminary toxicity and safety investigations of various drugs, herbs, and chemicals, animals commonly are used to predict risk in humans.

The suitability of experimental animal data is an important contemporary issue in toxicology [32,33]. The ultimate goal of toxicology assessments is to characterize toxicity in animal models and extrapolate to the human situation. There are basically two main guiding principles to experimental toxicology which are that animal assays, when properly qualified, serve as useful predictors of potential human effects, and that exposure of test animals to high doses is a valid determinant of human hazard [33,34]. Therefore, the use of adequate test systems is critical to the predictive ability of animal toxicity studies. However, drawbacks to animal testing do exist; the costs of the animals to be used can be prohibitive and subtle differences within species can affect the type of effects that are observed; and they are usually more tedious to arrange, in terms of duration of experiments [35].

In vitro studies have been used to complement whole-animal experiments. They are valuable in providing information on basic mechanistic processes in order to refine specific experimental questions to be addressed in the whole animal [31,33,36,37]. However, *in vitro* studies have limitations. Neurobehavioral effects such as loss of memory or sensory dysfunction, organ system interactions, and organ toxicity cannot be modeled with the current *in vitro* strategies [37].

The effects on human health of many plants and herbs used in traditional medicine systems are from experience or claims from ethnomedicinal usage. But little is known about both the exact pharmacological effects and the toxicological effects of these medicinal plants and herbs. WHO has been advocating recently for the integration with traditional medicine systems with conventional medical practices in resource-poor countries, and recent surveys in southern Africa have shown that the majority of the population relies on traditional medicines [3,10]. This unfortunately has exposed many people to potentially toxic plants with little or no documented evidence of efficacy. Therefore, WHO and many research institutes in African countries are placing emphasis on research on plants used in ethnomedical practices, particularly in the developing world.

Although it is necessary to have healthcare policies to regulate and encourage research on the safety and efficacy of herbal medicines, the methodologies for research and evaluation of traditional medicine should be based on the following basic principles. The methodologies should guarantee the safety and efficacy of herbal medicines and traditional procedure-based therapies. On the other hand, however, they should not become obstacles to the application and development of traditional medicine [38]. Standard guidelines for the conduct of animal toxicity tests have been harmonized by the Organization for Economic Cooperation and Development (OECD) as part of continuous efforts to internationally harmonize toxicity test guidelines [39]. The majority of studies on African herbal medicines toxicity have been on evaluating their acute systemic toxicity and research groups in countries such as Nigeria [40,41], Egypt [42], Cameroon [43], Zimbabwe [26,27], and South Africa [10,18,31] have been at the forefront of this research.

3.3   Acute Toxicity Testing

Acute toxicity is defined as that toxicity which arises soon after administration and unless death occurs, recovery is complete [32,34]. Test organisms range from simple systems such as brine shrimp to other animals such as mice, rats, guinea pigs, and rabbits. The median lethal dose (LD_{50}) of a substance is the amount required to kill 50% of a given test population of living organisms. The LD_{50} value for a substance will vary according to the species involved. The substance may be administered any number of ways, including orally, topically, intravenously, or through inhalation. The most commonly used species for these tests are rats, mice, rabbits, and guinea pigs [32,34].

Over the years acute toxicity testing has been taken to be synonymous with the lethal dose, yet the LD_{50} is only one of many indices used in defining acute toxicity. This misconception arose due to the relative ease of determination of and the frequent use of the LD_{50} [32]. However, there are two aspects of that are not addressed by the LD_{50} value which are symptomatology and pathological changes. In addition, the LD_{50} does not show the slope of the dose-response curve, meaning that it is difficult to tell exactly how toxic a substance is as you gradually increase the dose based solely on the LD_{50} [30]. Proper experimental design of an acute toxicity study should allow determination of nonlethal parameters together with the LD_{50} [30].

The significance of the LD_{50} value has been questioned by many toxicologists [30,44]. The determination of LD_{50} with 95% confidence limit has been described as wasteful in terms of number of animals sacrificed [44,45]. An inter-laboratory comparison study of LD_{50} by Hunter et al. [46] showed wide variations on the LD_{50} values. The LD_{50} has been shown to vary according to species, age, weight, sex, strain, nutrition, and environment factors [45]. Another weakness is that it does not take into account toxic effects that do not result in death but are nonetheless serious. However, estimation of the LD_{50} value could still provide valuable information about the toxicity of a compound. The US Environmental Protection Agency (EPA) and other international bodies now recommend the use of a limit test to estimate the LD_{50} [39].

3.4   Alternatives Methods for the Oral LD_{50} Test

A number of methods based on the concept of a Limit Test that use fewer animals as alternatives to estimate the acute lethal dose [47,48] have been developed over the years, and include the fixed-dose procedure (FDP), the acute toxic class (ATC) method, and the revised up-and-down procedure (UDP) [49−51]. All three methods emphasize humane treatment of animals undergoing testing and each assay utilizes a major endpoint other than lethality as its determining value [52].

3.4.1   Repeated Dose Toxicity Testing

Because some of the African traditional herbs are used for several days to weeks or even months, it is necessary to subject them to repeated dose toxicity tests.

The objective of repeated dose toxicity studies is to screen for potential adverse effects of substances using animal models as surrogates [33]. These studies are of varying duration, generally 1−4 weeks for short-term (subacute), 3 months for subchronic, and 6−12 months for chronic studies [33,53].

3.4.2 Subacute Studies

Short-term repeated dosing tests (1−4 weeks) are performed to obtain information on the toxicity of a substance after repeated administration and are generally required for the successful design of subchronic studies [33,54]. The major objective of short-term studies is to determine adverse effects at low doses, dose response, and sometimes to identify target organs [44].

3.4.3 Subchronic Studies

The principal goals of subchronic toxicity studies are to identify adverse effects not detected in subacute studies, to establish a nonobservable adverse effects level (NOAEL) and to further identify specific target organs or sites of action [33,44,54]. Subchronic studies data alone may be sufficient to predict the hazard of long-term exposure [33]. This is discussed in more detail in Chapter 6.

3.4.4 Chronic Studies

Chronic studies are performed in a similar manner to subchronic studies (see Chapter 6), except that the duration of exposure is prolonged. They are performed to assess the cumulative toxicity of substances and the majority of chronic effects. The design and conduct of chronic toxicity tests should allow for the detection of general toxic effects, including neurological, physiological, biochemical, and hematological effects, histopathological effects, and often include carcinogenic evaluation [53,54]. These studies usually are performed with rodent species, rats and mice, so as to extend over the lifetime of the species [54,55].

3.4.5 Genotoxicity and Mutagenicity Testing

Research in other regions of the world has shown that many plants that are used as food ingredients or in traditional medicine have *in vitro* mutagenic [19,22] or toxic and carcinogenic [23] properties. For this reason, it is also important to screen medicinal plants for their genotoxicity potential. This is increasingly more important for traditional medicines, because of the belief that because they have been in use for ages they are safe, yet it is known that cancers may not appear immediately, but over long periods that may stretch to decades. Thus it is conceivable that while a herb may not cause any immediate acute toxic effects (and thus be perceived as safe), it may actually be mutagenic or carcinogenic, with effects becoming evident after several years. Such toxic effects would be missed by the users who may not relate the cancers to earlier use of traditional herbs or medicines.

Various guidelines have been established to identify potential risk for carcinogenicity and heritable mutations [56,57]. The major challenge in genotoxicity testing resides in developing methods that can reliably and sensibly detect cellular response to genotoxic insult. It is recognized that no single test can detect every genotoxin; therefore, the concept of test batteries has been implemented in many regulatory guidelines [58].

The currently recommended approach is to carry out the following battery of tests, that is, an *in vitro* bacterial gene mutation test, an *in vitro* mammalian cell test, and an *in vivo* test for chromosomal damage using rodent hematopoietic cells [59,60]. However, a reduced battery one or two *in vitro* tests, can be acceptable in cases where physicochemical and metabolic properties indicate lack of genotoxicity activity [61]. Generally, the *in vitro* tests are used as screening tests before progressing to the hazard assessment *in vivo tests.* The most commonly used screening *in vitro* bacterial test is the Ames *salmonella* test due to its simplicity and relative inexpensive cost [61]. The comet assay is used for both *in vitro* and *in vivo* studies of assessment of chromosomal damage. Other tests that can be used to evaluate genotoxicity potential include the chromosome aberration assay, micronucleus assay, and *Saccharomyces cerevisiae* mutation assay [59,60]. The methods used in genotoxicity and mutagenicity assays are discussed in more detail in Chapters 9 and 10, respectively.

3.4.6   Clinical Testing Studies

After sufficient preliminary investigation showing the safety of a herbal product in preclinical studies, further studies can then be initiated in human participants. There are critical issues which must be considered to provide justification for the clinical trial of herbal products and guidelines to this effect have been provided by the WHO [35,38]. WHO also has provided guidelines for the pharmacovigilance on herbal products that are either available commercially or on informal markets [62]. Currently, there are no clinical trials in literature involving African herbal medicines and information on potential toxic exposure to herbal medicines can be gathered through WHO pharmacovigilance programs on herbal medicines [62].

3.4.6.1   Retrospective Case Reviews of Toxic Exposures

Retrospective cases reviews of human toxic exposure to herbal medicines are very important for getting a picture of the expected signs and symptoms of poisoning by a particular herb or herbal medicine. Although they have been used very rarely for herbal medicines in African countries, they have proved very useful with many poisoning agents all over the world [63−66]. Apart from providing information on the expected symptomatology, they sometimes aid in planning for animal studies to provide detailed analysis of mechanism of toxicity [26,27]. African countries such as Zimbabwe, South Africa, and Nigeria have been at the forefront of exploiting the usefulness of case reviews in determining the toxicology of herbal medicines [6,67−69].

3.5 Conclusion

In summary, the processes involved in the toxicological evaluation of herbal medicines in Africa has been focused mainly on preclinical evaluation of a relatively few herbal medicines that are used to treat a variety of ailments. However, despite an increased interest on research on the toxicology and efficacy of herbal medicines in Africa, very few countries led by South Africa and Nigeria and .others have made a great effort to set up the necessary research infrastructure to conduct research in this area. There is still a dearth of information on the toxicity of many popular herbs in Africa. It is very important that this research in these countries formulate research guidelines that are in line with international guidelines which can be adopted by other countries to improve the quality of toxicological research.

References

[1] Gilani AH, Rahman AU. Trends in ethnopharmacology. J Ethnopharmacol 2005;100:43−9.
[2] Saric-Kundalic B, Dobeš C, Klatte-Asselmeyer V, Saukel J. Ethnobotanical study on medicine use of wild and cultivated plants in middle, south and west Bosnia and Herzegovina. J Ethnopharmacol 2010;131:33−5520.
[3] World Health Organisation. Monographs on selected medicinal plants, vol. 1; 1999.
[4] Barnes J. Quality, efficacy and safety of complimentary medicines: fashions, facts and the future. Part 1. Regulation and quality. J Clin Pharmacol 2003;55:226−33.
[5] Makunga NP, Philander LE, Smith M. Current perspectives on an emerging formal natural products sector in South Africa. J Ethnopharmacol 2008;119:365−75.
[6] Tagwireyi D, Ball DE, Nhachi CFB. Traditional medicine poisoning in Zimbabwe: clinical presentation and management in adults. Hum Exp Toxicol 2002;21:579−86.
[7] Klopper RR, Chatelain C, Banninger V, Habashi C, Steyn HM, De Wet BC, et al. Checklist of the flowering plants of Sub-Saharan Africa. An index of accepted names and synonyms. South African Botanical Diversity Network Report No. 42. SABONET, Pretoria; 2006.
[8] Stafford GI, Pedersen ME, van Staden J, Jäger AK. Review on plants with CNS side effects used in traditional South African medicine against mental disease. J Ethnopharmacol 2008;119:513−37.
[9] van Wyk BE. A broad review of commercially important southern African medicinal plants. J Ethnopharmacol 2008;119:342−55.
[10] Fennell CW, Lindsey KL, McGaw LJ, Sparg SG, Stafford GI, Elgorashi EE, et al. Assessing African medicinal plants for efficacy and safety: pharmacological screening and toxicology. J Ethnopharmacol 2004;94:205−17.
[11] van Wyk BE, van Heerden FR, van Oudtshoorn B. Poisonous plants of South Africa. Pretoria: Briza Publications; 2002. p. 60.
[12] van Wyk BE, van Oudtshoorn B, Gericke N. Medicinal plants of South Africa. 1st ed. Pretoria: Briza Publications; 1997. p. 60−61.
[13] van Wyk BE, Gericke N. People's plants. Pretoria: Briza Publications; 2000. p. 156.
[14] Watt JM, Breyer-Brandwijk MG. The medicinal and poisonous plants of Southern Africa. Edinburgh: E & S Livingstone Ltd.; 1962. p. 23−35.

[15] Fennell CW, van Staden J. *Crinum* species in traditional and modern medicine. J Ethnopharmacol 2001;78:15−26.

[16] McGaw LJ, Eloff JN. Screening of 16 poisonous for antibacterial, antihelmintic and cytotoxic activity *in vitro*. S Afr J Bot 2005;71:302−6.

[17] Botha CJ, Penrith ML. Poisonous plants of veterinary and human importance in southern Africa. J Ethnopharmacol 2008;119:549−58.

[18] Elgorashi EE, Taylor JLS, Verschaeve L, Maes A, van Staden J, De Kimpe N. Screening of medicinal plants used in South African traditional medicine for genotoxic effects. Toxicol Lett 2003;143:195−207.

[19] Verschaeve L, van Staden J. Mutagenic and antimutagenic properties of extracts from South African traditional medicinal plants. J Ethnopharmacol 2008;119:575−87.

[20] Cardoso CRP, de Syllos Cólus IM, Bernardi CC, Sannomiya M, Vilegas W, Varanda EA. Mutagenic activity promoted by amentoflavone and methanolic extract of *Byrsonimacrassa niedenzu*. Toxicology 2006;225:55−63.

[21] Déciga-Campos M, Rivero-Cruz I, Arriaga-Alba M, Castañeda-Corral G, Angeles-López GE, Navarrete A, et al. Acute toxicity and mutagenic activity of Mexican plants used in traditional medicine. J Ethnopharmacol 2007;110:334−42.

[22] Mohd-Fuat AR, Kofi EA, Allan GG. Mutagenic and cytotoxic properties of three herbal plants from Southeast Asia. Trop Biomed 2000;24:49−59.

[23] De Sá Ferreira ICF, Ferrão Vargas VM. Mutagenicity of medicinal plant extracts in *Salmonella*/microsome assay. Phytother Res 1999;13:397−400.

[24] Sharad S, Manish P, Mayank B, Mitul C, Sanjay S. Regulatory status of traditional medicines in Africa region. Int J Res Ayuverda Pharm 2011;2:103−10.

[25] Moshi MJ, Mhame PP. Legislation on medicinal plants in Africa. Medicinal plant research in Africa. London: Elsevier Inc.; 2013. pp. 843−858.

[26] Gadaga LL, Tagwireyi D, Dzangare J, Nhachi CFB. Acute oral toxicity and neurobehavioural toxicological effects of hydroethanolic extract of *Boophone disticha* in rats. Hum Exp Toxicol 2011;30(8):972−80.

[27] Mutseura M, Tagwireyi D, Gadaga LL. Pre-treatment of BALB/C mice with a centrally acting serotonin antagonist (cyproheptadine) reduces mortality from *Boophone disticha* poisoning. Clin Toxicol 2013;51:16−22.

[28] Bandaranayake WM. Quality control, screening, toxicity, and regulation of herbal drugs. In: Ahmad I, Aqil F, Owais M, editors. Modern phytomedicine. Turning medicinal plants into drugs. Weinheim: WILEY-VCH Verlag GmbH and Co. KGaA; 2006. p. 25−53.

[29] Jordan SA, Cunningham DG, Marles RJ. Assessment of herbal medicinal products: challenges, and opportunities to increase the knowledge base for safety assessment. Toxicol Appl Pharmacol 2010;243:198−216.

[30] DiPasquale LC, Hayes AW. Acute toxicity and eye irritancy. In: Hayes AW, editor. Principles and methods of toxicology. 4th ed. Philadelphia, PA: Taylor and Francis; 2001. p. 853−906.

[31] World Health Organization. Neurotoxicity risk assessment for human health: principles and approaches. Geneva, Switzerland: WHO; 2001, Environmental Health Criteria 223. International Programme on Chemical Safety (IPCS).

[32] Brown VK. Pre-clinical testing. Acute and subacute toxicity testing. London: Arnold; 1988. p. 11−31.

[33] Wilson NH, Hardisty JF, Hayes JR. Short-term, subchronic and chronic toxicology studies. In: Hayes AW, editor. Principles and methods of toxicology. 4th ed. Philadelphia, PA: Taylor and Francis; 2001. p. 917−54.

[34] Wilson AB. Experimental design. In: Anderson D, Conning DM, editors. Experimental toxicology: the basic principles. 2nd ed. Cambridge: The Royal Society of Chemistry; 1990. p. 213−38.

[35] Ifeoma O, Oluwakanyinsola S. Screening of herbal medicines for potential toxicities. In: Gowder S, editor. New insights into toxicity and drug testing. Croatia: InTech Rijeka; 2013. Available from: < http://www.intechopen.com/books/new-insights-into-toxicity-and-drug-testing/screening-of-herbal-medicines-for-potential-toxicities >.

[36] Goldberg AM, Frazier JM. Alternatives to animals in toxicity testing. Sci Am 1989;261:24−30.

[37] Harry GJ, Billingsley M, Bruinink A, Campbell IL, Classen W, Dorman DC, et al. In vitro techniques for the assessment of neurotoxicity. Environ Health Perspect 1998;106:131−58.

[38] World Health Organization. General guidelines for methodologies on research and evaluation of traditional medicine. Geneva, Switzerland: WHO; 2001.

[39] Organisation for Economic Cooperation and Development. Guideline for testing of chemicals. OECD TG 425. Acute oral toxicity—revised up and down procedure. Paris: OECD; 2001.

[40] Awah FM, Uzoegwu PN, Ifeonu P, Oyugi JO, Rutherford J, Yao XJ, et al. Free radical scavenging activity, phenolic contents and cytotoxicity of selected Nigerian medicinal plants. Food Chem 2012;131:1279−86.

[41] Ashidi JS, Houghton PJ, Hylands PJ, Efferth. T. Ethnobotanical survey and cytotoxicity testing of plants of South-western Nigeria used to treat cancer, with isolation of cytotoxic constituents from Cajanus cajan Millsp. leaves. J Ethnopharmacol 2010;128:501−12.

[42] El-Seedi HR, Robert Burman R, Mansour A, Turki Z, Loutfy Boulos L, Gullbo J, et al. The traditional medical uses and cytotoxic activities of sixty-one Egyptian plants: discovery of an active cardiac glycoside from Urginea maritima. J Ethnopharmacol 2013;145:746−57.

[43] Kuete V, Krusche B, Youns M, Voukeng I, Fankam AG, Tankeo S, et al. Cytotoxicity of some Cameroonian spices and selected medicinal plant extracts. J Ethnopharmacol 2011;134:803−12.

[44] Kennedy Jr GL, Ferenz RL, Burgess BA. Estimation of acute toxicity in rats by determination of the approximate lethal dose rather than the LD_{50}. J Appl Toxicol 1986;6 (3):145−8.

[45] Rowan A. Shortcomings of LD_{50}: values and acute toxicity testing in animals. Acta Pharm Toxicol 1983;52(Suppl. II):53−64.

[46] Hunter WJ, Lingk W, Recht P. Intercomparison study on the determination of single administration toxicity in rats. J Ass Anal Chem 1979;62(4):864−73.

[47] Walum E. Acute oral toxicity. Environ Health Perspect 1998;106(Suppl. 2):497−503.

[48] Rispin A, et al. Alternative methods for the median lethal dose (LD_{50}) test: the up and down procedure for acute oral toxicity. ILAR J 2002;43(4):233−43.

[49] Organisation for Economic Cooperation and Development. Guideline for testing of chemicals. OECD TG 420. Acute oral toxicity—fixed dose procedure. Paris: OECD; 1992.

[50] Organisation for Economic Cooperation and Development. Guideline for testing of chemicals. OECD test guideline 471. Bacterial reverse mutation test. Paris: OECD; 1996.

[51] Organisation for Economic Cooperation and Development. Guideline for testing of chemicals. OECD TG 425. Acute oral toxicity—modified up and down procedure. Paris: OECD; 1998.

[52] Huggins J. Alternatives to animal testing: research, trends, validation, regulatory acceptance. ALTEX 2003;20(Suppl. 1/03):3−31.

[53] USEPA [United States Environmental Protection Agency]. Guidelines for neurotoxicity
      risk assessment. USEPA 630/R-95/001F. April 30, 1998. U.S. EPA, Risk Assessment
      Forum, Washington, DC; 1998. p. 90.

[54] Eaton DL, Klaasen CD. Principles of toxicology. In: Klaasen CD, editor. Casarett and
      Doull's toxicology: the basic science of poisons. New York, NY: McGraw-Hill; 2001.
      p. 11–34.

[55] Nielsen ND, Sandager M, Stafford GI, van Staden J, Jager AK. Screening of indige-
      nous plants from South Africa for affinity to the serotonergic reuptake transport pro-
      tein. J Ethnopharmacol 2004;94(1):159–63.

[56] Kirkland DJ, Galloway SM, Sofuni T. Report of the international workshop on stan-
      dardization of genotoxicity test procedures. Summary of major conclusions. Mut Res
      1994;312:205–9.

[57] Müller L, Kikuchi Y, Probst G, Schechtman L, Shimada H, Sofuni T, et al. ICH-
      harmonised guidances on genotoxicity testing of pharmaceuticals: evolution, reasoning
      and impact. Mut Res 1999;436:195–225.

[58] Abdelmigid HM. New trends in genotoxicity testing of herbal medicinal plants.
      In: Sivakumar Gowder, editor. New insights into toxicity and drug testing. InTech;
      2013. Available from: http://www.intechopen.com/books/new-insights-into-toxicity-
      and-drug-testing/new-trends-in-genotoxicity-testing-of-herbal-medicinal-plants

[59] Brambilla G, Martelli A. Failure of the standard battery of short-term tests in detecting
      some rodent and human genotoxic carcinogens. Toxicology 2004;196:1–19.

[60] Witte I, Plappert U, de Wall H, Hartmann A. Genetic toxicity assessment: employing
      the best science for human safety evaluation part III: the comet assay as an alternative
      to in vitro clastogenicity tests for early drug candidate selection. Toxicol Sci
      2007;97:21–6.

[61] Eisenbrand G, Pool-Zobel B, Baker V, Balls M, Blaauboer BJ, Boobis A, et al.
      Methods of in vitro toxicology. Food Chem Toxicol 2002;40:193–236.

[62] World Health Organization. The importance of pharmacovigilance- safety monitoring
      of medicinal products. Geneva, Switzerland: WHO; 2002, <http://apps.who.int/medi-
      cine-docs/en/d/Js4893e/1.html> [accessed 16.01.14].

[63] Spiller HA, Willias DB, Gorman SE, Sanftleban J. Retrospective study of mistletoe
      ingestion. Clin Toxicol 1996;34(4):405–8.

[64] Woolf AD. Safety evaluation and adverse events monitoring by poison control centers:
      a framework for herbs and dietary supplements. Clin Toxicol 2006;44:617–22.

[65] Kim EJY, Chen Y, Huang JQ, Li KM, Razmovski-Naumovski V, Poon J, et al.
      Evidence-based toxicity evaluation and scheduling of Chinese herbal medicines.
      J Ethnopharmacol 2013;146:40–61.

[66] Nyazema NZ. Herbal toxicity in Zimbabwe. Trans R Soc Trop Med Hyg
      1986;80:448–50.

[67] Nyazema NZ. Poisoning due to traditional remedies. Cent Afr J Med 1984;30:80–3.

[68] Kasilo OM, Nhachi CF. The pattern of poisoning from traditional medicines in urban
      Zimbabwe. S Afr Med J 1992;82:187–8.

[69] Kadiri S, Arije A, Salako BL. Traditional herbal preparations and acute renal failure in
      South West Nigeria. Trop Doct 1999;29:244–6.

4 Discordant Results in Plant Toxicity Studies in Africa: Attempt of Standardization

Roland E. Akhigbe

Department of Physiology, Ladoke Akintola University of Technology, Ogbomoso, Oyo state, Nigeria

4.1 Introduction

Plants and their natural products have been used in the management of human diseases from time immemorial [1,2]. Despite the growth of conventional medicine, the use of medicinal plants has been on the increase and become a part of human life [3−7], especially in African and Asian communities. The rise in the medicinal use of botanicals has been associated with factors such as their easy availability, inexpensiveness [8], and probably, cultural attachment of the community. Since it is a known fact that all drugs, conventional and/or orthodox, have side effects, it has become pertinent to evaluate the toxicological profile of commonly used medicinal plants to provide a safe and inexpensive alternative therapy [2]. This has stimulated and led to the toxicological survey of medicinal plants by interested researchers, particularly in Africa where an elaborate plant knowledge base is found [9].

It has been found that there are astonishingly high variations in the values of both physical and biochemical parameters of experimental animals in toxicological studies, including control animals. Some authors report an increase in a variable as a sign of toxicity, while others report a decrease in the same variable as a sign of toxicity. The aim of this chapter is to point out these variables and the causes of their differences, and suggest standard values which can be considered normal for each parameter.

4.2 Toxicological Evaluation: Parameters and Interpretation

Various parameters are important in evaluating the toxicity of medicinal plants. These include body and organ weight with their respective biomarkers, hematological profile, and redox status.

Toxicological Survey of African Medicinal Plants. DOI: http://dx.doi.org/10.1016/B978-0-12-800018-2.00004-2

4.2.1   Body and Organ Weight

Body and organ weights are important parameters in assessing the toxicity of substances, including plants. It is essential also to assess the body—organ weight ratio in evaluating the toxicity of plants as the organ weight is related to the body weight of the animal [10]. This reveals the weight of the organs in relation to the body weight. Studies have shown that toxicity can be revealed by reduction in the body and organ weights following exposure to a particular substance [10,11]. The reduction in weight has been associated with consequent depression of cellular metabolism and growth [12,13]. It also has been linked to lipid peroxidation. Cholesterol acyltransferase plays an essential role in the absorption of cholesterol in intestinal enterocytes and in the accumulation of lipid-laden foam cells [14]. Oxidation of polyunsaturated fatty acids in the mitochondria, erythrocytes, and platelets leads to the formation of lipid peroxide with consequent cell damage [15].

4.2.2   Hematological Profile

Assessment of hematological indices is crucial in evaluating the toxicity of exogenous compounds, including medicinal plants, in the system [2,16]. It also can be used to explain blood-related functions of plant extracts [17]. This is important in toxicity evaluation because hematological responses to botanicals have higher predictive values for humans when the data are translated from animal studies [2,16,18]. The blood accounts for the greatest percentage of total body fluid [16,19—21], hence it is the major route of food and drug transport [16,22]. A rise or decline in hematological variable could suggest excess or suppressed hemopoiesis [16]; both exaggerated and suppressed hemopoiesis could be a pointer to toxicity. Reduction of the blood cells suggests a suppression of bone marrow function [23]. Also, reduced red blood cells suggest suppression of the secretion of erythropoietin, an important hemopoietic factor, by the kidney [23]. A decline in red cell indices which include mean corpuscular volume (MCV), mean corpuscular hemoglobin (MCH), and mean corpuscular hemoglobin concentration (MCHC), following exposure to a botanical reveals the antihematinic potential (potential to induce anemia) of the extract [16].

Increase or decrease in total white blood cell and its differentials could suggest toxicity. White blood cells are indicators of the ability to fight infections [24]. Following the administration of a botanical, a rise in white cell counts reveals the stimulatory effect of the botanical on the immune system. This might be beneficial as such plants may be valuable in conditions associated with immunosuppression. On the other hand, a fall in white cell counts following exposure to a botanical shows the suppressive potentials of the plant on the immune system. However, it is important to note that a rise in cytokines (such as IL-1 and IL-8), basophiles, serotonin, and complement proteins following plant extract administration suggests inflammatory activity of the extract [25].

4.2.3 Plasma Enzyme Activities

Plasma enzymes of diagnostic importance are vital in toxicological survey. Because these enzymes are found in one or more tissues, changes in plasma activities may reflect damage to any of the tissues where the enzyme is present. Increase in these enzymes is proportional to the extent of tissue damage [13,26]. Also, increases in the plasma concentrations of these enzymes are indicative of loss of functional integrity of tissues cell membranes and consequent leakage into the blood stream [27].

Aminotransferases, which include alanine aminotransferase (ALT) and aspartate aminotransferase (AST), are involved in the transfer of an amino group from a 2-amino acid to a 2-oxoacid with the aid of pyridoxal phosphate as a cofactor for optimal activity [28]. A rise in any of these is an indicator of damage to cystolic and/or mitochondrial membranes [28]. AST is present in high concentrations in hepatic, renal, cardiac, and skeletal muscle cells and erythrocytes, hence damage to any of these tissues may increase plasma AST levels [13,28,29]. Similarly, ALT is present in high concentrations in hepatic, renal, skeletal muscle, and cardiac cells, thus damage to any of these tissues may increase plasma ALT levels [13,28,29].

Although hepatic cells contain more AST than ALT, ALT is confined to the cytoplasm in which its concentration is higher than AST [28,29]. The relative plasma activities of ALT and AST may indicate the type of cell damage; ALT is more specific for hepatic damage while AST may be present in skeletal muscle and it is more sensitive than ALT [28].

Alkaline phosphatases (ALPs) are hydrolyzed organic phosphates at high pH [28]. They are present in high concentrations in osteoblasts of bone and the cells of the hepatobiliary tract, intestinal wall, renal tubules, and placenta [28]. However, plasma ALP is derived from bone and liver in equal proportions in adults. Hence, damage to any of these tissues will cause a rise in plasma ALP.

Creatine kinase (CK) is another useful diagnostic enzyme. It is most abundant in cardiac and skeletal muscle cells, and in the brain [28]. It is also found in smooth muscle cells [28]. A rise in plasma creatine indicates damage to any of these tissues.

In addition, lactate dehydrogenase (LDH) is important in toxicological assessment of African plants. LDH catalyzes the reversible interconversion of lactate and pyruvate [28]. However, as it is widely distributed in the body with high concentrations in the cells of the liver, kidney, brain, erythrocytes, cardiac, and skeletal muscle, it is not a specific marker for cell damage [28]. The toxicological relevance of these enzymes is discussed in detail in Chapter 23.

Since most of these enzymes are found in more than one tissue, the rise in the plasma concentrations of the enzymes is not specific. Histomorphologic studies of these tissues are thus necessary to assess cyto-architectural distortion. For example, if following administration of a botanical, the plasma CK rises above expected when compared with the control, while histomorphologic analyses show normal histomorphology of all cells except the cardiac cells, it can be concluded that the rise in CK is a result of the toxic effect of the botanical on cardiac cells/heart.

4.2.4 Hepatic and Renal Function Tests

The liver performs a myriad of metabolic functions, thus accounting for the greatest percentage of metabolism of food, drugs, and foreign substances [30]. Besides the plasma enzymes that are suggestive of liver toxicity, the liver function test is very important in toxicological survey of medicinal plants. Liver function tests involve synthetic and excretory function tests. The synthetic function tests include measurement of plasma proteins (total plasma protein, plasma albumin, fibrinogen, and globulin) and prothrombin time [28]. These are factors secreted by the liver; hence, when the liver cells are damaged, these functions are impaired. A fall in the levels of plasma proteins and prolonged prothrombin time are indications of hepatotoxicity [28].

Renal function tests assess the functions of the kidneys. In evaluating the toxicity of medicinal plants, it is essential to assess renal functions alongside the liver functions. Although many tests are done to evaluate renal functions, the basic ones regularly reported are creatinine clearance, urea clearance, glomerular filtration rates (GFRs), and serum and urine electrolytes [19].

Creatinine clearance calculated from creatinine concentrations in urine and plasma samples and the urine flow rate, as well as the urea clearance, which is determined by urea concentrations in urine and plasma samples and the urine flow rate are used to measure GFR [19,31]. Although creatinine clearance is higher than inulin clearance because some creatinine is secreted by the tubule, while urea clearance is lower than inulin clearance because some urea is reabsorbed into the tubules [28], they still remain the common ways of determining GFR in Africa, possibly because creatinine and urea are already present in the body fluids and their plasma levels are quite constant [19]. Hence, the plasma concentrations of creatinine and urea could be used as indicators of nephrotoxicity [31−35]. Low creatinine clearance and/or low urea clearance suggests an impaired ability of the kidneys to filter these waste products from the blood and excrete them in urine [31]; this also impairs excretion of other serum electrolytes. As the clearance values decrease, their blood concentrations increase [31−35]. Thus, higher than normal plasma concentrations of creatinine and/or urea with or without alterations in serum electrolytes indicate nephrotoxicity.

4.2.5 Oxidative Stress

Oxidative stress is the presence of reactive oxygen species (ROS) in excess of the available antioxidant-buffering capacity [36,37], that is, an imbalance between pro-oxidant (free radical species) and antioxidant [38,39]. ROS causes damage to molecular target; proteins, lipids, and DNA, thus altering the structure and function of the cell, tissue, organ, or system [36,40]. Oxidative stress is assessed by determining malondialdehyde (MDA; also discussed in Chapter 23), which is the product of lipid peroxidation [36,39]. A rise in MDA level is associated with concomitant decline in one or more antioxidants (such as superoxide dismutase, catalase, reduced glutathione); this is suggestive of toxicity of a botanical. The serum levels of pro- and antioxidant are suggestive of whether or not a plant extract is toxic; it is nonspecific as to what organ(s) is/are affected. Hence, it is a necessary assay for oxidative indices both in the serum and other vital tissue (homogenates).

4.2.6 Gastrointestinal and Neurological Manifestations

Some qualitative variables also are necessary in assessing the toxicity of plant extracts. These signs of toxicity are made on observation. They include respiratory distress, nasal discharge, aggression, fearfulness, loss of response to touch, pain, or noise, salivation, loss of appetite, diarrhea, changes in hair appearance and fur loss, tremor, convulsion, hypotonicity or hypertonicity of the limbs, loss of reflex (pinna and cornea reflex), piloerection, writhing, as well as mortality [41,42].

4.3 Standard Values for Parameters in Toxicological Survey

Various studies reporting plant toxicity in Africa have documented surprisingly wide differences in the values of the toxicological indices determined. The variation in the values reported could be due to one or more of the following:

- Species of animal used; this affects how the botanicals are absorbed, transported, chemically altered, excreted, and bound to specific molecules within the body
- Diversity of the animal used as per genetic background, age, gender, and diet
- Sample size; smaller sample sizes are more prone to error
- Laboratory conditions; this is important to maintain room temperature
- Poor technique
- Analytical error

The differences observed across various studies have made experimental studies with animal models insufficient to draw a general conclusion. Since animal studies still remain accepted as part of the standard ways of predicting human toxicity, it is pertinent to extrapolate data from animal studies to lend credence to the nontoxic human use of African plants, as this is necessary for the protection of human health and the environment. Hence, a useful tool is provided as a reference document for African scientists for good research practice and interpretation of data.

Tables 4.1 and 4.2 show recommended range of values for toxicological variables. These values are means from documented studies.

4.4 Conclusion

Toxicological survey remains an important aspect of medicinal plants research. This chapter points out important toxicological variables, causes of the variations observed in different studies, and suggests standard values. This provides a standardized guide for researchers to enhance their studies.

Table 4.1 Recommended Values for Hematological Parameters [43]

Parameters	Animal Models and Values			
	Dog	**Rabbit**	**Rat**	**Mouse**
PCV (%)	29−55	30−50	36−54	39−49
Hemoglobin (g/dL)	14.2−19.2	10−15	11−19.2	10.2−16.6
Mean corpuscular volume (fl)	65−80	−	−	41−49
Mean corpuscular hemoglobin (pg)	12.2−25.4	−	−	−
Mean corpuscular hemoglobin Concentration (g/dL)	32−36	−	40	−
White blood cell (×1000)	5.9−16.6	7−13	6−18	6−15
Differentials (%)				
Lymphocytes	8−38	40−80	65−85	55−95
Monocytes	1−9	1−4	0−5	1−4
Eosinophils	0−9	0−4	0−6	0−4
Basophils	0−1	1−7	0−1	0−1
Platelet count (×1000)	160−525	125−270	500−1300	160−410
Fibrinogen (mg/dL)	200−400	263−572	−	−

Table 4.2 Recommended Values for Biochemical Parameters [43]

Parameters	Animal Models and Values			
	Dog	**Rabbit**	**Rat**	**Mouse**
Albumin (g/dL)	2.6−4.0	2.7−4.6	3.8−4.8	2.5−3
Alkaline phosphatase (U/L)	3−70	5−20	16−50	35−96
Alanine transaminase (U/L)	4−90	12−67	35−80	17−77
Amylase (U/L)	220−1070	−	−	−
Aspartate transaminase (U/L)	Less than 105	14−113	−	54−298
Total bilirubin (mg/dL)	0.2−0.8	0.2−0.7	0.2−0.5	0−0.9
Direct bilirubin	0−0.4	−	−	−
BUN (mg/dL)	7−24	10−25	10−21	8−33
Ca 2+(mg/dL)	9−11.4	5.6−12.5	8−13	7.1−10.1
Cholesterol	110−330	20−80	40−130	−
Creatinine (mg/dL)	0.7−1.4	0.8−1.8	0.5−1	0.2−0.9
GGT (U/L)	−	−	−	−
Glucose (g/dL)	80−120	75−150	50−160	62−175
Na+ (mEq/L)	140−165	130−150	140−150	140−160
K+ (mEq/L)	4.4−6.1	3.6−7.5	4.3−5.6	5−7.5
Chloride (mEq/L)	109−122	95−120	95−115	88−110

References

[1] Ajayi AF, Akhigbe RE. Antifertility activity of *Cryptolepis sanguinolenta* leaf ethanolic extract in male rats. J Hum Reprod Sci 2012;5:43–7.

[2] Ajayi AF, Akhigbe RE, Adewumi OM, Olaleye SB. Haematological evaluation of *Cryptolepis sanguinolenta* stem ethanolic extract in rats. Int J Med Biomed Res 2011;1:56–61.

[3] Sindiga I, Nyaigotti-Chacha C, Kanunah MP. Traditional medicine in Africa. Nairobi: East African Educational Publishers; 1995.

[4] Sindiga I, Kanunah MP, Aseka EM, Kiriga GW. Kikuyu traditional medicine. In: Sindiga I, Nyaigotti-Chacha C, Kanuna MP, editors. Traditional medicine in Kenya. Nairobi: East African Educational Publishers; 1995.

[5] Toledo BA, Galetto L, Colantonio S. Ethnobotanical knowledge in rural communities of Cordoba (Argentina): the importance of cultural and biogeographical factors. J Ethnobiol Ethnomed 2009;5:40.

[6] WHO. Armonización de los sistemas de salud indígenas y el sistema de salud convencional en las Américas. Washington, DC; 2003.

[7] Harun-Or-Rashid Y, Aminur Rashid Md, Nahar S, Sakamoto J. Perceptions of the muslim religious leaders and their attitudes on herbal medicine in Bangladesh: a cross-sectional study. BMC Res Notes 2011;4:366–95.

[8] Bussmann RW, Paul S, Aserat W, Paul E. Plant use in Odo-Bulu and Demaro, Bale region, Ethiopia. J Ethnobiol Ethnomed 2011;7:28–81.

[9] Barrow EGC. The dry lands of Africa: local participation in tree management. Nairobi: Initiatives Publishers; 1996.

[10] Anderson H, Larsen S, Splid H, Christenson ND. Multivariate statistical analysis of organ weights in toxicity studies. Toxicology 1999;136:67.

[11] Nadir R, Suat E. Oral administration of lycopene reverses cadmium-suppressed body weight loss and lipid peroxidation in rats. Biol Trace Elem Res 2007;118:175–83.

[12] Hispard F, Vaufleury A, Martin H, Devaux S, Cosson RP, Scheifler R, et al. Effects of subchronic digestive exposure to organic or inorganic cadmium on biomarkers in rat tissues. Ecotox Environ Safe 2007;24:45–50.

[13] Ige SF, Akhigbe RE, Edeogho O, Ajao FO, Owolabi OQ, Oyekunle OS, et al. Hepatoprotective activities of *Allium cepa* in cadmium-treated rats. Int J Pharm Pharm Sci 2011;3:60–3.

[14] Chang TY, Chang CC, Lin S, Yu C, Li BL, Miyazaki A. Role of acyl-coenzyme A: acyltransferase-1 and -2. Curr Opin Lipidol 2001;12:289.

[15] Gurr MI, James AT. Lipid biochemistry. 3rd ed. New York, NY: Sciences Publisher; 1980, p. 230.

[16] Saka WA, Akhigbe RE, Azeez OM, Babatunde TR. Effects of pyrethroid insecticide exposure on haematological and haemostatic profiles in rats. Pak J Biol Sci 2011;14:1024–7.

[17] Yakubu MT, Akanji MA, Oladiji AT. Haematological evaluation in male albino rats following chronic administration of aqueous extract of *Fadogia agrestis* stem. Pharmacog Mag 2007;3:34.

[18] Olson H, Betton G, Robinson D, Thomas K, Monro A, Kolaja G, et al. Concordance of toxicity of pharmaceuticals in humans and in animals. Reg Toxicol Pharmacol 2000;32:56–67.

[19] Sembulingam K, Sembulingam P. Essentials of medical physiology. 5th ed. New Delhi, India: Jaypee Brothers Medical Publishers (P) Ltd; 2010.

[20] Guyton AC, Hall JE. Textbook of medical physiology. 10th ed. New Delhi, India: Elsevier; 2001.

[21] Ganong W. Review of medical physiology. 18th ed. San Francisco, CA: Appleton and Lange; 1997.

[22] Katzung BG, Masters SB, Trever AJ. Basic and clinical pharmacology. 11th ed. USA: The McGraw-Hill Companies, Inc.; 2009 [International edition].

[23] Hoffbrand AV, Moss PAH, Pettit JE. Essential haematology. 5th ed. Malden, MA: Blackwell Publishing; 2006.

[24] Yakubu MT, Afolayan AJ. Effect of aqueous extract of *Bulbine natalensis* Baker stem on haematological and serum lipid profile of male Wistar rats. Indian J Exp Biol 2009;47:283−8.

[25] Kumar V, Abbas AK, Fausto N. Acute and chronic inflammation. Robbins and cotran pathologic basis of disease. 7th ed. Elsevier, Saunders; 2005.

[26] Chatterjea MN, Rana S. Liver function tests. Textbook of medical biochemistry. New Delhi, India: Jaypee Brothers Medical Publishers Ltd.; 2005.

[27] Lahon K, Das S. Hepatoprotective activity of *Ocimum sanctum* alcoholic leaf extract against paracetamol-induced liver damage in Albino rats. Pharmacogn Res 2011;3:13−8.

[28] Crook MA. Clinical chemistry and metabolic medicine. 7th ed. London, UK: Edward Arnold Publishers Ltd.; 2006.

[29] Saka WA, Akhigbe RE, Ishola OS, Ashamu EA, Olayemi OT, Adeleke GE. Hepatotherapeutic effect of *aloe vera* in alcohol-induced hepatic damage. Pak J Biol Sci 2011;14:742−6.

[30] Ajayi AF, Akhigbe RE. Implication of altered thyroid state on liver function. Thyroid Res Pract 2012;9:84−7.

[31] Saka WA, Akhigbe RE, Popoola OT, Oyekunle OS. Changes in serum electrolytes, urea, and creatinine in aloe vera-treated rats. J Young Pharm 2012;4:78−81.

[32] Pagna KD. Mosby's manual of diagnostic and laboratory tests. St. Louis, MO: Mosby Inc.; 1998.

[33] Brenner BM, Floyd C. The kidney. 6th ed. Philadelphia, PA: W.B. Saunders Company; 1999.

[34] Wallach J. Interpretation of diagnostic tests. 7th ed. Philadelphia, PA: Lippincott Williams and Wilkins; 2000.

[35] Henry JB. Clinical diagnosis and management by laboratory methods. 20th ed. Philadelphia, PA: W.B. Saunders Company; 2001.

[36] Azeez OM, Akhigbe RE, Anigbogu CN. Oxidative status in rats kidney exposed to petroleum hydrocarbons. J Nat Sci Biol Med 2013;4:149−54.

[37] Adly AA. Oxidative stress and disease: an updated review. Res J Immunol 2010;3:129−45.

[38] Agarwal A, Gupta S, Sharma RK. Role of oxidative stress in female reproduction. Reprod Biol Endocrinol 2005;3:28.

[39] Ige SF, Akhigbe RE. The role of *Allium cepa* on aluminium-induced reproductive dysfunction in experimental male rat models. J Hum Reprod Sci 2012;5:200−5.

[40] Robert JM, Hubel CA. Oxidative stress in preeclampsia. Am J Obstet Gynecol 2004;190:1177−8.

[41] Esimone CO, Akah PA, Nworu CS. Efficacy and safety assessment of T. Angelica herbal tonic, a phytomedicinal product popularly used in Nigeria. Evid Based Complement Alternat Med 2011;2011:123036.

[42] Yakubu MT, Adeshina AO, Oladiji AT, Akanji MA, Oloyede OB, Jimoh GA, et al. Abortifacient potential of aqueous extract of *Senna alata* leaves in rats. J Reprod Contracept 2010;21:163−77.

[43] Research animal resources, University of Minnesota. <www.ahc.umn.edu/rar/refvalues.html> [accessed 12.12.13].

5 Acute and Subacute Toxicities of African Medicinal Plants

Gerald Ngo Teke[1] and Victor Kuete[2]

[1]Department of Biomedical Sciences, Faculty of Health Sciences, University of Bamenda, Bambili, Cameroon, [2]Department of Biochemistry, Faculty of Science, University of Dschang, Dschang, Cameroon

5.1 Introduction

Plant poisoning in animals is usually accidental, and most frequently occurs during unfavorable conditions when pastures are poor due to drought, veldt fires, and overstocking and trampling of the grazing. In humans, it may be accidental or intentional. Accidental poisoning in humans may be due to confusing poisonous with edible plants, contamination of food with poisonous plants, or by the use of plants as remedies. Toxic principles in plants can affect the entire spectrum of organ systems; the dominant effect may depend on the condition, growth stage or part of the plant, the amount consumed, and the species and susceptibility of the victim. Considering the fact that herbal medicine as preparations derived from naturally occurring plants with medicinal or preventive properties are a major component in all indigenous peoples' traditional medicine across Africa and the world at large, pharmacological and toxicological evaluations of medicinal plants are essential for drug safety and development. The main types of toxicological evaluations include: acute toxicity, subacute toxicity, subchronic toxicity, and chronic toxicity studies. We will emphasize the acute and subacute toxicities of African medicinal plants.

Acute toxicity refers to those adverse effects occurring following oral or dermal (topical) administration of a single dose of a substance or medicinal plant extract, or multiple doses given within 24 h, or a single uninterrupted exposure by inhalation over a short period of time [1,2]. Subacute toxicity is an adverse effect occurring as a result of repeated daily dosing of a chemical, or exposure to the chemical, over a period of several days or weeks [3]. The study of the toxicities of medicinal plants is important in predicting their safety. Most of the methods used in the evaluation of acute and subacute toxicities of chemicals are based on the routes of administration/exposure to humans: oral; dermal; intraperitoneal (IP), inhalation; and the eye irritation assays. The inhalation and eye irritation tests are not very common in Africa [4–7]. In most acute toxicity tests, each test animal is administered a single (relatively high) dose of the test substance, observed for 1 or

Toxicological Survey of African Medicinal Plants. DOI: http://dx.doi.org/10.1016/B978-0-12-800018-2.00005-4

2 weeks for signs of treatment-related effects, then necropsied. Some acute toxicity tests (such as the "classical" LD_{50} test) are designed to determine the mean lethal dose of the test substance. The median lethal dose (or LD_{50}) is defined as the dose of a test substance that is lethal for 50% of the animals in a dose group.

5.2   The Acute and Subacute Effects of Medicinal Plants

Plant poisonings of humans in eastern and southern Africa have been documented in a comprehensive treatise by Watt and Breijer-Brandwyk [8]. Human poisoning from plants usually arises either from the unintentional use of toxic plants as food or from the use of poisonous plants for medicinal purposes. Accidental ingestion of plants resulting in acute poisoning is more common in preschool children [9]. Human intoxication due to plant exposure is far less important than poisonings involving paraffin, pesticides, pharmaceuticals, household cleaning chemicals, and cosmetics [9,10]. It is, however, likely to be higher in societies where plant-based traditional medicines are commonly used. A survey conducted at a large hospital in such an area revealed that poisoning with traditional medicines is the second most common cause of acute poisoning, representing 12.1%, mostly of plant origin [11]. Through the centuries, traditional medicines based on plants known and selected for their therapeutic effects have been used to good purpose and have undoubtedly cured many more people than they have killed. However, the levels of poisonous principles in plants are usually unpredictably variable, and occasional overdoses are likely. Furthermore, the cumulative effects of plants taken over long periods may be subtle and not well understood in traditional medicine. The toxic effects of poisonous plants, particularly chronic effects, are not always easy to reverse [10].

Poisonous plants, including those used in African Traditional Medicine (ATM), can affect the entire spectrum of organ systems, with some plants having several toxic principles that affect different systems. The dominant effect may depend on the condition, growth stage or part of the plant, the amount consumed, and the species and susceptibility of the victim. While the active principles and mode of action are known for many plants, many others are known to induce poisoning, but the mechanism of intoxication has yet to be elucidated.

In this chapter, some outstanding toxic effects, both acute and subacute, of medicinal plants worldwide will be examined. For simplicity, this will be based on the different plant parts.

5.2.1   Leaf Poisoning

The leaves of many plants used in traditional medicine can be an equal threat to the lives of humans or animals. *Manihot esculenta* (Euphorbiaceae) leaves are used in herbal medicine to treat measles, small pox, chicken pox, and/or skin rashes, gonorrhea, toothache, and as a purgative [12,13].

In Nigeria, they are used in the treatment of ringworm, tumor, conjunctivitis, sores, and abscesses. Despite the multiple applications in herbal medicine, acute ingestion exposes the individual to toxins such as prussic acid and cyanogenic glycoside characterized by konzo (cassava-associated spastic paraparesis/cassava-associated tropical ataxic myeloneuropathy) [14,15]. Other medicinal plants that contain cardiac glycosides, cardenolide, digoxin, (*Digitalis purpurea*; foxglove) are used for the treatment of congestive heart failure; and some of the bufadienolide-containing plants, in particular *Drimia sanguinea* (slangkop, sekanama) and *Bowiea volubilis* (climbing potato) are used by traditional healers in the treatment of various ailments and have been implicated in human poisoning [16,17]. *Callilepis laureola* (impila, oxeye daisy, wild-emagriet) is used in traditional medicine to treat tapeworm, snakebite, infertility, pregnancy tonics (known as *inembe* and *isihlambezo*), to kill maggots in cattle, and to treat whooping cough. Among the Swati, the macerated leaf is used as an external disinfectant [18]. An atractyloside (discussed in Chapter 19) has been isolated from the tuberous roots [17,19]. This plant has been reported to induce severe hepatotoxicity, nephrotoxicity, and hypoglycemia [20].

5.2.2 Flowers and/or Fruits

After prolonged administration of *Lathyrus sativa* (chickling pea), lathyrism (peripheral neuropathy) may result, caused by excitatory amino acids in this plant. The disease manifests as a spastic muscle weakness and is clinically very similar to konzo [14].

Datura stramonium and *Datura ferox* (moon flower, jimson weed, stinkblaar, oliebome) are cosmopolitan weeds that contain parasympatholytic alkaloids such as atropine and hyoscine. Humans are extremely susceptible to their effects and hallucinations may occur, and the proverb "blind as a bat, red as a beet, dry as a bone, and mad as a hatter" aptly describes atropine poisoning in humans. Consumption of these weeds is usually inadvertent. The plants may be mistaken for *Amaranthus* and prepared as "marog," resulting in severe poisoning, or the seeds may end up in harvested maize. On occasion, young children have been forced to swallow seeds ("malpitte") during initiation ceremonies at schools [8,21].

A plant like *Nicotiana glauca* (wild tobacco), of which the young seedlings may be mistakenly collected as "marog," contains a pyridine alkaloid, anabasine (discussed in Chapter 21), which is very similar to nicotine. Ingestion may result in nausea, vomiting, gait abnormalities, tremors, confusion, and convulsive seizures [22]. *Boophane disticha* (seeroogblom, bushman poison bulb) contains various alkaloids such as buphanidrine, buphanisine, and buphanamine. Poisoning usually occurs in humans that utilize the bulb for medicinal purposes. Acute poisoning induces vomiting, weakness, coma, and mortality [17]. Cardenolide-containing plants are highly toxic and may be very important in humans, as a number of garden/medicinal plants contain cardenolides. Yellow oleander (*Thevetia peruviana*) is also very toxic, highlighted by the fact that the fruit contains cardenolides (oleandrin) and is referred to as "Be-still-nut" [21].

Toxalbumins that occur principally in the seeds but sometimes also in other parts of plants are highly toxic plant lectins. Ricin, derived from the castor oil plant *Ricinus communis*, is one of the most toxic substances known, and likely to be used for bioterrorism. *Ricinus communis* (castor oil plant) is used traditionally as a purging agent against stomach upset and contains toxalbumin (ricin) [23].

5.2.3 Stem Bark

The stem bark of many plants used in traditional medicine can sometimes provoke deleterious effects in humans. *Spirostachys africana* used in healing headaches [24]; infantile cradle cap (*ishimca*); cough and fever [25], but is reported to provoke severe diarrhea, headache, and nausea [26,27]. In Cameroon, *Pteleopsis hylodendron* is used to treat measles, chicken pox, sexually transmitted diseases, female sterility, liver and kidney disorders, as well as dropsy [28,29]. It possibly alters liver and kidney functions [30]. Also, *Fagara macrophylla* used to treat malaria, diabetes, and hypertension provokes anemic and hyperglycemic effects. Also, hepatic and renal disturbances can occur after acute or subacute intoxication [31]. The plant *Erythrina senegalensis*, whose stem bark is used traditionally in the Western region of Cameroon against liver disorders such as hepatitis and jaundice [32], is known to cause gastrointestinal disorders, amenorrhea, dysmenorrhea, sterility, Onchocerciasis body pain, abdominal pain, and headache [33] following prolonged treatment. There is also a possibility of blood cell cancer and reduction in blood ion levels [34].

5.2.4 Roots

The roots of some medicinal plants contain certain chemicals that can cause harm to our systems. Exposure to pyrrolizidine alkaloids through the use of herbal remedies also may be a contributing factor to the high rates of liver cancer and cirrhosis seen in Africa [35]. Another pyrrolizidine alkaloid-containing plant that often is used as herbal medicine and has also been associated with poisoning is comfrey (*Symphyttum officinale*) [36]. The role of pyrrolizidine alkaloids in plant toxicity is discussed in detail in Chapter 21. The plant decoction *Annona senegalensis* is used in the treatment of sleeping sickness in Northern Nigeria [37]; cancer [38] and malarial infection [39]; against snake venom [40]; and is reported to have a possible effect on degeneration and necrosis of hepatocytes after acute/prolonged exposure [41].

5.3 Experimental Models for Toxicological Studies

Researchers can carry out the acute toxicity test or subacute test with any animal species, but they use rats or mice most often. Other species include dogs, hamsters, cats, guinea pigs, rabbits, and monkeys. In each case, the choice for the appropriate animal for a particular test is influenced by the test method, which is also linked to the route of product administration. Some methods in acute toxicity include: oral (fixed-dose procedure, acute toxic class method, up-and-down procedure), dermal, inhalation, IP, and eye irritation methods.

5.3.1 Animals Used in Acute and Subacute Toxicity Surveys

Animal studies have formed the cornerstone of toxicology and safety assessments despite the difficulties in interpreting animal test results with regard to human relevance. This is complex, considering the biological differences among species and the use of high experimental doses. Animal studies are used to assess a variety of toxicological outcomes: potential noncancer effects, and genotoxicity, carcinogenicity, as well as adverse reproductive and developmental outcomes. Rodents (males and females) are the most widely used animal models, but other types of animals with special sensitivities may be used for specific endpoints.

5.3.2 Animals Used in Acute Toxicity

The different types of animals used in acute toxicity may vary with respect to the route of product administration.

i. *Oral*: In Africa and elsewhere, the preferred rodent species is the albino rat (Wistar) and the test substance is administered in a single dose by gavage (if not possible, the dose may be given in smaller fractions over a period not exceeding 24 h) [42]. There is a good amount of research involving the mouse (Swiss mice) instead of the rat. Some reasons could be that the mouse is more sensitive and would easily express toxic symptoms, and its small size makes the assay more economical. Some years earlier, for most classes of chemical substances governmental regulatory agencies demanded single-dose toxicity on at least two rodents and one nonrodent species. Dogs and sometimes even monkeys were used [43].

ii. *Dermal*: The rat, rabbit, or guinea pig may be used. The albino rabbit is preferred because of its size, ease of handling, skin permeability, and extensive data base. Commonly used laboratory strains must be employed. If a species other than rats, rabbits, or guinea pigs is used, the tester must provide justification and reasoning for its selection. Young adult animals, rats between 8 and 12 weeks old, rabbits at least 12 weeks old, and guinea pigs between 5 and 6 weeks old at the beginning of dosing should be used. The weight variation of animals used in a test must be within 20% of the mean weight for each sex [44].

iii. *Inhalation*: A variety of test species can be used for acute inhalation tests. For comparison with data obtained in other acute tests, the rat is the preferred species. Commonly used laboratory strains of healthy young adult rats should be employed. The weight range within the test animal population should not exceed $\pm20\%$ of the mean weight. For each exposure time, at least ten animals (five females and five males) should be used. The females should be nulliparous and nonpregnant [1].

iv. *Intraperitoneal*: Commonly used laboratory animals are mice and rats [45].

   IP administration results in a faster absorption into the vasculature and is thus akin to intravenous (IV) administration. We think that mice and rats are good in this type of test because they have fewer hairs on their skin which facilitates access to the IP site for drug administration, compared to thick hairs on bigger mammals such as guinea pigs and rabbits. Any irritating compound, such as ketamine or pentobarbital, is less irritating if administered IP. A mouse or rat may be injected easily by one person, whereas a bigger mammal may require restraint by one person and injection by the other.

v. *Eye irritation test*: Although several mammalian test species may be used, bigger rodents such as rabbits usually are used [45]. It is obvious that the larger the organ under test (eye), the easier it is to observe and detect minute changes or symptoms of toxicity.

5.3.3   Animals Used in Subacute Toxicity

The animals used in subacute toxicity are bigger rodents such as rats. These animals are manipulated probably on a daily basis (according to designed protocol) and will have to resist test duration. For example, daily oral administration of test substance with an intubation needle or even IP (which are common exposure routes) for about 28 days may be stressful for mice [46–48]. Hence they are not very suitable. Even when the test product has to be mixed with feed or water for animals to freely consume orally or a product is meant for inhalation route, the use of mice should be strongly justified [49].

5.3.4   Methods in Acute and Subacute Toxicity Survey

In animal testing, a route of exposure that reflects probable human exposure is key. If there is no appropriate route of exposure, study interpretation can be difficult. This is because the toxicity (or detoxification) of a compound may be highly dependent on absorption into the gastrointestinal tract and subsequent metabolism. In addition, oral exposures usually do not reflect dermal toxicity and vice versa. In this light, the different methods in acute and subacute toxicity rely on the route of exposure. Animals are often observed for 2 or 4 weeks after the end of treatment for reversibility, persistence, or delayed occurrence of adverse effects.

5.3.4.1   Oral Toxicity

The various methods for oral acute toxicity rely on the end results expected. Usually animals are starved for less than 24 h prior to test in this toxicity type.

- *Fixed-dose procedure*: the aim of this test is to minimize the number of animals required to estimate the acute oral toxicity of a chemical. The method permits estimation of an LD_{50} with a two confidence interval and the results will permit the substance to be ranked and classified according to the Globally Harmonized System (GHS) [50]. Briefly, the test consists in dosing groups of animals (single sex, normally females) in a stepwise procedure using the fixed doses of 5, 50, 300, and 2000 mg/kg (exceptionally an additional dose of 5000 mg/kg may be considered). The initial dose level is selected as the dose expected to produce some signs of toxicity without causing severe toxic effects or mortality based on *in vivo/in vitro* data (if no information exists, the starting dose will be 300 mg/kg). Further groups of animals may be dosed at higher or lower fixed doses, depending on the presence or absence of signs of toxicity. This procedure continues until the dose causing evident toxicity is identified. The test substance may be given in smaller fractions over a period not exceeding 24 h if the volume is high for a single administration. This method is common with some researchers in Africa.

• *Acute toxic class method*: with similar objective as in fixed dose, the test consists of a stepwise procedure with the use of three animals of a single sex (normally females) per step. Absence or presence of compound-related mortality of the animals dosed at one step will determine the next step (i.e., no further testing is needed; dosing of three additional animals with the same dose; dosing of three additional animals at the next higher or lower dose level). The starting dose level is selected from one of four fixed levels (5, 50, 300, and 2000) and it should be that which is most likely to produce mortality in some of the dosed animals [51].

• *Up-and-down procedure*: the test consists of a single ordered dose progression in which animals of a single sex (normally females) are dosed, one at a time. The first animal receives a dose step below the level of the best estimate of the LD_{50} (when no information is available to make a preliminary estimate of the LD_{50}, the suggested starting dose is 175 mg/kg). If the animal survives, the dose for the next animal is increased by a factor of 3.2 times the original dose; if it dies, the dose for the next animal is decreased by a similar dose progression [52].

5.3.4.2 Inhalation Toxicity

The purpose for this test is to provide information on health hazards likely to arise from short-term exposure by the inhalation route. Data from an acute study (such as the estimation of LC_{50}, median lethal concentration) may serve as a basis for classification and labeling. Several groups of animals (at least ten animals, i.e., five females and five males) are exposed for a defined period to the test substance in graduated concentrations (at least three), one concentration being used per group. Animals should be tested with inhalation equipment designed to sustain a dynamic airflow of 12−15 air changes per hour, ensure adequate oxygen content of 19%, and an evenly distributed exposure atmosphere. Where a chamber is used, its design should minimize crowding of the test animals and maximize their exposure to the test substance. The animals are observed daily for a total of 14 days for any toxic symptoms [53].

5.3.4.3 Acute Dermal Toxicity

In the assessment and evaluation of the toxic characteristics of a substance, determination of acute dermal toxicity is useful where exposure by the dermal route is likely. Data from an acute study may serve as a basis for classification and labeling. It is an initial step in establishing a dosage regimen in prolonged administration and may provide information on dermal absorption and the mode of toxic action of a substance by this route [44].

Conventionally, the test substance is applied dermally in graduated doses to several groups of experimental animals, one dose being used per group. The doses chosen may be based on the results of a range-finding test. Subsequently, observations of effects and deaths are made. Animals that die during the test are necropsied, and at the conclusion of the test, the surviving animals are sacrificed and necropsied. Dosing test substances in a way known to cause marked pain and distress due to corrosive or irritating properties need not be carried out [54]. Guinea pigs in such tests are increasingly being replaced by mice (local lymph node assay). Here, the test material is applied to the ears of the mice. After an interval, the mice are euthanized and the early stages of sensitization are detected by measuring the level of induced DNA synthesis in the lymph nodes. This test provides more useful information, uses fewer animals than the guinea pig test, and causes substantially less pain and distress to the animals involved.

5.3.4.4   Acute IP Toxicity

This is one of the methods of dosing, which may occasionally provide information about local as well as systemic toxicity. To give drugs by IP dosing, the animal is laid on its back and the abdomen shaved. This area is thoroughly cleansed and, using an appropriate syringe and needle, the abdominal wall is punctured. To ensure minimal danger of perforation of abdominal viscera, the injection should be made rostral and lateral to the bladder at an angle of about $15°$ to the abdomen. The depth of penetration should not exceed 5 mm [55,56].

In this assay, the animal is taken out of its cage by its tail, placed on a flat surface and the palm of the free hand placed on its back. The thumb and the forefinger are passed around the neck, with the thumb going behind one forelimb ending up under the lower jaw. Once the animal is securely held and lifted, the injection using a 1-mL syringe is made into the lower left quadrant of the abdomen and observed for toxic symptoms.

5.3.4.5   Acute Eye Irritation Toxicity

According to Environmental Protection Agency (EPA) [57] and Organization for Economic Co-operation and Development (OECD) [58], the animals subjected to acute eye irritation test are pretreated with a systemic analgesic and induction of appropriate topical anesthesia. The substance to be tested is applied in a single dose to one of the eyes of the experimental animal; the untreated eye serves as the control. The degree of eye irritation/corrosion is evaluated by scoring lesions of conjunctiva, cornea, and iris, at specific intervals. Other effects in the eye and adverse systemic effects also are described to provide a complete evaluation of the effects. The duration of the study should be sufficient to evaluate the reversibility or irreversibility of the effects. Animals showing signs of severe distress and/or pain at any stage of the test or lesions (moribund and severely suffering animals) should be humanely killed, and the substance assessed accordingly. This method is not common in Africa, as most of the available literature on this assay indicates laboratories in America and Europe.

5.4   The Significance of LD_{50} in Acute Toxicity Screenings

In toxicology, the median lethal dose, LD_{50} (abbreviation for "lethal dose, 50%"), LC_{50} (lethal concentration, 50%), or LCt_{50} (lethal concentration and time) of a toxin, radiation, or pathogen is the dose required to kill half the members of a tested population after a specified test duration.

The LD_{50} value is expressed as the weight of chemical administered per kilogram body weight of the animal and it states the test animal used and route of exposure or administration; for example, LD_{50} (oral, rat)—5 mg/kg, LD_{50} (skin, rabbit)—5 g/kg.

If the lethal effects from breathing a compound are to be tested, the chemical (usually a gas or vapor) is first mixed in a known concentration in a special air chamber where the test animals will be placed. This concentration is usually quoted as parts per million (ppm) or milligrams per cubic meter (mg/m^3). In these experiments, the concentration that kills 50% of the animals is called an LC_{50} (lethal concentration 50) rather than an LD_{50}.

The LD_{50} is important for the prediction of human lethal dose and for the prediction of the symptomatology of poisoning after acute overdosing in humans. The LD_{50} value is a base from which other doses could be designed in subacute and chronic toxicity experiments.

Based on the value of LD_{50}, the degree of toxicity of substances can be classified: LD_{50} of 1 mg/kg or less (extremely toxic); 1−50 mg/kg (highly toxic); 50−500 mg/kg (moderately toxic); 500−5000 mg/kg (slightly toxic); 5000−15,000 mg/kg (practically nontoxic); more than 15,000 mg/kg (relatively harmless).

5.4.1   History of LD_{50} Determination

The concept of LD_{50} was first introduced in 1927 for establishing the toxic potency of biologically active compounds. It was desirable to have a test that could determine whether a chemical substance was very toxic, toxic, less toxic, or whether the toxic effects were significant. At that time, a toxic substance was one which is injurious after a specific mode of intake. Because the term "injuries" is highly subjective, an objective criterion was selected, namely death. Thus the means of determining the likelihood of death as exactly as possible was sought. Finally, the LD_{50} was recognized and justified of being the best parameter by Trevan [59]. Since the end of the 1970s, the LD_{50} test has been widely criticized for both scientific and animal welfare reasons, and the test procedure has been modified in various ways to reduce the number of animals required, and to reduce the suffering caused to any animal used. This modification to the classical LD_{50} test includes the fixed-dose procedure, the acute toxic class method, and the up-and-down procedure as described above.

5.4.2   Significance of LD_{50} in Toxicity Studies

The LD_{50} test was developed in 1927 for the biological standardization of dangerous drugs. It was incorporated into the routine toxicological protocol of other

Table 5.1 Classification of Toxicity of Substances Based on LD_{50} Dose Ranges

LD_{50} (mg/kg)	Category [60]	LD_{50} (mg/kg)	Category [61]
<5	Super-toxic	1 or less	Extremely toxic
5–50	Extremely toxic	1–50	Highly toxic
50–500	Very toxic	50–500	Moderately toxic
500–5000	Moderately toxic	500–5000	Slightly toxic
5000–15,000	Slightly toxic	5000–15,000	Practically nontoxic
>15,000	Practically nontoxic	More than 15,000	Relatively harmless

classes of chemical compounds and is now part of practically all governmental guidelines that regulate toxicological testing of chemicals. For scientific, economic, and ethical reasons it is necessary to periodically reassess all toxicological test procedures, including the LD_{50} test. Although the numerical value of the LD_{50} is influenced by many factors, such as animal species and strain, age and sex, diet, food deprivation prior to dosing, temperature, caging, season, experimental procedures, etc., its variability can be reduced but never fully eliminated. The LD_{50} is important for the prediction of human lethal dose and for the prediction of the symptomatology of poisoning after acute overdosing in humans [43]. The LD_{50} value is a base from which other doses could be designed in subacute and chronic toxicity experiments. The use of the LD_{50} parameter toxicity studies is generalized in Africa, where there is still a lot of research on plant extracts. Whether the test sample is a pure product or an extract, the classification/category and cutoff points are based on the corresponding range where the LD_{50} of the substance falls as indicated in Table 5.1.

Though the LD_{50} is a useful tool in classifying substances based on their degree of toxicity, it has some limitations. It cannot be considered a biological constant due to its variability. It provides less information except physiological functions; biochemical and histopathological examinations are incorporated to the assay. The LD_{50} is poorly suited in proving information on special risks for the human newborn and infant. For the appraisal of pharmacokinetic behavior and bioavailability, the LD_{50} test gives only semiquantitative, and often ambiguous results [43].

5.4.3 Toxic Symptoms in Acute and Subacute Toxicity Studies

Generally, animals under acute toxicity study are observed daily for a total of 14 days. Observations (irrespective of the mode of drug administration) include changes in skin and fur, eyes and mucous membranes, respiratory, circulatory, autonomic and central nervous systems, and somato-motor activity and behavior pattern. Attention should be given to tremors, convulsions, salivation, diarrhea, lethargy, writhing, edema, opisthotonos, exophthalmos, corneal opacity (visual acuity), eye swelling or reddening, sleep, and coma. The weight of each animal should be recorded before and after the test substance is administered.

All animals should be subjected to gross necropsy and pathological changes should be recorded [42,62−65]. Despite the multiple symptoms available for acute toxicity studies, most researchers communicate only a few of them at end of the assay. For a laboratory to permit the assessment of all these symptoms and more, then it should be well equipped. This is an ideal condition that almost no laboratory, especially in the developing countries, can claim. Some of these symptoms are correlated; the appearance of a symptom would request the verification of other symptoms that can arise from the direct or indirect effect of the former.

- A variation in body weight animal which is directly linked to appetite and the amount of food intake [66] would probably provoke a change the level of structural/ storage proteins/lipids/carbohydrates.
- The presence of edema, which is an indirect cause of decrease in plasma proteins [63].
- Respiratory depression, congestion, and edema of the trachea are indicative of wide circulation of toxic molecules in the respiratory system [67].
- Gross necropsy of the liver (damage)—variation in serum biochemistry—key enzymes [68−71]; total serum protein [48]; other organs such as the spleen/ bone marrow affect the level of blood cells [72].
- Gross necropsy of the kidney (damage)—variation in water intake [73].
- Gross necropsy of the lungs (damage)—variation in ventilation rate [74].
- Central nervous system (CNS)—sedative [75] tremors, convulsions, salivation, auditory defects, motor activity, sight defects.
- Diarrhea—lethargy [76].
- Cardiac activity—ventilation, motor activity, specific serum enzymes [77,78].

5.5 Limitations of the Study of the Acute and Subacute Toxicities of Medicinal Plants

In the past, a compound often was considered harmless if it was without immediate adverse health effects when administered in a large single dose. It has also been recognized that a short-term low-dose exposure to some types of toxicants (e.g., potent genotoxic agents) may be sufficient to induce delayed adverse effects such as malignant tumors [79−81]. One obvious example of a delayed effect from an acute exposure is lung cancer due to an acute inhalation of plutonium-239 (in the form of a Pu^{4+} salt). A less obvious, but still striking, example is the delayed neurotoxicity that was observed in humans who were exposed to the organophosphorus ester triorthocresylphosphate (TOCP) [79,80]. Now it is known that some toxicants accumulate in the body and that the "tissue doses" will eventually become critically high if the exposure to such agents continues for a sufficiently long time, even at rather low doses. Substances that accumulate in certain organs may cause slow progressive necrosis and eventual organ malfunction, which might be life threatening [79,80]. Important systemic parameters that relate to organ malfunction are missing after a toxicity assay lasting for some days like the subacute toxicity.

5.6   African Plants Screened for Their Acute and Subacute Toxicities

Several African medicinal plant were screened for their acute and subacute toxicity effects in different experimental models. The most prominent data are depicted in Table 5.2.

These were found to be either relatively nontoxic: (*Acanthus montanus, Albizia chevalieri, Anacardium occidentale, Annona senegalensis, Aspilia africana, Azadirachta indica, Berlina grandiflora, Bersama engleriana, Canthium mannii, Carissa edulis, Cassia occidentalis, Centaurium erythraea, Corrigiola telephiifolia, Coula edulis, Crassocephalum bauchiense, Cylicodiscus gabunensis, Drypetes gossweileri, Ficus exasperata, Glinus lotoides, Guiera senegalensis, Herniaria glabra, Ludwiguia abyssinica, Ocimum suave, Persea americana, Phyllanthus muellerianus, Phyllanthus niruri, Piptadeniastum africana, Polygala fruticosa, Sanseviera liberica, Sida rhombifolia, Spathodea campanulata, Syzigium aromaticum, Tabernaemontana crassa, Tridesmostemon omphalocarpoides, Turraeanthus africanus, Zanthoxylum xanthoxyloides*), fairly toxic (*Croton membranaceus, Dichapetalum barteri, Leonotis leonurus, Pteleopsis hylodendron, Pterocarpus soyauxii, Sacoglottis gabonensis, Vernonia guineensis*); or potentially toxic (*Ajuga iva, Butyrospermum paradoxum, Euphorbia kamerunica, Lawsonia inermis, Murraya koenigii, Musanga cecropioides*); and very toxic (*Ocimum gratissimum, Ozoroa insignis, Stachytarpheta indica*).

From the results compiled in Table 5.2, it appears that a limited number of medicinal plants have been screened for their toxicity profiles as compared to the high numbers of botanicals used in ATM or to the numbers of pharmacological data reported from phytochemicals from Africa. However, this chapter indisputably shows that efforts are being made in the toxicological survey of the medicinal plants throughout the continents.

5.7   Conclusion

Acute and subacute toxicity screenings of medicinal plants are the fastest ways to evaluate the toxicological profiles of medicinal plants. It helps to have a quick idea on the harmful or the safety potency of drugs. The acute and subacute toxicities of many African plants were screened in the past three decades. In this chapter, we reviewed more than 50 plants used in ATM for several purposes. The criteria of classification their toxicities are also summarized. It appears that, though most of the plants in ATM are safe, numbers of them are potentially toxic and caution should be taken when they are used for therapeutic purpose. Some of these plants include *Ocimum gratissimum, Pteleopsis hylodendron, Annona senegalensis, Syzigium aromaticum, Murraya koenigii, Sacoglottis gabonensis,* and *Spathodea campanulata*. However, in regard to the high numbers of plants used in ATM, this chapter also highlights the shortcomings of the toxicological studies. Efforts should be made throughout the continent to fill the gap and to place ATM in the standard of other world well-renowned traditional medicine.

Table 5.2 African Medicinal Plants Screened for Their Acute Toxicity Profiles

Plant (Family)	Traditional Use	Area of Plant Collection	Plant Constituents	LD$_{50}$ (g/kg Body Weight) and Toxic Syndrome (Biological System)
Acanthus montanus (Acanthaceae)	Treatment of urogenital infections, urethral pain, endometritis, urinary disease, cystitis, leukorrhea [82], bathing to relieve aches and pains [83]; epilepsy, dysmenorrhea, miscarriages, and false labor [84]	Cameroon; Nigeria	Saponins and the gammaceranes acanthusol and its 3-O-β-D-glucopyranoside [82]	I.P and Oral LD$_{50}$ > 5000 mg/kg [82].
Ajuga iva (Labiatea)	A panacea (cure-all) [85], and specifically for gastrointestinal disorders [86], hypertension and diabetes [87].	Morocco	Antileukemic sterol glycosides [88], hypoglycemic ecdysteroids [89], antibacterial and insect antifeedant neo-clerodane diterpenoids [90–93], insect antifeedant diglyceride [94], vasoconstrictor 8-O-acetyharpagide [95].	I.P LD$_{50}$ of 3.6 g/kg [96]; Symptoms: asthenia, anorexia, salivation, diarrhea, and syncope
Albizia chevalieri (Fabaceae)	Lumbago [97], diabetes mellitus [98].	Congo-Brazzaville Nigeria	4-Methoxypyridoxine [21]	Oral LD$_{50}$ > 3 g/kg [98] Hypersensitivity, convulsions
Anacardium occidentale (Anacardiaceae)	Treatment of diabetes and hypertension [99].	Cameroon	Unknown	Oral LD$_{50}$ of 16 g/kg i.p. LD$_{50}$ of 0.250 g/kg [99] Symptoms: asthenia, anorexia, diarrhea, and syncope

(Continued)

Table 5.2 (Continued)

Plant (Family)	Traditional Use	Area of Plant Collection	Plant Constituents	LD$_{50}$ (g/kg Body Weight) and Toxic Syndrome (Biological System)
Annona senegalensis (Annonaceae)	Chest pain, coughs, anemia, and urinary tract infection [39,100,101]	Nigeria	Unknown	Oral LD$_{50}$ of 1296 g/kg in rats
Aspilia africana, (Asteraceae)	Stop bleeding, remove corneal opacities, induce delivery and in the treatment of anemia and various stomachs complaints [102,103].	Cameroon	Saponines and tannins; flavonoids [104], sesquiterpenes and monoterpenes, precocene [105]	Oral LD$_{50}$ of 6.1 g/kg [106].
Azadirachta indica (Meliaceae)	Leprosy, intestinal worm, skin ulcers, cough, asthma, diabetes, and inflammatory diseases [107–110].	Nigeria	Tannins, phenolic compounds (salicylic acid and allocatechin) [110,111].	[a]I.M LD$_{50}$ of 6.2 mL/kg I.P LD$_{50}$ of 870 mg/kg [67] Symptoms: anorexia, dehydration, malaise, respiratory depression, coma, and death [111].
Berlina grandiflora (Leguminosae)	Treatment of gastrointestinal disorders [112].	Nigeria	Flavonoids, steroids, triterpenoids, and fatty acids. 1-sitoseterol, betulinic acid, apigenin, and mannitol [113].	LD$_{50}$ >2000 mg/kg [114].
Bersama engleriana (Melianthaceae)	Treatment of cancer, spasms, infectious diseases, male infertility, and diabetes [115,116].	Cameroon	Triterpenes and saponins [117]	Oral LD$_{50}$ > 5 g/kg [118]

Plant (Family)	Uses	Country	Constituents	Toxicity
Butyrospermum paradoxum (Sapotaceae)	Treatment of both animal and human trypanosomosis [74].	Nigeria	Alkaloids, steroids, tannins, saponins, and flavones [119,120].	I.P LD$_{50}$ of 240 mg/kg [120] and 820 mg/kg [74]. Symptoms: anorexia, dehydration, starry hair coat, depression, general malaise, prostration, difficulty in respiration, coma, and death [74].
Canthium mannii (Rubiaceae)	Intestinal worms [121].	Cameroon	Unknown	Oral LD$_{50}$ > 16 g/kg [121]
Carissa edulis (Apocynaceae)	Treatment of a variety of diseases including sickle cell anemia, toothache, ulcer, worm infestation, epilepsy, pain, and inflammation [122].	Nigeria	Saponins, flavonoids, tannins, anthraquinones, and cardiac glycosides [123].	Oral LD$_{50}$ > 5.0 g/kg [123]
Cassia occidentalis (Leguminosae)	Treat bacterial and fungal infections, liver disorders (jaundice, hepatitis, cirrhosis, detoxification, injury/failure, bile stimulant; intestinal worms, internal parasites, skin parasites [124].	Nigeria	Achrosine, emodin, anthraquinones, anthrones, apigenin, sitosterols, tannins and xanthones [124].	Oral LD$_{50}$ > 5 g/kg [124,125].
Centaurium erythraea (Gentianaceae)	Treatment of asthma, eczema, jaundice, intestinal parasitic infestation, rheumatism, wounds and sores, reduce blood pressure, gastrointestinal smooth muscle spasm, edema and digestive disorders (loss of appetite, stomach discomfort, bloating, indigestion), liver and gall bladder stimulant [126–130].	Morocco	Centauroside, centapicrin, flavonoids, gentiopicrin, gentiopicroside, isocoumarin, phenolic acids, swertiamarin, triterpenes, wertiamarine, and xanthones [131–135].	I.P LD$_{50}$ of 12.13 g/kg. mice [135]. Symptoms: anorexia, syncope, piloerection, asthenia, diarrhea, convulsions

(Continued)

Table 5.2 (Continued)

Plant (Family)	Traditional Use	Area of Plant Collection	Plant Constituents	LD$_{50}$ (g/kg Body Weight) and Toxic Syndrome (Biological System)
Corrigiola telephiifolia (Caryophyllaceae)	Treatment of flu, dermatological diseases, inflammation, ulcer, cough, and jaundice; also used as anesthetic and a diuretic; and also given to parturient women [136,137].	Morocco	Saponins and terpenes [136].	Oral LD$_{50}$ >14,000 mg/kg
Coula edulis (Olacaceae)	Stomachache and skin diseases, tonifiant, and kidney problems [102].	*Cameroon*	3-O-β-D-glucopyranoside of sitosterol, β-sitosterol, stigmasterol, *n*-hexadecanoid acid [138].	Oral LD$_{50}$ of 16.8 g/kg Symptoms: diarrhea, reduction in sensitivity [138].
Crassocephalum bauchiense (Asteraceae)	Treat epilepsy, pain, arthritis, intestinal pain, and colics [139], cerebral deficit, behavioral disturbances, inflammatory disorders and neuropathic pain [140]	Cameroon	Unknown	Oral LD$_{50}$ > 5.120 g/kg [140]
Croton membranaceus (Euphorbiaceae)	stem bark essential oil treats cough, fever, flatulence, diarrhea, and nausea [78].	Ghana	Furano-clerodane diterpenoid, crotomembranafuran, glutarimide alkaloid, (julocrotine), sitosterol, sitosterol-3-d-glucoside, labdane diterpenoid, gomojoside H and dithreitol. scopoletin and julocrotinecalcium oxalate crystals [78].	Oral LD$_{50}$ > 3 g/kg [78].

Plant (family)	Uses	Constituents	Country	Toxicity
Cylicodiscus gabunensis (Mimosaceae)	Against diarrhea and gastrointestinal disorder, rheumatism, filariasis, and against headache [103]	Saponins [141].	Cameroon	Oral LD_{50} of 11 g/kg [142] Symptoms: prostration, motionlessness, slow response to external stimuli and rapid breathing [142]
Dichapetalum barteri (Dichapetalaceae)	Unknown	Monofluoroacetate [143]	Nigeria	LD_{100} of 2 g/kg mice and 0.1 g/kg in rabbits Symptoms: restlessness, convulsion and death [143]
Drypetes gossweileri (Euphorbiaceae)	Treatment of wounds and toothache [144].	Steroids, triterpenoids, alkaloids, saponins [145].	Cameroon	Oral LD_{50} > 12 g/kg [145]
Erythrina senegalensis (Fabaceae)	Hepatitis and jaundice [32]. Gastrointestinal disorders, amenorrhea, dysmenorrhea, sterility, onchocerchosis, body pain, abdominal pain, and headache [33].	2, 3 dihydroauriculatin, erybraedin A, 6—8 diprenylgenistein [34]	Cameroon	LD_{50} unknown but possible blood cell cancer and reduction in blood ions level [34].
Euphorbia kamerunica (Euphorbiaceae)	Some herbalists use the latex either medicinally or as a poison [103].	Lectins and esters of certain diterpene, alcohols, phorbol derivatives, tigliane, daphnane, and ingenane diterpene ester toxins, triterpene saponins, rotenoids, sesquiterpenes, lignans, proanthocyanidin polymers, and polyacetylenic compounds [146].	Nigeria	[b]LC_{50} of 0.023 g/L (fish) Symptoms: skin inflammation with reddening edematous swellings. poisonous lattices in eyes cause inflammation of cornea and conjunctiva, leading to blindness. severe gastroenteritis, vomiting, and colicky diarrhea [147].
Euphorbia mauritanica (Euphorbiaceae)	Infections and arrow poison	Terpenoids	South Africa	Tremors (bloat and diarrhea) [21].

(Continued)

Table 5.2 (Continued)

Plant (Family)	Traditional Use	Area of Plant Collection	Plant Constituents	LD$_{50}$ (g/kg Body Weight) and Toxic Syndrome (Biological System)
Ficus exasperate (Moraceae)	Against cardiac arrhythmias, respiratory tract infections such as asthma, bronchitis, tuberculosis, and emphysema; chest pain [148,149], hastening childbirth; eye troubles and stomach pains; arrest bleeding [149]	Ivory Coast; Nigeria	Pheophytin/pheophorbide derivatives, flavonoids, fatty acids and glycerol derivatives [150].	I.P LD$_{50}$ of 0.54 g/kg [65].
Glinus lotoides (Molluginaceae)	Anthelmintic for the prevalent tapeworm infestation [151].	Ethiopia	Unknown	Oral LD$_{50}$ > 5000 mg/kg Symptoms: distended abdomen, miosis, soft stool, urination, abdominal colic and severe piloerection in both sexes [151]
Guiera senegalensis (Combretaceae)	Treatment of ulcer, leprosy and malaria, increase lactation, dried leaves are powdered and mixed with food and taken by women post-partum [152]. Remedy for diarrhea, dysentery and abdominal discomfort [153].	Nigeria	1,3-digalloylquinic acid, quinic acid gallates, 5-Methyldihydroflavasperone [154].	Oral LD$_{50}$ >5000 mg/kg [154].

Plant (Family)	Traditional uses	Country	Chemical constituents	Toxicity
Herniaria glabra (Caryophyllaceae)	Treatment of dropsy, catarrh of the bladder, cystitis, kidney stones, gout, hernias, jaundice, nerve inflammation, respiratory disorders, and removing excess of mucus in the stomach and to increase the flow of urine [155–157a].	Morocco	Unknown	Oral LD$_{50}$ of 8.50 g/kg Symptoms: asthenia, piloerection, ataxia, anorexia, urination, diarrhea, syncope [157b]
Khaya ivorensis (Meliaceae)	Management of malaria, emmenagogue, emetic, hypotensive, and antipyretic agent [158].	Nigeria	Tetranortriterpenoids, khivorin (limonoid) [159]; methyl-6-acetoxy angolensate, swietenolide, 3-deacetyl khivorin, 7-deacetyl khivorin, 6-hydroxymethyl angolensate, methyl ivorensate, mexicanolide, fissinolide, methyl ivorensate [160].	Unknown
Lawsonia inermis (Lythraceae)	Antihelminthic and astringent and in the control of perspiration and partly menstrual cycles [81].	Nigeria	Lawsoniaside and laliodise lawsone and gallic acid reducing sugar, tannins, flavanoids, saponins, alkaloids [81]	Oral LD$_{50}$ > 1600 mg/kg Symptoms: paralysis, total collapse of the body, weakness, laziness and loss of appetite [81]
Leonotis leonurus (Lamiaceae)	Relief of epilepsy, coughs, influenza, bronchitis, hypertension, and headaches [161]; treatment of diabetes, snake bites, and intestinal worms [162]; boils, eczema, skin disease, itching, and muscular cramps [18,27].	South Africa	Diterpenoid labdane lactones, tannins, quinones, saponins, alkaloids, and triterpene steroids [162].	Oral LD$_{50}$ > 3200 mg/kg Symptoms: Decreased respiratory rate, decreased motor activity loss of righting reflex Ataxia, respiratory failure, convulsions, paralysis of skeletal muscle, and coma [163].

(Continued)

Table 5.2 (Continued)

Plant (Family)	Traditional Use	Area of Plant Collection	Plant Constituents	LD$_{50}$ (g/kg Body Weight) and Toxic Syndrome (Biological System)
Ludwiguia Abyssinica (Onagraceae)	Treats syphilis and poulticing pimples, diarrhea, dysentery, flatulence, leukorrhea and as vermifuge [164]	Cameroon	Rich in oil, flavonoids, phenols tannin, alkaloids, saponins	Oral LD$_{50}$ of 21 g/kg [164]
Murraya koenigii (Rutaceae)	Tonic, febrifuge, stomachic, antivomiting, dysentery and diarrhea; kin eruptions and bites of venomous animals, stimulant, carminative, hypotensive, hypoglycemic, antiperiodic, and antifungal [165–167].	Nigeria	Carbazole alkaloids, carbazoles [168,169]; mahanimbilylacetate, Girinimbilylacetate, bicyclomahanimbiline [66].	Oral LD$_{50}$ of 316.23 mg/kg [66].
Musanga cecropioides (Cecropiaceae)	Treat hypertension, lumbago, rheumatism, leprosy, chest infections, trypanosomosis [100,170].	Nigeria	Saponins, flavonoids, alkaloids, tannins, phlobatannins, cardiac glycosides, reducing sugars and anthraquinones [170].	Oral LD$_{50}$ > 3000 mg/kg oral (rats) symptoms: irritation, restlessness, tachypnea, anorexia, bilateral narrowing of the eyelids and abnormal posture [170].
Ocimum gratissimum (Lamiaceae)	Leaf essential oil against seminal weakness, as aromatic baths, tonic expectorants, and antispasmodic [115].	Nigeria	Essential oil contains mostly thymol (47.0%), *p*-cymene (16.2%) and α-terpinene (6.2%) [171].	I.P LD$_{50}$ of 0.27 g/kg Oral LD$_{50}$ of 1.41 g/kg in mice [115]. Symptoms: weakness and fatigability, sedation and sleep

Plant (Family)	Uses	Country	Chemical constituents	Toxicity
Ocimum suave (Lamiaceae)	Treat ulcers and as an anticathartic in East Africa [8], menstrual problems, stomachache and broncho-pneumonic infections [172].	Cameroon	Phenols [8] triterpenes [46].	$LD_{50} > 5.00$ g/kg in mice [46].
Ozoroa insignis (Anacardiaceae)	Mixed with other plants to treat of peptic ulcers [173].	Tanzania	Tannins, saponins, steroids, cardiac glycosides, flavonoids and terpenoids	LC_{50} of 10.63 µg/mL Symptoms: hyperventilation
Persea americana (Lauraceae)	Chronic hypertension, wound healing [174] analgesic and anti-inflammatory [175], anticonvulsant [176].	Nigeria	Unknown	Oral LD_{50} >10 g/kg (rats) [177].
Phyllanthus muellerianus (Euphorbiaceae)	Various infectious ailments [103,178].	Cameroon	Triterpens, Sterols, coumarins, glycosides, Anthraquinons, cardiac glycosides [178].	Oral LD_{50} > 5 g/kg [178]
Phyllanthus niruri (Euphorbiaceae)	Treat mild malaria, problems of stomach, genitourinary system, liver, kidney (elimination of renal stones), and spleen [179].	Ghana	Lectins and esters of certain diterpene alcohols; phorbol derivatives, tigliane, daphnane, and ingenane diterpene ester toxins [180–183].	Oral LD_{50} > 5000 mg/kg b.w. [78]. Symptoms: skin inflammation, with reddening and formation of edematous swellings. Eye contact causes inflammation of the cornea and conjunctiva. gastroenteritis, vomiting, and colicky [147].
Piptadeniastum africana (Mimosaceae)	Various infectious ailments [103,178]	Cameroon	Alkaloids, triterpens, sterols, coumarins, glycosides [178]	Oral LD_{50} > 5 g/kg [178]
Polygala fruticosa (Polygalaceae)	Chronic ulcer, poor circulation, intestinal sores, gonorrhea and the snuff to improve sinusitis; Facilitate childbirth, blood purification and as a cure for tuberculosis [8,24,184]	South Africa	Unknown	Oral LD_{50} of 10.8 g/kg body [185]

(Continued)

Table 5.2 (Continued)

Plant (Family)	Traditional Use	Area of Plant Collection	Plant Constituents	LD$_{50}$ (g/kg Body Weight) and Toxic Syndrome (Biological System)
Preleopsis hylodendron (Combretaceae)	Treatment of measles, chicken pox, sexually transmitted diseases, female sterility, liver and kidney disorders as well as dropsy [28,29].	Cameroon	Unknown	LD$_{50}$ of 3–3.8 g/kg (stem bark) in mice; possible alterations of liver and kidney functions [30].
Pterocarpus soyauxii (Papilionaceae)	Treat hypertension, diabetes, intestinal parasitizes, renal and cutaneous diseases, used as diuretic and as food in Nigeria [186–189].	Cameroon	Biflavonoids, isoflavone quinone (claussequinone), isoflavones, tannins, triterpenes, xanthones [190–192].	LD$_{50}$ > 10.75 g/kg for the mouse Symptoms: hair bristling or piloerection, body writhing, hind limbs and body stretching, diarrhea, gasping and breathing difficulty [47].
Sacoglottis gabonensis (Humiriaceae)	Inebriating effect of raffia/palm wine, control of gastrointestinal helminthiasis [193].	Nigeria	Unknown	I.P LD$_{50}$—rats between 1600 and 3200 mg/kg, Symptoms: dehydration, congestion, edema of the lungs, bronchi and bronchioles [193].
Sanseviera liberica (Agavaceae)	Fever, headache and cold, as well as analgesic, antibiotic, and anti-inflammatory [8]	Nigeria	Unknown	^{c}i.p LD$_{50}$ of 668.3 g/kg Symptoms: weakens the defense system [194]
Sida rhombifolia (Malvaceae)	Antiseptic, wound healing, treatment of diarrhea, cough, ulcer, abscess, and furuncle [195]	Cameroon	Polyphenols, alkaloids, and steroids [196]	Oral LD$_{50}$ > 12 g/kg [196]

Plant (Family)	Uses	Country	Phytochemicals	Toxicity
Spathodea campanulata (Bignoniaseae)	Used for the control of epilepsy and convulsion [197].	Nigeria	Tannins and saponins [198].	Oral LD_{50} of 4466.84 mg/kg Symptom: anorexia, hyperventilation and diarrhea
Stachytarpheta indica	Unknown	Nigeria	Flavoniods, steroids, tannin, cardiac glycoside, carbohydrate, alkaloids, saponins [199]	I.P LD_{50} < 0.020 g/kg [199]
Syzigium aromaticum (Myrtaceae)	Relieve pain and promote healing [200].	Nigeria	Flavonoids [201,202].	i.p LD_{50} of 263.0 mg/kg and oral LD_{50} of 2500.0 mg/kg; Symptom of writhing [203] and hypoxia.
Tabernaemontana crassa (Apocynaceae)	Healing dressing to sores, furuncles and to anthrax pustules, filaria, ringworm and other fungal troubles, wound infections, coryza, headaches, abscesses, boils and carbuncles [100,153], contusions and sprained backs [204]	Cameroon	Indole alkaloids: pericyclivine, perivine, vobasine, tabersonine, conopharyngine, oxocoronaridine, 3-oxocoronaridine hydroxyindolenine, (−)-heyneanine, (−)-3-oxoheyneanine, (−)-ibogamine, isovoacangine, voacristine, crassanine, conoduramine [205], indole alkaloids: ibogaine, isovoacangine, conopharyngine, conodurine, conoduramine, coronaridine, and in the seeds, voacamidine, coronaridine-hydroxyindolenine, tabersonine andcoronaridine have been reported [206–208]	Oral LD_{50} of 6.75 g/kg convulsion and respiratory distress prostation, stretching and sluggishness, rapid breathing all punctuated by a diarrhea [209].

(Continued)

Table 5.2 (Continued)

Plant (Family)	Traditional Use	Area of Plant Collection	Plant Constituents	LD$_{50}$ (g/kg Body Weight) and Toxic Syndrome (Biological System)
Tridesmostemon omphalocarpoides (Sapotaceae)	Treatment of gastroenteritis and skin lesions [210].	Cameroon	alkaloids, phenols, polyphenols, saponins, tannins, triterpenes, anthraquinones, and steroids [210]	Oral LD$_{50}$ > 20 g/kg [210]
Turraeanthus africanus (Meliaceae)	Treat infertility, sexual weakness, fibroma and cancer, asthma, stomachache, intestinal worms, cardiovascular and inflammatory diseases [211].	Cameroon	Turraeanthin C, Sesamin and stigmasterol [212].	LD$_{50}$ of 7.2 g kg (I.P.-mice) [213].
Vernonia guineensis (Asteraceae)	combat stress and as a stimulant, antibiotic, anthelmintic, aphrodisiac, antidote against poison, and treatment for malaria and jaundice [102,214].	Cameroon	Vernoguinoside A, stigmasterol 3-O-β-D-glucoside and sitosterol 3-O-β-D-glucoside [215]	Oral LD$_{50}$ > 4 g/kg [216]
Zanthoxylum xanthoxyloides (Rutaceae)	Stomachache, toothache, coughs, urinary and venereal diseases, leprous ulcerations, rheumatism, lumbago [217,218].	Cameroon, Nigeria	Benzo-phenanthridines (fagaronine, dihydroavicine 6. chelerythrine); aporphines (berberine tembetarine, magnoflorine, *n*-methyl-corydine) [219].	Oral LD$_{50}$ of 5.8 g/kg (stem) [220]

[a]IM = intramuscular.
[b]LC = lethal concentration.
[c]I.P = intraperitoneal administration.

References

[1] OECD. Organization for economic co-operation and development: OECD guideline for testing of chemicals. Acute inhalation toxicity; 1981 (May 12).

[2] Phillip LW, Robert CJ, Stephen MR. Principles of toxicology: environmental and industrial applications. 2nd ed. New York, NY: John Wiley & Sons; 2000.

[3] Eaton DL, Klaassen CD. Principles of toxicology. In: Klassen CD, editor. Casarett and Doull's toxicology: the basic science of poisons. 6th ed. McGraw-Hill Medical Publishing Division; 2001. p. 11−34.

[4] Hussain SS, Gajanan AN. Hepatoprotective effect of leaves of *Erythroxylum monogynum* Roxb. On paracetamol induced toxicity. Asian Pac J Trop Biomed 2013;3 (11):877−81.

[5] Gadanya AM, Sule MS, Atiku MK. Acute toxicity study of "gadagi" tea on rats. Bayero J Pure Appl Sci 2011;4(2):147−9.

[6] Seidle T, Pilar P, Bulgheroni A. Examining the regulatory value of multi-route mammalian acute systemic toxicity studies. Altex 2011;28:2/11.

[7] Cotovio J, Marie-Hélène G, Damien L, Christelle B, Nicole F, Sophie LJ, et al. The use of the reconstructed Human Corneal Model (HCE) to assess *in vitro* eye irritancy of chemicals. Proc. 6th world congress on alternatives & animal use in the life sciences, 14. AATEX; 2000 [Special issue, p. 343−50].

[8] Watt JM, Breijer-Brandwyk MG. Medicinal and poisonous plants of Southern and Eastern Africa. 2nd ed. Edinburgh: E. & S. Livingstone; 1962.

[9] Van Wyk BE, Van Heerden FR, Van Oudtshoorn B. Poisonous plants of South Africa. Arcadia, Pretoria: Briza Publications; 2002. p. 8, 30, 108, 144, 180, 194, 196.

[10] Gaillard Y, Paquin G. Poisoning by plant material: review of human cases and analytical determination of main toxins by high-performance liquid chromatography-(tandem) mass spectrometry. J Chromatogr B Biomed Sci App 1999;733:181−229.

[11] Joubert PH, Mathibe L. Acute poisoning in developing countries. Adverse Drug React Acute Poisoning Rev 1989;8:165−78.

[12] Haxaire C. Phytothérapie et Médecine Familiale chez les Gbaya-Kara (République Centrafricaine). Thèse de Doctorat, Université de Paris, Fac. Pharmacie; 1979. p. 320.

[13] Abel AA, Banjo AD. Honeybee floral resources in Southwestern Nigeria. J Biol Life Sci 2012;3(1):2157−6076.

[14] Ellenhorn MJ. Ellenhorn's medical toxicology: diagnosis and treatment of human poisoning. 2nd ed. Baltimore, MD: Williams and Wilkins; 1997. p. 33.

[15] Tor-Agbidye J, Palmer VS, Laserev MR, Craig AM, Blythe LL, Sabri MI, et al. Bioactivation of cyanide to cyanate in sulphur amino acid deficiency: relevance to neurological disease in humans subsisting on cassava. Toxicol Sci 1999;50:228−35.

[16] Marx J, Pretorius E, Espag WJ, Bester MJ. *Urginea sanguinea*: medical wonder or death in disguise? Environ Toxicol Pharmacol 2005;20:26−34.

[17] Steenkamp PA. Chemical analysis of medicinal and poisonous plants of forensic importance. Ph.D. thesis, Department of Chemistry, University of Johannesburg, South Africa; 2005.

[18] Hutchings A, Scott AH, Lewis G, Cunningham A. Zulu medicinal plants: an inventory. Pietermaritzburg: University of Natal Press; 1996.

[19] Laurens JB, Bekker LC, Steenkamp V, Stewart MJ. Gas chromatographic mass spectrometric confirmation of atractyloside in a patient poisoned with *Callilepis laureola*. J Chromatogr B 2001;765:127−33.

[20] Wainwright J, Schonland MM, Candy HA. Toxicity of *Callilepis laureola*. S Afr Med J 1977;52:313–5.

[21] Botha CJ, Penrith ML. Poisonous plants of veterinary and human importance in southern Africa. J Ethnopharmacol 2008;119:549–58.

[22] Steenkamp PA, Van Heerden FR, Van Wyk BE. Accidental fatal poisoning by *Nicotiana glauca*: identification of anabasine by high performance liquid chromatography/photodiode array/mass spectrometry. Forensic Sci Int 2002;127:208–17.

[23] Farrell M. Poisons and poisoners. An encyclopedia of homicidal poisonings. London: Robert Hale Publishers; 1992. p. 51–2.

[24] Van Wyk B, van Oudshoorn B, Gericke N. Medicinal plants of South Africa. 1st ed. Arcadia, Pretoria: Briza Publications; 1997. p. 196–7.

[25] Mulaudzi RB, Ndhlala AR, Kulkarni MG, Van Staden J. Pharmacological properties and protein binding capacity of phenolic extracts of some Venda medicinal plants used against cough and fever. J Ethnopharmacol 2012;143(1):185–93.

[26] Palmer E. A field guide to the trees of Southern Africa. 2nd ed. London/Johannesburg: Collins; 1981. p. 191.

[27] Van Wyk BE, Gericke N. People's plants. Arcadia, Pretoria: Briza Publications; 2000.

[28] Ngounou NF, Atta-Ur-Rahman CM, Malik S, Zareen S, Ali R, Lontsi D. New saponins from *Pteleopsis hylodendron*. Phytochemistry 1999;52:917–21.

[29] Magnifouet HN, Ngono RN, Kuiate JR, Koanga LM, Tamokou JD, Ndifor F, et al. Acute and sub-acute toxicity of the methanolic extract of *Pteleopsis hylodendron* stem bark. J Ethnopharmacol 2011;137:70–6.

[30] Motso CP. Recensement de quelques plantes camerounaises à activité antivirale. Mémoire de Maîtrise de Biochimie. Université de Douala; 2007. p. 25–7.

[31] Ngogang J, Bernard AN, Lucie FB, Jean LE, Nole T, Louis Z, et al. Evaluation of acute and sub acute toxicity of four medicinal plants extracts used in Cameroon. Toxicol Lett 2008;180:S185–6.

[32] Moundipa FP, Njayou FN, Yanditoum S, Sonke B, Mbiapo TF. Medicinal plants used in the Bamoun region of the Western province of Cameroon against jaundice and other liver disorders. Cameroon J Biochem Sci 2002;12:39–46.

[33] Togola A, Austarheim I, Theis A, Diallo D, Paulsen BS. Ethnopharmacological uses of *Erythrina senegalensis*: a comparison of three areas in Mali, and link between traditional knowledge and modern biological science. J Ethnobiol Ethnomed 2008;4:6.

[34] Atsamo AD, Nguelefack TB, Datté JY, Kamanyi A. Acute and subchronic oral toxicity assessment of the aqueous extract from the stem bark of *Erythrina senegalensis* DC (Fabaceae) in rodents. J Ethnopharmacol 2011;134:697–702.

[35] Steenkamp V, Stewart MJ, Zuckerman M. Clinical and analytical aspects of pyrrolizidine poisoning caused by South African traditional medicines. Ther Drug Monit 2000;22:302–6.

[36] Betz JM, Page SW. Perspectives on plant toxicology and public health. In: Garland T, Barr AC, editors. Toxic plants and other natural Toxicants. Wallingford: CAB International; 1998. p. 367–72.

[37] Igweii AG, Onabanjo AO. Chemotherapeutic effects of *Annona senegalensis* in *Trypanosoma brucei brucei*. Ann Trop Med Parasitol 1989;83:527–34.

[38] Graham JG, Quinn ML, Fabricant DS, Farnsworth NR. Plants used against cancer: an extension of the work of Jonathan Hartwell. J Ethnopharmacol 2000;73:343–77.

[39] Ajaiyeoba E, Mofolusho F, Omonike LO, Dora A. *In vivo* antimalarial and cytotoxic properties of *Annona senegalensis* extract. Afr J Trad Complement Altern Med 2006;3 (1):137–41.

[40] Adzu B, Abubakar MS, Izebe KS, Akumka DD, Gamaniel KS. Effect of *Annona sene-galensis* root bark extracts on *Naja nigricotlis* venom in rats. J Ethnopharmacol 2005;96(3):507−13.

[41] Theophine CO, Peter AA, Adaobi CE, Maureen OO, Collins AO, Frankline N, et al. Evaluation of the acute and sub acute toxicity of *Annona senegalensis* root bark extracts. Asian Pac J Trop Med 2012;277−82.

[42] OECD. Guidance document on the recognition assessment and use of clinical signs as humane endpoints for experimental animals used in safety evaluation. Environmental health and safety monograph series on testing and assessment No. 19; 2000.

[43] Zbinden G, Flury-Roversi M. Significance of the LD50-test for the toxicological evaluation of chemical substances. Arch Toxicol 1981;47:77−99.

[44] OECD. Guideline for testing of chemicals. Guideline 401 "Acute Oral Toxicity" Adopted; February 24, 1987.

[45] Fakeye TO, Awe SO, Odelola HA, Ola-Davies OE, Itiola OA, Obajuluwa T. Evaluation of valuation of toxicity profile of an alkaloidal fraction of the stem bark of *Picralima nitida* (Fam. Apocynaceae). J Herbal Pharmacother 2004;4(3):37−45.

[46] Tan PV, Mezui C, Enow-Orock G, Njikam N, Dimo T, Bitolog P. Teratogenic effects, acute and sub chronic toxicity of the leaf aqueous extract of *Ocimum suave* Wild (Lamiaceae) in rats. J Ethnopharmacol 2008;115:232−7.

[47] Tchamadeu MC, Dzeufiet PD, Nana P, Kouambou CC, Ngueguim FT, Allard J, et al. Acute and sub-chronic oral toxicity studies of an aqueous stem bark extract of *Pterocarpus soyauxii* Taub (Papilionaceae) in rodents. J Ethnopharmacol 2011;133:329−35.

[48] Teke GN, Kemadjou NE, Kuiate JR. Chemical composition, antimicrobial properties and toxicity evaluation of the essential oil of *Cupressus lusitanica* Mill. leaves from Cameroon. BMC Complement Altern Med 2013;13(130):1−9.

[49] OECD. Guidelines for the testing of chemicals. Section 4: health effects. Paris, <http://www.oecd.org/env/testguidelines>; 2009.

[50] United Nations-Economic Commission for Europe (UN/ECE). Globally harmonised system of classification and labelling of chemicals (GHS). UN, New York and Geneva; 2003.

[51] OECD. Guideline for testing of chemicals. Guideline 423: acute oral toxicity: acute toxic class method. Adopted: December 17, 2001.

[52] OECD. Guideline for testing of chemicals. Guideline 425: acute oral toxicity: up-and-down procedure. Adopted: December 17, 2001.

[53] OECD. Guideline for testing of chemicals. Draft proposal for a new guideline: 433. Acute inhalation toxicity: fixed concentration procedure. 2nd version, June 8, 2004.

[54] TSCA. Toxic Substances Control Act: acute dermal toxicity; Protection of Environment (40 CFR 799.9120); 2000.

[55] Poole A, Leslie GB. 1st ed. A practical approach to toxicological investigations, 2. Cambridge: Cambridge University Press; 1989. p. 30−117.

[56] Waynforth HB. Experimental and surgical technique in the rat. London: Academic Press; 1980. p. 17−68.

[57] Environmental Protection Agency (EPA). Prevention, pesticides and toxic substances (7101). Health effects test guidelines; OPPTS 870.2400. Acute Eye Irritation; 1998.

[58] OECD. Organisation for economic Co-operation and development for the testing of chemicals. Guideline 405: Acute Eye Irritation/Corrosion. Adopted: October 2, 2012.

[59] Trevan JW. The error of determination of toxicity. Proceedings of the Royal Society (London), Series B 1927;101:483−514.

[60] Pascoe D. Toxicology. London: Edward Arnold Ltd; 1983. p. 1−60.

[61] Loomis TA, Hayes AW. Loomis's essentials of toxicology. 4th ed. San Diego, CA: Academic Press; 1996. p. 208—45.

[62] TSCA. Toxic substances control Act test guideline is the office of prevention. Pesticides, and Toxic Substances (OPPTS) harmonized test guideline 870.1200; 1998.

[63] Chawla R. Serum total proteins and albumin—globulin ratio. In: Chawla R, editor. Practical clinical biochemistry (methods and interpretations). New Delhi: Jaypee Brothers Medical Publishers; 1999. p. 106—18.

[64] OECD. Guidelines for the testing of chemicals: proposal for a new guideline 223— Avian acute oral toxicity test; 2002.

[65] Bafor EE, Igbinuwen O. Acute toxicity studies of the leaf extract of *Ficus exasperata* on haematological parameters, body weight and body temperature. J Ethnopharmacol 2009;123:302—7.

[66] Adebajoa AC, Ayoolab OF, Iwalewac EO, Akindahunsib AA, Omisorec NA, Adewunmid CO, et al. Anti-trichomonal, biochemical and toxicological activities of methanolic extract and some carbazole alkaloids isolated from the leaves of *Murraya koenigii* growing in Nigeria. Phytomedicine 2006;13:246—54.

[67] Mbaya AW, Ibrahim UI, God OT, Ladi S. Toxicity and potential anti-trypanosomal activity of ethanolic extract of *Azadirachta indica* (Maliacae) stem bark: an *in vivo* and *in vitro* approach using Trypanosoma brucei. J Ethnophamarcol 2010;128:495—500.

[68] Schmidt E, Schmidt FW. Determination of serum GOT and GPT. Enzym Clin Biol 1963;3:1—3.

[69] Baggot JD. Disposition and fate of drugs in the body. In: Booth WH, McDonald LE, editors. John's veterinary pharmacology and therapeutics. Ludhiana/New Delhi: Kalyani Publishers; 1984. p. 30—70.

[70] Hayes AW. Guidelines for acute oral toxicity testings. Principles and methods of toxicity. 2nd ed. New York, NY: Raven Press Ltd.; 1989, Table 4; p. 185.

[71] Olagunju JA, Oyedapo OO, Onasanya OO, Osoba OO, Adebanjo OO, Eweje O, et al. Effects of Isosaline extracts of *Tetrapleura tetraptera* and *Olax subscopiodes* on certain biochemical parameters of albino rats. Pharmaceut Biol 2000;38:187—91.

[72] Mitruka BM, Rawnsley HM. Clinical biochemical and haematological reference values in normal experimental animals. New York, NY: Masson; 1977.

[73] Tietz NW. Fundamentals of clinical chemistry with clinical correlation. 1st ed. London: Bailliere Tindall; 1994. p. 234.

[74] Mbayaa AW, Chukwunyere ON, Patrick AO. Toxicity and anti-trypanosomal effects of ethanolic extract of *Butyrospermum paradoxum* (Sapotaceae) stem bark in rats infected with *Trypanosoma brucei* and *Trypanosoma congolense*. J Ethnopharmacol 2007;111:526—30.

[75] Guillemain J, Rousseau A, Delaveau P. Neurodepressive effect of the essential oil of *Lavandular angustifolia* mill. Ann Pharm Fr 1989;47:337—43.

[76] Orafidiya OO, Elujoba AA, Iwalewa EO, Okeke IN. Evaluation of anti-diarrheal properties of *Ocimum gratissimum* volatile oil and its activity against enteroaggregative *Escherichia coli*. Pharm Pharmacol Lett 2000;10:9—12.

[77] Nanji AA, Blank D. Low serum creatine kinase activity in patients with alcoholic liver disease. Clin Chem 1981;27:1954.

[78] Asare GA, Phyllis A, Kwasi B, Ben G, Samuel A, Lydia SO, et al. Acute toxicity studies of aqueous leaf extract of *Phyllanthus niruri*. Interdiscip Toxicol 2011;4 (4):206—10.

[79] Klaasen CD. Casarett & Doull's toxicology. The basic science of poisons. 6th ed. New York, NY: McGraw-Hill; 2001.

[80] Hayes AW. Principles and methods of toxicology. 4th ed. Philadelphia, PA: Taylor and Francis; 2001.

[81] Mudi SY, Ibrahim H, Bala MS. Acute toxicity studies of the aqueous root extract of *Lawsonia inermis* Linn. in rats. J Med Plants Res 2011;5(20):5123−6.

[82] Okoli C, Akah P, Onuoha N, Okoye T, Nwoye A, Nworu C. *Acanthus montanus*: an experimental evaluation of the antimicrobial, anti-inflammatory and immunological properties of a traditional remedy for furuncles. BMC Complement Altern Med 2008;1(8):27.

[83] Ibe AE, Nwufo MI. Identification, collection and domestication of medicinal plants in southern Nigeria. Afr Dev 2005;30:66−77.

[84] Noumi E, Fozi FL. Ethnomedical botany of epilepsy treatment in Fongo-Tongo village, Western province, Cameroon. Pharmaceut Biol 2003;41(5):330−9.

[85] Hassar M. La phytothérapie au Maroc. Espérance Médicale 1999;47:83−5.

[86] Bellakhdar J, Claisse R, Fleurentin J, Younos C. Repertory of standard herbal drugs in the Moroccan pharmacopoeia. J Ethnopharmacol 1991;35:123−43.

[87] Ziyyat A, Legssyer A, Mekhfi H, Dassouli A, Serhrouchni M, Benjelloun W. Phytotherapy of hypertension and diabetes in oriental Morocco. J Ethnopharmacol 1997;58:45−54.

[88] Akbay P, Gertsch J, Calis I, Heilmann J, Zerbe O, Sticher O. Novel antileukemic sterol glycosides from *Ajuga salicifolia*. Helvet Chim Acta 2002;85:1930−42.

[89] Kutepova TA, Syrov VN, Khushbaktova ZA, Saatov Z. Hypoglycemic activity of the total ecdysteroid extract from *Ajuga turkestanica*. Khimiko-Farmatsevticheskii Zhurnal 2001;35:608−9.

[90] Chen H, Tan RX, Liu ZL, Zhang Y, Yang L. Antibacterial neoclerodane diterpenoids from *Ajuga lupulina*. J Nat Prod 1996;59:668−70.

[91] Bremner PD, Simmonds MSJ, Blaney WM, Veitch NC. Neo-clerodane diterpenoid insect antifeedants from *Ajuga reptans* ev Catlins Giant. Phytochemistry 1998;47:1227−32.

[92] Ben JH, Harzallah-Skhiri F, Mighri Z, Simmonds MSJ, Blaney WM. Responses of *Spodoptera littoralis* larvae to Tunisian plant extracts and to neo-clerodane diterpenoids isolated from *Ajuga pseudoiva* leaves. Fitoterapia 2000;71:105−12.

[93] Bondì ML, Al-Hillo MRY, Lamara K, Ladjel S, Bruno M, Piozzi F, et al. Occurrence of the antifeedant 14,15-dihydroajugapitin in the aerial parts of *Ajuga iva* from Algeria. Biochem Syst Ecol 2000;28:1023−5.

[94] Breschi MC, Martinotti E, Catalano S, Flamini G, Morelli I, Pagni AM. Vasoconstrictor activity of 8-O-acetylharpagide from *Ajuga reptans*. J Nat Prod (Lloydia) 1992;55:1145−8.

[95] Ben JH, Harzallah-Skhiri F, Mighri Z, Simmonds MSJ, Blaney WM. Antifeedant activity of plant extracts and of new diglyceride compounds isolated from *Ajuga pseudoiva* leaves against *Spodoptera littoralis* larvae. Ind Crop Prod 2001;14:213−22.

[96] Jaouad EH, Zafar HI, Badiâa L. Acute and chronic toxicological studies of *Ajuga iva* in experimental animals. J Ethnopharmacol 2004;91:43−50.

[97] Betti JL. An ethnobotanical study of medicinal plants among the Baka Pygmies in the Dja biosphere reserve, Cameroon. Afr Study Monogr 2004;25(1):1−27.

[98] Yusuf S, Lawal SB, Mansur L, Simeon AI, Rabiu AU. Hematotoxicity study of the leaf extract of *Albizia chevalieri* harms (Leguminosae). Biochem Med 2007;17(2):203−11.

[99] Tédong L, Djomeni PDD, Dimo T, Asongalem EA, Ndogmo SS, Flejou JF, et al. Acute and subchronic toxicity of *anacardium occidentale* linn (anacardiaceae) leaves hexane extract in mice. Afr J Trad Complement Altern Med 2007;4(2):140−7.

[100] Burkill HM. The useful plants of West Africa. Kew: Royal Botanical Gardens; 1985. p. 103—5.

[101] Muanza DN, Kim BW, Euler KL, Williams L. Antibacterial and antifungal activities of nine medicinal plants from Zaire. Int J Pharmacogn 1994;32:337—45.

[102] Iwu MM. Handbook of African medicinal plants. Boca Raton, FL: CRC Press; 1993. p. 121—2.

[103] Adjanohoun E, Aboubakar N, Dramane K, Ebot ME, Ekpere JA, Enow-Orock EG, et al. Contribution to ethnobotanical and floristic studies in Cameroon. Lagos: CSTR/OUA; 1996.

[104] Obadoni BO, Ochuko PO. Phytochemical studies and comparative efficacy of the crude extracts of some haemostatic plants in edo and delta states of Nigeria. Global J Pure Appl Sci 2002;8(2):203—8.

[105] Kuiate JR, Zollo PHA, Lamaty G, Menut C, Bessiere JM. Composition of the essential oils from the leaves of two varieties of *Aspilia africana* (Pers.) C. D. Adams from Cameroon. Flav Frag J 1999;14(3):167—9.

[106] Taziebou LC, Etoa F-X, Nkegoum B, Pieme CA, Dzeufiet DPD. Acute and subacute toxicity of *Aspilia africana* leaves. Afr J Trad Complement Altern Med 2007;4(2):127—34.

[107] Akah PA, Offiah VN, Onuagu E. Hepatotoxic effect of *A. indica* leaf extract in rabbits. Protozoa 1992;63:311—9.

[108] Charterjee A, Pakrashi S. Treatise on Indian medicinal plants. New Delhi: IBH Publishing Co.; 1994.

[109] Pakrashi S. Treatise on Indian medicinal plants. New Delhi: IBH Publishing Co.; 1994.

[110] Nwachukwu N, Iweala EJ. Influence of extraction methods on the hepatotoxicity of *Azadirachta indica* bark extract on albino rats global. J Pure Appl Sci 2009;15(3):369—72.

[111] Ashafa AOT, Latifat OO, Musa TY. Toxicity profile of ethanolic extract of *Azadirachta indica* stem bark in male Wistar rats. Asian Pac J Trop Biomed 2012;811—7.

[112] Asuzu IU, Nwelle OC, Anaga AO. The Pharmacological activities of the methanolic bark of *Berlina grandiflora*. Fitoterapia 1993;64:529—34.

[113] Enwerem NM. Aphytochemical and anthelmintic activity investigation of the stem bark extracts of *Berlina grandiflora* Dutch and Dalz. Ph.D. thesis, Department of chemistry, faculty of science, University of Ibadan, Nigeria; 1999.

[114] Aniagu SO, Nwiyi FCO, Lanubi B, Akumka DD, Ajoku GA, Izebe KS, et al. Is *Berlina grandiflora* (leguminosae) toxic in rats? Phytomedicine 2004;11:352—60.

[115] Orafidiya LO, Agbani EO, Iwalewa EO, Adelusola KA, Oyedapo OO. Studies on the acute and sub-chronic toxicity of the essential oil of *Ocimum gratissimum* L. leaf. Phytomedicine 2004;11:71—6.

[116] Makonnen E, Hagos E. Antispasmodic effect of *Bersama abyssinica* aqueous extract on guinea-pig ileum. Phytother Res 1993;7:211—2.

[117] Tapondjou LA, Miyamoto T, Lacaille-Dubois MA. Glucuronide triterpene saponins from *Bersama engleriana*. Phytochemistry 2006;67:2126—32.

[118] Kuete V, Armelle TM, Maurice T, Véronique PB, Etoa FX, Nkengfack AE, et al. Antitumor, antioxidant and antimicrobial activities of *Bersama engleriana* (Melianthaceae). J Ethnopharmacol 2008;115:494—501.

[119] Ogunwande IA, Bello MO, Olawore ON, Muili KA. Phytochemical and antimicrobial studies on *Butyrospermum paradoxum*. Fitoterapia 2001;72:54—6.

[120] Rabo JS. Toxicity and trypanossuppressive effects of stem bark extract of *Butyrospermum paradoxum* in laboratory animals. Ph.D. thesis, University of Maiduguri: Nigeria; 1998.

[121] Wabo PJ, Mpoame M, Bilong BCF. *In vivo* evaluation of potential nematicidal properties of ethanol extract of *Canthium mannii* (Rubiaceae) on *Heligmosomoides polygyrus* parasite of rodent. Vet Parasitol 2009;166:103–7.

[122] Ibrahim H, Abdurrahman EM, Shok M, Ilyas N, Musa KY, Ikandu I. Comparative analgesic activity of the root bark, stem bark, leaves, fruits and seed of *Carissa edulis* VAHL (Apocynaceae). Afr J Biotechnol 2007;6:1233–5.

[123] Ya'ua J, Chindo BA, Yaro AH, Okhale SE, Anuka JA, Hussaini IM. Safety assessment of the standardized extract of *Carissa edulis* root bark in rats. J Ethnopharmacol 2013;147:653–61.

[124] Sadiq IS, Shuaibu M, Bello AB, Tureta SG, Isah A, Izuagie T, et al. Phytochemistry and antimicrobial activities of *Cassia occidentalis* used for herbal remedies. J Chem Eng 2012;1(1):38–41.

[125] Mirtes GBS, Ticiana PA, Carlos FBV, Pablo AF, Bruno AA, Igor MA, et al. Acute and subacute toxicity of *Cassia occidentalis* L. stem and leaf in Wistar rats. J Ethnopharmacol 2011;136:341–6.

[126] Grieve M. A modern herbal, vol. 1. New York, NY: Dover Publications; 1971. p. 443.

[127] Capasso F, Mascolo N, Morrica P, Ramundo E. Phytotherapeutic profile of some plants used in folk medicine. Boll Soc Ital Biol Sper 1983;59:1398–404.

[128] Bisset NG. Herbal drugs and phytopharmaceuticals. New York, NY: CRC Press; 1994. p. 566.

[129] Beth M. Healing threads: traditional medicines of the highlands and islands. Polygon. Scotland: Edinburgh University Press; 1995. p. 1–304.

[130] Kultur S. Medicinal plants used in Kirklareli province (Turkey). J Ethnopharmacol 2007;111:341–64.

[131] Van Der SWG, Labadie RP. Secoiridoids and xanthones in the genus *Centaurium*. Part III: decentapicrins A, B and C, new m-hydroxybenzoyl esters of sweroside from *Centaurium littorale*. Planta Med 1981;41:150–60.

[132] Takagi S, Yamaki M, Yumioka E, Nishimura T, Sakina K. Studies on the constituents of *Centaurium erythraea* (Linne) Persoon. II. The structure of centauroside, a new bis-secoiridoid glucoside. Yakugaku Zasshi 1982;102:313–7.

[133] Kaouadji M, Vaillant I, Mariotte AM. Polyoxygenated xanthones from *Centaurium erythraea* roots. J Nat Prod 1986;49:359.

[134] Hatjimanoli M, Favre-Bonvin J, Kaouadji M, Mariotte A-M. Monohydroxy and 2,5-dihydroxy terephthalic acids, two unusual phenolics isolated from *Centaurium erythraea* and identified in other Gentianaceae members. Ann Pharm Fr 1988;35:107–11.

[135] Tahraouia A, Zafar HI, Badiaa L. Acute and sub-chronic toxicity of a lyophilised aqueous extract of *Centaurium erythraea* in rodents. J Ethnopharmacol 2010;132:48–55.

[136] Lakmichi H, Fatima ZB, Chemseddoha AG, Ezoubeiri A, Younes EJ, Abdellah EM, et al. Toxicity profile of the aqueous ethanol root extract of *Corrigiola telephiifolia* Pourr. (Caryophyllaceae) in rodents. Evid Based Complement Altern Med 2011;1–10.

[137] Faiz CA, Thami IA, Saïdi, N. (2007) Domestication of some MAP species. In: Faiz CA, editor. Biological diversity, cultural and economic value of medicinal, herbal and aromatic plants in Morocco. Annual report, p. 15–22.

[138] Tamokou JD, Kuiate JR, Gatsing D, Nkeng-Efouet AP, Njouendou AJ. Antidermatophytic and toxicological evaluations of dichloromethane-methanol extract, fractions and compounds isolated from *Coula edulis*. Iran J Med Sci 2011;36 (2):111.

[139] Arbonnier M. Arbres, arbustes et lianes des zones seches d'Afrique de l'Ouest Trees, shrubs and lianas of West Africa dry zones, Mali, Ouagadougou: Centre de Cooperation Internationale en Recherche Agronomique pour le developpement/ Museum national d'histoire naturelle/Union mondiale pour la nature. 1st ed. CIRAD/ MNHN/UICN; 2000.

[140] Taiwe GS, Elisabeth NB, Talla E, Theophile D, Neteydji S, Amadou D, et al. Evaluation of antinociceptive effects of *Crassocephalum bauchiense* Hutch (Asteraceae) leaf extract in rodents. J Ethnopharmacol 2012;141:234−41.

[141] Tchivounda HP, Koudogbo B, Besace Y, Casadevall E. Triterpene saponins from *Cylicodiscus gabunensis*. Phytochemistry 1991;30:2711−6.

[142] Kouitcheu LBM, Penlap VB, Kouam J, Oyono E, Etoa FX. Toxicological evaluation of ethyl acetate extract of *Cylicodiscus gabunensis* stem bark (Mimosaceae). J Ethnopharmacol 2007;111:598−606.

[143] Nwude N, Parsons LE, Adaudi AO. Acute toxicity of the leaves and extracts of *Dichapetalum barteri* (Engl.) in mice, rabbits and goats. Toxicology 1997;7: 23−9.

[144] Troupin G. Flore des plantes ligneuses du Rwanda. Musée royal de l'Afrique Centrale Turuven; 1983. p. 257−8.

[145] Ngouana V, Fokoua PVT, Foudjoa BUS, Ngouel SA, Boyom FF, Zollo PHA. Antifungal activity and acute toxicity of stem bark extracts of *Drypetes gossweileri*s. moore—Euphorbiaceae from Cameroon. Afr J Tradit Complement Altern Med 2011;8(3):328−33.

[146] Fai PBA, Fagade SO. Acute toxicity of *Euphorbia kamerunica* on *Oreochromis niloticus*. Ecotox Environ Safe 2005;62:128−31.

[147] Frohne D, Pfander HJK. A colour atlas of poisonous plants. London: Wolfe Publishing; 1984.

[148] Assi AL. Utilisation de diverses espèces de Ficus (Moraceae) dans la pharmacopée traditionnelle africaine de Côte d'Ivoire. Mitteilungen aus dem Institut für allgemeine Botanik in Hamburg 1990;23:1039−46.

[149] Ijeh II, Ukweni AI. Acute effect of administration of ethanol extracts of *Ficus exasperata* Vahl on kidney function in albino rats. J Med Plant Res 2007;1:027−9.

[150] Bafor EE, Lim CV, Rowan EG, Edrada-Ebel R. The leaves of *Ficus exasperata* Vahl (Moraceae) generates uterine active chemical constituents. J Ethnopharmacol 2013;145(3):803−12.

[151] Endale A, Kassa M, Gebre-Mariam T. *In vivo* anthelmintic activity of the extracts of *Glinus lotoides* in albino mice infested with *Hymenolepis nana* worms. Ethiop Pharmaceut J 1998;16:34−41.

[152] Olver-Bever B. Medicinal plants in tropical West Africa. Cambridge: Cambridge University Press; 1986.

[153] Dalziel JM. The useful plants of West tropical Africa. London: Crown Overseas Agents for the Colonies; 1937. p. 2−3.

[154] Aniagu SO, Binda LG, Nwinyi FC, Orisadipe A, Amos S, Wambebe C, et al. Antidiarrheal and ulcer-protective effects of the aqueous root extract of *Guiera senegalensis* in rodents. J Ethnopharmacol 2005;97:549−54.

[155] Grieve MM. A modern herbal. London, U.K: Penguin; 1984.

[156] Chopra RN, Nayar SL, Chopra IC. Glossary of Indian medicinal plants (including the supplement). New Delhi: Council of Scientific and Industrial Research; 1986.

[157a] Chevallier A. The encyclopedia of medicinal plants. London: Dorling Kindersley; 1996.

[157b] Rhiouani H, El-Hilaly J, Israili ZH, Lyoussi B. Acute and sub-chronic toxicity of an aqueous extract of the leaves of Herniaria glabra in rodents. J Ethnopharmacol 2008;118(3):378—86.

[158] Hutchinson J, Dalziel JM. Flora of West Africa. Crown agents for overseas government and administration. London; 1955. p. 488—97.

[159] Adesida GA, Adesogan EK, Okorie DA, Taylor DAH, Styles BT. The limonoid chemistry of the genus Khaya (Meliaceae). Phytochemistry 1971;10:1845.

[160] Agbedahunsi JM, Fakoyab FA, Adesanyac SA. Studies on the anti-inflammatory and toxic effects of the stem bark of Khaya ivorensis (Meliaceae) on rats. Phytomedicine 2004;11:504—8.

[161] Biennvenu E, Amabeoku GJ, Eagles PK, Scott G, Springfield EP. Anticovulsant activity of aqueous extract of Leonotis leonurus. Phytomedicine 2002;9:217—23.

[162] Mutsabisa M. Traditional medicines, <http://www.sahealthinfo.org/traditionalmeds/aboutuwc.htm>.

[163] Maphosa V, Masika PJ, Adedapo AA. Safety evaluation of the aqueous extract of Leonotis leonurus shoots in rats. Hum Exp Toxicol 2008;27:837.

[164] Siméon-Pierre CF, Donatien G, Gerald NT, Fabrice K, Sedric DT, Joseph T. In vivo antityphoid activity of Ludwiguia abyssinica aqueous extract and side effects induced by the treatment on infected rats. Int J Pharmacol Clin Trials 2013;46(2):1191—9.

[165] Chakraborty DP, Barma BK, Bose PK. On the constitution of murrayanine, a carbazole derivative isolated from Murraya koenigii Spreng. Tetrahedron 1965;21:681—5.

[166] Gupta GL, Nigam SS. Chemical examination of the leaves of Murraya koenigii. Planta Med 1971;19(1):83—6.

[167] Nutan MTH, Hasnat A, Rashid MA. Antibacterial and cytotoxic activities of Murraya koenigii. Fitoterapia 1998;69(2):173—5.

[168] Fiebig M, Pezzuto JM, Soejarto DD, Kinghorn AD. Koenoline, a further cytotoxic carbazole alkaloid from Murraya koenigii. Phytochemistry 1985;24(12):3041—3.

[169] Atta-ur-Rahman ZR, Firdous S. NMR studies on mahanine. Fitoterapia 1988;59 (6):494—5.

[170] Adeneye AA, Amole OO, Adeneye AK. Hypoglycemic and hypocholesterolemic activities of the aqueous leaf and seed extract of Phyllanthus amarus in mice. Fitoterapia 2006;77:511—4.

[171] Ekejiuba EC. Establishment of some pharmacopoeia standards for selected Nigerian medicinal plants. M. Phil. thesis, UNIFE, 1984.

[172] Bouquet A. Féticheurs et Médecine traditionnelles au Congo (Brazzaville). Mémoire O.R.S.T.O.M. No. 36. Paris; 1969. p. 49, 100, 198.

[173] Haule EE, Moshi MJ, Nondo RSO, Mwangomo DT, Rogasian LA. A study of antimicrobial activity, acute toxicity and cytoprotective effect of a polyherbal extract in a rat ethanol—HCl gastric ulcer model. BMC Res Notes 2012;5:546.

[174] Naya BS, Raju SS, Chalapathi RAV. Wound healing activity of Persea Americana (avocado) fruit: a preclinical study on rats. J Wound Care 2008;17:123—6.

[175] Adeyemi OO, Okpo SO, Ogunti OO. Analgesic and anti-inflammatory effects of the aqueous extract of leaves of Persea americana mill (Lauraceae). Fitoterapia 2002;73:375—80.

[176] Ojewole JA, Amabeoku GJ. Anticonvulsant effect of Persea americana Mill (Lauraceae) (Avocado) leaf aqueous extract in mice. Phytother Res 2006;20:696—700.

[177] Ozolua O, Anaka N, Stephen OO, Sylvester EI. Acute and sub-acute toxicological assessment of the aqueous seed extract of Persea americana mill (lauraceae) in rats. Afr J Trad Complement Altern Med 2009;6(4):573—8.

[178] Assob JC, Kamga HL, Nsagha DS, Njunda AL, Nde PF, Asongalem EA, et al. Antimicrobial and toxicological activities of five medicinal plant species from Cameroon traditional medicine. BMC Complement Altern Med 2011;11:70.

[179] Patel JR, Tripathi P, Sharma V, Chauhan NS, Dixit VK. *Phyllanthus amarus*: ethnomedicinal uses, phytochemistry and pharmacology: a review. J Ethnopharmacol 2011;138(2):286–313.

[180] Dagang W, Sorg B, Adolf W, Seip EH, Hecker E. Oligo- and macrocyclic diterpenes in Thymelaeaceae and Euphorbiaceae occurring and utilized in Yunnan (Southwest China): two ingenane type diterpene esters from *Euphorbia nematocypha*. Phytother Res 1992;26(5):237–40.

[181] Gundidza M, Sorg B, Hecker E. A skin irritant principle from *Euphorbia metabelensis* Pax. J Ethnopharmacol 1993;39(3):209–12.

[182] Milillo MA, Laffaldano D, Sasso G, De-Laurentis N. Poisoning by *Euphorbia characias* L.1. clinical and toxicological findings. Obiettivie-Documento-Veterianari 1993;14(5):35–7.

[183] Neuwinger HD. Fish poisoning plants in Africa. Bot Acta 1994;107(4):263–70.

[184] Iwalewa EO, McGaw L, Naidoo V, Eloff JN. Inflammation: the foundation of diseases and disorders. A review of phytomedicines of South African origin used to treat pain and inflammatory conditions. Afr J Biotechnol 2007;25: 2868–85.

[185] Mukinda JT, Peter FKE. Acute and sub-chronic oral toxicity profiles of the aqueous extract of *Polygala fruticosa* in female mice and rats. J Ethnopharmacol 2010;128: 236–40.

[186] Oteng-Gyang K, Mbachu JI. Changes in the ascorbic acid content of some tropical leafy vegetables during traditional cooking and local processing. Food Chem 1987;23:9–17.

[187] Kimpouni V. Etude de marché préliminaire sur les produits forestiers non ligneux commercialisés dans les marchés de Pointe-Noire (Congo-Brazaville). In: Recherches actuelles et perspectives pour la conservation et le développement Archives de Documents de la FAO, X2161/F, Département des Forêts; 1999.

[188] Okafor J. La contribution du savoir des exploitants agricoles à la recherche sur les produits forestiers non ligneux. In: Recherches Actuelles et perspectives pour la conservation et le développement. Archives de Documents de la FAO, X2161/F, Département des Forêts; 1999.

[189] Sarah AL. L'exploitation du bois d'oeuvre et des produits forestiers non ligneux (PFNL) dans les forêts d'Afrique Centrale. In: Recherches Actuelles et perspectives pour la conservation et le développement. Archives de Documents de la FAO, X2161/F, Département des Forêts; 1999.

[190] Arnone A, Camarda L, Merlini L, Nasini G, Taylor DAH. Colouring matters of the West African red woods *Pterocarpus osun* and *P. soyauxii*. Structures of santarubins A and B. Journal of Chemical Society 1977;1:2116–8 (London Perkin).

[191] Banerjee A, Mukherjee KA. Chemical aspects of santalin as a histological stain. Biotec Histochem 1981;56:83–5.

[192] Bezuidenhoudt BCB, Brandt EV, Ferreira EV. Flavonoid analogues from *Pterocarpus* species. Phytochemistry 1987;26:531–5.

[193] Chukwunyere ON, Tafarki AE, Patrick AO, Victor OO. Toxicity and anthelmintic efficacy of crude aqueous of extract of the bark of *Sacoglottis gabonensis*. Fitoterapia 2008;79:101–5.

[194] Muibat BA, Omoniyi KY, Olufunmilayo OA. Toxicological assessment of the aqueous root extract of *Sanseviera liberica* Gerome and labroy (Agavaceae). J Ethnopharmacol 2007;113:171−5.

[195] Noumi E, Yomi A. Medicinal plants used for intestinal diseases in Mbalmayo region Central province, Cameroon. Fitoterapia 2001;3:246−54.

[196] Assam AJP, Dzoyem JP, Pieme CA, Penlap VB. *In vitro* antibacterial activity and acute toxicity studies of aqueous-methanol extract of *Sida rhombifolia* Linn. (Malvaceae). BMC Complement Altern Med 2010;10:40.

[197] Ilodigwe EE, Akah PA, Nworu CS. Evaluation of the acute and subchronic toxicities of ethanol leaf extract of *Spathodea campanulata* P. Beauv. Int J Appl Res Nat Prod 2010;3(2):17−21.

[198] Kahnut AM, Probstle H, Rimpter R, Heinrich R. Biological and pharmacological activities and further constituents of *Hyptis suaveolens*. Planta Med 1995;6(3):227−37.

[199] Otimenyin SO, Uguru MO, Ojeka K. Acute toxicity studies and some pharmacological properties of *Stachyphata indica*. Int J Dev Neurosci 2006;24(8):146.

[200] Prashar A, Locke IC, Evans CS. Cytotoxicity of clove (*Syzigium aromaticum*) oil and its major components to human skin cells. Cell Prolif 2006;39:241−8.

[201] Agbaje EO. Gastrointestinal effects of *Syzigium aromaticum* (L). Merr. & Perry (Myrtaceae) in animal model. Nig Qt J Hosp Med 2008;18(3):137−41.

[202] Sanchez-Elsner T, Ramirez JR, Rodriguez-Sanz F, Varela E, Bernabew C, Botella LM. A cross talk between hypoxia and TGF-beta orchestrates erythropoietin gene regulation through SPI and smads. J Mol Biol 2004;36(1):9−24.

[203] Agbaje EO, Adeneye AA, Daramola AO. Biochemical and toxicological studies of aqueous extract of *Syzigium aromaticum* (l.) merr. & Perry (myrtaceae) in rodents. Afr J Trad Complement Altern Med 2009;6(3):241−54.

[204] Sandberg F. Etude sur les plantes medicinales et toxiques d'Afrique équatoriale. Cahiers de la Maboké 1965;3:5−49.

[205] Van Beek TA, Verpoorte R, Baerheim SA, Leeuwenberg AJM, Bisset NG. *Tabernaemontana* L. (Apocynageae): a review of its taxonomy, phytochemistry, ethnobotany and pharmacology. J Ethnopharmacol 1984;10:1−156.

[206] Dass B, Fellion E, Plat M. Alkaloids in *Conopharyngia durissima* seeds. Comptes Rendus de 1_Académie des Sci 1967;C264:1765−7.

[207] Van Beek TA, De Smidt C, Verpoorte R. Phytochemical investigation of *Tabernaemontana crassa*. J Ethnopharmacol 1985;14:315−8.

[208] Danieli B, Palmisano G. Alkaloids from *Tabernaemontana*. The alkaloids Chapter 1 Chem Pharmacol 1987;27:1−130.

[209] Kuete V, Manfouo RN, Beng VP. Toxicological evaluation of the hydroethanol extract of *Tabernaemontana crassa (Apocynaceae) stem bark*. J Ethnopharmacol 2010;130:470−6.

[210] Kuete V, Tangmouo JG, Beng VP, Ngounou FN, Lontsi D. Antimicrobial activity of the methanolic extract from the stem bark of *Tridesmostemon omphalocarpoides* (Sapotaceae). J Ethnopharmacol 2006;104:5−11.

[211] Ekwalla N, Tongo E. Nos plantes qui soignent. Cameroon: Douala; 2003, Ed. I.C

[212] Vardamides JC, Dongmo AB, Meyer M, Ndom JC, Azebaze AGB, Sahmeza ZMR, et al. Alkaloids from the stem bark of *Turraeanthus africanus* (Meliaceae). Chem Pharm Bull 2006;54:1034−6.

[213] Lembe DM, Sonfack A, Gouado I, Dimo T, Dongmo A, Demasse MFA, et al. Evaluations of toxicity of *Turraeanthus africanus* (Meliaceae) in mice. Blackwell Verlag GmbH Æ Andrologia 2009;41:341−7.

[214] Tchinda AT, Tsopmo A, Tane P, Ayafor JF, Connolly JD, Sterner O. Vernoguinosterol and vernoguinoside, trypanocidal stigmastane derivatives from *Vernonia guineensis* (Asteraceae). Phytochemistry 2002;59:371−4.

[215] Donfack ARN, Toyang NJ, Wabo HK, Tane P, Awoufack MD, Kikuchi H, et al. Stigmatane derivatives from the root extract of *Vernonia guineensis* and their antimicrobial activity. Phytochem Lett 2012;5:596−9.

[216] Toyang NJ, Wabo HK, Ateh EN, Davis H, Tane P, Kimbu SF, et al. *In vitro* antiprostate cancer and *in vivo* antiangiogenic activity of *Vernonia guineensis* Benth. (Asteraceae) tuber extracts. J Ethnopharmacol 2012;141:866−71.

[217] Olatunji OA. The biology of Zanthoxylum Linn (Rutaceae) in Nigeria. In Essien, Adebanjo, Adewunmi, Odebiyi editors. Anti infective agents of higher plants origin. In: Proceedings of the fifth international symposium on medicinal plants. Lagos, Nigeria: Medex Publications; 1983. p. 56−9.

[218] Oliver-Bever B. Medicinal plants in tropical West Africa II. Plants acting on the nervous system. J Ethnopharmacol 1983;7:1−93.

[219] Adesina SK. The Nigerian *Zanthoxylum*; chemical and biological values. Afr J Trad Compl Altern Med 2005;2(3):282−301.

[220] Ngono RAN, Ebelle Etame RM, Koanga M, Ndifor F, Biyiti L, Bouchet PH, et al. Fungicidal effect and acute toxicity of extracts from *Zanthoxylum xanthoxyloides* (Rutaceae). Int J Infect Dis 2008;12(Suppl. 1):e282.

6 Subchronic and Chronic Toxicities of African Medicinal Plants

Adejuwon Adewale Adeneye

Department of Pharmacology, Faculty of Basic Medical Sciences, Lagos State University College of Medicine, Ikeja, Lagos, Nigeria

6.1 Introduction

The use of medicinal plants as a therapy for disease condition is an ancestral medical practice which has formed an important component of the health care delivery system, particularly in regions that are abundantly rich in diverse flora [1]. Medicinal plants commonly used in traditional medicine are frequently promoted as natural and harmless. In addition, the affordability, reliability, availability, and low toxicity of medicinal plants in therapeutic use have made them popular and acceptable by all religions, for implementation in medical health care all over the world [2]. The wide acceptance and use of the medicinal plants are on the premise that: (i) being of natural source, the constituents are naturally balanced; (ii) their ancestral use has proven the efficacy and safety of these medicinal plants in disease prevention and treatment; (iii) they are easy accessed, readily available at little or no cost [3]. However, despite these beliefs, medicinal plants must be used with great caution, as their use could be associated with adverse reactions, especially when taken in high dose, over long periods even at tolerable dose, or when they are co-administered with synthetic drugs from herbal-drug interactions [4−7]. Also, in recent times there has been a growing concern about the safety of herbal and herbal-related products [8,9]. Consequently, in response to public health concerns, research focusing on the potential therapeutic and toxicities of medicinal plants is currently being encouraged by all stakeholders in traditional complementary and alternative medicine practice [10−14].

In order to determine the potential toxicities inherent in a medicinal plant, toxicity testing is often undertaken and these toxicity studies are broadly divided into three main categories, namely:

- *Descriptive toxicology*, which is directly concerned with toxicity testing providing much-needed information on the safety evaluation and regulatory requirements. This category of toxicology involves the design of appropriate toxicity tests in experimental animals that provide information that could be used to assess the potential human risks posed by exposure to specific drugs and herbal products [15]. However, there are two basic

Toxicological Survey of African Medicinal Plants. DOI: http://dx.doi.org/10.1016/B978-0-12-800018-2.00006-6

principles underlying all descriptive animal toxicity testing. First, drug effects in the laboratory animals tested, when properly quantified, can be applied to humans as on the basis of
dose per unit body surface, toxic effects in humans are often within the same range as
those obtainable in animals, while on the basis of body weight, humans are generally more
vulnerable than laboratory animals, probably by a factor of 10 [15]. Therefore, appropriate
safety factors can be applied to calculate relatively safe doses for humans. The second
main principle is that exposure of experimental animals to toxic agents in high doses is a
necessary and valid method of discovering possible hazards in humans. This principle is
based on the quantal dose-response concept that the incidence of an effect in a population
is greater as the dose or exposure increases.

- *Mechanistic toxicology*, which is primarily concerned with elucidating the mechanisms by
which drugs exert their toxic effects in animals. Results of this often lead to development of
sensitive predictive tests useful in risk assessment, design, and production of safer alternative drugs and in rational therapy for drug/chemical poisoning and disease treatment [15];
- *Regulatory toxicology*, which is saddled with the responsibility of deciding on the basis of
data provided by the descriptive toxicology if a drug possesses a sufficiently low risk to be
marketed for the stated purpose(s) for which such a drug was designed and developed [15].

Descriptive toxicology could further be classified into acute, subacute, subchronic and chronic oral, intraperitoneal, subcutaneous, dermal, and inhalational toxicities. Of these toxicity types based on the route of drug administration, the oral
route is the most common, especially for medicinal plants. However, definitions
and peculiarities of the different types of toxicity testing are discussed as follows:

- *Acute toxicity*: Acute toxicity refers to the adverse effects occurring following oral or dermal administration of a single high dose of a drug, or multiple doses given within 24 h or
an inhalation exposure of 4 h. The purposes of acute toxicity testing are to obtain information on the biological (biokinetic, cellular, and molecular) activity and possible mechanism of action of a drug [16]. The LD_{50} (defined as the statistically derived dose that
when administered in an acute toxicity test is expected to cause death in 50% of the treated animals in a given period) is often the determined outcome of this study [16].
- *Subacute toxicity*: This refers to adverse effects occurring as a result of repeated daily
dosing or exposure to a drug for a part of the animal's life span (usually not exceeding
10%). The importance of preclinical subacute toxicity testing is to identify the target
organs toxicity, determine the response, relationship, and potential toxicity reversibility
of these target organs to the administered dose levels of the drug. The information
obtained from this toxicity test is important for the estimation of baseline starting dose
for human studies and identification of target organs and cells for clinical monitoring for
potential untoward effects [17].
- *Subchronic toxicity*: This refers to the adverse effects occurring as a result of the repeated
daily dosing of a drug to experimental animals for more than 1 year but less than the lifetime of the exposed experimental animals. This study is aimed at (i) identifying toxicity
that develops only after a certain length of continuous exposure to the drug/chemical; (ii)
identifying the organs most affected by the drug; (iii) determining the doses at which each
effect occurs. The major difference between subacute (repeated dose) and subchronic toxicity studies is the duration as the repeated dose toxicity studies are conducted over a treatment period of 28 days, while that of subchronic toxicity are conducted over 90 days [18].
- *Chronic toxicity*: This is the ability of a drug to cause adverse effects over an extended
period following repeated or continuous exposure, sometimes lasting for the entire life

span of the exposed experimental animal. The importance and purpose of chronic toxicity is to provide a reliable set of information on the safe dose levels to be administered in the different phases of the clinical development of a drug [17]. There should be little individual variation between preclinical animal and human data due to their close phylogeny.

This chapter describes some of the conventional and scientifically acceptable subchronic and chronic animal toxicity methods, experimental toxicity animal types, measured end points, and interpretation of data generated on some selected African medicinal plants. In addition, a range of welfare considerations (including selection, dosing, and sampling methods) that may arise for experimental animals involved in subchronic and chronic toxicity testing are discussed.

6.2 Subchronic and Chronic Toxicity Effect of Medicinal Plants

Toxicological screening of medicinal plants is vital for the development of new drugs from and for promoting the continuous therapeutic use of the already existing medicinal plants. Thus, toxicity testing, particularly of subchronic and chronic toxicity, is mostly used to examine specific adverse events or specific end points such as the "no observed adverse effect level (NOAEL)" dose which usually forms the benchmark dose (BMD) for human risk assessment data in the clinical trials of these medicinal plants [19].

Subchronic and chronic oral toxicity profiles of some selected commonly used medicinal plants employed in the treatment of various human diseases worldwide are summarized below.

6.2.1   Hypericum perforatum *(Hypericaceae)*

Hypericum perforatum extract is an extract of the capsules, flowers, leaves, and stem heads of *Hypericum perforatum*, commonly called St. John's Wort. This plant is native to the temperate and subtropical regions of North America, Europe, Asia Minor, Russia, India, and China, but also could be found in hot temperate regions of Africa, particularly in sub-Saharan African countries such as Nigeria where it is known as "Afomo" among the Yoruba tribe (Southwest Nigeria). The extract of this plant commonly is used in the treatment of depression, insomnia, exhaustion, somatoform disorders, nervousness, and convalescence; in addition to its uses as a remedy for skin diseases, superficial injury, mucosal lesions, and gastrointestinal illness [20]. Preclinical studies showed that rats fed with *Hypericum perforatum* flowers for 2 weeks manifested significant signs of toxicity, including erythema, edema of the portion of the body exposed to light, alopecia, and changes in blood chemistry, while chronic oral toxicity study of *Hypericum perforatum* resulted in significant attenuation in weight gain in *Hypericum perforatum*-treated rats when compared to control rats [21]. Long-term clinical dosing with the polyphenol fraction of *Hypericum perforatum* resulted in enhanced immunostimulating activity in subjects orally administered with *Hypericum perforatum* extract, while administration of its lipophilic portion had immunosuppressing properties in another set of subjects. Other observed toxicities associated with *Hypericum perforatum* extract in the clinical treatment of depression include skin reddening and itching, dizziness, constipation, fatigue, anxiety, tiredness, photosensitivity, acute neuropathy, and even episodes of mania and serotonergic syndrome, particularly when administered simultaneously with other antidepressant drugs [21].

6.2.2   Rosmarinus officinalis *Linn. (Lamiaceae)*

In folk medicine, *Rosmarinus officinalis*, commonly called rosemary leaf extract, is used as an antispasmodic in renal colic and dysmenorrhea and for relieving respiratory disorders, while its essential oil is used internally as a carminative and as an appetite stimulant [22,23]. The plant was found also to grow in South Africa. Preclinical subchronic oral toxicity of doses as high as 14.1 g/kg of *Rosmarinus officinalis* hexane and ethanol leaf extracts as well as its leaf oil to rats for 90 days caused significant increases in absolute and relative liver-to-body weights, although these changes were reversible, and no other signs of toxicity were observed [24].

6.2.3   Erythrina mulungu *(Fabaceae)*

Erythrina mulungu, also known as Lucky Bean Tree (South Africa), is a branched tree native to Southern Brazil, growing in the rain forests of Amazonia. Its common name in English is "coral tree." In native Southern Brazilian and Peru herbal medicine, a leaf or bark decoction or tincture from *Erythrina mulungu* has long been used in folk medicine due to its tranquilizing effects and as natural sedative [25]. It is also has potent anxiolytic and antibacterial properties [26]. In both Brazil and Peru, *Erythrina mulungu* (Figure 6.3) is used for epilepsy [27]. Practitioners in the United States use *Erythrina mulungu* to quieten hysteria from trauma or shock, as a mild sedative to calm the nervous system, to treat insomnia, and promote healthy sleeping patterns [28]. *Erythrina mulungu* also decreases blood pressure and normalizes heart arrhythmia [29]. Chronic oral toxicity studies of the plant stem bark extract conducted in mice revealed alterations in the levels of biochemical markers of hepatic and cardiac damage. The alterations were significantly pronounced in mice fed with the highest inclusion of the extract (1.0 g extract/kg feed), showing significant increases in the serum activities of alanine aminotransferase, aspartate aminotransferase, and alkaline phosphatase. Hematological assessments of mice in this group showed significant decreases in the red blood cell count and the packed cell volume strongly indicates the anemic potential of the plant extract. The plant extract at the oral doses of 0.5 and 1 g/kg/day also increased the relative weights of the liver and heart, as well as increased the serum levels of lipid peroxidation marker, malondialdehyde, on the 84th day of the study. Significant histopathological changes of myocardial hemorrhages and degeneration of hepatocytes were observed in the heart and liver of the extract-treated mice, respectively [28].

6.2.4 Hydrastis canadensis *(Ranunculaceae)*

Goldenseal (*Hydrastis canadensis*), also called orangeroot or yellow puccoon, is a perennial herb belonging to the Ranunculaceae family. It is native to southeastern Canada and the northeastern United States. The root of the goldenseal plant is traditionally used to treat wounds, ulcers, digestive problems, and eye and ear infections. The herb is also used as a laxative, tonic, and diuretic. Goldenseal is used in feminine products such as vaginal douches and is claimed to help to regularize menstrual cycle and flow in women with irregular menstrual cycle and excessive bleeding. Hydrastine, one of the chief components of goldenseal, has been reported to elicit abortifacient effects and induce preterm labor in pregnant women when taken orally. Ingestion of high oral doses of goldenseal may cause convulsions and stomatitis, pharyngitis, gastritis, cutaneous tingling sensation, paralysis, respiratory failure, and possibly death. Its chronic use also inhibits intestinal absorption of vitamin B resulting into vitamin B deficiencies which may manifest as megaloblastic anemia, peripheral neuropathy, and encephalopathy. It is known also to induce hallucinations and delirium at high doses [30].

6.2.5 Symphytum officinale *(Bignoniaceae)*

Comfrey (*Symphytum officinale*) acts as an anti-inflammatory to promote healing of bruises, sprains, and open wounds when applied topically. The roots and leaves of this plant contain the protein allantoin, which stimulates cell proliferation and promote wound and bone healings. It is also ingested as a herbal tea to treat gastric ulcers, rheumatic pain, arthritis, bronchitis, and colitis. Despite its folkloric uses, comfrey primarily contains symphytine and echimidine, pyrrolizidine alkaloids, which have been linked to hepatic cancers and hepatotoxicity in chronic animal and clinical studies, respectively [31,32]. These toxic alkaloids have also been reported to induce lung cancers in rats [32−34].

6.2.6 Ginkgo biloba *(Ginkgoaceae)*

Ginkgo biloba, also known as the maidenhair tree, is one of the oldest species of trees on the planet. Its leaf extract is among the most widely sold herbal dietary supplements in the United States, with its purported medicinal benefits to include free radical scavenging activity, lowering of oxidative stress; reducing neural damages, reducing platelet aggregation, anti-inflammation, antitumor activity, and antiaging activities. It is clinically prescribed to treat central nervous system (CNS) disorders such as Alzheimer's disease and cognitive deficits, and there are still insufficient data supporting its clinical benefits in this respect [35]. Indeed, *Ginkgo biloba* extract is listed as the fifth or sixth most frequently used herbal dietary supplement in the United States, and the third best-selling herbal product in health food stores in the United States in 1997 [35]. The leaf extract of *Ginkgo biloba* when administered to humans at the oral dose of 120 mg/day for 52 weeks was associated with headaches, dizziness, palpitations, gastrointestinal disturbances, and allergic skin reactions [36]. It has been reported that 4-O-methylpyridoxine, a neurotoxic antivitamin B_6, also known as ginkgotoxin, found in ginkgo seeds and leaves, was responsible for convulsions, loss of consciousness, and death in the human subjects of Japanese clinical trials [37]. Ginkgotoxin inhibits the formation of 4-aminobutyric acid from glutamate in the brain and 4-aminobutyric acid deficiency in the brain induces seizures [38]. There are also reports that *Gingko biloba* leaf extract may reduce the ability of the blood to clot and prolong bleeding time, resulting in spontaneous bleeding [39]. Thus, concurrent use of *Gingko biloba* with oral anticoagulant medication should be avoided as the additive effects of ginkgolide B, a potent inhibitor of platelet-activating factor, in the *Gingko biloba* leaf extract could potentiate the anticoagulant effect of the oral anticoagulant medication [40,41]. Dugoua et al. [42] suggested that ginkgo must be avoided or used with caution during pregnancy.

6.2.7   Piper methysticum *Forster f. (Piperaceae)*

This is a tropical shrub belonging to the family Piperaceae. The common names for *Piper methysticum* extract include ava, ava pepper, awa, intoxicating pepper, kava, kava kava, kava pepper, kawa, kawa kawa, kew, rauschpfeffer, sakau, tonga, wur zelstock, and yangona [43]. It is widely cultivated for its rootstock in three geographic regions of the Pacific: Polynesia, Melanesia, and Micronesia [44], although it is also widely cultivated in North African countries such as Morocco, Tunisia, Egypt, etc., where it is used as a ritual beverage to promote relaxation and a sense of well-being. Approximately 2.2 million Americans used Kava Kava as a natural alternative to antianxiety drugs [45], used as diuretic and antiseptic [44], and it has been reported to help children with hyperactivity [46]. The recommended oral dose for usage of Kava Kava as an anxiolytic is 50−70 mg kavalactones two to four times a day and, as a hypnotic, 150−210 mg in a single oral dose before bedtime [47]. The active principles of Kava Kava rootstock are mostly contained in the lipid-soluble resin which is generally grouped into categories namely: arylethylene-apyrones, chalcones and other flavanones, and conjugated diene ketones. The compounds of greatest pharmacological interest are the substituted α-pyrones or Kava pyrones, commonly known as kavalactones. Fifteen lactones have been isolated from Kava Kava rootstock, of which six are present in the highest concentrations and account for approximately 96% of the lipid resin. These are yangonin, methysticin, dihydromethysticin, kavain, dihydrokavain, and desmethoxyyangonin [48,49]. Subchronic studies following oral exposure to Kavain (10−400 mg/kg) in dogs for 91 days revealed the presence of mild toxicity with multicentric liver necrosis in a high dose dog [50]. Similarly, subchronic and chronic toxicity studies of 1 and 2 g/kg of Kava Kava in rats and mice reported for 3 months and 2 years, respectively, was associated with weight loss, ataxia and lethargy, hepatomegaly, and varying degrees of hepatotoxicity and hepatic cholestasis in both sexes of the animal species used [51]. Similarly, in humans, the main concern with the use of Kava Kava is hepatotoxicity in humans [52−56]. However, evidence on the hepatotoxic effects of Kava Kava currently remain conflicting, with some reports indicating an association of Kava

Kava administration with hepatotoxicity including hepatitis, cirrhosis, and liver failure [57−59]. Additional reports describing adverse effects such as visual disturbances, lethargy, and disorientation and possible drug interactions following the use of various preparations of kava also have been reported [60].

6.2.8 Glycyrrhiza glabra *(Fabaceae)*

This is also known as true licorice and sweet licorice. *Glycyrrhiza glabra*, of the Leguminosae/Fabaceae family, is a perennial herb or subshrub with horizontal underground stems, native to Eurasia, northern Africa, and western Asia but now cultivated worldwide [61,62]. It is commonly used in treating coughs and colds, chronic fatigue, gastric and duodenal ulcers, and canker sores. It is also used in the local management of abdominal pain, inflammation, muscle spasms, bronchitis, and as a heart tonic [62,63]. Subchronic oral toxicity of licorice (10 g/kg/day), glycyrrhizin (160 mg/day), glycyrrhizin salts (100−2000 mg/kg), glycyrrhetic acid (α- and β-isomers (300 mg/kg)), and deglycyrrhizinated licorice extract (extract containing 3−4% glycyrrhizin, as compared to 20−25% in the original extract (800 mg/kg)] in rats for 50 days to approximately 24 weeks, showed the extract to elicit hypertensive and deoxycorticosteromimetic effects, cognitive and neuromotor toxicities [64]. In humans, prolonged ingestion of licorice and/or its active metabolites has been associated with mineralocorticoid syndrome which is characterized by sodium retention, potassium loss, elevated blood pressure, generalized peripheral edema, and suppression of the rennin-angiotensin-aldosterone system [65,66]. In some cases, hypokalemia can be so severe as to induce myopathy [67,68]. This mineralocorticoid syndrome has been attributed to glycyrrhetic acid, the active metabolite of glycyrrhizic acid resulting from glycyrrhizin de-glucuronation in the gastrointestinal tract [67]. Glycyrrhetic acid is known to inhibit the oxidation of cortisol via inhibition of the enzyme 11β-hydroxysteroid dehydrogenase (11β-HSD) [69] by competitively binding to 11β-dehydrogenase, which complexes with 11-oxoreductase to form 11β-HSD [67]. This complex is responsible for the interconversion of cortisol and cortisone, and as a result of its inhibition, cortisol is not degraded and thus, may exert its mineralocorticoid action in mineralocorticoid-selective tissues (e.g., kidney, colon, and parotid gland) [70]. Cortisol mimics aldosterone, stimulating the reabsorption of sodium from renal tubules and the secretion of potassium into the urine, causing a state of apparent mineralocorticoid excess. The increased sodium reabsorption, thus, depresses the renin-angiotensin-aldosterone axis, and as a reaction to increases in atrial stretch caused by fluid retention, the serum concentration of atrial natriuretic peptide (ANP) increases [71,72]. Occurrence of arterial hypertension and edema indicate that the compensatory mechanisms counteracting the glycyrrhetic acid-induced sodium retention are overwhelmed. However, discontinuation of licorice administration showed that recovery of the renin-angiotensin system after discontinuation of licorice is very slow, apparently due to the slow clearance of the drug, and its continued inhibition of 11β-HSD [73]. In addition to the classical symptoms of hypertension, hypokalemia, and suppression of the renin-aldosterone system, hypertensive encephalopathy, as well as the increased incidence of proteinuria, headaches, nausea, and vomiting have been associated with the regular daily intake of low doses of licorice [74]. Other adverse effects reportedly associated with licorice ingestion include rhabdomyolysis, myoglobinuria [75], and transient visual impairments such as binocular scintillating scotomas [76]. Also, the potential for herb-drug interactions is reported to exist between licorice and thiazide diuretics and digitalis glycosides. Deglycyrrhizinated licorice also has been reported to significantly decrease the renal excretion of nitrofurantoin (an antibiotic used for treating urinary tract infection) [77].

6.3 Methods, Parameters, and Data Interpretation in Subchronic and Chronic Toxicity Screenings Worldwide and in Africa

6.3.1 Conventional Subchronic and Chronic Oral Tests

In conducting descriptive toxicity in laboratory animals, there are different types of toxicity testing that could be undertaken based on the route of drug administration (oral, dermal, inhalational, subcutaneous, and intraperitoneal) and frequency and duration of exposure to the toxicant (acute, subacute, subchronic, and chronic). Again, for the purpose of this chapter, emphasis shall be placed on the methods of subchronic and chronic oral toxicity testing, which are further discussed in detail below.

6.3.1.1 Subchronic Oral Toxicity Test

This toxicity type often follows subacute toxicity test after appropriate doses have been reliably established. Its duration is often varies, but preferably between 28 and 90 days. However, 90 days remains the most common and acceptable test duration. The main principle underlying it is to establish a nonobservable effect level and to further identify and characterize the specific organ(s) affected by the test drug after prolonged exposure to it. The subchronic study often is conducted in two animal species (preferably rats and dogs) using at least three graded doses (a high dose that effects toxicity with not more than 10% of death ensuing; a low dose that effect no apparent toxicity, and a middle dose usually of therapeutic dose). The test is conducted with 15−30 rats and 4−6 dogs of each sex per dose. Each healthy test animal is uniquely identified with either a permanent marking or ear tag and housed individually in well-controlled and standard laboratory conditions with the animals observed at least once a day for signs of toxicity, including body weight changes, feeding pattern, changes in fur color or texture, respiratory or cardiovascular distress, motor and behavioral changes, and any obvious and palpable body masses or swelling. There should be little individual variation between animals, and the allowed weight variation range is ±20%. A satellite group may be included in the study protocol, and this group has both a control group and a high dose group [78]. All preterm deaths should be recorded and autopsies conducted. All moribund animals should be separated and euthanized before being terminated as soon as possible in order to minimize the suffering of the affected animal.

Weekly body weight variations, monthly biochemical and hematological parameters, and behavioral changes are observed. At the end of the 90-day study, all the remaining animals should be weighed, before being fasted of food overnight but with drinking water made available. The animals are then sacrificed under light anesthesia with blood samples collected for both hematological and biochemical analyses and body tissues preferably, vital organs weighed and then collected for gross and microscopic examinations. Hematology and blood chemistry samples usually are collected at the beginning, middle, and end of the toxicity testing. Hematology measurements include the full blood count parameters (as hematocrit, hemoglobin concentration, erythrocyte count, mean corpuscular volume (MCV), mean hemoglobin concentration, mean corpuscular hemoglobin concentration (MCHC), total and differential white blood cell count, and platelet count), clotting time and prothrombin time (PT). Clinical chemistry parameters commonly measured include glucose, renal function parameters (sodium, potassium, chloride, bicarbonates, calcium, phosphate, urea, and creatinine), liver function parameters (alanine aminotransferase, aspartate aminotransferase, gamma-glutamyltranspeptidase, alkaline phosphatase, sorbitol dehydrogenase, lactate dehydrogenase, bilirubin, albumin, total protein, globulin, triglyceride, total cholesterol, and its fractions). Also, urinalysis is usually performed at the middle and end of the testing period with parameters such as specific gravity, pH, glucose, ketones, nitrites, bilirubin, and urobilinogen performed, in addition to urine microscopic examination for oxalates crystals, epithelial and pus cells conducted. Overall, the essence of subchronic toxicity studies is not only to characterize the dose-response relationship of a test compound following repeated exposure to it, but also to provide detailed information on more reasonable prediction of appropriate doses for the chronic toxicity studies. The basic parameters for measuring subchronic oral toxicity are summarized in Table 6.1.

6.3.1.2 Chronic Oral Toxicity Study

This is otherwise known as the long-term toxicity studies. It is performed similarly to subchronic toxicity studies except that its duration ranges from 6 months to 2

Table 6.1 Basic Parameters of Subchronic Oral Toxicity

Parameters	Observations
Species	Rodents (usually rats) preferred while nonrodents (usually dogs) recommended as a second species for oral tests
Age	Young adults
Number of animals	15–30 of each sex for rodents; 4–6 of each sex for nonrodents per dose level
Dosage	Three dose levels plus a control group; include a toxic dose level plus NOAEL; exposure are for 90 days
Observational period	90 days (same as treatment period)

Table 6.2 Basic Parameters for Measuring Chronic Oral Toxicity Testing

Parameters	Observations
Species	Two species recommended, rodents and nonrodent (rat and dog)
Age	Young adults
Number of animals	20 of each sex for rodents, 4 of each sex for nonrodents per dose level
Dosage	Three dose levels recommended; includes a toxic dose level and NOAEL; exposures generally for 12 months
Observational period	12−24 months

years depending on the animal species being studied. A chronic toxicology study provides inferences about the long-term effect of a test substance in animals, and the results may be extrapolated to the human safety of the test substance. The report on chronic toxicity is essential for new drug entity. In rodents, the test duration is usually for 6 months while lasts for between 1 and 2 years in nonrodents. The weight variations within and between the test groups should not exceed ±20%. A satellite group may be included in the study protocol. This group has both a control group and high dose group. During the study period, the animals are observed for normal physiological functions, behavioral variations, and alterations in biochemical parameters. At the end of the study, organ tissues obtained from the animals and subjected to histological studies [79]. The essence of this toxicity is to assess the cumulative toxicity of a drug. The basic parameters for this test are summarized in Table 6.2.

6.3.2 OECD Recommendations for Subchronic and Chronic Toxicity Study

The original OECD Guideline 452 (OECD TG 452) designed for conducting animal chronic toxicity was first designed and adopted in 1981 after standardized methods of testing were jointly accepted in principle by all the 30 OECD member countries. However, this guideline is periodically revised in order to reflect recent developments in the field of animal welfare and regulatory requirements. Most often chronic toxicity studies are conducted in rodent species (and rarely in nonrodent species) via the oral, dermal, and inhalation routes of drug administration depending on the physical and chemical characteristics of the test substance and the predominant route of exposure of humans. The OECD guideline focuses on exposure via the oral route, the route most commonly used in chronic toxicity studies.

The objectives of chronic toxicity studies covered by this test guideline include: the identification of the hazardous properties of a drug/chemical, identification of target organs, characterization of the dose-response relationship, identification of a NOAEL or point of departure for establishment of a BMD, prediction of chronic toxicity effects at human exposure levels, and provision of data to test hypotheses regarding mode of action. Thus, chronic toxicity study provides information on the possible health hazards likely to arise from repeated exposure over a considerable part of the entire life span (in rodents). The study also provides information on the possible toxic effects of the test drug, and indicates target organs and the possibility of accumulation. It can also provide an estimate of the NOAEL which can be used for establishing safety criteria for human exposure.

6.3.2.1 OECD Procedure for Chronic Toxicity

The criteria and procedures to be considered for chronic toxicity studies using OECD guidelines [80] include:

6.3.2.1.1 Animal Species
Young healthy adult, nulliparous, and nonpregnant female rodent species (particularly rats and mice) of same strain are the preferred experimental animal for this study. These rodent species have a relatively short life span, are widely used in pharmacological and toxicological studies, are highly susceptible to tumor induction, and sufficiently characterized strains are readily available. As a consequence of these characteristics, a large amount of information is available on their physiology and pathology.

6.3.2.2 Housing and Feeding Conditions

Experimental animals are housed individually or caged in small groups of the same sex. Each cage is properly labeled, identified, and arranged in the laboratory such cage misplacements are minimized. The recommended laboratory ambient conditions are 22°C (±3°C) temperature, 50−60% but not exceeding 70% relative humidity, 12/12 h light/dark periodicity. The diets are standard, meeting all the nutritional requirements of the species tested. Feeds and potable drinking water which are made freely available to the treated rats must be free of dietary contaminants such as pesticide residues, persistent organic pollutants, phytoestrogens, heavy metals, and mycotoxins that might influence the outcome of the toxicity test.

6.3.2.3 Animal Preparation

Healthy and at most 8-week-old rodents, which have been previously acclimatized standard laboratory conditions for at least 7 days and not previously subjected to experimental procedures, are used. The test animals are characterized as to species, strain, source, sex, weight, and age. At the commencement of the study, the weight variations of animals are ensured not to exceed ±20% of the mean weight of all the animals within the study, separately for each sex. Animals are randomly assigned to the control and treatment groups. After randomization, the differences in the mean body weights within and between groups within each sex are ensured not to exceed ±20% of the mean weight of all the animals. If there are statistically significant differences, then the randomization step should be repeated. Each animal should be assigned a unique identification number, and permanently marked with this number by tattooing, microchip implant, or other suitable method.

6.3.2.4   Number and Sex of Experimental Animals

Both sexes of animal strain are used. At least 20 animals per sex per group are used at each dose level. In studies involving mice, additional animals may be needed in each dose group to conduct all required hematological determinations because of their small blood volume. An additional two satellite groups of twenty animals of each sex, one high dose group and one control group, should be included for evaluation of chronic toxicity and nonneoplastic pathology at 12 months. An additional dose group may be useful to better define the dose response for chronic toxicity.

6.3.2.5   Provision for Interim Kills, Satellite Groups, and Sentinel Animals

The study makes provision for interim kills of treated animals at least 6 months old, to provide information on progression of toxicological changes and mechanistic information, if scientifically justified. Satellite groups also may be included to monitor the reversibility of any toxicological changes induced by the chemical under investigation; these will normally be restricted to the highest dose level of the study plus control. An additional group of sentinel animals (typically five animals per sex) also is included for monitoring of disease status during the study. If interim kills or inclusion of satellite or sentinel groups are planned, the number of animals included in the study design is increased by the number of animals scheduled to be killed before the completion of the study.

These animals should normally undergo the same sham-handling and observations, including body weight, food/water consumption, hematological and clinical biochemistry measurements, and pathological investigations as the animals in the chronic toxicity phase of the main study.

6.3.2.6   Dose Groups and Dosage

At least three dose levels (high, middle, and low) and a concurrent control are used in the study. The dose levels are chosen from either the results of short-term repeated dose toxicity studies or range finding studies and should take into account any existing toxicological and toxicokinetic data available for the test substance or related materials. The highest dose level is chosen to identify the principal target organs and toxic effects while avoiding suffering, severe toxicity, morbidity, or death. The highest dose level is chosen so as to elicit evidence of toxicity, as evidenced by, for example, alterations in serum enzyme levels or depression of body weight gain (by approximately 10%). The highest dose, however, should not exceed 1000 mg/kg body weight/day (limit dose). Dose level spacing should be designed to demonstrate a dose response and to establish a NOAEL or other intended outcome of the study (e.g., a BMD) at the lowest dose level. Factors to be considered in the placement of lower doses include the expected slope of the dose-response curve, the doses at which important changes may occur in metabolism or mode of toxic action, where a threshold is expected, or where a point of departure for low dose extrapolation is expected. Two- to fourfold intervals are frequently optimal for setting the descending dose levels and addition of a fourth test group is often preferable to using very large intervals (e.g., a factor of about 6−10) between dosages but factors greater than 10-fold are avoided. The control is the untreated group or a vehicle control group if a vehicle is used in administering the test substance with the animals in this group under same sham-handling as other treatment groups.

6.3.2.7 Preparation of Doses and Routes of Drug Administration

The commonest route of drug administration is the oral route. For animal welfare considerations, oral gavage is normally selected only for those drugs for which this route reasonably represents potential human exposure. For dietary drugs or environmental chemicals, administration should be via the diet or drinking water. The choice of the route of administration depends on the physical and chemical characteristics of the test substance and the predominant route of exposure of humans. All the animals are dosed with the drug on a daily basis for a period of 12 months. The drug is administered by oral gavage to the animals using a stomach tube or a suitable intubation cannula, at similar times each day, preferably, as a single daily dose. The volume for drugs in the aqueous suspension should normally not exceed 1−2 mL/100 g body weight.

6.3.2.8 Duration of Study

The period of dosing and duration of the chronic phase of this study is normally 12 months with the high dose and control satellite groups terminated at this stage for evaluation of chronic toxicity pathology. Satellite groups included to monitor the reversibility of any toxicological changes induced by the test drug is, however, maintained without dosing for a period not less than 4 weeks and not more than one-third of the total study duration after cessation of exposure.

6.3.2.9 Observations During Toxicity Study

General clinical observations of the treated rats is made at least once a day, preferably at the same time of the day, taking into consideration the peak period of anticipated effects after dosing. The clinical condition of the animals is checked and recorded. All animals are also checked for morbidity or mortality, and for specific signs of toxicological relevance, usually at the beginning and the end of each day throughout the study period including weekends and holiday. Signs of toxicity to be watched out for include, but are not limited to, changes in skin, fur, eyes, mucous membranes, occurrence of secretions and excretions, and autonomic activity (e.g., lacrimation, piloerection, pupil size, and unusual respiratory pattern). Changes in gait, posture, and response to handling as well as the presence of clonic or tonic movements, stereotypes (e.g., excessive grooming, repetitive circling) or bizarre behavior (e.g., self-mutilation, walking backward) should also be noted.

6.3.2.10   Body Weights and Food/Water Consumption

All animals are weighed at the start of treatment, at least once a week for the first 13 weeks and monthly thereafter. Measurements of food consumption are made at least weekly for the first 13 weeks and monthly thereafter. Water consumption also may be measured during the study as the drinking activity may be altered, and this should be measured at least weekly when a drug is administered in drinking water.

6.3.2.11   Hematology and Clinical Biochemistry

At the end of the treatment period, blood samples are taken directly from the heart by cardiac puncture or bleeding from retro-orbital sinus. Blood samples are obtained from at least 5−10 male and 5−10 female animals per group, at 3, 6, and 12 months, using the same animals throughout. In mice, satellite animals may be required in order to conduct all required hematological determinations. The hematological parameters to be measured include total and differential leukocyte count, erythrocyte count, platelet count, hemoglobin concentration, hematocrit (packed cell volume), methemoglobin, MCV, mean corpuscular hemoglobin (MCH), and MCHC, PT, and activated partial thromboplastin time. In the same vein, blood samples are obtained in the same way as that for hematological studies. Clinical biochemical parameters to be measured include glucose, urea (urea nitrogen), creatinine, total protein, albumin, calcium, sodium, potassium, total cholesterol, hepatocellular enzymes (alanine aminotransferase, aspartate aminotransferase, glutamate dehydrogenase), and hepatobiliary markers (alkaline phosphatase, gamma glutamyltransferase, 5′-nucleotidase, total bilirubin, total bile acids). In addition, urinalysis is performed on at least five to ten male and five to ten female animals per group on samples collected at the same intervals as for hematology and clinical chemistry. Urinalysis parameters to be measured include appearance, urine volume, osmolality or specific gravity, pH, total protein, glucose, ketone, urobilinogen, bilirubin, and occult blood.

6.3.2.12   Pathology: Gross Necropsy

All treated animals in the study are normally subjected to a full, detailed gross necropsy which includes careful examination of the external surface of the body, all orifices, and the cranial, thoracic, and abdominal cavities and their contents after the animals have been lightly anesthetized and sacrificed humanely. Organs such as the adrenals, brain, epididymides, heart, kidneys, liver, ovaries, spleen, testes, thymus, thyroid (weighed postfixation, with parathyroids), and uterus of all animals (apart from those found moribund and/or intercurrently killed) are identified, trimmed of any adherent tissue, harvested, and their wet weights taken as soon as possible after dissection to prevent drying.

6.3.2.13   Histopathology

Histopathological examinations should be performed on organs/tissue from the high dose and control groups; all tissues from animals dying or killed during the study; all tissues showing macroscopic abnormalities; target tissues, or tissues which showed treatment-related changes in the high dose group, from all animals in all other dose groups.

The OECD approach has largely removed the need for testing according to different protocols to satisfy regulatory authorities in different countries, and as such, drastically reduced the total number of animals used for certain standard tests. It also has provided a focus for the introduction of new methods that replace, reduce or refine animal use. However, adoption of OECD guidelines has been limited by the slow process and protocols involved in reflecting the periodic revision and changes of these guidelines.

6.3.3   WHO Recommendations for Subchronic and Chronic Toxicity Study

The research guidelines for evaluating the safety of herbal medicine/medicinal plants is as described in the *WHO's Research guidelines for evaluating the safety and efficacy of herbal medicines* [81] are discussed below:

- *Animal species*: Many regulatory agencies require that at least two species be used, one a rodent and the other a nonrodent.
- *Sex*: Normally, the same number of male and female animals should be used.
- *Number of animals*: In the case of rodents, each group should consist of at least 10 males and 10 females. In the case of nonrodents, each group should consist of at least three males and three females. When interim examinations are scheduled, the number of animals should be increased accordingly.
- *Route of administration*: Normally, the expected clinical route of administration should be used, preferably, the oral route for most drugs especially medicinal plants.
- *Administration period*: The period of administration of the test drug to animals depends on the expected period of clinical use. The period of administration of the toxicity study may vary from country to country, according to the country's regulatory authorities. However, for subchronic and chronic oral toxicity testing, treatment period of 3−6 and 9−12 months are generally recommended.
- *Dose levels*: Groups receiving at least three different dose levels should be used. These dose levels are: a low dose level that will not cause toxic changes (no-effect dose) and a high dose level that produces overt toxic effects. Within this range is the addition of at least one more dose may enhance the possibility of observing a dose-response relationship for toxic manifestations. A vehicle control group of test animals should be included. As a rule, the test dose is administered 7 days a week including weekend and holidays. Administration periods for the toxicity study are also recorded in each result.
- *Observations and examinations*: The following observations and examinations should be routinely or regularly performed on the animals during the treatment period.
- *General signs, body weight, and food and water intake*: For all experimental animals, the general signs should be observed daily and body weight and food intake should be measured periodically. Water intake should also be determined. The frequency of measurements should normally be as (i) body weight (before the start of drug administration, at least once a week for the first 3 months of administration, and at least once every 4 weeks thereafter) and (ii) food intake (before the start of drug administration, at least once a week for the first 3 months of administration and at least once every 4 weeks thereafter. If the test drug is administered mixed in the food, the intake should be measured once a week).

- *Hematological examination*: For rodents, blood samples should be taken before autopsy. For nonrodents, blood samples should be taken before the start of drug administration, at least once during the administration period (for studies of longer than 1 month), and before autopsy. Full hematological parameters should be measured.
- *Renal and hepatic function tests*: Because the liver and kidneys are the usual organs of metabolism and excretion, they are easily affected by potentially toxic agents; their functions should be monitored in long-term toxicity studies. For rodents, a fixed number of animals from each group should be selected and urinalysis should be performed before the start of drug administration, and at least once during the administration period. Clinical chemistry tests in general include measurements of electrolyte balance, carbohydrate metabolism, and liver and kidney function. Serum enzyme levels indicative of hepatocellular function that are typically evaluated include alanine aminotransferase, aspartate aminotransferase, sorbitol dehydrogenase, and glutamate dehydrogenase. Assessment of hepatobiliary function may include measurements of serum alkaline phosphatase, total bilirubin, gamma glutamyl transpeptidase (GGT), 5′-nucleotidase and total bile acids. Markers of cellular function or change include albumin, calcium, chloride, creatinine, globulin (calculated), fasting blood glucose, triglycerides, total cholesterol and total protein, phosphorus, potassium, sodium, and urea nitrogen. Significant changes in these biochemical parameters may signal renal, cardiac, or hepatic toxicity. They may be particularly useful for interpretation of study results where there are changes in organ weight, such as liver or kidney, but no overt histopathological changes, as alterations in clinical chemistry parameters associated with organ function can be the first indication of toxicity.
- *Other functional tests*: Functional tests such as the electrocardiograph (ECG) and visual, auditory tests should be performed in nonrodents. For rodents, ophthalmological examination should be performed on a fixed number of animals from each group at least once during the administration period; for nonrodents, examination should be performed on all animals before the start of drug administration and at least once during the period of administration.
- *Animals found dead during the examination should be autopsied as soon as possible*: A macroscopic examination should be made of organs and tissues. In addition, where possible, organ weight measurements and histopathological examinations should be performed in an attempt to identify the cause of death and the nature (severity or degree) of the toxic changes present.
- *In order to maximize the amount of useful information that can be obtained during the administration period, all moribund animals should be sacrificed rather than allowed to die*: Prior to sacrifice, clinical observations should be recorded and blood samples collected for hematological and biochemical analyses. At autopsy, a macroscopic examination of organs and tissues and measurement of organ weights should be recorded. A full histopathological examination should be performed in an attempt to characterize the nature (severity or degree) of all toxic changes. All survivors should be autopsied at the end of the administration period or of the recovery period after taking blood samples for hematological (including blood chemistry) examinations; organs and tissues should be examined macroscopically and organ weights measured.
- *Histopathological examination*: Histopathological examination of the organs and tissues of animals receiving lower dosage should also be performed, if changes are found on gross or macroscopic examination of their organs and tissues of these animals, or if the highest dose group reveals significant changes. On the other hand, histopathological examination of all rodents will further improve the chances of detecting toxicity.
- *Recovery from toxicity*: In order to investigate the recovery from toxic changes, animals that are allowed to live for varying lengths of time after cessation of the treatment.

Usually, a minimum recovery period of 4 weeks is allowed following cessation and terminal of chronic oral toxicity and the experimental animals are subjected to same set of anthropological, full hematological, biochemical, and histopathological investigations as in the treatment groups.

The WHO guidelines like that of OECD also have harnessed all the different protocols of WHO member states to develop common satisfactory animal testing guidelines for the regulatory authorities in different countries. These guidelines have significantly reduced the total number of animals used for certain standard tests. It also has provided a focus for the introduction of new methods that replace, reduce, or refine animal use. However, these guidelines have not been uniformly adopted by the WHO member countries due to the slow process and protocols involved in reflecting the periodic revision and changes of these guidelines. The WHO guidelines are closely related but slightly different from the OECD in that the former accommodates the use of nonrodents such as dogs, hamster, swine, etc., unlike the latter, which advocates strictly for the use of rodents.

6.3.4 Other Recommendations

6.3.4.1 Nonanimal Alternative Methods

Expert working groups have observed that "interspecies differences limit the usefulness of animal studies for predicting long-term target-organ and target-system effects in humans" [82] and that *in vitro* approaches often provide more relevant information for hazard assessment than the animal tests [83]. Therefore, current efforts are geared toward replacing animal chronic toxicity models with technologically driven human cell-based models. These nonanimal alternative models for chronic toxicity testing include human and animal perfused organs; organ tissue slices; isolated, suspended cells; primary cultured cells; cultured cell lines; genetically engineered cell lines; reaggregating cell cultures; three-dimensional cell cultures and co-cultures; and (quantitative) structure activity relationship [(Q)SAR] computational systems [84,85].

In vitro and *in silico* methods as alternatives to repeated dose testing as described by Prieto et al. [86] include:

- *Liver models*: such as isolated human hepatocytes, liver slices, and perfused liver; perfused, collagen sandwich, and bioreactor test systems based on human liver cells
- *Kidney models*: renal epithelial primary cells and cell lines
- *CNS models*: neuronal primary cells and cell lines; reaggregating brain cell cultures; brain slices; astrocyte cell cultures; oligodendrocyte cell cultures; microglia cell cultures
- *Pulmonary models*: isolated, perfused rat and mouse lung; lung slices; human tracheal/bronchial epithelial cell cultures; human alveolar cell model of Skinethic; primary rat pneumocyte type II cell cultures; and other cells models in development
- *Hematopoietic models*: long-term bone marrow cultures; long-term culture initiating cells; myeloid-lymphoid initiating cell assay; human cobblestone area-forming cells; blast colony-forming cell assay; high proliferative potential colony-forming cell assay; colony-forming unit-A method
- *Novel long-term culture methods*: hollow fiber bioreactors; perfusion culture models.

However, these *in vitro* testing methods are without challenges which include: the challenge of maintaining and exposing the cells for the extended periods of time required to assess chronic toxic effects; identification and use of cell culture conditions that retain the fundamental functions of the cell/tissue/organ type; identification and integration of all of the various biological parameters and tissue toxicities that occur *in vivo* with appropriate *in vitro* models; and extrapolation of the *in vitro* results to the *in vivo* system.

In vitro models are developed using cells or tissues from the organs that are the typical targets of toxicity. The liver is the primary site for drug metabolism in the body as well as the primary site of potential toxic injury (hepatotoxicity). The liver is predominantly composed of hepatocytes, each of which contains phase 1 and 2 drug metabolizing enzymes that convert many drugs and chemical into metabolites that the body can easily excrete. Species differences in the activities of the liver's drug metabolizing enzymes are one of the major contributors to the species differences observed in drug and chemical toxicity. Therefore, a human liver model for the prediction of human toxicity is particularly important. A major drawback to using *in vitro* liver cells/models for chronic toxicity testing has been the limited time the cells maintain their normal functions *in vitro* [83]. Improvements in cryopreservation and culturing techniques for human hepatocytes that extend their time in culture and their predictive capabilities are therefore important.

The kidney is another organ susceptible to drug-induced toxicity. Like the liver, perfused whole kidney and kidney slices have been used to test drugs and chemicals for short periods of time [83]. Kidney tubule cell cultures and cell lines have been developed, and some can be maintained for extended periods [85]. Likewise, cells from other human organs such as lung and brain, as well as blood and immune cells have been cultured for toxicity testing. However, despite the commendable progress made in the development of predictive *in vitro* organ models, technological and scientific limitations remain strong barriers to scientific breakthroughs in this fast-emerging area of science. *In vitro* organ models can only satisfactorily replace animal testing for chronic toxicity testing when they are organized into a predictive integrated testing scheme. Thus, biokinetics of a chemical in the human or animal body must be determined in order to develop appropriate *in vitro* methods and models. Combinations of *in vitro* and/or (Q)SAR models to determine the biokinetics (i.e., the absorption/uptake, distribution, metabolism/biotransformation, and excretion, or ADME) of a chemical must be evaluated to fully understand its potential chronic toxic effects [87−89].

6.4 Animals Used in Subchronic and Chronic Toxicity Screenings

Laboratory animals are sensitive to toxic substances occurring in plants. Hence, the administration of the plant extracts in increasing amounts enables the evaluation of the subchronic and chronic toxicity limits. Therefore, the subchronic and chronic toxicity tests are carried out for three doses and for both sexes, taking into account other factors such as age, weight, species, diet, and environmental conditions [90]. In toxicity studies, there is no single species of animal that can be used for all toxicity tests, thus necessitating species selection. Experimental animals often used in evaluating the subchronic and chronic toxicity testing of African medicinal plants include the mice, rats, and rabbits while others commonly used in developed countries include dogs, guinea pigs, hamster, sheep, and monkeys. However, in the developed countries, these latter species also are frequently used. Preference for the former species is apparently due to their availability, low costs in breeding and housing, and past history in producing reliable results. Limitations to the routine use of monkeys and dogs in subchronic and chronic toxicities may be unconnected to their strict restriction to special toxicity testing, even though they represent the species that may react closest to humans.

6.5 Observable Symptoms and Signs in Subchronic and Chronic Toxicity Screenings

In the course of conducting subacute and subchronic oral toxicity studies, there could be manifestation of behavioral and systemic end points of toxicity. Possible toxicity manifestations include abdominal contractions, blood pressure changes, excessive salivation, pallor, aggression, muscle weakness, hair loss, piloerection (hair standing on end), vomiting (in dogs), tremors, diarrhea, coma, and occasionally death. Similarly, symptoms observable during chronic oral toxicity testing in experimental animals include abnormal body posture, dyskinesia, blood pressure changes, loss of appetite, aggression, restlessness, muscle weakness, weight loss, hair loss, excessive salivation, internal organ damage, piloerection, vomiting (in dogs), tremor and/or convulsion, bloody diarrhea, local and generalized body swelling, bleeding from body orifices (such as the mouth, nostrils, ear, and anus), coma, and death.

6.6 Limitations of the Study of the Subchronic and Chronic Toxicity of Medicinal Plants

Despite the wide use and acceptance of animal testing in subchronic and chronic toxicity studies, there are limitations militating against their use particularly in the face of fast-evolving advances in science and technology. These limitations among others include:

- *Failure of animal testing methods in keeping pace with fast-evolving scientific breakthroughs in biology and biotechnology*: Advances in cell culture and robotic technology have resulted in "high-throughput" technologies and *in vitro* test systems such as bioinformatics, genomics, proteomics, metabonomics, and *in silico* (computer-based) systems have provided potential alternatives to animal testing. Discovery of these "high-throughput" *in vitro* systems which are of potentially of human origin have resulted in a major paradigm shift in toxicity testing. These new approaches would generate more relevant data to evaluate the potential risks associated with drug exposure, expand the number of medicinal compounds or plants that could be screened at a time, reduce the time, money, and animals used in conventional animal testing.

- *Questionable reliability and relevance of current animal testing methods*: Animal toxicity testing is predicated on the assumption that an untoward effect of a drug observed in one animal species could also be observed in others. However, it is also known that different animal species do respond differently to same administered drug of same dose [90,91] due to interspecies differences resulting from genetic, biochemical, or metabolic factors or their combinations. Thus, it is practically impossible to predict accurately if the toxicity results in rodents, rabbits, or dogs will provide an accurate prediction of toxic effects in humans and as such, puts in question the accuracy, reliability, and relevance of such data [92−94]. More importantly, in recent times there have been heated debates on the relevance of extrapolating data emanating from administration of high drug doses to animals to realistic human or environmental exposure levels [95]. Also, despite efforts to standardize the toxicity testing procedures, the results emanating from such animal studies could be highly variable and difficult to reproduce making the animal data highly unreliable [96,97].
- *Animal welfare concerns*: In the last three decades, there have been growing concerns about the use of laboratory animals in toxicity testing. These concerns were borne out of the fact that the conventional animal toxicity testing of a drug often requires a large number of experimental animals, sometimes in hundreds and thousands [98]. Also, emerging figures from countries around the world showed that animal toxicity testing accounts for up to 70% of the most painful procedures to which test animals are subjected for experimental purposes (e.g., the continued use of fatality or moribundity (near death) as end point in acute systemic toxicity studies).
- *Time and cost*: The subchronic and chronic toxicity tests usually take months and even years to conduct, analyze, and interpret, often at exorbitant cost of hundreds of thousands and sometimes millions of US dollars.
- *Legal obligations*: As public opposition to the continued use of animal toxicity testing grows, some countries have openly prohibited animal testing where alternative methods are "reasonably and practically available." For example, since 2004 there has been a ban on animal testing of cosmetic ingredients as well as the marketing of cosmetic products whose ingredients have been tested on animals within the European Union.

6.7   African Plants Screened for Their Subchronic and Chronic Toxicity

Several African medicinal plants have been screened for their subchronic and chronic toxicity profiles. A few of these commonly used African medicinal plants with their oral toxicities are discussed below:

6.7.1   Sphenocentrum jollyanum *Pierre (Menispermaceae)*

Sphenocentrum jollyanum is a perennial plant that grows naturally along the west coast subregion of Africa with expanse from Cameroon across Nigeria to Sierra Leone. Traditionally, the leaf decoctions are used as vermifuge, wound dressings (particularly for chronic ulcers), fever, cough, aphrodisiac [99], jaundice, menstruation-related breast engorgement, swelling, and inflammation [99,100]. Pharmacological studies of the leaf extract of this plant have shown the plant to possess significant anti-inflammatory, antiangiogenic, and analgesic [99,101] and polio type-2 antiviral properties [102]. Chronic oral administration of the 50, 100, and 200 mg/kg of the ethanolic leaf extract of the plant suspended in 2% Tween 80 solution to Wistar rats for 120 days caused significant increase in red blood cells, packed cell volume, hemoglobin, and high density lipoprotein, while causing significant reductions in the fasting blood glucose and total cholesterol levels. However, the prolonged oral treatment caused no significant alterations in the serum alanine aminotransferase and aspartate aminotransferase, mean corpuscular, MCH, MCHC, and white blood cells [1]. The leaf extract also caused no significant effect on the pattern of body weight gain, relative vital organ weights, gross, and histological changes in the vital body organs examined [1].

6.7.2 Spathodea campanulata P. Beauv (Bignoniaseae)

Spathodea campanulata is commonly known as the African tulip tree. The plant is widely distributed in Nigeria and other West African countries and is reputedly used for epilepsy and convulsion control, against kidney disease, urethritis, and as antidote against animal poisons [103]. Decoctions of the plant stem are also employed against eczemas, fungal skin disease, herpes, stomach ache, and diarrhea [104]. Its leaf decoction is used in the treatment of pain, inflammation [105,106], constipation, and dysentery [107]. The plant is also reported to exhibit antiplasmodial activity [108] and analgesic and anti-inflammatory actions [109]. Its subchronic toxicity study showed that oral administration of 750, 1000, and 3000 mg/kg/day of the ethanol leaf extract of the plant to Sprague-Dawley rats for 90 days caused dose and time-dependent hepatotoxicity marked by elevations in the serum liver enzyme markers (aminotransferase and alkaline phosphatase). Its use also improved hemograms in the treated rats while causing no effects on other measured hematological and biochemical parameters. The extract also caused decreased food and water intake, emaciation, muscle weakness, and sluggishness in the extract-treated rats [110]. However, these changes showed complete recovery after 28 days posttreatment (oral toxicity reversibility study) [110]. Similarly, 42 days of subchronic toxicity of the ethanol leaf extract of the plant in Swiss albino mice at the oral dose of 200 mg/kg of the extract caused no significant effects on the hematological profile, biochemical, hepatorenal analyses, antioxidant activities, and histological status of the mice liver suggests nontoxicity of the extract although it slightly elevated the serum levels of blood glucose, total cholesterol, triglyceride, and low density lipoprotein-cholesterol [103].

6.7.3 Syzigium aromaticum (L.) Merr. and Perry (Myrtaceae)

Syzigium aromaticum (Figure 6.1) synonymous with *Eugenia caryophyllata*, and known as clove in English, is one of the most widely used plants worldwide for medicinal purposes. It is an evergreen tree in the Myrtle family and grows in warm climates but is cultivated commercially in Tanzania, Sumatra, the Maluku

Figure 6.1 *Syzigium aromaticum* (L.) Merr. and Perry plant (family: Myrtaceae).

Figure 6.2 A mature tree of *Hunteria umbellata* (K. Schum.) Hallier f. (Apocynaceae).

(Molucca) Islands, and South America [111]. Essential oils are obtained from the buds, stems, and leaves [112] are used as a topical application to relieve pain and promote healing in herbal medicine [113]. Clove oil has long been recognized as safe and is used in foods, beverages, and tooth pastes, but toxicities which could be life-threatening have been documented [113−115]. The systemic use of this oil has been restricted to three drops per day for an adult human, as excessive use can cause severe renal damage [116], disseminated intravascular coagulation, and renal toxicity [115]. Oral administration of 300 and 700 mg/kg/day of boiled aqueous extract of *Syzigium aromaticum* to White Albino rats for 90 days significantly increased hemograms and platelet counts while exhibiting irreversible weight loss, hepatotoxicity, renal damage, cerebral edema, and gastritis [117].

6.7.4   Hunteria umbellata *(K. Schum.) Hallier f. (Apocynaceae)*

Hunteria umbellata (Figure 6.2) is reputed for the folkloric management of labor pain and swellings, stomach ulcers, diabetes, obesity, and anemia [118,119]. Among the Yoruba and Binis (Southwest Nigeria), it is locally known as "Abeere." Oral administration of 100−500 mg/kg/day of the aqueous seed extract of the plant to adult Wistar rats for 90 days resulted in weight loss, organomegally, proliferation of the testicular spermatogenic series, hypoglycemia, and hematopoiesis [120].

Following 14 days of oral toxicity reversibility test, there was no significant reversal in the serum levels of the biochemical and hematological parameters investigated, including the extract-induced histological lesions [120].

6.7.5 Cnestis ferruginea *(CF) Vahl ex DC (Connaraceae)*

Cnestis ferruginea (Figure 6.3), is a shrub commonly found in deciduous forests and secondary scrublands (Burkill, 1985) [121]. It is abundant in West Africa and other tropical regions. Its local names include "Apose" Akan-Asante (Ghana), "Alabalu" Adyukru (Ivory Coast), "Sanhanguin" Kruguere (Liberia), "Treventiito" Balanta (Guinea Bissau), "Diakanare dire" Banyun (Senegal), "Selialiwo" Kissi (Sierra Leone), "Fupeleen" Diola (Gambia), and "Tangolo sebe" Manding-Dyula (Burkinafaso). In Nigeria, the plant is locally called "Fura amarya," "otito" (Hausa); "Okpu nkita," "amunkita" (Igbo); and "Akara oje", "Bonyin bonyin" (Yoruba) [121]. Its root decoctions are used to treat headache, migraine, toothache, dysmenorrhea, skin infections, gynecological troubles, dysentery, urethritis, and sinusitis [121]. Decoctions from the plant root are also employed as an appetite stimulant, purgative, skin ointment, aphrodisiac, and as remedy for snake bite [122]. Oral administration of 80, 400, and 1000 mg/kg/day of the methanolic root extract of the plant to Wistar rats for 90 days resulted in profound weight loss, reductions in testicular weight loss and sperm count and testicular inflammation and necrosis, reduction in relative liver weight and hepatic steatosis, and interstitial nephritis [122]. Despite withdrawal of the extract from the rats for 30 days, these changes were irreversible [122].

6.7.6 Corrigiola telephiifolia *Pourr. (Caryophyllaceae)*

Corrigiola telephiifolia is a Moroccan medicinal plant called "*Sarghina*" and belongs to the Caryophyllaceae family. It is a herbal plant that is widely distributed in Southern Europe and North Africa. The plant root is also used to treat flu, dermatological diseases, inflammation, ulcer, cough, and jaundice; it is also used as an anesthetic and a diuretic [123]. Oral administration of 70 and 2000 mg/kg of the hydroethanolic root extract of the plant to both sexes of Wistar rats for 40 days

Figure 6.3 A flowering *Cnestis ferruginea* (CF) Vahl ex DC plant.

resulted in dose-related abdominal contractions, inactivity, prostration, diarrhea, anorexia, polyuria, and functional nephrotoxicity and hepatotoxicity [124].

6.7.7 Carica papaya *Linn. (Caricaceae)*

Carica papaya (Figure 6.4) is a herbaceous succulent plant popularly known as paw-paw, and belongs to the Caricaceae family. It is native to the tropics of the Americas but now is widely cultivated in other tropical regions of the world for its edible melon-like fruit, which is available throughout the year [125]. Different parts of the plant are employed in the treatment of different human and veterinary diseases in various parts of the world. For example, in Asian folk medicine, the latex is employed as an abortificient, antiseptic for wound dressing, and as a cure for dyspepsia [126]; while in Africa, the root infusion is reputed for treating venereal diseases, piles, and yaws [127]. In Cuba, the latex is used in the treatment of psoriasis, ringworm, and cancerous growth [127]. Its fruit and seed extracts have been reported of possessing pronounced bactericidal activity against *Staphylococcus aureus*, *Bacillus cereus*, *Escherischia coli*, *Pseudomonas aeruginosa*, and *Shigella flexneri* [128]. The pulverized seeds are also documented to possess antiparasitic activities against *Entamoeba histolytica* and *Dirofilaria immitis* infections [129]. The sedative and muscle relaxant [130], reversible antifertility [131], and purgative properties [132] of the plant extract have also been widely reported. Preclinical subchronic toxicity of 250, 500, and 1000 mg/kg of the aqueous leaf extract of *Carica papaya* have been reported to include hypoglycemia, hypolipidemia, improved hematological parameters while those of its ethanolic leaf extract include hyperuricemia, glomerulosclerosis, and tubular clarification, hepatotoxicity, hypoglycemia, and hypolipidemia [133].

6.7.8 Allium cepa *Linn. (Liliaceae)*

Allium cepa (Figure 6.5), commonly known as onion belongs to the Liliaceae family, and is traditionally employed in the management of diabetes mellitus.

Figure 6.4 Fruiting pawpaw trees.

Figure 6.5 Whole fresh bulbs of *Allium cepa* Linn. (Onions).

Figure 6.6 Dried bulbs of *Allium sativum* Linn. (Garlic).

Various soluble and insoluble fractions of dried onion powder exhibited significant hypoglycemic, hypolipidemic, and antioxidant activities in several animal models and diabetic patients [134,135]. Subchronic oral treatments Swiss albino mice with 250 and 500 mg/kg/day of the plant extract for 30−90 days did not produce visible toxic symptoms, but the oral dose of 30 g/kg/day for the same period caused hypothermia, tachypnea, tachycardia, piloerection, and polyuria in the treated mice [9]. However, chronic oral treatment of male Swiss mice with 150 mg/kg/day of the plant extract for 8 weeks resulted in hypothermia, aggression, alopecia, and itching with manifestation of hypothermia and itching in the female mice treated with same dose of plant extract for 12 weeks [9].

6.7.9 Allium sativum *Linn. (Liliaceae)*

Allium sativum (Figure 6.6), otherwise known as garlic, is a perennial herb also belonging to the Liliaceae family. It is locally known as "Lahasun" in Hindi (India) where it is commonly used as a food condiment [136].

Repeated oral treatment of male Swiss mice with 75 mg/kg/day of the plant extract for 6−10 weeks resulted in hypothermia, tachypnea, tachycardia, aggression, excitation, itching, and alopecia in both the treated male and female mice [9].

6.7.10 Carum carvi Linn. (Apiaceae/Umbellifereae)

The aromatic dried seeds of *Carum carvi*, also known as black caraway and cumin seeds are widely used as a spice for culinary purposes and for flavoring confectionaries (such as bread, biscuits, cakes, and candies), cheese, curries, meat products, sausages [137].

In folk medicine, it is used as carminative for stomach ache and to induce diuresis [138]. Its seed has also been reported to give relief in patients suffering from lumbago, rheumatism, toothache, diarrhea, epilepsy, diabetes mellitus, and hypertension [139]. However, repeated oral treatment of both male and female Swiss albino mice for 10−12 weeks resulted in toxicity symptoms such as hyperthermia, fast and shallow breathing, tachycardia, and aggression of the treated mice [9]. Also, a significant pharmacokinetic interaction of some herbal products from cumin and caraway with antitubercular drugs has been reported. An aqueous extract derived from cumin seeds produced a significant enhancement of rifampicin levels in rat plasma leading to exaggerated toxicity of rifampicin [140].

6.7.11 Nigella sativa Linn. (Ranunculaceae)

Nigella sativa Linn. is an annual flowering plant, native to South and Southwest Asia. The seeds of *Nigella sativa*, commonly known as black seed or black cumin, are used in folk medicine worldwide in the local treatment and prevention of bronchial asthma, cough, diarrhea, abdominal pain, and dyslipidemia [141,142]. The plant seed is known to be rich in both fixed and essential oils, proteins, alkaloids, and saponin [143]. However, much of the biological activity of the seeds (including antihypertensive, nephroprotection, hepatoprotection, analgesic, antipyretic, antimicrobial, and antineoplastic activities) has been attributed to its thymoquinone, the major component of the essential oil (although present in lower quantity in the fixed oil) of *Nigella sativa* [143,144]. Toxicological studies of the acute oral toxicity of thymoquinone component of the volatile oil of *Nigella sativa* by Badary et al. [145] showed its LD_{50} value to be 2.4 g/kg, while the immediate behavioral toxicity signs of 2 and 3 g/kg of the compound were hypoactivity and difficulty in respiration. Delayed toxicities of the acute oral toxicity of thymoquinone include a significant reduction in the relative organ weight and glutathione concentrations of the liver, kidneys, and heart. Plasma urea and creatinine concentrations and the enzyme activities of alanine aminotransferase, lactate dehydrogenase, and creatine phosphokinase were significantly increased [145]. In the subchronic study, mice treated with 30, 60, and 90 mg/kg/day of thymoquinone in drinking water at concentrations of 0.01%, 0.02%, and 0.03% (translating to 30, 60, and 90 mg/kg/day, respectively) for 90 days resulted in no mortality or signs of toxicity but significant lowering of fasting blood glucose [145]. There were no changes of toxicological significance in body and organ weights, food and water intake, or urine and feces output. Tissue-reduced glutathione contents, plasma concentrations of total protein, urea, creatinine, and triglycerides, and enzyme activities of alanine aminotransferase, lactate dehydrogenase, and creatine phosphokinase also were not affected. Histological examination revealed no gross or microscopic tissue damage [145]. Similarly, oral treatment of Wistar rats with the seed extract for 12 weeks significantly improved the hemogram (packed cell volume and hemoglobin) and decreased the plasma concentrations of cholesterol, triglycerides, and glucose [141]. In human subjects, topical administration of the seed oil has been reported to cause contact dermatitis [141]. Another clinical trial conducted recruited patients in Karachi reported that oral administration of powdered *Nigella sativa* seed in capsule induced weight loss in 25% of the subjects administered *Nigella sativa* capsule through reduction in appetite [146].

6.7.12   Viscum album *L. (Santalaceae)*

Viscum album is a hemi-parasitic shrub that grows on the stems and crowns of other broad-leaved trees, especially of apple, lime, hawthorn, and poplar. *Viscum album*, also commonly known as European mistletoe and common mistletoe, is one of many species of mistletoe which is native to Europe, South-western Asia, and Nepal [147]. A toxic lectin protein, viscumin has been reported isolated from extracts of mistletoe [148]. *Viscumin* (mistletoe lectin-1 or ML-1) is reported to be effective in cancer treatment. *In vitro* (cell-free and cellular preparations) and animal studies have also shown *viscumin* to be cytotoxic to 3T3 cells, lethal to mice and protein synthesis inhibitor [149−152]. *Viscumin* induces apoptosis in murine lymphocytes (ML-3, ML-2, and ML-1), human peripheral blood lymphocytes and monocytes, murine thymocytes, human monocytic leukemia cell line (THP-1 cells), and human erthromyeloblastoid leukemia (K562) cell line [153], as well as enhance the activity of natural killer cells and granulocyte phagocytosis in patients with breast cancers [154,155]. Clinical evidence suggests mistletoe extract ingested at a high dose induced hepatic injury resulting in a phenomenon referred to as "mistletoe hepatotoxicity" [156,157].

6.8   Conclusion

In summary, this chapter has surveyed the various ways in which animal species (mostly rodents—rats, mice, and occasionally rabbits) are used in safety assessments of drugs including medicinal plants within the context of legal and regulatory requirements governing drug use. These are essential for drug discovery, development, and use. The preclinical subchronic and chronic oral toxicity testing on various biological systems reveals the species-, organ-, and dose-specific toxic effects of the African medicinal plants and products in experimental animal models using conventional, OECD-, and WHO-prescribed guidelines. The preclinical toxicity data generated from this review, thus, provide an in-depth insight into the possible adverse/untoward effects associated and/or could be associated with the ingestion of such medicinal products. In addition, the preclinical toxicity data provide the NOAEL data of the medicinal products which could be much needed to initiate the clinical evaluations/trials of these medicinal products.

References

[1] Mbaka GO, Adeyemi OO, Oremosu AA. Acute and sub-chronic toxicity studies of the ethanol extract of the leaves of *Sphenocentrum jollyanum* (Menispermaceae). Agric Biol J N Am 2010;1(3):265−72.

[2] Akharaiyi FC. Antibacterial, phytochemical and antioxidant activities of *Datura metel*. Int J Pharm Tech Res 2011;3(1):479−83.

[3] Inamul H. Safety of medicinal plants. Pak J Med Res 2004;43(4): <http://www.pmrc. org.pk/434/43413.pdf>.

[4] Janetzky K, Morreale AP. Probable interaction between warfarin and ginseng. Am J Health Syst Pharm 1997;54(6):692−3.

[5] Miller LG. Herbal medicinals: selected clinical considerations focusing on known or potential drug-herb interactions. Arch Intern Med 1998;158(20):2200−11.

[6] Ernst E. Harmless herbs? A review of the recent literature. Am J Med 1998;104 (2):170−8.

[7] Johne A, Brockmöller J, Bauer S, Maurer A, Langhcinrich M, Roots I. Pharmacokinetic interaction of digoxin with an herbal extract from St John's wort (*Hypericum perforatum*). Clin Pharmacol Ther 1999;66(4):338−45.

[8] Cao Y, Colegate SM, Edgar JA. Safety assessment of food and herbal products containing hepatotoxic pyrrolizidine alkaloids: interlaboratory consistency and the importance of N-oxide determination. Phytochem Anal 2008;19:526−33.

[9] Alqasoumi S, Khan TH, Al-Yahya M, Al-Mofleh I, Rafatullah S. Effect of acute and chronic treatment of common spices in Swiss albino mice: a safety assessment study. Int J Pharmacol 2012;8(2):80−90.

[10] World Health Organization. Traditional medicine, Fact Sheet, 134. World Health Organization. Geneva, Switzerland. <http://www.who.int/mediacentre/factsheets/fs134/ en>; 2003.

[11] World Health Organization. WHO guidelines on safety monitoring of herbal medicines in pharmacovigilance systems. Geneva, Switzerland: World Health Organization; 2004.

[12] Cooper EL. Complementary and alternative medicine, when rigorous, can be science. Evid Based Complement Alternat Med 2004;1:1−4.

[13] Suzuki N. Complementary and alternative medicine: a Japanese perspective. Evid Based Complement Alternat Med 2004;1:113−8.

[14] Mallikharjuna PB, Rajann LN, Seetharan YN, Sharanabasappa GK. Phytochemical studies of *Strychnos potatorum* L.F.: a medicinal plant. E-J Chem 2007;4(4):510−8.

[15] Klaassen CD, Eaton DL. Principles of toxicology. In: Amdur MO, Doull J, Klaassen CD, editors. Casarett and Doull's toxicology the basic science of poisons. 4th ed. New York, NY: Pergamon Press; 1992. p. 12−49.

[16] Walum E. Acute oral toxicity. Environ Health Perspect 1998;106(Suppl. 2):497−503.

[17] Renwick AG. Safety factors and establishment of acceptable daily intakes. Food Addit Contamin 1991;8:135.

[18] OECD Template #67. Repeated dose toxicity: Oral. Available from: <http://www. oecd.org/dataoecd/56/40/45616929.html>; 2010.

[19] Setzer RW, Kimmel CA. Use of NOAEL, benchmark dose, and other models for human risk assessment of hormonally active substance. Pure Appl Chem 2003;75:2151−8.

[20] Saller R, Melzer J, Reichling J. St. John's Wort (*Hypericum perforatum*): a plurivalent raw material for traditional and modern therapies. Forsch Komplementarmed Klass Naturheilkd 2003;10(Suppl. 1):33−40.

[21] Rodríguez-Landa JF, Contreras CM. A review of clinical and experimental observations about antidepressant actions and side effects produced by *Hypericum perforatum* extracts. Phytomedicine 2003;10(8):688−99.

[22] Leung AY, Foster S. Encyclopedia of common natural ingredients used in food, drugs, and cosmetics. 2nd ed. New York, NY: John Wiley & Sons, Inc.; 1996.

[23] Al-Sereiti MR, Abu-Amer KM, Sen P. Pharmacology of rosemary (*Rosmarinus officinalis* Linn.) and its therapeutic potentials. Indian J Exp Biol 1999;37:124−30.

[24] European Food Safety Authority (EFSA). Scientific opinion of the panel on food additives, flavorings, processing aids and materials in contact with food on a request from the commission on the use of rosemary extracts as a food additive. EFSA J 2008;721:1−29.

[25] Teixeira-Silva F, Santos FN, Sarasqueta DFO, Alves MFS, Neto VA, Moreira de Paula C, et al. Benzodiazepine-like effects of the alcohol extract from *Erythrina velutina* leaves: memory, anxiety, and epilepsy. Pharm Biol 2008;46:321−8.

[26] Garín-Aguilar MA, Luna JER, Soto-Hernández M, Valencia del Toro G, Vásquez MM. Effect of crude extracts of *Erythrina americana* mill. on aggressive behavior in rats. J Ethnopharmacol 2000;69:189−96.

[27] Vasconcelos SM, Lima NM, Sales GT, Cunha GM, Aguiar LM, Silveira ER, et al. Anticonvulsant activity of hydroalcoholic extracts from *Erythrina velutina* and *Erythrina mulungu*. J Ethnopharmacol 2007;110(2):271−4.

[28] Patocka J. Mulungu: rainforest anxiolytic. In: Szirmai Ã, editor. Anxiety and related disorders. InTech; 2011. ISBN: 978-953-307-254-8, InTech. Available from: <http://www.intechopen.com/books/anxiety-and-relateddisorders/mulungu-rainforest-anxiolytic>.

[29] Begossi A, Hanazaki N, Tamashiro JY. Medicinal plants in the Atlantic forest (Brazil): knowledge, use, and conservation. Hum Ecol 2004;30:281−99.

[30] Mills S, Bone K. Principles and practice of phytotherapy. Philadelphia, PA: Churchill Livingstone; 2000.

[31] Roitman JN. Comfrey and liver damage (letter). Lancet 1981;1(8226):944.

[32] Miskelly FG, Goodyear LI. Hepatic and pulmonary complications of herbal medicines. Postgrad Med 1992;68(805):935.

[33] Hirono I, Mori H, Haga M. Carcinogenic activity of *Symphytum officinale*. J Natl Cancer Inst 1978;61(3):865−9.

[34] Oberlies NH, Kim NC, Brine DR, Collins BJ, Handy RW, Sparacino CM, et al. Analysis of herbal teas made from the leaves of comfrey (*Symphytum officinale*): reduction of N-oxides results in order of magnitude increases in the measurable concentration of pyrrolizidine alkaloids. Public Health Nutr 2004;7(7):919−24.

[35] Chan P-C, Xia Q, Fu PP. *Ginkgo biloba* leaf extract: biological, medicinal and toxicological effects. J Environ Sci Health C Environ Carcinog Ecotoxicol Rev 2007;25:211−44.

[36] Le Bars PL, Katz MM, Berman N, Itil TM, Freedman AM, Schatzberg AF. A placebo-controlled, double-blind, randomized trial of an extract of *Ginkgo biloba* for dementia. North American EGb study group. JAMA 1997;278(16):1327−32.

[37] Arenz A, Klein M, Fiehe K, Gross J, Drewke C, Hemscheidt T, et al. Occurrence of neurotoxic 4-O-methylpyridoxine in *Ginkgo biloba* leaves, Ginkgo medications and Japanese Ginkgo food. Planta Med 1996;62(6):548−51.

[38] Gilbert GJ. Ginkgo biloba. Neurology 1997;48(4):1137.

[39] Bent S, Goldberg H, Padula A, Avins AL. Spontaneous bleeding associated with *Ginkgo biloba*: a case report and systematic review of the literature. J Gen Intern Med 2005;20(7):657−61.

[40] Rosenblatt M, Mindel J. Spontaneous hyphema associated with ingestion of *Ginkgo biloba* extract. N Engl J Med 1997;336(15):1108.

[41] Hauser D, Gayowski T, Singh N. Bleeding complications precipitated by unrecognized *Gingko biloba* use after liver transplantation. Transpl Int 2002;15(7):377−9.

[42] Dugoua JJ, Mills E, Perri D, Koren G. Safety and efficacy of ginkgo (*Ginkgo biloba*) during pregnancy and lactation. Can J Clin Pharmacol 2006;13(3):e277−84.

[43] Food and Drug Administration. Consumer advisory: Kava-containing dietary supplements may be associated with severe liver injury, <http://www.cfsanfdagov-dms/aadskvahtml>; 2002 [accessed 17.12.13].

[44] Norton SA, Ruze P. Kava dermopathy. J Am Acad Dermatol 1994;31:89−97.

[45] Gardiner P, Graham R, Legedza AT, Ahn AC, Eisenberg DM, Phillips RS. Factors associated with herbal therapy use by adults in the United States. Altern Ther Health Med 2007;13:22−9.

[46] Symmetry. Tranquility, <http://www.go-symmetry.com/catalog/tranquility.htm>; 1998.

[47] Bilia AR, Gallon S, Vincieri FF. Kava-kava and anxiety: growing knowledge about the efficacy and safety. Life Sci 2002;70:2581−97.

[48] Lebot V, Merlin M, Lindstrom L. Kava: the pacific drug. New Haven, CT: Yale University Press; 1992. p. 225.

[49] Dentali SJ. Herb safety review. Kava. *Piper methysticum* Forster f. (Piperaceae). Boulder, CO: Herb Research Foundation; 1997. p. 29.

[50] Hapke HJ, Sterner W, Heisler E, Brauer H. Toxicological studies with Kavaform. Farmaco Ed Prat 1971;26:692−720.

[51] Behl M, Nyska A, Chhabra RS, Travlos GS, Fomby LM, Sparrow BR, et al. Liver toxicity and carcinogenicity in F344/N rats and B6C3F1 mice exposed to Kava Kava. Food Chem Toxicol 2011;49(11):2820−9.

[52] Russmann S, Lauterburg BH, Helbling A. Kava hepatotoxicity. Ann Intern Med 2001;135:68−9.

[53] Russmann S, Barguil Y, Cabalion P, Kritsanida M, Duhet D, Lauterburg BH. Hepatic injury due to traditional aqueous extracts of kava root in New Caledonia. Eur J Gastroenterol Hepatol 2003;15:1033−6.

[54] Campo JV, McNabb J, Perel JM, Mazariegos GV, Hasegawa SL, Reyes J. Kava-induced fulminant hepatic failure. J Am Acad Child Adolesc Psych 2002;41:631−2.

[55] Clough AR, Bailie RS, Currie B. Liver function test abnormalities in users of aqueous Kava extracts. J Toxicol Clin Toxicol 2003;41:821−9.

[56] Ernst E. Herbal remedies for anxiety: a systematic review of controlled clinical trials. Phytomedicine 2006;13:205−8.

[57] Escher M, Desmeules J, Giostra E, Mentha G. Hepatitis associated with Kava, a herbal remedy for anxiety. BMJ 2001;322:139.

[58] Humberston CL, Akhtar J, Krenzelok EP. Acute hepatitis induced by Kava Kava. J Toxicol Clin Toxicol 2003;41:109−13.

[59] Teschke R, Gaus W, Loew D. Kava extracts: safety and risks including rare hepatotoxicity. Phytomedicine 2003;10:440−6.

[60] Robinson V, Bergfeld WF, Belsito DV, Klaassen CD, Marks Jr JG, Shank RC, , et al. Cosmetic Ingredient Review Expert Panel Final report on the safety assessment of *Piper methysticum* leaf/root/stem extract and Piper *methysticum* root extract. Int J Toxicol 2009;28(65):175S−188SS.

[61] Wren RC, Williamson EM, Evans FJ. Potter's newcyclopaedia of botanical drugs and preparations. Saffron Walden: The C.W. Daniel Company; 1988.

[62] Mabberley DJ. Mabberley's plant-book: a portable dictionary of plants, their classification and uses. 3rd ed. Cambridge: Cambridge University Press; 2008.

[63] Shibata S. A drug over the millennia: pharmacognosy, chemistry, and pharmacology of licorice. Yakugaku Zasshi 2000;120:849−62.

[64] Sobotka T, Spaid SL, Brodie RE, Reed GF. Neurobehavioral toxicity of ammoniated glycyrrhizin, a licorice component, in rats. Neurobehav Toxicol Teratol 1981;3(1):37−44.

[65] Shintani S, Murase H, Tsukagoshi H, Shiigai T. Glycyrrhizin (licorice)-induced hypokalemic myopathy. Eur Neurol 1992;32:44−51.

[66] Isbrucker RA, Burdock GA. Risk and safety assessment on the consumption of Licorice root (*Glycyrrhiza* sp.), its extract and powder as a food ingredient, with emphasis on the pharmacology and toxicology of glycyrrhizin. Reg Tox Pharm 2006;46(3):167−92.

[67] Størmer FC, Reistad R, Alexander J. Glycyrrhizic acid in liquorice-evaluation of health hazard. Food Chem Toxicol 1993;31(4L):303−12.

[68] Wang ZY, Nixon DW. Review. Licorice and cancer. Nutr Cancer 2001;39(1):1−11.

[69] Walker BR, Edwards CR. Licorice-induced hypertension and syndromes of apparent mineralocorticoid excess. Endocrinol Metab Clin North Am 1994;23:359−77.

[70] Edwards CRW, Stewart PM, Burt D, Brett L, McIntryre MA, Sutanto W-S, et al. Localization of 11 beta-hydroxysteroid dehydrogenase-tissue specific protector of the mineralocorticoid receptor. Lancet 1988;2:986−9.

[71] Forslund T, Fuhrquilst F, Froseth B, Tikkanen I. Effects of licorice on plasma atrial natriuretic peptide in healthy volunteers. J Int Med 1989;22595−9.

[72] Van Gelderen CEM, Bijlsma JA, van Dokkum W, Savelkoul TJF. Glycyrrhizic acid: the assessment of a no effect level. Hum Exp Toxicol 2000;19:434−9.

[73] Schambelan M. Licorice ingestion and blood pressure regulating hormones. Steroids 1994;59:127−30.

[74] Russo S, Mastropasqua M, Mosetti MA, Persegani C, Paggi A. Low doses of liquorice can induce hypertension encephalopathy. Am J Nephrol 2000;20:145−8.

[75] Heidermann HT, Kreuzfelder E. Hypokalemic rhabdomyolysis with myoglobinuria due to licorice ingestion and diuretic treatment. Klin Wochenschr 1983;61:303−5.

[76] Dobbins KRB, Saul RF. Transient visual loss after licorice ingestion. J Neuro-Ophthalmol 2000;20:38−41.

[77] Dalta R, Rao SR, Murthy KJR. Excretion studies of nitrofurantoin and nitrofurantoin with deglycyrrhizinated liquorice. Ind J Physiol Pharmac 1981;25:59−63.

[78] Muralidhara S, Ramanathan R, Mehta SM, Lash LH, Acosta D, Bruckner JV. Acute, subacute, and subchronic oral toxicity studies of 1,1-dicloroethane in rats: application to risk evaluation. Toxicol Sci 2001;64:135−45.

[79] Jaijoy K, Soonthornchareonnon N, Lertprasertsuke N, Panthong A, Sireeratawong S. Acute and chronic oral toxicity of standardized water extract from the fruit of *Phyllanthus emblica* Linn. Int J Appl Res Natl Prod 2010;3:48−58.

[80] OECD Guidelines for the Testing of Chemicals. TG 452, Chronic toxicity studies. Draft consultant's proposal 2008;8.

[81] World Health Organization. Research guidelines for evaluating the safety and efficacy of herbal medicine. Manila: WHO Regional Office for the Western Pacific; 1993. p. 35−40. <http://apps.who.int/medicinedocs/en/d/Jh2946e/>.

[82] ECVAM. Target organ and target system toxicity. Altern Lab Anim 2002;30 (Suppl. 1):71−82.

[83] Pfaller W, Balls M, Clothier R, Coecke S, Dierickx P, Ekwall B, et al. Novel advanced *in vitro* methods for long-term toxicity testing. ECVAM Workshop Report 45. Altern Lab Anim 2001;29:393−426.

[84] Spielmann H, Bochkov NP, Costa L, Gribaldo L, Guillouzo A, Heindel JJ, et al. 13th meeting of the scientific group on Methodologies for the Safety Evaluation of Chemicals (SGOMSEC): alternative testing methodologies for organ toxicity. Environ Health Perspect 1998;106(Suppl. 2):427−39.

[85] Prieto P, Baird AW, Blaauboer BJ, Castell-Ripoll JV, Corvi R, Dekant W, et al. The assessment of repeated dose toxicity *in vitro*: a proposed approach. The report and recommendations of ECVAM workshop 56. Altern Lab Anim 2006;34:315−41.

[86] Prieto P, Clemedson C, Meneguz A, Pfaller W, Sauer UG, Westmoreland C. Subacute and subchronic toxicity. Altern Lab Anim 2005;33(Suppl. 1):109−16.

[87] Blaauboer BJ, Balls M, Barratt M, Casati S, Coecke S, Mohamed MK, et al. 13th meeting of the Scientific Group on Methodologies for the Safety Evaluation of Chemicals (SGOMSEC): alternative testing methodologies and conceptual issues. Environ Health Perspect 1998;106(Suppl. 2):413−8.

[88] Blaauboer BJ, Forsby A, Houston JB, Beckman M, Combes RD, DeJongh J. An integrated approach to the prediction of systemic toxicity by using biokinetic models and biological *in vitro* test methods. In: Balls M, van Zeller A-M, Halder M, editors. Progress in the reduction, refinement and replacement of animal experimentation. Amsterdam: Elsevier; 2000. p. 525−36.

[89] ECVAM. Biokinetics. Altern Lab Anim 2002;30(Suppl. 1):55−70.

[90] Ekwali B, Barile FA, Castano A, Clemedson C, Clothier RH, Dierickx P, et al. MEIC evaluation of acute systemic toxicity. Part VI: the prediction of human toxicity by rodent LD_{50} values and results from 61 *in vitro* methods. Altern Lab Anim 1998;26 (2):617−58.

[91] Hurrt ME, Cappon GD, Browning A. Proposal for a tiered approach to developmental toxicity testing for veterinary pharmaceutical products for food-producing animals. Food Chem Toxicol 2003;41:611−9.

[92] Robinson MK, McFadden JP, Basketter DA. Validity and ethics of the human 4-h patch test as an alternative method to assess acute skin irritation potential. Contact Dermatitis 2001;45(1):1−12.

[93] Cohen SM. Bioassay bashing is bad science: Cohen's response. Environ Health Perspect 2002;110:A737.

[94] Cohen SM. Human carcinogenic risk evaluation: an alternative approach to the two-year rodent bioassay. Toxicol Sci 2004;80:225−9.

[95] ACSH (American Council on Science and Health). Of mice and mandates: animal experiments, human cancer risk and regulatory policies. New York, NY: ACSH; 1997.

[96] Gottmann E, Kramer S, Pfahringer B, Helma C. Data quality in predictive toxicology: reproducibility of rodent carcinogenicity experiment. Environ Health Perspect 2001;109(5):509−14.

[97] Bremer S, Pellizzer C, Hoffmann S, Seidie T, Hartung T. The development of new concepts for assessing reproductive toxicity applicable to large scale toxicological programmes. Curr Pharm Des 2007;13:3047−58.

[98] Cooper RL, Lamb JC, Barlow SM, Bentley K, Brady AM, Doerrer NG, et al. A tiered approach to life stages testing for agricultural chemical safety assessment. Crit Rev Toxicol 2006;36(1):69−98.

[99] Iwu MM. Handbook of African medicinal plants. New York, NY: CRC Press; 1993. p. 239.

[100] Odugbemi T. Medicinal plants by species names: outlines and pictures of medicinal plants from Nigeria. Lagos: Lagos University Press; 2006.

[101] Nia R, Paper DH, Essien EE, Iyadi KC, Bassey AIL, Antai AB, et al. Evaluation of the anti-oxidant and anti-angiogenic effects of *Sphenocentrum jollyanum* Pierre. Afr J Biomed Res 2004;7:129−32.

[102] Moody JO, Roberts VA, Connolly JD, Houghton PJ. Anti-inflammatory activities of the methanol extracts and an isolated furanoditerpene constituent of *Sphenocentrum jollyanum* Pierre (Menispermaceae). J Ethnopharmacol 2006;104:87−91.

[103] Akharaiyi FC, Boboye B, Adetuyi FC. Study of acute and subchronic toxicity of *Spathodea campanulata* P. Beav leaf. IPCBEE 2012;41:76−80.

[104] Adriana P, Jurandir PP, Dalva TF, Noemia KI, Raimundo BF. Iridoid glucose and antifungal phenolic compounds from *Spathodea campanulata* roots. Cien Agrar 2007;28:251−6.

[105] Oliver B. Medicinal plants in Nigeria. Ibadan: Nigerian College of Arts, Science and Technology; 1960. p. 23.

[106] Dalziel JM. The useful plants of west tropical Africa. London: The Crown Agents for Overseas Colonies; 1948. p. 397−99.

[107] Ainslie JR. A list of plants used in native medicine in Nigeria. Oxford: Imperial Forest Institute, Institute paper 7, 1937.

[108] Makinde JM, Adesogan EK, Amusan OG. The schizontocidal activity of *Spathodea campanulata* leaf extract on *Plasmodium berghei* in mice. Phytother Res 1987;1 (2):65−8.

[109] Ilodigwe EE, Akah PA. *Spathodea campanulata*: an experimental evaluation of the analgesic and anti-inflammatory properties of a traditional remedy. Asian J Med Sci 2009;1(2):35−6.

[110] Ilodigwe EE, Akah PA, Nworu CS. Evaluation of the acute and subchronic toxicities of ethanol leaf extract of *Spathodea campanulata*. Int J Appl Res Nat Prod 2010;3 (2):17−21.

[111] Bisset N. Herbal drugs and Phytopharmaceuticals. Stuttgart: CRC Press; 1994. p. 130−31.

[112] Lawless J. The illustrated encyclopedia of essential oils. Rockport, MA: Element Publishing Co; 1995.

[113] Prashar A, Locke IC, Evans CS. Cytotoxicity of clove (*Syzigium aromaticum*) oil and its major components to human skin cells. Cell Prolif 2006;39:241−8.

[114] Lane BW, Ellenhorn MJ, Hulbert TV, McCarron M. Clove oil ingestion in an infant. Hum Exp Toxicol 1991;10(4):291−4.

[115] Brown SA, Biggerstaff J, Savidge GF. Disseminated intravascular coagulation and hepatocellular necrosis due to clove oil. Blood Coagul Fibrinolysis 1992;3 (5):665−8.

[116] Bensky DS, Clavey S, Stoger E. Chinese herbal medicine: materia medica. Seattle, WA: Eastland Press; 2004.

[117] Agbaje EO, Adeneye AA, Daramola AO. Biochemical and toxicological studies of the aqueous extract of *Syzigium aromaticum* (L.) Merr and Perry (Myrtaceae) in rodents. Afr J Trad Complement Altern Med 2009;6(3):241−54.

[118] Adegoke EA, Alo B. Abere-amines: water soluble seed alkaloids from *Hunteria umbellata*. Phytochem 1986;25(6):1461−8.

[119] Falodun A, Nworgu ZAM, Ikponmwonsa MO. Phytochemical components of *Hunteria umbellata* (K. Schum.) and its effect on isolated non-pregnant rat uterus in oestrus. Pak J Pharm Sci 2006;19(3):256−8.

[120] Adeneye AA, Adeyemi OO, Agbaje EO, Banjo AAF. Evaluation of the toxicity and reversibility profile of the aqueous seed extract of *Hunteria umbellata* (K. Schum.) Hallier f. in rodents. Afr J Tradit Complement Altern Med 2010;7(4):350−69.

[121] Burkill HM. 2nd ed. The useful plants of West Tropical Africa, vol. 1. Kew: Royal Botanic Gardens; 1985. Families A−D.

[122] Ishola IO, Akindele JA, Adeyemi OO. Sub-chronic toxicity study of the methanol root extract of *Cnestis ferruginea*. Pharm Bio 2012;50(8):994−1006.

[123] Al-Faïz C, Thami-Alami I, Saïdi N. Domestication of some MAP species. In: Al-Faïz C, editor. Biological diversity, cultural and economic value of medicinal, herbal and aromatic plants in Morocco. Annual report; 2006−2007. p. 15−22.

[124] Lakmichi H, Bakhtaoui FZ, Ghadi CA, Ezoubeiri A, El-Jahiri Y, Mansouri A, et al. Toxicity profile of the aqueous ethanol root extract of *Corrigiola telephiifolia* Pourr. (Caryophyllaceae) in rodents. Evid Based Complement Alternat Med 2011; doi: 10.1155/2011/317090 Available online at: <http://www.pam-morocco.org/pdf/annual%20report%202007.pdf>.

[125] Banerjee A, Vaghasiya R, Shrivastava N, Padh H, Nivsarkar M. Antihyperlipidemic effect of *Carica papaya* L. in Sprague Dawley rats. Nig J Nat Prod Med 2006;10:69−72.

[126] Chinoy NJ, Patel KG, Sunita C. Antifertility investigations of alcoholic papaya seed extract in female rats. J Med Arom Plant Sci 1997;19(2):422−6.

[127] Duke JA. Borderline herbs. Boca Raton, FL: CRC Press; 1984.

[128] Emeruwa AC. Antibacterial substance from *Carica papaya* fruit extract. J Nat Prod 1982;45(2):123−7.

[129] Tona L, Kambu K, Ngimbi N, Cimanga K, Vlietinck AJ. Antiamoebic and phytochemical screening of some Congolese medicinal plants. J Ethnopharmacol 1998;61 (1):57−65.

[130] Gupta A, Wambebe CO, Parsons DL. Central and cardiovascular effects of the alcoholic extract of the leaves of *Carica papaya*. Int J Crude Drug Res 1990;28 (4):257−66.

[131] Harsha J, Chinoy NJ. Reversible antifertility effects of benzene extract of papaya seed on female rats. Phytother Res 1996;10(4):327−8.

[132] Akah PA, Oli AN, Enwerem NM, Gamaniel K. Preliminary studies on purgative effect of *Carica papaya* root extract. Fitoterapia 2007;68(4):327−31.

[133] Tarkang PA, Agbor GA, Armelle TD, Yemthe TLR, David K, Mengue-Ngadena YS. Acute and chronic toxicity studies of the aqueous and ethanol leaf extracts of *Carica papaya* Linn in Wistar rats. J Nat Prod Plant Resour 2012;2(5):617−27.

[134] Bailey CJ, Day C. Traditional plant medicines as treatments for diabetes. Diabetes Care 1989;12(8):553−64.

[135] Shane-McWhorter L. Biological complementary therapies: a focus on botanical products in diabetes. Diab Spectr 2001;14(4):199−208.

[136] Grover JK, Yadav S, Vats V. Medicinal plants of India with anti-diabetic potential. J Ethnopharmacol 2002;81(1):81−100.

[137] Kirtikar KR, Basu BD. Indian medicinal plants. 2nd ed. New Delhi: Periodical Expert Book Agency; 1984. p. 838.

[138] Ene AC, Nwankwo EA, Samdi LM. Alloxan-induced diabetes in rats and the effects of black caraway (*Carum carvi* L.) oil on their body weights. J Pharmacol Toxicol 2008;3:141−6.

[139] Johri RK. *Cuminum cyminum* and *Carum carvi*: an update. Pharmacogn Rev 2011;5 (9):63−72.

[140] Sachin BS, Monica P, Sharma SC, Satti NK, Tikoo MK, Tikoo AK, et al. Pharmacokinetic interaction of some antitubercular drugs with caraway: implications in the enhancement of drug bioavailability. Hum Exp Toxicol 2009;28:175—84.

[141] Ali BH, Blunden G. Pharmacological and toxicological properties of *Nigella sativa*. Phytother Res 2003;17(4):299—305.

[142] Ahmad A, Husain A, Mujeeb M, Khan SA, Namji AK, Siddique NA, et al. A review on the therapeutic potential of *Nigella sativa*: a miracle herb. Asian Pac J Trop Biomed 2013;3(5):337—52.

[143] Nickavar B, Mojab F, Javidnia K, Amoli MA. Chemical composition of the fixed and volatile oils of *Nigella sativa* L. from Iran. Z Naturforsch C 2003;58(9—10):629—31.

[144] Badary OA, Taha RA, Gamal el-Din AM, Abdel-Wahab MH. Thymoquinone is a potent superoxide anion scavenger. Drug Chem Toxicol 2003;26:87—98.

[145] Badary OA, Al-Shabanah OA, Nagi MN, Al-Bekairi AM, Elmazar MMA. Acute and subchronic toxicity of thymoquinone in mice. Drug Dev Res 1998;44:56—61.

[146] Qidwai W, Hamza HB, Qureshi R, Gilani A. Effectiveness, safety, and tolerability of powdered *Nigella sativa* (kalonji) seed in capsules on serum lipid levels, blood sugar, blood pressure, and body weight in adults: results of a randomized, double-blind controlled trial. J Altern Complement Med 2009;15(6):639—44.

[147] Zuber D. Biological flora of central Europe: *Viscum album* L. Flora 2004;1999:181—203.

[148] Olsnes S, Stirpe F, Sandvig K, Phil A. Isolation and characterization of *viscumin*, a toxic lectin from *Viscum album* L. (mistletoe). J Biol Chem 1982;257:13263—70.

[149] Kuttan G, Vasudevan DM, Kuttan R. Isolation and identification of a tumor reducing component from mistletoe extract (Iscador). Cancer Lett 1988;41(3):307—14.

[150] Kuttan G, Vasudevan DM, Kuttan R. Effect of a preparation from *Viscum album* on tumor development *in vitro* and in mice. J Ethnopharmacol 1990;29(1):35—41.

[151] Mannel DN, Becker H, Gundt A, Kist A, Franz H. Induction of tumor necrosis factor expression by a lectin from *Viscum album*. Cancer Immunol Immunother 1991;33 (3):177—82.

[152] Kuttan G, Kuttan R. Immunological mechanism of action of the tumor reducing peptide from mistletoe extract (NSC 635089) cellular proliferation. Cancer Lett 1992;66 (2):123—30.

[153] Hostanska K, Hajto T, Weber K, Fischer J, Lentzen H, Sütterlin B, et al. A natural immunity-activating plant lectin, *Viscum album* agglutin-1, induces apoptosis in human lymphocytes, monocytes, monocytic THP-1 cells and murine thymocytes. Nat Immun 1996—7;15(6):295—311.

[154] Antony S, Kuttan R, Kuttan G. Inhibition of lung metastasis by adoptive immunotherapy using Iscador. Immunol Invest 1999;28(1):1—8.

[155] Grossarth-Maticek R, Ziegler R. Randomized and non-randomized prospective controlled cohort studies in matched pair design for the long-term therapy of *corpus uteri* cancer patients with a mistletoe preparation (Iscador). Eur J Med Res 2008;13 (3):107—20.

[156] Harvey J, Colin-Jones DG. Mistletoe hepatitis. Br Med J 1981;282:186—7.

[157] Fletcher HF. Mistletoe hepatitis. Br Med J 1981;282:739.

7 Toxic Plants Used in African Traditional Medicine

Jean-de-Dieu Tamokou and Victor Kuete

Faculty of Science, Department of Biochemistry, University of Dschang, Dschang, Cameroon

7.1 Introduction

Traditional medicinal plants have been used for thousands of years and are a therapeutic resource used by the population worldwide and especially in Africa for healthcare [1]. Due to the structural diversity of plant metabolites and their previously demonstrated pharmacological properties, medicinal plants also can serve as starting materials for drugs. In fact, the therapeutic use of plants continues today because of their biomedical benefits and place in cultural beliefs in many parts of the world. The economic reality of the inaccessibility of modern medication for many African peoples also has played a key role in the broad use of herbal medicines [2]. The World Health Organization (WHO) has recognized the contribution and value of the herbal medicines used by a large segment of the world's population. Iwu et al. [3] reported that infectious diseases account for one-half of all deaths in the tropical countries. As a result, people of all continents have long applied poultices and imbibed infusions of indigenous plants dating back to prehistory for health purposes [4]. Many scientific studies have shown various pharmacological properties of medicinal plants [5−9]. Isolation of active compounds, in almost all cases, provided scientific explanation for the use of the plants in traditional medicine. WHO has defined traditional medicine (including herbal drugs) as comprising therapeutic practices that have been in existence, often for hundreds of years, before the development and spread of modern medicine and are still in use today [10]. The traditional preparations comprise medicinal plants, minerals, organic matter, etc. Herbal drugs constitute only those traditional medicines which primarily use medicinal plant preparations for therapy. According to WHO [11], "a medicinal plant" is any plant which in one or more of its organ contains substances that can be used for the therapeutic purposes or which are precursors for the synthesis of useful drugs.

Despite the benefits derived from medicinal plants, some of them have some unpleasant side effects which may be related to overdoses, toxic principles, or other factors. This may lead to acute toxicity and death of patients; scientific evidence of

Toxicological Survey of African Medicinal Plants. DOI: http://dx.doi.org/10.1016/B978-0-12-800018-2.00007-8

the safety of herbal preparation will help to harness the therapeutic potentials of medicinal and aromatic plants for further drug development in the future. Toxic plants can affect the entire spectrum of organ systems, with some plants having several toxic principles that affect different systems. The dominant effect may depend on the condition, growth stage, or part of the plant, the amount consumed, and the species and susceptibility of the victim [12]. For the majority of medicinal plants in use in Africa, there is a scarcity of toxicological surveys. Besides, in many African countries, herbal medicines are not subjected to the same regulatory standards as pharmaceuticals in terms of efficacy and safety. Unfortunately, the majority of the population does not pay attention, believing that, if these products have long been used, they should be devoid of toxicity [13−18]. Hopefully, many medicinal plant preparations have been shown not to be toxic [19,20]; but toxicological tests have also shown some other plants currently used as highly toxic at high doses or upon more or less prolonged consumption. This is the case of *Momordica charantia* (Cucurbitaceae) traditionally used as potent antidiabetic [21,22], *Urtica dioica* (Urticaceae) used as antidiuretic antihypertensive [23], *Crocus sativus* (Iridaceae) used as analgesic and anti-inflammatory [24], and *Erythrophleum guineese* (Fabaceae) [25]. This raises concern about their safety and implications for their use as medicines. Toxicity testing revealed some of the risks that may be associated with use of these plants, therefore avoiding potential harmful effects when used as medicine. Thus, the use of plants as remedies could produce prominent toxic effects either due to inherent toxicity or to contaminants: heavy metals, microorganisms, pesticides, toxic organic solvents, radioactivity, etc. [26]. The ability to identify toxic plants or toxic contaminants of medicinal herbs is an important challenge to protect human health. In traditional drug development, pharmaceutical companies evaluate the toxic effects of drugs through preclinical studies, including acute toxicity, safety pharmacology, and reproductive toxicity, to ensure the safety of new drugs before administration to humans.

This chapter aims to compile available toxicological data on African medicinal plants in order to warn about their use in traditional medicine.

7.2 Toxic Symptoms of Medicinal Plants

Generally, the action of a toxic substance can be divided into two major phases, as illustrated in Figure 7.1 [27]. The kinetic phase involves absorption, metabolism, temporary storage, distribution, and, to a certain extent, excretion of the toxicant or its precursor compound, called the protoxicant. In the most favorable scenario for an organism, a toxicant is absorbed, detoxified by metabolic processes, and excreted with no harm resulting. In the least favorable case, a protoxicant that is not itself toxic is absorbed and converted to a toxic metabolic product that is transported to a location where it has a detrimental effect. The dynamic phase is divided as follows: [1] the toxicant reacts with a receptor or target organ in the primary reaction step, [2] there is a biochemical response, and [3] physiological or

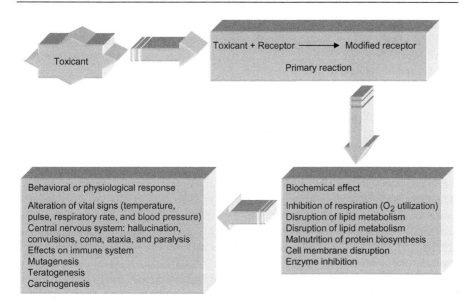

Figure 7.1 Major steps in the overall process leading to a toxic response [27].

behavioral manifestations of the effect of the toxicant occur. The final part of the overall toxicological process outlined in Figure 7.1 consists of behavioral and physiological responses, which are observable symptoms of poisoning. These are discussed here, primarily in terms of responses seen in humans and other animals after exposure to some toxic medicinal plants. This section divides these responses on the basis of responses in the major tissues and organs of animals, particularly humans. It should be noted that details of physical changes of some particularly toxic African plants are discussed in Chapter 22. The major systems considered here are the gastrointestinal tract system, liver, kidneys, skin, blood and cardiovascular system, nervous system, and reproductive system.

7.2.1 Gastrointestinal Tract Symptoms

The gastrointestinal tract responds to a number of toxic substances, usually by pain, vomiting, or paralytic ileus. Ricin, derived from the castor oil plant *Ricinus communis* (Euphorbiaceae) (see Chapter 22), is one of the most toxic substances known, and features very high on the list of substances likely to be used for bioterrorism. During the Cold War, an assassination using ricin was documented where a Bulgarian dissident, Georgi Markov, was eliminated by the implantation in his body of a perforated metal sphere containing ricin. The killer injected the sphere with a sharpened umbrella tip at a crowded bus stop, and had left by the time the victim collapsed and died [28]. Other members of the Euphorbiaceae family such as *Jatropha curcas* and *Jatropha multifida* also contain a toxalbumin, namely curcin, which can cause severe diarrhea. An indigenous tree of the Euphorbiaceae family, *Spirostachys africana* (tamboti), sometimes used for medicinal purposes, is nevertheless highly toxic, and use of the wood in fires over which meat is grilled has resulted in severe diarrhea after eating the meat. Even the smoke can cause headache and nausea [29,30]. Toxic diterpene esters found in *Euphorbia* species cause poisoning with severe gastroenteritis, vomiting, and colicky diarrhea when taken internally [31].

7.2.2   Liver Symptoms

The liver is often the first major metabolizing organ that an ingested toxicant encounters, and it has very high metabolic activity. Hepatotoxic plants may be consumed in traditional medicines. Animal experiments also have shown that some medicinal plants such as *Hollarrhena antidysenterica* (Apocynaceae), *Aegle marmelos* (Rutaceae), *and Terminalia chebula* (Combretaceae) used in traditional medicine for centuries in Pakistan have lately been reported to produce hepatic lesions [32]. There is a long list of medicinal plants used traditionally that are lately reported to be hepatotoxic [32,33]. The Asteraceae plant *Callilepis laureola* (impila, oxeye daisy, wildemagriet) is widely used as herbal medicine ("muti") and upon ingestion induces severe hepatotoxicity, nephrotoxicity, and hypoglycemic [34]. The diterpene atractyloside (discussed in Chapter 19) has been isolated from the tuberous roots [35,36]. Liver injury associated with the use of herbal medicine ranges from mild elevation of liver enzymes to fulminant liver failure often requiring a new transplant; and carcinogenesis [37]. Exposure to pyrrolizidine alkaloids (discussed in Chapter 21) through the use of herbal remedies also may be a contributing factor to the high rates of liver cancer and cirrhosis seen in Africa [38]. Another pyrrolizidine alkaloid-containing plant that often is used as herbal medicine and also has been associated with poisoning is comfrey (*Symphyttum officinale;* Boraginaceae) [39]. Established hepatotoxic phytochemicals include podophyllin, eugenol, neoclerodane diterpenes, among others [40−45].

7.2.3   Skin and Eye Symptoms

In many cases the skin and eye exhibit evidence of exposure to toxic substances. Generally, the two main skin characteristics observed as evidence of poisoning are skin color and degree of skin moisture. African poison ivy or pynboom, *Smodingium argutum* (Anacardiaceae) causes an allergic dermatitis and pruritus in sensitive individuals. It even occurs when the individual just passes near the tree [46]. Like animals, people who come into contact with the highly irritant milky latex of the candelabra tree or naboom, *Euphorbia ingens,* and the rubber hedge euphorbia, *Euphorbia tirucalli* (Euphorbiaceae), develop severe irritation and inflammation, especially when moist mucous membranes are affected [46]. *Pterocarpus soyauxii* (Fabaceae) wood dust is known to cause skin irritation, allergic contact dermatitis and sensitization [47].Toxic diterpene esters found in *Euphorbia* species cause inflammation of the skin, with reddening and formation of edematous swellings. If these poisonous lattices get into the eyes, they can cause inflammation of the cornea and conjunctiva, sometimes leading to blindness [31]. *Embelia ribes* (Myrsinaceae) had been reported to cause visual defects and eventually optic atrophy in animal experiments [48].

7.2.4 Blood and Cardiovascular System

The medicinal properties of the cardiac glycosides were known to the ancient Egyptians as well as the Romans who used it as an emetic, heart tonic, and diuretic [49]. Today, more than 200 cardiac glycoside-containing plants are used medicinally. For example, the foxglove *Digitalis purpurea* (Plantaginaceae) contains a cardenolide, digoxin (discussed in Chapter 19), which is used for the treatment of congestive heart failure. However, plants that contain cardiac glycosides may potentially impair the cardiovascular system [50]. Some of the bufadienolide-containing plants, in particular *Drimia sanguinea* commonly known as slangkop, sekanama (Asparagaceae), and the climbing potato *Bowiea volubilis* (Hyacinthaceae) are used by traditional healers in the treatment of various ailments and have been implicated in human poisoning [36,51]. The sap of *Acokanthera oppositifolia*, well-known as bushman poison bush (Apocynaceae), contains cardenolides and, as the name implies, that the sap has been used as arrow poisons for hunting by the San people. Humans may accidentally or intentionally become the target, resulting in fatal intoxication. *Senna alata* (Fabaceae) used traditionally for constipation, can have adverse effects on the heart because regular consumption is reported to deplete the body of potassium, causing fatalities. Other adverse reactions include grand mal seizures, circulatory failure, hypertension, and anaphylactic reaction [32].

Panax species (Araliaceae) is used as a general tonic and is claimed to increase the body's resistance to stress and builds up general vitality besides treating hypertension, diabetes, depression [32]. Recently the herb was reported to cause hypertension and mastalgia as documented side effects [32]. Taking *Panax species* may keep blood thinners from working correctly, resulting in problems with blood clotting [52]. Some recent reports also have linked ginseng to potential medication interactions with warfarin and digoxin. Taking *Panax species* and digoxin together may cause an increase in digoxin levels in the blood, leading to digoxin poisoning [32]. *Panax species* and antidiabetic drugs taken together may make blood sugar too low, leading to blurred vision, tremor (shaking), hunger, sweating, headache, skipped heart beats, confusion, nervousness, and extreme tiredness [32]. *Ginkgo biloba* (Ginkgoaceae) fruits and seeds have been used medicinally for thousands of years. Early reports have shown that the extract is a potent inhibitor of platelet-activating factor and long-term use has been associated with increased bleeding time, spontaneous hemorrhage, and subdural hematomas [32]. Adverse reactions that have been reported to WHO include hypertension, leucopoenia, thrombocytopenia, and hallucinations.

Several plants, for example, *Solanum tubersrosum, Solanum lycopersium, Solanum eleagnofolium, Solanum nigrum* (Solanaceae), *Mercurialis perennis*, and *Mercurialis annua* (Euphorbiaceae) have been reported to destroy red blood cells

(RBCs), thus leading to anemia [53−55]. The physiological consequence of anemia includes deficient level of circulating hemoglobin (Hb), the essential transport vehicle of oxygen, which is necessary for proper cell functions in the organs and body tissues. Administration of the juice of *Lantana camara* (Verbenaceae) leaves to rats resulted in a significant reduction in the total protein, globulin, absolute lymphocyte count, and percent lymphocyte count [56]. The effects of multiple doses of the petroleum ether extract of *Hygrophilia spinosa* (Acanthaceae) on the hematological and biochemical parameters as well as the hepatorenal functions of normal mice were evaluated [57]. The results showed that weekly moderate to high dose levels (above 40 mg/kg) and daily high dose (8 mg/kg) affected liver and kidney functions and metabolic and hematological parameters. Lower doses did not alter them. The polyherbal drug, Prostina [each capsule contains: Vapusha (*Juniperus Communis*, Cupressaceae) 200 mg; Pravala (*Corallium rubrum*, Coralliidae) 125 mg; Purified Silajit 40 mg; Kabab Chini (*Piper cubeba*, Piperaceae) q.s.; Ushira (*Vetiveria Zizanioides*, Poaceae) q.s.; Chandan Oil (Oil of *Santalum album*, Santalaceae) q.s.] recommended for use in benign prostatic hypertrophy showed no toxic effects as seen on morphological, gross behavior, body weight changes, and histopathological, biochemical, and hematological changes in rats up to doses of 450 mg/kg, which is 15 times higher than the recommended dose [58]. Many medicinal herbs known to lower blood sugar levels which include *Momordica charantia* (Cucurbitaceae), *Gymnema sylvestre* (Asclepiadaceae) and *Pterocarpus marsupium* (Fabaceae) when combined with Insulin could result in hypoglycemia [32]. Some most popular herbs such as *Alium sativum* (Alliaceae), *Ginkgo biloba* (Ginkgoaceae), *Zingiber officinalis* (Zingiberaceae), *Panax species* (Araliaceae), *Tanacetum parthenium* (Asteraceae) are known to delay the blood clotting time [32].

7.2.5 Nervous System Symptoms

The nervous system consists of the brain, spinal cord, and peripheral nerves. The nervous system is toxicologically important because of potential damage from neurotoxins that attack it. Nerve toxins from plants cause a variety of central nervous and peripheral nervous system effects. Several examples are mentioned herein. Plant-derived neurotoxic psychodysleptics affect peripheral neural functions and motor coordination, sometimes accompanied by delirium, stupor, trance states, and vomiting. Prominent among these toxins are the pyrollizidines from *Lophophora williamsii* (Cactaceae) [27]. A plant neurotoxin that is receiving much current publicity because of its effectiveness in the chemotherapeutic treatment of at least one form of cancer is taxol, a complex molecule that belongs to the class of taxine alkaloids. Taxol occurs in most tissues of the western yew tree, and is isolated from the bark (once considered a nuisance tree in forestry, but in short supply following discovery of the therapeutic value of taxol until alternate sources were developed). Ingestion of taxol causes a number of neurotoxic effects, including sensory neuropathy, nausea and gastrointestinal disturbances, and impaired respiration and cardiac function. It also causes blood disorders (leukopenia and thrombocytopenia) [27]. The mechanism of taxol neurotoxicity involves binding to tubulin, a protein involved in the assembly of microtubules, which assemble *Taxus breviofolia* (Taxaceae) and dissociate as part of cell function [27]. This binding of tubulin in nerve cell microtubules stabilizes the microtubules and prevents their dissociation, which can be detrimental to normal nerve cell function. Spotted hemlock contains the alkaloid nerve toxin coniine. Ingestion of this poison is followed within about 15 minutes by symptoms of nervousness, trembling, arrythmia, and bradycardia. Body temperature may decrease and fatal paralysis can occur [27].

The central nervous system generally responds to poisoning by exhibiting symptoms such as convulsions, paralysis, hallucinations, coma, and ataxia (lack of coordination of voluntary movements of the body). Other behavioral symptoms of poisoning include agitation, hyperactivity, disorientation, and delirium. *Boophane disticha* (seeroogblom, bushman poison bulb) contains various alkaloids such as buphanidrine, buphanisine, and buphanamine. Poisoning usually occurs in humans that utilize the bulb for medicinal purposes. Acute poisoning induces vomiting, weakness, coma, and mortality [36]. Cerpegin, a furopyridine alkaloid isolated from the chloroform extract of *Ceropegia juncea* (Apocynaceae) was found to be toxic at doses above 400 mg/kg and the mice showed excitation, irritability, convulsions and respiratory paralysis [27,59]. Two Solanaceae plants *Datura stramonium* (see more details in Chapter 22) and *Datura ferox* (moon flower, jimsonweed, stinkblaar, oliebome) are cosmopolitan weeds that contain parasympatholytic alkaloids such as atropine and hyoscine. Humans are extremely susceptible to their effects and hallucinations may occur, and the proverb "blind as a bat, red as a beet, dry as a bone, and mad as a hatter" aptly describes atropine poisoning in humans. Consumption of these weeds is usually inadvertent [27].

According to the National Academy of Sciences, Washington, DC (2009), *Azadirachta indica* (Meliaceae) seed oil is a strong spermicide and when it is used intravaginally, is effective in reducing the birth rate in laboratory animals [32]. This oil also has been reported to be toxic in large doses in human infants [32]. Thirteen infants who received 5−30 mL of the oil were reported to have been severely poisoned. Symptoms of the toxicity included metabolic acidosis, drowsiness, seizures, loss of consciousness, coma, and death in two infants. The infants exhibited Reye's syndrome-like symptoms, with death from hepatoencephalopathy.

7.2.6 Reproductive and Developmental Effects

The reproductive systems of both males and females are susceptible to adverse effects of toxic plants. Developmental toxic effects adversely influence the growth and development of an organism to adulthood. These may occur from exposure of either parent of the living organism to toxic plants even before conception. They may be the result of exposure of the embryo or fetus before birth. They can also include effects resulting from exposure during the growth of the juvenile organism from birth to adulthood. Teratology refers specifically to adverse effects of substances on an organism after conception up until birth. Artemisinin was found positive for teratogenicity on a Xenopus assay (FETAX) [60,61]; its derivative, dihydroartemisinin has been shown in a rat whole embryo culture (WEC) model to primarily affect primitive RBCs, causing subsequent tissue damage and dysmorphogenesis [61].

The chemical alteration of cell DNA that results in effects passed on through cell division is known as mutagenesis. Mutagenesis may occur in germ cells (female egg cells, male sperm cells) and cause mutations that appear in offspring. Mutagenesis also may occur in somatic cells, which are any body cells that are not sexual reproductive cells. Somatic cell mutagens are of particular concern because of the possibility that they will result in uncontrolled cell reproduction, leading to cancer. Somatic cell mutations are easier to detect than germ cell mutations through observation of chromosomal aberrations and other effects [27]. African plants such as *Crinum macowanii, Chaetacme aristata* Planch. (Celastraceae), *Plumbago auriculata* Lam. (Plumbaginaceae), *Catharanthus roseus* (L.) G.Don. (Apocynaceae), and *Ziziphus mucronata* Willd. (Rhamnaceae) were found to have mutagenic effects in the *Salmonella*/microsome assay [62]. A few well-characterized compounds include (i) 1−2 unsaturated pyrrolizidine alkaloid esters from many Boraginaceae, Asteraceae, and Fabaceae [63,64] that exhibit a large variety of genotoxicities, including DNA binding, DNA cross-linking, DNA−protein cross-linking, sister chromatid exchange, and chromosomal aberrations [65−68]; (ii) aristolochic acids (AA; see Chapter 19), nitro-polyaromatic compounds responsible for terminal nephropathies observed after intoxication by many *Aristolochia* species [64]; a series of studies confirmed that they are genotoxic in both bacterial and mammalian cells, yielding highly persistent and molecules nonrepaired DNA adducts; (iii) allylalkoxybenzenes (e.g., eugenol, methyleugenol, estragole), safrole (4−allyl−1,2−ethylenedioxybenzene), and β-asarone, potentially genotoxic components from some essential oils [69−72]. The notion of threshold for genotoxic insult is still a matter of heavy debate; consequently such compounds should be proscribed from herbal medicines or at least severely limited. In the *in vitro* comet assay, artesunate, a semisynthetic derivative from artemisinin (*Artemisia annua* L.), induced DNA breakage in a dose-dependent manner [60].

Polar and apolar extracts of more than 50 African plants were tested for potential genotoxic effects using *in vitro* bacterial and mammalian cell assays such as the Ames test (tester strains TA98 and TA100), VITOTOX test, micronucleus test, and comet assay. It was found that most of the plant species investigated caused either DNA damage, detected by the comet assay (at doses of 500 and 250 μg/mL); chromosomal aberrations and/or nondisjunction or chromosome lagging in human white blood cells (WBCs) as detected in the micronucleus test (at doses of 2500, 500, and 100 μg/mL) [73,74]. Among these, the genotoxicity of Copaiba oil *Copaifera langsdorffii* Desf. (Fabaceae) was demonstrated by *in vitro* micronucleus and comet assays [75]. *Punica granatum* (Punicaceae) whole extract and the enriched fractions of flavonoids and tannins from *Mouriri pusa* reacted positively to genotoxic test [76]. Metabonomics techniques have been used in the toxicity studies of Guan Mutong *Aristolochia manshuriensis* Kom (Aristolochiaceae) [77] and its toxic component aristolochic acid [78]; aristolochic acid I was suggested to possess genotoxic potency also by quantitative structure−activity relationship (QSAR) modeling [79−81].

Boehmeria nivea (Urticaceae) aqueous extract at 32 g/kg/day did not cause significant embryotoxicity or maternal toxicity in mice, although it might cause cytotoxicity in cultured embryonic stem cells (ESCs test) at a higher dose [82,83]. Many plant species investigated using human WBCs showed genotoxicity in the micronucleus test, in the form of structural and numerical chromosome aberrations. The included *Kigelia africana, Merwilla plumbea* (Lindl) Speta (syn. *Scilla natalensis* Planch., (Hyacinthaceae), *Boophane disticha* (L.f.) Herb. (Amaryllidaceae), *Celtis africana* Burm. f. (Celtidaceae), *Crinum macowanii, Erythrina caffra* Thunb. (Fabaceae), *Ochna serrulata* (Hochst.) Walp. (Ochnaceae), *Sclerocarya birrea* (Anacardiaceae), and *Tulbaghia violacea* (Alliaceae). Other plants such as *Acokanthera oblongifolia* (Hochst.) Codd (Apocynaceae), *Afzelia quanzensis* Welw. (Fabaceae), *Bersama lucens* (Hochst.) Szyszyl. (Melianthaceae), *Ocotea bullata* (Lauraceae), *Siphonochilus aethiopicus* and *Tetradenia riparia* (Hocsht.) Codd (Lamiaceae) caused DNA damage as detected in the comet assay [73,74]. The above-mentioned plant species should be taken with caution when their therapeutic used is involved.

7.2.7 *Kidney and Bladder Symptoms*

The kidney and bladder are very important in toxicology because they are the main route of elimination of hydrophilic toxicant metabolites and because damage to them in the form of impaired kidney function or bladder cancer is one of the major adverse effects of toxicants [27]. In Belgium, several preparations based on Chinese plants were reported to be responsible for 70 cases of renal impairment [32,84]. Horse chestnut *Aesculus hippocastanum* (Hippocastanaceae) extracts are used medicinally in patients with venous insufficiency. Its therapeutic use at high doses ($>340\,\mu$g/kg) is associated with renal failure or a lupuslike syndrome [85,86]. In addition to the gastrointestinal manifestations of vomiting, diarrhea, and dehydration, ricin, extracted from the castor bean (*Ricinus communis*), can cause cardiac, hematologic, hepatic, and renal toxicity. All contribute to death in humans and animals [87,88]. Consumption of colchicine from plant sources such as *Colchicum autumnale* (Colchicaceae) produces a spectrum of symptoms, including nausea, vomiting, watery diarrhea, hypotension, bradycardia, electrocardiographic abnormalities, diaphoresis, alopecia, bone marrow depression, renal failure, hepatic necrosis, hemorrhagic acute lung injury, convulsions, and death [89,90]. Despite the long history of *Aristolochia* use in herbal remedies, evidence of the plants' inherent danger emerged only recently. In the early 1990s, women who had received *Aristolochia* treatments at a weight-loss clinic in Belgium developed kidney problems that progressed to renal failure and, in later years, to abnormal growths in their upper urinary tracts [32,91]. More recently, *Aristolochia* contamination of local wheat crops was determined to be the cause of a high incidence of urothelial carcinomas of the upper urinary tract (UTUC) among rural communities on the banks of the Danube River in Europe. And in Taiwan, where recent prescription records reveal that approximately one-third of the population has taken *Aristolochia*-containing medicines, the incidence of UTUCs is the highest in the world [32,91]. Animal experiments have also shown that some medicinal plants like *Terminalia chebula* (Combretaceae) and *Withania somnifera* (Solanaceae) used in traditional medicine since centuries in Pakistan lately have been reported to produce marked renal lesions [32].

7.3 Toxicity of African Medicinal Plants

Toxic medicinal plants contain poisons or toxin that can induce adverse side effects in animal and humans upon consumption or administration for the therapeutic purposes. In general, plants considered poisonous to humans are likely to be poisonous to animals as well. Reactions to toxic plants can range from mild to serious, depending on the age and size of the patient, health factors such as allergies, and the amount of the exposure.

The earliest report of the toxicity of herbs originated from Galen, a Greek pharmacist and physician who showed that herbs do not contain only medicinally beneficial constituents, but may also be constituted with harmful substances [92]. The stories of Socrates' execution from being forced to drink an extract of the deadly poisonous spotted hemlock plant *Conium maculatum* (Apiaceae) and Cleopatra's (last pharaoh of Ancient Egypt) suicide at the fangs of a venomous asp are rooted in antiquity [27]. Many household poisonings result from children ingesting toxic plant leaves or berries [27]. High on this list is philodendron. Other plants that may be involved in poisonings include *Diefenbachia* spp., the jade plant *Crassula ovata* (Crassulaceae), the wandering Jew plants *Tradescantia* spp., the Swedish ivy *Plectranthus verticillatus* (Lamiaceae), the plants of the genus *Phytolacca* known as pokeweed, the string-of-pearls *Senecio rowleyanus* (Asteraceae), and the yew plant *Taxus baccata* (Taxaceae). Pollen from plants causes widespread misery from allergies, and reactions to toxins from plants such as the poison ivy *Toxicodendron radicans* (Anacardiaceae) can be severe [27]. A review of plant poisonings reported to American poison control centers over an 18-year period (1983−2000) revealed 30 fatalities, with 7 due to *Cicuta* species (water hemlock) and 5 due to *Datura* species (jimsonweed) [93]. These findings indicate that despite the benefits derived from plants, some of them can be toxic as mentioned above. This may lead to acute toxicity and death. By way of indication, 43% of poisoning cases recorded in a forensic database for Johannesburg from 1991 to 1995 were caused by traditional plant medicines [94]. *Peganum harmala* L. (Nitrariaceae) is commonly used in traditional medicine in Morocco for its sedative and emmenagogue properties but expose to the risk of overdose and poisoning. During a period of 24 years from January 1984 to December 2008, it was cited in 200 cases of poisoning collected in poison control and pharmacovigilance center of Morocco [95]. Therapeutic circumstance was found in 32.5%, followed by suicide (28.5%) and abortion (13.5%). The symptomatology was dominated by neurological, gastrointestinal, and cardiovascular signs, respectively, 34.4%, 31.9%, and 15.8%. The evolution has been specified in 114 cases; 7 deaths have been deplored with a fatality rate of 6.2%

[95]. Fifty-six medical records of patients (2−75 years old) admitted to the toxicological intensive care unit in Tunis, Tunisia during 1983−1998 following the ingestion of medicinal plants, were investigated and the results showed that the most common poisonings involved were *Atractylis gummifera* (32%), *Datura stramonium* (25%), *Ricinus communis* (9%), *Nerium oleander* (7%) and *Peganum harmala* (7%) [96]. Poisonings involved neurological (91%), gastrointestinal (73%), and cardiovascular systems (18%). The only lethal cases of liver failure involved 16 *Atractylis gummifera* poisonings [96].

However, many cases of poisoning remain unrecorded, and mortality from traditional plant medicines may be higher than currently known [97,98]. Similarly, morbidity caused by the use of traditional medicines may be significantly higher than presently estimated [74]. For example, *Callilepis laureola* DC. (Asteraceae) was identified as the primary cause of the high occurrence of liver necrosis recorded in the black population of KwaZulu-Natal in the 1970s [34,98,99]. Demographics indicate that the majority of traditional medicine-related poisoning cases are pediatric [98,100,101]. Due to a lack of regulation of the marketing and use of herbal medicines in most African countries, many potentially toxic plants are available over the counter from herbalist retailers and medicinal plant traders [98,102,103].

Some scientific researches have shown that many plants used in traditional medicine are potentially toxic, mutagenic, and carcinogenic [104−108]. The following medicinal plants are highly toxic because they cause both DNA damage and chromosomal aberrations: *Antidesma venosum* E. Mey. ex Tul. (Euphorbiaceae), *Balanities maughamii* Sprague (Balanitaceae), *Catharanthus roseus, Catunaregam spinosa* (Thunb.) Tirveng. (Rubiaceae), *Chaetacme aristata, Croton sylvaticus* Hochst. (Euphorbiaceae), *Diospyros whyteana* (Hiern) F.White (Ebenaceae), *Euclea divinorum* Hiern (Ebenaceae), *Gardenia volkensii* K.Schum. (Rubiaceae), *Heteromorpha arborescens* (Spreng.) Cham. & Schltdl. Var. *abyssnica* (A. Rich.) H. Wolff (syn. *Heteromorpha trifoliata* (H.L. Wend.) Eckl. and Zeyh.) (Apiaceae), *Hypoxis colchicifolia* Baker (Hypoxidaceae), *Ornithogalum longibractaetum* Jacq. (Hyacinthaceae), *Plumbago auriculata, Prunus africana* (Hook. f.) Kalkm. (Rosaceae), *Rhamnus prinoides* L'Hér. (Rhamnaceae), *Ricinus communis, Spirostachys africana* Sond. (Euphorbiaceae), *Trichelia emetica* Vahl subsp. Emetic (Meliaceae), *Turraea floribunda* Hochst. (Meliaceae), *Vernonia colorata* (Asteraceae), and *Ziziphus mucronata* (Rhamnaceae) [73,74]. Their therapeutic use should be treated with caution and rigorous toxicological and clinical studies are necessary before they are widely prescribed for use in traditional medicine. According to Fennell et al. [74], several plants used in South African traditional medicine (ATM) can cause damage to genetic material and therefore have the potential to cause long-term damage in patients when administered as medical preparations.

Cassia occidentalis L. (syn. *Senna occidentalis*; Fabaceae) has been used as natural medicine in rainforests and tropical regions as laxative, analgesic, febrifuge, diuretic, hepatoprotective, vermifuge, and colagogo. Toxicological studies have shown potent toxic effects of *C. occidentalis* seeds in animals [109]. Signs of *C. occidentalis* seeds poisoning in general include independently of animal species: ataxia, muscle weakness, stubbing, and body weight loss, eventually leading to death. Skeletal muscle degeneration is the predominant lesion found in the majority of animals intoxicated with this plant [110−112]. Mechanism of *C. occidentalis* toxicity has been described as being due to impairment of mitochondrial function, including swelling, loss of mitochondrial matrix, fragmented mitochondrial cristae, and glycogen depletion [111,113]. Haraguchi et al. [112] identified dianthrone, an anthraquinone-derived compound in *C. occidentalis* seeds, and demonstrated that these substances could cause the characteristic mitochondrial myopathy produced by this plant.

Euphorbia kamerunica (Euphorbiaceae) is a cactus-like plant that prefers dry, rocky country. Its distribution extends from Guinea to Cameroon [114]. Fishermen commonly use this plant to capture fish. In addition, some herbalists use the latex either medicinally or as a poison. This plant contains a skin irritant and has cocarcinogenic principles and other constituents responsible for health hazards to both humans and grazing livestock [115]. The main compounds responsible for the toxic properties of Euphorbiaceae are the lectins and esters of certain diterpene alcohols. In addition to phorbol derivatives, tigliane, daphnane, and ingenane diterpene ester toxins are found in *Euphorbia* species. The chemical identity and structural formulae of the more important of these compounds have been determined [115−117,118]. However, their mechanisms of action are incompletely understood.

Sharathchandra and Balakrishnamurthy [119] have studied the mode of action of *Cleistanthus collinus* (Euphorbiaceae), a toxic plant that is frequently implicated in poisoning. *Cleistanthus collinus* causes a depletion of thiol/thiol-containing enzymes in most organs, which results in its toxicity. Thiol compounds may act as antidotes. Significant increase in the relative weights of adrenals was also observed. High doses (1500 mg/kg) significantly inhibited granulomatous tissue formation in rats, similar to cyclophosphamide [56]. Extracts from *Ricinus communis* (Euphorbiaceae) can be used to isolate ricin, an extremely potent poison that is one of the leading candidates for use by terrorists. Plant-derived cocaine causes many deaths among those who use it or get into fatal disputes marketing it [27]. The toxic agents in plants have been summarized in a review chapter dealing with that topic [120]. Pyrrolizidine alkaloids in the plant *Senecio vernalis* (Asteraceae) have been implicated in the poisoning of cattle [121]. The toxic agents in this plant include three closely related alkaloids: senecionine, senkirkin, and seneciphyllin.

7.4 Toxic African Medicinal Plants as Identified by Their Medium Lethal Doses (LD$_{50}$) Values and Induced Physical Symptoms

With the current emphasis on research and development of phytomedicines in Africa, it is imperative to be aware of the side effects of some medicinal plants and to have information on common toxic plant occurring in humans (see Chapter 22). Over 120 toxicological studies were conducted on some African medicinal plants for the period of 1977−2013 and led to the identification of 49 toxic plants used in ATM based on their LD$_{50}$ values and physical symptoms (Table 7.1).

Plant (Family)	Traditional Use	Area of Plant Collection	Part Used; Extraction Solvent	LD$_{50}$	Physical Signs of Toxicity (Animal Model)	Toxin or Other Compounds (Toxin Class)
Acanthus montanus (Acanthaceae)	Pain, female infertility, and threatened abortion	Cameroon	Leaves; water		Kidneys revealed crystals resulting in glomerulosclerosis [122]	–
Ajuga iva (Labiateae)	Widely used as panacea (cure-all), anthelmintic and also to treat gastrointestinal disorders, hypertension, and diabetes	Morocco	Whole plant; water	3.6 g/kg IP (in mice)	Asthenia, anorexia, hypoactivity, piloerection, salivation, diarrhea, syncope, and death (mice) [123]	–
Anacardium occidentale (Anacardiaceae)	Diabetes and hypertension	Cameroon	Leaves; methanol	16 g/kg po	A reduction in the food intake and weight gain, asthenia, anorexia, diarrhea, and syncope were observed (mice). Histopathological studies revealed evidence of microscopic lesions in the liver and kidney [124]	Urushiol oleoresins (Terpenoid) [125]
Annona senegalensis (Annonaceae)	Sleeping sickness, malaria, diarrhea, intestinal troubles, stomach ache, anthelmintic, skin cancers, and leukemia	Nigeria	Root bark; hexane, ethyl acetate and CH$_2$Cl$_2$/MeOH	1265–3808 mg/kg/p.o.	Liver sections showed degeneration and necrosis of the hepatocytes [126]	–
			Root; 80% ethanol	11.16 μg/mL	Very toxic to brine shrimps [127]	

(Continued)

Table 7.1 (Continued)

Plant (Family)	Traditional Use	Area of Plant Collection	Part Used; Extraction Solvent	LD$_{50}$	Physical Signs of Toxicity (Animal Model)	Toxin or Other Compounds (Toxin Class)
Aspilia Africana (Asteraceae)	Used to stop bleeding, remove corneal opacities, induce delivery and treat anemia and various stomach complaints	Cameroon	Leaves; water	6.6 g/kg bw	Increased aggressiveness and motility (mice) Necrosis, edema, and inflammatory infiltrations in the liver and kidneys [128]	—
Azadirachta indica (Maliaceae)	Tuberculosis, skin infections, pests, nematodes of livestock, malaria, *Setaria cervi* infection of cattle, coccidiosis of chicken, and trypanosomosis	Nigeria	Stem bark; ethanol	870 mg/kg bw (IP)	Decrease in the spleen–body weight ratio, feed and water intakes as well as anorexia, dehydration, depression, malaise, respiratory depression, coma, and death (rats). Death animals showed varying degrees of dehydration, congestion and edema of the trachea, bronchi, bronchioles, lungs, and kidneys. There was also hepatomegally, nephritis and diffuse necrosis of hepatocytes, and glomeruli (rats) [129,130]	—

Plant (Family)	Uses	Country	Plant part; solvent	Dose	Toxicity	
Berlina grandiflora (Fabaceae)	Intestinal problems, purgative, chewing stick, enemas, and constipation	Nigeria	Stem bark; H₂O:MeOH (70:30)	>2000 mg/kg oral routes (po)	Reductions in the weight gain, food, and water intakes. Dose-dependent mortalities as well as mild sedative effects, anorexia, and respiratory distress (rats) [131]	—
Bryophyllum calycinum (Crassulaceae)	Infections, rheumatism, inflammation, hypertension, kidney stones, cancers, and cancer related	Nigeria	Whole plant; 80% ethanol	8.38 µg/mL	Very toxic to brine shrimps [127]	
Butyrospermum paradoxum (Sapotaceae)	Trypanosomosis and several human and animal diseases	Nigeria	Stem bark; ethanol	820 mg/kg, IP	Anorexia, dehydration, depression, prostration, coma, and death. At necropsy, the pathological lesions were mainly congestion and edema of the lungs, bronchi, bronchioles and the kidney, hepatomegally, and focal necrosis of the liver cells [132]	
Chenopodium ambrosioides (Chenopodiaceae)	Analgesic, purgative and vermifuge, cancers, and cancer-related problems	Nigeria	Leaves; water	/	Body weight loss, pathological features like congestion of the lungs, metaplastic changes in the mucosal surface of the stomach, and necrosis of the kidney tubules were noticed (rats) [133]	—

(Continued)

Table 7.1 (Continued)

Plant (Family)	Traditional Use	Area of Plant Collection	Part Used; Extraction Solvent	LD_{50}	Physical Signs of Toxicity (Animal Model)	Toxin or Other Compounds (Toxin Class)
Chrysophyllum albidum (Sapotaceae)	cancers and cancer-related problems	Nigeria	Whole plant; 80% ethanol	2746.34 μg/mL	Fairly toxic to brine shrimps [127]	
			Root bark. 80% ethanol	329.24 μg/mL	Fairly toxic to brine shrimps [127]	
Corrigiola telephiifolia (Caryophyllaceae)	Dermatological diseases, flu, inflammation, ulcer, cough, jaundice, anasthenie diuretic, and given to parturient women	Morocco	Roots; ethanol: water (75:25)		Decrease in the relative body weight [134]	—
Coula edulis (Olacaceae)	Tonifiant, kidney problems, stomach ache, and skin diseases	Cameroon	Stem bark; CH_2Cl_2: MeOH (1:1)	16.8 and 19.6 g/kg bw in male and female mice, respectively	Diarrhea and decrease of the reaction to pinch and noise (mice). Decreases in the body weight gain, food and water consumptions (rats). Gross anatomical analysis revealed white vesicles on the liver (rats) [135]	—

Plant name (family)	Ethnomedicinal use	Country	Part; solvent	Dose	Effect	Compound
Cylicodiscus gabunensis (Mimosaceae)	Diarrhea, gastrointestinal disorder, rheumatism, filariasis, and headache	Cameroon	Stem bark; ethyl acetate	14.5 and 11 g/kg bw po for male and female rats, respectively	Decrease in relative weight of the spleen. Toxic effect on liver, kidneys and lungs (rats) [136]	Monofluoroacetate (organofluorine compounds) [137]
Dichapetalum barteri (Dichapetalaceae)	Animal feed	Nigeria	Leaves; water	/	Depression, restlessness, convulsion, and death. The main lesions observed were acute vasculitis and congestion of the liver, lung, kidney, spleen as well as extensive edema, and congestion of the myocardium (rabbits, mice, and goats) [137]	
Diospyros canaliculata (Ebenaceae)	Whooping cough, leprosy, snakebites, skin eruptions, dysentery, menstrual troubles, abdominal pains, wounds, ulcers, eye, chest pains, and skin infections	Cameroon	Leaves; methanol	>8000 mg/kg bw	Prominent hepatic and renal tissue injuries [138]	–

(Continued)

Table 7.1 (Continued)

Plant (Family)	Traditional Use	Area of Plant Collection	Part Used; Extraction Solvent	LD$_{50}$	Physical Signs of Toxicity (Animal Model)	Toxin or Other Compounds (Toxin Class)
Entada abyssinica (Fabaceae)	Coughs, diarrhea, fever, gonorrhea, to prevent miscarriage, rheumatic and abdominal pains	Tanzania	Stem bark; 80% ethanol	/ 140.39 μg/mL	At doses >2000 mg/kg body wt, the mice exhibited increased respiratory rate and scruffy hair. Two mice died at the dose of 3000 mg/kg body wt. Fairly toxic effect to brine shrimp [139]	
Erythrina senegalensis (Fabaceae)	Malaria, hepatitis jaundice, gastrointestinal disorders, amenorrhea, dysmenorrhea, sterility, pains, onchocerchosis, and headache	Cameroon	Stem bark; water	>12.5 g/kg	Reduction in locomotion, exploration, aggressiveness, touch, sensibility, and pain sensibility (mice) [140]	—
Erythrophleum suaveolens (Fabaceae)	Pains, fever, headache, cancers, used in trial-by-ordeal ritual, as arrow and fish poison	Nigeria	Leaves; 80% ethanol	54.42 μg/mL	Toxic to brine shrimps [127]	

Plant (family)	Ethnomedicinal uses	Country	Plant part; solvent	Dose	Toxicity	
Ficus exasperata (Moraceae)	Cardiac arrhythmias, asthma, bronchitis, tuberculosis, emphysema, hastening the expulsion of placenta, arrest bleeding, eye troubles, stomach and chest pains	Nigeria	Leaves; water Leaves; 80% ethanol	0.54 g/kg IP in mice [141] 31,372.75 (µg/mL)	/ Fairly toxic to brine shrimps [127]	—
Garcinia kola (Guttiferae)	Bronchitis, throat infections, colic, head or chest colds, cough, liver disorders, cancer, and as a chewing stick, purgative, antiparasitic, antimicrobial	Nigeria	Root; 80% ethanol	20.98 µg/mL	Toxic to brine shrimps [127]	
Glinus lotoides (Molluginaceae)	Anthelmintic and tapeworm infestation	Ethiopia	Seeds; 60% methanol	4.5 g/kg (mice)	No physical sign of toxicity was observed (rats) [142]	—
Herniaria glabra (caryophyllaceae)	Astringent, diuretic, expectorant, dropsy, catarrh of the bladder, cystitis, kidney stones, gout,	Morocco	Leaves; water	8.50 g/kg bw po	Asthenia, piloerection, ataxia, anorexia, urination, diarrhea, syncope, and mortality (mice). Reduction in the body weight gain, centrolobular	—

(Continued)

Table 7.1 (Continued)

Plant (Family)	Traditional Use	Area of Plant Collection	Part Used; Extraction Solvent	LD_{50}	Physical Signs of Toxicity (Animal Model)	Toxin or Other Compounds (Toxin Class)
	hernias, jaundice, nerve inflammation, and respiratory disorders				sinusoidal congestion, disruption of the central vein, and hepatocellular necrosis in the liver as well as interstitial and intraglomerular congestion, tubular atrophy, and inflammation in the kidney (rats) [143]	—
Hibiscus sabdariffa (Malvaceae)	Diuretic, aphrodisiac, antiseptic, astringent, tonic, sedative, laxative, and infectious diseases	Nigeria	Calyces; water	5 g/kg p.o.	Decreases in the water intake, weights of the testis and epididymal sperm counts; distortion of tubules; disruption of normal epithelial organization; hyperplasia of testis with thickening of the basement membrane and disintegration of sperm cells (rats) [144]	
			Leaves; 80% ethanol	236.96 µg/mL	Fairly toxic to brine shrimps [127]	

Plant (family)	Traditional uses	Country	Part; solvent	Dose	Toxicity effect	Active compound
Hydnora johannis Becca. (Hydnoraceae)	Dysentery, diarrhea, cholera, and swelling tonsillitis.	Sudan	Roots; ethanol	/	Toxic effect on liver, kidney, and spleen (Wistar rats) [145]	—
Hymenocardia acida (Euphorbiaceae)	Cancers, eye infection, sickle cell anemia, fever, jaundice, muscular pains, diarrhea, dysentery, skin diseases, diabetes, urinary tract infections, sexual incapacity in males, as an agent for female genital hygiene	Nigeria	Stem bark; 80% ethanol	24.12 μg/mL	Toxic to brine shrimps [127]	
Jatropha curcas (Euphorbiaceae)	Widely used in traditional folk medicine in many parts of West Africa	Nigeria	Leaves; methanol	46.0 mg/kg bw po	Reduction in the food intake and mobility, asthenia and death (rats). Before rats died, they exhibited signs of depression, closing of eyes, languishment, loss of body mass, and black excreta [146]	Curcin (Protein, lectin, peptide, amino acid) [125]
Khaya ivorensis (Meliaceae)	Malaria, emmenagogue, emetic, hypotensive, and antipyretic	Nigeria	Stem bark; 90% ethanol	/	Morphological abnormalities within the white matter of the cerebral cortex (rats) [147]	—

(Continued)

Table 7.1 (Continued)

Plant (Family)	Traditional Use	Area of Plant Collection	Part Used; Extraction Solvent	LD$_{50}$	Physical Signs of Toxicity (Animal Model)	Toxin or Other Compounds (Toxin Class)
Lannea schimperi (Anacardiaceae)	Peptic ulcers, pains, diarrhea, herpes, anemia, and cough	Tanzania	Stem bark; 80% ethanol	/	The extract caused increased defaecation/diarrhea but it did not kill any mice up to 2000 mg/kg bw. Mortality to mice occurred at doses of 3000 mg/kg bw and above [139]	
				128.41 µg/mL	Fairly toxic effect to brine shrimp [139]	
Lawsonia inermis (Lythraceae)	Vaso-constriction of the uterine and abortion	Nigeria	Root; water	/	Dizziness, loss of appetite, partial paralysis, temporary amnesia, and spontaneous abortion (rats) [148]	—
Leonotis leonurus (Lamiaceae)	Epilepsy, coughs, influenza, bronchitis, hypertension, headaches, diabetes, snakebites, intestinal worms, boils, eczema, skin disease, itching,	Nigeria	Shoots; water	/	Decreased respiratory rate and motor activity, loss of righting reflex, ataxia and death (rats). The symptoms observed before death were respiratory failure, convulsions, paralysis of skeletal muscle, and coma. Marked hyperplasia of pulmonary arteries;	—

Plant (Family)	Traditional uses	Country	Part; extraction	Concentration	Toxicity
	and muscular cramps				glomerulonephritis; necrosis, and mild hemosiderosis in the liver; severe hemosiderosis in the spleen and decrease in the body weights (rats) [149]
Maytenus senegalensis (Celastraceae)	Peptic ulcers, dysentery, snakebites, wounds, respiratory diseases, also used as antimicrobial and anti-inflammatory agents	Tanzania	Stem bark; 80% ethanol	81.63 µg/mL	Toxic effect to brine shrimp [139]
Morinda lucida (Rubiaceae)	Cancers, malaria, diabetes, jaundice, fever, hypoglycemic, trypanocidal, bitter tonic and astringent for dysentery, abdominal colic, and intestinal worm infestation	Nigeria	Root bark; 80% ethanol	37.51 µg/mL	Toxic to brine shrimps [127]

(Continued)

Table 7.1 (Continued)

Plant (Family)	Traditional Use	Area of Plant Collection	Part Used; Extraction Solvent	LD$_{50}$	Physical Signs of Toxicity (Animal Model)	Toxin or Other Compounds (Toxin Class)
Murraya koenigii (Rutaceae)	Tonic, febrifuge, stomachic, antivomiting, diarrhea, used to treat dysentery, skin eruptions, and bites of venomous animals, stimulant, carminative, hypotensive, hypoglycemic, anti-periodic, and antifungal	Nigeria	Leaves; methanol	316.23 mg/kg bw p.o. on rats	Hepatotoxic and nephrotoxic effects [150]	—
Musanga cecropioides (Cecropiaceae)	Lumbago, hypertension, rheumatism, leprosy, chest infections, trypanosomosis	Nigeria	Stem bark; water	>3000 mg/kg bw/p.o.	Decrease in weight gain, irritation, restlessness, tachypnoea, anorexia, bilateral narrowing of the eyelids and abnormal posture which was characterized by tugging of the head in-between the hindlimbs) (rats) [151]	—

Plant (family)	Traditional uses	Country	Part; solvent	Dose/concentration	Effect
Nymphaea lotus (Nymphaeaceae)	Aphrodisiac, cancers, and cancer-related problems	Nigeria	Whole plant; 80% ethanol	58.29 μg/mL	Toxic to brine shrimps [127]
Ocimum gratissimum (Lamiaceae)	Seminal weakness, tonic expectorants and antispasmodic, febrifuge, and malaria	Nigeria	Leaves; water	0.27–1.41 g/kg (in mice), 0.43–2.29 g/kg (in rats)	A dose-dependent sedative effect and death (mice and rats) [152]
Ozoroa insignis (Anacardiaceae)	Peptic ulcers, diarrhea, venereal diseases, tapeworms, hookworms, schistosomiasis, and kidney ailments	Tanzania	Stem bark; 80% ethanol	/	The extract was toxic to mice at doses above 1000 mg/kg body wt; causing hyperventilation within 30 min of extract administration, increased defaecation and passage of loose stools. The mice showed scruffy hair mostly after 24–72 h. Two mice died and four exhibited diarrhea at the dose of 2000 mg/kg bw [139]
				10.63 μg/mL	Very toxic to brine shrimp [139]

(Continued)

Table 7.1 (Continued)

Plant (Family)	Traditional Use	Area of Plant Collection	Part Used; Extraction Solvent	LD_{50}	Physical Signs of Toxicity (Animal Model)	Toxin or Other Compounds (Toxin Class)
Picralima nitida (Apocynaceae)	Analgesic, antipyretic, anti-inflammatory, hypertension, and infectious diseases	Nigeria	Stem bark; water:ethanol (20:80)	/	Reduction in the activity and mobility, inflammation and necrosis of liver hepatocytes, reduction in the neutrophilic count and a corresponding increase in lymphocytic count (mice) [153]	Akuammine, akuammidine, akuammicine, akuammigine, akuammiline, and pseudo-akuammigine (alkaloids)
Pteleopsis hylodendron (Combretaceae)	Measles, dropsy, chickenpox, sexually transmitted diseases, female sterility, liver and kidney disorders	Cameroon	Stem bark; methanol	3.00 and 3.60 g/kg bw po on male and female mice, respectively.	Growth retardation, inflammation and vascular congestion in the liver and kidneys (rats) [154]	—
Rhynchosia recinosa (Fabaceae)	Peptic ulcers	Tanzania	Aerial parts; 80% ethanol	222.43 μg/mL	Fairly toxic to brine shrimp [139]	—
Sacoglottis gabonensis (Humiriaceae)	Gastrointestinal helminthiasis	Nigeria	Stem bark; water	Between 1600 and 3200 mg/kg, IP	Depression, drowsiness and unsteady gait, paralysis of the hindlimbs, dyspnea, coma and death (rats). Congestion and edema of the lungs, bronchi and bronchioles and hepatomegally with focal necrosis of liver cells [155]	—

Species (Family)	Ethnomedicinal use	Country	Part; solvent	Dose	Toxicity	Active compound [Ref.]
Senecio latifolius (Asteraceae)	Wounds, burns, and abortion	South Africa	Leaves; water	/	A dose-related toxicity of the cell hepatocytes, with apoptosis, abnormalities of the cytoskeleton, and necrosis was observed (rats) [156]	Pyrrolizidine alkaloids [156]
Spathodea campanulata (Bignoniaseae)	Epilepsy, painful inflammation, constipation, and dysentery	Nigeria	Leaves; water: ethanol (30:70)	4466.84 mg/kg po	Anorexia, weakness, sluggishness, reduction in the body weight, food and water intake (rats) [157]	—
Spondiathus preussii (Euphorbiaceae)	Cancers and cancer-related problems	Nigeria	Stem bark; 80% ethanol	26.09 µg/mL	Toxic to brine shrimps [127]	
Syzigium aromaticum (Myrtaceae)	Gastrointestinal tract diseases and also used as food spices	Nigeria	Buds; water	263 mg/kg/IP and 2500 mg/kg/p.o. in adult albino rats	Abdominal writhing and a decrease in mean weight gained and death (rats). Severe and irreversible toxicity was observed on the liver (steatosis; perivenular, periportal, and sinusoidal inflammatory infiltrates), brain (edema and congestion), kidney (tubular necrosis), and stomach (gastritis) [158]	—

(*Continued*)

Table 7.1 (Continued)

Plant (Family)	Traditional Use	Area of Plant Collection	Part Used; Extraction Solvent	LD$_{50}$	Physical Signs of Toxicity (Animal Model)	Toxin or Other Compounds (Toxin Class)
Tabernaemontana crassa (Apocynaceae)	Abscesses, furuncles, anthrax pustules, dermal infections, ringworm, coryza, headaches, contusions coryza, sinusitis, stomach disorders, hematuria, gonorrhea, wound infections	Cameroon	Stem bark; 70% ethanol	6.75 g/kg bw po	Slow response to external stimuli, prostation, stretching and sluggishness, rapid breathing all punctuated by a diarrhea. Toxic effects in the liver, lungs and kidneys, decrease in the body weight and death (rats). The apparent causes of death were convulsion and respiratory distress [159]	—
Tithonia diversifolia (Asteraceae)	Chronic malaria	Nigeria	Aerial parts; water: ethanol 30:70 (v/v)	>1600 mg/kg/day	A dose- and time-dependent toxic effect on the morphology of the kidney and liver was observed on (rats) [160]	—
Turraeanthus africanus (Meliaceae)	Infertility, sexual weakness, fibroma, cancer, asthma, stomach ache, intestinal worms, cardiovascular and inflammatory diseases	Cameroon	Stem bark; water	7.2 g/kg/IP on mice	Decrease in the feed consumption and body weight and a dose-dependent mortality. Morphological changes in the liver, kidneys, and lung through parenteral administration of the extract (mice). Not toxic through the oral route in mice [161]	—

(-) not available; p.o, oral administration; i.p, intraperitoneal administration

These parameters were determined through conventional methods using animal and cell toxicity studies. The symptomatology was dominated here by neurological (depression, restlessness, anorexia, amnesia, convulsions, paralysis, coma, etc.); gastrointestinal (diarrhea, black excreta, gastritis, abdominal writhing, etc.); reproductive (distortion of tubules; disruption of normal epithelial organization; hyperplasia of testis with thickening of the basement membrane and disintegration of sperm cells, spontaneous abortion, etc.); respiratory (respiratory distress and failure, tachypnoea, hyperplasia of pulmonary arteries, congestion and edema of the trachea, bronchi, bronchioles, lungs, etc.); hepatic (necrosis, vasculitis, congestion, hepatomegally, steatosis, etc.); renal (glomerulonephritis, urination, tubular atrophy and necrosis, congestion, etc.); and cardiovascular (acute vasculitis in various organs, edema and congestion of the myocardium, disruption of the central vein, reduction in the neutrophilic count, increase in lymphocytic count, etc.) signs. Our simplified presentation of one plant per LD_{50} value and symptoms group also allows us to note that reactions to toxic medicinal plant can be range from mild to serious, depending on factors such as type, part and amount of plant administrated, extraction solvent, animal species, duration, and route of administration, etc.

Severe toxicity may cause symptoms that could be life-threatening, for example, irregular heartbeat, breathing distress, seizures, shock, or paralysis. Moderate toxicity may cause symptoms such as hallucinations, severe stomach irritation, agitation, or severe dermatitis. In mild toxicity, symptoms are generally not life-threatening in nature, such as nausea, vomiting, diarrhea, or skin rashes [162]. These findings show that the medicinal plants listed below can be classified among the above three categories of toxic agents. Among the plants mentioned in Table 7.1, those that cause severe toxicity are mainly *Dichapetalum barteri, Anacardium occidentale, Jatropha curcas, Picralima nitida, Herniaria glabra, Pteleopsis hylodendron, Leonotis leonurus, Tabernaemontana crassa, Azadirachta indica, Sacoglottis gabonensis, Butyrospermum paradoxum, Ocimum gratissimum, Ozoroa insignis, Murraya koenigii, Ficus exasperate,* and *Syzigium aromaticum* (see Chapter 22 for more details on the toxicity of some of these plants).

The data depicted in Table 7.1 also show that undesirable side effects may result from a single ingestion of large amount of a poisonous plant (e.g., *Coula edulis, Hibiscus sabdariffa, Spathodea campanulata, Erythrina senegalensis, Herniaria glabra, Leonotis leonurus,* and *Aspilia Africana*); some plants are so toxic that very small amounts may result in severe disease or death. Among these are *Dichapetalum barteri, Jatropha curcas, Azadirachta indica, Butyrospermum paradoxum, Ocimum gratissimum, Ozoroa insignis, Murraya koenigii, Ficus exasperate,* and *Syzigium aromaticum.* Other plants cause chronic poisoning only after ingestion over week or months. For example, hepatotoxicity and nephrotoxicity of *Chenopodium ambrosioides, Senecio latifolius, Anacardium occidentale, Acanthus montanus, Cylicodiscus gabunensis,* and *Turraeanthus africanus* were observed following chronic administration of these plant extracts. The latter situation may result in delayed clinical signs after the exposure to the toxic plant material, and treatment may no longer be possible.

The level of toxicity of particular plant species is achieved through the determination of its lethal dose (LD). An LD_{50} is an indication of the lethality (ability to kill) of a given substance [163]. The result of acute oral toxicity recorded the LD_{50} higher than 2500 mg/kg body weight (bw) for the most of the studied plants which showed their relative safety as classified by the Organization for Economic Cooperation and Development (OECD, Paris, France) [164]. On the other hand, acute and chronic administration of the respective plants to the animal showed some adverse side effects including neurological, gastrointestinal, respiratory, hepatic, renal, reproductive, and cardiovascular signs. These findings have once more highlighted the limitations of acute toxicity LD_{50} testing and suggest that these plants may exert varied toxicological effects when administered orally in animal species. Furthermore, for *Syzigium aromaticum*, these values were low compared with their corresponding LD_{50} determined by intraperitoneal (IP) route. It could be conjectured that the herbal drug undergoes the first pass effect when administered orally. LD_{50} is not a biological constant because many variables such as animals' species and strain, age, gender, diet, bedding, ambient temperature, caging conditions, and time of the day can all affect its value; hence there are considerable uncertainties in extrapolating LD_{50} value obtained for a species to other species. Consequently, LD_{50} test provides at best, only a ballpark estimate of human lethality [165,166]. Based on the LD_{50} values, *Jatropha curcas* (46.0 mg/kg bw p.o. on rats), *Azadirachta indica* (870 mg/kg bw IP on rats), *Butyrospermum paradoxum* (820 mg/kg bw IP on mice), *Ocimum gratissimum* (0.27−1.41 g/kg bw in mice and 0.43−2.29 g/kg bw in rats), *Murraya koenigii* (316.23 mg/kg bw p.o. on rats), *Ficus exasperate* (0.54 g/kg bw IP in mice), *Syzigium aromaticum* (263 mg/kg bw IP in rats) can be considered to be the most toxic plants on these animal models in their respective administration route.

The results of the brine shrimp lethality test showed different patterns of toxicity (Table 7.1). LD_{50} values of extracts with ≤ 100 µg/mL was considered toxic and those with < 20 were considered to be very toxic [127]. Thus, *Hibiscus sabdariffa*, *Chrysophyllum albidum*, *Chenopodium ambrosioides*, *Entada abyssinica*, *Ficus exasperata*, *Lannea schimperi* and *Rhynchosia recinosa* were fairly toxic while *Bryophyllum calycinum*, *Ozoroa insignis* and *Annona senegalensis* were very toxic and the others (*Erythrophleum suaveolens*, *Garcinia kola*, *Hymenocardia acida*, *Maytenus senegalensis*, *Morinda lucida*, *Nymphaea lotus*, *Spondiathus preussii*) were moderately toxic.

Toxic effects of *Ozoroa insignis*, *Entada abyssinica* and *Lannea schimperi* also were observed on mice while the other plants (*Rhynchosia recinosa*, *Maytenus senegalensis*), and the combined extract of these five plants were well tolerated [139].

Plant chemistry is complex and presentation of one plant per toxin herein, overlooks the fact that plants contain multiple chemicals and chemical classes that work independently or in synergy. Besides, different plant families may contain similar toxic principles. In many cases, xenobiotics remain unidentified. Many of these cases lack clear links between toxin exposure and illness, and qualitative serum concentrations are unavailable. Uncertainty is compounded by the fact that plants themselves are inherently variable, and potency and type of toxin depend on the season, geography, local environment, plant part, and methods of processing [167,168]. For example, during the acute toxicity of the leaves and extracts of *Dichapetalum barteri* on mice, rabbits, and goats, Nwude et al. [137] showed that plants collected in the dry season were more toxic than those collected during the wet season. Monofluoroacetate was detected in this plant while pyrrolizidine alkaloids, urushiol oleoresins, *curcin*, and alkaloid fraction were detected in *Senecio latifolius, Anacardium occidentale, Jatropha curcas*, and *Picralima nitida*, respectively. These compounds are probably among the toxic principles of these plants.

7.5 Toxic African Medicinal Plants as Identified by Biochemical Indices

Acute, subchronic, and chronic toxicity tests of 120 plants as determined using animal models helped to identify 23 plants used in ATM which are able to affect negatively some hematological and biochemical parameters (Table 7.2). In most cases, the repeated oral administrations of the plant extracts induced alterations on hematological (packed cell volume (PCV), hemoglobin concentrations, hematocrit, platelets, RBCs and WBCs, etc.) and biochemical (aspartate aminotransferase (AST), alanine aminotransferase (ALT), alkaline phosphatase (ALP), creatine kinase (CK), γ-glutamyl transpeptidase (GGT), calcium, total bilirubin, cholesterol, urea, creatinine, total proteins, glucose, etc.) parameters on rats and mice. Details on physical and biochemical parameters altered by African medicinal plants are discussed in Chapters 22 and 23, respectively. However, an overview of the plant-induced biochemical changes is provided in this chapter, especially as indication of their potential toxic effects. Significant increases in serum AST, ALT, ALP, GGT, and total bilirubin levels were recorded in the groups receiving several medicinal plant extracts as compared to those of the control groups suggesting liver injuries. These liver damages are further characterized here into hepatocellular (predominantly initial AST and ALT elevations) and cholestatic (initial ALP, GGT, and total bilirubin rises) types [171,172]. It is inferred also from the results that some of the studied plants have toxic effects on the hematological parameters such as hematocrit, platelets, WBC, and RBC counts. Reduction in the levels of RBCs and hematocrit shows that use of these plants may predispose animals to anemia. The physiological consequence of anemia include deficient level of circulating hemoglobin (Hb), the essential transport vehicle of oxygen, which is necessary for proper cell functions in the organs and body tissues. The effect of the plant extracts on WBC counts indicates that these plants may interfere with the blood, immune systems, and spleen. Administration of the studied plants to rats resulted in a significant reduction in the serum protein suggesting chronic liver damage and alteration of the renal function. Equally, there was also a significant rise in creatinine and urea in group receiving the plant extract when compared to that of the control group. Indeed, creatinine is known as a good indicator of renal function [123]. Any rise in creatinine level is only observed, if there is a marked damage at the

Table 7.2 Toxic African Medicinal Plants as Identified by the Biochemical Indexes

Plant (Family)	Area of Plant Collection	Biochemical Parameters and Variation
Acanthus montanus (Acanthaceae)	Cameroon	Hypercreatinemic effects at a dose >500 mg/kg/day were observed [122].
Anacardium occidentale (Anacardiaceae)	Cameroon	Serum parameters like creatinine, transaminases, and urea were found to be significantly ($p < 0.01$) abnormal after repeated oral administration of the methanol leaf extract on mice [124].
Annona senegalensis (Annonaceae)	Nigeria	The subacute toxicity studies in rats indicated a significant ($p < 0.05$) increase in the total WBC count at 100 and 400 mg/kg/p.o. of root extract. The differential analysis showed a decrease in the neutrophils [126].
Aspilia Africana (Asteraceae)	Cameroon	Significant increases in the levels of kidneys and serum creatinine, serum glutathione, liver AST and serum ALT were noted on rats at the dose of 500 mg/kg bw [128].
Azadirachta indica (Meliaceae)	Nigeria	In subchronic toxicity, the WBC, platelets, serum triacylglycerol and HDL-cholesterol decreased significantly ($p < 0.05$). In contrast, serum globulins, total and conjugated bilirubin, serum cholesterol, LDL-cholesterol and computed atherogenic index increased significantly. The ALP, ALT, AST, sodium, potassium, and calcium were altered at specific doses [130].
Berlina grandiflora (Fabaceae)	Nigeria	Aqueous methanolic stem bark extract provoked the decreases in the PCV, hemoglobin concentrations, calcium, ALP and CK levels whereas total leukocytes counts, glutamate oxaloacetate, and GGT concentrations increased remarkably [131].
Corrigiola telephiifolia (Caryophyllaceae)	Morocco	In the 40-day study of rats, the dose of 2000 mg/kg/day of the ethanol–water (75:25) root extract significantly increased the serum concentrations of creatinine, ALP, GGT, and phosphorus ($p < 0.05$) all suggestive of functional nephrotoxicity and hepatotoxicity. At 70 mg/kg/day, the treated group differed from the control only by a significant decrease in serum concentrations of sodium and chloride ions [134].

(Continued)

Table 7.2 (Continued)

Plant (Family)	Area of Plant Collection	Biochemical Parameters and Variation
Coula edulis (Olacaceae)	Cameroon	In subchronic oral administration, CH_2Cl_2-MeOH (1:1 v/v) stem bark extract (at 200 mg/kg) induced significant ($p < 0.05$) changes on hematological (hematocrit, RBC, and WBC) and biochemical (AST, ALT, total cholesterol, creatinine, proteins, and glucose) parameters on rats [135].
Cylicodiscus gabunensis (Mimosaceae)	Cameroon	Ethyl acetate stem bark extract induces significant increases in hematocrit, serum AST, ALT, total cholesterol, and glucose levels at doses ≥ 3 g/kg bw. It also causes a significant reduction in hepatic malondialdehyde, renal urea, and creatinine levels [136].
Diospyros canaliculata (Ebenaceae)	Mont Eloumden, Center Region of Cameroon	Methanolic stem bark extract significantly ($p < 0.05$) increased concentrations of liver AST, serum ALP, AST, total protein, and urea at 2 g/kg bw (rats) [138].
Ficus exasperata (Moraceae)	Nigeria	In 14-day single-dose study, oral administration of the aqueous leaf extract showed significant increase in body temperature and a significant decrease in the RBC count, Hb count, and hematocrit values [141].
Herniaria glabra (Caryophyllaceae)	Morocco	At the dose 4 g/kg bw, the aqueous leaf extract causes on rats, a significant increase in erythrocytes, leukocytes, platelets, eosinophils, serum ALT, AST, and creatinine levels. The study also revealed the hypoglycemic activity of the extract in normoglycemic rats [143].
Hydnora johannis Becca. (Hydnoraceae)	Sudan	Alterations in the levels of AST, ALT, ALP, cholesterol, and urea were observed (rats) [145].
Khaya senegalensis (Meliaceae)	Nigeria	Aqueous stem bark extract induces on albino rats, a significant decrease in total protein, as well as significant increases in ALT, AST, ALP, and total bilirubin levels in comparison with control [169].

(Continued)

Table 7.2 (Continued)

Plant (Family)	Area of Plant Collection	Biochemical Parameters and Variation
Leonotis leonurus (Lamiaceae)	Nigeria	The aqueous shoots extract causes significant changes in RBCs, PCV, hemoglobin concentration, mean corpuscular volume, platelets, WBC, and its differentials at doses of 1600 mg/kg in subacute toxicity and in as low as 200 mg/kg in chronic toxicity. This extract also causes a significant decrease in the levels of urea and creatinine at 1600 mg/kg dose and a significant reduction in urea, total bilirubin, total protein, albumin, globulin, GGT, and ALT in the 400 mg/kg dose in chronic toxicity [149].
Murraya koenigii (Rutaceae)	Nigeria	Acute doses (≥ 500 mg/kg) of methanolic leaf extract on rats reduced significantly serum globulin, albumin, urea, glucose, total protein, AST, and increased cholesterol and ALT indicating hepatic injury. However, chronic administration for 14 days gave a significant reduction in the serum cholesterol, glucose, urea, bilirubin, ALT and AST showing that the plant has hypoglycaemic and hepatoprotective effects after prolonged use [150].
Musanga cecropioides (Cecropiaceae)	Nigeria	The aqueous stem bark extract also caused a significant ($p < 0.05$) decrease in differential eosinophil count and increase in serum creatinine [151].
Ocimum gratissimum (Lamiaceae)	Nigeria	Data analyses of blood biochemical, hematological, and histopathological findings showed significant differences between control and treated groups and revealed that *Ocimum* oil is capable of invoking an inflammatory response that transits from acute to chronic on persistent administration [152].
Pteleopsis hylodendron (Combretaceae)	Cameroon	The methanolic stem bark extract causes on rats a dose-dependent significant increase in liver enzymes (ALT and AST), proteins, and creatinine levels, which correlates the observed histopathological damages in the liver and kidneys.Hematological parameters showed a significant decrease in WBC count and significant increases RBC count. A dose-dependent significant increase in HDL was observed only in male rats [154].

(Continued)

Table 7.2 (Continued)

Plant (Family)	Area of Plant Collection	Biochemical Parameters and Variation
Sidarhombifolia (Malvaceae)	Cameroon	Significant increases of some biochemical parameters such as AST, ALT, ALP, and creatinine were found on Wistar rats after repeated oral administration of aqueous–methanol (1v:4v) extract [170].
Syzigium aromaticum (Myrtaceae)	Nigeria	Significant and prominent effects on ALP, AST, Hb, RBC, PCV, platelets, and granulocytes These findings were in consonance with the severe and irreversible toxicity observed on the liver (steatosis; perivenular, periportal, and sinusoidal inflammatory infiltrates), brain (edema and congestion), kidney (tubular necrosis), and stomach (gastritis) [158].
Tabernaemontana crassa (Apocynaceae)	Cameroon	70% (v/v) hydroethanol extract from the stem bark induces significant variation of the ALP, ALT, AST, total bilirubin, glutathione, malondialdehyde, urea, direct bilirubin, and creatinine at the dose of 6 g/kg bw suggesting nephrotoxic and hepatotoxic effects of this plant [159].
Turraeanthus africanus (Meliaceae)	Cameroon	In subacute toxicity in mice, the intraperitoneal administration of aqueous stem bark extract induced significant decreases in triglyceride, total cholesterol, HDL cholesterol, LDL cholesterol, serum, and tissue creatinine levels at doses ≥ 1.5 g/kg bw. Serum protein level of treated animal enhanced at doses of 1.5 and 6 g/kg bw with $p < 0.05$ while tissue creatinine level of treated animal enhanced with $p < 0.001$ [161].

AST, aspartate aminotransferase; ALT, alanine aminotransferase; ALP, alkaline phosphatse; GGT, gamma-glutamyl transpeptidase; Hb, hemoglobin; WBCs, white blood cells; RBCs, red blood cells; LDL, low density lipoproteins; HDL, high density lipoproteins; PCV, packed cell volume; CK, creatine kinase.

nephrons [173]. Thus, the results recorded with the biochemical parameters similarly suggest that the studied plants might have altered the renal function. The kidney damage was further characterized by a number of signs such proteinuria, hypoalbuminemia, and hyperlipidemia. The results on hematological and biochemical parameters are in consonance with the nephrotoxicity and hepatotoxicity already revealed by the histopathological analysis. Overall, the alterations in the hematological and biochemical parameters of toxicity have consequential effects on the normal functioning of the organs of the animals mainly liver and kidneys. Therefore, the administration of extract preparation from these plants in traditional medicine may not be completely safe as an oral remedy and should be taken with caution if absolutely necessary. Results of many subchronic and chronic toxicity tests of various plant extracts showed that the major organs usually affected are the liver and kidneys. Hepatotoxic and nephrotoxic effects are mostly to be expected, as the liver acts as the main detoxifying organ for chemical substances, while the kidney is a principal route of excretion for many chemical substances in their active and/or inactive forms [174].

7.6 Conclusion

The general observation is that a very low amount of African medicinal plants has been subjected to toxicological studies. In this chapter, of up to 120 African medicinal plants investigated, 49 were identified as potentially toxic, with reported adverse effects, serious allergic reactions. They include: *Acanthus montanus, Ajuga iva, Anacardium occidentale, Annona senegalensis, Aspilia Africana, Azadirachta indica, Berlina grandiflora, Bryophyllum calycinum, Butyrospermum paradoxum, Chenopodium ambrosioides, Chrysophyllum albidum, Corrigiola telephiifolia, Coula edulis, Cylicodiscus gabunensis, Dichapetalum barteri, Diospyros canaliculata, Entada abyssinica, Erythrina senegalensis, Erythrophleum suaveolens, Ficus exasperate, Garcinia kola, Glinus lotoides, Herniaria glabra, Hibiscus sabdariffa, Hydnora johannis, Hymenocardia acida, Khaya ivorensis, Lannea schimperi, Lawsonia inermis, Leonotis leonurus, Maytenus senegalensis, Morinda lucida, Murraya koenigii, Musanga cecropioides, Nymphaea lotus, Ocimum gratissimum, Ozoroa insignis, Picralima nitida, Pteleopsis hylodendron, Rhynchosia recinosa, Sacoglottis gabonensis, Senecio latifolius, Sida Rhombifolia, Spathodea campanulata, Spondiathus preussii, Syzigium aromaticum,* and *Tabernaemontana crassa, Tithonia diversifolia, Turraeanthus africanus.* Therefore, awareness of their use in therapy should be given while rigorous toxicological and clinical studies are necessary before they are widely prescribed. High-risk patients such as the elderly, expectant mothers, children, those taking several medications for chronic conditions, those with hypertension, depression, high cholesterol, or congestive heart failure, should be more cautious in taking these herbal medicines.

This report should not lead to the false conclusion that all toxic plants used in ATM are known or that the plant reported herein could not succesfully undergo clinical studies. Reassurance can be achieved by excluding exposure to these toxic plants particularly at the high doses. Clinical trials should be conducted to establish facts such as the average effective dose for any herbal preparation, as well as potential induced side effects at therapeutic doses. In short, these herbal drugs need to be analyzed in the same way as any modern drug, that is, with randomized controlled clinical trials. At the same time, legislators at the national level should continue to press for effective laws to protect consumers from potentially harmful herbal drugs.

Upon selecting plants for ethnopharmacological studies, researchers are encouraged to search available literature for known toxic properties of plants of interest, prior to conducting biological activity studies. Where toxic effects are unknown, parallel toxicity studies are useful in detecting potential toxicity when screening plant extracts or isolated natural products for anticancer, antibacterial, antifungal, antiviral, and antiparasitic activities [12,175].

This report and documentation of traditional medicines will go a long way toward protecting people from being poisoned by substances that were meant to cure them. It will also be important that the governments of African countries encourage their institutes or researchers to currently conduct toxicity studies in conjunction with the traditional practitioners on commonly used African traditional plants in order to define the margin between the therapeutic dose and the toxic dosage. Also, traditional practitioners should be aware of the potential risks associated with the use of some medicinal plants in ATM.

References

[1] Sofowora AE. Medicinal plants and traditional medicines in Africa. 2nd ed. Ibadan, Nigeria: Spectrum Books; 1993, p. 289.

[2] WHO. Research guidelines for evaluating the safety and efficacy of herbal medicines. Manila: WHO Regional Office for the Western Pacific; 1993.

[3] Iwu MM, Duncan AR, Okunji CO. New antimicrobials of plant origin. In: Janick J, editor. Perspectives in new crops and new uses. Alexandria, VA: ASHS Press; 1999, p. 457−62.

[4] Cowan MM. Plant products as antimicrobial agents. Clin Microbiol Rev 1999;12 (4):564−82.

[5] Tene M, Tane P, Kuiate JR, Tamokou JDD, Connolly JD. Anthocleistenolide, a new rearranged nor-secoiridoid derivative from the stem bark of *Anthocleista vogelii*. Planta Med 2008;74:80−3.

[6] Nyaa TBL, Tapondjou AL, Barboni L, Tamokou JDD, Kuiate JR, Tane P, et al. NMR assignment and antimicrobial/antioxidant activities of 1β-hydroxyeuscaphic acid from the seeds of *Butyrospermum parkii*. Nat Prod Sci 2009;15(2):76−82.

[7] Tamokou JDD, Mpetga Simo DJ, Lunga PK, Tene M, Tane P, Kuiate JR. Antioxidant and antimicrobial activities of ethyl acetate extract, fractions and compounds from the stem bark of *Albizia adianthifolia* (Mimosoideae). BMC Compl Altern Med 2012;12(99).

[8] Mouokeu RS, Ngono NRA, Tume C, Kamtchueng OM, Njateng GSS, Dzoyem JP, et al. Immunomodulatory activity of ethyl acetate extract and fractions from leaves of *Crassocephalum bauchiense* (Asteraceae). Pharmacologia 2013;4(1):38−47.

[9] Tamokou JDD, Chouna JR, Fischer-Fodor E, Chereches G, Barbos O, Damian G, et al. Anticancer and antimicrobial activities of some antioxidant-rich Cameroonian medicinal plants. PLoS One 2013;8(2):e0055880.

[10] WHO. In progress report by the director general, document no. A44/20. Geneva: World Health Organization; 1991.

[11] WHO. Resolution—promotion and development of training and research in traditional medicine. WHO document nos. 30−49; 1977.

[12] Botha CJ, Penrith M-L. Poisonous plants of veterinary and human importance in southern Africa. J Ethnopharmacol 2008;119:549−58.

[13] Cosyns JP, Jadoul M, Squifflet JP, De Plaen JF, Ferluga D, van Ypersele de Strihou C. Chinese herbs nephropathy: a clue to Balkan endemic nephropathy? Kidney Int 1994;45:1680−8.

[14] Tanaka A, Nishida R, Sawai K, Nagae T, Shinkai S, Ishikawa M, et al. Traditional remedy-induced Chinese herbs nephropathy showing rapid deterioration of renal function. Nippon Jinzo Gakkai Shi 1997;39:794−7.

[15] Tanaka A, Shinkai S, Kasuno K, Maeda K, Murata M, Seta K, et al. Chinese herbs nephropathy in the Kansai area: a warning report. Nippon Jinzo Gakkai Shi 1997;39:438−40.

[16] Stengel B, Jones E. End-stage renal insufficiency associated with Chinese herbal consumption in France. Nephrology 1998;19:15−20.

[17] Lord GM, Tagore R, Cook T, Gower P, Pusey CD. Nephropathy caused by Chinese herbs in the UK. Lancet 1999;354:481−2.

[18] Luyckx VA, Naicker S. Acute kidney injury associated with the use of traditional medicines. Nat Clin Pract Nephr 2008;4:664−71.

[19] Joshi CS, Priya ES, Venkataraman S. Acute and subacute toxicity studies on the polyherbal antidiabetic formulation Diakyur in experimental animal models. J Health Sci 2007;52:245−9.

[20] Lahlou S, Israili ZH, Lyoussi B. Acute and chronic toxicity of a lyophilized aqueous extract of *Tanacetum vulgare* leaves in rodents. J Ethnopharmacol 2008;117:221−7.

[21] Raman A, Lau C. Anti-diabetic properties and phytochemistry *Momordica charantia* L. (Cucurbitaceae). Phytomedicine 1996;2:349−62.

[22] Basch E, Gabardi S, Ulbricht C. Bitter melon (*Momordica charantia*): a review of efficacy and safety. Am J Health Syst Pharm 2003;60:356−9.

[23] Tahri A, Yamani S, Legssyer A, Mohammed A, Mekhfi H, Bnouham M, et al. Acute diuretic, natriuretic and hypotensive effects of a continuous perfusion of aqueous extract of *Urtica dioica* in the rat. J Ethnopharmacol 2000;73:95−100.

[24] Hosseinzadeh H, Younesi HM. Antinociceptive and anti-inflammatory effects of *Crocus sativus* L. stigma and petal extracts in mice. BMC Pharmacol 2002;2:7.

[25] Adeoye BA, Oyedapo OO. Toxicity of *Erythrophleum guineese* stem-bark: role of alkaloidal fraction. Afr J Tradit Complement Altern Med 2004;1:45−54.

[26] Efferth T, Li PC, Konkimalla VS, Kaina B. From traditional Chinese medicine to rational cancer therapy. Trends Mol Med 2007;13:353−61.

[27] Manahan SE. Toxicology in toxicological chemistry and biochemistry. 3rd ed. Boca Raton, FL: Lewis Publishers/CRC Press; 2003, p. 134−57.

[28] Farrell M. Poisons and poisoners, an encyclopedia of homicidal poisonings. London: Robert Hale Publishers; 1992, p. 51−2.

[29] Palmer E. A field guide to the trees of Southern Africa. 2nd ed. London/Johannesburg: Collins; 1981, p. 191.

[30] Van Wyk B, Van Wyk P, Van Wyk B-E. Photographic guide to trees of Southern Africa. Arcadia, Pretoria: Briza Publications; 2000, p. 16, 290.

[31] Frohne D, Pfander HJK. A colour Atlas of poisonous plants. London: Wolfe Publishing Company; 1984.

[32] Haq I. Safety of medicinal plants. Pak J Med Res 2004;43:4.

[33] Newall C, Anderson LA, Phillipson JD. Herbal medicines: A guide for health-care professionals. London, England: Pharmaceutical Press; 1996.

[34] Wainwright J, Schonland MM, Candy HA. Toxicity of *Callilepis laureola*. S Afr Med J 1977;52:313—5.

[35] Laurens JB, Bekker LC, Steenkamp V, Stewart MJ. Gas chromatographic mass spectrometric confirmation of atractyloside in a patient poisoned with *Callilepis laureola*. J Chromatogr B 2001;765:127—33.

[36] Steenkamp PA. Chemical analysis of medicinal and poisonous plants of forensic importance in South Africa. Ph.D. thesis, University of Johannesburg; 2005. p. 65, 80, 168.

[37] Maurer HH. Toxicokinetics- variations due to genetics or interactions: basics and examples. <www.gtfch.org/cms/images/stories/media/tb/tb2007/s153-155.pdf>; 2013 [accessed 20.09.13].

[38] Steenkamp V, Stewart MJ, Zuckerman M. Clinical and analytical aspects of pyrrolizidine poisoning caused by South African traditional medicines. Ther Drug Monit 2000;22:302—6.

[39] Betz JM, Page SW. Perspectives on plant toxicology and public health. In: Garland T, Barr AC, editors. Toxic plants and other natural toxicants. Wallingford: CAB International; 1998. p. 367—72.

[40] Kao WF, Hung ZZ, Tsai WJ, Lin KP, Deng JF. Podophyllotoxin intoxication: toxic effect of Bajiaolian on herbal therapeutics. Hum Exp Toxicol 1992;11:480—7.

[41] Larrey D, Vial T, Pauwels A, Castot A, Biour M, David M, et al. Hepatitis after germander (*Teucrium chamaedrys*): another instance of herbal medicine hepatotoxicity. Ann Intern Med 1992;117:129—32.

[42] Farrell GC, Weltman M. Drug-induced liver disease. In: Ginick G, editor. Current hepatology, vol. 16. Chicago, IL: Mosby-year Book Medical Publishers; 1996. p. 143—208.

[43] Chitturi S, Farrel GC. Herbal hepatotoxicity: an expanding but poorly defined problem. J Gastroen Hepatol 2000;15:1093—9.

[44] Pak E, Esrason KT, Wu VH. Hepatotoxicity of herbal remedies: an emerging dilemma. Progr Transplant 2004;14(2):91—6.

[45] Seeff LB. Herbal hepatotoxicity. Clin Liver Dis 2007;11(3):5777—96, vii.

[46] Van Wyk B-E, Van Heerden FR, Van Oudtshoorn B. Poisonous plants of South Africa. Arcadia, Pretoria: Briza Publications; 2002. p. 8, 30, 108, 144, 180, 194, 196.

[47] Kiec-Swierczynska M, Krecisz B, Swierczynska-Machura D, Palczynski C. Occupational allergic contact dermatitis caused by padauk wood (*Pterocarpus soyauxii* Taub.). Contact Dermatitis 2004;50:384—5.

[48] Vanherweghem JL, Depierreux M, Tielemans C, Abramowicz D, Dratwa M, Jadoul M, et al. Rapidly progressive interstitial renal fibrosis in young women: association with slimming regimen including Chinese herbs. Lancet 1993;341:387—91.

[49] Gilman AG, Rall TW, Nies AS, Taylor P, editors. Goodman and Gilman's the pharmacological basis of therapeutics. New York, NY: Macmillan Publishing Company; 1990.

[50] Kellerman TS, Naudé TW, Fourie N. The distribution, diagnoses and estimated economic impact of plant poisonings and mycotoxicoses in South Africa. Onderstepoort J Vet Res 1996;63:65—90.

[51] Marx J, Pretorius E, Espag WJ, Bester MJ. *Urginea sanguinea*: medical wonder or death in disguise? Environ Toxicol Pharmacol 2005;20:26—34.

[52] Rowin J, Lewis SL. Spontaneous bilateral subdural hematomas associated with chronic Ginkgo biloba ingestion. Neurology 1996;46(6):1775—6.

[53] Blood DC, Radostits OM. Veterinary medicine. London: Balliere Tindall; 1989. p. 127—30.

[54] Adedapo AA, Abatan MO, Olorunsog OO. Toxic effects of some plants in the genus *Euphorbia* on haematological and biochemical parameters of rats. Vet Arhiv 2004;74:53−62.

[55] Adedapo AA, Abata MO, Olorunsog OO. Effects of some plants of the spurge family on the haematological and biochemical parameters of rats. Vet Arhiv 2007;77:29−38.

[56] Garg SK, Shah MA, Garg KM, Farooqui MM, Sabir M. Antilymphocytic and immuno-suppressive effects of *Lantana camara* leaves in rats. Indian J Exp Biol 1997;35:1315−8.

[57] Mazumdar UK, Gupta M, Maiti S. Effect of petroleum ether extract from *Hygrophila spinosa* on hematological parameters and hepatorenal functions in mice. Indian J Exp Biol 1996;34:1201−3.

[58] Biswas NS, Sen S, Singh S, Gopal N, Pandey RM, Giri D. Sub-acute toxicity study of a polyherbal drug (Prostina®) in rats. Indian J Pharmacol 1998;30:239−44.

[59] Sukumar E, Hamsaveni GR, Bhima RR. Pharmacological actions of cerpegin, a novel pyridine alkaloid from *Ceropegia juncea*. Fitoterapia 1995;66:403−6.

[60] Efferth T, Li PCH, Lam E, Roos WP, Zdzienicka MZ, Kaina B. Artesunate derived from traditional Chinese medicine induces DNA damage and repair. Cancer Res 2008;68:4347−51.

[61] Longo M, Zanoncelli S, Della, Torre P, Rosa F, Giusti A, Colombo PM, et al. Investigations of the effects of the antimalarial drug dihydroartemisinin (DHA) using the Frog Embryo Teratogenesis Assay-Xenopus (FETAX). Reprod Toxicol 2008;25:433−41.

[62] Elgorashi EE, Taylor JLS, Verschaeve L, Maes A, van Staden J, De Kimpe N. Screening of medicinal plants used in South African traditional medicine for genotoxic effects. Toxicol Lett 2003;143:195−207.

[63] Prakash AS, Pereira TN, Reilly PEB, Seawright AA. Pyrrolizidine alkaloids in human diet. Mutat Res 1999;443:53−67.

[64] Fang Z-Z, Zhang Y-Y, Wang X-L, Cao Y-F, Huo H, Yang L. Bioactivation of herbal constituents: simple alerts in the complex system. Expert Opin Drug Metab Toxicol 2011;7:1−19.

[65] Roeder E. Medicinal plants in China containing pyrrolizidine alkaloids. Pharmazie 2000;55:711−26.

[66] Fu PP, Chou MW, Xia Q, Yang YC, Yan J, Doerge DR, et al. Genotoxic pyrrolizidine alkaloids and pyrrolizidine alkaloid N-oxides-mechanisms leading to DNA adduct formation and tumorigenicity. Environ Carcinog Ecotoxicol Rev 2001;19:353−86.

[67] Fu PP, Xia Q, Lin G, Chou MW. Genotoxic pyrrolizidine alkaloids—mechanisms leading to DNA adduct formation and tumorigenicity. Int J Mol Sci 2002;3:948−64.

[68] Fu PP, Xia Q, Lin G, Chou MW. Pyrrolizidine alkaloids-genotoxicity, metabolism enzymes, metabolic activation, and mechanisms. Drug Metab Rev 2004;36:1−55.

[69] Liu T-Y, Chunga Y-T, Wang P-F, Chi C-W, Hsieh L-L. Safrole-DNA adducts in human peripheral blood-an association with areca quid chewing and CYP2E1 polymorphisms. Mutat Res 2004;559:59−66.

[70] Munerato MC, Sinigaglia M, Reguly M, Rodriguez-de-Andrade HH. Genotoxic effects of eugenol, isoeugenol and safrole in the wing spot test of *Drosophila melanogaster*. Mutat Res 2005;582:87−94.

[71] Zhang F, Xu Q, Fu S, Ma X, Xiao H, Liang X. Chemical constituents of the essential oil of *Asarum forbesii* Maxim (Aristolochiaceae). Flavour Frag J 2005;20:318−20.

[72] Smith B, Cadby P, Leblanc J-C, Setzer RW. Application of the margin of exposure (MoE) approach to substances in food that are genotoxic and carcinogenic example: methyleugenol, CASRN: 93-15-2. Food Chem Toxicol 2010;48:89−97.

[73] Taylor JLS, Elgorashi EE, Maes A, Van Gorp U, De Kimpe N, van Staden J, et al. Investigating the safety of plants used in South African traditional medicine: testing for genotoxicity in the micronucleus and alkaline comet assays. Environ Mol Mutagen 2003;42:144−54.

[74] Fennell CW, Lindsey KL, McGaw LJ, Sparg SG, Stafford GI, Elgorashi EE, et al. Assessing African medicinal plants for efficacy and safety: pharmacological screening and toxicology. J Ethnopharmacol 2004;94:205−17.

[75] Rao VSN, Cavalcanti BC, Costa-lotufo LV, Moraes MO, Burbano RR, Silveira ER, et al. Genotoxicity evaluation of kaurenoic acid, a bioactive diterpenoid present in Copaiba oil. Food Chem Toxicol 2006;44:388−92.

[76] Ouedraogo M, Baudoux T, Stévigny C, Nortier J, Efferth T, Qu F, et al. Review of current and "omics" methods for assessing the toxicity (genotoxicity, teratogenicity and nephrotoxicity) of herbal medicines and mushrooms. J Ethnopharmacol 2012;140:492−512.

[77] Zhao JY, Yan XZ, Peng SQ. Metabonomics study on nephrotoxicity of *Aristolochia manshuriensis*. Chin Tradit Herbal Drugs 2006;37:725−30.

[78] Jia W, Chen MJ, Su MM, Zhao LP, Jiang J, Liu P, et al. Metabonomic study of aristolochic acid-induced nephrotoxicity in rats. J Proteome Res 2006;5:995−1002.

[79] Hashimoto K, Higuchi M, Makino B, Sakakibara I, Kubo M, Komatsu Y, et al. Quantitative analysis of aristolochic acids, toxic compounds, contained in some medicinal plants. J Ethnopharmacol 1999;64:185−9.

[80] Arvidson KB, Valerio LG, Diaz M, Chanderbhan RF. In silico toxicological screening of natural products. Toxicol Mech Methods 2008;229−42.

[81] Chen XW, Serag ES, Sneed KB, Zhou SF. Herbal bioactivation, molecular targets and the toxicity relevance. Chem Biol Interact 2011;192:161−76.

[82] Huang KL, Lai YK, Lin CC, Chang JM. Inhibition of hepatitis B virus production by *Boehmeria nivea* root extract in HepG2 2.2.15 cells. World J Gastroenterol 2006;12:5721−5.

[83] Tian XY, Xu M, Deng B, Leung KS, Cheng KF, Zhao ZZ, et al. The effects *Boehmeria nivea* (L.) Gaud. on embryonic development: *in vivo* and *in vitro* studies. J Ethnopharmacol 2011;134:393−8.

[84] D'Arcy PF. Adverse reactions and interactions with herbal medicines, Part I, adverse reactions. Adverse Drug React Toxicol Rev 1991;10(4):189−208.

[85] Grob PJ, Muller-Schoop JW, Hacki MA, Joller-Jemelka HI. Drug-induced pseudolupus. Lancet 1975;2:144−8.

[86] Hellberg K, Ruschewski W, de Vivie R. Drug induced acute renal failure after heart surgery. Scanning Microsc 1975;23:396−9.

[87] Ansford AJ, Morris H. Fatal oleander poisoning. Med J Aust 1981;1:360−1.

[88] Challoner KR, McCarron MM. Castor bean intoxication. Ann Emerg Med 1990;19:1177−83.

[89] Furet Y, Ernouf D, Brechot JF, Autret E, Breteau M. Collective poisoning by flowers of laburnum. Presse Med 1986;15:1103−4.

[90] Klintschar M, Beham-Schmidt C, Radner H, Henning G, Roll P. Colchicine poisoning by accidental ingestion of meadow saffron (*Colchicum autumnale*): pathological and medicolegal aspects. Forensic Sci Int 1999;106:191−200.

[91] Williams R. Cancer-Causing herbal remedies: a potent carcinogen lurks within certain traditional Chinese medicines. The Scientist Magazine®. <http://www.the scientist.com/?articles.view/articleNo/36919/title/Cancer-Causing-Herbal Remedies/.Retrieved on 16th October 2013.

[92] Cheng ZF, Zhen C. The Cheng Zhi-Fan collectanea of medical history. Beijing, China: Peking University Medical Press; 2004.

[93] Krenzelok EP. Lethal plant exposures reported to poisons centres: prevalence, characterization and mechanisms of toxicity. J Toxicol Clin Toxicol 2002;2002:40303—4. [abstract].

[94] Stewart MJ, Moar JJ, Steenkamp P, Kokot M. Findings in fatal cases of poisoning attributed to traditional remedies in South Africa. Forensic Sci Interview 1999;101:77—183.

[95] Achour S, Rhalem N, Khattabi A, Lofti H, Mokhtari A, Soulaymani A, et al. L'intoxication au *Peganum harmala* L. au Maroc: à propos de 200 cas. Thérapie 2012;67(1):53—8.

[96] Hamouda C, Amamou M, Thabet H, Yacoub M, Hedhili A, Bescharnia F, et al. Plant poisonings from herbal medication admitted to a Tunisian toxicologic intensive care unit, 1983—1998. Vet Hum Toxicol 2000;42(3):137—41.

[97] Thomson SA. South African government genocide and ethnopiracy. South Africa: The Gaia Research Institute; 2002.

[98] Popat A, Shear NH, Malkiewicz I, Stewart MJ, Steenkamp V, Thomson S, et al. The toxicity of *Callilepis laureola*, a South African traditional herbal medicine. Clin Biochem 2001;34:229—36.

[99] Wainwright J, Schonland MM. Toxic hepatitis in black patients in Natal. S Afr Med J 1977;51:571—3.

[100] Joubert P, Sebata N. The role of prospective epidemiology in the establishment of a toxicology service for a developing community. South Afr Med J 1982;62:853—4.

[101] Savage A, Hutchings A. Poisoned by herbs. Brit Med J 1987;295:1650—1.

[102] Bodenstein JW. Observations on medicinal plants. S Afr Med J 1973;47:336—8.

[103] Cunningham AB. An investigation of the herbal medicine trade in Natal/KwaZulu. Pietermaritzburg: University of Natal, Institute of Natural Resources; 1988.

[104] Schimmer O, Haefele F, Kruger A. The mutagenic potencies of plant extracts containing quercetin in *Salmonella typhimurium* TA98 and TA100. Mutat Res 1988;206:201—8.

[105] Schimmer O, Kruger A, Paulini H, Haefele F. An evaluation of 55 commercial plant extracts in the Ames mutagenicity test. Pharmazie 1994;49:448—51.

[106] Higashimoto M, Purintrapiban J, Kataoka K, Kinouchi T, Vinitketkumnuen U, Kimoto S, et al. Mutagenicity and antimutagenicity of extracts of three species and a medicinal plant in Thailand. Mutagen Res 1993;303:135—42.

[107] Kassie F, Parzefall W, Musk S, Johnson I, Lamprecht G, Sontag G, et al. Genotoxic effects of crude juices from *Brassica* vegetables and juices and extracts from phytopharmaceutical preparations and spices of cruciferous plants origin in bacterial and mammalian cells. Chem Biol Interact 1996;102:1—16.

[108] De Sã Ferrira ICF, Ferrão Vargas VM. Mutagenicity of medicinal plant extracts in *Salmonella*/microsome assay. Phytother Res 1999;(13)397—400.

[109] Tasaka AC, Weg R, Calore EE, Sinhorini IL, Dagli MLZ, Haraguchi M, et al. Toxicity testing of *Senna occidentalis* seed in rabbits. Vet Res Commun 2000;24:573—82.

[110] O'Hara PJ, Pierce KR, Read WK. Degenerative myopathy associated with the ingestion of *Cassia occidentalis* L.: clinical and pathologic features of the experimental disease. Am J Vet Res 1969;30:2173—80.

[111] Calore EE, Cavaliere MJ, Haraguchi M, Górniak SL, Dagli MLZ, Raspantini PCF, et al. Experimental mitochondrial myopathy induced by chronic intoxication by *Senna occidentalis* seeds. J Neurol Sci 1997;146:1—6.

[112] Haraguchi M, Górniak SL, Dagli MLZ, Raspantini PCF. Determinaç ão dos constituintes químicos das fraç ões tóxicas de fedegoso (*Senna occidentalis* L.). In: Proceedings of annual meeting of the Brazilian chemical society, Poçosde Caldas, Brazil; 1996.

[113] O'Hara PJ, Pierce KR. A toxic cardiomyopathy caused by *Cassia occidentalis*. I. Morphological studies in poisoned rabbits. Vet Pathol 1974;11:97—109.

[114] Keay RWJ. Trees of Nigeria. New York, NY: Oxford University Press; 1989. p. 146—51.

[115] Gundidza M, Sorg B, Hecker E. A skin irritant principle from *Euphorbia metabelensis* Pax. J Ethnopharmacol 1993;39(3):209—12.

[116] Dagang W, Sorg B, Adolf W, Seip EH, Hecker E. Oligo- and macrocyclic diterpenes in Thymelaeaceae and Euphorbiaceae occurring and utilized in Yunnan (Southwest China): two ingenane type diterpene esters from *Euphorbia nematocypha*. Phytother Res 1992;6(5):237—40.

[117] Milillo MA, Laffaldano D, Sasso G, De-Laurentis N. Poisoning by *Euphorbia characias* L.1.Clinical and toxicological findings. Obiettivi-e-Documento-Veterianari 1993;14(5):35—7.

[118] Neuwinger HD. Review of plants used for poison fishing in tropical Africa. Toxicon 2004;44(4):417—30.

[119] Sarathchandra G, Balakrishnamurthy P. Perturbations in glutathione and adenosine triphosphatase in acute oral toxicosis of *Cleistanthus collinus*: an indigenous toxic plant. Indian J Pharmacol 1997;29:82—5.

[120] Norton S. Toxic effects of plants. In: Klaassen CD, editor. Casarett and Doull's toxicology: the basic science of poisons. 6th ed. New York, NY: McGraw-Hill; 2001. p. 965—76. [Chapter 27].

[121] Skaanild MT, Friis C, Brimer L. Interplant alkaloid variation and *Senecion vernalis* toxicity in cattle. Vet Hum Toxicol 2001;43:147—51.

[122] Djami TAT, Asongalem EA, Nana P, Choumessi A, Kamtchouing P, Asonganyi T. Subacute toxicity study of the aqueous extract from *Acanthus montanus*. Electron J Biol 2011;7(1):11—5.

[123] El Hilaly J, Israili HZ, Lyoussi B. Acute and chronic toxicological studies of *Ajuga iva* in experimental animals. J Ethnopharmacol 2004;9:43—50.

[124] Tédong L, Dzeufiet DPD, Dimo T, Asongalem EA, Sokeng NS, Flejou JF, et al. Acute and subchronic toxicity of *Anacardium occidentale* Linn. (Anacardiaceae) leaves hexane extract in mice. Afr J Trad CAM 2007;4(2):140—7.

[125] Palmer M, Betz JM. Plants. In: 9th ed. Nelson LS, Lewin NA, Howland MA, Hoffman RS, Goldfrank LR, Flomenbaum NE, editors. Goldfrank's toxicologic emergencies, 2011. New York, NY: McGraw-Hill; 2011. p. 1537.

[126] Okoye CT, Akah AT, Ezike CA, Okoye OM, Onyeto AC, Ndukwu F, et al. Evaluation of the acute and sub acute toxicity of *Annona senegalensis* root bark extracts. Asian Pac J Trop Med 2012;277—82.

[127] Sowemimo AA, Fakoya FA, Awopetu I, Omobuwajo OR, Adesanya SA. Toxicity and mutagenic activity of some selected Nigerian plants. J Ethnopharmacol 2007;113:427—32.

[128] Taziebou LC, Etoa F-X, Nkegoum B, Pieme CA, Dzeufiet DPD. Acute and subacute toxicity of *Aspilia africana* leaves. Afr J Trad CAM 2007;4(2):127—34.

[129] Mbaya WA, Ibrahim IU, God TO, Ladi S. Toxicity and potential anti-trypanosomal activity of ethanolic extract of *Azadirachta indica* (Maliacea) stem bark: an *in vivo* and *in vitro* approach using *Trypanosoma brucei*. J Ethnopharmacol 2010;128:495—500.

[130] Ashafa TOA, Orekoya OL, Yakubu TM. Toxicity profile of ethanolic extract of *Azadirachta indica* stem bark in male Wistar rats. Asian Pac J Trop Biomed 2012;811−7.

[131] Aniagu SO, Nwinyi FC, Olanubi B, Akumka DD, Ajoku GA, Agala P, et al. Is *Berlina grandiflora* (Leguminosae) toxic in rats? Phytomedicine 2004;11:352−60.

[132] Mbaya WA, Nwosu OC, Onyeyili AP. Toxicity and anti-trypanosomal effects of ethanolic extract of *Butyrospermum paradoxum* (Sapotaceae) stem bark in rats infected with *Trypanosoma brucei* and *Trypanosoma congolense*. J Ethnopharmacol 2007;111:526−30.

[133] Amole OO, Izegbu MC. Chronic toxicity of *Chenopodium ambrosioides* in rats. Biomed Res 2005;16(2):111−3.

[134] Lakmichi H, Bakhtaoui ZF, Gadhi AC, Ezoubeiri A, El Jahiri Y, El Mansouri A, et al. Toxicity profile of the aqueous ethanol root extract of *Corrigiola telephiifolia* Pourr. (Caryophyllaceae) in rodents. Evid Based Complement Altern Med 2011; Article ID 317090, 10 pages.

[135] Tamokou JDD, Kuiate JR, Gatsing D, Nkeng-Efouet AP, Njouendou AJ. Antidermatophytic and toxicological evaluations of dichloromethane-methanol extract, fractions and compounds isolated from *Coula edulis*. Iran J Med Sci 2011;36 (2):111−21.

[136] Mabeku KBL, Beng PV, Kouam J, Essame O, Etoa F-X. Toxicological evaluation of ethyl acetate extract of *Cylicodiscus gabunensis* stem bark (Mimosaceae). J Ethnopharmacol 2007;111:598−606.

[137] Nwude N, Parsons LE, Adaudi AO. Acute toxicity of the leaves and extracts of *Dichapetalum barteri* (Engl.) in mice, rabbits and goats. Toxicology 1977;7:23−9.

[138] Dzoyem JP, Nkegoum B, Kuete V. A 4-week repeated oral dose toxicity study of the methanol extract from *Diospyros canaliculata* in rats. Comp Clin Pathol 2013;22:75−81.

[139] Haule EE, Moshi JM, Mahunnah LAR, Nondo SOR, Mwangomo TD. A study of antimicrobial activity, acute toxicity and cytoprotective effect of a polyherbal extract in a rat ethanol-HCl gastric ulcer model. BMC Res Notes 2012;5:546.

[140] Atsamo DA, Nguelefack BT, Datté YJ, Kamanyi A. Acute and subchronic oral toxicity assessment of the aqueous extract from the stem bark of *Erythrina senegalensis* DC (Fabaceae) in rodents. J Ethnopharmacol 2011;134:697−702.

[141] Bafor EE, Igbinuwen O. Acute toxicity studies of the leaf extract of *Ficus exasperata* on haematological parameters, body weight and body temperature. J Ethnopharmacol 2009;123:302−7.

[142] Demma J, Gebre-Mariam T, Asres T, Ergetie W, Engidawork E. Toxicological study on *Glinus lotoides*: a traditionally used taenicidal herb in Ethiopia. J Ethnopharmacol 2007;111:451−7.

[143] Rhiouani H, El-Hilaly J, Israili HZ, Lyoussi B. Acute and sub-chronic toxicity of an aqueous extract of the leaves of *Herniaria glabra* in rodents. J Ethnopharmacol 2008;118:378−86.

[144] Orisakwe EO, Husaini CD, Afonne JO. Testicular effects of sub-chronic administration of *Hibiscus sabdariffa* calyx aqueous extract in rats. Reprod Toxicol 2004;18:295−8.

[145] Yagi S, Yagi AI, AbdelGadir EH, Henry M, Chapleur Y, Laurain-Mattar D. Toxicity of *Hydnora johannis* Becca. dried roots and ethanol extract in rats. J Ethnopharmacol 2011;137:796−801.

[146] Igbinosa OO, Igene O, Oviasogie FE, Igbinosa HI, Igbinosa OE, Idemudia GO. Effects of biochemical alteration in animal model after short-term exposure of *Jatropha curcas* (Linn) leaf extract. Sci World J 2013;2013: Article ID 798096, 5 pages <http://dx.doi.org/10.1155/2013/798096>.

[147] Agbedahunsi JM, Fakoya FA, Adesanya SA. Studies on the anti-inflammatory and toxic effects of the stem bark of *Khaya ivorensis* (Meliaceae) on rats. Phytomedicine 2004;11:504—8.

[148] Mudi SY, Ibrahim H, Bala M. SAcute toxicity studies of the aqueous root extract of *Lawsonia inermis* Linn. in rats. J Med Plants Res 2011;5(20):5123—6.

[149] Maphosa V, Masika PJ, Adedapo AA. Safety evaluation of the aqueous extract of *Leonotis leonurus* shoots in rats. Hum Exp Toxicol 2008;27:837—43.

[150] Adebajo AC, Ayoola OF, Iwalewa EO, Akindahunsi AA, Omisore NOA, Adewunmi CO, et al. Anti-trichomonal, biochemical and toxicological activities of methanolic extract and some carbazole alkaloids isolated from the leaves of *Murraya koenigii* growing in Nigeria. Phytomedicine 2006;13:246—54.

[151] Adeneye AA, Ajagbonna OP, Adeleke TI, Bello SO. Preliminary toxicity and phyto-chemical studies of the stem bark aqueous extract of *Musanga cecropioides* in rats. J Ethnopharmacol 2006;105:374—9.

[152] Orafidiya LO, Agbani EO, Iwalewa EO, Adelusola KA, Oyedapo OO. Studies on the acute and sub-chronic toxicity of the essential oil of *Ocimum gratissimum* L. leaf. Phytomedicine 2004;11:71—6.

[153] Fakeye TO, Awe SO, Odelola HA, Ola-Davies OE, Itiola OA, Obajuluwa T. Evaluation of valuation of toxicity profile of an alkaloidal fraction of the stem bark of *Picralima nitida* (Fam. Apocynaceae). J Herbal Pharmacother 2004;4(3):37—45.

[154] Manifouet NH, Ngono GAR, Mouokeu RS, Koanga MM, Tiabou TA, Tamokou JD, et al. Acute and subacute toxicities of ethyl acetate extract from *Crassocephalum bauchiense* (Hutch.) Milne-Redh (Asteraceae). J Ethnopharmacol 2011;137(1):70—6.

[155] Nwosu OC, Eneme AT, Onyeyili AP, Ogugbuaja OV. Toxicity and anthelmintic effi-cacy of crude aqueous of extract of the bark of *Sacoglottis gabonensis*. Fitoterapia 2008;79:101—5.

[156] Zuckerman M, Steenkamp V, Stewart MJ. Hepatic veno-occlusive disease as a result of a traditional remedy: confirmation of toxic pyrrolizidine alkaloids as the cause, using an *in vitro* technique. J Clin Pathol 2002;55:676—9.

[157] Ilodigwe EE, Akah PA, Nworu CS. Evaluation of the acute and subchronic toxicities of ethanol leaf extract of *Spathodea campanulata* P. Beauv. Int J Applied Res Nat Prod 2010;3(2):17—21.

[158] Agbaje EO, Adeneye AA, Daramola AO. Biochemical and toxicological studies of aqueous extract of *Syzigium aromaticum* (l.) merr. & perry (myrtaceae) in rodents. Afr J Trad CAM 2009;6(3):241—54.

[159] Kuete V, Manfouo NR, Beng PV. Toxicological evaluation of the hydroethanol extract of *Tabernaemontana crassa* (Apocynaceae) stem bark. J Ethnopharmacol 2010;130:4706.

[160] Elufioye TO, Alatise OI, Fakoya FA, Agbedahunsi JM, Houghton PJ. Toxicity studies of *Tithonia diversifolia* A. Gray (Asteraceae) in rats. J Ethnopharmacol 2009;122:410—5.

[161] Lembè MD, Sonfack A, Gouado I, Dimo T, Dongmo A, Demasse MFA, et al. Evaluations of toxicity of *Turraeanthus africanus* (Meliaceae) in mice. Andrologia 2009;41:341—7.

[162] Dolan R, Welch-Keesey M. A guide to poisonous & non-poisonous plants in Indiana. <www.iuhealth.org/poisoncontrol>. <http://iuhealth.org/images/met-doc-upl/plant-guide.pdf>; 2013. Consulted on September 17, 2013.

[163] Fulder S. Complementary medicine. J Clin Herbal Med 1988;4(7):12.

[164] Walum E. Acute oral toxicity. Environ Health Perspect 1998;106:497−503.

[165] Zbinden G, Flury-Rovers M. Significance of the LD_{50} test for toxicological evaluation of chemical substances. Arch Toxicol 1981;49(1):99−103.

[166] Oduola T, Adeniyi FAA, Ogunyemi EO, Bello IS, Idowu TO, Subair HG. Toxicity studies on an unripe *Carica papaya* aqueous extract: biochemical and haematological effects in wistar albino rats. J Med Plant Res 2007;1(1):1−4.

[167] Frohne D, Pfander HJ. A colour Atlas of poisonous plants. A handbook for pharmacists, doctors, toxicologists, biologists. London: Wolfe Publishing Company; 1983.

[168] Ameri A. The effects of Aconitum alkaloids on the central nervous system. Prog Neurobiol 1998;56:211−35.

[169] Abubakar MG, Lawal A, Usman MR. Hepatotoxicity studies of sub-chronic administration of aqueous stem bark of *Khaya senegalensis* in albino rats. BAJOPAS 2010; 3(1):26−8.

[170] Assam Assam JP, Dzoyem JP, Pieme CA, Penlap VB. *In vitro* antibacterial activity and acute toxicity studies of aqueous-methanol extract of *Sida rhombifolia* Linn. (Malvaceae). BMC Complement Altern Med 2010;10:40.

[171] Mumoli N, Cei M, Cosimi A. Drug-related hepatotoxicity. N Engl J Med 2006;354 (20):2191−3.

[172] Bénichou C. Criteria of drug-induced liver disorders. Report of an international consensus meeting. J Hepatol 1990;11(2):272−6.

[173] Lameire NH, Van Biesen W, Vanholder R. Acute renal failure. Lancet 2005;365:417−30.

[174] Abdulrahman FI, Onyeyili PA, Sanni S, Ogugbuaja VO. Toxic effect of aqueous root-bark extract of *Vitex doniana* on liver and kidney functions. Int J Biol Chem 2007;1:184−95.

[175] Cos P, Vlietinck AJ, Van den Berghe D, Maes L. Anti-infective potential of natural products: how to develop a stronger *in vitro* 'proof-of-concept'. J Ethnopharmacol 2006;106:290−2.

8 Cytotoxicity of African Medicinal Plants Against Normal Animal and Human Cells

Lyndy J. McGaw[1], Esameldin E. Elgorashi[2] and Jacobus N. Eloff[1]

[1]Phytomedicine Programme, Department of Paraclinical Sciences, Faculty of Veterinary Science, University of Pretoria, Onderstepoort, South Africa, [2]Toxicology and Ethnoveterinary Medicine, ARC-Onderstepoort Veterinary Research Institute, Onderstepoort, South Africa

8.1 Introduction

The word "cytotoxicity" has a broad and sometimes vague meaning [1]. With regard to *in vitro* cell culture systems, a test substance is considered to be cytotoxic if it interferes with attachment of cells, significantly alters morphology, adversely affects the rate of cell growth, or causes cells to die [2].

There has been much attention devoted to cytotoxicity studies as a first step in evaluating the toxicity of test substances. This is especially valid in connection with screening biological activity of plant extracts and active compounds isolated from plants. Cytotoxicity assays are widely performed because the test compounds, including plant-derived extracts and purified compounds, may be intended for use as pharmaceuticals or cosmetics, in which case minimal to no toxicity is important. Alternatively the chemicals may be designed to act as anticancer drugs, in which case selective cytotoxicity to cancerous cells is essential. A low cytotoxic profile exerted in noncancerous cells indicates that these extracts do not act as indiscriminate cellular toxins, but have a specific cell type-based cytotoxicity. In the same way, cytotoxicity testing allows the researcher investigating antimicrobial or other bioactivities of plant extracts and compounds to detect at an early stage selective activity against certain microbes or other organisms, and not merely extracts which are generally toxic to cells of all organisms. This process allows the identification and prioritization of test substances useful for further biological activity studies.

The basal cytotoxicity concept [3] may be useful in explaining the relationship between acute toxicity and cytotoxicity. This concept originates from classifying chemical toxicity to humans into three categories, namely organizational

Toxicological Survey of African Medicinal Plants. DOI: http://dx.doi.org/10.1016/B978-0-12-800018-2.00008-X

(extracellular) toxicity, organ-specific cytotoxicity, and basal cell toxicity or chemical injury to structures and functions common to all human cells [4]. This classification is integral to understanding the potential of cellular methods in toxicology testing. Basal cytotoxic mechanisms are studied using undifferentiated finite or continuous cell lines, and organ-specific cytotoxicity is evaluated in primary cultures with well-differentiated cells from different organs [5]. Organizational toxicity on the other hand can be indirectly observed in cell cultures by analyzing the substrates or products of cell metabolism [5].

The mechanism of action of most toxic chemicals is associated with basic biochemical processes common to all cells and it has therefore been proposed that a correlation is likely to exist between toxic concentrations determined *in vitro* to those seen *in vivo* [4]. Examples of basic metabolic functions common to all cells include maintenance of cell membrane integrity, mitochondrial activity and protein synthesis [5]. Different organs may reflect different levels of activity depending on the metabolic rate and other mechanisms. It has been hypothesized that in some cases, target organ toxicity in mammals reflects basal cytotoxicity of chemicals distributed to the affected organ [5].

In the international program known as the Multicenter Evaluation of *In Vitro* Cytotoxicity (MEIC), organized by the Scandinavian Society of Cell Toxicology, the relevance of *in vitro* toxicity assays for human acute toxicity was evaluated [6–9]. In this program, 29 laboratories tested 50 reference chemicals known to be toxic to humans in 61 different *in vitro* assays [10], comprising a selection of animal and human cell lines, primary cultures, fish cell cultures, and other ecotoxicological systems, as well as cell-free systems. *In vitro* IC_{50} values were compared to human mean lethal serum concentrations. To investigate the predictive value of *in vitro* assays to animal tests, rodent and human toxicity data were compared [11] and a comparison of rodent LD_{50} values and the human acute lethal dose was done [10]. Rat and mouse LD_{50} values predicted human acute lethal doses with correlation coefficients (r^2) of 0.61 and 0.65 respectively while the *in vitro* assay predictions of human lethal blood concentrations was $r^2 = 0.69$ (all human cell lines combined). This demonstrated that for the 50 reference chemicals investigated, *in vitro* tests using human cell lines were comparable to rodent LD_{50} determinations in predicting human lethal blood concentrations.

This basic approach does not account for metabolic and pharmacokinetic phenomena in the living organism, but the results support the further consideration of the use of human cell lines for *in vitro* cytotoxicity assays to predict human acute toxicity, as alternatives to acute lethality studies in rodents [4]. It is difficult to concur on *in vitro* replacement methods for acute toxicity testing, and therefore some scientists have proposed a battery of *in vitro* tests to use in a strategy toward replacing animal acute toxicity tests [12].

8.1.1 *Important Factors to Consider in Cytotoxicity Experiments*

Cell viability or cytotoxicity assay results may be heavily influenced by the design of the experiment, so it is useful to take into consideration a number of factors when comparing results obtained using similar methods.

When designing cell culture experiments, it is critically important to note the growth state of a culture and its kinetic parameters. As many of the properties of cell cultures vary significantly between the lag phase, the log phase (logarithmic or exponential growth phase), and the stationary phase, it is necessary to determine the status of the culture when the experiment begins and at the time of sampling [13]. It is generally found that cultures in the log phase are more consistent and uniform, and when sampling is performed at the end of the log phase, reproducibility is high. Cells in the stationary phase may be more differentiated, have a different morphology, and may become polarized, often secreting more extracellular matrix [13].

The duration of an experiment is also an important consideration. Addition of a test substance at a certain time in the growth phase and then assaying when the cells are still undergoing exponential growth, or when their growth has plateaued, will influence cytotoxicity results [13]. Detailed knowledge of the growth cycle, including population doubling time (PDT) of the cells during exponential growth, is necessary in planning cytotoxicity experiments.

8.2 Necessity for Cytotoxicity Testing

Toxicity is a complicated process in human and animal systems, potentially involving direct cellular damage (e.g., with a cytotoxic anticancer agent), physiological effects (such as membrane transport in the kidney or neurotoxicity in the brain), inflammation, and other systemic effects. As it is difficult to measure systemic and physiological effects *in vitro*, most assays investigate effects at the cellular level [13]. Cells may die by necrosis, apoptosis, or autophagy (self-digestion), or they may cease proliferating (cytostasis) or become terminally differentiated.

A cytocidal, or cell killing, effect is required to demonstrate efficacy of an anticancer agent, but verifying a lack of toxicity of a pharmaceutical substance is far more complicated and may require analysis of specific targets such as alteration of gene transcription, cell signaling, or cell–cell interactions [13]. More intricate assays focusing on metabolic pathway regulation and signaling are necessary to supplement standard cytotoxicity assays, such as the induction of an allergic or inflammatory response. Cytotoxicity tests are therefore still useful, and new drugs, cosmetics, and food additives must be subjected to extensive cytotoxicity testing, usually involving many animal experiments. In Europe, new legislation governs these experiments [14] for topical application from 2009 and for systemic application from 2013. Motivation for performing as many cytotoxicity tests as possible *in vitro* originates from an economic as well as an ethical perspective. *In vitro* tests are carried out widely in current ethnopharmacological research, partly because of the aforementioned ethical and financial constraints of using animals or animal tissue, and also because they facilitate bioassay-guided isolation of "active" compounds, that is, those responsible for any activity shown as a result of the bioassay using the total extract. The significantly lower quantity of material required for testing is also an important aspect of using *in vitro* assays [15].

The selectivity index (SI) is defined as the ratio of cytotoxicity to biological activity. It is generally considered that biological efficacy is not due to *in vitro* cytotoxicity when SI ≥ 10 [16]. This means that there is a safety factor of >10 built into the therapeutic use. A lack of selectivity in cytotoxic effect between cancer cell lines and noncancerous cell lines minimizes the prospect that these plants contain compounds which could serve as leads for novel anticancer drugs [17]. In antimicrobial investigations of plant extracts in our research group we have found SI values to be a useful means of distinguishing highly active but toxic extracts from those that are selectively active against certain test organisms. This provides good leads for continuing research on promising extracts that may be sources of interesting biologically active and therapeutically useful compounds with excellent activity and low toxicity. It is sometimes found that fractionation of plant extracts may result in fractions with higher activity and lower toxicity owing to removal of toxic compounds. Such fractions may have better activity or lower toxicity than the purified compound resulting from bioassay-guided fractionation for isolation of the active compound(s). Synergistic effects between different compounds in a plant extract may also have a substantial outcome on activity or toxicity of fractions.

8.3   Limitations of Cytotoxicity Testing

Limitations associated with extrapolating results obtained using *in vitro* assays to the *in vivo* situation should be kept in mind. Toxicity *in vivo* may occur as a result of a tissue response, possibly caused by an inflammatory reaction or kidney failure, or a systemic response. As toxic responses *in vitro* are generally measured by changes in the survival or metabolism of cells, models relating more closely to tissue or systemic toxicity should be considered. An example involves utilizing organotypic cultures prepared from different cell types and maintained in medium containing hormones and other appropriate supplements.

The pharmacokinetics of bioactive compounds is a crucial consideration. The time of exposure to the test substance and the rate of change of the drug concentration will vary between the *in vitro* and *in vivo* tests. Other factors not able to be accounted for when relying upon *in vitro* assays are the ability of the drug to penetrate tissue, as well as clearance and excretion. A critical factor in toxicology is metabolism *in vivo*, as some substances lacking toxicity initially may produce toxic metabolites after being exposed to liver enzymes, while other substances that are toxic *in vitro* may become detoxified. If *in vitro* tests are to become an accepted alternative to animal tests, the cells should be exposed to the same form of the substance without alterations caused by metabolism. To address this issue, some alternatives have been suggested, such as incorporating an additional step of incubating the test substance with purified liver microsomal enzymes [18], coculturing with activated hepatocytes [19], or using hepatoma-derived cells such as HepG2 [13]. It also has been proposed to modify target cells genetically by introducing genes for enzyme metabolism which can be controlled with regulatable promoters [20].

For mixtures of compounds, such as plant extracts, it is exceptionally challenging to predict the potential interactions that may take place between the different components. The composition of constituents reaching the target site may differ from that present at the exposure site because of differences in bioavailability of the various compounds [21]. The components of a plant mixture may interact toxicodynamically, displaying additive synergistic or antagonistic effects, or toxicokinetically during absorption or transport to the target site [22]. Certain constituents of the extract may enhance penetration or influence the absorption of other compounds [23,24].

8.4 Different Types of Cytotoxicity Assay

The choice of assay to use when investigating cytotoxicity depends on the nature of the test substance, the expected response, and the target cell [13]. Cell growth or survival, the traditional measure of cytotoxicity, can be measured in various ways, including evaluating the net change in population size, a change in cell mass (total protein or DNA) or metabolic activity (e.g., DNA, RNA, or protein synthesis, MTT reduction). General viability criteria for cell cultures can also be divided in terms of the indicator and method of evaluation used, and some examples of these are given in Table 8.1. The use of biochemical parameters as an indication of toxicity is discussed in Chapter 23.

The specific research goals will impact on the choice of a particular viability or cytotoxicity assay. Niles and coworkers [1] classify the assays which are the most practical, useful and popular for drug discovery into three main classes, namely (i) viability by metabolism reductase activities; (ii) viability by bioluminescent ATP (adenosine triphosphate) assays; and (iii) cytotoxicity by enzymes released into culture medium. Each of these methods has its limitations as well as merits, but the inherent shortcomings of single assays may be mitigated by employing multiparametric methods using various viability and cytotoxicity markers [1]. In metabolism and ATP-based viability assays, the scientific premise is that activity is

Table 8.1 Indicators and Methods for Assessment of Cell Viability and Cytotoxicity [5,13]

Indicator	Method for Toxicity Determination
Cell differentiation	PCR analysis for gene markers
Cell division	Plating efficiency, clonal formation
Cell membrane	Enzyme leakage (e.g., LDH), uptake of dye to which cell is normally impermeable (e.g., trypan blue, erythrosine, naphthalene black, propidium iodide)
Cell metabolism	Uptake of fluorescent or isotope-labeled precursors, assays for enzymatic activity
Lysosomes	Vital staining, neutral red uptake assay
Mitochondria	Mitochondrial reduction (MTT) assay

proportional to viable cell number, inferring that reduction in activity after treatment when compared with the control results from cytotoxicity. On the other hand, cytotoxicity assays measure parameters proportional to the degree of cell death. Viability and cytotoxicity measures in most cases are inversely proportional, but differences between the approaches may be emphasized depending on the length of compound exposure [25]. For practical purposes, most researchers refer to cytotoxicity when conducting what are better classified as viability assays. Brief descriptions are given in the following sections of some of the most commonly used bioassays for detection of cytotoxic effects of plant extracts and purified plant compounds.

8.4.1 Metabolism Reductase Viability Assays

8.4.1.1 Tetrazolium Assays

In recent years, many assay methods have been developed for assessing cellular viability based on cell metabolism using tetrazolium and resazurin reduction. The first tetrazolium salt to be employed in a multi-well viability assay for mammalian cells was MTT, or 3-(4,5-dimethylthiazol-2-yl)-2,5-diphenyltetrazolium bromide [26]. In this assay, the MTT is reduced by metabolically active cells (via mitochondrial and cytosolic enzymes) to a blue-purple formazan product. This product accumulates within cells as it cannot pass through intact cell membranes, but upon removal of the aqueous medium and solubilizing in DMSO (dimethyl sulfoxide) or isopropanol, the colored product can be measured spectrophotometrically. Only viable cells can reduce MTT, so the amount of reduced MTT formazan is proportional to the intensity of color and the greatest cell viability [5].

As a substitute for MTT, XTT (2,3-bis(2-methoxy-4-nitro-5-sulfophenyl)-5-carboxanilide-2H-tetrazolium) can be used as it is soluble in most aqueous media. The XTT assay is also suitable for cells in suspension, which is a useful advantage. The assay is based on the extracellular reduction of XTT by NADH (a form of NAD, or nicotinamide adenine dinucleotide) produced in the mitochondria via trans-plasma membrane electron transport and an electron mediator [27]. Third-generation tetrazoliums with good solubility and better stability in the form of stock solutions are MTS (5-(3-carboxymethoxy-phenyl)-2-(4,5-dimethylthiazoly)-3-(4-sulfophenyl)tetrazolium, inner salt) and WST-1 ((4-[3-4-iodophenyl]-2-(4-nitrophenyl)-2H-5-tetrazolio)-1,3-benzene disulfonate) [1].

There have been reports that some plant extracts and compounds, such as flavonoids, have the ability to reduce MTT nonspecifically in the absence of cells [28−30]. It is therefore recommended to include steps to wash the cells with phosphate-buffered saline between removal of the plant extract or test substance and dissolving of the MTT crystals. Cell-free controls should also be included.

The MTT assay is well characterized and widely used, and is often a gold standard to which new viability/cytotoxicity assays are compared [1]. In our research group, the Phytomedicine Programme (www.up.ac.za/phyto), we have made extensive use of the MTT assay for evaluating toxic effects of plant extracts and purified compounds to various cell lines in efforts to gain indications of selective antimicrobial activity of test substances.

8.4.1.2 Resazurin Assays

Various formulations of resazurin are becoming more commonly used as viability reagents in drug screening. The original patented commercial product, Alamar Blue, contained resazurin and possibly also a mixture of stabilizer salts to minimize spontaneous background reduction in the absence of cells [1]. Resazurin assays are based on the reduction by living cells of the oxidized blue dye to a pink fluorescent resorufin product. The assay also can be monitored by absorbance although there is a slight loss of sensitivity [1]. Reduction of resazurin is believed to be accomplished by reductase or diaphorase-type enzymes from mitochondria and the cytosol [31].

A more recently available resazurin-based colorimetric agent is PrestoBlue, which is flexible in that the results of an assay can be measured visually, using absorbance, or by fluorescence reading of the reduced resorufin product. PrestoBlue is a fast, live assay for measuring cell viability with incubation steps as short as 10 min. It is also sensitive, being able to detect as few as 12 cells per well [32]. The time of color development following addition to treated cells is, however, dependent upon the metabolic rates of various bacteria and cell lines being tested [33].

8.4.2 Bioluminescent ATP Assays

Measuring the ATP concentration in cells *in vitro* is widely accepted as a valid marker of the number of cells present [34]. Eukaryotic cells growing in culture have a constant amount of ATP that is regulated to maintain homeostasis. When cells die, they lose the ability to synthesize ATP, and endogenous ATPases remove remaining ATP [1]. The most popular method for measuring ATP is based on the ability of firefly luciferase to generate a luminescent signal [1]. It is purported to be the most sensitive microplate assay for detecting viable cells growing in culture, as it can detect fewer than 10 cells per well, mainly because of the low background luminescence present in biological samples [1].

8.4.3 Enzyme Release-Based Cytotoxicity Assays

A reliable method for evaluating cell death is to detect and quantify leakage of cellular constituents from affected cells into the culture medium. Lactate dehydrogenase (LDH) is a preferred marker of cell death for *in vitro* systems [35,36]. LDH activity can be indirectly measured by subjecting the sample to a coupled enzymatic chemistry reagent containing lactate, NAD + , diaphorase, and an appropriate redox dye such as resazurin, which yields a change in absorbance or a shift in the fluorescence profile [1].

Other highly sensitive bioluminescent assays for cytotoxicity measure ATP generation through the activity of released ATP cycling enzymes. For example, adenylate kinase (AK) activity can be measured in the culture medium by providing ADP (adenosine diphosphate) to serve as a substrate for the production of ATP. In a similar way, GAPDH (glyceraldehyde-3-phosphate dehydrogenase) activity can be measured by adding coupled glycolytic pathway enzymes and other constituents necessary for the generation of ATP [1]. As these assays are based on luminescence, they have excellent signal-to-noise ratios from relatively few dead cells, but they are limited in terms of net signal decay owing to limiting reagents and luciferase inactivation [1].

8.4.4 Other Commonly Used Assays

The neutral red uptake assay uses the weak cationic dye neutral red (3-amino-7-dimethylamino-2-methylphenazine hydrochloride), that is preferentially absorbed into lysosomes [5]. The premise underlying the assay is that a cytotoxic chemical, regardless of site or mechanism of action, interferes with normal lysosomal uptake, representing the number of viable cells. Intact lysosomal processes are only maintained in viable cells, so the degree of inhibition of viability in proportion to the concentration of test compounds indicates the toxicity of the test substance [5]. A disadvantage of the assay is the induction of precipitation of the neutral red dye into visible needle-like crystals by some chemicals, which affects the accuracy of the absorbance readings used to detect the presence of the dye.

The antiproliferative sulforhodamine B (SRB) assay is widely used to assess growth inhibition or cytotoxicity of cells by a colorimetric assay which estimates cell number indirectly by staining total cellular protein with the dye SRB [37]. The SRB assay relies on the uptake of the negatively charged pink aminoxanthine dye, SRB, by basic amino acids in the cells. The greater the number of cells, the more dye is taken up, and when the cells are lysed following fixing, the released dye results in a more intense color and greater absorbance [37,38]. The SRB assay is not suitable for suspension cells, as they cannot be fixed easily.

8.4.5 Brine Shrimp Assay

A simple test which has been widely used as an indication of cytotoxicity is based on lethality to the larvae of the brine shrimp *Artemia salina* [39−42]. The brine shrimp lethality bioassay is a rapid and inexpensive test requiring a relatively small amount of sample (2−20 mg). This bioassay has been proposed to have a good correlation with cytotoxic activity in some human solid tumors and with pesticidal activity, leading to the discovery of annonaceous acetogenins as a new class of natural pesticides and antitumoral agents [43]. However, the value of a crustacean model in drawing conclusions regarding mammalian cytotoxicity is likely to be less than that of mammalian cell culture models.

8.5 xCELLigence Real-Time Cell Analyzer and SpectraMax i3 System

The relatively recently introduced xCELLigence real-time cell analysis system (Roche Applied Science) is used for real-time, label-free monitoring of cell viability or cytotoxicity. The system measures impedance via interdigitated gold microelectrode sensors integrated into the bottom of each well of tissue culture plates, whereby an electric field can be generated [44,45]. This electric field is affected by the number of cells attached to the electrodes and also by changes in cellular morphology. Where there are no or few cells, electrical impedance is minimal, but when cells are present in higher numbers, they attach to the electrode sensor surfaces, forming a barrier to the electric field, thus leading to increased impedance [44]. An impedance read-out known as the cell index (CI) value digitally represents the morphological changes and number of cells in a given microelectronic well, supplying information about the status of the cells, including cell number, cell viability, and cell morphology. Therefore, the greater the cytotoxicity of a compound, the lower the number of cells growing on the electrodes, and the lower the CI value of the electrode impedance. As it is a real-time analyzer, the xCELLigence system is able to isolate optimum time points, such as when cytotoxic compounds achieve their maximum effect, as indicated by the lowest CI values in cytotoxicity assays.

The SpectraMax i3 System (Molecular Devices) with the MiniMax imaging cytometer option allows images to be acquired to provide phenotypic information in addition to concentration response of cell viability of cells in the same wells. Cell toxicity effects are assessed by measuring the area of the well covered by the cells after treatment with various compounds. After image analysis, the built-in software is able to plot results and fit the curves to determine IC_{50} values. The use of this new technology has not been reported extensively in the literature as yet, but it is anticipated that such real-time analysis with StainFree™ Cell Detection Technology will become more widely used in the future.

8.6   Cytotoxicity of African Medicinal Plants

A literature search was conducted using Scopus and Pubmed in pursuit of reported cytotoxicity of extracts and compounds derived from African plant species. Keywords such as "cytotoxicity" and "plant extracts" together with each of the 53 African countries were used during the search. Results where a single concentration was used or where fractions of a crude extract were investigated without successful isolation of the active principles were not included. The results are presented in Table 8.2, which provides information on the plant species and family, IC_{50} values, and cell lines used for cytotoxicity testing. A variety of plant parts have been investigated for cytotoxicity, and this is of course largely dependent on the traditional use of the species of interest. Plants used in traditional medicine to treat diseases of considerable economic burden to the African continent especially malaria, leishmania, and sleeping sickness received much attention. Relatively little information is available on the cytotoxicity of purified plant compounds against normal cells compared to cancerous cells. Several different cell lines and assays have been used, with many researchers relying on the easily established MTT assay. Cell lines employed range from human monocytes and fibroblasts to mouse macrophages and

Table 8.2 African Plant Species Tested for Cytotoxicity

Family	Species	Extract or Compound Tested	Cell Line	IC$_{50}$[a] (μg/mL)	References
Acanthaceae	*Adhatoda latibracteata* R. Be	Stem	MRC-5 (human normal lung fibroblast)	11.5	46
	Ermomastax speciosa (Hochst) Cufod	Stem with leaves	AML 12 (mouse hepatocyte)	>40.0	47
	Justicia gendarussa Burm f.	Aerial parts	Human skin fibroblast (WS1)	>200.0	48
	Thomandersia hensii De Wild. and T. Durand	Leaves	MRC-5	>64.0	49
	T. laurifolia Petit	Stem bark	MRC-5	>64.0	50
Alliaceae	*Tulbaghia violacea* Harv.	Rhizome	Brine shrimp	18.0–19.0	51
Amaranthaceae	*Amaranthus caudatus* L.	Seed	RAW264.7 (murine macrophage)	>100.0	52
	A. hybridus L.	Whole plant	Brine shrimp	116.1	53
	Cyathula cylindrica Moq.	Leaves	Brine shrimp	153.3	54
		Root		137.2	
	C. polycephala Bak.	Leaves	Brine shrimp	2.9	54
		Root		8.4	
	Sericocomopsis hildebrandtii Schinz	Aerial parts	Vero cells	93.9– >500.0	55
		Root		367.1– >500.0	
	Pupalia lappacea Juss.	Aerial parts	Macrophage-like murine cells (J774), human normal fibroblasts (W138)	(66.5– >100), (68.3– >100)	56
Amaryllidaceae	*Crinum bulbispermum* (Bum. F.) Milne-Redh. and Schweick	Bulb	Human kidney epithelial cells (Graham 293)	8.7–177.6	57
		Leaves		11.1–45.2	
		Root		2.4–171.1	
	C. zeylanicum Linn.	Whole plant	AML 12	>40.0	47

Family	Species	Plant part	Assay	Value	References
	Gethyllis gregoriana D. Müll.-Doblies	Bulb	Brine shrimp	0.2–120.7	58
	G. multifolia L. Bolus	Bulb	Brine shrimp	6.2–139.6	58
	G. villosa Thunb.	Bulb	Brine shrimp	4.2–1956.4	58
	Hymenocallis senegambica Willem Reuter	leaves	MRC-5	>64.0	50
Anacardiaceae	*Lannea acida* A. Rich	Bark	THP-1 monocytoid human cells	Less toxic (LDH), No IC50	59
	L. schweinfurthii (Engl.) Engl.	Roots	Vero E6	76.0	60
	Mangifera indica L.	Bark	Brine shrimp	2456.0	61
		Leaves		3079.1	62
		Bark		365.0	
	Ozoroa engleri R.A. Fernandes	Stem bark	Monkey kidney cells	35.0	63
	O. sphaerocarpa R. Fern and A. Fern	Bark	Human kidney epithelial cells	8.11	64
	Pseudospondias microcarpa (A. Rich.) Engl.	Root bark	L6 (rat skeletal myoblast)	>90.0	65
		Stem bark		>90.0	
	Rhus lancea L.f.	Bark	Brine shrimp	3900.0	66
		Leaves		600.0–1000.0	
	R. pyroides Burch.	Leaves	Graham 293	7.2–942.5	57
		Seeds		22.9–230.2	
		Stem		5.3–445.5	
Annonaceae	*Annickia kummeriae* (Engl. and Diels) Setten and Maas	Leaves	L6	30.0–72.0	67
		Root bark		10.5–61.4	65
		Stem bark		22.0–58.2	

(Continued)

Table 8.2 (Continued)

Family	Species	Extract or Compound Tested	Cell Line	IC$_{50}$[a] (μg/mL)	References
		Lysicamine		1.6	
		Trivalvone		45.3	
		Palmatine		>90.0	
		Jatrorrhizine		>90.0	
	Annona senegalensis Rolyns andGh	Leaves	Brine shrimp	6811.0	61
		Bark	Vero cells	71.0	68
	A. squamosa L.	Seed	Brine shrimp	232.0	62
	Anonidium mannii Engl. and Diels	Stem bark	MRC-5	>64.0	49
	Enantia chlorantha Oliv.	Stem bark	MRC-5	5.66	49
		Bark	Brine shrimp	214.3	61
	Isolona hexaloba Engl. and Diels	Root bark	MRC-5	32.2	49
	Monodora myristica (Gaertn.) Dunal.)	Stem	MRC-5	95.7–170.5	46
	Polyalthia suaveolens Engl. and Diels	Leaves	MRC-5	>64.0	49
			Human monocytes	8.5–12.4	50
		Stem bark		4.4–10.5	69
	Uvaria acuminata Oliv.	Roots	Vero E6	2.37	60
	Xylopia aethiopica (Dunal) A. Rich.	Seeds	HepG2 (human hepatocellular carcinoma cells)	18.28	47
			AML 12	>40.0	
			Human umbilical vein endothelial cells (HUVECs)	>80.0	70

Family	Species	Plant part	Cell line/organism	Value	Ref.
Aphloiaceae	*Aphloia theiformis* (Vahl) Benn.	Bark	WS1	61.0	48
Apiaceae	*Heteromorpha trifoliata* (Wendl.) Eckl. and Zeyh.	Leaves	Vero cells	43.0	71
Apocynaceae	*Acokanthera schimperi* (A.DC.) Schweinf.	Leaves	Green monkey kidney (GMK)	>200.0	72
			Madin–Darby canine kidney (MDCK)	>200.0	73
			Vero cells	0.4–36.6	73
	Alstonia boonei De Wild.	Stem bark	MRC-5	>64.0	49
	A. congensis Engl.	Stem bark	MRC-5	>64.0	50
	Carissa edulis (Forssk.) Vahl	Root bark	Vero E6	480.0	74
	Landolphia owarensis P. Beauv.	Leaves	Vero cells	0.37–111.7	73
		Leaves	MRC-5	>64.0	50
	Nerium oleander L.	Leaves	Brine shrimp	398.0	62
	Ochrosia borbonica J.F. Gmel.	Leaves	WS1	149–194	48
	Picralima nitida Th. and H. Dur.	Stem bark	MRC-5	>64.0	49
	Rauvolfia macrophylla Ruiz and Pav	Stem bark	Rat skeletal muscle myoblasts (l-6)	>90.0	75
	R. vomitoria Afzel.	Leaves	MRC-5	32.0	50
	Tabernaemontana elegans Stapf	Roots	THP-1	<4.0	76
	Thevetia peruviana (Pers.) K. Schum.	Leaves	Brine shrimp	641.0	62
	Vinca rosea (L.) G. Don		Brine shrimp	326.0	62

(*Continued*)

Table 8.2 (Continued)

Family	Species	Extract or Compound Tested	Cell Line	IC$_{50}$[a] (µg/mL)	References
Araceae	*Anchomanes difformis* (Blme) Engl.	Root	J774, WI38	(2.2–22.0), (12.7–22.0)	56
	Culcasia lancifolia N.E.Br.	Stem	MRC-5	10.3–44.3	46
Araliaceae	*Cussonia spicata* Thunb.	Root	Brine shrimp	2600.0	66
Arecaceae	*Eramospatha haulleleana* De Wild	Leaves	MRC-5	42.0	50
		Whole plant			
Aristolochiaceae	*Aristolochia bracteolata* Lam.	Seeds	Brine shrimp	185.0	62
		Whole plant	3T3 (mouse fibroblast)	50.0	77
				44.6	
Asclepiadaceae	*Calotropis procera* (Aiton) W.T. Aiton	Latex	Brine shrimp	110.0	62
		Leaves		159.0	
		Root		393.0	
	Gongronema latifolia (Benth)	Leaves	Brine shrimp	1175.2	78
		Stem		2238.5	
	Periploca lineariifolia (Dill and Rich)	Stem bark	Brine shrimp	231.3	79
Asteraceae	*Acanthospermum hispidum* DC	Aerial parts, 15-acetoxy-8β-[(2-methylbutyryloxy)]-14-oxo-4, 5cis-acanthospermolide; 9α-acetoxy-15-hydroxy-8β-[(2-methylbutyryloxy)]-14-oxo-4, 5-trans-acanthospermolide	J774, WI38	(43.2– >100), (1,1), (13.8) (34.9– >100), (5.7), (17.3)	56,80 81
			Normal melanocytes	79.8	62
			THP-1	60.2	
			Malignant melanoma cells	100.8	
		Seeds	Brine shrimp	126.0	
		Whole plant		1224.0	

Species	Plant part / Compound	Test system	Value	Ref.
Ageratum conyzoides L.	Aerial parts	WS1	99.0—>200.0	48
Anisopapus chinensis Hook. and Arn	Whole plant	W138	98.3–126.3	82
Artemisia afra Jacq. ex Willd.	Whole plant	McCoy fibroblast cells		83
	Acacetin		16.9	
	Betulinic acid		35.4	
	Scopoletin		30.9	
			132.5	
A. annua L.	Leaves	L6	>90.0	65
	Stem bark		>90.0	
A. gorgonum Webb	Gorgonolide		87.9	84
	1β, 10α-dihydroxy-1, 10-deoxygorgonolide		>100.0	
	1, 10-dioxo-1, 10-deoxy-1, 10-secogorgonolide		63.8	
	3β, 4β-epoxy-1β, 10β-epiarborescin		87.5	
	5α-hydroxy-leukodin		>100.0	
	Arborescin		71.0	
	2α-hydroxyarborescin		>100.0	
	1β, 10β-epoxy-2α-hydroxykauniolide		44.3	
	Ridentin		67.4	
	Hanphyllin		26.4	
Athrixia elata Sond.	Aerial parts	Brine shrimp	283.0	85
		Vero cells	102.0	
A. phylicoides DC.	Aerial parts	Brine shrimp	394.0	85
		Vero cells	252.0	
Brachylaena discolour DC.	Leaves	Vero cells	4.0	71

(Continued)

Table 8.2 (Continued)

Family	Species	Extract or Compound Tested	Cell Line	IC$_{50}$[a] (μg/mL)	References
	Conyza aegyptica L. Aiton	Leaves	WI38	81.9	86
	C. bonariensis (L.) Cronquist	Whole plant	Brine shrimp	40%	87
	C. canadensis (L.) Cronquist	Aerial parts	MRC-5	250.0	88
	C. dioscoridis (L.) Desf.	Whole plant	Brine shrimp	30%	
	Crassocephalum crepidioides (Benth.) S. Moore	Leaves	Brine shrimp	901.0	89
	C. rubens (Juss. ex Jacq.) S. Moore	Leaves	Brine shrimp	374.0	89
	Echinops giganteus var. *lelyi* (C.D. Adams) A. Rich.	Roots	HepG2	14.3	47
			AML 12	>40.0	
		2-(penta-1,3-diynyl)-5-(4-hydroxybut-1-ynyl)-thiophene	HepG2	37.4	
			AML 12	>40.0	
		Tubers	HUVECs	>80.0	47
	Eupatorium riparium Regel	Leaves	WS1	50.0– >200.0	48
	E. triplinerve Vahl	Aerial parts	WS1	>200.0	48
	Dicoma tomentosa Cass.	Whole plant	WI38	17.0, (5.9–47.2)	90, 91
		Urospermal A 15-O-acetate		3.03	
	Inula confertiflora A. Rich.	Leaves	GMK	>200.0	72
			MDCK	>150.0	
	Lagera alata L.	Leaves	MRC-5	>64.0	50
		Leaves	Brine shrimp	2367.0	79

Species	Plant part	Test system	Value	Ref.
Microglossa pyrifolia (Lam.) Kuntze		WI38	4.7	86
Psiadia arguta Voiget	Leaves	WS1	24.0–43.0	48
P. dentata DC.	Aerial parts	WS1	55.0–125.0	48
Schkuhria pinnata (Lam.) Cabrera	Aerial parts	Brine shrimp	2500.0–3400.0	66
Solanecio mannii (Hook.f.) C. Jeffrey	Leaves	WI38	122.3	86
Tarchonanthus camphoratus L.	Bark	Brine shrimp	73.7–161.0	92
	Leaves		205.0–1165.6	
		Human embryonic kidney cells	385.7–572.2 291.0–899.1	
Tithonia diversifolia A. Gary	Leaves	Brine shrimp	24304.0	61
	Fruit	WI38	6.5	86
	Leaves		1.0	
Tridax procumbens L.	Leaves	MRC-5	>64.0	50
Vernonia amygdalina Delile	Leaves	Brine shrimp	300.0	62
V. auriculifera Hiem	Leaves	L6	84.8	93
V. clorata (Willd) Drake	Leaves/twigs	Human dermal fibroblast (HDF)	562.2	94
V. lasiopus O. Hoffim	Root bark	L6	>90.0	93
Xanthium brasilicum Waller.		3T3	74.5	77

(Continued)

Table 8.2 (Continued)

Family	Species	Extract or Compound Tested	Cell Line	IC$_{50}$[a] (µg/mL)	References
Balanitaceae	*Balanites aegyptiaca* L.		3T3	80.9	77
Besellaceae	*Besella alba* Linn	Leaves	Brine shrimp	1200.4	78
		Stem		1887.8	
Bignoniaceae	*Kigelia africana* (Lam.) Benth	Bark	LLC/MK2 monkey kidney epithelial cells	125.0	95
		Atranorin		27.78	
		2β, 3β, 19α-trihydroxy-urs-12-en-28-oic acid		9.37	
		Specicoside		>125.0	62
		p-hydroxycinnamoic acid	Brine shrimp	>125.0	
		Fruits and seeds		124.0	
		Root		593.0	
	Newbouldia laevis (P. Beauv.)	Leaves (essential oils)	Vero cells	>50.0	96
		2-acetyl-furo-1,4-naphthoquinone	AML 12	>20.0	70
	Stereospermum acuminatissimum K. Schum.	Stem bark	Rat skeletal muscle myoblasts (L-6)	>90.0	75
	S. zenkeri K. Schum. Ex De Wild	Stem bark	Rat skeletal muscle myoblasts (L-6)	>90.0	75
Bobaceae	*Adansonia digitata* L.	Leaf	RAW264.7	>100.0	52
Boraginaceae	*Heliotropium indicum* L.	Aerial parts	J774, W138	(>100), (>100)	56
	Cassia didymobotrya Fres.	Leaves	Brine shrimp	230.9	53

Family	Species	Plant part	Cell line	Value	Ref
Burseraceae	Commiphora edulis (Koltzsch) Engl.	Leaves	Graham 293	99.5	97
		Stem		194.0	
	C. grandulosa Schinz	Leaves	Graham 293	106.5	97
		Stem		30.5	
	C. marlothii Engl.	Leaves	Graham 293	97.5	97
		Stem		123.0	
	C. mollis (Oliv.) Engl.	Stem	Graham 293	172.0	97
	C. neglecta I. Verd	Leaves	Graham 293	111.5	97
	C. pyracanthoides Engl.	Leaves	Graham 293	104.0	97
		Stem		101.5	
	C. schimperi (O.Berg) Engl.	Stem	Graham 293	136.5	97
	C. viminea Burtt Davy	Leaves	Graham 293	141.5	97
Caesalpiniaceae	Cassia sieberiana DC	Leaves, Twigs	J774	(19–167), (19–167)	98
	Guilbourtia demensei (Hams) J. Léonard	Stem bark	MRC-5	51.0	50
	Poliostigma thonningii (Schum.) Milne-Readhead	Bark	AML 12	>40.0	47
	Scorodophoeus zinkeri Harms	Stem bark	MRC-5	51.0	50
Canellaceae	Warburgia stuhlmannii Engl.	Stem bark	Vero cells	233.0–3337.8	99
Capparaceae	Maerua edulis (Gilg and Gilg-Ben.)	Root	THP-1	>125	76

(Continued)

Table 8.2 (Continued)

Family	Species	Extract or Compound Tested	Cell Line	IC$_{50}$^{a} (µg/mL)	References
Capparidaceae	*Boscia senegalensis* (Pers) Lam. Ex Poir	Seeds	Brine shrimp	384.0	62
Capparidaceae	*Physena sessiliflora* Tul.	Leaves	Brine shrimp	50.4	100
		Physenoside S1		>500.0	
		Physenoside S2		>500.0	
		Physenoside S3		>500.0	
		Physenoside S4		>500.0	
		Physenoside S5		394.0	
		Physenoside S6		>500.0	
		Physenoside S7		8.5	
		Physenoside S8		22.1	
	Warburgia ugandensis Sprague	Stem bark	IEC-6 (Intestinal epithelial cells)	<50.0	101
Cecropiaceae	*Masanga cecropioides* R. Br. apud Tedlie	Stem bark	MRC-5	>64.0	49
Celastraceae	*Maytenus heterophylla* Eckl. And Zeyh.	Leaves	Graham 293	7.4–64.4	57
		Seeds		5.4–6259.3	
		Stem		34.2–230.8	
	M. peduncularis (Sond) Loes	Leaves	Vero cells	89.4	102
	M. procumbens (L.f.) Loes	Leaves	Vero cells	187.7	102
	M. putterlickioides (Loes.) Excell and Mendoca	Root bark	Vero cells	112.4–380.8	99
	M senegalensis (Lam.) Loes	Leaves	Vero cells	87.6	102
		Root bark	L6	>90.0	65

Family	Species	Part	Cell line	Activity	Ref
	M. undata (Thunb.) Blakelock	Leaves	Vero cells	251.3–3645.7	99
Chrysobalanceae	Parinari curatellifolia Planch. ex Benth	Bark	Vero cells	99.2	102
				92.3	68
Clusiaceae	Allanblackia monticola Staner L.C.	Fruit	Rat skeletal muscle myoblasts (1-6)	55.6	75
	Garcinia kola Heckel	Fruit	MRC-5	>64.0	50
	G. punctata Oliv.	Stem bark	MRC-5	>64.0	49
	Harungana madagascariensis Lam exPoir.	Seed	Rat skeletal muscle myoblasts (1-6)	28.1	75
	Mammea Africana Sab.	Stem bark	MRC-5	20.1	49
	Psorospermum senegalense Spach.	Stem bark	MRC-5	38.1	49
		Leaves	WI38	36.4	90
	Symphonia globulifera L.F.	Leaves	Rat skeletal muscle myoblasts (1-6)	>90.0	75
Cochlospermaceae	Cochlospermum planchonii Hook. f. Planch.	Root	Rat skeletal muscle myoblasts (L-6)	67.3	16
Combretaceae	Anogeissus leiocarpus (DC) Guill. And Perr.	Leaves	K562S (Human monocytes cell)	21.4–>125	103
			Rat skeletal muscle myoblasts (L-6)	71.9	16
	Combretum comosum G. Don	Leaves	Human monocytes	63.1–>100	69
	C. cuspidatum Planch. ex Benth.	Leaves	Human monocytes	8.8–74.5	69
		Stem bark		25.3–>100	

(Continued)

Table 8.2 (Continued)

Family	Species	Extract or Compound Tested	Cell Line	IC$_{50}$^{a} (μg/mL)	References
	C. molle R. Br. Ex G Don	Leaves	K562S (Human monocytes cell)	36.5–>125	103
	C. erythrophyllum (Burch.) Sond.	Leaves	Graham 293	32.9–925.7	57
		Stems		2.7–70.5	
	Guiera senegalensis (Gmel)	Harman	THP-1	22.0	104
		Tetrahydroharman		75.0	62
		Seeds	Brine shrimp	289.0	
	Terminalia bentzoe L.	Bark	WS1	102.0–>200.0	48
	T. catappa Linn.	Leaves	L6	159.9–377.6	105
	T. glaucescens Planch.	Leaf, stem	Human fibroblasts (HeLa)	8.8–95.5	106
	T. ivorensis A. Chev.	Root	Bovine aorta endothelial cells	153.5	107
		Stem bark		13750.0	
	T. kilimandscharica Engl.	Bark	Brine shrimp	164.1	53
	T. latifolia Engl.	Leaves	Brine shrimp	272.9	61
	T. mollis M.A. Lawson	Stem bark	WI38	18.9	86
	T. sericea Burch. ex. DC	Root	Monkey kidney cells	24.0	108
	T. spinosa Engl.	Root	Brine shrimp	126.4–380.1	109
		Stem bark		75.8–126.4	
	T. stenostachya Engl. and Diels	Leaves	Brine shrimp	142.7–545.5	109
		Stem bark		133.3	
	T. superba Engl. and Diels	Root	Bovine aorta endothelial cells	526.3	107
Connaraceae	*Byrsocarpus coccineus* Schomach. and Thonn.	Aerial parts	J774, WI38	(>100), (>100)	56

Family	Species	Plant part	Test system	Value	Ref
	Chnestis ferruginea Vahl ex DC.	Leaves	Chinese hamster Ovary cells (CHOK1)	280.0–610.0	110
Convolvulaceae	*Manotes pruinosa* Gilg.	Stem bark	MRC-5	27.0	50
	Calycobolus sp.	Stem bark	MRC	>64.0	49
	Ipomea aquatica Forsk	Leaves	Brine shrimp	160.9	111
		Stem		111.4	
Cucurbitaceae	*Bryonia dioica* Jacq.	Root	BL41 Burkitt's lymphoma	15.63	112
	Citrullus colocynthis (L) Schrad.	Fruit and seeds	Brine shrimp	189.0	62
			3T3	78.0	77
	Luffa cylindrica (L.)	Leaves	MRC-5	34.0	50
	Momordica charantia L.	Whole plant	MRC-5	3.0	50
	Telfairia occidentalis Hook f.	Leaves	Brine shrimp	1392.6	78
		Stem		1577.7	
Cupressaceae	*Cupressus lusitanica* Mill.	Bark	Vero cells	1580.0	113
		Leaves		1008.6	
		Root		>3200.0	
Curtisiaceae	*Curtisia dentata* (Burm.f) C.A. Sm.	Stem bark	Brine shrimp	302.0	114
Dilleniaceae	*Tetracena alinifolia* Wild.	Leaves	MRC-5	38.0	50
Dioscoreaceae	*Dioscorea bulbifera* L	Rhizome	AML 12	>40.0	47
	D. preussii Pax	Leaves	Human monocytes	2.4–5.9	69
Dracaenaceae	*Sansevieria liberica* Hort ex.Gérôme and Labroy	Aerial parts	J774, WI38	(59.0–>100), >100.0	56
Ebenaceae	*Euclea divinorum* Hiern	Bark, leaves	Vero cells	142.3	68

(Continued)

Table 8.2 (Continued)

Family	Species	Extract or Compound Tested	Cell Line	IC$_{50}$^{a} (μg/mL)	References
	E. natalensis A.DC.	Leaves	Vero cells	285.1	68
	E. schimperi (A.DC.) F. White	Leaves	GMK	>200.0	72
			MDCK	>200.0	
Euphorbiaceae	Alchornea cordifolia (Schumach.) Muell. Arg.	Leaves, stem	Human fibroblasts (HeLa)	55–146.1	106
		Leaves	MRC-5	>64.0	49
		Leaves	Bovine aorta endothelial cells	220.7	107
	A. floribunda Muell.Arg.	Root bark	MRC-5	>64.0	49
	Bridelia micrantha Benth.	Leaves	Brine shrimp	>90,000.0	61
	Chrozophora senegalensis (Lam) A. Juss	Leaves	HUVECs	15.0	115
	Croton macrostachyus Del.	Leaf Root	Brine shrimp	387.1	53
	C. pseudopulchellus Pax	Stem bark	Monkey kidney cells	64.0	63
	Dichostemma glaucescens Pierre	Stem bark	MRC-5	>64.0	50
	Drypetes gossweileri S. Moore	Stem bark	MRC-5	>64.0	49
	Drypetes natalensis (Harv.) Hutch	Root bark Stem bark	L6	19.0 88.9	65
	Elaoephorbia drupifera (Thonn.) Stapf.	Leaves	AML 12	>40.0	47
	Euphorbia hirta Linn.	Leaves	L6	65.8–450.0	105
	Flueggea virosa (Willd.) Voigt	Leaves	Vero cells	682.6–2990.5	99

Family	Species	Part	Cell line/assay	Value	Ref.
	Jatropha curcas L.	Root bark	MRC-5	>64.0	105
		Leaves	L6	126.5—>450.0	49
	Manniophyton fulvum Muell. Arg.	Leaves	MRC	>64.0	
		Root bark			
		Stem bark			
	Margaritaria discoidea (Baill) Webster	Leaves	Monkey kidney epithelial cells	31.3	116
	Phyllanthus amarus Schum and Thunn	Root		93.8	
		Leaves	L6	77.7–427.5	105
	P. niruri L.	Leaves	Vero cells	>100.0	117
		Root		>100.0	
		Stem		>100.0	
	Ricinus communis L.	Leaves/stem	Brine shrimp	1400.0–1500.0	66
	Sapium cornutum Pax.	Stem bark	MRC-5	37.0	50
	Sebastiania chamaelea (l.) Müll. Arg.	Whole plant	HUVECs	50.0	115
	Spirostachys africana Sond.	Stem bark	Vero cells	102.8	118
Fabaceae	*Abrus schimperi* Hochst. Ex Baker	Leaves	Vero	60.0	119
		Amorphaquinone		11.8	
		Pendulone		11.0	
		Leaves	Pig kidney epithelial cells (LLC-PK11)	30.0	
		Amorphaquinone		4.8	
		Pendulone		4.5	
	Acacia artaxacantha DC.	Root	Bovine aorta endothelial cells	438.7	107
	A. tortilis (*Hayne*)	Seeds	Vero cells	>100.0	
	A. letestui Pellegr	Stem bark	Human monocytes	6.2–7.2	55
		Leaves		4.5–8.5	69

(Continued)

Table 8.2 (Continued)

Family	Species	Extract or Compound Tested	Cell Line	IC$_{50}^{a}$ (µg/mL)	References
	Bauhinia bowkeri Harv	Leaves	Vero cells	17.9	120
	B. galpinii N.E. Br	Leaves	Vero cells	35.7	120
	B. petersiana Bolle	Leaves	Vero cells	40.7	120
	B. variegata Linn	Leaves	Vero cells	76.4	120
	Cajanus cajan Millsp.	Leaves	Brine shrimp	988.5	61
	Cassia fistula L.	Leaves	WS1	>200.0	48
	C. floribunda L.	Root bark	MRC-5	>64.0	50
	C. hirsuta L.	Root bark	MRC-5	>64.0	
	C. siamea Lam.	Stem bark	Vero cells	56.9–>100.0	121
		Leaves	L6	14.0–105.0	105
		Bark	L6	808.8	61
		Leaves	Brine shrimp	8232.2	
	Copaifera religiosa	Bark	MRC-5	4.8	122
	Dialium lopense Breteler	Leaves	Human monocytes	12.3–42.2	69
		Stem bark		6.2–36.1	
	Erythrina lysistemon Hutch	Stem bark	Monkey kidney cells	69.0	63
	Piliostigma thonnigii Schum	Leaves	Brine shrimp	7958.0	61
	Prosopis juliflora	Bulb	RAW264.7	>100.0	52
	Pterocarpus angolensis DC.	Leaves	Brine shrimp	3600.0–3800.0	66
		Bark		1400.0–1500.0	
	Senna petersiana (Bolle) Lock	Seeds	Monkey kidney cells	24.0	108

Family	Species	Part	Assay/Cell line	Value	Ref
	Schotia brachypetala Sond.	Leaves	Brine shrimp	3300.0	66
	Trigonella foenum-graecum L.	Seeds	Brine shrimp	60.0	62
Flacourtiaceae	*Flacourtia indica* (Burm. F.) Merr.	Pyrocatechol Homaloside Poliothrysoside	THP-1	<10.0 / >25.0 / 26.0	123
	Homalium africanum (Hook. f) Benth	Leaves	Monkey kidney epithelial cells	46.9–250	116
Geraniaceae	*Geranium incanum*	Leaves	Brine shrimp	0.43–4156.2	58
Gramineae	*Cymbopogon citratus* L.	Leaves	MRC-5	>64.0	50
	C. densiflorus (Stend.) Stapf.	Whole plant	MRC-5	>64.0	50
	Imperata cylindrica Beauv. Var. Koenigii Durand et Schinz	Root	HepG2 / AML 12 / HUVECs	33.4 / >40.0 / 47.7	47
Guttiferaceae	*Harungana madagascariensis* Poir	Leaves	Vero cells	461.5–3569.2	99
Guttiferaceae	*Mammea africana* G.	Stem bark	MRC-5	>64.0	50
	Psorospermum febriugum Spach.	Leaves	MRC-5	>64.0	50
Huaceae	*Afrostyrax lepidophyllus* Mildbr.	Stem bark	MRC-5	>64.0	49
Hymenocardiaceae	*Hymenocardia acida* Tul.	Leaves, twigs	J774, L6	(28– ≥167), (19–167), L6 (12.2, leaves)	98
		Leaves	L6	65.4–71.8	16

(Continued)

Table 8.2 (Continued)

Family	Species	Extract or Compound Tested	Cell Line	IC$_{50}$[a] (µg/mL)	References
Hypericaceae	*Hypericum lanceolatum Lam.*	Bark	LLC-MK2 monkey epithelial cells	>1000.0	124
		Betulinic acid		25.0	
		2, 2′, 5, 6′-Tetrahydroxybenzophenone		>100.0	
		5-hydroxy-3-methoxyxanthone		>100.0	
		3-hydroxy-5-methoxyxanthone		125	
Icacinaceae	*Vismia laurentii* De Wild	Xanthone V$_1$	AML 12	>20.0	70
	Apodytes dimidiata	Leaves	Vero cells	3.0	71
	Pyrenacantha klaineana Pierre ex Exell and Mendonça var. klaineana	Leaves	MRC-5	>64.0	49
Labietaceae	*Hoslundia opposite* Vahl	Roots	Vero E6	37.0	60
Lamiaceae	*Ajuga remota* Benth	Leaves	Brine shrimp	61.6	53
	Clerodendrum glabrum E. Mey.	Leaves	Vero cells	172.0	71
	Fuerstia africana T.C.E. Friers	Aerial parts	Vero cells	64.7–>500	55
		Roots		23.8–366.4	86
		Leaves and Stem	WI38	13.0	
	Leonotis africana (P. Beauv.) T. Durand and H. Durand	Stem	MRC-5	439.1	46
	Ocimum basilicum L.	Leaves	Brine shrimp	>1000.0	62
	O. gratissimum L.	Leaves	MRC-5	>64.0	49
		Whole plant		12.0	50
		Leaves	L6	34.7–424.2	105
	Salvia repens Burch. ex Benth.	12-methoxycarnosic acid (Whole plant)	L6	6.0 (17.3 µM)	125

Family	Species	Plant part	Cell line	Value	Ref
Lecythidaceae	Napoleona vogelii Hook and Planch.	Stem bark	MRC-5	>64.0	49
	Albizia schimperiana Oliv.	Stem bark	Vero cells	22.0	126
		5, 14-dimethylbudmunchiamine K		1.8	
		5-normethylbudmunchiamine K		5.3	
		5, 14-dimethylbudmunchiamine K	LLC-PK$_{11}$	1.7	
		6-hydroxybudmunchiamine K		5.3	
		5-normethylbudmunchiamine K		2.5	
		6-hydroxy-5-normethylbudmunchiamine K		6.0	
Leguminosae	Caesalpinia volkensii Harms	Leaves	L6	82.4	93
	Dalhousiea africana S. Moore	Leaves	MRC-5	>64.0	49
	Dialium guineense Willd.	Aerial parts	J774, WI38	(>100), (73.3—>100)	56
	Guibourtia tessmannii Harms Leonard	Bark	Vero cells	1580	113
		Leaves		>3200	
		Root		>3200	
	Melilotus elegans Salzm. ex Ser.	Leaves	GMK	>200	72
			MDCK	>200	
	Pericopsis laxiflora (Benth. ex Baker) Meeuwen	Leaves	L6	58.5—>450	105
	Piptadeniastrum africanum (Hook.f.) Brenan	Stem bark	MRC-5	8.8	49
	Scorodophloeus zenkeri Harms	Stem bark	MRC-5	>64.0	49

(Continued)

Table 8.2 (Continued)

Family	Species	Extract or Compound Tested	Cell Line	IC$_{50}$[a] (µg/mL)	References
	Tephrosia villosa (L) Pers	Fruit	Brine shrimp	9.7	127
		Leaves		134.0	
		Root		4.5	
		Twigs		238.2	
	Tetrapleura tetraptera Traub.	Fruit	MRC-5	29.3	49
		Stem bark		32.0–89.0	46
Liliaceae	*Aloe greatheadii* Schonl. var. *davjana* (Schonl.) Glen and Hardy	Leaves	Graham 293	4.7–392.9	57
Loganiaceae	*Buddleja salviifolia* (L.) Lam	Leaves	WS1	>200.0	48
		Bark		162.0	
	Geniostoma borbonicum Spreng.	Bark	WS1	68.0	48
		Leaves		41.0	
	Nuxia verticillata Lam.	Leaves	WS1	125.0	48
	Strychnos heningsii Gilg	Twigs	L6	>90.0	93
	S. spinosa Lam.	Leaves	J774, W138	(>100), (>100)	56
Maesaceae	*Maesa lanceolata* Forssk.	Leaves	Vero cells	104.0	71
Magnoliidae (or Canellaceae)	*Warburgia ugandensis* Sprague	Stem bark	Vero E6	>250.0	128
			L6	0.34	93
Malpighiaceae	*Acridocarpus chloropterus* Oliv.	Leaves	L6	77.3– >90.0	65
		Root bark		64.7– >90.0	
		Stem bark		82.5–88.1	
	Hiptage benghalensis Kurz	Leaves	WS1	90.0	48
Malvaceae	*Gossypium arboretum* L.	Leaves	Brine shrimp	94.1	61
	G. barbadense L.	Leaves	Brine shrimp	3585.0	61

	H. trifurca L.	Aerial parts (leaves, stem, and flowers)	Human kidney epithelial cells	75.6	129
	Hermannia saccifera (Turcz.) K. Schum	Aerial parts (leaves, stem, and flowers)	Human kidney epithelial cells	61.4	129
Meliaceae	G. hirsutum L.	Leaves	Brine shrimp	257.0	61
	Azadirachta indica A. Juss.	Leaves	Brine shrimp	21.0	62
		Seeds		45.0	
	Ekebergia capensis Sparrm	Leaves	Vero cells	13.1→27.3	73
	Entandophragma caudatum (Sprague) Sprague	Stem bark	Monkey kidney cells	100.0	63
	Khaya anthotheca (Welv.) C.D.C.	Seeds	L6	90.0	130
		Grandifolione		44.7	
		7-deacetykhivorin		14.9	
	Melia azedarach L.	Leaves	Vero cells	145.0	71
	Trichilia emetica Vahl subsp. suberosa J.J.F.E. Dewilde	Leaves	J774, W138	(83.7→ >100)	56
				(59.2→ >100)	63
	T. giliana Harms	Stem bark	Monkey kidney cells	50.0	50
	T. gilletii (De Wild.) Stanner	Stem bark	MRC-5	>64.0	50
		Stem bark	MRC-5	34.0	50
Melianthaceae	Bersama engleriana Gürke	Bark	Vero cells	748.2	113
		Leaves		745.0	
		Root		1300.0	
	Turraea robusta Guerke	Root bark	L6	14.3	93
Mimosaceae	Acacia nilotica L.	Leaves	3T3	65.2	77
	Albizia gummifera (G.F. Gmel.)	Leaves	Brine shrimp	274.4	54
		Root		86.9	
	A. zygia J.F. Macbr.	Stem bark	L6	4.5	75

(Continued)

Table 8.2 (Continued)

Family	Species	Extract or Compound Tested	Cell Line	IC$_{50}$[a] (μg/mL)	References
	Entada abyssinica Steud. ex A. Rich.	Bark	AML 12	>40.0	47
	E. leptostachya Harms	Tuber	Brine shrimp	421.3	53
	Pentaclethra etveeldeana Benth.	Root bark	MRC-5	27.0	50
	Piptadenia africanum Hoof. f.	Stem bark	MRC-5	<0.25	50
	Schrankia leptocarpa DC.	Leaves, Twigs	J774, WI38	(83.7->100), (88.6->100)	56
Menispermaceae	*Cissampelos capensis* L.	Aerial parts Root	Brine shrimp	6.2–2446.3 3.8–3526.2	58
	C. mucronata A. Rich	Aerial Parts Root	Brine shrimp	114.2–125.7 59.6–143.0	127
	Penianthus longifolius Miers	Root bark	MRC-5	>64.0	49
	Synclisia scabrida (Meirs)	Root	Brine shrimp	36.6–202.1	131
	Triclisia dictyophylla Diels	Leaves	MRC-5	>64.0	49
Moraceae	*Dorstenia klaineana* Pierre ex Heckel and Schlagd	Stem	MRC-5	0.43–29.3	46
	D. psilirus Welwitch	Roots	HUVECs	>80	70
	Ficus exasperate Vahl.	Leaves	Human monocytes	8.6–39.1	69
	Ficus sur Forssk.	Fruit	RAW264.7	>100	52
Moringaceae	*Moringa oleifera* Lam.	Leaves Seeds	Brine shrimp	>1000.0 900.0	62
	M. stenopetala	Leaf	RAW264.7	>100	52

Family	Species	Part	Assay/Cell line	Value	Ref
Myristicaceae	Myristica fragrans Houtt	Kernel	Brine shrimp	4359.7	132
	Staudtia gabonensis Warb.	Stem	MRC-5	46.6–123.2	46
	S. kamerunensis Warb. var. gabonensis (Warb.) Fouilly	Stem bark	MRC-5	>64.0	49
Myrothamnaceae	Myrothamnus flabellifolia Welw.	Aerial parts	Vero cells	50.0	133
Myrtaceae	Eugenia uniflora L.	Leaf	Bovine aorta endothelial cells	149,4	107
	Psidium guajava L.	Bark	Brine shrimp	707.2	61
		leaves	MRC	32.9	49
	Syzygium cordatum Hochst ex C Krauss	Bark	Human kidney epithelial cells	26.8	64
Ochnaceae	Campylospermum flavum (Schum.) Farron.	Bark	Brine shrimp	60%	134
		Leaves		70%	
		4'''-O-methylagathisflavone		72%	
		Flavumchalcone		100.0	
		Flavumindole		90.0	
Olacaceae	Olax subscorpioidea var. subscorpioidea Oliv.	Seeds	HUVECs	>80.0	70
	Ongokea gore (Hua) Pierre	Stem bark	MRC-5	12.0	50
Oleaceae	Schrebera alata (Hochst.) Welw.	Leaves	Vero	30.4–243.7	73
Pandaceae	Microdesmis puberula Hoof. f.	Leaves	MRC-5	>64.0	50
Papaveraceae	Argemone mexicana L.	Leaves	3T3	67.9	77
Papilionaceae	Indigofera frutescens L.f.	Leaves	Vero cells	52.0	71

(Continued)

Table 8.2 (Continued)

Family	Species	Extract or Compound Tested	Cell Line	IC$_{50}$[a] (µg/mL)	References
	Milletia grandis (E.Mey.) Skeels	Leaves	Vero cells	21.0	71
	Neurautanenia mitis	Tuber	L6	22.8	65
	Pericopsis elata Harms	Leaves	Brine shrimp	601.8	61
	P. laxiflora (Bentham ex Baker) Van Meeuwen	Leaves, Twigs	J774, L6	(19-—≤167), (32.2, L6)	98
Passifloraceae	*Adenia cissampeloides* Forssk.	Stem	Bovine aorta endothelial cells	240.7	107
Pedaliaceae	*Dicerocarym eriocarpum* (Decne.) Abels	Whole plant	Brine shrimp	800.0–2800.0	66
	D. senecioides (Klotzsch) Abels.	Roots	Vero cells	122.1	68
Phytolaccaceae	*Phytolacca dodecandra* l'Herit.	Leaves	MRC-5	>64.0	50
Piperaceae	*Piper capense* L.f.	Seeds	HepG2	16.7	47
			AML 12	>40.0	70
			HUVECs	>80	
	P. guineense (Schum and Thonn	Seeds	HUVECs	70.0	70
		Leaves	MRC-5	>64.0	49
Pittosporaceae	*Pittosporum lanatum* Hutch. and Bruce	Leaves	Brine shrimp	27.4	54
Plumbaginaceae	*Plumbago zeylanica* L.	Leaves	MRC-5	>64.0	50
			GMK	96.2	72
			MDCK	20.8	73
			Vero cells	6.7–243.3	

Family	Species	Part	Assay	Value	Ref
Poaceae	*Imperata cylindrical* P. Beauv.	Leaves	L6	169.1–>450	105
Podocarpaceae	*Podocarpus henkelii* Stapf ex Dallim. and Jacks.	Leaves	Vero cells	107.3	73
Polygalaceae	*Carpolobia lutea* G. Donn	Aerial parts	J774, WI38	(38.7–>100), (65.4–>100)	56
	Securidaca longepedunculata Fresen.	Roots	THP-1	>125.0	76
Polygonaceae	*Calligonum comosum* L.	Aerial parts	Brine shrimp (mg/mL)	41.8–174.8	135
		(+)-catechin		61.1	
		Dehydrodicatechin A		35.1	
		Kaempferol-3-O-rhamnopyranoside		80.1	
		Quercitrin		145.3	
		B-sitosterol-3-Oglucoside		261.0	
		Isoquercitrin		105.5	
		Kaempferol-3-O-glucuronide		106.6	
		Mequilianin		103.8	
	Polygonium glabrium (Willd.)	Leaves	Brine shrimp	354.0	62
Polypodiaceae	*Rumex abyssinicus* Jacq.	Root	WI38 WI-38	13.3	86
	Polypodium polycarpon Cav.	Leaves	Human monocytes	5.7–8.2	69
Rhamnaceae	*Berchemia zeyheri* (Sond.) Grubov	Bark	Brine shrimp	3800.0	66

(Continued)

Table 8.2 (Continued)

Family	Species	Extract or Compound Tested	Cell Line	IC$_{50}$[a] (μg/mL)	References
	Maesopsis eminii Engl.	Leaves	Human monocytes	12.4–86.4	69
		Stem bark		23.1–33.1	
	Scutia commersonii Brongn.	Leaves	WS1	116.0	48
	Ziziphus mucronata Willd.	Leaves	Brine shrimp	900.0	66
	Z. spina-christi L.	Leaves	Brine shrimp	300.0	62
Rosaceae	*Leucosidea sericea* Eckl. and Zeyh.	Leaves	Vero cells	16.0	71
Rosaceae	*Prunus africana* (Hook.f.) Kalkman	Stem bark	Vero E6	104.1	128
	Rubus rosifolius Sm.	Aerial parts	WS1	119.0	48
Rubiaceae	*Breonadia salicina* (Vahl) Hepper and J.J.R. Wood	Bark	Human kidney epithelial cells	>200.0	64
	Canthium addonii L.	Stem bark	MRC-5	>64.0	50
	Canthium henriquesianum (K. Schum)	Leaves/twigs	HDF	615.4	94
	Corynanthe mayumbensis (R.D. Good)Raym.-Hamet ex N. Hallé	Leaves	Human monocytes	36.4– >100	69
		Stem bark		86.4– >100	
	Crossopteryx febrifugum Benth.	Leaves	MRC-5	>64.0	50
	Feretia apodanthera (Del.)	Methanol fraction	THP-1	293.0	104

Gardenia sokotensis Hutch.	Leaves/twigs	HDF	(12, 222.0)	90
				94
Keetia leucantha (K. Krause) Bridson	Twigs: Extract, ursolic acid, oleanolic acid, leaves	W138	>100, 6.7, 59.6, (65.6->100)	56, 136
Massularia acuminate (G. Don) Bullock ex Hoyle	Stem bark	MRC-5	>64.0	49
Mitracarpus scaber Zucc.	-			
Mitragyna inermis (Willd.) O. Kuntze	Total alkaloids	Vero	10.0	137
		THP-1	100.0	104
	Ursolic acid		1.0	
Morelia senegalensis A. Rich. ex DC	Leaves	Human monocytes	12.4–35.4	69
	Stem bark		5.4–26.1	
Morinda citrifolia L.	Stem bark	MRC-5	>64.0	50
	Fruit	WS1	>200.0	48
	Leaves		>200.0	
M. lucida Benth.	Bark	Brine shrimp	2.6	61
	Leaves		383.9	
Nauclea diderrichii (De Wild. and T. Durand) Merrill	Bark	CHOK1	67.4–>100.0	138
N. latifolia (Sm.)	Bark	Brine shrimp	9368.0	61
	Stem	MRC-5	11.5	46
	Total alkaloids	THP-1	5.0	104
	Bark	CHOK1	12.3–35.7	138
N. pobeguinii (Pob. Ex. Pell.)	Stem bark	MRC-5 SV$_2$	>64.0	139
	(5S)-5-carboxystrictosidine		>64.0	
	19-O-methylangustoline		>64.0	138
	3-O-β-fucosyl-quinovic acid		>64.0	
	Bark	CHOK1	54.2–>100.0	138
N. vanderguthii	Bark	CHOK1	>100.0	138

(Continued)

Table 8.2 (Continued)

Family	Species	Extract or Compound Tested	Cell Line	IC$_{50}$^{a} (μg/mL)	References
	Pavetta crassipes (K. Schum)	Leaves	THP-1	62.7	81
			Malignant melanoma cells	46.1	
				100.8	
	P. gardenifolia A. Rich.	Leaves	Graham 293	25.9–729.7	57
		Seeds		153.4–495.9	
		Stem		50.7–1065.9	
	Pentas lanceolata (Forssk.) Defleurs	Aerial parts	Vero cells	>500.0	55
		Roots		>500.0	
	P. longiflora Oliv.	Leaves	Brine shrimp	12.3	54
		Root		6.4	
	Uncaria africana G. Don	Leaves	Human monocytes	24.6–25.4	69
		Stem bark		15.3–51.2	
	Sarcocephalus latifolius (J.E. Smith) E.A. Bruce	Leaves	K562S (Human monocytes cell)	52.9–>125.0	103
Rutaceae	*Citrus auranthus* (Chr.) SVV.	Leaves	MRC-5	>64.0	50
	Clausena anisata (Willd) Hook	Leaves	Vero cells	53.0	71
	Teclea trichocarpa (Engl.) Engl.	Leaves	L-6	>90.0	140
		Melicopicine		>90.0	
		Normelicopicine		>90.0	
		Arborinine		12.2	
		Skimmianine		38.6	
		B-Sitosterol		>90.0	
		α-Amyrin		>90.0	
		Fruit	Vero cells	>100.0	141

Species	Plant part	Cell line	Compound	Value	
Toddalia asiatica (L) Lam.	Leaves			>100.0	
Zanthoxylum calybeum Engl.	Root bark	WI38		80.0	
	Root bark			1.9	86
Z. capense Thunb.	Root	THP-1	Decarine	45.7	76
			Norchelerythrine	66.0	142
			Dihydrochelerythrine	75.4	
			6-Acetonyldihydrochelerythrine	68.7	71
			Tridecanoncheleyrthrine	57.1	
			6-Acetonyldihydronitidine	94.5	
			Zanthocapensine	1.7	
			Rutaecarpine	40.3	
			Skimmianine	58.2	
			(-)-sesamin	50.3	
			(-)-episesamin	49.8	
			(-)-savinin	48.9	
		Vero cells		3.7	
			Zanthocapensol	72.3	
			Zanthocapensate	52.1	
			N-isobutyl-(2E,4E)-2,4-tetradecadienamide	61.2	
			Lupeol	60.0	
			Leaves	8.0	
Z. zanthoxyloides (Lam.) B. Zepernick and Timler	Seeds (essential oils)	Vero cells		>50	96
	Bark	K562S (Human monocytes cell)		4.7->125	103

(Continued)

Table 8.2 (Continued)

Family	Species	Extract or Compound Tested	Cell Line	IC$_{50}$[a] (μg/mL)	References
Salvadoraceae	*Salvadora persica* L.	Leaves	Brine shrimp	>1000.0	62
Sapindaceae	*Blighia sapida* K.D. Koenig	Leaves	Brine shrimp	114.9	143
	B. unijugata Baker	Leaves	Brine shrimp	803.8	143
	Ganophyllum giganteum (Chev.)Hauman	Leaves	Human monocytes	1.3–8.6	69
		Stem bark		5.5–11.2	
Sapotaceae	*Autranella congolensis* (De Wild.) A.Chev.	Stem bark	MRC-5	40.3	49
	Englerophytum magalismontanum (Sond.) T.D. Penn.	Bark	Vero cells	98.8	68
	Manilkara discolor (Sond.) J.H. Hemsl.	Leaves	Vero cells	>100.0	55
		Stem bark		>100.0	
	Minusops kummel Bruce	Pulp	RAW264.7	>100.0	52
	Omphalocarpum glomerata P. Beauv.	Root bark	MRC-5	9.0	50
	Vitellaria paradoxa C. F. Gaertn	Bark	Vero cells	350.0	113
		Leaves		2391.2	
		Root		183.3	
Simaroubaceae	*Harrisonia abyssinica* Oliv	Stem Bark	L6	32.8	93
		Leaves, Bark	Brine shrimp	392.4	53
	Irvingia grandifolia Engl.	Leaves	Human monocytes	5.4–>100.0	69
		Stem bark		6.2–8.4	
	Quassia africana Baill.	Root bark	MRC-5	6.3	49
		Stem Bark		17.3	

Family	Species	Part	Test system	Value	Ref
Solanaceae	*Datura stramonium* L.	Leaves	Brine shrimp	39.0	62
Solanaceae	*Physalis angulata* L.	Leaves	WI38	7.8–15.7	82
Sterculiaceae	*Solanum nigrum*	Leaves	RAW264.7	>100	52
	Cola lizae N. Hallé	Leaves / Stem bark	Human monocytes	3.8–42.1 / 2.5–32.5	69
Strychnaceae	*Strychnos henningsii* (Gilc.)	Root bark	Brine shrimp	822.2	79
	S. mitis S. Moore	Leaves	Vero cells	43.0	71
Thymelaeaceae	*Gnidia capitata* L.f.	Root	Brine shrimp	700.0	66
Umbelliferae	*Centella asiatica* Urban	Whole plant	L6	82.6	93
Ulmaceae	*Trema guineensis* (Schumm. and Thom.) Fic.	Root bark	MRC-5	>64.0	50
	T. orientalis (L.) Blume	Bark / Leaves	L6	46.2–129.4 / 32.5–409.0	105
Urticaceae	*Pouzolzia mixta* Solms	Leaves stem	Brine shrimp	5000.0 / 4500.0	66
	Urtica simensis (Steud.)	Leaf	RAW264.7	>100.0	52
Verbenaceae	*Clerodendrum capitata* (Willd)	Root	Bovine aorta endothelial cells	259.8	107
	C. eriophyllum Guerke	Root bark	L6	7.9	93
	C. myricoides (Hochst) Vatke	Root bark	L6	>90.0	93
	Lantana camara L.	Leaves	MRC-5	38.0	50
	Vitex trifolia L	Aerial parts	Brine shrimp	0.01	144
			Brine shrimp	140–180 mg/mL	145
	Lippia javanica (Burm F.)	Root	Brine shrimp	1138.0	79

(Continued)

Table 8.2 (Continued)

Family	Species	Extract or Compound Tested	Cell Line	IC$_{50}$[a] (μg/mL)	References
	L. multiflora Moldenke	Aerial parts	Brine shrimp	1.1	61
	Vitex doniana Sweet	Bark	L6	335.4—>450	105
		Leaves		122.6—>450.0	
Violaceae	*Allexis cauliflora* (Oliv.)	22-hydroxyclerosterol	MRC-5 fibroblasts	513.6	146
	Rinorea subintegrifolia	Stem	MRC-5	520.0	46
Vitaceae	*Cissus quadrangularis* L.	Stem	Brine shrimp	1300.0	66
Zingiberaceae	*Aframomum melegueta*	Seeds	HUVECs	21.0	70
	(Roscoe) K. Schum.	Stem	Brine shrimp	749.7	132
		Stem	MRC-5	47.2—168.5	46
	Zinziber officinale Roscoe	Rhizomes	HUVECs	>80.0	70
			Brine shrimp	187.9	132
Zygophyllaceae	*Zygophyllum geslini* Coss.	Root/ 3β-(3,4-dihydroxycinnamoyl)-erythrodio	KB cells (carcinoma)	2.5	147

[a]Where a range of IC$_{50}$ values is reported this represents a range of different extracting solvents used to prepare various extracts.

Vero African GMK cells. Other widely used cell lines are HUVECs (human umbilical vein endothelial cells) and MRC-5 (normal human lung fibroblast) cells. As IC_{50} values can vary quite significantly depending on which cell line is used, it would be advisable to select a commonly used and widely available cell line, which is relevant to the particular study, so that results may be compared with those of other researchers to obtain meaningful interpretations of data. Another popular assay used to indicate cytotoxicity of African medicinal plants is the brine shrimp assay, which is rapid and inexpensive. However, manuscripts reporting correlation of results obtained in the brine shrimp assay to those resulting from tests using mammalian cells are sparse.

Although a wide range of families have been represented in terms of cytotoxicity testing of plant species, some plant families have been focused on to a greater extent than others. This potentially reflects their popularity in traditional African medicine, translating to heightened interest in biological activity and toxicity studies of such species. The Asteraceae is a large family with many members of medicinal value, and 34 species belonging to this family have been reported in available literature to have been screened for cytotoxicity. Other relatively well-represented families include the Rubiaceae, Fabacaeae, Euphorbiaceae, and Combretaceae.

8.7 Conclusion

Medicinal plants constitute an integral part of traditional healing customs in African culture. A large number of extracts of African plants, as well as some pure compounds isolated from these plants, have been screened for cytotoxicity against a wide range of cell lines. Such studies are necessary to determine whether biological activity is a specific phenomenon or whether it should be merely attributed to nonspecific toxicity. A list of African medicinal plants tested for cellular toxicity has been collated from available scientific literature, and the cell lines used and IC_{50} values obtained were summarized. Several different cell lines have been used, with a focus on certain cell types such as macrophages and monocytes, as well as easily available cell lines such as Vero and MRC-5 cells.

It should be kept in mind that almost any chemical substance has the potential to be a toxin. Paracelsus, considered by some to be the father of toxicology stated "Poison is in everything, and no thing is without poison. The dosage makes it either a poison or a remedy." (http://www.brainyquote.com/quotes/authors/p/paracelsus. html). Factors such as the concentration (or dose), the duration, and route of exposure will affect the toxicity of a chemical. Additionally, the response of a human or animal to a particular chemical involves many physiological targets, and complex pharmacokinetics. Chemicals may induce harm through a variety of mechanisms, so it is impossible to address the complexity of general toxicity in a single method. Rapid, simple, and economic *in vitro* cytotoxicity assays, especially those correlated with basal cytotoxicity in general and local toxicity in particular, serve an important purpose in evaluating medicinal plants for potential pharmaceutical and cosmetic application.

Acknowledgments

The National Research Foundation (NRF), Medical Research Council (MRC) of South Africa, and the University of Pretoria Institutional Research Theme for Animal and Zoonotic Diseases provided financial assistance.

References

[1] Niles AL, Moravec RA, Riss TL. Update on *in vitro* cytotoxicity assays for drug development. Expert Opin Drug Discov 2008;3:655−69.

[2] Horvath S. Cytotoxicity of drugs and diverse chemical agents to cell cultures. Toxicology 1980;16:59−66.

[3] Ekwall B. The basal cytotoxicity concept. In: Goldberg AM, van Zupthen AFM, editors. Alternative methods in Toxicology and the life sciences. Vol 11, the world congress on alternatives and animal use in the life sciences, education, research, testing. New York, NY: Mary Ann Liebert; 1994.

[4] DiPasquale LC, Hayes AW. Acute toxicity and eye irritancy. In: Hayes AW, editor. Principles and methods of toxicology. 4th ed. Philadelphia, PA: Taylor and Francis; 2001. p. 853−916.

[5] Barile FA. Principles of toxicology testing. Boca Raton, FL: CRC Press (Taylor and Francis Group); 2008.

[6] Bondesson I, Ekwall B, Hellberg S, Romert L, Stenberg K, Walum E. MEIC. A new international multicenter project to evaluate the relevance to human toxicity of *in vitro* cytotoxicity tests. Cell Biol Toxicol 1989;5:331−47.

[7] Ekwall B. Features and prospects of the MEIC cytotoxicity evaluation project. AATEX 1992;1:231−7.

[8] Ekwall B, Bondesson I, Castell JV, Gomez-Lechon MJ, Hellberg S, Hogberg J, et al. Cytotoxicity evaluation of the first ten MEIC chemicals: acute lethal toxicity in man predicted by cytotoxicity in five cellular assays and by oral LD_{50} tests in rodents. ATLA 1989;17:83−100.

[9] Walum E, Clemedsen C, Ekwall B. Principles for the validation of *in vitro* toxicology test methods. Toxicol In Vitro 1994;8:807−12.

[10] Ekwall B, Walum E, Clemedsen C, Barile FA, Castano A, Clothier RA, et al. MEIC evaluation of acute systemic toxicity. Part IV: prediction of human toxicity by rodent LD_{50} values and results from 61 *in vitro* tests. ATLA 1998;26(Suppl. 2):617−58.

[11] Ekwall B, Clemedsen C, Crafoord B, Ekwall B, Hallander S, Walum E, et al. MEIC evaluation of acute systemic toxicity. Part V: rodent and human toxicity data for the 50 reference chemicals. ATLA 1998;26(Suppl. 2):571−616.

[12] Nardone RM. The LD_{50} test and *in vitro* toxicology strategies. Acta Pharmacol Toxicol (Copenhagen) 1989;52(Suppl. 2):65−79.

[13] Freshney RI. Culture of animal cells: a manual of basic technique and specialised applications. 6th ed. Hoboken, NJ: John Wiley & Sons, Inc.; 2010.

[14] Cox P, Chrisochoidis M. Directive 2003/15/EC of the European Parliament and the Council of 27 February 2003. Off J Eur Union 2003;L66:26−35.

[15] Houghton PJ, Howes M-J, Lee CC, Steventon G. Uses and abuses of *in vitro* tests in ethnopharmacology: visualizing an elephant. J Ethnopharmacol 2007;110:391−400.

[16] Vonthron-Sénécheau C, Ouattara M, Trabi F, Kamenan A, Anton R, Weniger B. *In vitro* antiplasmodial activity and cytotoxicity of extracts of ethnobotanically selected Ivorian plants. J Ethnopharmacol 2003;87:221−5.

[17] Ashidi JS, Houghton PJ, Hylands PJ, Efferth T. Ethnobotanical survey and cytotoxicity testing of plants of South-western Nigeria used to treat cancer, with isolation of cytotoxic constituents from *Cajanus cajan* Millsp. leaves. J Ethnopharmacol 2010;128:501−12.

[18] McGregor DB, Edwards I, Riach CJ, Cattenach P, Martin R, Mitchell A, et al. Studies of an S9 based metabolic activation system used in the mouse lymphoma L51768Y cell mutation assay. Mutagenesis 1988;3:485−90.

[19] Guillouzo A, Guguen-Guillouzo C. Evolving concepts in liver tissue modeling and implications for *in vitro* toxicology. Expert Opin Drug Metab Toxicol 2008;4:1279−94.

[20] Macé K, Gonzalez FJ, McConnell IR, Garner RC, Avanti O, Harris CC, et al. Activation of promutagens in a human bronchial epithelial cell line stably expressing human cytochrome P450 1A2. Mol Carcinog 1994;11:65−73.

[21] Nielsen JB. What you see may not always be what you get: bioavailability and extrapolation from *in vitro* tests. Toxicol In Vitro 2008;22:1038−42.

[22] Lambert JD, Hong J, Yang G, Liao J, Yang CS. Inhibition of carcinogenesis by polyphenols: evidence from laboratory investigations. Am J Clin Nutr 2005;81:284S−91S.

[23] El-Kattan AF, Asbill CS, Kim N, Michniak BB. The effects of terpene enhancers on the percutaneous permeation of drugs with different lipophilicities. Int J Pharm 2001;215:229−40.

[24] Narishetty STK, Panchagnula R. Transdermal delivery of zidovudine: effect of terpenes and their mechanism of action. J Control Release 2004;95:367−79.

[25] Riss T, Moravec R. Use of multiple assay endpoints to investigate the effects of incubation time, dose of toxin, and plating density in cell-based cytotoxicity assays. Assay Drug Dev Technol 2004;2:51−62.

[26] Mosmann T. Rapid colorimetric assay for cellular growth and survival: application to proliferation and cytotoxicity assays. J Immunol Methods 1983;65:55−63.

[27] Berridge MV, Herst PM, Tan AS. Tetrazolium dyes as tools in cell biology: new insights into their cellular reduction. Biotechnol Annu Rev 2005;11:127−52.

[28] Bruggisser R, Von Daeniken K, Jundt G, Schaffner W, Tullberg-Reinert H. Interference of plant extracts, phytoestrogens and antioxidants with the MTT tetrazolium assay. Planta Med 2002;68:445−8.

[29] Shoemaker M, Cohen I, Campbell M. Reduction of MTT by aqueous herbal extracts in the absence of cells. J Ethnopharmacol 2004;93:381−4.

[30] Peng L, Wang B, Ren P. Reduction of MTT by flavonoids in the absence of cells. Colloids Surf B Biointerfaces 2005;45:108−11.

[31] O'Brien J, Wilson I, Orton T, Pognan F. Investigation of the Alamar Blue (resazurin) fluorescent dye for the assessment of mammalian cell cytotoxicity. Eur J Biochem 2000;267:5421−6.

[32] Boncler M, Różalski M, Krajewska U, Podsędek A, Watala C. Comparison of PrestoBlue and MTT assays of cellular viability in the assessment of anti-proliferative effects of plant extracts on human endothelial cells. J Pharmacol Toxicol Methods 2014;69:9−16.

[33] Lall N, Henley-Smith CJ, De Canha MN, Oosthuizen CB, Berrington D. Viability reagent, PrestoBlue, in comparison with other available reagents, utilized in cytotoxicity and antimicrobial assays. Int J Microbiol 2013; Article ID 420601, 5 pages.

[34] Crouch SP, Kozlowski R, Slater KJ, Fletcher J. The use of ATP bioluminescence as a measure of cell proliferation and cytotoxicity. J Immunol Methods 1993;160:81−8.

[35] Korzeniewski C, Callewaert D. An enzyme-release assay for natural cytotoxicity. J Immunol Methods 1983;64:211−24.

[36] Decker T, Lohmann-Matthes M. A quick and simple method for the quantification of lactate dehydrogenase release in measurements of cellular cytotoxicity and tumour necrosis factor (TNF) activity. J Immunol Methods 1988;115:61−9.

[37] Skehan P, Storeng R, Scudiero D, Monks A, McMahon J, Vistica D, et al. New colorimetric cytotoxicity assay for anticancer-drug screening. J Natl Cancer Inst 1990;82:1107−12.

[38] Houghton PJ, Fang R, Techatanawat I, Steventon G, Hylands PJ, Lee CC. The sulforhodamine (SRB) assay and other approaches to testing plant extracts and derived compounds for activities related to reputed anticancer activity. Methods 2007;42:377−87.

[39] Meyer BN, Ferrigni NR, Putnam JE, Jacobsen LB, Nichols DE, McLaughlin JL. Brine shrimp: a convenient general bioassay for active plant constituents. Planta Med 1982;45:31−4.

[40] McLaughlin JL. Assays related to cancer drug discovery. In: Hostettmann K, editor. Methods in plant biochemistry, assays for bioactivity, vol. 6. London: Academic Press; 1991. p. 1−32.

[41] Solís P, Wright C, Anderson M, Gupta M, Phillipson JD. A microwell cytotoxicity assay using Artemia salina (brine shrimp). Planta Med 1993;59:250−2.

[42] Desmarchelier C, Mongelli E, Coussio J, Ciccia G. Studies on the cytotoxicity, antimicrobial and DNA-binding activities of plants used by the Ese'ejas. J Ethnopharmacol 1996;50:91−6.

[43] McLaughlin JL, Rogers LL, Anderson JE. The use of biological assays to evaluate botanicals. Drug Inf J 1998;32:513−24.

[44] Ke N, Wang X, Xu X, Abassi YA. The xCELLigence system for real-time and label-free monitoring of cell viability. Methods Mol Biol 2011;740:33−43.

[45] Solly K, Wang X, Xu X, Strulovici B, Zheng W. Application of real-time cell electronic sensing (RT-CES) technology to cell-based assays. Assay Drug Dev Technol 2004;2:363−72.

[46] Lekana-Douki JB, Bongui JB, Liabagui SLO, Edou SEZ, Zatra R, Bisvigou U, et al. In vitro antiplasmodial activity and cytotoxicity of nine plants traditionally used in Gabon. J Ethnopharmacol 2011;133:1103−8.

[47] Kuete V, Sandjo LP, Wiench P, Efferth T. Cytotoxicity and modes of action of four Cameroonian dietary spices ethno-medically used to treat cancers: Echinops giganteus, Xylopia aethiopica, Imperata cylindrical, and Piper capense. J Ethnopharmacol 2013;149:245−53.

[48] Jonville MC, Kodja H, Strasberg D, Pichette A, Ollivier E, Fédérich M, et al. Antiplasmodial, anti-inflammatory and cytotoxic activities of various plant extracts from the Mascarene Archipelago. J Ethnopharmacol 2011;136:525−31.

[49] Muganza DM, Fruth BI, Lami JN, Mesia GK, Kambu OK, Tona GL, et al. In vitro antiprotozoal and cytotoxic activity of 33 ethnopharmacologically selected medicinal plants from democratic republic of Congo. J Ethnopharmacol 2012;141:301−8.

[50] Mesia GK, Tona GL, Nanga TH, Cimanga RK, Apers S, Cos P, et al. Antiprotozoal and cytotoxic screening of 45 plant extracts from Democratic Republic of Congo. J Ethnopharmacol 2008;115:409–15.

[51] Olorunnisola OS, Bradley G, Afolayan AJ. Antioxidant properties and cytotoxicity evaluation of methanolic extract of dried and fresh rhizomes of *Tulbaghia violacea*. Afr J Pharm Pharmacol 2011;5:2490–7.

[52] Ayele Y, Kim J, Park E, Kim Y, Retta N, Dessie G, et al. A methanol extract of *Adansonia digitata* L. leaves inhibits pro-inflammatory iNOS possibly via the inhibition of NF-κB activation. Biomol Ther 2013;21:146–52.

[53] Cyrus WG, Daniel GW, Nanyingi MO, Njonge FK, Mbaria JM. Antibacterial and cytotoxic activity of Kenyan medicinal plants. Mem Inst Oswaldo Cruz Rio de Janeiro 2008;103:650–2.

[54] Wanyoike GN, Chhaba SC, Lang'at-Thoruwa CC, Omar SA. Brine shrimp toxicity and antiplasmodial activity of five Kenyan medicinal plants. J Ethnopharmacol 2004;90:129–33.

[55] Kigondu EVM, Rukunga GM, Gathirwa JW, Irungu BN, Mwikwabe NM, Amalemba GM, et al. Antiplasmodial and cytotoxicity activities of some selected plants used by the Maasai community, Kenya. S Afr J Bot 2011;77:725–9.

[56] Bero J, Ganfon H, Jonville M-C, Frédérich M, Gbaguidi F, DeMol P, et al. *In vitro* antiplasmodial activity of plants used in Benin traditional medicine to treat malaria. J Ethnopharmacol 2009;122:439–44.

[57] Van Dyk S, Griffiths S, van Zyl RL, Malan SF. The importance of including toxicity assays when screening plant extracts for antimalarial activity. Afr J Biotechnol 2009;8:5595–601.

[58] Babajide OJ, Mabusela WT, Green IR, Ameer F, Weitz F, Iwuoha EI. Phytochemical screening and biological activity studies of five South African indigenous medicinal plants. J Med Plants Res 2010;2:1924–32.

[59] Ouattara L, Koudou J, Karou DS, Giaco L, Capelli G, Simpore J, et al. *In vitro* anti-*Mycobacterium tuberculosis* H37Rv activity of *Lannea acida* A. Rich from Burkina Faso. Pak J Biol Sci 2011;14:47–52.

[60] Gathirwa JW, Rukunga GM, Mwitari PG, Mwikwabe NM, Kimani CW, Muthaura CN, et al. Traditional herbal antimalarial therapy in Kilifi district, Kenya. J Ethnopharmacol 2011;134:424–34.

[61] Ajaiyeoba EO, Abiodun OO, Falade MO, Ogbole NO, Ashidi JS, Happi CD, et al. *In vitro* cytotoxicity studies of 20 plants used in Nigerian antimalarial ethnomedicine. Phytomedicine 2006;13:295–8.

[62] Gadir SA. Assessment of bioactivity of some Sudanese medicinal plants using brine shrimp (*Artemia salina*) lethality assay. J Chem Pharm Res 2012;4:5145–8.

[63] Prozesky EA, Meyer JJM, Louw AI. *In vitro* antiplasmodial activity and cytotoxicity of ethnobotanically selected South African plants. J Ethnopharmacol 2001;76:239–45.

[64] Sibandze G, van Zyl RL, van Vuuren SF. The anti-diarrhoeal properties of *Breonadia salicina*, *Syzygium cordatum*, and *Ozora sphaerocarpa* when used in combination in Swazi traditional medicine. J Ethnopharmacol 2010;132:506–11.

[65] Malebo HM, Tanja W, Cal M, Swaleh SAM, Omolo MO, Hassanali A, et al. Antiplasmodial, anti-trypanosomal, anti-leishmanial and cytotoxic activity of selected Tanzanian medicinal plants. Tanzania J Health Res 2009;11:226–34.

[66] McGaw LJ, Van der Merwe D, Eloff JN. *In vitro* anthelmintic, antibacterial and cyto-toxic effects of extracts from plants used in South African ethnoveterinary medicine. Vet J 2007;173:366–72.

[67] Malebo HM, Wenzler T, Cal M, Swaleh SM, Omolo MO, Hassanali A, et al. Anti-protozoal activity of aporphine and protoberberine alkaloids from *Annickia kummeriae* (Engl. & Diels) Setten & Maas (Annonaceae). BMC Complementary Altern Med 2013;13:48.

[68] More G, Tshikalange TE, Lall N, Botha F, Meyer JJM. Antimicrobial activity of medicinal plants against oral microorganisms. J Ethnopharmacol 2008;119:473–7.

[69] Lamidi M, DiGiorgio C, Delmas F, Favel A, Mve-Mba CE, Rondi ML, et al. *In vitro* cytotoxic, antileishmanial, and antifungal activities of ethnopharmacologically selected Gabonese plants. J Ethnopharmacol 2005;102:185–90.

[70] Kuete V, Krusche B, Youns M, Voukeng I, Fankam A, Tankeo S, et al. Cytotoxicity of some Cameroonian spices and selected medicinal plants. J Ethnopharmacol 2011;134:803–12.

[71] Adamu M, Naidoo V, Eloff JN. Some Southern African plant species used to treat hel-minth infections in ethnoveterinary medicine have excellent antifungal activity. BMC Complementary Altern Med 2012;12:213.

[72] Gebre-Mariam T, Neubert R, Schmidt PC, Wutzler P, Schmitke M. Antiviral activities of some Ethiopian medicinal plants used for the treatment of dermatological disorders. J Ethnopharmacol 2006;104:182–7.

[73] Bagla VP, McGaw LJ, Eloff JN. The antiviral activity of six South African plants tradition-ally used against infections in ethnoveterinary medicine. Vet Microbiol 2012;155:198–206.

[74] Tolo FM, Rukunga GM, Muli FW, Njagi ENM, Njue W, Kumon K, et al. Anti-viral activity of the extracts of a Kenyan medicinal plant *Carissa edulis* against herpes sim-plex virus. J Ethnopharmacol 2006;104:92–9.

[75] Lenta BN, Sénécheau CV, Soh RF, Tantangmo F, Ngouela S, Kaiser M, et al. *In vitro* antiprotozoal activities and cytotoxicity of some selected Cameroonian medicinal plants. J Ethnopharmacol 2007;111:8–12.

[76] Luo X, Pires D, Aínsa JA, Gracia B, Duarte N, Mulhovo S, et al. Antimycobacterial evaluation and preliminary phytochemical investigation of selected medicinal plants traditionally used in Mozambique. J Ethnopharmacol 2011;137:114–20.

[77] Koko WS, Osman EE, Galal M. Antioxidant and antiglycation potential of some Sudanese medicinal plants and their isolated compounds. Boletin Latinoamericano ydel Caribo de Plantas Medicinales y Aromaticas 2009;8:402–11.

[78] Omale J, Nnacheta OP, Okpara M. Cytotoxicity and antioxidant screening of some selected Nigerian medicinal plants. Asian J Pharm Clin Res 2009;2:48–53.

[79] Ayuko TA, Njau RN, Cornelius W, Leah N, Ndiege IO. *In vitro* antiplasmodial activity and toxicity assessment of plant extracts used in traditional malaria therapy in the Lake Victoria region. Mem Inst Oswaldo Cruz Rio de Janeiro 2009;104:689–94.

[80] Ganfon H, Bero J, Tchinda AT, Gbaguidi F, Gbenou J, Moudachirou M, et al. Antiparasitic activity of two sesquiterpenic lactone isolated from *Acanthospermum his-pidum* D.C. J Ethnopharmacol 2012;141:411–7.

[81] Sanon S, Azas N, Gasquet M, Ollivier E, Mahiou V, Barro N, et al. Antiplasmodial activity of alkaloid extracts from *Pavetta crassipes* (K. Schum) and *Acanthospermum hispidum* (DC) two plants used in traditional medicine in Burkina Faso. Parasitol Res 2003;90:314–7.

[82] Lusakibanza M, Mesia G, Tona G, Karemere S, Lukuka A, Tits M, et al. *In vitro* and *in vivo* antimalarial and cytotoxic activity of five plants used in Congolese traditional medicine. J Ethnopharmacol 2010;129:398−402.

[83] More G, Lall N, Hussein A, Tshikalange TE. Antimicrobial constituents of *Artemisia afra* Jacq.ex Willd. against periodontal pathogens. Evidence-Based Complementary Altern Med 2012; Article 252758, 7 pages.

[84] Ortet R, Prado S, Mouray E, Thomas OP. Sesquiterpene lactones from the endemic Cape Verdean *Artemisia gorgonum.* Phytochemistry 2008;69:2961−5.

[85] McGaw LJ, Steenkamp V, Eloff JN. Evaluation of *Athrixia* bush tea for cytotoxicity, antioxidant activity, caffeine content and presence of pyrrolizidine alkaloids. J Ethnopharmacol 2007;110:16−22.

[86] Muganga R, Angenot L, Tits M, Frédérich M. Antiplasmodial and cytotoxic activities of Rwandan medicinal plants used in the treatment of malaria. J Ethnopharmacol 2010;128:52−7.

[87] El Zalabani SM, Hetta MH, Ismail AS. Genetic profiling, chemical characterization and biological evaluation of two *Conyza* species growing in Egypt. J Appl Pharm Sci 2012;2:54−61.

[88] Edziri HL, Laurent G, Mahjoub A, Mastouri M. Antiviral activity of *Conyza canadensis* (L.) Cronquist extracts grown in Tunisia. Afr J Biotechnol 2011;10:9097−100.

[89] Adjatin A, Dansi A, Badoussi E, Loko YL, Dansi M, Azokpota P, et al. Phytochemical screening and toxicity studies of *Crassocephalum rubens* (Juss. ex Jacq.) S. Moore and *Crassocephalum crepidioides* (Benth.) S. Moore consumed as vegetable in Benin. Int J Curr Microbiol Appl Sci 2013;2:1−13.

[90] Jansen O, Angenot L, Tits M, Nicolas JP, De Mol P, Nikiéma J-B, et al. Evaluation of 13 selected medicinal plants form Burkina Faso for their antiplasmodial properties. J Ethnopharmacol 2010;130:143−50.

[91] Jansen O, Tits M, Angenot L, Nicolas J-P, De Mol P, Nikiema J-P, et al. Antiplasmodial activity of *Dicoma tomentosa* (Asteraceae) and identification of urospermal A-15-*O*-acetate as the main active compound. Malar J 2012;11:289.

[92] Nanyonga SK, Opoku AR, Lewu FB, Oyedeji OO, Singh M, Oyedeji AO. Antioxidant activity and cytotoxicity of the leaf and bark extracts of *Tarchonanthus camphoratus.* Trop J Pharm Res 2013;12:377−83.

[93] Irungu BN, Rukunga GM, Mungai GM, Muthaura CN. *In vitro* antiplasmodial and cytotoxicity activities of 14 medicinal plants from Kenya. S Afr J Bot 2007; 73:204−7.

[94] Ilboudo DP, Basilico N, Parapini S, Corbett Y, D'Alessandro S, Dell'Agli M, et al. Antiplasmodial and anti-inflammatory activities of *Canthium henriquesianum* (K. Schum), a plant used in traditional medicine in Burkina Faso. J Etnopharmacol 2013;148:763−9.

[95] Zofou D, Kengne ABO, Tene M, Ngemenya MN, Tane P, Titanji VPK. *In vitro* antiplasmodial activity and cytotoxicity of crude extracts and compounds from the stem bark of *Kigelia africana* (Lam.) Benth (Bignoniaceae). Parasitol Res 2011;108: 1383−90.

[96] Olounladé PA, Azando EVB, Hounzangbé-Adoté MS, Ha TBT, Leroy E, Moulis C, et al. *In vitro* antihelmintic activity of the essential oils of *Zanthoxylum zanthoxyloides* and *Newbouldia laevis* against *Strongyloides ratti.* Parasitol Res 2012; 110:1427−33.

[97] Paraskeva MP, van Vuuren SF, van Zyl RL, Davids H, Viljoen AM. The *in vitro* biological activity of selected South African *Commiphora* species. J Ethnopharmacol 2008;119:673−9.

[98] Hoet S, Opperdoes F, Brun R, Adjakidjé V, Quetin-Leclercq J. *In vitro* antiplasmodial activity of ethnopharmacologically selected Beninese plants. J Ethnopharmacol 2004;91:37−42.

[99] Muthaura CN, Rukunga GM, Chhaba SC, Omar SA, Guantai AN, Gathirwa JW, et al. Antimalarial activity of some plants used in treatment of malaria in Kwale district of Kenya. J Ethnopharmacol 2007;112:545−51.

[100] Inoue M, Ohtani K, Kasai R, Okukubo M, Andriantsiferana M, Yamasaki K, et al. Cytotoxic 16-β-((-xylopyranosyl)oxy)oxohexadecanyl triterpene glycosides from a Malagasy plant, *Physena sessilifolia*. Phytochemistry 2009;70:1195−202.

[101] Mwitara PG, Ayeka PA, Ondicho J, Matu EN, Bii CC. Antimicrobial activity and probable mechanisms of action of medicinal plants of Kenya: *Withania somnifera*, *Warbugia ugandensis*, *Prunus africana* and *Plectrunthus barbatus*. PLoS ONE 2013;8:e65619.

[102] Ahmed AS, McGaw LJ, Eloff JN. Evaluation of pharmacological activities, cytotoxicity and phenolic composition of four *Maytenus* species used in southern African traditional medicine to treat intestinal infections and diarrhoeal diseases. BMC Complementary Altern Med 2013;13:100.

[103] Gansane A, Sanon S, Ouattara PL, Hutter S, Ollivier E, Azas N, et al. Antiplasmodial activity and cytotoxicity of semi purified fractions from *Zanthoxylum zanthoxyloides* Lam. bark of trunk. Int J Pharmacol 2010;6:921−5.

[104] Azas N, Laurencin N, Delmas F, Di Giorgio C. Synergistic *in vitro* antimalarial activity of plant extracts used as traditional herbal remedies in Mali. Parasitol Res 2002;88:165−77.

[105] Abiodun OO, Gbotosho GO, Ajaiyeoba EO, Brun R, Oduola AM. Antitrypanosomal activity of some medicinal plants from Nigerian ethnomedicine. Parasitol Res 2012;110:521−6.

[106] Mustafa, Valentin A, Benoit-Vical F, Pélissier Y, Koné-Bamba D, Mallié M. Antiplasmodial activity of plant extracts used in West African traditional medicine. J Ethnopharmacol 2000;73:145−51.

[107] Adewunmi CO, Agbbedahunsi JM, Adebajo AC, Aladesanmi AJ, Murphy N, Wando J. Ethno-veterinary medicine: screening of Nigerian medicinal plants for trypanosomal properties. J Ethnopharmacol 2001;77:19−24.

[108] Tshikalange TE, Meyer JJM, Hussein AA. Antimicrobial activity, toxicity and isolation of a bioactive compound from plants used to treat sexually transmitted diseases. J Ethnopharmacol 2005;96:515−9.

[109] Mbwambo ZH, Erasto P, Nondo RSO, Innocent E, Kidukuli AW. Antibacterial and cytotoxic activities of *Terminalia stenostachya* and *Terminalia spinosa*. Tanzania J Health Res 2011;13:1−8.

[110] Garon D, Chosson E, Rioult J-P, de Pecoulas PE, Brasseur P, Vérite P. Poisoning by *Gnestis ferruginea* in Casamance (Senegal): an etiological approach. Toxicon 2007;50:189−95.

[111] James O, Nnacheta OP, Wara HS, Aliyu UR. *In vitro* and *in vivo* studies on the antioxidative activities, membrane stabilization and cytotoxicity of water spinach (*Ipomoea aquatic* Forsk) from Ibaji Ponds, Nigeria. Int J Pharm Tech Res 2009;1:474−82.

[112] Benarba B, Meddah B, Aoues A. *Bryonia dioica* aqueous extract induces apoptosis through mitochondrial intrinsic pathway in BL41 Burkitt's lymphoma cells. J Ethnopharmacol 2012;141:510–6.

[113] Mbaveng A, Kuete V, Mapunya BM, Beng VP, Nkengfack AE, Meyer JJM, et al. Evaluation of four Cameroonian medicinal plants for anticancer, antigonorrheal and antireverse transcriptase activities. Environ Toxicol Pharmacol 2011;32:162–7.

[114] Oyedemi SO, Oyedemi BO, Arowosegbe S, Afolayan AJ. Phytochemical analysis and medicinal potentials of hydroalcoholic extract from *Curtisia dentata* (Burm.f) C.A. Sm stem bark. Int J Mol Sci 2012;13:6189–203.

[115] Garcia-Alvarez M-C, Moussa I, Soh PN, Nongonierma R, Abdoulaye A, Nicolau-Travers M-L, et al. Both plants *Sebastiana chamaelea* from Niger and *Crozophora senegalensis* from Senegal used in African traditional medicine in malaria treatment share a same active principle. J Ethnopharmacol 2013;149:676–84.

[116] Cho-Ngwa F, Abongwa M, Ngemenya MN, Nyongbela KD. Selective activity of extracts of *Margaritaria discoidea* and *Homalium africanum* on *Onchocerca ochengi*. BMC Complementary Altern Med 2010;10:62.

[117] Soh PN, Banzouzi JT, Mangoombo H, Lusakibanza M, Bulubulu FO, Tona L, et al. Antiplasmodial activity of various parts of *Phyllanthus niruri* according to its geographical distribution. Afr J Pharm Pharmacol 2009;3:598–601.

[118] Mathabe MC, Hussein AA, Nikolova RV, Basson AE, Meyer JJM, Lall N. Antibacterial activities and cytotoxicity of terpenoids isolated from *Spirostachys africana*. J Ethnopharmacol 2008;116:194–7.

[119] Rahman AA, Samoylenko V, Jain SK, Tekwani BL, Khan SI, Jacob MR, et al. Antiparasitic and antimicrobial isoflavanquinones from *Abrus schimperi*. Nat Prod Commun 2011;6:1645–50.

[120] Ahmed AS, Elgorashi EE, Moodley N, McGaw LJ, Naidoo V, Eloff JN. The antimicrobial, antioxidative, anti-inflammatory activity and cytotoxicity of different fractions of four South African *Bauhinia* species used traditionally to treat diarrhea. J Ethnopharmacol 2012;143:826–39.

[121] Ntandou NGF, Banzouzi JT, Mbatchi B, Elion-Itou RDG, Etou-Ossibi AW, Ramos S, et al. Analgesic and anti-inflammatory effects of *Cassia siamea* Lam. stem bark extracts. J Ethnopharmacol 2010;127:108–11.

[122] Lekana-Douki JB, Liabagui SLO, Bongui JB, Zatra R, Lebibi J, Toure-Ndouo F. *In vitro* antiplasmodial activity of crude extracts of *Tetrapleura tetraptera* and *Copaifera religiosa*. BMC Res Notes 2011;4:506.

[123] Kaou AM, Mahiou-Leddet V, Canlet C, Debrauwer L, Hutter S, Laget M, et al. Antimalarial compounds from the aerial parts of *Flacourtia indica* (Flacourtiaceae). J Ethnopharmacol 2010;130:272–4.

[124] Zofou D, Kowa TK, Wabo HK, Ngemenya MN, Tane P, Titanji VPK. *Hypericum lanceolatum* (Hypericaceae) as a potential source of new anti-malarial agents: a bioassay guided-fractionation of the stem bark. Malar J 2011;10(167).

[125] Mokoka TA, Peter XA, Fouche G, Moodley N, Adams M, Hamburger M, et al. Antileishmanial activity of 12-methoxycarnosic acid of *Salvia repens* Burch.ex Benth. (Lamiaceae). S Afr J Bot 2014;90:93–5.

[126] Samoylenko V, Jacob MR, Khan SI, Zhao J, Tekwani BL, Midiwo JO, et al. Antimicrobial, antiparasitic and cytotoxic spermine alkaloids from *Albizia schimperiana*. Nat Prod Commun 2009;4:791–6.

[127] Nondo RSO, Mbwambo ZH, Kidukuli AW, Innocent EM, Mihale MJ, Erasto P, et al. Larvicidal, antimicrobial and brine shrimp activities of extracts from *Cissampelos mucronata* and *Tephrosia villosa* from coast region, Tanzania. BMC Complementary Altern Med 2011;11:33.

[128] Karani LW, Tolo FM, Karanja SM, Khayeka-Wandabwa C. Safety of *Prunus africana* and *Warburgia ugandensis* in asthma treatment. S Afr J Bot 2013;88:183–90.

[129] Essop AB, van Zyl RL, van Vuuren SF, Mulholland D, Viljoen AM. The *in vitro* pharmacological activities of 12 South African *Hermannia* species. J Ethnopharmacol 2008;119:615–9.

[130] Obbo CJD, Makanga B, Mulholland DA, Coombes PH, Brun R. Antiprotozoal activity of *Khaya anthotheca* (Welv.) C.D.C. a plant used by chimpanzees for self-medication. J Ethnopharmacol 2013;147:220–3.

[131] Okoli S, Iroegbu CU. *In vitro* antibacterial activity of *Synclisa scabrida* whole root extracts. Afr J Biotechnol 2005;4:946–52.

[132] Kazeem MI, Akanji MA, Hafizur Rahman M, Choudhary MI. Antiglycation, antioxidant and toxicological potential of polyphenol extracts of alligator pepper, ginger and nutmeg from Nigeria. Asian Pac J Trop Biomed 2012;2:727–32.

[133] Gescher K, Kühn J, Lorentzen E, Hafezi W, Derksen A, Deters A, et al. Proanthocyanidin-enriched extract from *Myrothamnus flabellifolia* Welw. exerts antiviral activity against herpes simplex virus type 1 by inhibition of viral adsorption and penetration. J Ethnopharmacol 2011;134:468–74.

[134] Ndongo JT, Shaaban M, Mbing JN, Bikobo DN, Atchadé AT, Pegnyemb DE, et al. Phenolic dimers and an indolealkaloid from *Campylospermum flavum* (Ochnaceae). Phytochemistry 2010;71:1872–8.

[135] Badria FA, Ameen M, Akl MR. Evaluation of cytotoxic compounds from *Calligonum comosum* L. growing in Egypt. Z Naturforsch 2007;62:656–60.

[136] Bero J, Hérent M-F, Schmeda-Hirschmann G, Frédérich M, Leclercq JQ. *In vivo* antimalarial activity of *Keetia leucantha* twigs extracts and *in vitro* antiplasmodial effect of their constituents. J Ethnopharmacol 2013;149:176–83.

[137] Benoit-Vical F, Soh PN, Saléry M, Harguem L, Poupat C, Nongonierma R. Evaluation of Senegalese plants used in malaria treatment: focus on *Chrozophora senegalensis*. J Ethnopharmacol 2008;116:43–8.

[138] Liu W, Di Giorgio C, Lamidi M, Elias R, Ollivier E, De Méo MP. Genotoxic and clastogenic activity of some saponins extracted from *Nauclea* bark as assessed by the micronucleus and the comet assays in Chinese Hamster Ovary cells. J Ethnopharmacol 2011;137:176–83.

[139] Mesia K, Cimanga RK, Dhooghe L, Cos P, Apers S, Totté J, et al. Antimalarial activity and toxicity evaluation of a quantified *Nauclea pobeguinii* extract. J Ethnopharmacol 2010;131:10–6.

[140] Mwangi ESK, Keriko JM, Machocho AK, Wanyonyi AW, Malebo HM, Chhabra SC, et al. Antiprotozoal activity and cytotoxicity of metabolites from leaves of *Teclea trichocarpa*. J Med Plants Res 2010;4:726–31.

[141] Orwa JA, Ngeny L, Mwikwabe NM, Ondicho J, Jondiko IJO. Antimalarial and safety evaluation of extracts from *Toddalia asiatica* (L) Lam. (Rutaceae). J Ethnopharmacol 2013;145:587–90.

[142] Luo X, Pires D, Aínsa JA, Gracia B, Duarte N, Mulhovo S, et al. *Zanthoxylum capense* constituents with antimycobacterial activity against *Mycobacterium tuberculosis in vitro* and *ex vivo* within human macrophages. J Ethnopharmacol 2013;146:417−22.

[143] Sonibare MA, Oloyede GK, Adaramola TF. Antioxidant and cytotoxicity evaluations of two species of *Blighia* providing clues to species diversity. Electronic J Environ Agric Food Chem 2011;10:2960−71.

[144] Sonibare OO, Effiong I. Antibacterial activity and cytotoxicity of essential oil of *Lantana camara* L. leaves from Nigeria. Afr J Biotechnol 2008;7:2618−20.

[145] El-Kousy S, Mohamed M, Mohamed S. Phenolic and biological activities of *Vitex trifolia* aerial parts. Life Sci J 2012;9:670−7.

[146] Nganso YOD, Ngantchou IEW, Nkwenoua E, Nyasse B, Denier C, Hannert V, et al. Antitrypanosomal and cytotoxic activities of 22-hydroxyclerosterol, a new sterol from *Allexis cauliflora* (Violaceae). Sci Pharm 2011;79:137−44.

[147] Smati D, Longeon A, Guyot M. 3,β-(3, 4-Dihydroxycinnamoyl)-erythrodiol, a cytotoxic constituent of *Zygophyllum geslini* collected in the Algerian Sahara. J Ethnopharmacol 2004;95:405−7.

9 Genotoxicity and Teratogenicity of African Medicinal Plants

Armel Jackson Seukep, Jaures A.K. Noumedem, Doriane E. Djeussi and Victor Kuete

Faculty of Science, Department of Biochemistry, University of Dschang, Dschang, Cameroon

9.1 Introduction

Genotoxicity is the ability of chemicals to damage the genetic information within a cell resulting in mutations, which may lead to malignancies. The genotoxic substances induce damage to the genetic material in the cells through interactions with the DNA sequence and structure. Most often, genotoxicity is confused with mutagenicity as all mutagenic chemicals are genotoxic; however, not all genotoxic compounds are mutagenic. Teratogenic substances induce abnormalities of physiological development and the study involves human congenital abnormalities, but it is much broader than that, taking in other nonbirth developmental stages, including puberty, and other nonhuman life forms, including plants. The developmental toxicity takes in account all manifestations of abnormal development by toxic substances and can either be caused toxic substances (xenobiotics including plant-based drugs), transmitted infection, lack of nutrients (e.g., lack of folic acid in nutrition during pregnancy for humans can result in spina bifida), physical restraint (e.g., Potter syndrome due to oligohydramnios in humans), and genetic disorders. Worldwide and especially in Africa, the population does not pay enough attention to the toxic potencies of medicinal plants contrary to other chemicals, believing that, if these products have been used so far, they should be devoid of toxicity [1−6]. However, compiled data highlighting the genotoxicity and teratogenicity of African medicinal plants are scarce, though there are several scientific publications documenting the effects of individual plants. This chapter aims not only to review the genotoxicity and teratogenicity of the medicinal plants used in African traditional medicine (ATM), but also to bring together information regarding the ability of plant used in the continent to display genoprotective effects.

Toxicological Survey of African Medicinal Plants. DOI: http://dx.doi.org/10.1016/B978-0-12-800018-2.00009-1

9.2   Methods Used in Genotoxicity Assays

The investigation of DNA damage in cells exposed to the toxic substrates is the main way to assess the genotoxicity of a chemical. This DNA damage can be in the form of single- and double-strand breaks, loss of excision repair, cross-linking, alkali-labile sites, point mutations, and structural and numerical chromosomal aberrations [7]. Pronounced alteration or a compromised integrity of chromosome can lead to various types of cancers. Therefore, many sophisticated techniques including the Ames assay, *in vitro* and *in vivo* toxicology tests, and the Comet assay have been developed to assess the chemicals' potential to cause DNA damage that may lead to cancer. Table 9.1 below summarizes the different validated nonanimal alternatives used in genotoxicity screening of chemicals as retrieved from AltTox.org database [7].

The genotoxicity testing is performed in bacterial, yeast, and mammalian cells and is aimed at determining if a substrate will influence genetic material or may cause cancer [7]. In laboratories, the test for gene mutation is based on the Bacterial Reverse Mutation Assay, also known as the Ames assay. In this assay, different bacterial strains are used in order to compare the different changes in the genetic material. This has led to the detection of the majority of genotoxic carcinogens and genetic changes and the types of mutations detected are frame shifts and base substitutions [8].

The *in vitro* toxicology testing is used to determine whether a substrate, product, or environmental factor induces genetic damage. This technique detects chromatid

Table 9.1 Validated Nonanimal Alternatives Methods in Genotoxicity Study of Chemicals [7]

Methods	Test Purpose
In vitro mammalian cell micronucleus test	Alternative to the *in vitro* chromosome aberration assay for genotoxicity testing
Bacterial reverse mutation test (Ames Test)	Genotoxicity testing—using bacterial cells
Genetic toxicology: *Saccharomyces cerevisiae* gene mutation assay	Genotoxicity testing—using bacterial cells
Genetic toxicology: *Saccharomyces cerevisiae* mitotic recombination assay	Genotoxicity testing—using bacterial cells
In vitro mammalian cell gene mutation test	Genotoxicity testing—using mammalian cells
Genetic toxicology: *in vitro* sister chromatid exchange assay in mammalian cells	Genotoxicity testing—using mammalian cells
Genetic toxicology: DNA damage and repair, unscheduled DNA synthesis in mammalian cells *in vitro*	Genotoxicity testing—using mammalian cells

and chromosome gaps, chromosome breaks, chromatid deletions, fragmentation, translocation, complex rearrangements, and many more [8].

The *in vivo* testing determines the potential of DNA damage that can affect chromosomal structure or disturb the mitotic apparatus that changes chromosome number. Most often, the Comet assay is used *in vivo* to detect the genotoxicity of a substance. The technique involves cells lysis using detergents and salts. The DNA released from the lysed cells is electrophoresed in an agarose gel under neutral pH conditions. Cells containing DNA with an increased number of double-stranded breaks will migrate more quickly to the anode [9,10].

9.3   Methods Used in Teratogenicity Assays

Teratogenesis is the disturbed growth process involved in the production of a malformed neonate. There are six principles of teratology as defined by Wilson since 1959 [11]. These principles guide the study and understanding of teratogenic agents and their effects on developing organisms:

- Susceptibility to teratogenesis depends on the genotype of the conceptus and the manner in which this interacts with adverse environmental factors.
- Susceptibility to teratogenesis varies with the developmental stage at the time of exposure to an adverse influence. There are critical periods of susceptibility to agents and organ systems affected by these agents.
- Teratogenic agents act in specific ways on developing cells and tissues to initiate sequences of abnormal developmental events.
- The access of adverse influences to developing tissues depends on the nature of the influence. Several factors affect the ability of a teratogen to contact a developing conceptus, such as the nature of the agent itself, route, and degree of maternal exposure, rate of placental transfer and systemic absorption, and composition of the maternal and embryonic/fetal genotypes.
- There are four manifestations of deviant development (death, malformation, growth retardation, and functional defect).
- Manifestations of deviant development increase in frequency and degree as dosage increases from the no observable adverse effect level (NOAEL) to a dose producing 100% lethality (LD_{100}).

The screenings of the potential teratogenicity of chemicals or environmental agents use animal model systems such as rat, mouse, rabbit, dog, and monkey. In the past, pregnant animals were exposed to the chemical and their fetuses were observed for gross visceral and skeletal abnormalities. Though this is still part of the teratological evaluation procedures today, the field of teratology is moving to a more molecular level, seeking the mechanism(s) of action by which these agents act. Animals such as genetically modified mice are commonly used nowadays. The study of the teratogenicity of xenobiotics is important in preventing congenital abnormalities and also has the potential for developing new safe therapeutic drugs for use with pregnant women.

9.4   Medicinal Plants with Genotoxic Effects

Pyrrolizidine alkaloids (PAs) are a well-known example of a genotoxic substance causing DNA damage. The transition metal chromium interacts with DNA in its high-valent oxidation state so to incur DNA lesions, leading to carcinogenesis. Medicinal plants worldwide containing PAs and several other metabolites were found to be genotoxic. Taylor et al. [12] demonstrated genotoxicity of many medicinal plants from South Africa. They used samples made of methanol and dichloreomethane extracts of the following plants: leaves of *Balanites maughamii*, *Catharanthus roseus* (Apocyanaceae), *Catunaregam spinosa*, *Chaetacme aristata* (Rubiaceae), *Plumbago auriculata* (Plumbaginaceae), *Turraea floribunda* (Meliaceae), *Vernonia colorata* (Asteraceae), *Trichelia emetica* (Meliaceae) and *Rhamnus prinoides* (Rhamnaceae), bark of *Ocotea bulata* (Lauraceae), *Prunus africana* (Rosaceae), *Boophane disticha* (Amaryllidaceae), *Combretum mkhzense* (Combretaceae), *Crinum macowanii* (Amaryllidaceae), *Erythrina caffra* (Leguminosae), *Hypoxis colchicifolia*, *Hypoxis hemerocallidea* and *Ornithogalum longibracteatum* (Hyacinthaceae), *Ricinus comminus* (Euphorbiaceae), *Scilla natalensis* (Liliaceae), *Sclerocarya birrea* (Anacardiaceae), *Trichelia emetic* (Meliaceae); twigs and leaves of *Diospyros whyteana* (Ebenaceae); twigs and bark of *Spirostachys africana*, *Croton sylvaticus* (Euphorbiaceae), and *Gardenia volkensii* (Rubiaceae). They demonstrated in Alkaline Comet/Single Cell Gel Electrophoresis (SCGE) Assay that these plants exert DNA damage in cells at concentration of 250 and 500 ppm. Some of the extracts also displayed genotoxicity by increasing the number of micronuclei in dividing human blood cells. It has been shown that *Arictum lappa* root extract induces chromosomal aberrations, micronuclei (MNC) formations in *Allium cepa* root tip cells. It also induced mitotic spindle disturbance and presented mitodepressive effects in *Allium cepa* by decreasing the mitotic index with more pronounced effects at 62.5 and 125 mg/mL [13]. It was shown that the aqueous extracts of *Azadirachta indica* (A. Juss), *Morinda lucida* (Benth.), *Cymbopogon citratus* (DC Stapf.), *Mangifera indica* (Linn.), and *Carica papaya* (Linn.) displayed mitodepressive effects on cell division and induced mitotic spindle disturbance in *Allium cepa*. *Senna alata*, already known for its toxicity on liver, intestine, and blood components [14], was shown recently as a potent mutagen [15] as it increased (in presence of metabolic activation system, namely S9 mixture and made up of microsome fraction of rats liver) the number of revertant strains of Salmonella *typhimrium* TA98 and TA1537 in the mutagenicity assay [10], while genotoxicity of *Jatropha multifida*, *Sterculia africana*, and *Spirostachys Africana* were demonstrated by vititox assay in absence of S9 [16]. A synopsis of African medicinal plants screened for their genotoxicity is summarized in Table 9.2.

Some African medicinal plants also are known for their protective effects of the genome. These include *Acacia salicina* (Mimosoideae), *Marrubium deserti* de Noé (Lamiaceae), *Myrtus communis* (Myrtaceae), *Pistacia lentiscus* (Anacardiaceae), *Pituranthos chloranthus* (Apiaceae), *Rosa roxburghii* (Rosaceae), and *Teucrium ramosissimum* (Lamiaceae) (Table 9.3).

Table 9.2 African Medicinal Plants Screened for their Genotoxicity

Plant (Family)	Traditional Use	Area of Plant Collection	Chemical Constituents	Effects of the Plant
Antidesma venosum (Euphorbiaceae)	Abdominal pain, enema [17,18].	South Africa	Alkaloids, phenols [19]; saponins, tannins, flavonoids [20]; 3,5,7,3′,4′-pentahydroxyflavanol (epicatechin), 5,7,8,2′,3′,4′-hexahydrodihydroflavanol-4-O-β-ᴅ-glucoside, 5,7,4′-trihydroxyl-isoflavone (genistein) [21]; γ-lactone ((3R,4R,5S)-4-hydroxy-5-methyl-3-tetradecanyl γ-lactone), β-sitosterol, triterpenoids (friedelin, lupeol), phytosterols, stigmasterol [22].	DNA damage: The DCM leaf extract showed positive result in micronucleus test while the twigs extract was positive in micronucleus and Comet tests without S9 (human white blood cells *in vitro*) [12].
Balanites maughamii (Balanitacea)	Mulluscicidal [17,18].	South Africa	Saponins (diogenin, sapogenins), steroidal glycosides [23], steroids, terpenoids, terpenes, flavonoids, organic acids (vanilic acid, syringic acid), furanocoumarins (marmesin, bergapten), hydroxy cholestenones (cryptopgenin) [24].	The DCM and MeOH extracts of leaves and twigs induced DNA damage according to micronucleus and Comet tests without S9 in human lymphocyte cultures [12].

(Continued)

Table 9.2 (Continued)

Plant (Family)	Traditional Use	Area of Plant Collection	Chemical Constituents	Effects of the Plant
Boophane disticha (Amaryllidaceae)	Dressing for cuts, boils, septic wounds, headaches, abdominal pain, weakness, eye conditions, sedative [17,18].	South Africa	Alkaloids (buphanadrine, buphanamine, distichamine [25], 6α-hydroxycrinamine, 6β-hydroxycrinamine) [26].	The MeOH extract of bark induced micronuclei while the DCM extract was toxic suggesting an indirect effect on cell division, or other toxic processes according to micronucleus test without S9 in human lymphocyte cultures [12].
Catha edulis (Celastraceae)	Antalgic [27].	Ethiopia	Alkaloids (kathine, cathinine, cathidine, D-norpseudoephedrine, cathinone) [28], tannins (tannic acid), mannitol, flavonoids, volatile oils [29].	The MeOH leaves extract induced a significant increase in sister chromatid exchanges. Moreover, it induced various types of chromosomal aberrations in mice bone marrow cells which include: broken, sticky and ring chromosomes and disturbed metaphase and anaphase [30].
Catharanthus roseus (Apocyanaceae)	Diabetes, rheumatism [17,18] anti-inflammatory, antimalarial, antimitotic, antihypertensive, antifertility, antihypercholesterolemic, antimutagenic, antidiuretic,	South Africa	DCM-MeOH whole plant extract: alkaloids (vincristine vinblastine, ajmalicine, catharanthine, vindolinine), carbohydrates [32,33], saponins, flavonoids, anthraquinone glycosides [33].	The MeOH leaf extract induced micronuclei in micronucleus and Comet tests without S9 while the DCM extract was found toxic in micronucleus test suggesting an (in)direct effect on cell division, or other

Plant species (family)	Ethnomedicinal uses	Country	Phytochemicals	Toxicity
	antifungal, antispasmodic, antiviral, cardio tonic, central nervous system depressant, antitumor, cytotoxic, antispermatogenic, anticancer [31].			toxic processes in human lymphocyte cultures [12]. The MeOH and DCM leaf extracts induced mutant strains in Ames test with metabolic activation S9 (*Salmonella typhimurium* strain TA98) [34].
Catunaregam spinosa (Rubiaceae)	Emetic, fever, aphrodisiac, gonorrhea, headaches, nausea, respiratory and febrile complaints, gynecological ailments, epilepsy, arthritis [17,18].	South Africa	Triterpenoid, alkaloids, glycosides, flavonoids, saponins, carbohydrates, tannins, phenols [35]. Triterpenoid saponins (Catunaroside, Catunaroside F, Catunaroside H, Mussaendoside J) [36].	The DCM leaves extract induced micronuclei in micronucleus without exogenous metabolic activation S9 indicating DNA damage (human lymphocyte cultures) [12].
Celtis africana (Ulmaceae)	Cancer, syphilis, rhumatism, pains, magical properties [17,18].	South Africa	Trans-*N*-coumaroyltyramine, trans-*N*-feruloyltyramine, trans-*N* caffeoyltyramine, lauric acid, oleic acid, palmitic acid, lupeol, β-sitosterol, oleanolic acid [37], polyphenols, flavonoids [38].	The MeOH root extract induced micronuclei in micronucleus test without S9 in human lymphocyte cultures [12].
Chaetacme aristata (Ulmaceae)	Hemorrhoids [17,18].	South Africa	/	DNA damage in micronucleus and Comet tests without S9 in human lymphocyte cultures [12]. The DCM leaves extract showed mutagenicity in Ames test with and without metabolic activation (*Salmonella typhimurium* strain TA98) [34].

(Continued)

Table 9.2 (Continued)

Plant (Family)	Traditional Use	Area of Plant Collection	Chemical Constituents	Effects of the Plant
Cleome amblyocarpa (Capparidaceae)	Colic, diabetes, pains [39].	Tunisia	Dammarane triterpenes, cleocarpanol [40], cabraleahydroxy lactone [41], stigma-4-en-3-one, lupeol, taraxasterol [42].	Single and double strand DNA breaks as well as alkali-labile sites at 0–1 µg/mL concentrations of the plant extract according to the alkaline Comet assay [39].
Crinum macowanii (Amaryllidaceae)	Scrofula, rheumatic fever, kidney and bladder diseases, fever, sores, glandular swellings [17,18].	South Africa	Lycorine, crinamine [43]; macowine, cherylline, pratorimine [44] buphanidrine [45].	The MeOH bark extract induced micronuclei in micronucleus test without metabolic activation S9 in human lymphocyte cultures [12]. The DCM bark extract increased the number of His$^+$ revertants in Ames test (*Salmonella typhimurium* strain TA98) with and without metabolic activation [34].
Croton sylvaticus (Euphorbiaceae)	Abdominal and internal inflammations, uterine disorders, tonic, febrile conditions, purgative, pleurisy, indigestion, TB, rheumatism [17,18].	South Africa	/	The DCM and MeOH (leaves, twigs and bark) extracts induced micronuclei in micronucleus test without exogenous metabolic activation S9 in human lymphocyte cultures [12].

Plant species (family)	Medicinal uses	Phytochemical compounds	Region	Genotoxicity / mutagenicity
Diospyros whyteana (Ebenaceae)	Dysmenorrhea, irritating rashes, antibacterial [17,18].	Isodiospyrin [46].	South Africa	Mutagenic effect of DCM twigs extract in Comet test and DNA damage with the MeOH extracts in micronucleus and Comet tests without S9 (human lymphocyte cultures) [12]. The DCM leaf extract induced mutant strains in Ames test with metabolic activation in *Salmonella typhimurium* strain TA98 [34].
Erythrina caffra (Fabaceae)	Sores, wounds, arthritis, sprains, aches [17,18].	Alkaloids [47], triterpènes [48], coumarins, steroids, flavonoids (isoflavones, pterocarpanes, flavanones and isoflavanones) [49].	South Africa	The MeOH bark extract induced micronuclei in micronucleus test without exogenous metabolic activation S9 in human lymphocyte cultures [12].
Euclea divinorum (Ebenacea)	Purgative, for headache, toothache, constipation, antihelmintics, tonics, chest pain, pneumonia, stomach pain [17,18].	alkaloids, saponins, diterpenes, tannins, phytosterols [50], DCM extract: triterpenoids, resins tannins, amino acids. DCM/MeOH extract: alkaloids, reducing sugar, saponins, triterpenoids, resins, phenols, tannins, amino acids, diterpenes. Water extract: alkaloids, reducing sugar, saponins, triterpenoids, phenols, flavonoids, amino acids [51].	South Africa	The extract induced micronuclei in micronucleus test without S9 (human lymphocyte cultures) [12].

(Continued)

Table 9.2 (Continued)

Plant (Family)	Traditional Use	Area of Plant Collection	Chemical Constituents	Effects of the Plant
Gardenia volkensii (Rubiacea)	Emetic, sore eyes, headaches, asthma, dysmenorrhea, infertility, epilepsy, convulsions, earache [17,18].	South Africa	Irridoids (4-(2N-gardenamide)n-butanoic acid, genipin, genipin gentiobioside, pterocarpin medicarpin), coumarins, phenylpropanoids, benzenoids, triterpenes [52].	Toxic effect with DCM leaves extract and DNA damage with DCM twigs/bark extracts in micronucleus test without S9. DNA damage of DCM and MeOH twigs/bark extracts in Comet assay (human white blood cells cultures) [12].
Heteromorpha trifoliata (Apiaceae)	Enemas, abdominal disorders, mental/nervous disorders, intestinal worms, headaches, antiscabies [17,18].	South Africa	Fulcarindiol, sarison [53].	The DCM twigs/bark extracts and MeOH leaves extract induced micronuclei in micronucleus test without metabolic activation S9. Mutagenic effect of DCM leaves extract in Comet test without S9 in human lymphocyte cultures [12].
Hypoxis colchicifolia (Hyacinthaceae)	Tonic, anti-HIV, anti-inflammatory [17,18].	South Africa	/	DNA damage of DCM and MeOH bark extracts in Comet and micronucleus tests, respectively (human lymphocyte cultures) [12].
Kigelia africana (Bignoniaceae)	Ulcers, sores, syphilis, rheumatism, enema [17,18].	South Africa	Irridoids (Verminoside, Specioside, Minecoside, Norviburtinal); flavonoids	The DCM fruits extract induced micronuclei in micronucleus test without exogenous

			(Luteolin, quercetin); naphtoquinone (Pinnatal, Isopinnatal, Kigelinone, Kigelinol, Isokigelinol, Lapachol); coumarins (Kigelin); phenylpropanoid (Caffeic acid, p-coumaric acid, Ferulic acid); fatty acids (Palmatic acid) [12].	metabolic activation S9 indicating DNA damage (human lymphocyte cultures) [12].
Marrubium alysson L (Lamiaceae)	Hypertension, rheumatics, cough, burns, intestinal troubles [54].	Tunisia	MeOH extract: alkaloids, diterpenoids, saponosids [55]	The ethyl acetate and MeOH extracts of aerials parts induced DNA damage according to the alkaline Comet assay in C3A cells [54].
Peganum harmala L (Zygophylaceae)	Cancer, hypothermia, hallucinogen factor, antidepressant, antimicrobial, antinociceptive [56].		alkaloids (harmaline, harmine), flavonoids, saponins, reducing compounds, tannins, volatile oils, triterpenes, sterols, anthraquinone [56].	The intercalation of P. harmala alkaloids into DNA has led to its mutagenic property which causes genotoxic effects [57].
Plumbago auriculata (Plumbaginaceae)	Headaches, emetics, warts, fractures, scrofula, edema, malaria, skin lesions [17,18].	South Africa	Hexane:ethyl acetate:MeOH:water (40:10:10:2, v/v): naphthoquinones (plumbagin, epi-isoshinanolone), steroids (sitosterol, 3-O-glucosylsitosterol), plumbagic, palmitic acids [58].	DCM, MeOH extracts of leaves, fruits, twigs cause dependent DNA damage in micronucleus and Comet assays without S9 (human lymphocyte cultures) [12]. The DCM foliage extract induced mutant strains in Ames test with and without metabolic activation (Salmonella typhimurium tester strain TA98) [34].

(Continued)

Table 9.2 (Continued)

Plant (Family)	Traditional Use	Area of Plant Collection	Chemical Constituents	Effects of the Plant
Prunus Africana (Rosaceae)	Intercostal pain, prostate hypertrophy, hair tonics [17,18].	South Africa	Amygdalin, β-sitosterol [59].	DNA damage with DCM twigs extract in micronucleus and Comet assays without exogenous metabolic activation (S9). Mutagenic effect of MeOH leaves extract in Comet test (human lymphocyte cultures) [12].
Rhamnus prinoides (Rhamnaceae)	Sprains, blood purifiers, pneumonia, emetics, purgative, colic, stimulants [17,18].	South Africa	Naphthalene glycoside (Geshoidin) [12].	Toxic effect of DCM twigs extract and DNA damage of DCM bark/twigs extracts in micronucleus assay without exogenous metabolic activation (S9). Mutagenic effect of MeOH extracts and DCM leaves extract in Comet test (human lymphocyte cultures) [12].
Scilla natalensis (Liliaceae)	Enema, sprains, fractures, purgative, boils, sores, infertility [17,18].	South Africa	/	The DCM and MeOH bark extracts induced micronuclei in micronucleus test without S9 (human lymphocyte cultures) [12].
Sclerocarya birrea (Anacardiaceae)	Diarrhea, dysentery, stomach problems, fever, malaria, tonic, diabetes [17,18].	South Africa	Flavonol glycoside, quercetin 3-O-α-L-(5''-galloyl)-arabinofuranoside, phenols, epicatechin derivatives; sesquiterpenes [60].	The MeOH bark induced micronuclei in micronucleus test without exogenous metabolic activation S9 indicating DNA damage (human lymphocyte cultures) [12].

Plant (Family)	Country	Traditional uses	Phytochemicals	Toxicity
Senecio serratuloides (Asteraceae)	South Africa	Cuts, swellings, burns, sores, blood purifiers, headaches [17,18].	/	The DCM leaves exhibit toxic effect in micronucleus test suggesting an (in)direct effect on cell division, or other toxic processes without exogenous metabolic activation S9 (human lymphocyte cultures) [12].
Senna alata (Fabaceae)	Nigeria	Constipation, malaria, stomach pains, diabetes, purgative, skin, infectious diseases [15].	Flavonoid glycoside [61], alkaloids, Saponins, tannins, anthracionones, carbohydrates [62].	The extract dose-dependently increased the number of revertants in *Salmonella typhymurium* TA98 and 1537 strains according to the Ames Test [15].
Spirostachys Africana (Euphorbiaceae)	South Africa	Wood: stomach ulcers, acute gastritis, eye washes, headaches, rashes, boils, emetic, renal ailment, purgative, blood purifiers, diarrhea, dysentery [17,18].	Diterpenes, 2-hydroxyketones, 2-Stachenone, stachenol, diosphenol [63].	The DCM and MeOH leaves/ bark/twigs extracts induced micronuclei in micronucleus test without addition of a metabolizing enzyme solution (S9 mix) at 500 and 100 mg/ mL (human lymphocyte cells) [12]. Mutagenic effect of MeOH leaves extract and DCM bark/ twigs extracts in Comet test without S9 (human lymphocyte cultures) [12].

(Continued)

Table 9.2 (Continued)

Plant (Family)	Traditional Use	Area of Plant Collection	Chemical Constituents	Effects of the Plant
Trichelia emetica (Meliaceae)	Stomach and intestinal complaints, dysentery, kidney problems, indigestion, parasites, fever, purgative, bruises, rheumatism [17,18].	South Africa	Steroids, triterpenoids, coumarins [64].	Appearance of micronuclei with DCM and MeOH bark extracts in micronucleus assay. DNA damage of MeOH leaves extract in Comet test without S9 (human lymphocyte cultures) [12].
Tulbaghia violaceae (Alliaceae)	Fever, colds, asthma, enemas, tuberculosis, stomach problems, rheumatism, paralysis [17,18].	South Africa	2,4-Dithiapentane, *p*-Xylene, Chloromethylmethyl sulfide, O-Xylene, Thiodiglycol, and *p*-xylol [65], Alliin [66].	The MeOH leaves/bark extracts induced micronuclei according to micronucleus assay without S9 (human lymphocyte cultures) [12].
Turraea floribunda (Meliaceae)	Emetic, rheumatism, dropsy, heart disease, swollen and painful joints [17,18].	South Africa	Limonoids and limonoid derivatives (turraflorins D–I, turraflorins An and B) [67].	Mutagenic effect in Comet test without exogenous metabolic activation (S9). Appearance of micronuclei with DCM leaves/bark extracts and MeOH leaves extract in micronucleus assay without S9 (human lymphocyte cultures) [12].

Plant species (Family)	Traditional uses	Region	Phytochemicals	Toxicity
Vernonia colorata (Asteraceae)	Abdominal pain, colic, rheumatism, dysentery, diabetes, ulcerative colitis [17,18].	South Africa	Aqueous and ethanolic extracts: reducing sugars, saponins, polyphenols, tannins, phlobatannins, alkaloids, sterols, Triterpènes [68]; coumarins, cardenolids, anthracenosids [69].	Toxic effect of DCM leaves/roots extracts in micronucleus test suggesting an (in)direct effect on cell division, or other toxic processes. DNA damage with MeOH leaves extract in Comet test without exogenous metabolic activation (S9) (human lymphocyte cultures) [12].
Ziziphus mucronata (Rhamnaceae)	Boils, sores, glandular swellings, diarrhea, dysentery, expectorant, emetic for coughs, chest problems, boils, sores, glandular swellings [17,18].	South Africa	Anthocyanins, quinones, tanins [70] Cyclopeptides alkaloids (mucronine), isoquinolone alkaloids (coclaurine, juziphine), porphine alkaloids (laurelliptine, asimilobine), flavonoids (swertish, apigenin glycosides), anthraquinones, triterpenoids (jujubogenin, betulinic acid) [71].	Appearance of micronuclei with DCM and MeOH leaf extracts in micronucleus and Comet tests without exogenous metabolic activation (S9). DNA damage of DCM extract in Comet test (human lymphocyte cultures) [12]. The 90% MeOH leaves extract induced mutant strains in Ames test with metabolic activation (*Salmonella typhimurium* strain TA98) [34].

(/): not reported; MeOH: methanol; DCM: dichloromethane.

Table 9.3 African Medicinal Plants Screened for their Genoprotective Effects

Plant (Family)	Traditional Use	Area of Plant Collection	Chemical Constituents	Effects of the Plant
Acacia salicina (Mimosoideae)	Inflammatory diseases, febrifuge, cancer, promote human fertility [72]	Tunisia	Chloroform and petroleum ether extracts: polyphenols, flavonoids, tannins, sterols [73].	The extracts may adsorb the mutagens in a way similar to the carcinogen adsorption which has been associated with pyrrole pigments (hemin and chlorophyllin) according to *Salmonella typhimurium* TA1535 and TA98 assay (Ames Test) [74,75]. They can induce also DNA glycosylase enzymes which are capable of repairing alkylating DNA bases [72].
Marrubium deserti de Noé (Lamiaceae)	Digestive disorders, scorpion stings, allergy [76].	Algeria	Dichloromethane extract: β-stigmasterol 1 [77,78], diterpenes, labdane diterpenes (cyllenin A 2), 15-epi-cyllenin A 3 [79], marrubiin [80,81], marrulactone, marrulibacetal 7 [82]. Ethyl acetate and n-BuOH phases from methanol extract: Apigenin-7-O-b neohesperidoside, acteoside [83], forsythoside B [84], apigenin-7-O-glucoside, terniflorin (apigenin-7-O-(600-E-p-coumaroyl)-glucoside), apigenin [85], apigenin-7-O-glucuronide [86].	The isolated compounds protect directly DNA strands from the electrophilic metabolite of the mutagen. The protective effect may correspond to a synergic participation of several of each component [76].

Myrtus communis (Myrtaceae)	Candidiasis, lung disorders [87].	Tunisia	Aqueous, ethyl acetate and methanol extracts: flavonoids, coumarins, tannins [87].	The extracts, according to Nifuroxazide and aflatoxine B assays, inhibit microsomal enzyme activation or protect DNA strands from the electrophilic metabolites of the mutagens [87].
Pistacia lentiscus (Anacardiaceae)	/	Tunisia	Flavonols; galloyl derivatives (galloyl-glucosides, ellagitannins and galloyl-quinic acids) [88].	An isolated compound (from fruits) Digallic Acid (DGA) is able to interact and neutralize electrophiles such as nitrofurantoine or may inhibit microsomal activation of B[a]p to electrophilic metabolite. It may act by inhibiting microsomal activation or by directly protecting DNA strands from the electrophilic metabolite of the mutagen. They may inhibit also several metabolic intermediates and Reactive Oxygen Species (ROS) formed during the process of microsomal enzyme activation which is capable of breaking DNA strands [89,90].
Pituranthos chloranthus (Apiaceae)	/	Tunisia	Essentials oils (monoterpenoid: thymol and carvacrol; sesquiterpenoid), alkanes, alcohols and aldehydes [91].	Essential oils showed a protective effect against damages induced by radicals, obtained from the photolysis of H_2O_2, on DNA plasmid through free radical scavenging mechanisms [91].

(Continued)

Table 9.3 (Continued)

Plant (Family)	Traditional Use	Area of Plant Collection	Chemical Constituents	Effects of the Plant
Rosa roxburghii (Rosaceae)	Detoxification and restoration of the liver, colon, kidney, lungs and skin, supports healthy cardiovascular function and reduces the risk of heart attacks and stroke [92].	South Africa	Ascorbic acid, polyphenols [92].	The fruit extract dose-dependently protected against the metabolically activated carcinogens, 2-AAF and AFB1 in *Salmonella typhimurium* T98 and TA100 strains, respectively [92].
Teucrium ramosissimum (Lamiaceae)	Gastric ulcer, intestinal inflammation, cicatrisant.	Tunisia	Tannins, coumarins, sterols, flavonoids [93], sesquiterpene hydrocarbons, oxygenated sesquiterpenes, δ-Cadinene, α-cadinol, germacradien-4-α-1-ol [94].	The total oligomer flavonoids (TOF), petroleum ether and aqueous extracts reduced the number of frameshift mutations induced by the direct genotoxicant NOPD (10 μg/plate) and the indirect genotoxicant B [a] P (5 μg/plate) in *S. typhymurium* TA98 strain, as well as the base pair substitution induced by the direct acting agent SA (1.5 μg/plate) and indirect acting agent AFB1 (0.5 μg/plate) in strains TA100 and TA1535 [93].

9.5 Medicinal Plants with Teratogenic Effects

Several phytochemicals among which some are commonly used as food and medicinal plants are potential teratogenic agents. The use of laxatives containing anthraquinones should be avoided during pregnancy and especially in the first trimester (the period during which organogenesis occurs and during which malformations can arise from brief exposure), as they can induce uterine contractions [95], increase blood flow to the uterus and its attachments, increase the risk of fetal loss, and may pass into breast milk and cause unwanted effects such as spasms in the infant [96]. Caffeine, considered as the most widely consumed stimulant in the world, could also be the cause of teratogenic effects of plants. It crosses the placental barrier and decreases blood flow to the placenta, where it can be a cause of weight loss in newborns, though there is no consensus on this activity [97,98]. It is estimated that 75% of pregnant women in the United States consume drinks containing caffeine, including those from plants such as black tea (*Camellia sinensis*), green tea (made from the leaves from *Camellia sinensis* that have undergone minimal oxidation during processing), and guarana (*Paullinia cupana*) [97]. *Zingiber officinalis* (Zingiberaceae) commonly known as ginger, is effective in the treatment of nausea and vomiting during pregnancy [99,100] and the prevention of motion sickness. It is controversial, as there are many differences in relation to its teratogenic potential. It was reported to cause embryonic loss when administered to rats during pregnancy, but also to increase the weight of the remaining fetuses [101]. Ginger components are indicated as potential inducers of apoptosis in human cells [102], a process involved in the remodeling of the fetal brain and other organs. In contrast, Weidner and Sigwart [103] found that an ethanol extract of ginger administered to pregnant rats caused no harm to either mother or developing fetus.

Cyanide contained in cassava (*Manihot esculenta*) could be responsible for its teratogenic effects as this was demonstrated in pregnant rats receiving cassava powder as 80% of their total diet during the first 15 days of gestation. Singh [104] demonstrated that there was a growth retardation of fetuses, embryonic death in 19%, and malformations in 28% of the total implantations, limb defects in 39% and microcephaly with open eyes in 5.5% of cases due to cassava. *Garcinia kola* (Guttiferae) was also shown to produce duration-dependent teratogenicity in fetal rats at 200 mg/kg body weight (bw) during the first 5 days of gestation, inducing a decrease in the weight and malformations of left upper limb in 7% of the fetuses from pregnant rats. The teratogenesis of *Acanthus montanus* (Acanthaceae) was demonstrated in another study where the following changes were observed on rat fetuses: decrease in body weight, crown-rump and tail length, placental weight, and poor ossification of bones extremities (forelimbs and hindlimbs) suggesting intrauterine growth retardation after the administration of 500 and 1000 mg/kg bw per day of the leaves methanol/methylene chloride (1:1) extract for 5 days [105]. African medicinal plants screened for their teratogenic potential are reported in Table 9.4.

Table 9.4 African Medicinal Plants Screened for their Teratotoxicity

Plant (Family)	Traditional Use	Area of Plant Collection	Chemical Constituents	Effects of the Plant
Rauwolfia vomitoria (Apocynaceae)	Snake and insect bites and stings, insomnia and insanity [106].	Nigeria	Alkaloids (rauwolfine, reserpine, rescinnamine, serpentine, ajmaline), steroid-serposterol and saponin [107].	Ethanolic root bark and leaf extracts induced adverse effects on developing liver of fetal rats [106].
Acanthus montanus (Acanthaceae)	Cough, epilepsy, dysmenorrhea, pain, miscarriages, rheumatism, hypertension, skin infections [105].	Cameroon	β-sitosterol from MeOH/CH₂Cl₂ extract of leaves [105].	The MeOH/CH₂Cl₂ extract had an embryotoxic effect at higher doses (500 and 1000 mg/kg day)) when it's administered orally to Wistar pregnant rats during organogenesis. This effect disappeared within the first 5 days after parturition [105].
Garcinia kola (Guttiferae)	Cough, purgative, antiparasitic, antimicrobial, bronchitis, throat infections, liver disorders, hypoglycemic, aphrodisiac, antioxidant [108].	Nigeria	Biflavonoid (Kolaviron) [1–3,8–11], benzophenones, *Garcinia* biflavonones (GB-1, GB-2), kolaflavonone [109]. Apigenin (flavonoids) [110].	The seed extract at 200 mg/kg bw administered alters estrous cycle in rats, partly inhibits ovulation and may produce duration-dependent teratogenicity in fetal female Sprague–Dawley rats [108].

Manihot (esculenta), and Manihot dulcis (Euphorbiaceae)	/	Nigeria	Cyanide [104].	The milled cassava powder on albino rats at 80% of the diet showed a low incidence of limb defects, open eye, microcephaly, and growth retardation [104].
Treculia africana (Moraceae)	Malaria, worms, coughs and digestive disorders [111].	Nigeria	Polyphenols [111]. Methanolic extracts: 4-hydroxybenzoic acid, caffeic acid, vanillic acid, syringic acid, p-coumaric acid, syringaldehyde, ferulic acid [112].	The extract induced significant numbers of malformations (gross and skeletal) in pregnant female rats treated by polyphenols obtained from outer coat of the fruit [111].

Table 9.5 Teratogenic Plants Reported in Other Parts of the World but also Found in Africa

Plant (Family)	Traditional Use	Area of Plant Collection	Chemical Constituents	Effects of the Plant
Luffa operculata L. Cogn. (Cucurbitaceae)	Purgative effect, treatment of parasitic diseases [113].	Brazilia	Ceramides, triterpenoid, steroids [114].	The decoction administered to female mice during the implantation of embryos caused a reduction in birth rate [115].
Artemisia absinthium L (Asteraceae)	It is used for their antiparasitic effects, and to treat anorexia and indigestion [116].	Brazilia	Thujone (terpenoid) [117], terpenes (limonene, myrcene, α and β thujone [118], sesquiterpene, caryophellene [119], sabinyl acetate, chrysanthenyl acetate [120], sesquiterpene lactone endoperoxide (Artemisinin) [121].	When ingested in large amounts, it can cause epileptiform seizures and even abortion due to the presence of toxic isolated terpenoid compound thujone [119].
Oxytropis lambertii Pursh. (Fabaceae)	/	India	Alkaloid (swansonine) [122,123].	The isolated alkaloid compound swansonine is an alphamannosidase inhibitor which induces vascular resistance and vasoconstriction in the fetus of sheep and cattle. [122,123].
Nicotiana glauca Grahm. (Solanaceae)	/	India	Alkaloids (piperidine Anatabine, anabaseine) [122,123]. Nicotine [124].	Contains piperidine alkaloids and cause congenital contracture-type skeletal malformations and cleft palates in the fetus. Anatabine and anabaseine are involved in the teratogenic effects on swine [122,123]. Moreover, these induced multiple congenital contractures

			(MCC) and palatoschisis in goat kids when their dams were gavaged with the plant during gestation. The skeletal abnormalities included fixed extension of the carpal, tarsal and fetlock joints, scoliosis, lordosis, torticollis and rib cage problems. The clinical signs of toxicity in sheep, cattle, and pigs included, ataxia, incoordination, muscular weakness, prostration, and death [125].
Lupinus formosus (Fabaceae)	/	India	Alkaloids (piperidine) [122,123].	Contains piperidine alkaloids and cause congenital contracture-type skeletal malformations and cleft palates in the fetus [122,123]. Moreover, these induced MCCs and palatoschisis in goat kids when their dams were gavaged with the plant during gestation. The skeletal abnormalities included fixed extension of the carpal, tarsal and fetlock joints, scoliosis, lordosis, torticollis and rib cage problems. The clinical signs of toxicity in sheep, cattle, and pigs included, ataxia, incoordination, muscular weakness, prostration, and death [125].

(Continued)

Table 9.5 (Continued)

Plant (Family)	Traditional Use	Area of Plant Collection	Chemical Constituents	Effects of the Plant
Veratrum californicum Durr. (Melanthiaceae)	Treatment of high blood pressure and cancer [126].	India	Steroidal alkaloids (a cyclopamine, jervine) [123,126,127].	Contains steroidal alkaloids that cause craniofacial malformations including cyclopia in the fetus of sheep [123,127].
Conium maculatum L. (Umbelliferae)	Antispasmodic, sedative, analgesic [123].	India	Piperidine alkaloids (coniine and gamma-coniceine) [122,123], 2-Pentylpiperidine, (conmaculatin) [128].	Contains piperidine alkaloids (coniine and gamma-coniceine) that cause congenital contracture-type skeletal malformations and cleft palates, restricted fetal movement, arthrogrypotic limb deformities in calves (cow) [122,123]. Moreover, these induced MCC and palatoschisis in goat kids when their dams were gavaged with the plant during gestation. The skeletal abnormalities included fixed extension of the carpal, tarsal and fetlock joints, scoliosis, lordosis, torticollis and rib cage problems. The clinical signs of toxicity in sheep, cattle and pigs included, ataxia, incoordination, muscular weakness, prostration and death [125].

Descurainia sophia / L. (Brassicaceae)	India	Hyoscyamine (alkaloid), quercetin-3-*O*-β-D-glucopyranosyl-7-*O*-β-gentiobioside, kaempferol-3-*O*-β-D-glucopyranosyl-7-*O*-β-gentiobioside, isorhammetin-3-*O*-β-D-glucopyranosyl-7-*O*-β-gentiobioside, quercetin-7-*O*-β-gentiobioside, kaempferol-7-*O*-β-gentiobioside, isorhammetin-7-*O*-β-gentiobioside, quercetin-3,7-di-*O*-β-D-glucopyranoside, kaempferol-3, 7-di-*O*-β-D-glucopyranoside, isorhammetin-3, 7-di-*O*-β-D-glucopyranoside, kaempferol-3-*O*-β-D-glucopyranosyl-7-*O*-[(2-*O*-*trans*-sinnapoyl)-β-D-glucopyranosyl (1-->6)]-β-D-glucopyranoside), sinapic acid ethyl ester, 3, 4, 5-trimethoxyl-cinnamic acid [129]. Coumarins, flavonoids, terpenes, fatty acids, amino acids [130].	Contains hyoscyamine (alkaloid) that cause deformed foals in the fetus [131].
Astragalus mollisimus Torr. (Fabaceae)	India	Swainsonine alkaloids [131].	Contains swainsonine alkaloids that cause neurological defects in the fetus [131].
Astragalus pubentissimus Torr. (Fabaceae)	India	Alkaloids (swainsonine, indolizidine), [132,133].	Contains swainsonine, and indolizidine alkaloid that caused generalized stunting in the fetus of sheep and cattle [132,133].

(Continued)

Table 9.5 (Continued)

Plant (Family)	Traditional Use	Area of Plant Collection	Chemical Constituents	Effects of the Plant
Solanum tuberosum L. (Solanaceae)	/	India	Solanidanes and spirosolanes	Contains solanidanes and spirosolanes those cause brain defects and cleft palate in the fetus of sheep and cattle [134,135].
Solanum melongena L. (Solanaceae)	Ani-diabetic, liver complaints [136].	India	Solanidanes and spirosolanes [132–134], alkaloids, saponins, steroids, tannins/ phenolics, flavonoids, proteins, carbohydrates, glycoalkaloids (solasodine), anthocyanin, glycosides (cardiac and cyanogenic glycosides). [136].	Contains solanidanes and spirosolanes those cause brain defects and cleft palate in the fetus of sheep and cattle [132–134].
Lycopersicon esculentum Mill. (Solanaceae)	Chronic degenerative disease [137].	India	Solanidanes and spirosolanes [132–134], lycopenes [137].	Contains solanidanes and spirosolanes those cause brain defects and cleft palate in the fetus of sheep and cattle [132–134].
Voacanga globosa (Blco.) Merr. (Apocynaceae)	/	India	Alkaloids, saponins, 2-deoxysugars, and hydrolysable tannins [138].	Causes hypoxia in fetus showed by *in vitro* assay [139].
Baccaurea tetrandra (Baill.) Muell.Arg. (Euphorbiaceae)	/	India	/	Causes hypoxia in fetus showed by *in vitro* assay [139].
Ficus septica Burm.f. (Moraceae)	Purgative, emetic effects [140].	India	Flavonoids, coumarins, alkaloids, steroids, ceramides, triterpenes, tannins [141].	Causes hypoxia in fetus showed by *in vitro* assay [139].

Plant name (family)	Traditional use	Country	Chemical compounds	Teratogenic effect
Uncaria perrottetii (A.Rich) Merr. (Rubiaceae)	/	India	Alkaloids, tannins and leucoanthocyanin [142].	Causes hypoxia in fetus showed by *in vitro* assay [139].
Aglaia loheri (Blco.) (Meliaceae)	insecticidal, antifungi, anti-inflammatory, neuroprotector, cardioprotector, anticancerous properties [143].	India	Phenolic ester [144], spinasterol, trilinolein, phytyl fatty acid ester [145], flavaglines [143].	Causes hypoxia in fetus showed by *in vitro* assay [139].
Veratrum viride Aiton. (Melanthiaceae)	Nervous palpitations, epileptics convulsions, neuralgia, paralysis, whooping-cough, drop, hydropisy [146].	India	Steroidal alkaloids cyclopamine and jervine [147,148].	The isolated compounds are teratogenic, studied *in vitro* by ELISA test [147,148].
Caulophyllum thalictroides L. Michx. (Berberidaceae)	Abortive and contraceptive effects	United States of America	Alkaloids (*N*-methylcytisine, baptifoline, anagyrine, magnoflorine) [149].	The isolated compound *N*-methylcytisine exhibit teratogenicity, studied *in vitro* rat embryo culture [149].
Cortex cinnamom L. (Lauraceae)	/	India	Essentials oil, Cinnamaldehyde [133].	Cinnamaldehyde is teratogenic in chick embryos [133].
Cannabis sativa L. (Cannabaceae)	Antimicrobial, anti-inflammatory and antiageing balances skin disorder, rheumatism, inflammation, diabetes, excessive epidermal water loss [150].	India	Terpenes (cannabinoids) [151], 9-trans-Tetrahydrocannabivarin, sterols (campesterol, stigmasterol, and β-sitosterol), amino acid (L-proline) [150].	Causes brain defects in the fetus [152].

(Continued)

Table 9.5 (Continued)

Plant (Family)	Traditional Use	Area of Plant Collection	Chemical Constituents	Effects of the Plant
Podophyllum hexandrum L. (Berberidaceae)	/	India	/	Causes absent right thumb and radius, a supernumerary left thumb, a probable septal defect of the heart, a defect of the right external ear, and skin tagsin the fetus in humans [153].
Nicotiana tobacum L. (Solanaceae)	Hair treatment, dysmenorrheal, vermiguge, rheumatism, hoarseness, fungal diseases of the skin, wounds, ulcers, bruises, sores, mouth lesions, stomartitis and mucosa; leaf is orally taken for kidney diseases, bronchitis and pneumonia [125].	India	Anatabine and anabaseine [122], nicotine, anabasine (an alkaloid similar to the nicotine but less active), glucosides (tabacinine, tabacine), 2,3,6-trimethyl-1,4-naphthoquinone, 2-methylquinone, 2-napthylamine, propionic acid, anatalline, anthalin, anethole, acrolein, anatabine, cembrene, choline, nicotelline, nicotianine, and pyrene [125].	The compounds anatabine and anabaseine cause arthrogryposis in the fetus and skeleton defects in pigs and also in swines [122].
Goniothalamus amuyon (Annonaceae)	Edema, rheumatism [128].	India	Styrylpyrones ((6R,7R,8R)-8-methoxygoniodiol, (6R,7R,8R)-8-chlorogoniodiol) [128], goniothalesacetate, goniothalesdiol A, goniodiol-7-monoacetate, goniodiol-8-monoacetate, leiocarpin C,	It induced morphological abnormalities in the fetus of mice [139].

Plant	Use	Country	Compounds	Teratogenic effect
			liriodenine, griffithazanone A, 4-methyl-2,9,10-(2H)-1-azaanthracencetrione, velutinam, aristolactam BII, isoquinolones [128].	
Espinheira santa Klen. (*Maytenus ilicifolia* Mart. ex Reissek) (Celastraceae)	Wound, cancer [154]	Brazilia	Maytansine [154].	The isolated compound maytansine shows teratogenic activity in mice [154].
Alstonia macrophylla Wall. (Apocynaceae)	/	India	Bisindole alkaloid (macralstonine, thungfaine), secoiridoid glycoside (sweroside, naresuanoside) [155].	It induced morphological abnormalities in the fetus of mice [156].
Asparagus racemosus Willd. (Asparagaceae)	Gastric ulcers, dyspepsia and as a galactogogue [157].	India	Polycyclic alkaloid (Asparagamine A), steroidal saponins (shatavaroside A, shatavaroside B, shatavarins VI-X), saponin (filiasparoside C), shatavarin I, shatavarin IV (or asparinin B), shatavarin V, immunoside and schidigerasaponin D5 (or asparanin A), [158,159]. Isoflavone, 8-methoxy-5,6,4'-trihydroxyisoflavone 7-O-β-D-glucopyranoside [160].	Cause reabsorption of fetuses and intrauterine growth retardation in rats [156].
Lupinus argenteus Pursh. (Fabaceae)	/	India	Anagyrine, quinolizidine [161].	The isolated compound anagyrine shows teratogenic effect in calf [161].

(Continued)

Table 9.5 (Continued)

Plant (Family)	Traditional Use	Area of Plant Collection	Chemical Constituents	Effects of the Plant
Lupinus caudatus L. (Fabaceae)	/	India	Anagyrine [161].	The isolated compound anagyrine shows teratogenic effect in calf [161].
Lupinus nootkatensis Donn ex sims (Fabaceae)	/	India	Anagyrine [161].	The isolated compound anagyrine shows teratogenic effect in calf [161].
Thermopsis montana Nutt. (Fabaceae)	/	India	Anagyrine [162].	The isolated compound anagyrine shows teratogenic effect in calf [162].
Trigonella foenum-graecum (Fabaceae)	It's used to ease childbirth and to increase milk flow, for menstrual pain and as hilba tea to ease stomach problems [163].	India	Alkaloids: Trimethylamine, Neurin, Trigonelline, Choline, Gentianine, Carpaine and Bin. Amino acids: Isoleucine, 4-Hydroxyisoleucine, Histidine, Leucine, lysine, L-tryptophan, Argenine. Saponins: Graecunins, fenugrin B, fenugreekine, trigofoenosides A-G. Steroidal sapinogens: Yamogenin, diosgenin, smilagenin, sarsasapogenin, tigogenin, neotigogenin, gitogenin, neogitogenin, yuccagenin, saponaretin. Flavonoids:	Cause decrease bone marrow cell proliferation and increase fetal mortality rate in rats [164].

Botanical name (family)	Uses	Country	Chemical constituents	Toxicity
Trifolium hybridum L. (Fabaceae)	/	India	Quercetin, rutin, vetixin isovetixin. Fibers: Gum, neutral detergent. Other: Coumarin, lipids, vitamins, minerals [163].	Contains mycotoxin which is teratogenic [131].
Indigofera spicata Forssk. (Fabaceae)	/	India	Indospicine, (+)-5''-deacetylpurpurin, (+)-5-methoxypurpurin, (2S)-2,3-dihydrotephroglabrin, (2S)-2,3-dihydrotephroapollin C, flavanones, rotenoids, chalcone [165].	Contains indospicine which is teratogenic [131].
Sorghum bicolori L. (Poaceae)	/	India	cyanogenic glycosides	Contains cyanogenic glycosides which are teratogenic [131].
Datura stramonium Linn. (Solanaceae)	Antispasmodiques, sedatifs, asthma, neuralgia [131].	India	Alkaloid (hyoscyamine, scopolamine) [131].	Alkaloid hyoscyamine is responsible for teratogenicity. [131].
Boerhaavia diffusa Linn. (Nyctaginaceae)	Dyspepsia, abdominal pain, inflammation, cancer, diabetes [166].	India	alkaloids (punarnavine), rotenoids (boeravinones A to J), flavones [167].	It induced congenital malformation in albino rats [168].
Rauwolfia serpentina Benth. Ex Kurz. (Apocynaceae)	Cancer, liver disease, or mental illness [169].	India	yohimbine, reserpine, ajmaline, deserpidine, rescinnamine, serpentinine [169].	Teratogenicity in Wistar rats [170].
Aspilia africana (Pers.) C.D. Adams (Asteraceae)	Contraceptive and antifertility properties, antimalarial, rheumatic pain, gonorrhea [171].	India	Alkoloids, tannins, saponins, flavonoids, phenols [171].	Teratogenicity in Wistar rats. [170].

MCCs, multiple congenital contractures.

The teratogenicity of some plants found in Africa but collected in other parts of the world was also reported. Some of them include *Goniothalamus amuyon* (Annonaceae), *Rauwolfia serpentina* Benth. Ex Kurz. *Voacanga globosa* (Blco.) Merr., and *Alstonia macrophylla* Wall. (Apocynaceae), *Aspilia africana* (Pers.) C. D. Adams, and *Artemisia absinthium* L (Asteraceae), *Asparagus racemosus* Willd. (Asparagaceae), *Podophyllum hexandrum* L., and *Caulophyllum thalictroides* L. Michx. (Berberidaceae), *Descurainia sophia L.* (Brassicaceae), *Cannabis sativa* L. (Cannabaceae), *Espinheira santa* Klen. (*Maytenus ilicifolia* Mart. ex Reissek) (Celastraceae), *Luffa operculata* L. Cogn. (Cucurbitaceae), *Baccaurea tetrandra* (Baill.) Muell.Arg. (Euphorbiaceae), *Oxytropis lambertii* Pursh., *Lupinus formosus*, *Astragalus mollisimus* Torr. and *Lupinus argenteus* Pursh., and *Lupinus caudatus* L., *Lupinus nootkatensis* Donn ex sims, *Thermopsis montana* Nutt. (Fabaceae), *Trigonella foenum-graecum*, *Trifolium hybridum* L., *Indigofera spicata* Forssk. and *Astragalus pubentissimus* Torr. (Fabaceae), *Cortex cinnamom* L. (Lauraceae), *Veratrum californicum* Durr., and *Veratrum viride* Aiton. (Melanthiaceae), *Aglaia loheri* (Blco.) (Meliaceae), *Ficus septica* Burm.f. (Moraceae), *Boerhaavia diffusa* Linn. (Nyctaginaceae), *Sorghum bicolori* L. (Poaceae), *Nicotiana glauca* Grahm., *Solanurn tuberosum* L., *Solanum melongena* L., *Lycopersicon esculentum* Mill., *Nicotiana tobacum*, and *Datura stramonium* Linn. (Solanaceae), *Conium maculatum* L. (Umbelliferae), and *Uncaria perrottetii* (A.Rich) Merr. (Rubiaceae) (Table 9.5).

9.6 Conclusion

In this chapter, we reviewed African medicinal plants with ability to induce genotoxic or teratogenic effects. We also pointed out some of the plants available in the continent with such effects even though they were collected in other parts of the world, in order to warn African therapists of their potential harmful effects. Though most of the results relied on animal or bacterial systems, it is important to have this information in order to prevent as well as possible some cancers or congenital abnormalities by using or developing new drugs which can safely be used by pregnant women, as well as other patients.

References

[1] Cosyns JP, Jadoul M, Squifflet JP, De Plaen JF, Ferluga D, van Ypersele de Strihou C. Chinese herbs nephropathy, a clue to Balkan endemic nephropathy? Kidney Int 1994;45:1680−8.
[2] Lord GM, Tagore R, Cook T, Gower P, Pusey CD. Nephropathy caused by Chinese herbs in the UK. Lancet 1999;354(9177):481−2.
[3] Luyckx VA, Naicker S. Acute kidney injury associated with the use of traditional medicines. Nat Clin Pract Nephrol 2008;4:664−71.
[4] Stengel B, Jones E. End-stage renal insufficiency associated with Chinese herbal consumption in France. Nephrologie 1998;19:15−20.

[5] Tanaka A, Nishida R, Sawai K, Nagae T, Shinkai S, Ishikawa M, et al. Traditional remedy-induced Chinese herbs nephropathy showing rapid deterioration of renal function. Nihon Jinzo Gakkai Shi 1997;39:794−7.

[6] Tanaka A, Shinkai S, Kasuno K, Maeda K, Murata M, Seta K, et al. Chinese herbs nephropathy in the Kansai area, a warning report. Nihon Jinzo Gakkai Shi 1997; 39:438−40.

[7] <http://www.alttox.org/ttrc/existing-alternatives/genotoxicity.htmL>; 2011. Genotoxicity, validated non-animal alternatives. AltTox.org. 2011-06-20. [accessed on 12.12.13].

[8] Furman G. Genotoxicity testing for pharmaceuticals current and emerging practices, <http://www.pharmatek.com/pdf/PTEKU/Apr172008.pdf>; 2008 [accessed on 16.12.13].

[9] Ames BN, Durston WE, Yamasaki E, Lee FD. Proc Natl Acad Sci 1973;70(8): 2281−5.

[10] Tice RR, Agurell E, Anderson D, Burlinson B, Hartmann A, Kobayashi H, et al. Single cell gel/Comet assay, guidelines for *in vitro* and *in vivo* genetic toxicology testing. Environ Mol Mutagen 2000;35(3):206−21.

[11] Wilson JG. Environment and birth defects. Environmental science series. London: Academic Press; 1973.

[12] Taylor JLS, Elgorashi EE, Maes A, Gorp UV, Kimpe ND, Staden JV, et al. Investigating the safety of plants used in south African traditional medicine, testing for genotoxicity in the micronucleus and alkaline Comet assays. Environ Mol Mutagen 2003;42:144−54.

[13] Fatemeh K, Khosro P. Cytotoxic and genotoxic effects of aqueous root extract of *Arctium lappa* on *Allium cepa* Linn root tip cells. Intl J Agron Plant Prod 2012; 3(12):630−7.

[14] Hennebelle T, Weniger B, Joseph H, Sahpaz S, Bailleuil F. Senna alata. Fitoterapia 2009;80:385−93.

[15] Hong C-E, Lyu S-Y. Genotoxicity detection of five medicinal plants in Nigeria. J Toxicol Sci 2011;36(1):87−93.

[16] van den Bout-van den Beukel CJ, Hamza OJ, Moshi MJ, Matee MI, Mikx F, Burger DM, et al. Evaluation of cytotoxic, genotoxic and CYP450 enzymatic competition effects of Tanzanian plant extracts traditionally used for treatment of fungal infections. Basic Clin Pharmacol Toxicol 2008;102(6):515−26.

[17] Hutchings A, Scott AH, Lewis G, Cunningham AB. Zulu medicinal plants, an inventory. Natal, South Africa: University of Natal Press; 1996.

[18] Van Wyk B-E, van Oudtshoorn B, Gericke N. Medicinal plants of South Africa. Pretoria, South Africa: Briza; 1997.

[19] Fawole OA. Pharmacology and phytochemistry of South African traditional medicinal plants used as antimicrobials. Master of science in the research centre for plant growth and development school of biological and conservation sciences. Pietermaritzburg: University of KwaZulu-Natal; 2009. p. 135.

[20] Tor-Anyiin TA, Terseer YD. Phytochemical screening and antimicrobial activity of stem bark extracts of *Antidesma Venosum*. J Nat Prod Plant Res 2012;2(3):427−30.

[21] Gitu LM. Biological and phytochemical studies of medicinal plants, *Antidesma venosum* (Euphorbiaceae) and *Kotschya africana* (Fabaceae) used in traditional medicine in Kenya. JKUAT Abstracts of PostGraduate Thesis; 2009.

[22] Magadula JJ, Mwangomo DT, Moshi MJ, Heydenreich M. A novel γ-lactone and other constituents of a Tanzanian *Antidesma venosum*. Spatula DD Peer Rev J Complementary Med Drug Dis 2013;3(1):7−12.

[23] Yadov JP, Panghal M. *Balanites aegyptiaca* (L.) Del. (Hingot), a review of its traditional uses, phytochemistry and pharmacological properties. Int J Green Pharm 2010; 4(3):140−6.

[24] van Wyck B-E, Van Oudtshoorn MCB, Gericke N. Medicinal plants of South Africa. Pretoria: Briza Publications; 2009. p. 320.

[25] Sandager M, Nielsen ND, Stafford GI, van Staden J, Jager AK. Alkaloids from *Boophane disticha* with affinity to the serotonin transporter in rat brain. J Ethnopharmacol 2005;98:367−70.

[26] Adewusi EA, Fouche G, Steenkampa V. Cytotoxicity and acetylcholinesterase inhibitory activity of an isolated crinine alkaloid from *Boophane disticha* (Amaryllidaceae). Department of Pharmacology, Faculty of Health Sciences, University of Pretoria, Arcadia, South Africa; 2012. p. 1−30.

[27] Hautefeuille M, Véléa D. Les drogues de synthèse. Paris: Presses Universitaires de France, coll; 2002.

[28] Al-Ahdal MN, McGarry TJ, Hannan M. Cytotoxicity of Khat (*Catha edulis*) extract on cultured mammalian cells. Mutat Res 1988;204(3):17−322.

[29] Krikorian AD, Getahun A. Chat, Coffee's Rival from Harar, Ethiopia. II. Chemical composition. Econ Bot 1973;27:378−89.

[30] Abderrahman SM, Modallal N. Genotoxic effects of *Catha edulis* (Khat) Extract on Mice Bone Marrow cells. JJBS 2008;1(4):165−72.

[31] Huxley A. New RHS dictionary of gardening. London: Macmillan; 1992. ISBN 0-333-47494-5.

[32] Ibrahim M, Mehjabeen SS, Mangamoori LN. Pharmacological evaluation of *Catharanthus roseus*. Int J Pharm Appl 2011;2(3):165−73.

[33] Yadav PD, Bharadwaj NSP, Yedukondalu M, Methushala CH, Ravi KA. Pytochemical evaluation of *Nyctanthes arbortristis, Nerium oleander* and *Catharathnus roseus*. IJRPB 2013;1(3):333−8.

[34] Elgorashi EE, Taylor JLS, Maes A, Van Staden J, De Kimpe N, Verschaeve L. Screening of medicinal plants used in South African traditional medicine for genotoxic effects. Toxicol Lett 2003;143:195−207.

[35] Senthamarai R, Kirubha SV, Gayathri S. Pharmacognostical and Phytochemical studies on fruits of *Catunaregam spinosa* Linn. J Chem Pharm Res 2011;3(6):829−38.

[36] Guang-Chun G, Zhong-Xian L, Shu-Hong T, Si Z, Fa-Zuo W, Qing-Xin L. Triterpenoid saponins from the stem bark of *Catunaregam spinosa*. Can J Chem 2011;89(10):1277−82.

[37] Al-Taweel AM, Shagufta P, Azza ME-S, Ghada AF, Abdul M, Nighat A, et al. Bioactive phenolic amides from *Celtis africana*. Molecules 2012;17:2675−82.

[38] Adedapo AA, Jimoh FO, Afolayan AJ, Masika PJ. Antioxidant properties of the methanol extracts of the leaves and stems of *Celtis africana*. RNP 2009;3(1):23−31.

[39] Edziri H, Mastouri M, Aouni M, Anthonissen R, Verschaeve L. Investigation on the genotoxicity of extracts from *Cleome amblyocarpa* Barr. and Murb, an important Tunisian medicinal plant. S Afr J Bot 2013;84:102−3.

[40] Tsichritzis F, Abdel-Mogip M, Jakupovic J. Dammarane triterpenes from *Cleome africana*. Phytochemistry 1993;33:423−5.

[41] Cascon SC, Brown KS. Biogenetically significant triterpenes in a species of meliaceae, *Cabralea polytricha* A. Juss. Tetrahedron 1972;28:315−23.

[42] Jente R, Jakupovic J, Olatunji GA. A Cembranoid diterpene from *Cleome viscosa*. Phytochemistry 1990;29:666−7.

[43] Davey MW, Persiau G, De Bruyn A, van Damme J, Bauw G, Van Montau M. Purification of the alkaloid lycorine and simultaneous analysis of ascorbic acid and lycorine by micellar electrokinetic capillary chromatography. Anal Biochem 1998;257:80−8.

[44] Waller GR, Nowachi EK. Alkaloids biology and metabolism in plants. Warsaw, Poland: Polish Academy of Sciences; 1978. p. 34−67.

[45] Neuwinger HD. African ethnobotany poisons and drugs. Germany: Chapman & Hall; 1996. p. 10−6.

[46] Mallavadhani UV, Panda AK, Rao YR. Review article number 134 pharmacology and chemotaxonomy of Diospyros. Phytochemistry 1998;49(4):901−51.

[47] Ghosal S, Singh S, Bhattacharya SK. Alkaloids of Mucuna pruriens, chemistry and pharmacology. Planta Med 1971;19:279.

[48] Nkengfack AE, Vouffo W, Vardamides JC, Kouam J, Fomum ZT, Meyer M, et al. Phenolic metabolites from Erythrina species. Phytochemistry 1997;46(3):573−8.

[49] Chacha M, Bojase-Moleta G, Majinda RRT. Antimicrobial and radical scavenging flavonoids from the stem wood of Erythrina latissima. Phytochemistry 2005;66:99−104.

[50] Mbanga J, Ncube M, Magumura A. Antimicrobial activity of Euclea undulata, Euclea divinorum and Diospyros lycioides extracts on multi-drug resistant Streptococcus mutans. J Med Plants Res 2013;7(37):2741−6.

[51] Ngari FW, Gikonyo NK, Wanjau RN, Njagi EM. Safety and Antimicrobial properties of Euclea divinorum Hiern, ChewiSticks used for management of oral health in Nairobi County, Kenya. AJBPS 2013;3(3):1−8.

[52] Juma BF, Majinda RRT. Constituents of Gardenia volkensii, their brine shrimp lethality and DPPH radical scavenging properties. Nat Prod Res 2007;21(2):121−5.

[53] Villegas M, Vargas D, Msonthi JD, Marston A, Hostettmann K. Isolation of the antifungal compounds falcarindiol and sarisan from Heteromorpha trifoliata. Planta Med 1988;54:36−7.

[54] Edziri H, Mastouri M, Mahjoub A, Anthonissen R, Mertens B, Cammaerts S, et al. Toxic and mutagenic properties of extracts from Tunisian traditional medicinal plants investigated by the neutral red uptake, VITOTOX and alkaline Comet assays. S Afr J Bot 2011;77:703−10.

[55] Edziri H, Samia A, Groh P, Mahjoub MA, Mastouri M, Gutmann L, et al. Antimicrobial and cytotoxic activity of Marrubium alysson and Retama raetam grown in Tunisia. Pakistan J Biol Sci 2007;10:1759−62.

[56] Benbott A, Bahri L, Boubendir A, Yahia A. Study of the chemical components of Peganum harmala and evaluation of acute toxicity of alkaloids extracted in the Wistar albino mice. J Matern Environ Sci 2013;4(4):558−65.

[57] Moloudizargari M, Mikaili P, Aghajanshakeri S, Asghari MH, Shayegh J. Pharmacological and therapeutic effects of Peganum harmala and its main alkaloids. Pharmacogn Rev 2013;7(14):199−19212.

[58] De Paiva SR, Figueiredo MR, Kaplan MAC. Isolation of secondary metabolites from roots of Plumbago auriculata Lam. by countercurrent chromatography. Phytochem Anal 2005;16(4):278−81.

[59] Van Wyk BE, Gericke N. People's plants. A guide to useful plants of Southern Africa. Pretoria: Briza Publications; 2000. p. 351.

[60] Kpoviessi DSS, Gbaguidi FA, Kossouoh C, Agbani P, Yayi-Ladekan E, Sinsin B, et al. Chemical composition and seasonal variation of essential oil of Sclerocarya birrea (A. Rich.) Hochst subsp birrea leaves from Benin. J Med Plants Res 2011;5(18):4640−6.

[61] Aowoyale JA, Olatunji GA, Oguntoye SO. Antifungal and antibacterial activities of an alcoholic extract of *Senna alata* leaves. J Appl Sci Environ Manage 2005;9(3):105−7.

[62] Okonko IO, Sule WF, Joseph TA, Ojezele MO, Nwanze JC, Alli JA, et al. *In vitro* antifungal activity of *Senna Alata* Linn. Crude leaf extract. Adv Appl Sci Res 2010;1 (2):14−26.

[63] Baarschers WH, Horn DH, Johnson LRF. The structure of some diterpenes from tambooti wood, *Spirostachys africana* Sond. J Chem Soc 1962;4046−55.

[64] Traore M, Zhai L, Chen M, Olsen CE, Odile N, Pierre GI. Cytotoxic kurubasch aldehyde from *Trichilia emetica*. Nat Prod Res 2007;21:13−7.

[65] Soyingbe OS The chemical composition, antimicrobial and antioxidant properties of the essential oils of *Tulbaghia violacea* Harv. and *Eucalyptus grandis* W. Hill ex Maiden. Thesis of Master of Science in the Department of Biochemistry and Microbiology, Faculty of Science and Agriculture, University of Zululand, KwaDlangezwa, South Africa; 2012. p. 108.

[66] Van Wyk B-E, Van Oudtshoorn B, Gericke N. Medicinal plants of South Africa. Pretoria, South Africa: Briza; 1997. pp. 304.

[67] McFarland K, Mulholland DA, Fraser L-A. Limonoids from *Turraea floribunda* (Meliaceae). Phytochemistry 2004;65(14):2031−7.

[68] Adebayo OL, Asamoah B, Mills-Robertson E, Charles F. *In vitro* evaluation of aqueous and ethanolic extracts of *Vernonia colorata* as an antibacterial agent. Int J Curr Res Rev 2012;04(01):21−7.

[69] Guenné S, Hilou A, Ouattara N, Nacoulma OG. Anti-bacterial activity and phytochemical composition of extracts of three medicinal Asteraceae species from Burkina Faso. Asian J Pharm Clin Res 2012;5(2):0974−2441.

[70] Karola DO The ethnobotany and chemistry of South African traditional tonic plants. Thesis of Doctorate in Botany, University of Johannesburg; 2012. p. 481.

[71] Dictionary of natural products. London: Chapman & Hall; 1996.

[72] Steele VE, Kellov GJ. Development of cancer chemopreventive drugs based on mechanistic approaches. Mutat Res 2005;591:16−23.

[73] Chatti IB, Jihed B, Skandrani I, Bhouri W, Ghedira K, Ghedira LC. Antioxidant and antigenotoxic activities in *Acacia salicina* extracts and its protective role against DNA strand scission induced by hydroxyl radical. Food Chem Toxicol 2011;49:1753−8.

[74] Ferguson RL, Philpott M, Karunasinghe N. Dietary cancer and prevention using antimutagens. Toxicology 2004;198:147−59.

[75] Ikuma NEM, Passoni HM, Biso IF, Longo M. Investigation of genotoxic and antigenotoxic activities of *Melampodium divaricatum* in *Salmonella typhimurium*. Toxicol In Vitro 2006;20:361−6.

[76] Zaabat N, Hay A-E, Michalet S, Darbour N, Bayet C, Skandrani I, et al. Antioxidant and antigenotoxic properties of compounds isolated from *Marrubium deserti* de Noé. Food Chem Toxicol 2011;49:3328−35.

[77] Marsan MP, Warnock W, Muller I, Nakatani Y, Ourisson G, Milon A. Synthesis of deuterium-labeled plant sterols and analysis of their side-chain mobility by solid state deuterium NMR. J Org Chem 1996;61:4252−7.

[78] Pereira PS, Franca SC, Anderson O, Paulo VB, Camila MS, Pereira SIV, et al. Chemical constituents from *Tabernaemontana catharinensis* root bark, a brief NMR review of indole alkaloids and *in vitro* cytotoxicity. Quim Nova 2008;31:20−4.

[79] Karioti A, Heilmann J, Skaltsa H. Labdane diterpenes from *Marrubium velutinum* and *Marrubium cylleneum*. Phytochemistry 2005;66:1060−6.

[80] Knöss W, Reuter B, Zapp J. Biosynthesis of the labdane diterpene marrubiin in *Marrubium vulgare via* a non-mevalonate pathway. Biochem J 1997;326:449−54.

[81] Hussein AA, Meyer MJJ, Rodriguez B. Complete H and C NMR assignments of three labdane diterpenoids isolated from *Leonotis ocymifolia* and six other related compounds. Magn Reson Chem 2003;41:147.

[82] Rigano D, Aviello G, Bruno M, Formisano C, Rosselli S, Capasso R, et al. Antispasmodic effects and structure-activity relationships of labdane diterpenoids from *Marrubium globosum ssp. Libanoticum*. J Nat Prod 2009;72:1477−81.

[83] Li L, Tsao R, Liu Z, Liu S, Yang R, Young JC, et al. Isolation and purification of acteoside and isoacteoside from Plantago psyllium L. by high-speed counter-current chromatography. J Chromatogr A 2005;1063:161−9.

[84] Endo K, Takahashi K, Abe T, Hikino H. Structure of forsythoside B, an antibacterial principle of *Forsythia koreana* stems. Heterocycles 1982;19:261−4.

[85] Van LP, De Bruyne T. Reinvestigation of the structural assignment of signals in the 1 H and 13 C NMR spectra of the flavone apigenin. Magn Reson Chem 1986;24:879−82.

[86] Agrawal PK. Carbon-13 NMR of Flavonoids. Studies in organic chemistry, Carbon-13 NMR of flavonoids. Amsterdam: Elsevier; 1989. [Magn Reson Chem 28(6):562−63].

[87] Hayder NA, Abdelwahed S, Kilani R, Ben Ammar A, Mahmoud K, Ghedira L, et al. Anti-genotoxic and free-radical scavenging activities of extracts from (Tunisian) *Myrtus communis*. Mutat Res 2004;564:89−95.

[88] Bhouri W, Derbel S, Skandrani I, Boubaker J, Bouhlel I, Mohamed B, et al. Study of genotoxic, antigenotoxic and antioxidant activities of the digallic acid isolated from *Pistacia lentiscus* fruits. Toxicol In Vitro 2010;24:509−15.

[89] Shon MY, Choi SD, Kahng GG, Nam SH, Sung NJ. Antimutagenic antioxidant free radical scavenging activity of ethyl acetate extracts from white yellow and red onions. Food Chem Toxicol 2004;42:659−66.

[90] De flora S. Mechanisms of inhibitors of mutagenesis and carcinogenesis. Mutat Res 1998;402:151−8.

[91] Neffati I, Bouhlel M, Sghaiera B, Boubakera J, Limema I, Kilani S, et al. Antigenotoxic and antioxidant activities of *Pituranthos chloranthus* essential oils. Environ Toxicol Phar 2009;27:187−94.

[92] Westhuizen FH, Van D, Catharina S, van Rensburg J, George S, Rautenbach JL, et al. *In vitro* antioxidant, antimutagenic and genoprotective activity of *Rosa roxburghii* fruit extract. Phytother Res 2008;22:376−83.

[93] Sghaier MB, Bhouri W, Bouhlel I, Skandrani I, Boubaker J, Chekir-Ghedira L, et al. Inhibitory effect of *Teucrium ramosissimum* extracts on aflatoxin B1, benzo[a]pyrene, 4-nitro-o-phenylenediamine and sodium azide induced mutagenicity, Correlation with antioxidant activity. S Afr J Bot 2011;77:730−40.

[94] Hachicha SF, Skanji T, Barrek S, Ghrabi ZG, Zarrouk H. Composition of the essential oil of *Teucrium ramosissimum* Desf. (Lamiaceae) from Tunisia. Flavour Frag J 2007;22(2):101−4.

[95] Conover EA. Herbal agents and over-the-counter medications in pregnancy. Best Pract Res Clin En 2003;17(2):237−51.

[96] Shulz V, Hänsel R, Tyler VE. Fitoterapia Racional, um guia de fitoterapia para as Ciências da Saúde. Barueri, Ed. Manole, 2002. p. 386.

[97] Bicalho GC, Filho AAB. Peso ao nascer e influência do consumo de cafeína. Rev Saúde Públ 2002;36(2):180−7.

[98] Moreira LMA, Dias AL, Ribeiro HBS, Falcão CL, Felício TD, Stringuetti C, et al. Associação entre o uso de abortificantes e defeitos congênitos. Rev Bras Ginecol Obstet 2001;23(8):517—21.

[99] Marcus DM, Snodgrass WR. Do no harm, avoidance of herbal medicine during pregnancy. Obstet Gynecol 2005;105(5):1119—22.

[100] Vutyavanich T, Kraisarin T, Ruangsri R. Ginger for nausea and vomiting in pregnancy, randomized, double-masked, placebo-controlled trial. Obstet Gynecol 2001; 97(4):577—82.

[101] Wilkinson JM. Effect of ginger tea on the fetal development of Sprague-Dawley rats. Reprod Toxicol 2000;14(6):507—12.

[102] Myioshi N, Yoshimasa N, Yasuhiro U, Masako A, Yoshio O, Koji U, et al. Dietary ginger constituents, galanals A and B, are potent apoptosis inducers in human T lymphoma Jurkat cells. Cancer Lett 2003;199(2):113—9.

[103] Weidner MS, Sigwart K. Investigation of the teratogenic potential of a *Zingiber officinale* extract in the rat. Reprod Toxicol 2000;15(1):75—80.

[104] Singh JD. The teratogenic effects of dietary Cassava on the pregnant albino rat: a preliminary report. Teratology 1981;24:289—2917.

[105] Nana P, Asongalem EA, Foyet HS, Folefoc GN, Dimo T, Kamtchouing P. Maternal and developmental toxicity evaluation of *Acanthus montanus* leaves extract administered orally to Wistar pregnant rats during organogenesis. J Ethnopharmacol 2008;116:228—33.

[106] Eluwa MA, Ekanem TB, Udoh PB, Akpantah AO, Ekong MB, Asuquo OR, et al. Teratogenic effect of crude ethanolic root bark and leaf extracts of *Rauwolfia vomitoria* (Apocynaceae) on the liver of albino Wistar rat fetuses. Asian J Med Sci 2013; 4(1):30—4.

[107] Gill LS. Ethnomedical uses of plants in Nigeria. Benin: UNIBEN Press; 1992. p. 204.

[108] Akpantah AO, Oremosu AA, Noronha CC, Ekanem TB, Okanlawon AO. Effects of *Garcinia kola* seed extract on ovulation, oestrous cycle and foetal development in cyclic female Sprague-Dawley rats. Niger J Physiol Sci 2005;20(1—2):58—62.

[109] Cotterih P, Scheinmenn F, Stenhuise I. Composition of *G. kola* seeds. J Chem Soc 1978;1:532—3.

[110] Iwu MM, Igboko O. Flavonoids of *Garcinia kola* seeds. J Nat Prod 1982;45:650—1.

[111] Lawal RO. Effects of dietary protein on teratogenicity of polyphenols obtained from the outer coat of the fruit of *Treculia Africana*. Food Chem 1997;60(4):495—9.

[112] Smith JL, Stanley DW. Toughening in blanched asparagus, identification of phenolic compounds. Food Chem 1989;1314:271—87.

[113] Sanseverino MTV, Spritzer DT, Schuler-Faccini L. Manual de Teratogênese. Porto Alegre, Ed. da UFRGS; 2001. p. 423—50.

[114] Feitosa CRD-S, Da Silva RC, Braz-Filho R, De Menezes JESA, Siqueira SMC, Monte FJQ. Characterization of chemical constituents of *Luffa operculata* (Cucurbitaceae). Am J Anal Chem 2011;2:989—95.

[115] Barilli SLS, Santos ST, Montanari T. Efeito do decocto dos frutos de buchinha-do-norte (*Luffa operculata* Cogn.) sobre a reprodução feminina e o desenvolvimento embrionário e fetal. In, salão de iniciação científica (17, 2005, Porto Alegre). Livro de resumos. Porto Alegre, UFRGS; 2005. p. 539.

[116] Lachenmeier DW. Wormwood (*Artemisia absinthium* L.), a curious plant with both neurotoxic and neuroprotective properties. J Ethnopharmacol 2010;131:224—7.

[117] Robbers JE, Tyler VE. Tyler's Herbs of choice, the therapeutic use of phytomedicinals. New York, NY: Haworth Herbal; 1999. p. 287.

[118] Vostrowsky O, Brosche T, Ihm H, Zintl R, Knobloch K. The essential components from *Artemisia absinthium*. Z Naturforsch 1981;36:369−77.

[119] Tucker AO, Maciarello MJ. The essential oil of Artemisia "Powis Castle" and its putative parents, *A. absinthium and A. arborescens*. J Essent Oil Res 1993;5: 224−39.

[120] Chilava FP, Lidlle AP, Doglia G. Chemotaxonomy of wormwood (*Artemisia absinthium* L.). A. Lebensm Unters Forech 1983;176:363−6.

[121] Klayman DL, Lin AJ, Acton N, Scovill N, Hock JM, Milhous WK, et al. Isolation of artemisinin (qinghaosu) from *Artemisia annua* growing in the United States. J Nat Prod 1984;47:715−7.

[122] Bush LP, Crowe MW. Nicotiana alkaloids. Toxicants of plant origin. Alkaloids, vol. I. Boca Raton, FL: CRC Press Inc.; 1989. 335, p. 87−107.

[123] Schmidt SP, Forsythe WB, Cowgill HM, Myers RK. A case of congenital occipito atlantoaxial malformation (OAAM) in a lamb. J Vet Dia Inv 1993;5:458−62.

[124] MacMahon JA. (1997). *Deserts*, New York, NY, National Audubon Society Nature Guides, Knopf A.A. Inc, 9e éd.

[125] Panter KE, Keeler RF, Bunch TD, Callan RJ. Congenital skeletal malformations and cleft palate induced in goats by ingestion of *Lupinus, Conium* and *Nicotiana* species, USDA/ARS/Poisonous Plant Research Laboratory, Logan, UT. Toxicology 1990;28 (12):1377−85.

[126] Beachy PA, Chen JK, Cooper MK, Wang B, Mann RK, Milenkovic L, et al. Effects of oncogenic mutations in smoothened and patched can be reversed by cyclopamine. Nature 2000;406(6799):1005−9.

[127] Kimberling CY. Jensen and Swift's diseases of sheep. 3rd ed. Philadelphia/Paris: Lea and Febiger; 1988. p. 372−73.

[128] Lan Y-H, Chang F-R, Yang Y-L, Wu Y-C. New constituents from stems of *Goniothalamus amuyon*. Chem Pharm Bull 2006;54(7):1040−3.Changwichit K, Khorana N, Suwanborirux K, Waranuch N, Limpeanchob N, Wisuitiprot W, et al. Bisindole alkaloids and secoiridoids from *Alstonia macrophylla* Wall. ex G. Don. Fitoterapia 2011;82(6):798−804.

[129] Xue Y, Xue B. Isolation and structure identification of chemical constituents from the seeds of *Descurainia sophia* (L.) Webb ex Prantl. Yao Xue Xue Bao 2004;39(1):46−51.

[130] Mohamed NH, Mahrous AE. Chemical constituents of *Descurainia sophia* L. and its biological activity. Rec Nat Prod 2009;3(1):58−67.

[131] Hanson G. The toxicity of plants in equines, a modern three-point approach to disseminating information. A thesis presented in partial fulfillment of the requirements for the degree of Master of Science in Rangeland ecology and management in the college of graduate studies. Moscow: University of Idaho; 2008. p. 83−9.

[132] James LF, Keeler RF, Binns W. Sequence in the abortive and teratogenic effects of locoweed fed to sheep. Am J Vet Res 1969;30:377.

[133] Keller K. *Cinnamomum* species, adverse reactions of herbal drugs. Berlin: Springer-Verlag; 1992. p. 105−14.

[134] Warkany J. Congenital malformations. Chicago, IL: Year Book Medical Publishers Inc.; 1971. p. 1309.

[135] Gaffield W, Keeler RF. Structure-activity relations of teratogenic natural products. Pure Appl Chem 1994;66:2407−10.

[136] Tiwari A, Rajesh SJ, Tiwari P, Nayak S. Phytochemical investigations of crown of *Solanum melongena* fruit. Int J Phytomed 2009;1:9−11.

[137] Cano A, Acosta M, Arnao M. Hydrophilic and lipophilic antioxidant activity changes during on-vine ripening of tomatoes (*Lycopersicon esculentum* Mill). Postharvest Biol Tec 2003;28:59−65.

[138] Vital PG, Rivera WL. Antimicrobial activity, cytotoxicity, and phytochemical screening of *Voacanga globosa* (Blanco) Merr. leaf extract (Apocynaceae). Asian Pac J Trop Med 2011;4(10):824−8.

[139] Herrera AA. *In vivo* evaluation of the potent angiosuppressive activity of some indigenous plants from Bataan, Philippines. Asia Life Sci 2010;19(1):183−90.

[140] Liao JC. In Flora of Taiwan, vol. 2 (2nd ed.), Taipei: Editorial Committee of the Flora of Taiwan, 1996. p. 177−80.

[141] Simo CCF, Kouam SF, Poumale HMP, Simo KI, Ngadjui BT, Green RI, et al. Biochem Syst Ecol 2008;36:238−43.

[142] Pierangeli GV, Rivera WL. Antimicrobial activity and cytotoxicity of *Chromolaena odorata* (L. f.) King and Robinson and Uncaria perrottetii (A. Rich) Merr. Extracts. J Med Plants Res 2009;3(7):511−8.

[143] Ebada SS, Lajkiewicz N, Porco JA, Li-Weber M, Proksch P. Chemistry and biology of rocaglamides (=flavaglines) and related derivatives from aglaia species (meliaceae). Fortschr Chem Org Naturst 2011;94:1−58.

[144] Dapat E, Jacinto S, Efferth T. A phenolic ester from *Aglaia loheri* leaves reveals cytotoxicity towards sensitive and multidrug-resistant cancer cells. BMC Complement Altern Med 2013;13:286.

[145] Ragasa CY, Torres OB, Shen C-C, Mejia MGR, Ferrer RJ, Jacinto SD. Chemical constituents of *Aglaia loheri*. Pharmacogn J 2012;4(32):29−31.

[146] Durand, L., Morissette, F. and Lamoureux, G. (1981). Plantes sauvages comestibles, Le groupe Fleurbec.

[147] Campbell MA, Brown KS, Hassell JR, Horigan EA, Keeler RF. Inhibition of limb chondrogenesis by a Veratrum alkaloid, temporal specificity *in vivo* and *in vitro*. Dev Biol 1985;111:464−70.

[148] Lee ST, Panter KE, Gaffield W, Stegelmeier BL. Development of an enzyme-linked immunosorbent assay for the veratrum plant teratogens, cyclopamine and jervine. J Agr Food Chem. 2003;51(3):582−6.

[149] Dugoua JJ, Perri D, Seely D, Mills E, Koren G. Safety and efficacy of blue cohosh (*Caulophyllum thalictroides*) during pregnancy and lactation. Can J Clin Pharmacol 2008;15(1):e66−73.

[150] Mole MLJ, Turner CE. Phytochemical screening of *cannabis sativa* L. I, constituents of an Indian variant. J Pharm Sci 1974;63(1):154−6.

[151] Mechoulam R, Gaoni Y. Recent advances in the chemistry of hashish. Fortschr Chem Org Naturst 1967;25:175−213.

[152] Reece AS. Chronic toxicology of cannabis. Clin Toxicol (Phila) 2009;47(6):517−24.

[153] Cullis JE. Congenital deformities and herbal slimming tablets. Lancet 1962;2:5112.

[154] Sieber SM, Whang-Peng J, Botkin C, Knutsen T. Teratogenic and cytogenetic effects of some plant-derived antitumor agents (vincristine, colchicine, maytansine, VP-16-213 and VM-26) in mice. Teratology 1978;18(1):31−47.

[155] Changwichit K, Khorana N, Suwanborirux K, Waranuch N, Limpeanchob N, Wisuitiprot W, et al. Bisindole alkaloids and secoiridoids from *Alstonia macrophylla* Wall. ex G. Don. Fitoterapia 2011;82(6):798−804.

[156] Goel RK, Prabha T, Kumar MM, Dorababu M, Prakash, Singh G. Teratogenicity of *Asparagus racemosus* Willd. Root, an herbal medicine. Indian J Exp Biol 2006;44:570−3.

[157] Goyal RK, Singh JLH. *Asparagus racemosus*--an update. [Review]. Indian J Med Res 2003;57(9):408−14.

[158] Sharma PC, Yelne MB, Dennis TJ. New Delhi, central council for research in Ayurveda and Siddha, department of ISM&H, Ministry of Health and Family Welfare (Govt. of India). Database on medicinal plants used in Ayurveda 2005;3:76−87.

[159] Hayes PY, Jahidin AH, Lehmann R, Penman K, Kitching W, De Voss JJ. Steroidal saponins from the roots of *Asparagus racemosus*. Phytochemistry 2006;69(3): 796−804.

[160] Saxena VK, Chourasia SA. New isoflavone from the roots of *Asparagus racemosus*. Fitoterapia 2001;72(3):307−9.

[161] Keeler RF. Teratogens in plants. J Anim Sci 1984;58(4):1029−39.

[162] Chase RL, Keeler RF. Mountain thermopsis toxicity in cattle. Utah Sci 1983;44:28−31.

[163] Yadav R, Kaushik R, Ggupta D. The health benefits of *Trigonella foenum-graecum*: a review. IJCEA 2009;1(1):032−5.

[164] Araee M, Norouzi M, Habibi G, Sheikhvatan M. Toxicity of *Trigonella foenum-graecum* (fenugreek) in bone marrow cell proliferation in rat. Pak J Pharm Sci 2009;22(2):126−30.

[165] Pérez LB, Li J, Lantvit DD, Pan L, Ninh TN, Chai H-B, et al. Bioactive constituents of *Indigofera spicata*. J Nat Prod 2013;76(8):1498−504.

[166] Kapoor LD. Handbook of Ayurvedic medicinal plants. Boca Raton, FL: CRC Press; 1990. p. 200.

[167] Singh A, Singh RG, Singh RH, Mishra N, Singh N. An experimental evaluation of possible teratogenic potential in *Boerhaavia diffusa* in Albino rats. Planta Med 1991;57(4):315−6.

[168] Leyon PV, Lini CC, Kuttan G. Inhibitory effect of *Boerhaavia diffusa* on experimental metastasis by B16F10 melanoma in C57BL/6 mice. Life Sci 2005;76:1339−49.

[169] Oudhia, P. and Tripathi, R.S. (2002). Identification, cultivation and export of important medicinal plants. In: Proc. National Seminar on Horticulture Development in Chhattisgarh, Vision and Vistas. Indira Gandhi Agricultural University, Raipur (India), p. 78−85.

[170] Eweka AO. Histological studies of the teratogenic effects of oral administration of *Aspilia africana* (Asteraceae) leaf extract on the developing kidney of Wistar rats. Int J Toxicol 2008;4:2.

[171] Abii TA, Onuoha EN. The chemical constituents of the leaf of *Aspilia africana* as a scientific backing to its tradomedical potentials. Ag J 2011;6(1):28−30.

10 Mutagenicity and Carcinogenicity of African Medicinal Plants

Jean-de-Dieu Tamokou and Victor Kuete

University of Dschang, Faculty of Science, Department of
Biochemistry, Dschang, Cameroon

10.1 Introduction

The uses of medicinal plants have been part of human culture beginning from the era of early civilization, as evidenced by the earliest recorded uses found in Babylon (1770 BC) and in ancient Egypt (1550 BC) [1]. It is estimated that approximately 80% of the world's population rely primarily on traditional medications for their primary healthcare needs [2]. A greater part of these traditional medications involves the use of plant extracts and/or their active ingredients [3]. In developing countries where the availability of health facilities and basic medicines is limited, herbal medications play significant roles in the healthcare delivery of the people. In Africa, the use of crude herbal remedies is still popular among the rural population, mainly because of the peoples' social-cultural heritage and the affordability of medicinal plants. In South Africa for example, up to 60% of the population consult one of an estimated 200,000 traditional healers [4], especially in rural areas where traditional healers are more numerous and accessible than Western physicians.

All over the world, interest has increased in studying the biological effects of traditional medicinal plants as a prerequisite for using them as such or isolating their active components for treatment of illness [5–12]. Literature retrieved from scientific web sites such as Pubmed, Web-of-knowledge, Scopus, scirus, and Scholar google revealed that comprehensive screening programs have been established in different parts of the world for biological screenings of medicinal plants [5,13–16]. Already an estimated 122 drugs from 94 plant species have been discovered through ethnobotanical leads [17], among which were anticancer drugs such as taxanes and vincristine [18,19]. Cancer is characterized by a rapid and uncontrolled formation of abnormal cells which may mass together to form a growth or tumor, or proliferate throughout the body, indicating abnormal growth at other sites. Cancer is considered one of the most fearsome causes of morbidity and mortality in all over the world. Local herbalists have been treating various cancers and cancer-related conditions for ages [20] and many plants have been reported as useful in the management of such conditions.

Toxicological Survey of African Medicinal Plants. DOI: http://dx.doi.org/10.1016/B978-0-12-800018-2.00010-8

Despite the therapeutic advantages of the medicinal plants, some phytochemicals have been shown to be potentially toxic, mutagenic, and carcinogenic [21]. On the other hand, most of the traditionally used medicinal plants have never been subjected to exhaustive toxicological tests as required for modern pharmaceuticals. As a result, poisoning due to mutagenic/carcinogenic plants is not well documented because of the unwillingness of people to admit using traditional medicine derived from plant material and because of the fear that the cultural heritage of the people will be put under strong laws and regulations [22]. Damage of the genetic material by mutagenic plants can lead to mutations in many organisms, including humans. Mutations are associated with the development of most cancers and various degenerative disorders and genetic defects in offspring [23]. This concern has driven most of the mutagenicity testing programs [24]. Many plants are known to contain mutagenic, carcinogenic, or cocarcinogenic compounds such as furocoumarins [25], tannins, anthraquinones [26], and flavonoids [27]. The enzymes responsible for the activation of the promutagens/carcinogens are present in different cells of mammals, and such activation happens frequently [28] and in many cases, even a very low exposure to the mutagenic/carcinogenic agent may be enough to induce a genotoxic effect. To prevent mutagenic/carcinogenic risk, it is important to identify the involved mutagenic/carcinogenic plants and minimize human exposure to them. Short-term genetic bioassays such as the bacterial Ames test [24,29] and Vitotox® test [30], which detect gene mutations and/or SOS-repair events, as well as the micronucleus [31] and alkaline Comet assays [32], which respectively allow detection of genome and/or chromosome mutations and single-strand DNA breakage and alkali labile sites have been used as important tools in mutagenic/carcinogenic studies of medicinal plants [33]. As a result, the possible mutagenic effects of 51 South African species used frequently in traditional medicine were reported [34] while many others across the world have mutagenic [17,35−45] and carcinogenic [46−50] properties. Plants exhibiting clear mutagenic/carcinogenic properties should be considered as potentially unsafe and certainly require further testing before their continued use can be recommended in traditional medicine. The aim of this chapter is to report the mutagenic/carcinogenic plants used in African traditional medicine to treat different health conditions in order to warn about their use.

10.2   Medicinal Plants with Mutagenic Effects

In developing countries, there is increasing interest and research in the area of herbal medicines as an approach to reducing costs of healthcare [51]. About 25% of the medicaments are prescribed worldwide and 11% of these are considered essential by the World Health Organization (WHO) are products derived from plants [52]. However, the chemically complex nature of these medicinal preparations results in a significant increased risk of toxicity, including mutagenicity. A total of 138 medicinal plant preparations used in the Philippines have been examined for mutagenicity using various short-term bacterial and mammalian tests [51]. Of these plants, 12 exhibited detectable mutagenicity. These include *Allium sativum* (Amaryllidaceae), *Aloe barbadensis* (Liliaceae), *Archangelisa flava* (Lamiaceae), *Canarium luzonicum* (Bursecraceae), *Capsicum frutescens* (Solanaceae), *Entada phaseoloides* (Fabaceae), *Moringa oleifera* (Moringaceae), *Nerium indicum* (Apocynaceae), *Piper betle* (Piperaceae), *Pithecellobium dulce* (Fabaceae), *Pittosporum pentandrum* (Pittosporaceae), and *Plantaqo major* (Plantaginaceae). Little is known about the chemical nature of the mutagenic agents in these preparations. Some plants also contain substances which reduce genotoxicity either by acting directly on the mutagen (desmutagens) or by acting on the affected organism (antimutagens).

According to Ames [53], plants utilized in diets and employed in medical treatments possess chemical compounds that are capable of inducing mutations. Studies performed recently have demonstrated that extracts and enriched fractions obtained from Brazilian medicinal plants were mutagenic *in vitro* and *in vivo* [54,55]. Dantas de Carvalho et al. [56] evaluated an extract of *Schinus terebinthifolius* (Anacardiaceae) in a series of cell-free and bacterial assays in order to determine its mutagenic potential. The effect of the extract was negative in a cell-free plasmid DNA test, indicating that it did not directly break DNA. Positive results, however, were obtained in the SOS chromotest, in a forward mutagenesis assay employing CC104 and CC104*mut*M*mut*Y strains of *Escherichia coli*, and in the *Salmonella* reversion assay, using strains TA97, TA98, TA100, and TA102. All the bacterial tests were performed without exogenous metabolic activation due to the topical use of this preparation. The results indicate that *Schinus terebinthifolius* (Anacardiaceae) stem bark extract produces DNA damage and mutation in bacteria, and that oxidative damage may be responsible for the mutagenicity. Some mutagenic and genotoxic activities of *Morinda lucida* root (Rubiaceae) using *Escherichia coli* (0157:H7) mutagenic test (modified Ames test) and the *Allium cepa* (Amaryllidaceae) were reported [57]. Ethanolic extracts of 40 Jordanian medicinal plants were examined for cytotoxicity, mutagenicity, and antimicrobial activity [58]. Regarding the mutagenic effect, 10 plants namely *Daphne linearifolia* (Thymelaeaceae), *Peganum harmala* (Zygophyllaceae), *Achillea falcata* (Asteraceae), *Teucrium polium* (Lamiaceae), *Mentha longifolia* (Lamiaceae), *Citrullus colocynthis* (Cucurbitaceae), *Balanites aegyptiaca* (Balanitaceae), *Beta vulgaris* (Chenopodiaceae), *Ceratonia siliqua* (Fabaceae), and *Senecio vernalis* (Asteraceae) were active on *Salmonella typhimurium* strains TA98 and TA100 histidine deficient. The most active extract was *Peganum harmala* (Zygophyllaceae) which was mutagenic against both TA98 and TA100, while *Ceratonia siliqua* (Fabaceae) showed the lowest activity against both tested strains. Khan and Awasthy [59] reported that leaf extract of *Azadirachta indica* (Meliaceae) induced structural and numerical changes in the spermatocyte chromosomes as well as synaptic disturbances in murine germ cells at their first metaphase. They also observed a significant increase in the frequency of sperm with abnormal head morphology and a decrease in mean sperm count.

Aqueous extracts of three species used in Brazilian popular medicine (*Sambucus australis*: Adoxaceae, *Bauhinia forficata*: Compositae, *Mimosa bimucromata*: Fabaceae) were screened for the presence of mutagenic activity in the Ames test extracts, using TA98, TAIOO, and TA102 strains of *Salmonella typhimurium* with (+S9) and without (−S9) metabolization [60]. The extracts showed frameshift mutagenic activity after metabolization. In addition, *Mimosa bimucromata* (Fabaceae) presented positive results in the TA100 strain, which detects a base pair substitution, with and without metabolization. The metabolites of *Bauhinia forficata* (Compositae) extract also showed mutagenic activity in the TA102 strain. The presence of flavonoids and tannins in the extracts was correlated to the positive mutagenic activity [60]. It already has been shown that the flavones with certain hydroxylation patterns (5,7-hydroxy substitution) like quercetin and kaempferol cause different genetic injuries such as chromosomal aberrations, sister chromatid exchanges, single-strand DNA breaks, and weak cellular transformation [61−64]. Furthermore, the plant extracts of *Achyrocline satureioides* (Compositae) and *Luehea divaricata* (Malvaceae) showed mutagenic activity after metabolization in the Ames test related to the presence of flavonoids [63,64]. *Thermopsis turcica* (Fabaceae) is an endemic species of Turkey popularly known as Eber Sarisi or Piyam; it is a member of the *Thermopsis* genus, which contains plants considered throughout the world to be medicinal. The mutagenic effect of aqueous extracts obtained from the leaves, root, stem, and flowers of this plant species were evaluated by the Ames test, using *Salmonella typhimurium* TA97, TA98, TA100, and TA102 strains with and without an S9 activation system. The results further demonstrated that, in relation to the concentration, leaf extracts of *Thermopsis turcica* increased the frequency of reverse mutation in *Salmonella typhimurium* TA102 with S9 mix [65]. Positive results also have been reported from the Ames test using extracts of *Crinum macowanii* (Amaryllidaceae), *Catharanthus roseus* (Apocynaceae), *Combretum mkhzense* (Combretaceae), *Diospyros whyteana* (Ebenaceae), *Plumbago auriculata* (Plumbaginaceae), *Ziziphus mucronata* (Rhamnaceae), and *Chaetacme aristata* (Ulmaceae) [34], as well as the genus *Helichrysum* Mill. [66]. The hydroalcoholic extract of *Ocotea duckei* (Lauraceae) leaves was found to be mutagenic for the *Salmonella typhimurium* TA97a, TA100, and TA102 strains, with or without S9 mix [67]. Déciga-Campos et al. [36] found that *Gnaphalium* sp. and *Valeriana procera* (Valerianaceae) extracts induced mutations of *Salmonella typhimurium* TA98 with or without S9 mix and of TA100 with S9 mix, respectively. The tubers of *Gloriosa superba* (Colchicaceae) were found to contain potent mutagenic properties in an Ames mutagenicity test on *Salmonella* [68]. It was also shown that compounds present in the methanolic extracts of the leaves of *Alchornea castaneaefolia* (Euphorbiaceae) and *Alchornea glandulosa* (Euphorbiaceae) were mutagenic in an Ames test [69].

Six widely used medicinal plants, namely *Nerium odorum* (Apocynaceae), *Andrographis paniculata* (Acanthaceae), *Nyctanthes arbortristis* (Oleaceae), *Phlogacanthus thyrsiflorus* (Acanthaceae), *Solanum indicum* (Solanaceae), and *Kaempferia galanga* (Zingiberaceae), were examined for their mutagenicity using plant cytogenetic assay [70]. Exposure of *Vicia faba* (Fabaceae) root tips to different doses of the plant aqueous extracts, using acetocarmine squash technique revealed that the extracts of *Nyctanthes odorum* (Oleaceae) and *Solanum indicum* (Solanaceae) were mitodepressant and induced higher frequencies of traditional chromosome aberration, abnormal chromosome behavior and micronucleus, indicating their clastogenic potentials, while the other four plant extracts produced no significant change in all the testing protocols, indicating that they were not clastogenic.

10.3   Medicinal Plants with Carcinogenicity Effects

Carcinogenicity is the ability or tendency of a chemical to induce tumors (benign or malignant), increase their incidence or malignancy, or shorten the time of tumor occurrence when it is inhaled, ingested, dermally applied, or injected. Worldwide, many scientists have documented the carcinogenic properties of several medicinal plants. For example, 12 medicinal herbs used as home remedies and beverages in South Carolina, USA, were evaluated in outbred NIH Black rats following repeated subcutaneous injections to correlate a high incidence of esophageal carcinoma in natives of different places with their habitual consumption of these products [47]. The tannin-rich plant extracts from *Areca catechu* (Arecaceae) and *Rhus copallina* (Anacardiaceae) produced local tumors in 100% and 33% of the experimental animals, respectively. Other plants extract not rich in tannins namely *Sassafras albidum* (Lauraceae), *Diospyros virginiana* (Ebenaceae), *Chenopodium ambrosiodes* (Chenopodiaceae), *Hamamelis virginiana* (Hamamelidaceae), and *Psidium guajava* (Myrtaceae) were tumorigenic in 66%, 56%, 53%, 10%, and 6% of the treated animals, respectively. The carcinogenicity of uncured raw betel nuts (*Areca catechu*) combined with active shell lime and chewing tobacco (l:1:1) has been studied in mice by Reddy and Anguli [71]. Vaginal epithelial papillomatous growths and vaginal mucosal thickening with changes of epithelia and submucosa were observed. Metastases to lungs, kidneys, and intraperitoneal regions were also found. Dunham et al. [72] studied the carcinogenicity of arecoline, an alkaloid from *Areca catechu* (Arecaceae). Proliferative lesions were developed in the esophagi of two hamsters in a group of nine that received calcium hydroxide applied to the cheek pouch followed by painting with arecoline. *Sassafras albidum* is known to contain safrole, isosafrole, anethole, and eugenol. The propenyl benzene derivatives eugenol and anethole are known to be weak hepatotoxins [73,74]. Safrole has been reported to be a hepatocarcinogen [75−77].

   The decoctions of 14 plants commonly taken internally by esophageal cancer patients on the island of Curacao were assayed for carcinogenic activity in rats and mice [78]. *Krameria ixina* (Krameriaceae) induced subcutaneous sarcomas at the site of injection in 100% of the rats. *Annona muricata* (Annonaceae), *Heliotropium ternatum* (Boraginaceae), *Citrus aurantium* (Rutaceae), and *Gliricidia sepium* (Fabaceae) each caused subcutaneous sarcomas in one or two rats, of 15 tested. No tumors were induced in mice given the same decoctions orally. Plants of the *Aristolochia* genus have been used for centuries in Chinese herbal remedies, but they contain a naturally carcinogenic compound (aristolochic acid) that causes mutations in the cells of people who consume them [49,50]. The hurtful potential of aristolochic acid is discussed in Chapter 19. More recently, *Aristolochia* contamination of local wheat crops was determined to be the cause of a high incidence of upper tract urothelial carcinoma (UTUC) among rural communities on the banks of the Danube River in Europe [49,50]. The total aqueous extracts and the tannin-containing fractions of *Krameria ixina* (*Krameriaceae*), *Krameria triandra* (*Krameriaceae*), and *Aristolochia villosa* (Aristolochiaceae) produced malignant fibrous histiocytomas after varying numbers of injections in NIH Black rats [46]. The tannin-free fractions had little carcinogenicity, while the tannin-containing fractions contained most of the carcinogenic material. The extracts/fractions from *Aristolochia villosa* (Aristolochiaceae) produced tumors in the shortest time and were the most potent.

10.4   Methods in Mutagenic Study of Medicinal Plants

Mutagenicity refers to the induction of transmissible changes in the structure of the genetic material of cells or organisms [79] involving a single gene or a group of genes. To evaluate mutagenicity, various end points must be taken into consideration: (i) beside point mutations induction; (ii) changes in chromosomal number (polyploidy or aneuploidy); or (iii) in chromosome structure (breaks, deletions, rearrangements). Thus, there is currently no single validated test that can provide information on all three end points involved in determining mutagenic potential of a given chemical. Consequently, a battery of tests is needed to determine the mutagenic profile of a compound. This includes conventional and toxicogenomics methods.

10.4.1   Conventional Methods

Conventional methods are *in vitro* and *in vivo* methods as well as those able to detect phytochemical compounds bearing structural alerts for mutagenicity activity. The Organization for Economic Cooperation and Development (OECD) [80] and the European Centre for the Validation of Alternative Methods [81] have investigated the validation of mutagenicity assays and should be referred to for more details.

10.4.1.1   Detection of Natural Compounds Bearing Toxicophores for Mutagenic Activity

Toxicophores or structural alerts are molecular functionalities that are associated with toxicity. Their presence in compounds alerts the investigator to their potential toxicity [82]. Analytical methods based on thin layer chromatography, spectrophotometry (the Ehrlich reagent for pyrrolizidine alkaloids), liquid chromatography/mass spectroscopy, and gas chromatography/mass spectroscopy [83−85] have been developed for the characterization and quantification of many known molecules or molecular functionalities associated with mutagenicity. They are now being adopted in official pharmacopeias. Several other methods for predicting mutagenicity, namely the *in silico* methods, have been developed. Such methods generally refer to a computational experiment, mathematical calculation, and scientific analysis of substances data through a computer-based analysis [86]. Computer models used for mutagenicity prediction can be grouped into three principal categories [87]: (i) quantitative structure−activity relationship models (QSAR) such as TOPKAT that uses "electro-topological" descriptors (atom-type, bond-type, and group-type E-state) rather than chemical structure to predict mutagenic reactivity with DNA [87]; (ii) rule-based expert systems such as DEREK that estimates the presence of a DNA-reactive moiety in a given molecule [88]; and (iii) three-dimensional computational DNA-docking model to identify molecules capable of noncovalent DNA interaction [89]. In general, *in silico* methods are rapid, cheaper, have higher reproducibility and low compound synthesis requirements, can undergo constant optimization, and have potential to reduce or replace the use of animals [86]. However, the computation of mutagenicity is complex, and predictive capability has been limited.

10.4.1.2   In Vitro *Methods*

10.4.1.2.1   Bacterial Reverse Mutation Test (Ames test)

The bacterial reverse mutation test, also known as the Ames test, utilizes prokaryotic cells, which differ from mammalian cells in factors such as uptake, metabolism, chromosome structure, and DNA repair processes. As the test is investigated on nonmammalian cells, it can be performed both in the absence and in the presence of an exogenous metabolizing system, often a rat liver S9 (microsomal) suspension. The test uses amino acid requiring strains of *Salmonella typhimurium* and *Escherichia coli* to detect point mutations, which involve substitution, addition, or deletion of one or a few DNA base pairs [29,90]. The principle of this test is that it detects mutations which revert mutations present in the test strains and restore the functional capability of the bacteria to synthesize an essential amino acid. The revertant bacteria are detected by their ability to grow in the absence of the amino acid required by the parent test strain. The Ames test is rapid, inexpensive, and relatively easy to perform. However, it does not provide direct information on the mutagenic and carcinogenic potency of a substance in mammals.

10.4.1.2.2   *Aspergillus nidulans* Assays

Due to its parasexual cycle, the filamentous fungus *Aspergillus nidulans* constitutes an excellent system for studying mitotic crossing-over, because its cells spend a substantial part of their cell cycle in the G2 phase during the germination period. Two copies of each chromosome during that period of the cell cycle significantly favor a mitotic recombination event, which can be visually detected by simple plating tests [91,92].

10.4.1.2.3   *Saccharomyces cerevisiae* Gene Mutation Assay

A variety of haploid and diploid strains of the yeast *Saccharomyces cerevisiae*, a eukaryotic microorganism, can be used to measure the production of gene mutations induced by chemical agents, with and without metabolic activation. Forward mutation systems in haploid strains, such as the measurement of mutation from red, adenine-requiring mutants (*ade*1, *ade*2) to double adenine-requiring white mutants, and selective systems, such as the induction of resistance to canavanine and cyclo-heximide, have been used. The most extensively validated reverse mutation system involves the use of the haploid strain, XV185−14C, which carries the ochre non-sense mutations *ade*2-1, *arg*4-17, *lys*1-1, and *trp*5-48, which are reversible by base substitution mutagens that induce site-specific mutations or ochre-suppressor muta-tions. XV 185-14C also carries the *his*1-7 marker, a mis-sense mutation reverted mainly by second site mutations, and the marker, *hom*3-10, which is reverted by frameshift mutagens. In diploid strains, the only widely used *Saccharomyces cere-visiae* assay, proposed by Zimmermann et al. in 1975 [93,94], relies on the D7 strain, which is homozygous for *ilv*1-92.

10.4.1.2.4 The *Saccharomyces cerevisiae* Mitotic Recombination Assay
Mitotic recombination (gene conversion or crossing-over) in *Saccharomyces cere-visiae* can be detected between genes (or more generally between a gene and its centromere) and within genes following treatment with a mutagen. These recombi-nations are essentially DNA exchanges between segments of homologous chroma-tids and this assay gives an indication of non-specific DNA damage. The former event is called mitotic crossing-over and generates reciprocal products, whereas the latter event is most frequently nonreciprocal and is called gene conversion. The most commonly used strains for the detection of mitotic gene conversion are D_4 (heteroallelic at *ade*2 and *trp*5); D_7 (heteroallelic at *trp*5); BZ_{34} (heteroallelic at *arg*4), and JD1 (heteroallelic at *his*4 and *trp*5). Mitotic crossing-over producing red and pink homozygous sectors can be assayed in D_5 or in D_7 (which also measures mitotic gene conversion and reverse mutation at *ilv*1-92) both strains being hetero-allelic for complementing alleles of *ade*2.

10.4.1.2.5 *In Vitro* Mammalian Chromosome Aberration Test
The purpose of the *in vitro* chromosome aberration test is to identify agents that cause structural chromosome aberrations in cultured mammalian cells [95,96]. Structural aberrations may be of two types, chromosome or chromatid.

Cell cultures are exposed to the test substance both with and without metabolic activation.

At predetermined intervals after exposure of cell cultures to the test substance, they are treated with a metaphase-arresting substance (e.g., Colcemid® or colchi-cine), harvested, stained, and metaphase cells are analyzed microscopically for the presence of chromosome aberrations [97].

10.4.1.2.6 *In Vitro* Mammalian Micronucleus Assay
The purpose of the *in vitro* micronucleus assay is to identify agents that cause structural and numerical chromosome changes and is an alternative to *in vivo* micronucleus and chromosome aberration testing. The test employs cultures of established cell lines or primary cell cultures. Micronuclei are acentric chromo-somal fragments or whole chromosomes left behind during mitotic cellular divi-sion, appearing in the cytoplasm of interphase cells as small additional nuclei [98]. Cells treated with tried sample are grown in the presence of cytochalasin-B to pre-vent the cytoplasmic division after nuclear division, fixed, stained, and scored for

binucleated and micronucleated cells [31,93]. The manual or automated detection of micronuclei [99,100] provides a readily measurable index of chromosome breakage and loss. In addition to the general limitations of *in vitro* systems, apoptosis may interfere with the scoring of micronuclei, giving rise to false-positive results.

10.4.1.2.7 Sister Chromatid Exchange Test
Numerous cytomolecular protocols have been used to perform the sister chromatid exchanges (SCEs) assay [101−104]. The exchange process presumably involves DNA breakage and reunion, although little is known about its molecular basis. The detection of SCEs requires some means of differentially labeling sister chromatids, and this can be achieved by the incorporation of bromodeoxyuridine (BrdU) into chromosomal DNA for two cell cycles.

10.4.1.2.8 Unscheduled DNA Synthesis in Mammalian Cells
The unscheduled DNA synthesis (UDS) test measures DNA repair after the excision of a stretch of DNA containing the region of damage induced by chemical and physical agents. The test is based on the incorporation of tritium labeled thymidine (^{3}HTdR) into the DNA of mammalian cells which are not in the S-phase of the cell cycle. ^{3}HTdR uptake can be determined by autoradiography or by liquid scintillation counting. Unless metabolically competent primary rat hepatocytes are used, the cells are treated with the test agent, with and without an exogenous metabolic activation system. A core limitation of the UDS assay is inability to indicate if a xenobiotic is mutagenic; indeed, it provides no information regarding the fidelity of DNA repair and it does not identify DNA lesions handled by mechanisms other than excision repair [105].

10.4.1.2.9 *In Vitro* Comet Assay
The Comet assay or single-cell gel electrophoresis assay developed by Singh et al. [106] is a method for measuring DNA strand breaks and can serve as an alternative to *in vivo* testing for DNA damage. In this test, isolated cells embedded in agarose are lysed, washed to remove membranes and proteins, briefly electrophoresed, stained, and examined under epifluorescence microscopy; strand breaks, coming from either strand breakage or excision repair, result in DNA extending toward the anode in a structure resembling a "comet" [107,108]. A broad spectrum of DNA damage can then be detected either by visual classification of comet morphologies ("visual scoring") [109,110] or from morphological parameters obtained by image analysis and integration of intensity profiles using in-house or commercially available systems. The Comet assay has a number of advantages: DNA strand breaks in individual cells can be detected; only a small number of cells are necessary; proliferating cells are not required; and the assay can be performed on virtually any cell line or tissue [79]. There are only few limitations of the single-cell gel electrophoresis assay with regard to its application and interpretation. Short-lived primary DNA lesions such as single-strand breaks, which may undergo rapid DNA repair, could be missed when using inadequate sampling times. Furthermore, indirect mechanisms related to cytotoxicity (e.g., DNA fragmentation in apoptosis) can lead to positive effects [111].

10.4.1.2.10   In Vitro *Mammalian Cell Gene Mutation Test*

This test allows the detection of gene mutations induced by chemical substances. Suitable cell lines include L5178Y mouse lymphoma cells, the CHO, CHO-AS52, and V79 lines of Chinese hamster cells, and TK6 human lymphoblastoid cells [112]. In these cell lines, the most commonly used genetic end points measure mutation at thymidine kinase (TK) [113] and hypoxanthine-guanine phosphoribosyl transferase (HPRT), and a transgene of xanthine-guanine phosphoribosyl transferase (XPRT) [114]. The TK, HPRT, and XPRT mutation tests detect different spectra of genetic events. The autosomal location of TK and XPRT can detect some genetic events (e.g., large deletions) that cannot be detected at the HPRT locus on X-chromosomes [115,116]. Most of the substances that are positive in this mammalian gene mutation test also induce clastogenic effects [117].

10.4.1.3   In Vivo *Methods*

In order to overcome some limitations of the *in vitro* evaluations, *in vivo* assays have been also developed, not to replace them, but to complete their information on a whole organism.

10.4.1.3.1   *In Vivo* UDS Assay

This test is generally assessed in the hepatocytes of treated animals following the same detection systems as its corresponding *in vitro* assay.

10.4.1.3.2   *In Vivo* Micronucleus Assay

Animals are exposed to the test substance by an appropriate route. Peripheral blood and bone marrow are collected by venipuncture and removed from the femur, respectively; they are then smeared, stained, and scored as described for the *in vitro* assay [118,119]. For studies with peripheral blood, as little time as possible should elapse between the last exposure and cell harvest. Preparations are analyzed for the establishment of the frequency of micronuclei containing lagging chromosome fragments or whole chromosomes [94,120].

10.4.1.3.3   Transgenic Rodent Mutation Assay

The transgenic rodent (TGR) mutation assays have been reviewed extensively [121−123]. Transgenic animals carry multiple copies of chromosomally integrated plasmid and phage shuttle vectors that contain reporter genes for the detection of various types of mutations induced *in vivo* by test chemicals. Mutations arising in a rodent are scored by recovering the transgene and analyzing the phenotype of the reporter gene in a bacterial host deficient for the reporter gene. TGR gene mutation assays measure mutations induced in genetically neutral genes recovered from virtually any tissue of the rodent. Some deletions and insertion mutations may however not be detected in phage-based TGR. The assay does not involve a large number of animals and a major advantage is that mutations can be evaluated in any tissue; the protocol is reproducible but requires well-trained experts, is not yet automated, and the assay cost is superior to most of the other genotoxicity assays [105].

10.4.1.3.4   Mouse Spot Test

This is an *in vivo* assay in mice in which developing embryos are exposed to a compound. The test is based on the observation that mutagenic compounds can induce color spots on the coat of mice exposed *in utero*. The color spots arise when mouse melanoblasts, heterozygous for several recessive coat color mutations, lose dominant allele through a gene mutation, chromosomal aberration or reciprocal recombination, allowing the recessive gene to be expressed [105,124]. The frequency of such spots in treated groups is compared with their frequency in the control group.

10.4.1.3.5   *In Vivo* Comet Assay

The *in vivo* comet assay is now well established as a supplementary assay to the standard battery of mutagenicity tests and can be used to assist in evaluating chemical which have given equivocal results in other *in vivo* mutagenicity tests or to investigate the potential mechanisms of mutagenic responses [125,126]. Following treatment and sacrifice of animals, blood lymphocytes and/or cells, dissociated from organs by mincing a small piece into very fine fragments, are treated as per the same protocols as *in vitro* studies [110,127]. The *in vivo* Comet test is more suited to detect clastogens and also to detect low levels of DNA damage, requiring small numbers of cells per sample [111].

10.4.1.3.6   Somatic Mutation and Recombination Test

The somatic mutation and recombination test is based on the loss of heterozygosity for two recessive markers [128,129] and aimed to evaluate gene mutations, chromosome aberrations, and rearrangements related to mitotic recombination [130]. In the few past years, the Comet assay has been adapted to be used *in vivo* in *Drosophila melanogaster* [129,131].

10.4.2   Toxicogenomics Methods

Toxicogenomics is a comparatively new scientific subdiscipline that combines the emerging technologies of genomics and bioinformatics to identify and characterize mechanisms of action of toxicants. Changes in production of small metabolites as well as in gene and protein expression as a result of exposure to a toxic chemical or physical agent can be measured in virtually any tissue (*in vitro* or *in vivo*). The rapidly developing field of toxicogenomics is expected to have a large impact on both the fields of genetic toxicology and carcinogenicity, as a result of increased understanding of these processes, which will promote the development of better tools for assessing these end points. This new approach has the potential to reduce the amount of testing normally required to define a mechanism or mode of action [79]. This discipline is based on the concept that the toxic effects of xenobiotics on biological systems are generally reflected at cellular level by their impact on the expression of genes (transcriptomics) and proteins (proteomics) and on the production of small metabolite (metabonomics) [132–136]. However, toxicogenomics methods are at the stage of research and development and most of the experimental results are not sufficiently well established to be suitable for regulatory decision making. The laboratory techniques and test procedures are not well validated. In addition, the test systems used are variable, so the results obtained may not be consistent and cannot be generalized or extrapolated across species or to different human populations [79].

10.5 Methods in Carcinogenicity Survey of Medicinal Plants

Carcinogenic substances are those that induce tumors (benign or malignant), increase their incidence or malignancy, or shorten the time of tumor occurrence when they are inhaled, ingested, dermally applied, or injected [137]. Carcinogens are classified according to their mode of action as genotoxic or nongenotoxic carcinogens. Genotoxic carcinogens initiate carcinogenesis by direct interaction with DNA, resulting in DNA damage or chromosomal aberrations that can be detected by genotoxicity tests [138]. Nongenotoxic carcinogens are agents that, at least initially, directly interact with DNA. These indirect modifications to DNA structure, amount, or function may result in altered gene expression or signal transduction [138]. The main phases which play a role in the development of tumors are initiation, promotion, and progression. Due to the complexity of this process, numerous *in vivo* and *in vitro* methods were developed to assess or predict the carcinogenicity of a chemical agent [139].

10.5.1 In Vivo *Methods*

10.5.1.1 *Long-Term Rodent Carcinogenicity Bioassay*

The long-term rodent carcinogenicity bioassay is the conventional test for carcinogenicity described in OECD Test Guideline (TG) 451 [140]. The aim of this assay is to observe test animals for a major portion of their life span for the development of neoplastic lesions during or after exposure to various doses of a test substance by an appropriate route of administration. Two end points in animal bioassays, carcinogenicity and chronic toxicity, can be combined to reduce animal use, as described in OECD TG453 [141]. The test substance is administered daily in graduated doses to several groups of test animals (rats or mice) for the majority of their life span, mainly by the oral, dermal, or inhalation routes, based upon the expected type of human exposure. The animals are observed closely for signs of toxicity and for the development of neoplastic lesions. Animals that die or are killed during the test are necropsied and, at the conclusion of the test, surviving animals are killed and necropsied.

10.5.1.2 *TGR Models*

The transgenic mouse lines are chosen with the objective of developing a short-term assay *in vivo* that could be used to differentiate carcinogenic from noncarcinogenic chemicals, and mutagenic from nonmutagenic carcinogens [137].

Three mouse models are considered to have the greatest potential usefulness: Tg. AC, heterozygous $(+/-)$ p53-deficient, and TgHras 2 [142–144]. Regulators at the 2003 International Life Sciences Institute (ILSI) Health and Environmental Science Institute's (HESI) workshop on the use of transgenic animals for carcinogenicity testing concluded that these assays should be integrated with traditional test methods. Tg.AC Mice carry a v-Ha-ras oncogene fused to the promoter of the zeta globin gene [144]. The v-Ha-ras transgene has point mutations in codons 12 and 59, and the site of integration of the transgene confers, on these mice, the characteristic of genetically initiated skin as a target for tumorigenesis [137]. Mice that are heterozygous for suppressor gene p53 have an increased risk of tumor development [142]. The heterozygous p53-deficient mice used in rapid (26-weekexposure) studies for the identification of mutagenic carcinogens have one functional wild-type allele and one inactivated null allele, and they usually remain free of sporadic neoplastic disease during short-term studies [137]. CBF1-Tg-Hras2 mice carry five or six copies of a human c-Ha-ras proto-oncogene, which is expressed both in spontaneous tumors and normal tissue [143]. However, none of the above three models is considered sufficient as standalone assays. Most could detect genotoxic compounds that a genotoxicity test battery would already detect, but better detection of nongenotoxic carcinogens is still needed [145].

10.5.2   In Vitro *Methods*

The *in vitro* methods include cell-based assays and computational prediction models.

10.5.2.1   Cell Transformation Assays

The most widely used of cell transformation assay include the Syrian hamster embryo (SHE) assay, the low-pH SHE assay, the Balb/c 3T3 assay, and the C3H10T1/2 assay [146–148], which rely on changes in mammalian cell colony morphology and focus formation following exposure to chemical substances [137]. The SHE test is believed to detect early steps of carcinogenesis, and the Balb/c and C3H10 assays later carcinogenic changes [138]. Cytotoxicity of test substances can be determined by measuring the effects on morphology, colony-forming ability, and/or growth rate [149]. "Accumulated evidence strongly supports the assumption that cellular and molecular processes involved in cell transformation *in vitro* are similar to those of *in vivo* carcinogenesis" [138,149]. The OECD and ECVAM have started to consider formal validation of SHE and Balb/3T3 assays.

10.5.2.2   Gap Junction Intercellular Communication Method

The gap junction intercellular communication is a method for the identification and characterization of cancer causing substances without genotoxic activity [150]. The assay relies on the disruption of the intercellular exchange of low-molecular-weight molecules through the gap junctions of adjacent cells; this disruption can result in abnormal cell growth and behavior [137]. This method could be a candidate end point for use in screening assays for the identification of nongenotoxic carcinogens and tumor promoters not detected by conventional genetic toxicology tests, but it still needs to be standardized and validated.

10.5.2.3   Mutagenicity/Genotoxicity Assays

The mutagenicity/genotoxicity assays are the most commonly used *in vitro* test systems to predict carcinogenicity. According to UN Globally Harmonized System (GHS) guidance, chemical-induced tumorigenesis involves genetic changes; thus, chemicals that are mutagenic in mammals may warrant being classified as carcinogens [139]. A strong correlation between DNA adduct formation and tumor incidence has been reported by Ottender and Lutz [151] in their review of 27 genotoxic carcinogens. There was a 100% correlation reported between adduct formation and tumors. Other investigators have found that the micronucleus test is a useful *in vivo* assay for chemicals that cause neoplasms in the hematopoietic system or when the lung is the target tissue for carcinogenesis [152]. Dean et al. [153] reviewed the data from transgenic mouse mutation assays with 14 potent site-of-contact carcinogens that are also mutagenic. They reported a good correlation between mutations in specific tissues and primary tumors at the same site. The alkaline single-cell electrophoresis assay (Comet assay) has been used by several investigators to compare genetic toxicity in multiple organs of mice and rats with carcinogenicity. Sasaki et al. [154] showed that the direct acting mutagen and carcinogen, ethyl-nitrosourea, induced DNA damage in liver, lung, kidney, spleen, and bone marrow cells of mice 3 and 24 h after treatment, while the liver carcinogen, *p*-dimethylaminoazobenzene, was only positive in the mouse liver.

10.5.2.4   Computational Prediction Models

The quantitative/qualitative structure−activity relationship models (QSARs and SARs) and expert systems have been developed to predict carcinogenicity. Early reports have reviewed models such as TOPKAT, CASE, and DEREK, used by regulatory authorities [138,139,155]. Generally, it has been demonstrated that the best models tend to be those that can integrate mechanism-based reasoning with biological data [155]. The OECD recently released a report on the regulatory uses and applications of (Q)SAR models and expert systems by member countries in chemical assessment, including their use in predicting carcinogenic/genotoxic substances [138]. The aim of this report was "enhancing the regulatory acceptance of (Q)SAR models and expanding the opportunities for future applications of the models." US FDA scientists used the Carcinogenic Potency Database (CPDB) to develop a predictive model for organ-specific carcinogenicity [156]. They added molecular structures to the CPDB to generate a database of SAR analyses for predicting organ-specific carcinogenicity to use in their reviews of new chemicals submitted for approval. They reported a preliminary analysis for liver-specific carcinogenicity.

10.6 Interpretation of Data in Mutagenicity and Carcinogenicity Studies and Significance of Test Results

The use of mutagenicity and carcinogenicity data has been proposed and widely accepted as a relatively fast and inexpensive means of predicting long-term risk to man (i.e., cancer in somatic cells, heritable mutations in germ cells). This view is based on the universal nature of the genetic material, the somatic mutation model of carcinogenesis, and a number of studies showing correlations between mutagenicity and carcinogenicity. Interpretation of data in mutagenicity/carcinogenicity studies should include: interpretation of any modeling approaches, dose−response relationships, historical control data, consideration of any mode of action information and relevance for humans [80]. "False-positive" (relative to carcinogenicity) and "false-negative" mutagenicity results occur, often with rational explanations (e.g., high threshold, inappropriate metabolism, inadequate genetic end point), and thereby confound any straightforward interpretation of mutagenicity test results [157]. *In vitro* mutagenicity assays ignore whole animal protective mechanisms, may provide unphysiological metabolism, and may be either too sensitive (e.g., testing at orders of magnitude higher doses than can be ingested) or not sensitive enough (e.g., short-term treatments inadequately model chronic exposure in bioassay). Test battery composition affects both the proper identification of mutagens and, in many instances, the ability to make preliminary risk assessments [157]. Bacterial systems, particularly the Ames assay, cannot in principle detect chromosomal events which are involved in both carcinogenesis and germ line mutations in humans. Some substances induce only chromosomal events and little or no detectable single-gene events (e.g., caffeine, methapyrilene, and acyclovir). *In vivo* mutagenicity assays are more physiological but appear to be relatively insensitive due to the inability to achieve sufficiently high acute plasma levels to mimic cumulative long-term effects. Examination of the mutagenicity of naturally occurring analogs may indicate the irrelevance of a test substance's mutagenicity (e.g., deoxyguanosine and the structurally related antiviral drug, acyclovir, have identical mutagenicity patterns) [157].

If tests or batteries of tests are to be used in a program to screen unknown chemicals, the way in which the results of the tests will be interpreted should be determined before the program is started [158]. Chankong et al. [159] have developed the Carcinogenicity Prediction and Battery methodology which uses Bayes' theorem to interpret test results, based on the sensitivity and specificity of the tests.

10.7 African Medicinal Plants Screened for Their Mutagenicity

Over 42 mutagenic studies were conducted on some African medicinal plants for the period of 2003−2013 and led to the identification of 64 mutagenic plants and 4 mutagenic commercial herbal mixtures used in African traditional medicine based on their capacity to induce some genetic changes (Table 10.1). The mutagenic

Table 10.1 African Medicinal Plants Screened for Their Mutagenicity

Plant (Family)	Traditional Use	Area of Plant Collection	Part Used; Extraction Solvent	Observed Effects
Acokanthera oblongifolia (Apocynaceae)	Emetics for snakebite, severe gastrointestinal irritation, digitalis-like cardiac effects, irritant, convulsions.	South Africa	Bark/twigs/leaves; dichloromethane/ methanol	The dichloromethane and methanol extracts induce DNA damage in the comet assay without metabolic activation S9 (human white blood cells *in vitro*). The dichloromethane extract was also found toxic in micronucleus test [160].
Acokanthera oppositifolia (Apocynaceae)	Headaches, abdominal pains, convulsions, and septicemia	South Africa	Root; 50% methanol	Exhibited mutagenic effect in the Ames assay by increasing the number of His$^+$ revertants without metabolic activation at 5000 µg/ml (*Salmonella typhimurium* strain TA1535) [161].
Alstonia boonei (Apocynaceae)	Malaria, fever, intestinal worms, stomach disorder, aphrodisiac, diabetes, antipyretic, and wounds	Nigeria	Bark; distilled water	Induced mutant strains showing positive results to citrate (modified Ames assay; *Escherichia coli* 0157: H7). Occurrences of stickiness, multipolar anaphase, bridges and fragments, and vagrant chromosomal aberrations were observed (*Allium cepa* Linn. model) [57].
Annona senegalensis (Annonaceae)	Malaria, diarrhea, intestinal troubles, stomach ache, anthelmintic, skin cancers, and leukemia	Nigeria	Root; 80% ethanol	Dose-dependent (50, 100, 200 mg/kg) chromosomal aberrations *in vivo* (rat lymphocytes). Aberrations

Species (Family)	Traditional uses	Country	Plant part; solvent	Genotoxicity findings
Antidesma venosum (Euphorbiaceae)	Abdominal pain, enema	South Africa	Leaves/twigs; dichloromethane	observed in cells arrested at the metaphase stage of cell division using colchicine in treated rats include breakages, loss of chromosomes, condensation, and rounding up of the long arm of some chromosomes as well as loss of short arms [162]. DNA damage: The leaf extract showed positive result in micronucleus test while the twigs extract was positive in micronucleus and Comet tests without S9 (human white blood cells *in vitro*) [160].
Azadirachta indica (Meliaceae)	Cancer, tonic, refrigerant, skin diseases, ulcers, syphilis, rheumatic disorders, jaundice, cathartic, anthelmintic, leprosy, diabetes, malaria, emmenagogue, fever, filariasis, diabetes, gingivitis, wounds	Nigeria	Leaf; distilled water	Induced mutant strains showing positive results to citrate, urease, H_2S and unable to ferment lactose (modified Ames assay; *Escherichia coli* O157:H7). Absolute root growth inhibition at 10 mg/ml. The results also showed the occurrences of stickiness, bridges and fragments, c-mitosis and vagrant chromosomal aberrations (*Allium cepa* Linn. model) [57].

(Continued)

Table 10.1 (Continued)

Plant (Family)	Traditional Use	Area of Plant Collection	Part Used; Extraction Solvent	Observed Effects
Balanites maughamii (Balanitaceae)	Molluscicidal properties	South Africa	Leaves/twigs; dichloromethane/methanol	The dichloromethane and methanol extracts were positive in micronucleus and comet tests without S9 indicating DNA damage (human lymphocyte cultures) [160].
Bauhinia galpinii (Fabaceae)	Tuber: pneumonia, veneral disease, diarrhea	South Africa	leaves; methanol	Mutagenic effect by increasing the number of His$^+$ revertants in Umu-C test without S9 (*Salmonella typhimurium* strain) [163].
Boophane disticha (Amaryllidaceae)	Dressing for cuts, boils, septic wounds, headaches, abdominal pain, weakness, eye conditions, sedative	South Africa	Bark; dichloromethane/methanol	In micronucleus test without S9, the methanol extract induced micronuclei while the dichloromethane extract was toxic suggesting an (in)direct effect on cell division, or other toxic processes (human lymphocyte cultures) [160].
Bryophyllum calycinum (Crassulaceae)	Infections, rheumatism, inflammation, hypertension, kidney stones, cancers, and cancer-related problems	Nigeria	Whole plant; 80% ethanol	Dose-dependent (50, 100, 200 mg/kg) chromosomal aberrations *in vivo* (rat lymphocytes). The aberrations observed in cells arrested at the metaphase stage of cell division using colchicine in treated rats include breakages, loss of chromosomes, slightly condensed chromosomes, centrally located openings "holes" giving cells a ring appearance. Cotton wool appearance and loss of arms [162].

Plant species (Family)	Traditional use	Country	Plant part; extract	Findings
Catharanthus roseus (Apocynaceae)	Diabetes, rheumatism	South Africa	Leaves; dichloromethane/methanol	The methanol extract induced micronuclei in micronucleus and comet tests without S9 while the dichloromethane extract was found toxic in micronucleus test suggesting an (in)direct effect on cell division, or other toxic processes (human lymphocyte cultures) [160]. The dichloromethane and dichloromethane leaf extracts induced mutant strains in Ames test with metabolic activation (*Salmonella typhimurium* strain TA98) [34].
Catunaregam spinosa (Rubiaceae)	Emetic, fever, aphrodisiac, gonorrhea, headaches, nausea, respiratory and febrile complaints, gynecological ailments, epilepsy, arthritis	South Africa	Leaves; dichloromethane	The extract induced micronuclei in micronucleus test without exogenous metabolic activation S9 (human lymphocyte cultures) [160].
Celtis africana (Cannabaceae)	Cancer, syphilis, rhumatism, pains, magical properties	South Africa	Root; methanol	The extract induced micronuclei in micronucleus test without S9 (human lymphocyte cultures) [160].
Chaetacme aristata (Ulmaceae)	Hemorrhoids	South Africa	Twigs/leaves; methanol	DNA damage in micronucleus and comet tests without S9 (human lymphocyte cultures) [160]. The dichloromethane leaves extract showed mutagenicity in Ames test with and without metabolic activation (*Salmonella typhimurium* strain TA98) [34].

(Continued)

Table 10.1 (Continued)

Plant (Family)	Traditional Use	Area of Plant Collection	Part Used; Extraction Solvent	Observed Effects
Combretum mkhzense (Combretaceae)	Stomach disorders, enemas	South Africa	Bark; dichloromethane	The dichloromethane bark extract showed mutagenic effect by increasing the number of His$^+$ revertants in Ames test with metabolic activation S9 (*Salmonella typhimurium* strain TA98) [34].
Crinum macowanii (Amaryllidaceae)	Scrofula, rheumatic fever, kidney and bladder diseases, fever, sores, glandular swellings	South Africa	Bark; methanol/ dichloromethane	The methanol bark extract induced micronuclei in micronucleus test without metabolic activation S9 (human lymphocyte cultures) [160]. The dichloromethane bark extract increased the number of His$^+$ revertants in Ames test with and without metabolic activation (*Salmonella typhimurium* strain TA98) [34].
Croton sylvaticus (Euphorbiaceae)	Abdominal, internal inflammations, uterine disorders, tonic, febrile conditions, purgative, pleurisy, indigestion, TB, rheumatism	South Africa	Leaves/twigs/bark; dichloromethane/ methanol	The extracts induced micronuclei in micronucleus test without exogenous metabolic activation S9 (human lymphocyte cultures) [160].
Datura stramonium (Solanaceae)	Relieve asthma, reduce pain, hypnotic, aphrodisiac; Fruit: toothache, sore throat, tonsillitis;	South Africa	Seeds/leaves; dichloromethane/ methanol	Mutagenic effect by increasing the number of His$^+$ revertants with Umu-C test in the absence of S9

Plant species (family)	Traditional uses	Country	Part; solvent	Toxicity findings
				(*Salmonella typhimurium* strain) [163].
Diospyros whyteana (Ebenaceae)	Leaf: rheumatism, gout, boils, abscesses, wounds, bronchitis Dysmenorrhea, irritating rashes, antibacterial	South Africa	Leaves/twigs; dichloromethane/methanol	Mutagenic effect of dichloromethane twigs extract in comet test and DNA damage with the methanol extracts in micronucleus and comet tests without S9 (human lymphocyte cultures) [160]. The dichloromethane leaf extract induced mutant strains in Ames test with metabolic activation (*Salmonella typhimurium* strain TA98) [34].
Ekebergia capensis (Meliaceae)	Gastritis, fever, heartburn, ulcers, dysentery, scabies, epilepsy, acne, dysentery, gonorrhea, kidney problems, heartburn, headache, respiratory complaints vermifuge, abscesses, boils, pimples, and itching skin	South Africa	Bark; water/dichloromethane	The water bark extract weakly increases the number of His$^+$ revertant colonies with and without S9 + activation in Ames test (*Salmonella typhimurium* tester strain TA98) [164]. Toxic effect in micronucleus test without S9 suggesting an (in)direct effect on cell division, or other toxic processes was observed with the dichloromethane extract (human lymphocyte cultures) [160].
Elephantorrhiza burkei (Fabaceae)	Diarrhea, dysentery, stomachache, painful menstruation, heart troubles, hemorrhoids, skin diseases, acne, infertility, and aphrodisiac	South Africa	Roots; water	Weakly increases the number of His$^+$ revertant colonies with and without S9+ activation in Ames test (*Salmonella typhimurium* tester strain TA98) [164].

(Continued)

Table 10.1 (Continued)

Plant (Family)	Traditional Use	Area of Plant Collection	Part Used; Extraction Solvent	Observed Effects
Enantia chlorantha (Annonaceae)	Malaria, ulcers, jaundice, and urinary tract infections	Nigeria	Bark; distilled water	The extract induced mutant strains with positive results to urease, H_2S and unable to ferment lactose (Modified Ames assay; *Escherichia coli* 0157:H7). The occurrences of stickiness, bridges and fragments, and vagrant chromosomal aberrations were observed (*Allium cepa* Linn. model) [57].
Erythrina caffra (Leguminosae)	Sores, wounds, arthritis, sprains, aches	South Africa	Bark; methanol	The extract induced micronuclei in micronucleus test without exogenous metabolic activation S9 (human lymphocyte cultures) [160].
Erythrophleum suaveolens (Caesalpinaceae)	Pains associated with fever and headache, cancers and cancer-related problems, use in trial-by-ordeal ritual, also used as arrow and fish poison.	Nigeria	Leaves; 80% ethanol	About 5% chromosomal damage involving condensation of chromosomes were observed at 100 mg/kg while 15% of chromosomal damage observed at 200 mg/kg involving condensation of chromosomes (rat lymphocytes *in vivo*) [162].
Euclea divinorum (Ebenacea)	Purgative, for headache, toothache, constipation, antihelmintics, tonics, chest pain, pneumonia, stomach pain	South Africa	Root; dichloromethane	The extract induced micronuclei in micronucleus test without S9 (human lymphocyte cultures) [160].

Plant (Family)	Ethnomedicinal uses	Country	Part; solvent	Findings
Garcinia kola (Guttiferae)	Bronchitis, throat infections, colic, head or chest colds, cough, liver disorders, cancer, and as a chewing stick, purgative, antiparasitic, antimicrobial	Nigeria	Root; 80% ethanol	Dose-dependent (50, 100, 200 mg/kg) chromosomal aberrations *in vivo* (rat lymphocytes). These chromosomal damages involving rounding up of lower arms of chromosomes and loss of short arms of chromosomes [162].
Gardenia volkensii (Rubiacea)	Emetic, sore eyes, headaches, asthma, dysmenorrhea, infertility, epilepsy, convulsions, earache	South Africa	Leaves/twigs/bark; dichloromethane/ methanol	Toxic effect with dichloromethane leaves extract [160] and DNA damage with dichloromethane twigs/bark extracts in micronucleus test without S9 [160,165]. DNA damage of dichloromethane and methanol twigs/bark extracts in comet assay (human lymphocyte cultures) [160].
Gymnosporia Senegalensis (Celastraceae)	Pneumonia, tuberculosis, menstruation, uterine cramps, aphrodisiac, headache, hemoptysis, snakebites	South Africa	Root bark; dichloromethane/ methanol	Mutagenic effect by increasing the number of revertant colonies in Umu-C test without metabolic activation S9 (*Salmonella typhimurium* strain) [163]. The dichloromethane and methanol extracts induced micronuclei in micronucleus test without S9 (human white blood cells) [163].
Helichrysum herbacea (Asteraceae)	Coughs, colds, fever, infections, headache, and menstrual pain. Leaves burnt to invoke goodwill of ancestors	South Africa	Whole plant; methanol	2.9-fold increase in revertants compared to the solvent control with Ames test in the presence of S9 (*Salmonella typhimurium* strain TA100) [163].

(Continued)

Table 10.1 (Continued)

Plant (Family)	Traditional Use	Area of Plant Collection	Part Used; Extraction Solvent	Observed Effects
Helichrysum rugulosum (Asteraceae)	Coughs, colds, fever, infections, headache, and menstrual pain	South Africa	Whole plant; methanol	1.7-fold increase in revertants compared to the solvent control with Ames test in the presence of S9 (*Salmonella typhimurium* strain TA100) [163].
Helichrysum simillimum (Asteraceae)	Coughs, colds, fever, infections, headache, and menstrual pain	South Africa	Whole plant; methanol	Mutagenic effect by increasing the number of revertant colonies with Ames test in the absence or presence of S9 (*Salmonella typhimurium* strains TA98 and TA100) [163].
Heteromorpha trifoliata (Apiaceae)	Scrofula, enemas, abdominal disorders, mental/nervous disorders, intestinal worms, headaches, antiscabies	South Africa	Leaves/twigs/bark; dichloromethane/ methanol	The dichloromethane twigs/bark extracts and methanol leaves extract induced micronuclei in micronucleus test without metabolic activation S9. Mutagenic effect of dichloromethane leaves extract in comet test without S9 (human lymphocyte cultures) [160].
Hymenocardia acida (Euphorbiaceae)	Cancers, eye infection, sickle cell anemia, fever, jaundice, muscular pains, diarrhea, dysentery, skin diseases, diabetes, urinary tract infections, sexual incapacity in males	Nigeria	Stem bark; 80% ethanol	Dose-dependent (50, 100, 200 mg/kg) chromosomal aberrations *in vivo* (rat lymphocytes). The aberrations observed in cells arrested at the metaphase stage of cell division using colchicine in treated rats include breakages resulting in various chromosome fragments, rounding up of the chromosomes, presence of a ring chromosome [162].

Species (Family)	Uses	Country	Part; solvent	Results
Hypoxis colchicifolia (Hyacinthaceae)	Tonic, anti-HIV, antiinflammatory	South Africa	Bark; dichloromethane/methanol	DNA damage of dichloromethane and methanol bark extracts in comet and micronucleus tests respectively (human lymphocyte cultures) [160].
Hypoxis hemerocallidea (Hyacinthaceae)	Dizziness, bladder and urinary disease, tonic, burns	South Africa	Bark; methanol	The extract induced micronuclei in micronucleus test without S9 (human lymphocyte cultures) [160].
Kigelia africana (Bignoniaceae)	Ulcers, sores, syphilis, rheumatism, enema	South Africa	Fruits; dichloromethane	The extract induced micronuclei in micronucleus test without exogenous metabolic activation S9 (human lymphocyte cultures) [160].
Marrubium Alysson (Lamiaceae)	Hypertension, cough, burns, rheumatics, intestinal troubles	Tunisia	Aerial parts; ethylacetate/methanol	DNA damage according to the alkaline comet assay (human C3A cells) [166].
Moricandia arvensis (Cruciferae)	Syphilis, scorbut	Tunisia	Leaves; water, methanol, total oligomers flavonoids, chloroform, ethyl acetate	The aqueous extract produced a mutagenic effect by increasing the number of revertants in Ames test after metabolic activation S9 (*Salmonella typhimurium* TA100). The methanol, total oligomers flavonoids, chloroform, petroleum ether, ethyl acetate extracts also increased mutant strains in Ames test after metabolic activation S9 (*Salmonella typhimurium* tester strain TA1535) [167].

(Continued)

Table 10.1 (Continued)

Plant (Family)	Traditional Use	Area of Plant Collection	Part Used; Extraction Solvent	Observed Effects
Morinda lucida (Rubiaceae)	Cancers, malaria, diabetes, jaundice, fever, hypoglycemic, trypanocidal, bitter tonic and astringent for dysentery, abdominal colic, and intestinal worm infestation	Nigeria	Root bark/root; 80% ethanol/ distilled water	The 80% ethanol root bark extract showed dose-dependent (50, 100, 200 mg/kg) chromosomal aberrations *in vivo* (rat lymphocytes). The aberrations observed in cells arrested at the metaphase stage of cell division using colchicine in treated rats include breakages, loss of chromosomes, cotton wool appearance (plebing) of cells, super condensation [162]. The water roots extract induced mutant strains showing positive results to urease, H$_2$S. These mutant strains altered the normal motility characteristics of the organism to nonmotile state (modified Ames assay; *Escherichia coli* 0157:H7). Mitostatic effect at 5 and 10 mg/ml. The results also showed the occurrences of stickiness, bridges and fragments, c-mitosis, and vagrant chromosomal aberrations (*Allium cepa* Linn. model) [57].

Plant (family)	Traditional uses	Country	Part; solvent	Findings
Newbouldia laevis (Bignoniaceae)	Elephantiasis, malaria, peptic ulcer, otalgia, skin ulcer, epilepsy, hemorrhoids, constipation, convulsion, orchitis (gonococcal), pain, and several inflammatory conditions	Nigeria	Leaf; distilled water	The extract induced mutant strains with positive results to urease, H_2S (*Escherichia coli* 0157:H7). Turbagenic effect: no mitotic cell division was observed at 5 and 10 mg/ml. The results also showed the occurrences of stickiness, bridges and fragments, and vagrant chromosomal aberrations (*Allium cepa* Linn. model) [57].
Nymphaea lotus (Nymphaceae)	Aphrodisiac, cancers, and cancer-related problems	Nigeria	Whole plant; 80% ethanol	Ten percent of chromosomal damage involving fragmentation of chromosomes was observed only at the highest dose: 200 mg/kg (rat lymphocytes) [162].
Ochna serrulata (Ochnaceae)	Enema, gangrenous rectitis, bone diseases of children	South Africa	Leaves; methanol	The extract induced micronuclei in micronucleus test without exogenous metabolic activation S9 (human lymphocyte cultures) [160].
Ocotea bulata (Lauraceae)	Snuff, headaches, urinary disorders, stomach problems, infantile diarrhea	South Africa	Bark; methanol	DNA damage in comet assay without exogenous metabolic activation S9 (human lymphocyte cultures) [160].
Ornithogalum longibracteatum (Hyacinthaceae)	Charm, irritant	South Africa	Leaves/fruits/bark; dichloromethane/ methanol	Organ- and solvent-dependant DNA damage in micronucleus and Comet assays without exogenous metabolic activation S9 (human lymphocyte cultures) [160].

(*Continued*)

Table 10.1 (Continued)

Plant (Family)	Traditional Use	Area of Plant Collection	Part Used; Extraction Solvent	Observed Effects
Plumbago auriculata (Plumbaginaceae)	Headaches, emetics, warts, fractures, scrofula, edema, malaria, skin lesions	South Africa	Leaves/fruits/twigs; dichloromethane/ methanol	Organ- and solvent-dependant DNA damage in micronucleus and comet assays without S9 (human lymphocyte cultures) [160]. The dichloromethane foliage extract induced mutant strains in Ames test with and without metabolic activation (*Salmonella typhimurium* tester strain TA98) [34].
Plumbago zeylanica (Plumbaginaceae)	Malaria, fever, diarrhea, dyspepsia, piles, skin diseases including leprotic lesions, antiviral, anti-*Helicobacter*, gastrointestinal complaints, parasitic diseases, scabies, and ulcers	Nigeria	Root; distilled water	The occurrences of stickiness, bridges and fragments, c-mitosis, and vagrant chromosomal aberrations were observed (*Allium cepa* Linn. model) [57].
Prunus Africana (Rosaceae)	Intercostal pain, prostate hypertrophy, hair tonics	South Africa	Leaves/twigs; dichloromethane/ methanol	DNA damage with dichloromethane twigs extract in micronucleus and Comet assays without exogenous metabolic activation (S9). Mutagenic effect of methanol leaves extract in comet test (human lymphocyte cultures) [160].
Rauvolfia vomitoria (Apocynaceae)	Malaria, typhoid, jaundice, antipsychotic, arbotifacient, purgative, and emetic	Nigeria	Root; distilled water	Occurrences of stickiness, bridges, and fragments, multipolar anaphase, c-mitosis, and vagrant chromosomal aberrations were observed (*Allium cepa* Linn. model) [57].

Species (Family)	Traditional uses	Country	Plant part; Solvent	Toxicity
Retama Raetam (Fabaceae)	Hypertension, diabetes	Tunisia	Flowers; Ethylacetate	Mutagenic effect with the Vitotox® test in the absence of S9 (*Salmonella typhimurium* strain) [166].
Rhamnus prinoides (Rhamnaceae)	Sprains, blood purifiers, pneumonia, emetics, purgative, colic, stimulants	South Africa	Leaves/bark/twigs; dichloromethane/methanol	Toxic effect of dichloromethane twigs extract and DNA damage of dichloromethane bark/twigs extracts in micronucleus assay without exogenous metabolic activation (S9). Mutagenic effect of methanol extracts and dichloromethane leaves extract in comet test (human lymphocyte cultures) [160].
Ricinus communis (Euphorbiaceae)	Purgatives, enemas, stomach ache; root and leaf applied to wounds, sores, boils	South Africa	Roots; dichloromethane/methanol	Toxic effect of dichloromethane extract and DNA damage of methanol extract in micronucleus assay without exogenous metabolic activation S9 (human lymphocyte cultures) [160].
Scilla natalensis (Liliaceae)	Enema, sprains, fractures, purgative, boils, sores, infertility	South Africa	Bark; dichloromethane/methanol	The extracts induced micronuclei in micronucleus test without S9 (human lymphocyte cultures) [160].
Sclerocarya birrea (Anacardiaceae)	Diarrhea, dysentery, stomach problems, fever, malaria, tonic, diabetes	South Africa	Bark; methanol	Induced micronuclei in micronucleus test without exogenous metabolic activation S9 (human lymphocyte cultures) [160].

(Continued)

Table 10.1 (Continued)

Plant (Family)	Traditional Use	Area of Plant Collection	Part Used; Extraction Solvent	Observed Effects
Senecio serratuloides (Asteraceae)	Cuts, swellings, burns, sores, blood purifiers, headaches	South Africa	Leaves; dichloromethane	Toxic effect in micronucleus test suggesting an (in)direct effect on cell division, or other toxic processes without exogenous metabolic activation S9 (human lymphocyte cultures) [160].
Spirostachys Africana (Euphorbiaceae)	Wood: stomach ulcers, acute gastritis, eye washes, headaches, rashes, boils, emetic, renal ailment, purgative, blood purifiers, diarrhea, dysentery	South Africa	Leaves/bark/ twigs; dichloromethane/ methanol	The extract induced micronuclei in micronucleus test without addition of a metabolizing enzyme solution (S9 mix) at 500 and 100 mg/ml (human lymphocyte cells) [160,165]. Mutagenic effect of methanol leaves extract and dichloromethane bark/ twigs extracts in comet test without S9 (human lymphocyte cultures) [160].
Spondiathus preussii (Euphorbiaceae)	Cancers and cancer-related problems	Nigeria	Stem bark; 80% ethanol	Fifteen percent of chromosomal damage observed involving loss of the shorter arms and loss of the centromere in chromosome 7 at dose of 25 mg/kg. Rats administered extracts at doses of 50, 100, 200 mg/kg convulsed and died 5–10 min after being injected (rat lymphocytes *in vivo*) [162].

Species (Family)		Country	Part used; solvent	Findings
Tetrapluera tetraptera (Fabaceae)		Nigeria	Fruit; distilled water	Induced mutant strains showing positive results to urease (modified Ames assay; *Escherichia coli* 0157:H7). Mitostatic effect at 5 and 10 mg/ml: constant mitotic cell division. The results also showed the occurrences of stickiness, bridges and fragments, c-mitosis, and vagrant chromosomal aberrations (*Allium cepa* Linn. model) [57].
Tetradenia riparia (Lamiaceae)		South Africa	Fruits; methanol	DNA damage in Comet test without exogenous metabolic activation S9 (human lymphocyte cultures) [160].
Trichilia emetica (Meliaceae)		South Africa	Leaves/bark; dichloromethane/ methanol	Appearance of micronuclei with dichloromethane and methanol bark extracts in micronucleus assay. DNA damage of methanol leaves extract in Comet test without S9 (human lymphocyte cultures) [160].
Tulbaghia violaceae (Alliaceae)		South Africa	Leaves/bark; methanol	Induced micronuclei in micronucleus assay without S9 (human lymphocyte cultures) [160].

The first column also lists traditional uses (separate column):

Species	Traditional uses
Tetrapluera tetraptera (Fabaceae)	Convulsions, leprosy, hypertension, inflammation, and rheumatic pains
Tetradenia riparia (Lamiaceae)	Coughs, sore throats, malaria, dengue fever, dropsy, fever, diarrhea, hemoptysis, boils, mumps, induce drowsiness
Trichilia emetica (Meliaceae)	Stomach and intestinal complaints, dysentery, kidney problems, indigestion, parasites, fever, purgative, bruises, rheumatism
Tulbaghia violaceae (Alliaceae)	Fever, colds, asthma, enemas, tuberculosis, stomach problems, rheumatism, paralysis

(*Continued*)

Table 10.1 (Continued)

Plant (Family)	Traditional Use	Area of Plant Collection	Part Used; Extraction Solvent	Observed Effects
Turraea floribunda (Meliaceae)	Emetic, rheumatism, dropsy, heart disease, swollen and painful joints	South Africa	Leaves/bark; dichloromethane/methanol	Mutagenic effect in comet test without exogenous metabolic activation (S9). Appearance of micronuclei with dichloromethane leaves/bark extracts and methanol leaves extract in micronucleus assay without S9 (human lymphocyte cultures) [160].
Vernonia colorata (Asteraceae)	Abdominal pain, colic, rheumatism, dysentery, diabetes, ulcerative colitis	South Africa	Leaves/roots; dichloromethane/methanol	Toxic effect of dichloromethane extracts in micronucleus test suggesting an (in)direct effect on cell division, or other toxic processes. DNA damage with methanol leaves extract in comet test without exogenous metabolic activation (S9). (human lymphocyte cultures) [160].
Xylopia aethiopica (Annonaceae)	Bronchitis, dysentery, toothaches, asthma, stomach aches, rheumatism, biliousness, and febrile pain	Nigeria	Fruit; distilled water	Induced mutant strains showing positive results to citrate (modified Ames assay; *Escherichia coli* 0157:H7). Absolute root growth inhibition at 5 and 10 mg/ml. The results also showed the occurrences of stickiness, and vagrant chromosomal aberrations (*Allium cepa* Linm. model) [57].

Ziziphus mucronata (Rhamnaceae)	Boils, sores, glandular swellings, diarrhea, dysentery, expectorant, emetic for coughs, chest problems, boils, sores, glandular swellings	South Africa	Leaves/twigs; dichloromethane/ methanol	Appearance of micronuclei with dichloromethane and methanol leaf extracts in micronucleus and comet tests without exogenous metabolic activation (S9). DNA damage of dichloromethane extract in comet test (human lymphocyte cultures) [160]. The 90% methanol leaf extract induced mutant strains in Ames test with metabolic activation (*Salmonella typhimurium* strain TA98) [34].

effects of these plants were performed through *in vitro* and/or *in vivo* methods with human white blood cells, human C3A cells, rat lymphocytes, *Salmonella typhimurium* tester strains, or *Allium cepa* as models. These methods comprising the micronucleus test, Comet test, chromosomal aberration assay, Vitotox[®] test, and Umu-C test were applied with or without S9 metabolic activation. Our simplified presentation of one plant per mutagenic effect allowed us to note that the mutagenic effects of plants, depended on factors such as type, part, and amount of plant administrated, extraction solvent, animal model, methods used, etc. The observed mutagenic effects are mainly characterized here by the appearance of micronuclei, increase of number of His[+] revertants, DNA damage and chromosomal aberrations (breakages, loss of chromosomes, nondisjunction or incomplete separation of chromosomes, super condensation, loss of arms, cotton wool appearance, occurrences of stickiness, etc.) involving genetic changes induced by the studied plants (Table 10.1). Plant extracts, especially polar extracts, are complex mixtures of phytochemicals; thus, the response of an organism to this particular extract depends on the action of this complex extract, which includes interactions due to synergism, additivity, and antagonism. This may in some cases explain the lack of anecdotal information on the toxicity of those plants (and their continued use in traditional medical preparations), which have been shown to be potentially mutagenic after laboratory experiments.

Most of the plant extracts were tested using the *Salmonella* microsome assay based on the plate-incorporation procedure with *Salmonella typhimurium* tester strains TA98, TA100, TA102, TA1535, and TA1537 with and without S9 metabolic activation [24,29]. TA98 strain has a −1 frameshift mutation at *his*D3052 which affects the reading frame of a repetitive base pair −C−G−C−G−C−G−C−G−sequence; TA100 and TA1535 have a *his*G46 marker resulting from the substitution of a leucine (GAG/CTC) by a proline (GGG/CCC) [168]. The *his*G46 mutation can be reverted to a wild type by mutations that cause base pair substitution at the GC site. TA102 contains AT base pairs at the *his*G428 mutant site. The *his*G428 can be reverted by mutagens that cause oxidative damage. TA1537 carries a +1 frameshift mutation *his*C3076 located near the repetitive site −C−C−C−sequence and can be reverted by frameshift mutagens that are not readily detected by the *his*D3052 (TA98).

Extracts from *Acokanthera oblongifolia* and *Catharanthus roseus* (Apocynaceae), *Boophane disticha* (Amaryllidaceae), *Ekebergia capensis* (Meliaceae), *Gardenia volkensii* (Rubiaceae), *Rhamnus prinoides* (Rhamnaceae), *Ricinus communis* (Euphorbiaceae), *Senecio serratuloides*, and *Vernonia colorata* (Asteraceae) from Africa displayed toxic effects in micronucleus test without exogenous metabolic activation S9 suggesting an (in) direct effect on cell division (Table 10.1) [160].

All Umu-C tests were performed on dichloromethane and methanol extracts in the presence and absence of S9 metabolic activation [163]. The dichloromethane and methanol extracts of *Datura stramonium*, methanol extract of *Bauhinia galpinii*, and dichloromethane extract of *Gymnosporia senegalensis* were found mutagenic in this test in the absence of S9. This is in contradiction with the results obtained for the Ames and Vitotox[®] tests. This finding highlights the fact that the mutagenic effect of chemicals can be varied according to the methods employed, explaining the importance of multiple screening techniques for its detection.

The toxic effects of *Spondiathus preussii* (Euphorbiaceae) can be considered highest compared with those of *Annona senegalensis* (Annonaceae), *Bryophyllum calycinum* (Crassulaceae), *Erythrophleum suaveolens* (Caesalpinaceae), *Garcinia kola* (Guttiferae), *Hymenocardia acida* (Euphorbiaceae), *Morinda lucida* (Rubiaceae), *Nymphaea lotus* (Nymphaceae), as rats administered with this plant extract at doses of 50, 100, 200 mg/kg convulsed and died 5−10 min after being injected [162]. Furthermore, only 25 mg/kg of this plant extract induced 15% of chromosomal damage in rat lymphocytes *in vivo* [162].

Most of the extracts reported in this chapter were from plants with recognized toxicity, but widely used medicinally, explaining why they should be carefully examined from a safety perspective and cautions attached to their use. Herbal mixtures are concoctions of two or more plant species with the same or different medicinal uses [169]. In Africa, commercialization of such products has resulted in the production of complex mixtures for many conditions as well as energy boosters, detoxifiers, immune boosters, and aphrodisiacs [168]. Mutagenic effects of 13 commercial herbal mixtures sold in KwaZulu-Natal, South Africa, were evaluated using the Ames test [168]. The herbal mixtures showed no mutagenic effects against *Salmonella typhimurium* tester strains TA98, TA100, TA102, TA1535, and TA1537 when the assay was done without S9 metabolic activation. However, four herbal mixtures, namely Umpatisa inkosi, Imbiza ephuzwato, Vusa umzimba, and Stameta™ BODicare®, showed mutagenic effects in an Ames test against *Salmonella typhimurium* tester strains TA98 after using S9 metabolic activation. *Salmonella typhimurium* TA98 has a −1 frameshift mutation *his*D3052 which affects the reading frame of a repetitive −C−G−C−G−C−G−C−G−sequence [24]. Thus, the four herbal mixtures (Umpatisa inkosi, Imbiza ephuzwato, Vusa umzimba, and Stameta™ BODicare®) cause a reversion of the *his*D3052 mutation back to the wild-type state. It was suggested that the herbal mixtures could contain various aromatic nitroso-derivatives of amine carcinogens that have been shown to cause such a reversal of *his*D3052 mutation back to the wild-type state [168]. Umpatisa inkosi also exhibited weak mutagenic activity against TA1535 after metabolic activation. The remaining mixtures did not show mutagenic effects against all the tester strains after S9 metabolic activation [168]. Imbiza ephuzwato, a multipurpose Zulu herbal tonic, consists of a mixture of 21 plant species belonging to 17 families. Most of the 21 plant species that constitute Imbiza ephuzwato are used by traditional people to treat various conditions. These plant species include *Scadoxus puniceus* and *Cyrtanthus obliquus* (Amaryllidaceae), *Acokanthera venenata* (Apocynaceae), *Asclepias fruticosa* (Asclepiadaeceae), *Aster bakeranus* and *Hypericum aethiopicum* (Asteraceae), *Momordica balsamina* (Cucurbitaceae), *Eriosema cordatum* (Fabaceae), *Gunnera perpensa* (Gunneraceae), *Ledebouria* sp. (Hyacinthaceae), *Drimia robusta* and *Urginea physodes* (Hyacinthaceae), *Watsonia densiflora* (Iridaceae), *Tetradenia riparia* (Lamiaceae), *Lycopodium clavatum* (Lycopodiaceae), *Stephania abyssinica* (Menispermaceae), *Rubia cordifolia* (Rubiaceae), *Zanthoxylum capense* (Rutaceae), *Vitellariopsis marginata*

(Sapotaceae), *Gnidia kraussiana* (Thymelaeaceae), *Corchorus asplenifolius* (Tiliaceae) [170]. The Ames test results revealed that all the water extracts of the 21 plant species used to make Imbiza ephuzwato were nonmutagenic toward the *Salmonella typhimurium* TA98 strain for the assay with and without S9 metabolic activation [170]. Thus, the reported mutagenicity in Imbiza ephuzwato could be a result of interaction of biomolecules in the heterogeneous mixture, yielding compounds that are converted to mutagenic agents by xenobiotic metabolizing enzymes. It is therefore important to carry out further studies aimed at identifying and eliminating the sources of the mutagenic compounds in the heterogeneous mixture. The ingredients of Stameta™ BODicare® as listed on the label include *Hypoxis rooperi* (Hypoxidaceae), *Mentha piperita* (Lamiaceae), *Pimpinella anisum* (Apiaceae), *Aloe* (unspecified); fortified with multivitamins (unspecified), calcium, magnesium, potassium, phosphorus, and iron [171]. At present, the compositions of Umpatisa inkosi and Vusa umzimba are not provided.

Until now, no carcinogenic studies have been conducted in order to evaluate directly the ability of African medicinal plants to induce cancers. However, the mutagenicity assays are the most commonly used *in vitro* test systems to predict carcinogenicity. According to UN GHS guidance, chemical-induced tumorigenesis involves genetic changes; thus, chemicals that are mutagenic in mammals may warrant being classified as carcinogens [139].

10.8 Conclusion

From this chapter review of the mutagenic and carcinogenic plants used in African traditional medicine, it can be noted that little mutagenic studies and no carcinogenic works from African medicinal plants have been performed so far. Nonetheless, from 42 published scientific works, we identified 64 plant species and 4 commercial herbal mixtures widely used in African traditional medicine with mutagenic effects. These plant species and commercial herbal mixtures include *Acokanthera oblongifolia, Acokanthera oppositifolia, Alstonia boonei, Catharanthus roseus*, and *Rauvolfia vomitoria* (Apocynaceae); *Annona senegalensis, Enantia chlorantha*, and *Xylopia aethiopica* (Annonaceae); *Antidesma venosum, Ricinus communis, Croton sylvaticus, Hymenocardia acida, Spirostachys africana*, and *Spondiathus preussii* (Euphorbiaceae); *Azadirachta indica, Ekebergia capensis, Turraea floribunda*, and *Trichilia emetica* (Meliaceae); *Balanites maughamii* (Balanitaceae); *Bauhinia galpinii, Elephantorrhiza burkei, Erythrina caffra, Retama raetam*, and *Terapluera tetraptera* (Fabaceae); *Boophane disticha* and *Crinum macowanii* (Amaryllidaceae); *Bryophyllum calycinum* (Crassulaceae); *Catunaregam spinosa, Gardenia volkensii*, and *Morinda lucida* (Rubiaceae); *Celtis africana* (Cannabaceae); *Chaetacme aristata* (Ulmaceae); *Combretum mkhzense* (Combretaceae); *Datura stramonium* (Solanaceae); *Diospyros whyteana* and *Euclea divinorum* (Ebenaceae); *Erythrophleum suaveolens* (Caesalpinaceae); *Garcinia kola* (Guttiferae); *Gymnosporia Senegalensis* (Celastraceae); *Helichrysum herbacea,*

Helichrysum simillimum, Helichrysum rugulosum, Senecio serratuloides, and *Vernonia colorata* (Asteraceae); *Heteromorpha trifoliata* (Apiaceae); *Hypoxis colchicifolia, Hypoxis hemerocallidea,* and *Ornithogalum longibracteatum* (Hyacinthaceae); *Kigelia africana* (Bignoniaceae); *Marrubium alysson* and *Tetradenia riparia* (Lamiaceae); *Moricandia arvensis* (Cruciferae); *Newbouldia laevis* (Bignoniaceae); *Nymphaea lotus* (Nymphaceae); *Ochna serrulata* (Ochnaceae); *Ocotea bulata* (Lauraceae); *Plumbago auriculata* and *Plumbago zeylanica* (Plumbaginaceae); *Prunus africana* (Rosaceae); *Rhamnus prinoides* (Rhamnaceae); *Scilla natalensis* (Liliaceae); *Sclerocarya birrea* (Anacardiaceae); *Tulbaghia violaceae* (Alliaceae); *Ziziphus mucronata* (Rhamnaceae) and Umpatisa inkosi, Imbiza ephuzwato, Vusa umzimba, Stameta™ BODicare®.

The observed mutagenic effects were dominated by the appearance of micronuclei, increase of number of His⁺ revertants, DNA damage, and chromosomal aberrations involving genetic changes induced by the identified plants. It should be noted that, at this point, most of the extracts from these mutagenic plants presented in this chapter have only been screened using *in vitro* methods with or without S9 metabolic activation. These investigations should therefore be considered preliminary and more in-depth analyses, eventually involving mammalian *in vivo* studies, are needed before final conclusions can be reached. The likelihood of side effects increases when the production and sale of such products is largely uncontrolled and or unregulated and the consumer is not adequately informed about their proper uses. While in developed countries, most of herbal medicines are regulated through official controls and rigorous manufacturing standards, this is not the case in most of the African countries. Finally, the herbal drugs need to be analyzed in the same way as any modern drug, that is, with randomized controlled clinical trials. At the same time, legislators at the national level should continue to press for effective laws to protect consumers from potentially harmful herbal drugs in order to increase the standard of African traditional medicine.

References

[1] Veilleux C, King SR. In: Morganstein L, editor. An introduction to ethnobotany. Available at: <http://www.accessexcellence.org/RC/Ethnobotany/>; 1996 [accessed 09.07].

[2] Farnsworth NR, Akerele OO, Bingel AS, Soejarta DD, Eno Z. Medicinal plants in therapy. Bull World Health Organ 1985;63:961−5.

[3] Akerele O. Nature's medicinal bounty: don't throw it away! World Health Forum 1993;14:390−5.

[4] Van Wyk B-E, van Outshoorn B, Gericke N. Medicinal plants of South Africa. Pretoria: Briza Publications; 1997.

[5] Koltb FT. Medicinal plants in Libya. Arab Encyclopedia House, Beirut, Lebanon. University, Irbid, Jordan; 1983.

[6] Meyer BN, Ferrigni NR, Putnam JE, Jacobsen LB, Nichols DE, Mclaughlin JL. Brine shrimp: a convenient general. Bioassay for active plant constituents. Planta Med 1982;45:31−4.

 [7] Sofowora A. Medicinal plants and traditional medicine in Africa. New York, NY: John Wiley & Sons Limited; 1982.

 [8] Tene M, Tane P, Kuiate JR, Tamokou JDD, Connolly JD. Anthocleistenolide, a new rearranged nor-secoiridoid derivative from the stem bark of *Anthocleista vogelii*. Planta Med 2008;74:80−3.

 [9] Tamokou JDD, Kuiate JR, Tene M, Tane P. Antimicrobial clerodane diterpenoids from *Microglossa angolensis* Oliv. Et Hiern. Indian J Pharmacol 2009;41(2):60−3.

[10] Tamokou JDD, Mpetga-Simo DJ, Lunga PK, Tene M, Tane P, Kuiate JR. Antioxidant and antimicrobial activities of ethyl acetate extract, fractions and compounds from the stem bark of *Albizia adianthifolia* (Mimosoideae). BMC Complement Altern Med 2012;12:99.

[11] Noumedem KJA, Tamokou JDD, Teke NG, Momo RC, Kuete V, Kuiate JR. Phytochemical analysis, antimicrobial and radical-scavenging properties of *Acalypha manniana* leaves. SpringerPlus 2013;2:503.

[12] Tamokou JDD, Chouna JR, Fischer-Fodor E, Chereches G, Barbos O, Damian G, et al. Anticancer and antimicrobial activities of some antioxidant-rich Cameroonian medicinal plants. PLoS One 2013;8(2):e0055880.

[13] Azzam MS. Phytochemical investigation of certain plants used in Egyptian folk-medicine as antidiabetic drugs. *Ph.D. Thesis*, Faculty of Pharmacy, Cairo University, Cairo, Egypt; 1984.

[14] Boulos L. Medicinal plants of North Africa. Algonac, MI: Reference Publications, Inc.; 1983.

[15] Nyaa Tankeu BL, Tapondjou AL, Barboni L, Tamokou JDD, Kuiate JR, Tane P, et al. NMR assignment and antimicrobial/antioxidant activities of 1β-hydroxyeuscaphic acid from the seeds of *Butyrospermum parkii*. Nat Prod Sci 2009;15(2):76−82.

[16] Tamokou JDD, Tala FM, Wabo KH, Kuiate JR, Tane P. Antimicrobial activities of methanol extract and compounds from stem bark of *Vismia rubescens*. J Ethnopharmacol 2009;124:571−5.

[17] Fabricant DS, Farnsworth NR. The value of plants used in traditional medicine for drug discovery. Environ Health Perspect 2001;109:69−75.

[18] Noble RL, Beer CT, Cutts JH. Further biological activities of vincaleukoblastin—an alkaloid isolated from *Vinca rosea* (L.). Biochem Pharmacol 1959;347−8.

[19] Wani MC, Taylor HL, Wall ME, Coggon P, McPhail AT. Plant antitumour agents VI. Isolation and structure of taxol, a novel antileukemic and antitumour agent from *Taxus brevifolia*. J Am Chem Soc 1971;93:2325−7.

[20] Sofowora A. Medicinal plants and traditional medicine in Africa. New York, NY: John Wiley & Sons; 1984. p. 66−256.

[21] Gadano AB, Gumi AA, Carballo MA. Argentine folk medicine: genotoxic effects of Chenopodiaceae family. J Ethnopharmacol 2006;103:246−51.

[22] Steenkamp PA, Harding NM, Van Heerden FR, Van Wyk B-E. Identification of atractyloside by LC−ESI−MS in alleged herbal poisonings. Forensic Sci Int 2006; 163:81−92.

[23] Cariño-Cortés R, Hernández-Ceruelos A, Torres-Valencia JM, GonzálezAvila M, Arriaga-Alba M, Madrigal-Bujaidar E. Antimutagenicity of *Stevia pilosa* and *Stevia eupatoria* evaluated with the Ames test. Toxicol In Vitro 2007;21:691−7.

[24] Mortelmans K, Zeiger E. The Ames *Salmonella*/microsome mutagenicity assay. Mutat Res 2000;455:29−60.

[25] Varanda EA, Pozetti GL, Lourenço MV, Vilegas W, Raddi MSG. Genotoxicity of *Brosimum gaudichaudii* measured by the *Salmonella*/microsome assay and chromosomal aberrations in CHO cells. J Ethnopharmacol 2002;81:257−64.

[26] Ferreira ICDF, Vargas VMF. Mutagenicity of medicinal plant extracts in *Salmonella*/ microsome assay. Phytother Res 1999;13:397−400.

[27] Rietjens IMCM, Boersma MG, Woude H, Jeurissen SMF, Schutte ME, Alink GM. Flavonoids and alkenylbenzenes: mechanisms of mutagenic action and arcinogenic risk. Mutat Res 2005;574:124−78.

[28] Goldstein JA, Faletto MB. Advances in mechanisms of activation and deactivation of environmental chemicals. Environ Health Perspect 1993;100:169−76.

[29] Maron DM, Ames BN. Revised methods for the *Salmonella* mutagenicity test. Mutat Res 1983;113:173−215.

[30] Verschaeve L, Van Gompel J, Thilemans L, Regniers L, Vanparys P, van der Lelie D. Vitotox® bacterial genotoxicity and toxicity test for the rapid screening of chemicals. Environ Mol Mutagen 1999;33:240−8.

[31] Fenech M. The *in vitro* micronucleus technique. Mutat Res 2000;455:81−95.

[32] Singh NP. Microgels for estimation of DNA strand breaks, DNA protein crosslinks and apoptosis. Mutat Res 2000;455:111−27.

[33] Mathur N, Bhatnagar P, Mohan K, Bakre P, Nagar P, Bijarnia M. Mutagenicity evaluation of industrial sludge from common effluent treatment plant. Chemosphere 2007;67:1229−35.

[34] Elgorashi EE, Taylor JLS, Maes A, Van Staden J, De Kimpe N, Verschaeve L. Screening of medicinal plants used in South African traditional medicine for genotoxic effects. Toxicol Lett 2003;143:195−207.

[35] Cardoso CR, De Syllos Colus IM, Bernardi CC, Sannomiya M, Vilegas W, Varanda EA. Mutagenic activity promoted by amentoflavone and methanol extract of *Byrsonima crassa* Niedenzu. Toxicology 2006;225:55−63.

[36] Déciga-Campos M, Rivero-Cruz I, Arriaga-Alba M, Castañeda-Corral G, Angeles-López GE, Navarrete A, et al. Acute toxicity and mutagenic activity of Mexican plants used in traditional medicine. J Ethnopharmacol 2007;110(2):334−42.

[37] Mohd-Fuat AR, Kofi EA, Allan GG. Mutagenic and cytotoxic properties of three herbal plants from Southeast Asia. Trop Biomed 2007;24:49−59.

[38] Ahn HS, Jeon TI, Lee JY, Hwang SG, Lim Y, Park DK. Antioxidative activity of persimmon and grape seed extract: *in vitro* and *in vivo*. Nutr Res 2002;22(11):1265−73.

[39] Higashimoto M, Purintrapiban J, Kataoka K, Kinouchi T, Vinitketkumnuen U, Akimoto S, et al. Mutagenicity and antimutagenicity of extracts of three spices and a medicinal plant in Thailand. Mutat Res 1993;303(3):135−42.

[40] Kassie F, Parzefall W, Musk S, Johnson I, Lamprecht G, Sontag G, et al. Genotoxic effects of crude juices from *Brassica* vegetables and juices and extracts from phytopharmaceutical preparations and spices of cruciferous plants origin in bacterial and mammalian cells. Chem Biol Interact 1996;102(1):1−16.

[41] Liu RH. Health benefits of fruit and vegetables are from additive and synergistic combinations of phytochemicals. Am J Clin Nutr 2003;78(Suppl. 3):517S−20S.

[42] McGaw LJ, Jager AK, van Staden J. Antibacterial, anthelmintic and antiamoebic activity in South African medicinal plants. J Ethnopharmacol 2000;72(1−2):247−63.

[43] Rabe T, Van Staden J. Screening of *Plectranthus* species for antibacterial activity. S Afr J Botany 1998;64(1):62−5.

[44] Schimmer O, Hafele F, Kruger A. The mutagenic potencies of plant extracts containing quercetin in *Salmonella typhimurium* TA98 and TA100. Mutat Res 1988;206 (2):201−8.

[45] Schimmer O, Kruger A, Paulini H, Haefele F. An evaluation of 55 commercial plant extracts in the Ames mutagenicity test. Pharmazie 1994;49(6):448−51.

[46] Pradhan SN, Chung EB, Ghosh B, Paul BD, Kapadia GJ. Potential carcinogens. I. Carcinogenicity of some plant extracts and their tannin-containing fractions in rats. J Natl Cancer Inst 1974;52(5):1579−82.

[47] Kapadia GJ, Chung EB, Ghosh B, Shukla SYN, Basak SP, Morton JF, et al. Carcinogenicity of some folk medicinal herbs in rats. J Natl Cancer Inst 1978;60:683−6.

[48] De Sa Ferreira IC, Ferrao Vargas VM. Mutagenicity of medicinal plant extracts in *Salmonella*/microsome assay. Phytother Res 1999;(13)397−400.

[49] Hoang ML, Chen C-H, Sidorenko VS, He J, Dickman KG, Yun BH, et al. Mutational signature of aristolochic acid exposure as revealed by whole-exome sequencing. Sci Transl Med 2013;5: 197ra102.

[50] Poon SL, Pang S-E, McPherson JR, Yu W, Huang KK, Guan P, et al. Genome-wide mutational signatures of aristolochic acid and its application as a screening tool. Sci Transl Med 2013;5: 197ra101.

[51] Lim-Sylianco CY, Shier WT. Mutagenic and antimutagenic activities in Philippine medicinal and food plants. Toxin Rev 1985;4(1):71−105.

[52] Rates SMK. Plants as source of drugs. Toxicon 2001;39:603−13.

[53] Ames BN. Dietary carcinogens and anticarcinogens. Science 1983;221:1256−64.

[54] Santos FV, Colus IMS, Silva MA, Vilegas W, Varanda EA. Assessment of DNA damage by extracts and fractions of *Strychnos pseudoquina*, a Brazilian medicinal plant with antiulcerogenic activity. Food Chem Toxicol 2006;44:1585−9.

[55] Santos FV, Tubaldini FR, Colus IMS, Andreo MA, Bauab TM, Leite CQF, et al. Mutagenicity of *Mouriri pusa* Gardner and *Mouriri elliptica* Martius. Food Chem Toxicol 2008;46:2721−7.

[56] Dantas de Carvalho MCR, Barca FNTV, Agnez-Lima LF, Batistuzzo de Medeiros SR. Evaluation of mutagenic activity in an extract of pepper tree stem bark (*Schinus terebinthifolius* Raddi). Environ Mol Mutagen 2003;42:185−91.

[57] Akintonwa A, Awodele O, Afolayan G, Coker H. Mutagenic screening of some commonly used medicinal plants in Nigeria. J Ethnopharmacol 2009;125:461−70.

[58] Alkofahi AS, Abdelaziz A, Mahmoud I, Abuirjie M, Hunaiti A, El-Oqla A. Cytotoxicity, mutagenicity and antimicrobial activity of forty Jordanian medicinal plants. Pharm Biol 1990;28(2):139−44.

[59] Khan PK, Awasthy KS. Cytogenetic toxicity of neem. Food Chem Toxicol 2003;41 (10):1325−8.

[60] Bresolin S, Ferrão Vargas VM. Mutagenic potencies of medicinal plants screened in the Ames test. Phytother Res 1993;7(3):260−2.

[61] Ravanel P, Leclercq M, Mariotte AM. La genotocixité des flavonoides. Plantes medicinales et Phytoterapie 1987;XXI:63−78.

[62] MacGregor JT, Wilson RE. Flavone mutagenicity in *Salmonella typhimurium*: dependence on the pKM101 plasmid and excision-repair deficiency. Environ Mol Mutagen 1988;11:315−22.

[63] Vargas VMF, Motta VEP, Leitlo AC, Henriques JAP. Mutagenic and genotoxic effects of aqueous extracts of *Achyrocline satureioides* in procaryotic organisms. Mutat Res 1990;240:13−8.

[64] Vargas VMF, Guidobono RR, Henriques JAP. Genotoxicity of plant extracts. Memorias do Instituto Oswaldo Cruz 1991;86:67—70.

[65] Liman R, Eren Y, Akyil D, Konuk M. Determination of mutagenic potencies of aqueous extracts of *Thermopsis turcica* by Ames test. Turk J Biol 2012;36:85—92.

[66] Reid KA, Maes J, Maes A, van Staden J, de Kimpe Ugent N, Mulholland DA, et al. Evaluation of the mutagenic and antimutagenic effects of South African plants. J Ethnopharmacol 2006;106:44—50.

[67] Marques RCP, Batistuzzo-de-Medeiros SR, Agnez-Lima LF, Dias CS, Barbosa-Filho JM. Evaluation of the mutagenic potential of yangambin and of the hydroalcoholic extract of *Ocotea duckei* by the Ames test. Mutat Res 2003;536:117—20.

[68] Hemaiswarya S, Raja R, Anbazhagan C, Thiagarajan V. Antimicrobial and mutagenic properties of the root tubers of *Gloriosa superba* Linn. (Kalihari). Pak J Bot 2009;41:293—9.

[69] dos Santos F, Calvo TR, Colus IMS, Vilegas W, Varanda EA. Mutagenicity of two species of the genus *Alchornea* measured by *Salmonella* microsome assay and micronucleus test. Rev Bras Farmacogn 2010;20:382—9.

[70] Sobita K, Bhagirath Th. Effects of some medicinal plant extracts on *Vicia faba* root tip chromosomes. Caryologia 2005;58(3):255—61.

[71] Reddy DG, Anguli VC. Experimental production of cancer with betel nut, tobacco, and slaked lime mixture. J Indian Med Assoc 1967;49:315—8.

[72] Dunham LJ, Sheets RH, Morton JF. Proliferative lesions in cheek pouch and esophagus of hamsters treated with plants from Curacao, Netherland Antilles. J Natl Cancer Inst 1974;53:1259—69.

[73] Taylor JM, Jenner PM, Jones WI. A comparison of the toxicity of some allyl, propenyl and propyl compounds in rat. Toxicol Appl Pharmacol 1964;6:378—87.

[74] Hagan EC, Jenner PM, Jones WI, Fitzhugh OG, Long EL, Brouwer JG, et al. Toxic properties of compounds related to safrole. Toxicol Appl Pharmacol 1965;7:1824.

[75] Homburger F, Kelley TJR, Friedler G, Russfield AB. Toxic and possible carcinogenic effects of 4-allyl-1,2-methylenedioxybenzene (Safrole) in rats on deficient diets. Med Exp 1961;4:1—11.

[76] Homburger F, Kelley Jr T, Baker TR, Russfield AB. Sex effect on hepatic pathology from deficient diet and safrole in rats. Arch Pathol 1962;73:118—25.

[77] Long EL, Nelson AA, Fitzhugh OG, Hansen WH. Liver tumors produced in rats by feeding safrole. Arch Pathol 1963;75:595—604.

[78] O'Gara RW, Lee C, Morton JF. Carcinogenicity of extracts of selected plants from curacao after oral and subcutaneous administration to rodents. J Natl Cancer Inst 1971;46 (6):1131—7.

[79] Maurici D, Aardema M, Corvi R, Kleber M, Krul C, Laurent C, et al. Genotoxicity and mutagenicity. Altern Lab Anim 2005;33:117—30.

[80] OECD, 2013.<http://www.oecd-ilibrary.org/environment/oecd-guidelines-for-thetesting-of-chemicals-section-4TAB-health-effects-20745788> [retrieved on 18.10.13].

[81] ECVAM, 2013.<http://ihcp.jrc.ec.europa.eu/our_labs/eurl-ecvam> [retrieved on 18.10.13].

[82] Jacobson-Kram D, Contrera JF. Genetic toxicity assessment: employing the best science for human safety evaluation. Part I: early screening for potential human mutagens. Toxicol Sci 2007;96:16—20.

[83] Fu PP, Yang Y-C, Xia Q, Chou MW, Cui YY, Lin G. Pyrrolizidine alkaloids tumorigenic components in Chinese herbal medicines and dietary supplements. J Food Drug Anal 2002;10:198—211.

[84] Zenga Y-X, Zhao C-X, Liang Y-Z, Yang H, Fang H-Z, Yi L-Z, et al. Comparative analysis of volatile components from *Clematis* species growing in China. Anal Chim Acta 2007;595:328−39.

[85] Napoli EM, Curcuruto G, Ruberto G. Screening the essential oil composition of wild Sicilian fennel. Biochem Syst Ecol 2010;38:213−23.

[86] Valerio Jr LG. *In silico* toxicology for the pharmaceutical sciences. Toxicol Appl Pharm 2009;241:356−70.

[87] Votano JR, Parham M, Hall LH, Kier LB, Oloff S, Tropsha A, et al. Three new consensus QSAR models for the prediction of Ames genotoxicity. Mutagenesis 2004;19:365−77.

[88] Greene N. Computer systems for the prediction of toxicity: an update. Adv Drug Deliv Rev 2002;54:417−31.

[89] Snyder RD, Smith MD. Computational prediction of genotoxicity: room for improvement. Drug Discovery Today 2005;10.

[90] Gatehouse D, Haworth S, Cebula T, Gocke E, Kier L, Matsushima T, et al. Recommendations for the performance of bacterial mutation assays. Mutat Res 1994;312:217−33.

[91] Osman F, Tomsett B, Strike P. The isolation of mutagen-sensitive *nuv* mutants of *Aspergillus nidulans* and their effects on mitotic recombination. Genetics 1993; 134:445−54.

[92] Souza-Júnior SA, Goncalves EAL, Catanzaro-Guimarãe SA, Castro-Prado MAA. Loss of heterozygosity by mitotic recombination in diploid strain of *Aspergillus nidulans* in response to castor oil plant detergent. Braz J Biol 2004;64:885−90.

[93] Sanchez-Lamar A, Fuentes JL, Fonseca G, Capiro N, Ferrer M, Alonzo A, et al. Assessment of the potential genotoxic risk of *Phyllantus orbicularis* HBK aqueous extract using *in vitro* and *in vivo* assays. Toxicol Lett 2002;136:87−96.

[94] Nohynek GJ, Kirkland D, Marzin D, Toutaina H, Leclerc-Ribaud C, Jinnai H. An assessment of the genotoxicity and human health risk of tropical use of kojic acid [5-hydroxy-2-(hydroxymethyl)-4H-pyran-4-one]. Food Chem Toxicol 2004;42:93−105.

[95] Ishidate Jr M, Sofuni T. The *in vitro* chromosomal aberration test using Chinese hamster lung (CHL) fibroblast cells in culture. In: Ashby J, et al., editors. Progress in mutation research, vol. 5. Amsterdam, New York, Oxford: Elsevier Science Publishers; 1985, p. 427−32.

[96] Galloway SM, Armstrong MJ, Reuben C, Colman S, Brown B, Cannon C, et al. Chromosome aberration and sister chromatid exchanges in Chinese hamster ovary cells: evaluation of 108 chemicals. Environ Mol Mutagen 1987;10(Suppl. 10):1−175.

[97] Evans HJ. Cytological methods for detecting chemical mutagens. In: Hollaender A, editor. Chemical mutagens, principles and methods for their detection, vol. 4. New York, London: Plenum Press; 1976, p. 1−29.

[98] Bolognesi C. Genotoxicity of pesticides: a review of human biomonitoring studies. Mutat Res 2003;543:251−72.

[99] Diaz D, Scott A, Carmichael P, Shi W, Costales C. Evaluation of an automated *in vitro* micronucleus assay in CHO-K1 cells. Mutat Res 2007;630:1−13.

[100] Westerink WM, Schirris TJ, Horbach GJ, Schoonen WG. Development and validation of a high-content screening *in vitro* micronucleus assay in CHO-k1 and HepG2 cells. Mutat Res 2011;724:7−21.

[101] Djelic N, Spremo-Potparevic B, Bajic V, Djelic D. Sister chromatid exchange and micronuclei in human peripheral blood lymphocytes treated with thyroxine *in vitro*. Mutat Res 2006;604:1−7.

[102] Kaya F, Topaktas M. Genotoxic effects of potassium bromate on human peripheral lymphocytes *in vitro*. Mutat Res 2007;626:48−52.

[103] Bakkali F, Averbeck S, Averbeck D, Idaomar M. Biological effects of essential oils—a review. Food Chem Toxicol 2008;46:446−75.

[104] Hseu Y-C, Chen S-C, Chen Y-L, Chen J-Y, Lee M-L, Lu F-J, et al. Humic acid induced genotoxicity in human peripheral blood lymphocytes using comet and sister chromatid exchange assay. J Hazard Mater 2008;153:784−91.

[105] Lambert IB, Singer TM, Boucher SE, Douglas GR. Detailed review of transgenic rodent mutation assays. Mutat Res 2005;590:1−280.

[106] Singh NP, McCoy MT, Tice RR, Schneider EL. A simple technique for quantification of low levels of damage in individual cells. Exp Cell Res 1988;175:184−91.

[107] Speit G, Vasquez M, Hartmann A. The comet assay as an indicator test for germ cell genotoxicity. Mutat Res 2009;681:3−12.

[108] Berthelot-Ricou A, Perrin J, Di Giorgio C, De Meo M, Botta A, Courbiere B. Comet assay on mouse oocytes: an improved technique to evaluate genotoxic risk on female germ cells. Fertil Steril 2011;95:1452−7.

[109] Ramos A, Rivero R, Victoria MC, Visozo A, Piloto J, Garcia A. Assessment of mutagenicity in *Parthenium hysterophorus* L. J Ethnopharmacol 2001;77:25−30.

[110] Cavalcanti BC, Ferreiraa JRO, Moura DJ, Rosa RM, Furtado GV, Burbano RR, et al. Structure−mutagenicity relationship of kaurenoic acid from *Xylopia sericeae* in Kirchner (Annonaceae). Mutat Res 2010;(701)153−63.

[111] Speit G, Brendler-Schwaab S, Hartmann A, Pfuhler S. The *in vivo* comet assay: use and status in genotoxicity testing. Mutagenesis 2005;20:245−54.

[112] Moore MM, DeMarini DM, DeSerres FJ, Tindall KR, editors. Banbury report 28: mammalian cell mutagenesis. New York, NY: Cold Spring Harbor Laboratory; 1987.

[113] Clive D, Flamm WG, Machesko MR, Bernheim NJ. A mutational assay system using the thymidine kinase locus in mouse lymphoma cells. Mutat Res 1972;16(7):77−87.

[114] Chu EHY, Malling HV. Mammalian cell genetics. II. Chemical induction of specific locus mutations in Chinese hamster cells *in vitro*. Proc Natl Acad Sci USA 1968;61:1306−12.

[115] Aaron CS, Stankowski Jr LF. Comparison of the AS52/XPRT and the CHO/HPRT assays: evaluation of six drug candidates. Mutat Res 1989;223:121−8.

[116] Aaron CS, Bolcsfoldi G, Glatt HR, Moore M, Nishi Y, Stankowski L, et al. Mammalian cell gene mutation assays working group report. Report of the international workshop on standardisation of genotoxicity test procedures. Mutat Res 1994;312:235−9.

[117] Kirkland D, Aardema M, Henderson L, Müller L. Evaluation of the ability of a battery of three *in vitro* genotoxicity tests to discriminate rodent carcinogens and non-carcinogens. I. Sensitivity, specificity and relative predictivity. Mutat Res 2005; 584:1−256.

[118] Mavournin KH, Blakey DH, Cimino MC, Salamone MF, Heddle JA. The *in vivo* micronucleus assay in mammalian bone marrow and peripheral blood. A report of the U.S. environmental protection agency Gene-Tox program. Mutat Res 1990;239:29−80.

[119] MacGregor JT, Wehr CM, Henika PR, Shelby ME. The *in vivo* erythrocyte micronucleus test: measurement at steady state increases assay efficiency and permits integration with toxicity studies. Fundam Appl Toxicol 1990;14:513−22.

[120] Leopardi P, Villani P, Cordelli E, Siniscalchi E, Veschetti E, Crebelli R. Assessment of the *in vivo* genotoxicity of vanadate: analysis of micronuclei and DNA damage induced in mice by oral exposure. Toxicol Lett 2005;158:39−49.

[121] OECD. Detailed review paper on transgenic rodent mutation assays, series on testing and assessment, N° 103, ENV/JM/MONO[2009]7. Paris: OECD; 2009.

[122] OECD. Retrospective performance assessment of OECD test guideline on transgenic rodent somatic and germ cell gene mutation assays, series on testing and assessment, N° 145. Paris: OECD; 2011.

[123] OECD. Transgenic rodent somatic and germ cell gene mutation assays. Test no. 488; 2013.

[124] Wahnschaffe U, Bitsch A, Kielhorn J, Mangelsdorf I. Mutagenicity testing with transgenic mice. Part II: comparison with the mouse spot test. J Carcinog 2005;4:4.

[125] Sekihashi K, Yamamoto A, Matsumura Y, Ueno S, Watanabe-Akanuma M, Kassie F, et al. Comparative investigation of multiple organs of mice and rats in the comet assay. Mutat Res 2002;517:53−74.

[126] Hartmann A, Schumacher M, Plappert-Helbig U, Lowe P, Suter W, Mueller L. Use of alkaline in-vivo comet assay for mechanistic genotoxicity investigations. Mutagenesis 2005;19(1):51−9.

[127] Chiu SW, Wang ZM, Leung TM, Moore D. Nutritional value of ganoderma extract and assessment of its genotoxicity and antigenotoxicity using comet assays of mouse lymphocytes. Food Chem Toxicol 2000;38:173−8.

[128] Idaomar M, El-Hamss R, Bakkali F, Mezzoug N, Zhiri A, Baudoux D, et al. Genotoxicity and antigenotoxicity of some essential oils evaluated by wing spot test of Drosophila melanogaster. Mutat Res 2002;513:61−8.

[129] Carmona ER, Creus A, Marcos R. Genotoxicity testing of two lead compounds in somatic cells of Drosophila melanogaster. Mutat Res 2011;724:35−40.

[130] Munerato MC, Sinigaglia M, Reguly M, Rodriguez-de-Andrade HH. Genotoxic effects of eugenol, isoeugenol and safrole in the wing spot test of Drosophila melanogaster. Mutat Res 2005;582:87−94.

[131] Sharma A, Shukla AK, Mishra M, Chowdhuri DK. Validation and application of Drosophila melanogaster as an in vivo model for the detection of double strand breaks by neutral Comet assay. Mutat Res 2011;721:142−6.

[132] Aardema MJ, MacGregor JT. Toxicology and genetic toxicology in the new era of "toxicogenomics": impact of "-omics" technologies. Mutat Res 2002;499:13−25.

[133] Marchant GE. Toxicogenomics and toxic torts. Trends Biotechnol 2002;20:329−32.

[134] Heijne WH, Kienhuis AS, van Ommen B, Stierum RH, Groten JP. Systems toxicology: applications of toxicogenomics, transcriptomics, proteomics and metabolomics in toxicology. Expert Rev Proteomics 2005;767−80.

[135] Marques A, Lourenco HM, Nunes ML, Roseiro C, Santos C, Barranco A, et al. New tools to assess toxicity, bioaccessibility and uptake of chemical contaminants in meat and seafood. Food Res Int 2011;44:510−22.

[136] Borner FU, Schutz H, Wiedemann P. The fragility of omics risk and benefit perceptions. Toxicol Lett 2011;201:249−57.

[137] Maurici D, Aardema M, Corvi R, Kleber M, Krul C, Laurent L, et al. Carcinogenicity. Altern Lab Anim 2005;33(Suppl. 1):177−82.

[138] OECD. Detailed review paper on cell transformation assays for detection of chemical carcinogens. Series on testing and assessment number 31; 2007.

[139] AltTox. Non-animal methods for toxicity testing, <http://www.alttox.org/ttrc/toxicity-tests/carcinogenicity/>; 2013 [retrieved on 24.10.13].

[140] OECD. Draft OECD guideline for the testing of chemicals. Test guideline 451: Carcinogenicity studies, <http://www.oecd.org/chemicalsafety/testing/41753121.pdf>; 2013 [retrieved on 14.10.13].

[141] OECD. OECD guideline for the testing of chemicals. Test guideline 453: combined chronic toxicity\carcinogenicity studies. Adopted: 7 September 2009.

[142] Harvey M, McArthur MJ, Montgomery Jr CA, Butel JS, Bradley A, Donehower LA. Spontaneous and carcinogen-induced tumorigenesis in p53-deficient mice. Nat Genet 1993;5:225−9.

[143] Yamamoto S, Hayashi Y, Mitsumori K, Nomura T. Rapid carcinogenicity testing system with transgenic mice harbouring human prototype c-HRAS gene. Lab Anim Sci 1997;47:121−6.

[144] MacDonald J, French JE, Gerson RJ, Goodman J, Inoue T, Jacobs A, et al. The utility of genetically modified mouse assays for identifying human carcinogens: a basic understanding and path forward. The Alternatives to Carcinogenicity Testing Committee ILSI HESI. Toxicol Sci 2004;77:188−94.

[145] Goodman JI. A perspective on current and future uses of alternative models for carcinogenicity testing. Toxicol Pathol 2001;29:173−6.

[146] Berwald Y, Sachs L. *In vitro* transformation of normal cells to tumour cells by carcinogenic hydrocarbons. J Natl Cancer Inst 1963;35:641−61.

[147] Aaronson SA, Todaro GJ. Development of 3T3-like line from BALB/c mouse embryo culture: transformation susceptibility to SV40. J Cell Physiol 1968;72:141−8.

[148] LeBoeuf RA, Kerchaert GA. Enhanced morphological transformation of early passage Syrian hamster embryo cells cultured in medium with a reduced bicarbonate concentration and pH. Carcinogenesis 1987;8:689−97.

[149] Combes R, Balls M, Curren R, Fischbach M, Fusenig N, Kirkland D, et al. Cell transformation assays as predictors of human carcinogenicity—the report and recommendations of ECVAM workshop 39. Altern Lab Anim 1999;27:745−67.

[150] Yamasaki H, Ashby J, Bignami M, Jongen W, Linnainmaa K, Newbold RF, et al. Nongenotoxic carcinogens: development of detection methods based on mechanisms: a European project. Mutat Res 1996;353:47−63.

[151] Ottender M, Lutz W. Correlation of DNA adduct levels with tumor incidence: carcinogenicity potency of DNA adducts. Mutat Res 1999;424:237−47.

[152] Morita T, Asao N, Awogi T, Sasaki YF, Sato S, Shimada H, et al. Evaluation of the rodent micronucleus assay in the screening of IARC carcinogens (Groups 1, 2A and 2B). The summary report of the 6th collaborative study by CSGMT MMS. Mutat Res 1997;389:3−122.

[153] Dean SW, Brooks TM, Burlinson B, Mirsalis J, Myhr B, Recio L, et al. Transgenic mouse mutation assay systems can play an important role in regulatory mutagenicity testing *in vivo* for the detection of site-of-contact mutagens. Mutagenesis 1999;14:141−51.

[154] Sasaki YF, Fujikawa K, Ishida K, Kawamura N, Nishikawa Y, Ohta S, et al. The alkaline single cell gel electrophoresis assay with mouse multiple organs: results with 30 aromatic amines evaluated by the IARC and U.S. NTP. Mutat Res 1999;440:1−18.

[155] Cronin MT, Jaworska JS, Walker JD, Comber MH, Watts CD, Worth AP. Use of QSARs in international decision-making frameworks to predict health effects of chemical substances. Environ Health Perspect 2003;111:1391−401.

[156] Young JF, Tong W, Fang H, Xie Q, Pearce B, Hashemi R, et al. Building an organ-specific carcinogenic database for SAR analyses. J Toxicol Environ Health A 2004;67:1363−89.

[157] Clive D. Mutagenicity in drug development: interpretation and significance of test results. Regul Toxicol Pharmacol 1985;5(1):79−100.

[158] Ennever FK, Andreano G, Rosenkranz HS. The ability of plant genotoxicity assays to predict carcinogenicity. Mutat Res 1988;205:99−105.

[159] Chankong V, Haimes YY, Rosenkranz HS, Pet-Ed-wards J. The carcinogenicity prediction and battery selection (CPBS) method: a Bayesian approach. Mutat Res 1985;153:135−66.

[160] Taylor JLS, Elgorashi EE, Maes A, Van Gorp U, De Kimpe N, van Staden J, et al. Investigating the safety of plants used in South African traditional medicine: testing for genotoxicity in the micronucleus and alkaline comet assays. Environ Mol Mutagen 2003;42:144−54.

[161] Aremu AO, Moyo M, Amoo SO, Van Staden J. Mutagenic evaluation of 10 long-term stored medicinal plants commonly used in South Africa. S Afr J Bot 2013;87:95−8.

[162] Sowemimo AA, Fakoya FA, Awopetu I, Omobuwajo OR, Adesanya SA. Toxicity and mutagenic activity of some selected Nigerian plants. J Ethnopharmacol 2007;113:427−32.

[163] Verschaeve L, Van Staden J. Mutagenic and antimutagenic properties of extracts from South African traditional medicinal plants. J Ethnopharmacol 2008;119:575−87.

[164] Mulaudzi RB, Ndhlala AR, Kulkarni MG, Finnie JF, Van Staden J. Anti-inflammatory and mutagenic evaluation of medicinal plants used by Venda people against venereal and related diseases. J Ethnopharmacol 2013;146:173−9.

[165] Verschaeve L, Kestens V, Taylor JLS, Elgorashi EE, Van Puyvelde L, De Kimpe N, et al. Investigation of the antimutagenic effects of selected South African medicinal plant extracts. Toxicol In Vitro 2004;18:29−35.

[166] Edziri H, Mastouri M, Mahjou A, Anthonissen R, Mertens B, Cammaerts S, et al. Toxic and mutagenic properties of extracts from Tunisian traditional medicinal plants investigated by the neutral red uptake, VITOTOX and alkaline comet assays. S Afr J Bot 2011;77:703−10.

[167] Skandrani I, Bouhlel I, Limem I, Boubaker J, Bhouri W, Neffati A, et al. *Moricandia arvensis* extracts protect against DNA damage, mutagenesis in bacteria system and scavenge the superoxide anion. Toxicol In Vitro 2009;23:166−75.

[168] Ndhlala AR, Anthonissen R, Stafford GI, Finnie JF, Verschaeve L, Van Staden J. *In vitro* cytotoxic and mutagenic evaluation of thirteen commercial herbal mixtures sold in KwaZulu-Natal, South Africa. S Afr J Bot 2010;76:132−8.

[169] Cano JH, Volpato G. Herbal mixtures in the traditional medicine of Eastern Cuba. J Ethnopharmacol 2004;90:293−316.

[170] Ndhlala AR, Finnie JF, Van Staden J. Plant composition, pharmacological properties and mutagenic evaluation of a commercial Zulu herbal mixture: Imbiza ephuzwato. J Ethnopharmacol 2011;133:663−74.

[171] Ndhlala AR, Stafford GI, Finnie JF, Van Staden J. Commercial herbal preparations in KwaZulu-Natal, South Africa: the urban face of traditional medicine. S Afr J Bot 2011;77:830−43.

11 Hepatotoxicity and Hepatoprotective Effects of African Medicinal Plants

Faustin Pascal Tsagué Manfo[1], Edouard Akono Nantia[2] and Victor Kuete[3]

[1]Department of Biochemistry and Molecular Biology, Faculty of Science, University of Buea, Cameroon, [2]Department of Biochemistry, Faculty of Science, University of Bamenda, Cameroon, [3]Department of Biochemistry, Faculty of Science, University of Dschang, Cameroon

11.1 Introduction

Complementary and alternative medicine (CAM) is the most ancient therapeutic practice in the world, which remains very common in Africa [1]. CAM has evolved, to give rise to modern therapy, whereby manufactured drugs are produced as a result of stringent research protocols, which enables substantial reduction of side/toxic effects. However, original CAM, basically consisting of the use of original materials in the natural way, is still commonly practiced nowadays by populations of both industrialized and developing countries. This practice is even increasing among all populations, probably motivated by the beliefs from populations, who consider medicinal plants (i.e., plants that have at least one of their chemical components and/or structural parts -flowers, leaves, stem, seeds, barks, or roots- used for therapeutic purposes [2]) as natural and therefore less "harmful" than manufactured/synthesized drugs used in conventional medicine. In fact, the use of synthetic drugs in order to relieve or treat multiple diseases has been associated with many side effects and undesirable hazards. People have thus gone back to CAM, and particularly to extracts from medicinal plants, which are culturally acceptable, easily available, and economically viable [3]. Indeed, in rural areas of developing countries, the medicinal plants continue to be used as the primary source of medicine. About 80% of the people in developing countries use traditional medicines for their healthcare [4].

An impressive number of medicinal plants are currently used worldwide against diseases, and hepatoprotective and hepatocurative medicinal plant preparations have unique importance. In fact, liver diseases are among the most serious health

Toxicological Survey of African Medicinal Plants. DOI: http://dx.doi.org/10.1016/B978-0-12-800018-2.00011-X

problems worldwide. Liver diseases may be classified as acute or chronic hepatitis (inflammatory liver diseases), hepatosis (noninflammatory diseases), and cirrhosis (degenerative disorder resulting in fibrosis of the liver) [3,5].

The liver is one of the most vital organs; it plays a key function in digestion, storage, and metabolism of nutrients [3], and its dysfunction is thus a major problem to the whole organism. It is the major organ involved in metabolism, detoxification, and excretion of xenobiotics [6], and this function makes it one of the most exposed organs to xenobiotics. Although liver metabolic process is aimed at converting xenobiotics into water-soluble metabolites in order to facilitate further elimination in the urine, it may also lead to bioactivation of the xenobiotics into highly hepatotoxic compounds. The prevention and treatment options for liver diseases remain scarce nowadays despite tremendous advances in modern medicine [3,7]. Except for vaccines and interferon α-2b, which concern only viral infections, modern medicine is quite limited in treating liver diseases, and only a few available drugs include cholagogues, choleretics, drugs for cholesterolic lithiasis, N-acetylcysteine (NAC), alkaloid mixture from *Enantia chlorantha* (Hepazor), and flavolignans from *Silybum marianum* [8−10]. This encourages frequent use of medicinal plants as hepatoprotective or hepatocurative. Several reports and reviews do exist on modulation of liver function by medicinal plants from Asia, Europe, and America. Potential toxicity of medicinal plants from the latter continents has also been illustrated [3,11−19]. However, data on toxicity and protection of the liver by medicinal plants from Africa are still scanty, and the few available ones remain scattered. This chapter summarizes data on hepatotoxicity and hepatoprotective medicinal plants that are used in folkloric medicine in Africa.

11.2   The Liver and the Metabolism of Xenobiotics

11.2.1   Anatomy and Physiology of the Liver

The liver is the largest organ in the human body (about 2% of the total body weight) [20]. It consists of two lobes, each containing multiple lobules and sinusoids. The liver receives 75% of its blood supply from the portal vein, which carries blood returning to the heart from the small intestine, stomach, pancreas, and spleen. The remaining 25% of the liver's blood supply is arterial, carried to the liver by the hepatic artery. Blood from both the portal vein and hepatic artery empty into sinusoids, which are expandable vascular channels that run through the hepatic lobules. The content from the portal vein and hepatic artery is mixed upon entry in the liver sinusoids [21−23] and will further reach hepatocytes by leaking through the endothelial cells lining (the membrane of the liver capillary sinusoids is so permeable that even plasma proteins pass freely through these walls, almost as easily as water and other substances) [20]. After perfusion of the liver, the blood then exits from the organ by the hepatic veins, which merge into the inferior vena cava and return blood to the heart [24].

Hepatocytes are the primary functional cells of the liver [20]. Hepatocytes are also known as the hepatic parenchymal cells, and form the liver lobules, while the remaining cells (nonparenchymal cells) constitute the lining cells of the walls of the sinusoids. The lining cells comprise the endothelial cells, Kupffer cells (reticuloendothelial cells), hepatic stellate cells (perisinusoidal or Ito cells), and intrahepatic lymphocytes (e.g., pit cells, which are liver-specific natural killer cells) [25].

The liver is also the largest exocrine gland in the body, secreting large amounts of bile. The hepatocytes secrete bile into the bile canniculus. The canniculi empty into the bile ducts, and the ducts then fuse to form the common bile duct which releases bile into the duodenum [20,25]. The entire liver surface is covered by a capsule of connective tissue that branches and extends throughout the liver to subdivide the lobes into the smaller lobules. The capsule also provides support for blood vessels, lymphatic vessels, and bile ducts that permeate the liver [23,25]. The liver is responsible for some variety of functions including digestion, the body's immune defense, and metabolism.

11.2.2   Role of Liver the Metabolism of Xenobiotics

The liver is primarily responsible for the metabolism or biotransformation of a plethora xenobiotics and endogenous agents, although metabolizing enzymes are also present at other sites, such as the gastrointestinal and respiratory tracts, kidney, lung, brain, and skin [26].

The liver plays a crucial role in eliminating and detoxifying substances that are harmful to the body, including alcohol, drugs, solvents, pesticides, and heavy metals. Toxins are delivered to the liver by the portal vein, and the liver provides an effective barrier that prevents xenobiotics from entering the systemic circulation. The liver processes these chemicals, thanks to varieties of drug-metabolizing enzymes, and excretes them in the bile [27−29]. Biotransformations in the liver also may improve solubility of the xenobiotics, in order to ease elimination in the urine. In fact, most pharmacologically active molecules are lipophilic and, to be excreted, must be enzymatically converted into water-soluble metabolites. These metabolites often are less active than the parent drug, but sometimes biotransformation may enhance activity or toxic effects. The most common routes of drug metabolism are hydrolysis, oxidation, reduction, and conjugation. A drug can be biotransformed by several competing pathways, and resulting metabolites may undergo further metabolism. Oxidation, reduction, and hydrolysis often occur first and commonly are referred to as phase I or functionalization reactions, with conjugations commonly referred to as phase II reactions [26,30].

11.2.3   Phase I and Phase II Reactions

Many of the xenobiotics are lipophilic and are oxidized, hydroxylated, or hydrolyzed by enzymes in phase I reactions, prior to conjugation in phase II reactions. Phase I reactions introduce/expose hydroxyl groups or other reactive sites that can be used for conjugation reactions, the phase II reactions. These reactions include microsomal monooxygenations, cytosolic and mitochondrial oxidations, co-oxidations in the prostaglandin synthetase reaction, reductions, hydrolyses, and epoxide hydration [25,31]. Phase I oxidative reactions are catalyzed by the cytochrome P450 (CYP)-dependent monooxygenase system or by flavin-containing monooxygenases (FMOs), which are all located in the endoplasmic reticulum of the cell. The major reactions of phase I are hydroxylation in addition to deamination, dehalogenation, desulfuration, epoxidation, peroxygenation, and reduction [25,31,32]. From phase I reactions, xenobiotics are generally converted to more polar, hydroxylated derivatives. The monooxygenases incorporate one atom of oxygen molecule into the substrate while the other atom is reduced to water. Because the electrons involved in the reduction of CYPs or FMOs are derived from NADPH, the overall reaction can be written as follows:

$$RH + O_2 + NADPH + H^+ \rightarrow R - OH + H_2O + NADP^+$$

In phase II, products of phase I metabolism and other xenobiotics containing functional groups such as hydroxyl, amino, carboxyl, epoxide, or halogen undergo conjugation reactions with endogenous metabolites (e.g., glucuronic acid, sugars, amino acids, glutathione (GSH), sulfate, and acetate). Conjugation products, with only rare exceptions, are more polar, less toxic, and more readily excreted than are their parent compounds into the bile and urine [31−35].

However, conjugated metabolites excreted by the liver into the bile can be subjected to microbial degradation activity (e.g., hydrolysis of glucuronides by β-glucuronidase) in the gastrointestinal tract. The degradation can result in regeneration of the parent toxicant that is more lipophilic than the conjugate and will be reabsorbed into blood system by the gastrointestinal tract. This phenomenon called "enterohepatic circulation," will thus prolong the presence of the drug or xenobiotic in the systemic circulation [20,25].

11.2.4 Xenobiotic-Induced Hepatotoxicity

Xenobiotics include toxins in food, peroxides, drugs, environmental pollutants, etc. Their biotransformation has as purpose to increase their water solubility or polarity and thus excretion from the body. However, in certain cases, phase I metabolic reactions convert xenobiotics from inactive or less active to biologically highly active toxic compounds [32]. For example, acetaminophen and ethanol that are metabolized in phase I reactions into the toxic intermediates N-acetyl-p-benzoquinoneimine (NAPQI) and acetaldehyde, respectively, which can lead to the depletion of GSH and consequently to the damage of cellular proteins and the death of hepatocytes (hepatotoxicity) [25,32]. The compound vinyl chloride used in the synthesis of plastics is activated in phase I reactions to a reactive epoxide by the hepatic P450 isozyme (CYP2E1), which can react with guanine in DNA or other cellular molecules causing angiosarcoma in the liver [25]. Aflatoxin is metabolically activated to its

8,9-epoxide by two different isozymes of cytochrome P450. The epoxide modifies DNA by forming covalent adducts with guanine residues. In addition, the epoxide can combine with lysine residues within proteins and thus is also a hepatotoxin [25].

Many drugs including anti-HIV molecules and certain herbal remedies are processed by the liver and can cause liver damage through formation of highly reactive metabolites such as peroxides, epoxides, and other radicals [27,29].

The toxic effects of xenobiotics cover a wide spectrum, including:

- *Cell injury (cytotoxicity)*: Cytotoxicity can be severe enough to result in cell death through many mechanisms; the major one being the covalent binding of xenobiotic-produced reactive oxygen species (ROS) to cell macromolecules. ROS can initiate protein and lipid peroxidation, and deplete antioxidant defenses like reduced GSH, and modify sulfhydryl (SH) groups on cellular components. These macromolecular targets include DNA, RNA, lipid, and protein [22].
- *The reactive species of a xenobiotic may bind to a protein, altering its antigenicity*: The xenobiotic is said to act as a hapten, that is, a small molecule that by itself does not stimulate antibody synthesis but will combine with antibody once formed. The resulting antibodies can then damage the cell by several immunologic mechanisms that grossly perturb normal cellular biochemical processes [32].
- *Reactions of activated species of chemical carcinogens with DNA are thought to be of great importance in chemical carcinogenesis*: Some chemicals (e.g., benzo(α)pyrene) require activation by monooxygenases in the endoplasmic reticulum to become carcinogenic. They are thus called indirect carcinogens. Other chemicals such as alkylating agents can react directly (direct carcinogens) with DNA without undergoing intracellular chemical activation [22,31].
- *The stress in hepatocytes induced by reactive metabolites, including ROS, can activate several signal transduction pathways*: Several endogenous signal substances modify signal transduction including various cytokines. The pathways activate pro-death or survival proteins that determines the fate of the cell [36]. The important cytokine for liver injury is c-Jun N-terminal protein kinases (JNKs). JNK can be activated with many stresses including ROS and represents a common mechanism of cellular injury associated with several diseases [36].
- *Xenobiotics and/or their metabolites may attack the organelle directly or through above pathways and induce mitochondrial permeability transition (MPT) through opening of MPT pore in the membranes of the organelle, leakage of Ca^{2+} or depletion of intracellular ATP and activation of signals, death receptors, and proapoptotic pathways*: As a result acute necrosis, apoptosis, and autophagic cell death can occur [22,37a]. Xenobiotics can produce a variety of biological effects, including pharmacologic responses, toxicity, immunologic reactions, and cancers.

11.3 Methods in Hepatotoxic and Hepatoprotective Screenings of Medicinal Plants

Hepatotoxic and/or hepatoprotective medicinal plants are plants that can induce any structural or functional change on liver. Several experimental models are used in scientific research for investigation of the modulatory effect of xenobiotics on liver function, in order to design new efficient treatment options of liver diseases. These experimental models may involve the whole animal (*in vivo* models), the isolated liver, hepatocytes, organelles, and even purified compounds such as enzymes (*ex vivo* and *in vitro* studies). Reference chemicals with proven activity on liver are also integrated into protocols.

11.3.1 Reference Hepatotoxic/Hepatoprotective Compounds Currently Used for Assessment of Hepatoprotective Effect of Medicinal Plants

11.3.1.1 Hepatotoxic Compounds

The reference hepatotoxic compounds are chemicals with proven deleterious effects on the liver. The respective action mechanisms are also documented, in order to specify the type of hepatotoxicity model and ease interpretation of data obtained. These compounds include carbon tetrachloride (CCl_4), paracetamol (*N*-acetyl-para-aminophenol), and alcohol (methanol and ethanol), among others.

Methanol, the simplest alcohol from the formula, is a common adulterant in liquors locally manufactured in developing countries. This increases the chance of accidental methanol poisoning among humans [7]. Ethanol is a common drink worldwide. Induction of oxidative stress appears to be one mechanism by which ethanol causes liver injury [37b]. Chronic ethanol treatment induces increases of hepatic TBARS level, and markedly decreases GSH level, superoxide dismutase (SOD), and catalase (CAT) activities. Ethanol also induces liver and plasma aminotransferases (ALT and AST) [38].

Carbon tetrachloride (CCl_4) is the most used chemical compound for hepatotoxicity induction in animal models, and this may be due to elucidation of its action mechanism. CCl_4 is converted in the liver into $CCl_3{}^-$, which further reacts with O_2 to yield $CCl_3O_2{}^-$. The latter radicals ($CCl_3{}^-$ and $CCl_3O_2{}^-$) are highly reactive and will thereafter bind to proteins and lipids [3,39,40]. The radicals may also induce lipid peroxidation through removal of a hydrogen atom from unsaturated lipids. These changes lead to modifications in the endoplasmic reticulum and reduction in protein synthesis. Peroxidation of the cytoplasmic membrane lipids by CCl_4 leads to leakage of cytoplasmic enzymes such as aspartate aminotransferase (AST), alanine aminotransferase (ALT), and alkanine phosphatase (ALP). Leakage of these enzymes from the hepatocytes thus leads to significant increase of their activities in blood/serum, and this justifies the use of these enzymes as biochemical markers for hepatotoxicity [3,39]. Decrease in the activity of these enzymes has been observed in serum of CCl_4-treated rats following administration of known antioxidants or heptocurative efficient drugs [41,42].

Paracetamol (also called acetaminophen) is a commonly used analgesic and antipyretic available over the counter. Paracetamol is used for induction of experimental hepatotoxicity in laboratory animals [43]. It is mainly metabolized in hepatocytes to yield a highly toxic metabolite, NAPQI. NAPQI in turn is rapidly conjugated by cysteine and mercaptate compounds, leading to depletion of GSH reserves in the liver. NAPQI not only reacts with GSH, but also causes a loss in protein thiol groups [43,44]. The antioxidant system is thus weakened, and highly reactive metabolites alter hepatocyte molecules/components, leading to liver cirrhosis. Paracetamol has also been used in combination with caffeine, and coadministration of the two compounds to experimental animals has been shown to cause more inducement in the hepatic toxicity [45a,45b].

Lipopolysaccharide (LPS) is a substance produced by gram-negative bacteria. LPS stimulates nitric oxide (NO) production by the activation of macrophages, and the released NO serves as a signal molecule in the acute inflammatory response [46]. LPS is both hepatotoxic and nephrotoxic and has been used for induction of hepatotoxicity in experimental animals [47]. Another type of hepatotoxicity model, galactosamine-induced hepatotoxicity, is due to galactosamine metabolism in the liver. The hepatic metabolism of galactosamine induces decrease of uracil nucleotides levels. This results in inhibition of RNA synthesis and leads to cell necrosis [48,49].

Heavy metals are also hepatotoxic, and have been used in assessment of hepatoprotective medicinal plants. Cadmium for example induces liver injury by enhancing ROS production and lipid peroxidation, and via inhibition of antioxidant enzymes [50−53].

Thioacetamide (TAA: CH_3-$C(S)NH_2$) is an organosulfur compound that serves as a source of sulfide ions in the synthesis of organic and inorganic compounds. It induces liver cirrhosis in rats after prolonged exposure [54]. When administered in the diet, TAA is converted to TAA-S-oxide by hepatic microsomal cytochrome P450 2E1, then transformed to toxic TAA S-dioxide, which thereafter will damage liver biomolecules and lead to cirrhosis [54−56].

Other reference compounds that have been used for experimental hepatotoxicity include therapeutical drugs such as Cisplatin (cisplatinum or carboplatin, which reduces liver GSH levels and catalase activity), antibiotics (tetracycline, erythromycin) [57−59], pesticides (diazinon) [28], and petroleum products (diesel, kerosene, and hydraulic oil). The use of gamma irradiation for induction of hepatotoxicity also has been reported [60].

In general, administration of the hepatotoxic compounds leads to elevation in the level of ALT, AST, and ALP in serum. These enzymes are thus referred to as biomarkers of hepatotoxicity [41,61]. Activity of gammaglutamyl transferase (γGT) is elevated in animals exposed to other toxic compounds such as petroleum products and may also serve as an indication of liver lesion [61,62]. However, ALT appears to be more specific to the liver, and thus serves as the most sensitive marker of heatotoxicity [41,42]. Serum ALP and γ-GT are also increased (with/without bilirubin) when bile drainage is impaired and may be referred to as bioindicators of liver/hepatocytes functioning [63]. Tissue histological assay also are known to be useful in assessing the functional integrity of the liver [64].

11.3.1.2 Reference Hepatoprotective Compounds

Xenobiotics induce liver damage mainly through induction of oxidative stress. Efficiency of the antioxidant system may thus be of great importance in prevention or treatment of liver toxicity. GSH and NAC are well-known antioxidants. GSH is a tripeptide synthesized by mammalian cells, including hepatocytes. GSH is one of the most important intracellular peptides, playing a multifunctional role ranging from antioxidant defense to modulation of immune function. It is a non-enzymatic antioxidant that serves as a substrate for the enzymes glutathione peroxidase (GPx) and glutathione S-transferase (GST) [31]. GSH neutralizes cellular/tissue free radical species (hydrogen peroxide, superoxide radicals, alkoxy radicals) and reactive metabolites of xenobiotics, thereby maintaining membrane lipids and protein thiols [3,31]. GSH prevents covalent binding of reactive metabolite of paracetamol to liver proteins. For example, GSH protects hepatocytes by combining with the reactive metabolite of paracetamol, thus preventing their covalent binding to liver proteins. GSH supplementation is thus integrated in the clinical treatment of paracetamol-induced toxicity, and several research studies have used this tripeptide as a reference hepatoprotective compound, especially in the paracetamol-induced hepatotoxicity model [44]. NAC is an antioxidant and anti-inflammatory molecule used for treatment of paracetamol toxicity [43,65]. It is reported to replenish the depleted stores of GSH and can also conjugate directly with the highly reactive paracetamol metabolite, NAPQI, by acting as a GSH substitute [43,66].

Vitamins A, C, and E exhibit considerable antioxidant effects and have been demonstrated to alleviate oxidation induced on liver by toxicants [66]. Silymarin is the active extract from *Silybum marianum* (L.) Gaertn. (Asteraceae), a plant commonly known as milk thistle, blessed milk thistle, etc. Silymarin is a complex of seven flavonolignans and polyphenols, with the compound silibinin (Figure 11.1) regarded as the most active component [67]. Silymarin has been extensively used for efficient treatment of liver disorders associated with alcohol consumption, acute and chronic viral hepatitis, and toxin-induced hepatic failures [68]. Silymarin protects the liver mainly by acting as antioxidant (inhibition of oxidative stress) and stimulator of tissue regeneration. It also exhibits anti-apoptotic and anti-inflammatory activities [69]. Likewise, Hepazor (a protoberberine alkaloid mixture from *Enantia chlorantha*) is used for treatment of viral hepatitis [8−10]. The aforementioned properties make Silymarin, vitamins (A, C, and E), and hepazor good reference drugs for hepatoprotective testing procedures.

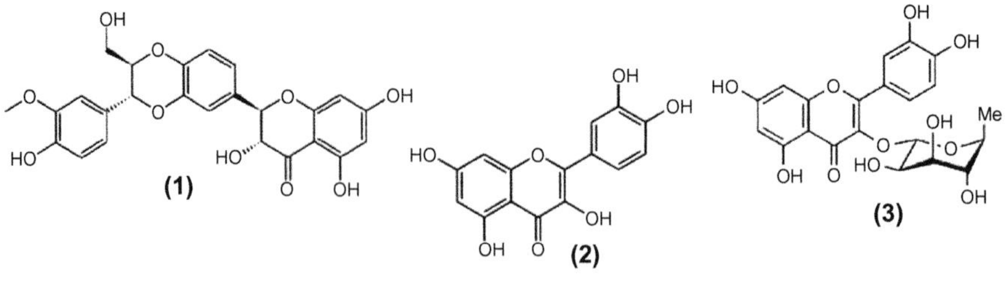

Figure 11.1 Chemical structures of hepatoprotective compounds isolated from medicinal plants. Silibinin (**1**) from *Silybum marianum*; quercetin (**2**) and quercitrin (**3**) from *Ficus gnaphalocarpa*.

11.3.2 Experimental Models

Experimental models for hepatotoxicity/hepatoprotective testing generally consist of animals (mainly rodents), liver tissue/slices, cells or cellular fractions, or tissue homogenates. For assessment of hepatoprotective effects, the experimental model receives the test compound/drug and a known hepatotoxic chemical, followed by evaluation of markers of hepatotoxicity. The test compound(s) may be administered concomitantly with the hepatotoxic compound(s). However, in other protocols, intoxication is performed before administration of the test drug(s), and this may be referred to as "recuperative effect."

In vitro experiments are commonly used for investigations of hepatotoxic/hepatoprotective effects of medicinal plants. *In vitro* experimental models represent an advantage of enabling testing several concentrations of the chemicals/extracts at once, with use of small quantity of the tested material, and have been useful for bioguided fractionation schemes [9]. These models involve the use of cells (primary cell cultures or cell lines), liver slices such as precision-cut liver slices (PCLS), and liver microsomes [9,70]. Antioxidant capacity of natural compounds is also an interesting parameter, as it enables prediction of the potential hepatoprotective effect of the compounds. Screening for antioxidant activity has been done *in vitro* using 2,4-dinitrophenyl-1-picryl hydrazyl radical activity, β-carotene-linoleic acid model system, ferric-reducing antioxidant power assay, and microsomal lipid peroxidation [70−72].

In vivo models or animal models for medicinal testing are mainly represented by rodents. The results recorded in *in vivo* experimentations are more likely to occur in other living organisms, compared to *in vitro* findings. Hepatoprotective effects of extracts have been conducted *in vivo* using a broad range of hepatotoxic compounds, including ethanol, CCl_4, Fe-NTA (ferric nitroprussiate), acetaminophen, caffeine, and cadmium [9,72]. The compounds generally are administered (with or without the reference hepatotoxin), and the degree of protection/toxicity recorded using biochemical parameters, such as serum ALT, AST, and ALP. In addition to these parameters, the *in vivo* experimental model has enabled evaluation of medicinal plants effects on bile flow and bile solids [73,74].

11.4 Hepatotoxic Medicinal Plants of Africa

Many indigenous plants are used by humans without the actual knowledge of their toxic potentials [75]. The safety of herbal medicines used in low-income countries is a major concern although not frequently addressed in the literature [76]. Liver toxicity is one of the common adverse effects. Hepatotoxicity is defined as injury to the liver or impairment of the liver function caused by exposure to xenobiotics such as drugs, food additives, alcohol, chlorinated solvents, peroxidized fatty acids, fungal toxins, radioactive isotopes, environmental toxicants, and even some medicinal plants [6,77,78]. Hepatotoxicity is very common, due to the fact that liver represents the main metabolizing site for xenobiotics entering the animal living system. The liver contains many enzymes that contribute for biodegradation of compounds in order to facilitate the elimination. However, the biotransformation process bioactivates some chemicals to render them more hepatotoxic. Medicinal plants, although widely assumed to be safe, are potentially toxic. Although some poisoning from medicinal plants are related to misidentification of medicinal plants used for extract(s) preparation, inadequate preparation, and inappropriate administration [1,59,79], other cases are related to the presence of toxic substances in the medicinal plants. Several medicinal plants from Africa have been implicated in the etiology of liver diseases, and hepatotoxicity related to medicinal plants may be higher than actual expectations.

Pteleopsis hylodendron Mildbr is a tree belonging to the family Combretaceae, commonly found in West and Central Africa [80]. It is used in Cameroonian traditional medicine for the treatment of several diseases, including measles, chickenpox, sexually transmitted diseases, female sterility, and liver disorders [81]. However, this medicinal plant induced hepatotoxicity, as shown by vascular congestion and inflammation in the liver tissue, as well as increased serum activities of aminotransferase enzymes (ALT and AST) when administered to rats at 85 mg/kg/day, for 28 days [82]. Increase of the latter enzymes was also observed in rats treated with the dichloromethane−methanol extract of the plant *Coula edulis* Bail (Olacaceae) at a dose of 200 mg/kg bwt/day, for 28 days, which is traditionally used in West Africa for the treatment of stomach ache and skin diseases [83,84]. Another plant found in Central and West Africa is *Morinda lucida* (L.) Benth. (Rubiaceae) [85], widely used in the treatment of malaria, diabetes, typhoid fever, jaundice, analgesic, laxative, etc. [85−88]. Oduola et al. [89] and Ashafa and Olunu [85] concluded from subchronic toxicological studies of the extracts from this plant that its use was unlikely to be hepatotoxic. However, the aqueous extract from roots of the plant (100 mg extract/kg bwt/day) increased serum AST and liver body-weight ratio in rats [85,89]. This suggests a possible hepatotoxicity of the plant and emphasizes the need of further toxicological studies on *Morinda lucida* in order to clarify the adverse effect on the liver. In the meantime, awareness should be raised among the populations who are currently using this medicinal plant at a daily basis for prevention of several aliments.

Tetrapleura tetraptera (Mimosaceae), *Khaya senegalensis* (Desv) A. Juss (Maliacceae), and *Jatropha tanjorensis* (Euphorbiaceae) are other potential hepatotoxic plants used in Nigeria. *Tetrapleura tetraptera* is a flowering plant native to West Africa [90]. *Tetrapleura tetraptera* produces fruits that are used as spice in Nigeria and also find application in folk medicine for management of convulsions, leprosy, inflammation, rheumatism, fever, and even jaundice [90−93]. Odesanmi et al. [94] administered the fruit ethanolic extract of the plant to male rabbits/rats and noticed elevation of AST activity in serum of the animals. Elevation of liver and heart ALT activity was also reported in rats following treatment with the plant extract [90], suggesting liver toxicity. *Khaya senegalensis* is highly reputed for its numerous medicinal uses in Nigeria. It is used traditionally as antisickling, antimicrobial, anthelmintic, and for treatment of malaria [95,96]. Its stem bark and leaves have been reported to be effective against *Plasmodium falciparum* in *in vitro* studies [97]. In a 28-day subchronic study, administration of its aqueous stem bark extract to albino rats (20 and 40 mg/kg bwt/day) resulted in a significant increase in serum activities of ALT, AST, ALP, and total bilirubin level [96]. Increased activities of the aminotransferases (AST, ALT) and ALP enzymes were also reported in liver of rats after oral administration of the plant ethanolic extract (2 mg/kg bwt/day) for 18 days [98]. These observations indicate that extracts from *Khaya senegalensis* may possess hepatotoxic potential, at the dose levels studied. *Jatropha tanjorensis* is widely grown in Southern Nigeria, where it is commonly called "hospital-too-far" or "iyana ipaja." It is used as a popular remedy for the treatment of malaria and control of hypertension and diabetes [99]. Administration of the *Jatropha tanjorensis* methanol leaf extract to male rats (0.5−2.0 g/kg bwt/day) for 21 days induced elevation of serum biomarkers of hepatic function (bilirubin, ALT, and AST activities) in the rats. This indicates that the plant leaf extract is hepatotoxic, despite its use as a popular remedy for the treatment of different ailment [100]. The possible adverse effect should thus be carefully considered when administering the extract from this plant to patients.

Guizotia scabra (Vis) Chiov (Asteraceae), *Microglossa pyrifolia* (Lam.) Kuntze (Asteraceae), and *Vernonia lasiopus* O. Hoffm (Asteraceae) are used in folk medicine in Rwanda for treatment of hepatitis. Mukazayire et al. [9] investigated their pharmacological effect *in vitro* on rat PCLS model. Methanolic extracts from leaves of these plants were found hepatotoxic and unable to prevent acetaminophen toxicity on the PCLS. Considering these results, awareness should be raised in the consumption/use of the extracts from the aforementioned plants, which found applications in treatment of several illnesses in folk medicine. *Vernonia lasiopus* is used for treatment of children's constipation and wound healing. *Guizotia scabra* is used against helminthes and malaria. *Microglossa pyrifolia* also is used as an antimalarial plant, in addition to treatment of rheumatism, gastric ulcer, abdominal pain, wounds, and eye diseases. The use of extracts from these plants may thus induce hepatotoxicity in patients who use it for treatment of several aliments and even worsen the health of patients with hepatic disorders [9,101−103].

Tithonia diversifolia A. Gray (Asteraceae), commonly known as Mexican sun flower, has been introduced in Africa, where it is used to treat several ailments, including malaria. Its efficiency for malaria treatment (effective control of *Plasmodial* infection) has even been proven scientifically [104,105]. Although its LD_{50} was found to be >1600 mg/kg, a decrease in liver weights and fatty degeneration of the hepatocytes was noticed 30 min postadministration of the extract, even at 400 mg/kg bwt, a dose that was recommended for effective control of *Plasmodial* infection [104,105]. Indeed, the plant extract administered at 400 mg/kg bwt/day for 4 days was more efficient on early malaria infection than the reference drug chloroquine (5 mg/kg bwt/day, 4 days) [104]. Occurrence of hepatotoxicity at 400 mg/kg thus has safety implications in the use of the extract for malaria control. Another plant of ethnobotanical importance imported in Africa is *Murraya koenigii* (L.) Spreng. (Rutaceae). *Murraya koenigii* is native to India and is currently grown in tropical countries [106,107]. Leaves of the latter plant were harvested in Nigeria, and its methanol extract prepared and tested in rats. Acute treatment of rats with the methanol extract induced increase of relative weights of liver and activity of serum ALT 72 h after administration (≥ 0.5 g/kg bwt, single dose). Stimulation of liver weight also was noticed after subchronic treatment (≥ 0.25 g/kg bwt/day, 14 days treatment) [107].

A South African traditional herbal medicine, *Callilepis laureola* DC. (Asteraceae), is a herbaceous perennial medicinal plant found commonly in grassland habitats of eastern South Africa and used by the Zulu in South Africa to treat tapeworm, snakebite, infertility, etc. It is commonly known as "Impila" in the Zulu language, which means "health" [108]. *Callilepis laureola* is very toxic and has been responsible for several deaths among the Zulu [109]. The cytotoxicity of its aqueous extract was observed *in vitro* in human hepatoblastoma Hep G2 cells after 6 h of incubation (100% toxicity observed at a concentration of 6.7 mg/mL), and these findings confirmed the observed liver necrosis in the black population of KwaZulu-Natal in the 1970s [76]. *Hippobromus pauciflorus* (L.f.) Radlk (Sapindaceae) is another medicinal plant widely used in South Africa. Its leaves are used in the Eastern Cape of South Africa for the treatment of malaria, dysentery, diarrhea, etc. The leaves were collected in Sikusthwana village (in the Eastern Cape of South Africa), extracted with water, and screened for toxicity in male Wistar rats [110,111]. Administration of the aqueous extract (200 mg/kg bwt/day) for 14 days significantly increased the serum AST activity, indicating that its use as oral remedy may induce liver toxicity [111].

African medicinal plants, although efficient in healing a broad range of diseases, are potentially toxic. The toxicity may arise from inappropriate use (usually an overdose), inappropriate identification of the plant material, or be related to the presence of specific hepatotoxic compounds in the plant material. It is also noteworthy to emphasize the eventual contamination of medicinal plants by environmental toxicants with proven hepatotoxicity. An illustration comes from a study conducted by Awodele et al. [112], who reported high levels of heavy metals (Pb, Cd, Cr, Ni, and Zn) leaves and roots of five medicinal plants (*Ageratum conyzoides*, *Aspilia africana*, *Alchornea cordifolia*, *Amaranthus brasiliensis*, and *Chromolaena odorata*) collected in polluted areas in Nigeria.

11.5 Hepatoprotective Medicinal Plants of Africa

Many medicinal plants from Africa are traditionally used for liver protection against intoxication. Some of these plant extracts have been investigated scientifically, and their efficiency proven.

Hepatoprotective plants from Africa include *Aloe vera* (Aloaceae), a plant that has been widely used in African and Asian folk medicine to prevent and treat many ailments, including liver diseases [113]. The hepatoprotective effect of its extracts has been demonstrated in several *in vitro* studies. For example, Sultana and Najam [114] reported its significant inhibitory effect on serum ALT and AST activities in rabbits after 7 days of administration. Single-day treatment of adult male Wistar rats (4—5 months, 150—250 g) with the aqueous extract of *Aloe vera* (250 and 500 mg/kg bwt) alleviated paracetamol-induced hepatotoxicity, by decreasing the serum transaminase activities (AST and ALT) and restoring the depleted liver thiol levels significantly [43]. The hepatoprotective effect of *Aloe vera* was also demonstrated using other animal models, such as CCl_4-, lindane-, diesel-, petrol/kerosene-, hydraulic oil-, and sodium arsenite (Na_2AsO_2)-induced hepatotoxicity in rodents [61,73,115,116]. The reported hepatoprotetive effect was attributed to some bioactive compounds of *Aloe vera* which are responsible to protect liver from oxidative stress [117,118]. Extracts from *Aloe vera* have been reported to exhibit significant antioxidant activity [119]. Although treatment with *Aloe vera* extract significantly ($p < 0.05$) reduced mean liver and serum aminotransferase enzyme activities when compared with groups of animals administered only the hepatotoxic compounds (hepatoprotective effect of *Aloe vera* against the toxicants), groups of mice treated only with the plant extract (300 mg gel/kg bwt/day for 7 days) recorded aminotransferase activities significantly higher than the group that received distilled water alone. This is an indication of liver toxicity in the treated animals, and may be related to high dose, as oral toxicity of *Aloe vera* at high dose was reported earlier [120]. Definition and use of appropriate therapeutic dose of the extract is thus needed, in order to guarantee an efficient and safe use among humans.

Treatment with *Murraya koenigii* (L.) Spreng. (≥ 0.25 g/kg bwt/day) for 14 days seems to protect the liver, as demonstrated by reduction of serum aminotransferase (ALT and AST) activities [107]. The possible hepatoprotective effect of this plant might be through inhibition of lipid peroxidation and induction of protective antioxidant enzymes (i.e., SOD, CAT, and GSH) [121]. *Hibiscus sabdariffa* Linn (Malvaceae) or Red Sorrel plant is another medicinal plant used in Nigeria. *Hibiscus sabdariffa* is eaten as a diuretic or purgative and used as a remedy for the treatment of liver disorders and hypertension. Its infusion was demonstrated to exhibit anti-inflammatory effects [57,122]. In addition, administration of the *Hibiscus sabdariffa* extract (300 mg/kg bwt/day, 5 days) ameliorated cisplatin-induced tissue damage in albino rats (*Rattus norvegicus*) [57]. The plant *Gongronema latifolium* (Asclepiadaceae), commonly called "utazi" and "arokeke" in the southwestern and southeastern parts of Nigeria, has potentials for use as food formulation operations in view of its amino acid profile and fatty acid contents [123,124]. It is also used in traditional folk medicine, and scientific studies have established its hypoglycemic, hypolipidemic, antimicrobial, and antioxidant effects. Aqueous extract from its leaves also offers protection against acetaminophen and caffeinated acetaminophen toxicity in Wistar rats when its leaf extract is coadministered by oral gavages to the animals (200 and 400 mg/kg bwt, respectively) with the toxicants [45b,123]. Oral administration of the ethyl acetate leaf extract of the medicinal plant "coast gold" or *Bridelia micrantha* (Euphorbiaceae), used in West African countries for purging, antisterility, gastrointestinal disorders, and to expel the guinea worm, is beneficial in scavenging free radicals and its ethyl acetate extracts (300 mg/kg bwt) for 14 days was found to exhibit better hepatoprotective capacity in acetaminophen-exposed rats than Silymarin (25 mg/kg) [125].

Telfairia occidentalis Hook F. (Curcurbitaceae) is grown in West Africa. It is one of the green leafy vegetables widely consumed in Nigeria, where it serves as food and herbal medicine [126,127]. Several beneficial effects of the plant on health have been illustrated, such as protection against cancer, antidiabetic activity [126,128]. In addition to these pharmacological effects, soluble free and bound polyphenols extracted from *Telfairia occidentalis* leaves showed antioxidant and hepatoprotective properties in acetaminophen-intoxicated rats, with soluble free polyphenols having significantly higher pharmacological properties than the bound polyphenols [127]. Hepatoprotective effect of the lyophilized plant aqueous extract also was observed in rats intoxicated with cyanide (3 mg/kg bwt) [129]. Indeed, the damaging effect of cyanide on the liver was ameliorated by *Telfairia occidentalis* aqueous extract, as revealed by histopathological (liver necrosis) and biochemical (ALT, AST) data following cyanide and/or extract treatment [129].

The leaves of *Ocimum lamiifolium* Hochst ex. Benth (Lamiaceae) and *Crassocephalum vitellinum* (Benth) S. Moore (Asteraceae) are used traditionally in Rwanda for treatment of hepatitis and other diseases. This traditional use was scientifically justified by investigations of Mukazayire et al. [9], using an *in vitro* PCLS model. Incubation of the PCLS from rats and methanolic extracts from the respective plants significantly reduced the acetaminophen-induced toxicity on liver slices. The leaves of *Crassocephalum vitellinum* and *Ocimum lamiifolium* are used as food (edible vegetables) in Tanzania [130] and Ethiopia [131], respectively, and this suggests that medical use of the extracts may be safe [9]. Another medicinal plant from the family Asteracease is *Ageratum conyzoides* L., an herb widely utilized in traditional medicine wherever it grows. *Ageratum conyzoides* is a widespread uncultivated plant largely available in Africa. It is employed in folk medicine as a purgative, febrifuge, antiasthmatic, antispasmodic, analgesic, antidiarrhoeic, and anti-inflammatory remedy [132]. It is used in Cameroon for treatment of malaria and liver diseases. The use of this herb as a bactericide, antidysenteric, anti-inflammatory, purgative, antipyretic, antirheumatic, and antibiotic has been reported [71,112,133−135]. Its antioxidant and hypoglycaemic activities scientifically [136,137]. Daily administration of the ethanol leaf extract of *Ageratum conyzoides* (250 and 500 mg/kg bwt) for 14 days was shown to offer protection against acute acetaminophen (600 mg/kg bwt, intraperitoneal injection) and caffeinated acetaminophen overdose (acetaminophen plus 100 mg caffeine/kg bwt, oral gavage) in rat liver [45b]. For, the increase in ALT activity in serum of the intoxicated rats was significantly inhibited when the animals were cotreated with the plant extract. The acetaminophen-induced hepatotoxicity model was also used to demonstrate the protective effect of *Zingiber officinale* Roscoe (Zingiberaceae), commonly known as ginger. This familiar dietary spice has several medicinal properties. It is used in Africa for treatment of respiratory diseases, nausea, cardiovascular diseases, etc. [138], in addition to its proven pharmacological effect (e.g., immunomodulatory, anti-inflammatory, and antioxidant effects) [139−141].

Moringa oleifera Lam. (Moringaceae) is rich in minerals (Ca^{2+}, Fe^{3+}), proteins, vitamins (Vitamin A), and its leaves have been used to combat malnutrition [142−144]. It has an impressive range of medicinal uses. Various parts of the plant (leaves, roots, seed, bark, fruit, and flowers) possess antitumor, antipyretic, antiepileptic, anti-inflammatory, antiulcer, diuretic, antihypertensive, antioxidant, antidiabetic, antibacterial, antifungal, and hepatoprotective effects and are used with great dietary importance and for the treatment of different ailments in the African indigenous system [144−146]. The hydroalcoholic (80% ethanol: 20% distilled water) extract of the plant (200 and 800 mg/kg bwt/day, 14 days) prevented acetaminophen-induced liver injury in male Sprague-Dawley rats. The hepatoprotective activity was observed through significant reduction of the level of ALT, AST, and ALP in serum, and restoration of GSH level in rats pretreated with *Moringa oleifera* compared to those treated with acetaminophen alone [147]. *Moringa oleifera* therapeutic action in protecting liver damage in rats given an over-dosage of acetaminophen was also reported [144], and Hamza [148] illustrated its action against CCl_4-induced liver injury and fibrosis. Coadministration of methanolic extract of the plant (500 mg/kg bwt/2 days, oral route, for 3 weeks) and cadmium (2.5 mg/kg bwt, subcutaneous injection) to adult male Wistar rats alleviates cadmium-induced liver and kidney dysfunction in the rats [53]. The hepatoprotective properties of *Moringa oleifera* were attributed to its antioxidant potentials, anti-inflammatory effects, as well as its ability to attenuate hepatic stellate cells activation [148]. The plant *Hibiscus rosa-sinensis* (Malvaceae) extract also exhibited an antioxidant effects (through inhibition of superoxide anions generation and hydroxyl radicals formation) [149], which could explain the activity of the plant against CCl_4-induced hepatotoxicity. Indeed, the antioxidant compounds anthocyanins [150] were obtained from the petals of *Hibiscus rosa-sinensis* and their activity in preventing CCl_4-induced acute liver damage confirmed in the rat (5 days/week for 4 weeks before 0.5 mL/kg CCl_4 treatment) [151]. *Hibiscus rosa-sinensis* has also been used in folk medicine in Africa and elsewhere. The plant is used in fatigue, skin disease, gonorrhea, menorrhagia, epilepsy, leprosy, etc. [14]. It is also used to control diabetes mellitus in Cameroon and elsewhere [149,152], and hypoglycemic and antidyslipidemic activities of its extracts (ethanol and aqueous extracts) have been demonstrated in diabetic rats [149,152−154].

The apple guava or common guava (*Psidium guajava* Linn.; family: Myrtaceae) is native to Central and South America but has been widely cultivated in Africa for its fruit [64,155,156]. Its leaves are also used in Nigeria and Cameroon for treatment of gastrointestinal disorders (e.g., diarrhea, dysentery) and in South African folk medicine to manage, control, and/or treat hypertension and diabetes mellitus [11]. Its leaf extract (200−500 mg/kg bwt) significantly reduced the elevated serum levels of AST, ALT, ALP, and bilirubin in the acute liver damage induced by different hepatotoxins CCl_4 and acetaminophen in male and female Wistar rats [156], and this pharmacological effect was supported by histological examination of the liver tissues in the animals [157]. The safe use of *Psidium guajava* for prevention/ treatment of liver diseases was previously reported by Uboh et al. [64], who noticed no significant change in liver biochemical markers of toxicity (ALT, AST, ALP, bilirubin) after administration of the aqueous extract (200 mg/kg bwt/day) to nonintoxicated rats for 30 days. The treated animals also showed improvement in their hematological parameters (increased red blood cell counts, hematocrit, and hemoglobin concentrations) [64].

Rhoicissus tridentata (L.F.) Wild and Drum subspecies cuneifolia (Vitaceae), known as *Nwazi* in Zulu, is one of the many plants that are commonly used by traditional healers in southern Africa. Extracts are prepared from the plant for the treatment of epilepsy, kidney and bladder complaints, birth complications, headaches, stomach ache, sores, kidney complications, etc. [158,159]. Several pharmacological actions of *Rhoicissus tridentata* extracts have been reported, including the antiproliferative, antimicrobial, and anti-inflammatory activities [160,161]. In addition, administration of the plant aqueous extract to male Sprague-Dawley rats (0.05 mg/kg bwt/weeks old, 80−100 g) after CCl_4 intoxication (1.0 mL/kg bwt CCl_4 dissolved in olive oil, 1:1) resulted in a significant reduction of TBARS in liver tissue and microsomes, as well as the serum activities of aminotransferases (ALT and AST), indicating the hepatoprotective effects of the plant extract [162]. The savannah tree *Ficus cordata* Thunb. (Moraceae) is another medicinal plant found in South Africa. It is also grown in Angola, Senegal and Cameroon. In Cameroon, tradional traditional healers from the West Region use its stem bark for treatment of jaundice [163]. This medicinal plant was harvested in Cameroon and screened for hepatoprotective properties by Donfack et al. [70,163,164] along with two other plants from the same family Moraceae (*Ficus chlamydocarpa* and *Ficus gnaphalocarpa*). Five chemical compounds exhibiting hepatoprotective properties against CCl_4-induced toxicity in hepatoma cells were isolated from *Ficus cordata* stem bark; that is, β-amyrin acetate, lupeol, catechin, epiafzelechin, and stigmasterol. The latter compounds prevented liver cell death and LDH leakage from the cells during CCl_4 intoxication [163]. When rats were pretreated with the methanol extract of *Ficus chlamydocarpa* before CCl_4 administration, the extract significantly prevented CCl_4-induced liver toxicity [164]. The methanol extract and three compounds isolated from *Ficus gnaphalocarpa* (Miq.) Steud. ex A. Rich (3-methoxyquercetin, quercetin, and quercitrin) (Figure 11.1) prevented cytotoxicity (LDH leakage) of hepatoma cell line BS TCL 41 intoxicated by CCl_4 *in vitro* [70]. The chemical investigation of the twigs of *Morus mesozygia* resulted in the isolation of three compounds (two flavonoids and one 2-arylbenzofurans) which showed significant hepatoprotective effect [165]. The analyzed plant extracts and some isolated compounds (moracin LM from *Morus mesozygia*; α-amyrin acetate, and luteolin isolated from *Ficus chlamydocarpa*; β-amyrin acetate, lupeol, catechin, epiafzelechin, and stigmasterol from *Ficus cordata*) also displayed antioxidant activity. This suggests that the observed hepatoprotective/pharmacological effect possibly resulted from the antioxidant molecules present in the investigated medicinal plants [70,164,166,167]. *Alchornea cordifolia* Mull. Arg (Euphorbiaceae), which is used in African traditional medicine as topical anti-inflammatory, antibacterial, and antifungal agent, exhibits antioxidant as well as hepatoprotective activities against CCl_4, which reside mainly in the ethyl acetate fraction of methanol leaf extract [72]. Njayou et al. [168,169] also illustrated the hepatoprotective effect of *Erythrina senegalensis* DC (Fabaceae), *Entada africana* Guill. et Perr. (Fabaceae), and *Khaya grandifoliola* (Welw) CDC (Meliaceae) stem bark extracts, which are traditionally used as source of medicines against liver related diseases in Cameroon. Changes in liver function (increased serum ALT and liver TBARS) effected by CCl_4 intoxication in rats were significantly reversed by *Erythrina senegalensis* extract. The methylenechloride—methanol (1:1 v/v) extracts from *Entada africana* and *Khaya grandifoliola* (≥ 100 and ≥ 10 μg/mL, respectively) inhibited paracetamol (30 mM)-induced LDH leakage in liver slices *in vitro* [169]. Likewise, *Boerhaavia diffusa* Linn. (Nyctaginaceae) aqueous and ethanolic extracts decreased the activities of ALP, LDH, ALT, AST, and the level of bilirubin in the serum of rats that were elevated following paracetamol intoxication. The extracts also attenuated the elevation in the activities of the enzymes in the liver, and protected against acetaminophen-induced TBARS levels, suggesting an action mechanism through stimulation of antioxidant defense system [170]. The paracetamol-induced liver toxicity (500 mg/kg bwt) was prevented in rats by pretreatment of the animals with *Kohautia grandiflora* DC. (Rubiaceae) aqueous extract (300 mg/kg bwt/day, 7 days) [171].

Vernonia amygdalina (Astereaceae) and *Byrsocarpus coccineus* Schum. and Thonn. (Connaraceae) hepatoprotective effect was proven in CCl_4-treated rats. The plant *Vernonia amygdalina* is widely consumed as vegetable (e.g., bitter-leaf soup) [172] and used in tropical Africa in the treatment of fever, jaundice, stomach disorders, and diabetes. Administration of the leaf methanolic extract (500—1000 mg/kg, five times a week for 2 weeks) to male rats treated with CCl_4 (1.2 g/kg bwt, three times a week for 3 weeks) inhibited the CCl_4-induced serum transaminases (ALT, AST), ALP and γ-GT, and attenuated CCl_4-induced lipid peroxidation. Similarly, administration of the extract stimulated the antioxidant system in intoxicated rats (increased activities of SOD, GST, CAT) [173]. Aqueous leaf extract of *Byrsocarpus coccineus* (200, 400, and 1000 mg/kg) decreased serum ALT, AST and ALP levels dose dependently (with peak effect produced at the highest dose), produced reversal of CCl_4-diminished activity of the antioxidant enzymes, and reduced CCl_4-elevated level of TBARS. This suggests that *Byrsocarpus coccineus* possesses hepatoprotective and *in vivo* antioxidant effects, and thereby justifies the use of its extracts for the treatment of liver disease in West African traditional medicine [174].

Chromolaena odorata (L.) R. King and H. Robinson (formerly *Eupatorium odoratum* L.) (Asteraceae or Compositae) is a perennial scrambling shrub found in southern Asia and western Africa [40]. Pretreatment with *Chromolaena odorata* leaf ethanolic extract significantly prevented the elevation of serum aminotransferase enzymes (AST and ALT), LDH, γ-GT, total bilirubin, and TBARS resulting from CCl_4 intoxication. The extract also improved the antioxidant system (stimulation of SOD, CAT, GST activities, and GSH levels) in the liver of intoxicated animals [40]. The ethanol extract of the plant thus protected the hepatocytes against injury and prevented loss of functionality, probably through stimulation of the antioxidant system, which was attributed to secondary plant metabolites flavonoids, phenolic compounds, saponins, alkaloids, glycosides, and tannins present in the extract [40,175,176].

Musanga cecropioides R.Br. Apud Tedlie (Cecropiaceae) also is used in the folkloric medicine. The Yorubas in south-west Nigeria utilize its stem bark aqueous extract for treatment of hepatic injuries resulting from acute gastric poisonings, infective hepatitis, or other liver diseases. The beneficial pharmacological effect of this extract was proven on acute hepatocellular injuries induced by intraperitoneal CCl_4 (20% CCl_4 in olive oil, 1.5 mL/kg bwt) and acetaminophen (800 mg/kg bwt) in male Wistar rats. This was illustrated by the attenuation of the hepatotoxin-induced histopathological lesions in the rat livers and decrease in the acute elevation of the liver enzymes [177]. Extracts from *Musanga cercropioides* also exhibit hypotensive, antidiabetic, and antidiarrheal effects [178–180]. In the CCl_4-induced toxicity model, ethanolic extract from *Lagenaria breviflora* (Cucurbitaceae) leaves enhanced the recovery from CCl_4-induced hepatic damage and oxidative stress via its antioxidant and hepatoprotective properties [42]. *Lagenaria breviflora* is widespread in tropical Africa (especially from Senegal to Cameroon), where its seeds and fruits found application in folk medicine as painkiller (to treat headache), vermifuge, schistosomiasis, etc. [181], and its antifertility and hemantic properties were proven scientifically [182,183].

Garcinia kola Heckel (Guttiferae) is a tropical plant which grows in moist forest and commonly called bitter kola because of the bitter seeds it produces. It is used traditionally for the treatment of liver diseases and cough, and contains various components, such as kolaviron, xanthones, garcinoic acid, garcinal, and tocotrienol [47,184]. Methanolic extracts from *Garcinia kola* (100–250 mg/kg bwt/day, 14 days) significantly reduced the tissue damage induced by LPS (1 mg/kg bwt) in male albino rats (*Rattus norvegicus*) [47].

A typical Mediterranean plant, *Ceratonia siliqua* (Fabaceae) is mainly used in food and Tunisian traditional folk medicine. Its ethyl acetate fraction contains phenolic and flavonoid compounds, and prevents CCl_4-induced disorders, as indicated by the decrease in levels of hepatic markers (ALT, AST, ALP, LDH, γ-GT), when rats were exposed to the fraction (250 mg/kg bwt) prior to CCl_4 intoxication. The biochemical changes were in accordance with histopathological observations in the liver (suppression of CCl_4-induced sinusoidal dilation, vacuolization, inflammation, and congestion) suggesting a marked hepatoprotective effect of the ethyl acetate fraction [185]. Likewise, the nonstarch polysaccharides extracted from the mushroom *Pleurotus ostreatus* (Pleurotaceae) mycelia suppressed the CCl_4-induced toxicity in liver, when administered at 100–200 mg/kg bwt for 7 days [39]. *Silybum marianum* (L.) Gaertn. (Asteraceae) is an annual to biennial herb found in all North African countries. It is used in folk medicine for treatment of several diseases, including cancer, diabetes, hemorrhoids, malaria, constipation, asthma, duodenal ulcer, amenorrhea, and liver diseases [186]. The protective effects of its polyphenolic compounds extracted from the plant alleviated TAA-induced hepatotoxicity in rats [187]. It is the source of Silymarin, a well-known drug for treatment of liver diseases, which has been used as a reference drug for investigations of hepatoprotective active principles. For much scientific research has illustrated its action in alleviating hepatotoxicity induced by several xenobiotics, including CCl_4, acetaminophen, etc. [68,125,168,188]. Silymarin is a mixture of polyphenolic flavonoids, which include isosilyins, silibinins, silydianin silychristins, etc. Silymarin has a good safety profile, although little data are available on its potential for drug interaction [189].

The northern African flora also comprises mushrooms. Although some species of the mushrooms are extremely toxic, other species such as *Pleurotus ostreatus* (Pleurotaceae) are generally known as important source of nutrients, and also exhibit potential therapeutic effects. *Pleurotus ostreatus* (Pleurotaceae) has been shown to exhibit hepatoprotective, antitumor, and hypoglycemic effects [190,191]. The plant *Basella alba* (Basellaceae) also contains nutrients and is used in the Ayurvedic treatment in India for treatment of cancer [192a]. Its leaves are used in Nigeria for treatment of hypertension, and in Cameroon as an antimalarial and anti-fertility remedy [135,192b,193]. Previous studies revealed its androgenic effect in male rats [192b,194]. It is a potential hepatoprotective plant, given its antioxidant properties [193], and the effect of its aqueous extract in reducing the activity of the liver enzymes (ALP, ALT, and AST) in albino rats [195].

Several African medicinal plants showed hepatoprotective protective effects in hepatotoxic animal models induced by alcohol. These include *Phyllanthus amarus* (Thonn and Schum) (Euphorbiaceae), and *Ficus carica* (Moraceae), among others. The leaf methanolic extract of *Phyllanthus amarus* (250 and 500 mg/kg bwt/day, 4 weeks) inhibited lipid peroxidation in rats previously exposed to chronic ethanol treatment (5 g/kg/day, 3 weeks), with a concomitant marked reduction in the plasma activity of the transaminases (ALT, AST) in the ethanol-challenged rats. These results stipulated that the plant extract could protect the liver against ethanol-induced oxidative damage by possibly increasing the antioxidant defense mechanism in the intoxicated rats [38]. A study conducted by Saoudi and El Feki [7] showed that oral (in drinking water) administration of the aqueous extracts from stems of *Ficus carica* (Moraceae) reduced methanol-induced hepatotoxicity in adult male rats, as shown by normalization of serum ALT, AST, ALP, and lactate dehydrogenase (LDH) activities, as well as inhibition of lipid peroxidation in the liver of animals co-exposed to methanol and plant extract. The hepatoprotective effect also correlated with improvement of the antioxidant system (increasing antioxidant enzyme activities). The hepatoprotective effect of the extract from *Aframomum melegueta* (Zingiberaceae) (commonly known as Grains of Paradise, Melegueta pepper, alligator pepper) [196,197] against ethanol-induced damage (elevation of serum ALT and decrease in reduced GSH histological changes, and SOD activity were attenuated in rats coadministered ethanol and the plant extract) was also reported by Nwozo and Oyinloye [37b].

Enantia chlorantha (Annonaceae) is a tree that grows in Cameroon, Nigeria, and several other African countries. Several parts of the tree—especially the bark—have been used in folkloric medicine for the treatment of skin, gastric and duodenal ulcers, and as an antimalarial [198]. Efficiency of the plant extract in malaria, convulsions, inflammatory disease, and infections has been proven scientifically [199—201]. Several research studies have also illustrated the therapeutic effect of *Enantia chlorantha* on liver diseases in animal models [202—204]. Moreover, studies conducted in Cameroon have enabled isolation of the active compounds from the bark of *Enantia chlorantha*, that exhibit the hepatoprotective effect. The active compounds are protoberberines (alkaloid mixture), and are commercially available under the name Hepazor. In addition to the hepatoprotective effect in chemically induced traumatization and promotion of the healing process in the hepatic injury models, Hepazor has also been demonstrated to be efficient for treatment of viral hepatitis in a clinical trial [8,10,202,204].

The species including onion, garlic, and guava, are edible plants. Apart from culinary purposes, onion and garlic (*Allium cepa* L. and *Allium sativum* L., respectively) have received considerable attention for their functional health benefits. The two plants are used for the treatment of diabetes, stimulation of immune function, antimicrobial, and antioxidant effects, etc. [205−207]. Aqueous extracts from these plants were effective in alleviating Cd-induced oxidative damage in rat liver, possibly via reduced lipid peroxidation and enhanced antioxidant defense system [205], suggesting the use of onion and garlic in the prevention or treatment of hepatotoxicity related to heavy metals exposure. This activity may be related to the presence of active compounds in the plants, such as Quercetin (Figure 11.1), a well-known flavonoid and a strong antioxidant which was isolated from *Allium cepa* [208,209]. Quercetin administration to diabetic rats (15 mg/kg) prevented liver cells damages of the animals (by reducing the number of apoptotic cells in the liver), suggesting its use for treatment of liver diseases [210].

Administration of *Psidium guajava* aqueous extract (200 mg/kg bwt/day) to non-intoxicated rats for 30 days, also restored oxidative stress markers (TBARS and GSH levels, SOD, and CAT) in arsenic-induced hepatotoxic animal model [211]. Moreover, *Psidium guajava* extract protected the liver when coadministered to rats with the antibiotic erythromycin [58]. Tetracycline has a broad-spectrum antibiotic, malaria prophylactic, and animal growth promoter, but also induces liver toxicity. The tetracycline (1000 mg/kg/48 h, for 1 days)-induced hepatotoxicity was shown to be alleviated in rabbits by aqueous crude leaf extract from *Senna occidentalis* (25−100 mg/kg bwt, for 14 days). Interestingly, the LD_{50} of the orally administered extract was found to be >2000 mg/kg, and this suggests its safe use [59].

Most of the reviewed hepatoprotective plants herein have been used in folk medicine without prior scientific assessment of the efficiency. Despite the utilization of the latter plants in humans, the biochemical basis of action (mechanism) which is necessary for efficiency and safety of drugs is lacking in almost all the cases, and attempts to clarify this issue have been focused almost on antioxidant potentials of the extracts or derived/isolated chemicals only [3].

11.6 Conclusion

The African pharmacopoeia comprises many plant species, of which a good number exhibit hepatoprotective effects. Some medicinal plants used either for treatment of hepatic diseases or control of aliments unrelated to the liver are potential inducers of hepatotoxicity. The wide assumption of the safety of medicinally used plants thus represents a risk for the African population, which relies mainly on this treatment method. This emphasizes a need to raise awareness on the potential toxicity of medicinal plants. The toxicity that may result from interaction between compounds from different medicinal plants, plant products, and other xenobiotics (even from synthetic origin) should also be emphasized. Luckily, other investigated plant extracts showed promising results in alleviating the burden of toxicants on liver, and may be used as hits for further development of hepatoprotective drugs. The existing efficient hepatoprotective drugs discovered from African medicinal plants (e.g., Hepazor isolated from *Enantia chlorantha* in Cameroon) give more hope for further establishment of other active principles from the latter plants. Oxidative stress appears to be an important mechanism that contributes to the initiation and progression of hepatotoxicity, and most of the identified hepatoprotective medicinal plants could exhibit antioxidant properties as well. The phytochemical antioxidants therefore represent a logical therapeutic strategy for treatment of liver diseases. This may constitute the basis for further clarification of the action mechanism through which these extracts exert their pharmacological effects, in order to properly define their rational use as therapeutics. Modulation of the antioxidant system should thus be the main focus for further elucidation of the routes through which extracts/compounds from the medicinal plants exert their hepatoprotective effects, though other possible action mechanisms should be defined. Indeed, the biochemical basis of action (mechanism) is necessary for the rational development of a safe and potent drug. Moreover, the number of medicinal plants reviewed is by far less than existing plants used in African pharmacopeia, and the liver, which is the main site of xenobiotics metabolism, should be protected through proper assessment of the potential hepatotoxicity of all the medicinal plants used in Africa.

References

[1] Fennell CW, Lindsey KL, McGaw LJ, Sparg SG, Stafford GI, Elgorashi EE, et al. Assessing African medicinal plants for efficacy and safety: pharmacological screening and toxicology. J Ethnopharmacol 2004;94(2−3):205−17.

[2] Bruneton J. Plantes medicinales: phytochimie, pharmacognosie. 2nd ed. New York, NY: Lavoisier; 1993.

[3] Kumar A, Rai N, Kumar N, Gautam P, Kumar JS. Mechanisms involved in hepatoprotection of different herbal products: a review. Int J Res Pharm Sci 2013;4(2):112−7.

[4] Nagpal M, Sood S. Role of curcumin in systemic and oral health: an overview. J Nat Sci Biol Med 2013;4(1):3−7.

[5] Alshawsh AM, Abdulla AM, Ismail S, Amin AZ. Hepatoprotective effects of *Orthosiphon stamineus* extract on thioacetamide-induced liver cirrhosis in rats. Evid Based Complement Alternat Med 2011;2011:103039. Available from: http://dx.doi.org/doi:10.1155/2011/103039.

[6] Navarro VJ, Senior JR. Drug related hepatotoxicity. N Engl J Med 2006;354:731−9.

[7] Saoudi M, El Feki A. Protective role of *Ficus carica* stem extract against hepatic oxidative damage induced by methanol in male Wistar rats. Evid Based Complement Alternat Med 2012;2012:150458. Available from: http://dx.doi.org/doi:10.1155/2012/150458.

[8] Pousset JL. Place des médicaments traditionnels en Afrique. Med Trop 2006;66 (6):606–9.

[9] Mukazayire MJ, Allaeys V, Buc Calderon P, Stévigny C, Bigendako MJ, Duez P. Evaluation of the hepatotoxic and hepatoprotective effect of Rwandese herbal drugs on *in vivo* (guinea pigs barbiturate-induced sleeping time) and *in vitro* (rat precision-cut liver slices, PCLS) models. Exp Toxicol Pathol 2010;62(3):289–99.

[10] Fokunang CN, Ndikum V, Tabi OY, Jiofack RB, Ngameni B, Guedje NM, et al. Traditional medicine: past, present and future research and development prospects and integration in the national health system of Cameroon. Afr J Tradit Complement Altern Med 2011;8(3):284–95.

[11] Gutiérrez RM, Mitchell S, Solis RV. *Psidium guajava*: a review of its traditional uses, phytochemistry and pharmacology. J Ethnopharmacol 2008;117(1):1–27.

[12] Moro PA, Cassetti F, Giugliano G, Falce MT, Mazzanti G, Menniti-Ippolito F, et al. Hepatitis from Greater celandine (*Chelidonium majus* L.): review of literature and report of a new case. J Ethnopharmacol 2009;124(2):328–32.

[13] Gori L, Galluzzi P, Mascherini V, Gallo E, Lapi F, Menniti-Ippolito F, et al. Two contemporary cases of hepatitis associated with *Teucrium Chamaedrys* L. decoction use. Case reports and review of literature. Basic Clin Pharmacol Toxicol 2011;109 (6):521–6.

[14] Kumar A, Singh A. Review on *Hibiscus rosa sinensis*. Int J Pharm Biomed Res 2012;3(2):534–8.

[15] Abdualmjid JR, Sergi C. Hepatotoxic botanicals—an evidence-based systematic review. J Pharm Pharm Sci 2013;16(3):376–404.

[16] Stickel F, Egerer G, Seitz HK. Hepatotoxicity of botanicals. Public Health Nutr 2000;3113–24.

[17] Stickel F, Baumüller HM, Seitz K, Vasilakis D, Seitz G, Seitz HK, et al. Hepatitis induced by kava (*Piper methysticum* rhizoma). J Hepatol 2003;39(1):62–7.

[18] Stickel F, Poschl G, Seitz H, Waldherr R, Hahn E, Schuppan D. Acute hepatitis induced by Greater Celandine (*Chelidonium majus*). Scand J Gastroenterol 2003;38(5):565–8.

[19] Wang J, Ji L, Liu H, Wang Z. Study of the hepatotoxicity induced by *Dioscorea bulbifera* L. rhizome in mice. Bio Sci Trends 2010;4(2):9–85.

[20] Guyton AC, Hall JE. In: Guyton AC, Hall JE, editors. Textbook of medical physiology. 11th ed. Philadelphia, PA: Elsevier Inc.; 2006.

[21] Johnson EK. Histology and cell biology. Baltimore, MD: Williams & Wilkins; 1991.

[22] Duffus HJ, Worth GJH. Fundamental toxicology. Norfolk, UK: RSC Publishing; 2006.

[23] Postlethwart HJ, Hopson LJ. Modern biology. New York, NY: Holt, Rinehart and Winston; 2006.

[24] Hodgson E, Levi PE. Hepatotoxicity. In: Hodgson E, editor. A textbook of modern toxicology. 3rd ed. Hoboken, New Jersey, USA: John Wiley & Sons, Inc.; 2004.

[25] Smith C, Marks DA, Lieberman M. Liver metabolism. Marks' basic medical biochemistry: a clinical approach. 2nd ed. Mishawaka, IN, USA: Lippincott Williams & Wilkins publishing; 2005.

[26] Meyer UA. Overview of enzymes of drug metabolism. J Pharmacokinet Biopharm 1996;24:449–59.

[27] Mroueh M, Saab Y, Rizkallah R. Hepatoprotective activity of *Centaurium erythraea* on acetaminophen-induced hepatotoxicity in rats. Phytother Res 2004;18:431–3.

[28] Hajigholamali M, Jafari M, Hajihossaini R, Abasnezhad M, Salehi M. Antioxidant defences in Brown Norway rats in response to acute doses of diazinon. Abstracts/ Toxicol Lett 2010;196S:S207. P206-010.

[29] Highleyman L, Franciscus A. An introduction to the liver. HCSP Fact Sheet, Hepatitis C Support Project, Version 4, March 2012; <http://www.hcvadvocate.org/hepatitis/factsheets_pdf/The%20Liver.pdf>; 2012 [accessed 13.04.14].

[30] Rowland M, Tozer TN. Clinical pharmacokinetics, concepts and applications. 3rd ed. Baltimore, MD: Williams & Wilkins; 1995.

[31] Rose RL, Hodgson E. Metabolism of toxicants. In: Hodgson E, editor. A textbook of modern toxicology. 3rd ed. Hoboken, New Jersey, USA: John Wiley & Sons, Inc.; 2004.

[32] Murray KR. Metabolism of xenobiotics. Harper's illustrated biochemistry. 26th ed. New York, NY: Lange Medical Books/McGraw-Hill, Medical Publishing Division; 2003.

[33] Lüllmann H, Mohr K, Ziegler A, Bieger D. Color atlas of pharmacology 2nd ed., revised and expanded. Stuttgart, New York: Thieme; 2000.

[34] Karasch ED. Anesthetic Pharmacology: Physiologic Principles and Clinical Practice. In: Evers AS, Maze M. (editors). Churchill Livingstone, Philadelphia; 2004.

[35] Schonborn JL. The role of the liver in drug metabolism. Anaesthesia tutorial of the week 179, 17th May 2010, ATOTW; 2010.

[36] Han D, Shinohara M, Ybanez MD, Saberi B, Kaplowitz N. Signal transduction pathways involved in drug-induced liver injury. In: Uetrecht J, editor. Adverse drug reactions. Handbook of experimental pharmacology. Heidelberg, Germany: Springer; 2010.

[37a] Kass GEN. Mitochondrial involvement in drug-induced hepatic injury. Chem Biol Interact 2006;163(1−2):145−59.

[37b] Nwozo SO, Oyinloye BE. Hepatoprotective effect of aqueous extract of Aframomum melegueta on ethanol-induced toxicity in rats. Acta Biochim Pol 2011;58(3):355−8.

[38] Faremi TY, Suru SM, Fafunso MA, Obioha UE. Hepatoprotective potentials of Phyllanthus amarus against ethanol-induced oxidative stress in rats. Food Chem Toxicol 2008;46:2658−64.

[39] Nada SA, Omara EA, Abdel-Salam OM, Zahran HG. Mushroom insoluble polysaccharides prevent carbon tetrachloride-induced hepatotoxicity in rat. Food Chem Toxicol 2010;48(11):3184−8.

[40] Alisi CS, Onyeze GO, Ojiako OA, Osuagwu CG. Evaluation of the protective potential of Chromolaena odorata Linn. extract on carbon tetrachloride-induced oxidative liver damage. Int J Biochem Res Rev 2011;1(3):69−81.

[41] Fleurentin J, Joyeux M. Les tests in vivo et in vitro dans l'évaluation des propriétés antihépatotoxiques des substances d'origine naturelles. In: Fleurentin J, Calaillon P, Masarj G, Santos JD, Younos C, editors. Ethnopharmacologie: sources, méthodes, objectifs. Actes du 1P^erP colloque Européen d'Ethnopharmacologie-Metz (France)-ORSTOM; 1990.

[42] Saba AB, Onakoya OM, Oyagbemi AA. Hepatoprotective and in vivo antioxidant activities of ethanolic extract of whole fruit of Lagenaria breviflora. J Basic Clin Physiol Pharmacol 2012;23(1):27−32.

[43] Nayak V, Gincy TB, Prakash M, Chitralekha J, Soumya SR, Somayaji SN, et al. Hepatoprotective activity of Aloe vera gel against paracetamol induced hepatotoxicity in albino rats. Asian J Pharm Biol Res 2011;1(2):94−8.

[44] Albano E, Rundgren M, Harvison PJ, Nelson SD, Moldéus P. Mechanisms of N-acetyl-p-benzoquinone imine cytotoxicity. Mol Pharmacol 1985;28(3):306−11.

[45a] Johnkennedy N, Adamma E. The protective role of Gongronema latifolium in acetaminophen induced hepatic toxicity in Wistar rats. Asian Pac J Trop Biomed 2011; S151−4.

[45b] Ita S, Akpanyung EO, Umoh BI, Ben EE, Ukafia SO. Acetaminophen induced hepatic toxicity: protective role of *Ageratum conyzoides*. Pak J Nutr 2009;8(7): 928−32.

[46] Park YC, Kimback G, Salion C, Valacchi G, Parcker L. Activity of monomeric, dimeric, and trimeric flavonoids on NO production, TNF-α secretion, and NF-kB dependent gene expression in RAW264.7 macrophages. FEBS Lett 2000;465:93−7.

[47] Okoko T, Ndoni S. The effect of *Garcinia kola* extract on lipopolysaccharide-induced tissue damage in rats. Trop J Pharm Res 2009;8(1):27−31.

[48] Decker K, Keppler D. Galactosamine hepatitis: key role of the nucleotide deficiency period in the pathogenesis of cell injury and cell death. Rev Physiol Biochem Pharmacol 1974;71:77−106.

[49] Palanisamy D, Syamala, Kannan E, Bhojraj S. Protective and therapeutic effects of the Indian medicinal plant *Pterocarpus santalinus* on D-galactosamine-induced liver damage. Asian J Tradit Med 2007;2(2):51−7.

[50] Jurczuk M, Brzoska M, Moniuszko-Jakoniuk J, Galazyn-Sidorszuk M, Eulikowska-Karpinska E. Antioxidant enzymes activity and lipid peroxidation in liver and kidney of rats exposed to cadmium and ethanol. Food Chem Toxicol 2004;42:429−38.

[51] Alhazza IM. Cadmium-induced hepatotoxicity and oxidative stress in rats: protection by selenium. Res J Environ Sci 2008;2(4):305−9.

[52] Yadav N, Khandelwal S. Therapeutic efficacy of Picroliv in chronic cadmium toxicity. Food Chem Toxicol 2009;47(4):871−9.

[53] Ajilore BS, Atere TG, Oluogun WA, Aderemi VA. Protective effects of *Moringa oleifera* Lam. on cadmium-induced liver and kidney damage in male Wistar rats. Int J Phytother Res 2012;2(3):42−50.

[54] Salama SM, Abdulla MA, AlRashdi AS, Ismail S, Alkiyumi SS, Golbabapour S. Hepatoprotective effect of ethanolic extract of *Curcuma longa* on thioacetamide induced liver cirrhosis in rats. BMC Complement Altern Med 2013;13:56. Available from: http://dx.doi.org/doi:10.1186/1472-6882-13-56.

[55] Ramaiah SK, Apte U, Mehendale HM. Diet restriction as a protective mechanism in noncancer toxicity outcomes: a review. Int J Toxicol 2000;19(6):413−24.

[56] Chilakapati J, Shankar K, Korrapati MC, Hill RA, Mehendale HM. Saturation toxicokinetics of thioacetamide: role in initiation of liver injury. Drug Metab Dispos 2005;33 (12):1877−85.

[57] Okoko T, Oruambo IF. The effect of *Hibiscus sabdariffa* calyx extract on cisplatin-induced tissue damage in rats. Biokemistri 2008;20(2):47−52.

[58] Sambo N, Garba SH, Timothy H. Effect of the aqueous extract of *Psidium guajava* on erythromycin-induced liver damage in rats. Niger J Physiol Sci 2009;24 (2):171−6.

[59] Abongwa M, Rageh GA, Arowolo O, Dawurung C, Oladipo O, Atiku A, et al. Efficacy of *Senna occidentalis* in the amelioration of tetracycline induced hepato- and nephro-toxicities in rabbits. Abstracts/Toxicol Lett 2010;196S:S207. P206-009.

[60] Mansour HH, Hafez HF, Fahmy NM, Hanafi N. Protective effect of *N*-acetylcysteine against radiation induced DNA damage and hepatic toxicity in rats. Biochem Pharmacol 2008;75(3):773−80.

[61] Gbadegesin MA, Odunola OA, Akinwumi KA, Osifeso OO. Comparative hepatotoxicity and clastogenicity of sodium arsenite and three petroleum products in experimental Swiss Albino Mice: the modulatory effects of *Aloe vera* gel. Food Chem Toxicol 2009;47(10):2454−7.

[62] Lum G, Gambino SR. Serum gamma-glutamyl transpeptidase activity as an indicator of disease of liver, pancreas, or bone. Clin Chem 1972;18:358−62.

[63] Dooley JS. Gallstones and Benign Biliary Diseases. In: Dooley JS, Lok ASF, Burroughs AK, Heathcote EJ, editors. Sherlock's diseases of the liver and biliary system. 12th ed. Blackwell Publishing Ltd.; 2011. p. 257−93.

[64] Uboh FE, Okon IE, Ekong MB. Effect of aqueous extract of *Psidium guajava* leaves on liver enzymes, histological integrity and hematological indices in rats. Gastroenterol Res 2010;3(1):32−8.

[65] Keays R, Harrison PM, Wendon JA, Forbes A, Gove C, Alexander GJ, et al. Intravenous acetylcysteine in paracetamol induced fulminant hepatic failure: a prospective controlled trial. BMJ 1991;303:1026−9.

[66] Schilling A, Corey R, Leonard M, Eghtesad B. Acetaminophen: old drug, new warnings. Cleve Clin J Med 2010;77(1):19−27.

[67] Comelli MC, Mengs U, Schneider C, Prosdocimi M. Toward the definition of the mechanism of action of silymarin: activities related to cellular protection from toxic damage induced by chemotherapy. Integr Cancer Ther 2007;6(2):120−9.

[68] Borah A, Paul R, Choudhury S, Choudhury A, Bhuyan B, Das Talukdar A, et al. Neuroprotective potential of silymarin against CNS disorders: insight into the pathways and molecular mechanisms of action. CNS Neurosci Ther 2013;19(11):847−53.

[69] Raza SS, Khan MM, Ashafaq M, Ahmad A, Khuwaja G, Khan A, et al. Silymarin protects neurons from oxidative stress associated damages in focal cerebral ischemia: a behavioral, biochemical and immunohistological study in Wistar rats. J Neurol Sci 2011;309(1−2):45−54.

[70] Donfack JH, Adamou D, Ngueguim TF, Kapche DWFG, Tchana NA, Buonocore D, et al. *In vitro* hepatoprotective and antioxidant activities of crude extract and isolated compounds from *Ficus gnaphalocarpa*. Inflammopharmacology 2011;19(1):35−43.

[71] Njayou FN, Moundipa PF, Tchana AN, Ngadjui BT, Tchouanguep FM. Inhibition of microsomal lipid peroxidation and protein oxidation by extracts from plants used in Bamun folk medicine (Cameroon) against hepatitis. Afr J Tradit Complement Altern Med 2008;5(3):278−89.

[72] Osadebe PO, Okoye FBC, Uzor PF, Nnamani NR, Adiele IE, Obiano NC. Phytochemical analysis, hepatoprotective and antioxidant activity of *Alchornea cordifolia* methanol leaf extract on carbon tetrachloride-induced hepatic damage in rats. Asian Pac J Trop Med 2012;5(4):289−93.

[73] Chandan BK, Saxena AK, Shukla S, Sharma N, Gupta DK, Suri KA, et al. Hepatoprotective potential of *Aloe barbadensis* Mill. Against carbon tetrachloride induced hepatotoxicity. J Ethnopharmacol 2007;111:560−6.

[74] Chandan BK, Sharma AK, Anand KK. *Boerhaavia diffusa*: a study of its hepatoprotective activity. J Ethnopharmacol 1991;31:299−307.

[75] Musa TY, Adebayo OJ, Egwim EC, Owoyele VB. Increased liver alkaline phosphatase and amino transferases activities following administration of ethanolic extract of *Khaya senegalensis* stem bark to rats. Biochemistry 2005;17(1):27−32.

[76] Popat A, Shear NH, Malkiewicz I, Stewart MJ, Steenkamp V, Thomson S, et al. The toxicity of *Callilepis laureola*, a South African traditional herbal medicine. Clin Biochem 2001;34:229−36.

[77] Evans WC. An overview of drugs with antihepatotoxic and oral hypoglycaemic activities. In: Evans WC, editor. Trease and evans pharmacognosy. Edinburh, TX: W.B. Saunders; 2002.

[78] Onaolapo AY, Onaolapo OJ, Mosaku TJ, Akanji OO, Abiodun O. A histological study of the hepatic and renal effects of subchronic low dose oral monosodium glutamate in Swiss albino mice. Br J Med Med Res 2013;3(2):294−306.

[79] Xia Q, Chiang HM, Zhou YT, Yin JJ, Liu F, Wang C, et al. Phototoxicity of kava— formation of reactive oxygen species leading to lipid peroxidation and DNA damage. Am J Chin Med 2012;40(6):1271−88.

[80] Ngounou NF, Atta-Ur-Rahman, Choudhary MI, Malik S, Zareen S, Ali R, et al. New saponins from Pteleopsis hylodendron. Phytochemistry 1999;52:917−21.

[81] Motso CPR. Recensement de quelques plantes camerounaises à activité antivirale. Mémoire de Maîtrise de Biochimie. Université de Douala; 2007. pp. 25−27.

[82] Nana HM, Ngane RA, Kuiate JR, Mogtomo LM, Tamokou JD, Ndifor F, et al. Acute and sub-acute toxicity of the methanolic extract of Pteleopsis hylodendron stem bark. J Ethnopharmacol 2011;137:70−6.

[83] Iwu MM. Handbook of African medicinal plants. Boca Raton, FL, London, Tokyo: CRC Press; 1993.

[84] Tamokou JD, Kuiate JR, Gatsing D, Nkeng Efouet AP, Njouendou AJ. Antidermatophytic and toxicological evaluations of dichloromethane-methanol extract, fractions and compounds isolated from Coula edulis. Iran J Med Sci 2011;36(2): 111−21.

[85] Ashafa AOT, Olunu OO. Toxicological evaluation of ethanolic root extract of Morinda lucida (L.) Benth. (Rubiaceae) in male Wistar rats. J Nat Pharm 2011;2(2): 108−14.

[86] Makinde M, Obi PO. Screening of Morinda lucida leaf extract for antimalarial action on Plasmodium berghei in mice. Afr J Med Sci 1985;14:59−63.

[87] Kamanyi A, Njamen D, Nkeh B. Hypoglycaemic properties of the aqueous root extract of Morinda lucida (Benth.) (Rubiaceae): studies in the mouse. Phytother Res 1994;8:369−71.

[88] Akinmoladun AC, Obuotor EM, Farombi EO. Evaluation of antioxidant and free radical scavenging capacities of some Nigerian indigenous plants. J Med Food 2010;13:444−51.

[89] Oduola T, Bello I, Adeosun G, Ademosun AW, Raheem G, Avwioro G. Hepatotoxicity and nephrotoxicity evaluation in Wistar albino rats exposed to Morinda lucida leaf extract. N Am J Med Sci 2010;2(5):230−3.

[90] Ajayi OB, Fajemilehin AS, Dada CA, Awolusi OM. Effect of Tetrapleura tetraptera fruit on plasma lipid profile and enzyme activities in some tissues of hypercholesterol-emic rats. J Nat Prod Plant Resour 2011;1(4):47−55.

[91] Bouquet AAP. Plantes médicinales du Congo-Brazaville (III). Plantes Médicinales et Phytothérapie Tome 1971;5(2):154.

[92] Ojewole JAO, Adesina SK. Cardiovascular and neuromuscular actions of scopoleptinn from fruit of Tetrapleura tetraptera. Planta Med 1983;48:99−102.

[93] Nwawu JI, Akali PA. Anticonvulsant activity of the volatile oil from the fruit of Tetrapleura tetraptera. J Ethnopharmacol 1986;18:103−7.

[94] Odesanmi SO, Lawal RA, Ojokuku SA. Effects of ethanolic extract of Tetrapleura tetraptera on liver function profile and histopathology in male Dutch white rabbits. Int J Trop Med 2009;4(4):136−9.

[95] Full AB, Vanhaelen-Faster R, Vanhaelen M, Lo I, Toppet M, Ferster A, et al. In vitro antisickling activity of a rearranged Limonoid isolated from Khaya senegalensis. Plant Med 1999;65(3):209−12.

[96] Abubakar MG, Lawal A, Usman MR. Hepatotoxicity studies of sub-chronic administration of aqueous stem bark of *Khaya senegalensis* in albino rats. Bayero J Pure Appl Sci 2010;3(1):26−8.

[97] Egwim EC, Badru AA, Ajiboye KO. Testing pawpaw (*Carica papaya*) leaves and African Mahogany (*Khaya senegalensis*) bark for antimalaria activities. Niger Soc Exp Biol J 2002;2.

[98] Yakubu MT, Adebayo OJ, Egwim EC, Owoyele VB. Increased liver alkaline phosphatase and aminotransferase activities following administration of ethanolic extract of *Khaya senegalensis* stem bark to rats. Biokemistri 2005;17(1):27−32.

[99] Orhue ES, Idu M, Ataman JE, Ebite LE. Haematological and histopathological studies of *Jatropha tanjorensis* (J.L. Ellis and Soroja) leaves in rabbits. Asian J Biol Sci 2008;1:84−9.

[100] Oyewole OI, Oladipupo OT, Atoyebi BV. Assessment of renal and hepatic functions in rats administered methanolic leaf extract of *Jatropha tanjorensis*. Ann Biol Res 2012;3(2):837−41.

[101] Rucker G, Kehrbaum SK, Sakulas H, Lawong B, Goeltenboth F. Medicinal plants from Papua New Guinea. Part III. Acetylated aurone glucosides from *Microglossa pyrifolia*. Planta Med 1994;60:288−9.

[102] Johri RK, Singh C, Kaul BL. *Vernonia lasiopus* and *Vernonia galamansis*: a medicinal perspective. Res Ind 1995;40:327−8.

[103] Köhler I, Jenett-Siems K, Kraft C, Siems K, Abbiw D, Bienzle U, et al. Herbal remedies traditionally used against malaria in Ghana: bioassay-guided fractionation of *Microglossa pyrifolia* (Asteraceae). J Biosci 2002;57:1022−7.

[104] Elufioye TO, Agbedahunsi JM. Antimalaria activities of *Tithonia diviersifolia* (Asteraceae) and *Crossopteryx febrifuga* (Rubiaceae) on mice *in vivo*. J Ethnopharmacol 2004;93:167−71.

[105] Elufioye TO, Alatise OI, Fakoya FA, Agbedahunsi JM, Houghton PJ. Toxicity studies of *Tithonia diversifolia* A. Gray (Asteraceae) in rats. J Ethnopharmacol 2009;122(2):410−5.

[106] Devatkal SK, Thorat PR, Manjunatha M, Anurag RK. Comparative antioxidant effect of aqueous extracts of curry leaves, fenugreek leaves and butylated hydroxytoluene in raw chicken patties. J Food Sci Technol 2012;49(6):781−5.

[107] Adebajo AC, Ayoola OF, Iwalewa EO, Akindahunsi AA, Omisore NO, Adewunmi CO, et al. Anti-trichomonal, biochemical and toxicological activities of methanolic extract and some carbazole alkaloids isolated from the leaves of *Murraya koenigii* growing in Nigeria. Phytomedicine 2006;13(4):246−54.

[108] Bye SN, Dutton MF. The inappropriate use of traditional medicines in South Africa. J Ethnopharmacol 1991;34(2−3):253−9.

[109] Marimuthoo D. *Callilepis laureola* DC., South African National Biodiversity Institute, Biodiversity for Life. KwaZulu-Natal Herbarium, March 2008. <http://www.plantzafrica.com/plantcd/callilepislaur.htm>; 2008 [accessed 29.12.13].

[110] Clarkson C, Vinesh JM, Neil RC, Olwen MG, Pamisha P, Motlalepula GM, et al. *In vitro* antiplasmodial activity of medicinal plants native to or naturalised in South Africa. J Ethnopharmacol 2004;92:177−91.

[111] Pendota SC, Yakubu MT, Grierson DS, Afolayan AJ. Effect of administration of aqueous extract of hippobromus pauciflorus leaves in male Wistar rats. Afr J Tradit Complement Altern Med 2010;7(1):40−6.

[112] Awodele O, Popoola TD, Amadi KC, Coker HA, Akintonwa A. Traditional medicinal plants in Nigeria—remedies or risks. J Ethnopharmacol 2013;150(2):614−8.

[113]   The Wealth of India. A dictionary of Indian raw materials and industrial products, vol. 1. New Delhi: National Institutes of Science Communication, Council of Scientific and Industrial Research; 2000. A-Ci (Revised), p. 47−9.

[114]   Sultana N, Najam R. Gross toxicities and hepatoprotective effect of *Aloe vera* L. IRJP 2012;3(10):106−10.

[115]   Etim OE, Farombi EO, Usoh IF, Akpan EJ. The protective effect of *Aloe vera* juice on lindane induced hepatotoxicity and genotoxicity. Pak J Pharm Sci 2006;19 (4):337−40.

[116]   Kim SH, Cheon HJ, Yun N, Oh ST, Shin E, Shim KS, et al. Protective effect of a mixture of *Aloe vera* and *Silybum marianum* against carbon tetrachloride induced acute hepatotoxicity and liver fibrosis. J Pharmacol Sci 2009;109:119−27.

[117]   Gupta AK, Misra N. Hepatoprotective activity of aqueous ethanolic extract of *Chamomile capitula* in paracetamol intoxicated albino rats. Am J Pharmacol Toxicol 2006;1:17−20.

[118]   Arunkumar S, Muthselvam M. Analysis of phytochemical constituents and antimicrobial activities of *Aloe vera* L. against clinical pathogens. World J Agric Sci 2009;5:572−6.

[119]   Hu Y, Xu J, Hu Q. Evaluation of antioxidant potential of *Aloe vera* (*Aloe barbadensis* miller) extracts. J Agric Food Chem 2003;51:7788−91.

[120]   Bottenberg MM, Wall GC, Harvey RL, Habib S. Oral *Aloe vera*-induced hepatitis. Ann Pharmacother 2007;41:1740−3.

[121]   Patil R, Dhawale K, Gound H, Gadakh R. Protective effect of leaves of *Murraya koenigii* on reserpine-induced orofacial dyskinesia. Iran J Pharm Res 2012; 11(2):635−41.

[122]   Dafallah AA, Al-Mustafa Z. Investigation of the anti-inflammatory activity of *Acacia nilotica* and *Hibiscus sabdariffa*. Am J Clin Med 1996;24:263−9.

[123]   Eleyinmi AF. Chemical composition and antibacterial activity of *Gongronema latifolium*. J Zhejiang Univ Sci B 2007;8(5):352−8.

[124]   Morebise O, Fafunso MA. Antimicrobial and phytotoxic activities of saponin extracts from two Nigerian edible medicinal plants. Biokemistri 1998;8(2):69−77.

[125]   Nwaehujor CO, Udeh NE. Screening of ethyl acetate extract of *Bridelia micrantha* for hepatoprotective and anti-oxidant activities on Wistar rats. Asian Pac J Trop Med 2011;4(10):796−8.

[126]   Adaramoye OA, Achem J, Akintayo OO, Fafunso MA. Hypolipidemic effect of *Telfairia occidentalis* (fluted pumpkin) in rats fed a cholesterol-rich diet. J Med Food 2007;10(2):330−6.

[127]   Nwanna EE, Oboh G. Antioxidant and hepatoprotective properties of polyphenol extracts from *Telfairia occidentalis* (fluted pumpkin) leaves on acetaminophen induced liver damage. Pak J Biol Sci 2007;10(16):2682−7.

[128]   Nwozo SO, Adaramoye OA, Ajaiyeoba EO. Anti-diabetic and hypolipidemic studies of *Telifairia occidentalis* on alloxan-induced diabetic rabbits. Niger J Natl Prod Med 2004;8:37−9.

[129]   Bolaji OM, Olabode OO. Modulating effect of aqueous extract of *Telfairia occidentalis* on induced cyanide toxicity in rats. Niger J Physiol Sci 2011;26(2):185−91.

[130]   Copeland SR. Paleoanthropological implications of vegetation and wild plant resources in modern savanna landscapes, with applications to Plio-Pleistocene Olduvai Gorge, Tanzania. Graduate School—New Brunswick Rutgers, the State University of New Jersey, New Brunswick; 2004. p. 423.

[131] Demissew S, Asfaw N. Some useful indigenous labiates from Ethiopia. Lamiales
       Newslett (Kew) 1994;3:5−6.
[132] Harel D, Khalid SA, Kaiser M, Brun R, Wünsch B, Schmidt TJ. Encecalol angelate,
       an unstable chromene from *Ageratum conyzoides* L.: total synthesis and investigation
       of its antiprotozoal activity. J Ethnopharmacol 2011;137(1):620−5.
[133] Menut C, Sharma S, Luthra C. Aromatic plants of tropical central Africa, part X
       chemical composition of essential oils of *Ageratum houstonianum* Mill. and
       Ageratum conyzoides L. from Cameroon. Flavour Fragr J 1993;8(1):1−4.
[134] Lima MAS, Barros MCP, Pinheiro SM, do Nascimento RF, Matos FJD, Silveira ER.
       Volatile compositions of two Asteraceae from the North-east of Brazil: *Ageratum
       conyzoides* and Acritopappus confertus (Eupatorieae). Flavour Fragr J
       2005;20:559−61.
[135] Titanji VPK, Denis Z, Moses MN. The antimalarial potential of medicinal plants
       used for the treatment of malaria in Cameroonian folk medicine. Afr J Trad CAM
       2008;5(3):302−21.
[136] Jagetia GC, Shirwaikar A, Rao SK, Bhilegaonkar PM. Evaluation of the radioprotec-
       tive effect of *Ageratum conyzoides* Linn. extract in mice exposed to different doses
       of gamma radiation. J Pharm Pharmacol 2003;55(8):1151−8.
[137] Agunbiade OS, Ojezele OM, Ojezele JO, Ajayi AY. Hypoglycaemic activity of
       Commelina africana and *Ageratum conyzoides* in relation to their mineral composi-
       tion. Afr Health Sci 2012;12(2):198−203.
[138] Tapsell LC, Hemphill I, Lobiac L, Parch CS, Sullivan DR, Fenech M, et al. Health bene-
       fits of herbs and spices: the past, the present, the future. Med J Aust 2006;185
       (4):54−524.
[139] Abdel-Azeem AS, Hegazy AM, Ibrahim KS, Farrag AR, El-Sayed EM. Hepatoprotective,
       antioxidant, and ameliorative effects of ginger (*Zingiber officinale* Roscoe) and vitamin E
       in acetaminophen treated rats. J Diet Suppl 2013;10(3):195−209.
[140] Grzanna R, Lindmark L, Frondoza CG. Ginger an herbal medicinal product with
       broad anti-inflammatory actions. J Med Food 2005;8:125−32.
[141] Sharma P, Singh R. Dichlorvos and lindane Induced oxidative stress in rat brain: pro-
       tective effects of ginger. Pharmacognosy Res 2012;4(1):27−32.
[142] Fuglie LJ. The miracle tree: *Moringa oleifera*: natural nutrition for the tropics.
       Dakar: Church World Service; 1999. p. 68.
[143] Verma KS, Nigam R. Nutritional Assessment of Different parts of *Moringa oleifera*
       Lamm collected from Central India. J Nat Prod Plant Resour 2014;4(1):81−6.
[144] Sharifudin SA, Fakurazi S, Hidayat MT, Hairuszah I, Moklas MAM, Arulselvan P.
       Therapeutic potential of *Moringa oleifera* extracts against acetaminophen-induced
       hepatotoxicity in rats. Pharm Biol 2013;51(3):279−88.
[145] Kar A, Choudhary BK, Bandyopadhyay NG. Comparative evaluation of hypoglycaemic
       activity of some Indian medicinal plants in alloxan diabetic rats. J Ethnopharmacol
       2003;84(1):105−8.
[146] Anwar F, Latif S, Ashraf M, Gilani AH. *Moringa oleifera*: a food plant with multiple
       medicinal uses. Phytother Res 2006;21:17−25.
[147] Fakurazi S, Hairuszah I, Nanthini U. *Moringa oleifera* Lam prevents acetaminophen
       induced liver injury through restoration of glutathione level. Food Chem Toxicol
       2008;46:2611−5.
[148] Hamza AA. Ameliorative effects of *Moringa oleifera* Lam seed extract on liver fibro-
       sis in rats. Food Chem Toxicol 2010;48:345−55.

[149] Kumar V, Mahdi F, Khanna AK, Singh R, Chander R, Saxena JK, et al. Antidyslipidemic and antioxidant activities of *Hibiscus rosa-sinensis* root extract in alloxan induced diabetic rats. Indian J Clin Biochem 2013;28(1):46−50.

[150] Khoo HE, Azlan A, Nurulhuda MH, Ismail A, Abas F, Hamid M, et al. Antioxidative and cardioprotective properties of anthocyanins from defatted dabai extracts. Evid Based Complement Alternat Med 2013;2013:434057. Available from: http://dx.doi. org/doi:10.1155/2013/434057.

[151] Obi FO, Usenu IA, Osayande JO. Prevention of carbon tetrachloride-induced hepato-toxicity in the rat by *H. rosasinensis* anthocyanin extract administered in ethanol. Toxicology 1998;131(2−3):93−8.

[152] Nantia EA, Manfo FPT, Telefo PB, Zambou FN, Tchouanguep FM. Study of the hypoglycaemic activity of the aqueous extract of *Senna alata* L. (Caesalpinaceae), *Eucalyptus saligna* Hook (Myrtaceae) and *Hibiscus rosa-sinensis* L. (Malvaceae) leaves in adult male rats. Biosci Proc 2005;11:30−9.

[153] Sachdewa A, Khemani LD. Effect of *Hibiscus rosa-sinensis* Linn. ethanol flower extract on blood glucose and lipid profile in streptozotocin induced diabetes in rats. J Ethnopharmacol 2003;89(1):61−6.

[154] Kumar V, Singh P, Chander R, Mahdi F, Singh S, Singh R, et al. Hypolipidemic activity of *Hibiscus rosa-sinensis* root in rats. Indian J Biochem Biophys 2009;46 (6):507−10.

[155] USDA, ARS, National Genetic Resources Program. Germplasm Resources Information Network—(GRIN) [Online Database]. National Germplasm Resources Laboratory, Beltsville, MD. 17 October 1995, <http://www.ars-grin.gov/cgi-bin/ npgs/html/taxon.pl?30205> [accessed 04.01.14].

[156] Roy CK, Kamath JV, Asad M. Hepatoprotective activity of *Psidium guajava* Linn. leaf extract. Indian J Exp Biol 2006;44(4):305−11.

[157] Roy CK, Das AK. Comparative evaluation of different extracts of leaves of *Psidium guajava* Linn. for hepatoprotective activity. Pak J Pharm Sci 2010;23(1):15−20.

[158] Hutchings A, Scott AH, Lewis G, Cunningham A. Zulu medicinal plants: an inventory. Scottsville, South Africa: University of Natal Press; 1996. p. 195−6.

[159] Katsoulis LC. *Rhoicissus tridentata* subsp. cuneifolia: the effect of geological distribution and plant storage on rat uterine contractive activity. S Afr J Bot 1999;65:1−4.

[160] Lin J, Opoku AR, Geheeb-Keller M, Hutchings AD, Terblanche SE, Jäger AK, et al. Preliminary screening of some traditional Zulu medicinal plants for anti-inflammatory and antimicrobial activities. J Ethnopharmacol 1999;68:267−74.

[161] Opoku AR, Geheeb-Keller M, Lin J, Terblanche SE, Hutchings A, Chuturgoon A, et al. Preliminary screening of some traditional Zulu medicinal plants for anti-neoplastic activities versus the HepG2 cell line. Phytother Res 2000;14:534−7.

[162] Opoku AR, Ndlovu IM, Terblanche SE, Hutchings AH. *In vivo* hepatoprotective effects of *Rhoicissus tridentata* subsp. cuneifolia, a traditional Zulu medicinal plant, against CCl$_4$-induced acute liver injury in rats. S Afr J Bot 2007;73:372−7.

[163] Donfack JH, Kengap RT, Ngameni B, Chuisseu PDD, Tchana AN, Huonocore D, et al. *Ficus cordata* Thumb. (Moraceae) is a potential source of hepatoprotective and antioxidant compounds. Pharmacologia 2011;2(5):137−45.

[164] Donfack JH, Simo CC, Ngameni B, Tchana AN, Kerr PG, Finzi PV, et al. Antihepatotoxic and antioxidant activities of methanol extract and isolated compounds from *Ficus chlamydocarpa*. Nat Prod Commun 2010;5(10):1607−12.

[165] Kapche GD, Amadou D, Waffo-Teguo P, Donfack JH, Fozing CD, Harakat D, et al. Hepatoprotective and antioxidant arylbenzofurans and flavonoids from the twigs of *Morus mesozygia*. Planta Med 2011;77(10):1044—7.

[166] Kuete V,. Ngameni B, Simo CC, Tankeu RK, Ngadjui BT, Meyer JJ, et al. Antimicrobial activity of the crude extracts and compounds from *Ficus chlamydocarpa* and *Ficus cordata* (Moraceae). J Ethnopharmacol 2008;120(1):17—24.

[167] Kuete V, Fozing DC, Kapche WFGD, Mbaveng AT, Kuiate JR, Ngadjui BT, et al. Antimicrobial activity of the methanolic extract and compounds from *Morus mesozygia* stem bark. J Ethnopharmacol 2009;124:551—5.

[168] Njayou FN, Moundipa PF, Donfack JH, Chuisseu DPD, Tchana AN, Ngadjui BT, et al. Hepato-protective, antioxidant activities and acute toxicity of a stem bark extract of *Erythrina senegalensis* DC. Int J Biol Chem Sci 2010;4(3):738—47.

[169] Njayou FN, Aboudi ECE, Tandjang MK, Tchana AK, Ngadjui BT, Moundipa PF. Hepatoprotective and antioxidant activities of stem bark extract of *Khaya grandifoliola* (Welw) CDC and *Entada africana* Guill. et Perr. J Nat Prod 2013;6:73—80.

[170] Olaleye MT, Akinmoladun AC, Ogunboye AA, Akindahunsi AA. Antioxidant activity and hepatoprotective property of leaf extracts of *Boerhaavia diffusa* Linn against acetaminophen-induced liver damage in rats. Food Chem Toxicol 2010;48:2200—5.

[171] Garba SH, Sambo N, Bala U. The effect of the aqueous extract of *Kohautia grandiflora* on paracetamol induced liver damage in albino rats. Niger J Physiol Sci 2009;24(1):17—23.

[172] Aregheore EMK, Makkar HPS, Becker K. Feed value for some browse plant from the central zone of Delta state. Niger Trop Sci 1998;38(2):97—104.

[173] Adesanoye OA, Farombi EO. Hepatoprotective effects of *Vernonia amygdalina* (astereaceae) in rats treated with carbon tetrachloride. Exp Toxicol Pathol 2010;62 (2):197—206.

[174] Akindele AJ, Adeyemi OO. Anxiolytic and sedative effects of *Byrsocarpus coccineus* Schum. and Thonn. (Connaraceae) extract. Int J Appl Res Nat Prod 2010;3 (1):28—36.

[175] Alisi CS, Onyeze GOC. Biochemical mechanisms of wound healing using extracts of *Chromolaena odorata*-Linn. Niger J Biochem Mol Biol 2009;24(1):22—9.

[176] Igboh MN, Ikewuchi CJ, Ikewuchi CC. Chemical profile of *Chromolaena odorata* L. (King and Robinson). Pak J Nutr 2009;8(5):521—4.

[177] Adeneye AA. Protective activity of the stem bark aqueous extract of *Musanga cecropioides* in carbon tetrachloride- and acetaminophen-induced acute hepatotoxicity in rats. Afr J Tradit Complement Altern Med 2009;6(2):131—8.

[178] Kamanyi A, Bopelet M, Lontsi D, Noamesi BK. Hypotensive effects of some extracts of the leaves of *Musanga cecropioides* (Cecropiaceae). Studies in the cat and the rat. Phytomedicine 1996;2(3):209—12.

[179] Adeneye AA, Ajagbonna OP, Ayodele OW. Hypoglycemic and antidiabetic activities on the stem bark aqueous and ethanol extracts of *Musanga cecropioides* in normal and alloxan-induced diabetic rats. Fitoterapia 2007;78(7—8):502—5.

[180] Owolabi OJ, Ayinde BA, Nworgu ZA, Ogbonna OO. Antidiarrheal evaluation of the ethanol extract of *Musanga cecropioides* stem bark. Methods Find Exp Clin Pharmacol 2010;32(6):407—11.

[181] Elujoba AA, Olagbende SO, Adesina SK. Anti-implantation activity of the fruit of *Lagenaria breviflora* Robert. J Ethnopharmacol 1985;13:281—8.

[182]   Saba AB, Oridupa OA, Oyeyemi MO, Osanyigbe OD. Spermatozoa morphology
        and characteristics of male Wistar rats administered with ethanolic extract of
        Lagenaria breviflora Roberts. Afr J Biotechnol 2008;8:1170−5.
[183]   Saba AB, Oridupa OA, Ofuegbe SO. Evaluation of haematological and serum elec-
        trolyte changes in Wistar rats administered with ethanolic extract of whole fruit of
        Lagenaria breviflora Robert. J Med Plants Res 2009;3:758−62.
[184]   Terashima K, Takawa Y, Niwa M. Powerful antioxidant agents based on garcinoic
        acid from *Garcinia kola*. Bioorg Med Chem 2002;10:619−1625.
[185]   Hsouna AB, Saoudi M, Trigui M, Jamoussi K, Boudawara T, Jaoua S, et al.
        Characterization of bioactive compounds and ameliorative effects of *Ceratonia sili-
        qua* leaf extract against CCl_4 induced hepatic oxidative damage and renal failure in
        rats. Food Chem Toxicol 2011;49:3183−91.
[186]   Hammouda FM, Ismail SI, Abdel-Azim NS, Shams KA. *Silybum marianum* (L.)
        Gaertn. Asteraceae (Compositae). In: Batanouny KH, editor. A guide to medicinal
        plants in North Africa. International Union for Conservation of Nature (IUCN),
        Graficas la Paz, Torredonjimeno, Jaén, Spain; 2013. pp. 255−8.
[187]   Madani H, Talebolhosseini M, Asgary S, Nader GH. Hepatoprotective activity of
        Silybum marianum and Ci-chorium intybus against thioacetamide in rat. Pak J Nutr
        2008;7(1):172−6.
[188]   PDR for herbal medicines. 1st ed., Montvale, New Jersey; 1998. pp.1177−8.
[189]   Wu JW, Lin LC, Tsai TH. Drug-drug interactions of silymarin on the perspective of
        pharmacokinetics. J Ethnopharmacol 2009;121(2):185−93.
[190]   Manzi P, Marconi S, Guzzi A, Pizzoferrato L. Commercial mushrooms: nutritional
        quality and effect of cooking. Food Chem 2004;84:201−6.
[191]   Soares AA, de Sá-Nakanishi AB, Bracht A, da Costa SM, Koehnlein EA, de Souza
        CG, et al. Hepatoprotective effects of mushrooms. Molecules 2013;18(7):7609−30.
[192a] Shruthi SD, Roshan A, Naveen KHN A. Review on Medicinal Importance of
        Basella alba L. International J Pharmaceut Sci Drug Res 2012;4(2):110−4.
[192b] Manfo FPT, Chao W-F, Moundipa PF, Pugeat M, Wang PS. Effects of maneb on
        testosterone release in male rats. Drug Chem Toxicol 2011;34(2):120−8.
[193]   Nantia EA, Manfo FPT, Beboy ESN, Moundipa FP. *In vitro* the antioxidant activity
        of the methanol extract of *Basella alba* L. (Basellaceae) in rat testis homogenate.
        Oxid Antioxid Med Sci 2013;2(2):131−6.
[194]   Manfo FPT, Nantia EA, Déchaud H, Tchana AN, Zabot M-T, Pugeat M, et al.
        Protective effect of *Basella alba* and *Carpolobia alba* extracts against maneb-
        induced male infertility. Pharm Biol 2014;52(1):97−104.
[195]   Bamidele O, Akinnuga AM, Olorunfemi JO, Odetola OA, Oparaji CK, Ezeigbo N.
        Effects of aqueous extract of *Basella alba* leaves on haematological and biochemical
        parameters in albino rats. Afr J Biotechnol 2010;9(41):6952−5.
[196]   Doherty VF, Olaniran OO, Kanife UC. Antimicrobial activities of *Aframomum mele-
        gueta* (Alligator Pepper). Int J Biol 2010;2(2):126−31.
[197]   Ilic N, Schmidt BM, Poulev A, Raskin I. Toxicological evaluation of grains of para-
        dise (*Aframomum melegueta*) [Roscoe] K. Schum. J Ethnopharmacol 2010;127
        (2):352−6.
[198]   Siminialayi IM, Agbaje EO. Gastroprotective effects of the ethanolic extract of
        Enantia chlorantha in rats. West Afr J Pharmacol Drug Res 2005;20(1&2):35−8.
[199]   Agbaje EO, Onabanjo AO. The effects of extracts of *Enantia chlorantha* in malaria.
        Ann Trop Med Parasitol 1991;85(6):585−90.

[200] Agbaje EO, Tijani AY, Braimoh OO. Effects of *Enantia chlorantha* extracts in laboratory-induced convulsion and inflammation. Orient J Med 2013;15 (1&2):68–71.

[201] Odoh UE, Okwor IV, Ezejiofor M. Phytochemical, trypanocidal and anti-microbial studies of *Enantia chlorantha* (Annonaceae) root. J Pharm Allied Sci 2010;7:4.

[202] Virtanen P, Lassila V, Njimi T, Mengata ED. Regeneration of D-galactosamine-traumatized rat liver with natural protoberberine alkaloids from *Enantia chlorantha*. Acta Anat (Basel) 1988;132(2):159–63.

[203] Virtanen P, Lassila V, Njimi T, Mengata DE. Effect of splenectomy on hepasor treatment in allyl-alcohol-traumatized rat liver. Acta Anat (Basel) 1989;134(4):301–4.

[204] Virtanen P, Lassila V, Söderström KO. Protoberberine alkaloids from *Enantia chlorantha* therapy of allyl-alcohol- and D-galactosamine-traumatized rats. Pathobiology 1993;61(1):51–6.

[205] Obioha UE, Suru SM, Ola-Mudathir KF, Faremi TY. Hepatoprotective potentials of onion and garlic extracts on cadmium-induced oxidative damage in rats. Biol Trace Elem Res 2009;129(1–3):143–56.

[206] Bhanot A, Shri R. A comparative profile of methanol extracts of *Allium cepa* and *Allium sativum* in diabetic neuropathy in mice. Pharmacognosy Res 2010;2 (6):374–84.

[207] Bakht J, Khan S, Shafi M. *In vitro* antimicrobial activity of *Allium cepa* (dry bulbs) against Gram positive and Gram negative bacteria and fungi. Pak J Pharm Sci 2014;27(1):139–45.

[208] Khaki A, Fathiazad F, Nouri M, Khaki AA, Jabbari-kh H, Hammadeh M. Evaluation of androgenic activity of *Allium cepa* on spermatogenesis in rat. Folia Morphol 2009;68(1):45–51.

[209] Khaki A, Fathiazad F, Nouri M, Khaki AA, Maleki N, Jabbari-Khamnei H, et al. Beneficial effects of quercetin on sperm parameters in streptozotocin-induced diabetic male rats. Phytother Res 2010;24(9):1285–91.

[210] Bakhshaeshi M, Khaki A, Fathiazad F, Khaki AA, Ghadamkheir E. Anti-oxidative role of quercetin derived from *Allium cepa* on aldehyde oxidase (OX-LDL) and hepatocytes apoptosis in streptozotocin-induced diabetic rat. Asian Pac J Trop Biomed 2012;2(7):528–31.

[211] Tandon N, Roy M, Roy S, Gupta N. Protective effect of *Psidium guajava* in arsenic-induced oxidative stress and cytological damage in rats. Toxicol Int 2012; 19(3):245–9.

12 Nephrotoxicity and Nephroprotective Potential of African Medicinal Plants

Martins Ekor

Department of Pharmacology, School of Medical Sciences, University of Cape Coast, Cape Coast, Ghana

12.1 Introduction

The kidney is the major organ of excretion of xenobiotics and their metabolites. The high metabolic activity and transporter capacity of this organ enable it to also carry out metabolic inactivation of compounds and increase their water solubility, to facilitate or enhance elimination via the urine [1,2]. This role of the kidney as the primary eliminator of exogenous drugs and toxins, in addition to the fact that it is characterized by a large volume of blood supply (20–25% of the cardiac output) which ensures a high level of toxicant delivery over a period of time, predisposes this important organ to nephrotoxicity and enhances its vulnerability to developing various forms of injury [2–4]. The kidney's susceptibility to injury due to the adverse effects of a wide range of chemicals is becoming more apparent, especially over the last three and a half decades. Exposure level to various toxicants varies from minute quantities to very high doses, may be chronic or limited to a single event, and injury may result from exposure to a single substance or to multiple chemicals. The circumstances of exposure may be inadvertent, accidental, or intentional overdose or of therapeutic necessity. While some chemicals cause acute kidney injury (AKI), others produce chronic renal changes that may lead to end-stage renal failure and renal malignancies [5].

Previous epidemiological evidence indicated that nephrotoxicity leading to acute renal failure (ARF) and/or chronic renal failure (CRF) represented a substantial financial burden to society [5,6]. Although the full extent of the economic impact is difficult to define because diagnosis of early injury and the documentation of the cascade of secondary degenerative changes have not been adequately identified, the extent and cost of clinically relevant nephrotoxicity is now very apparent. According to the recent UK National Institute for Health and Care Excellence [7] clinical guidelines, the costs of AKI (excluding costs in the community) to the National Health Service are estimated to be between £434 and £620 million annually. This is more than the costs associated with breast cancer or lung and skin cancer combined [7].

Toxicological Survey of African Medicinal Plants. DOI: http://dx.doi.org/10.1016/B978-0-12-800018-2.00012-1

Drug-induced nephrotoxicity is believed to contribute to 8–60% of cases of AKI in hospitalized patients [8–11]. The incidence has been shown to range between 1% and 23% in the intensive care unit setting [12–14]. Age-related decline in glomerular filtration rate or renal blood flow, which generally leads to decrease in renal drug clearance as well as decline in hepatic clearance and altered free drug concentration, are factors which further increase the susceptibility of the elderly patients to AKI from nephrotoxic agents [15]. Chemotherapeutic agents, analgesics, angiotensin-converting enzyme inhibitors, and contrast agents, to mention a few, are part of the major culprits in drug-associated kidney damage [16]. This adverse effect significantly limits the therapeutic usefulness of these agents in clinical practice, as clearly demonstrated with many cancer chemotherapeutic agents and antibiotics [17–20].

In Africa, the incidence of AKI is on the increase and fast becoming a common health problem in almost all subregions. Recent evidence indicates that the burden of disease (especially the human immunodeficiency virus-related kidney injury in sub-Saharan Africa, diarrheal disease, malaria, etc.) has made AKI one of the challenging problems currently plaguing the continent [21]. Although there are no reliable statistics about the incidence of this disease in Africa, the incidence has been estimated at 150 per million population based on sporadic regional publications [22]. In addition to hemodynamic causes, infectious diseases, obstetric and surgical complications, nephrotoxins (either as therapeutic orthodox or traditional medications, industrial and environmental chemicals) are known to play a major etiologic role in the pathogenesis of AKI in all regions of Africa [21,23].

Most of the methods used in assessment of nephrotoxicity and nephroprotection of African medicinal plants involve the traditional measurements of blood urea nitrogen (BUN), plasma creatinine concentration, glomerular filtration rate, and creatinine clearance in animal models of renal injury/dysfunction induced by gentamicin, cisplatin, carbon tetrachloride, adriamycin, cyclosporine, etc. [24–34]. In most of the studies, histopathological examinations are usually carried out to validate observations from biochemical analyses. While measurement of these biomarkers is still generally accepted and widely employed in assessment of drug or xenobiotic-induced nephrotoxicity, they are less specific and useful only when there is advanced kidney damage and severe impairment of renal function [20,35]. This limitation provides the basis for recent efforts geared toward discovering and developing novel biomarkers that can detect kidney dysfunction or injury at the early stage and with improved specificity and accuracy [11].

Various experimental studies and efforts directed toward the screening of African medicinal plants have revealed the potential of many African folklore medicinal plants in attenuating nephrotoxic injury associated with exposure to therapeutic as well as industrial and environmental toxicants. Similar studies also have demonstrated the potential of some others to induce renal dysfunction and injury. In many of these studies, only experimental or preclinical data are available for the tested plants or compounds. Extended preclinical and clinical studies are, therefore, required to establish their safety and efficacy in humans. This chapter provides an overview of the important role of the kidney in xenobiotic metabolism and how this function increases its vulnerability to nephrotoxicity. Particular attention is paid to the role or contribution of African medicinal plants in nephrotoxic renal failure and other related injuries, as well as their potential as useful source of nephroprotective agents. Some of the documented efforts from studies in different laboratories across the continent in the screening of African medicinal plants for either nephrotoxic and/or nephroprotective potential, as well as the prominent methods of assessment are also reviewed.

12.2 The Kidney and the Metabolism of Xenobiotics

The kidney is a paired, bean-shaped organ found along the posterior muscular wall of the abdominal cavity with the left kidney located slightly more superior than the right because of the larger size of the liver on the right side of the body. The kidneys, unlike other abdominal organs, are considered to be retroperitoneal organs because they lie behind the peritoneum that lines the abdominal cavity. The ribs and muscles of the back protect them from external damage and they are also surrounded by perirenal fat which acts as protective padding.

12.2.1 Structure and Function of the Kidney

Macroscopic Structure: Each kidney has a concave and convex surface. The convex side of each organ is located laterally and the concave side medial. The indentation on the concave side (the renal hilus) provides a space for the renal artery, renal vein, and ureter to enter the kidney. A thin layer of fibrous connective tissue forms the renal capsule which surrounds each kidney, and this provides a stiff outer shell to maintain the shape of the soft inner tissues. Morphologically, the kidney can be divided into the lighter, superficial area of the renal cortex and the darker, deep medulla. Both structures together build up the renal lobes (Figure 12.1). Deep to the renal capsule is the soft, dense, vascular renal cortex,

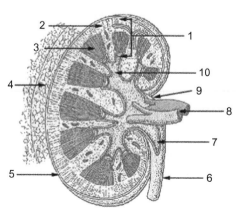

Figure 12.1 A cross-section of the kidney. 1, parenchyma; 2, cortex; 3, medulla; 4, perirenal fat; 5, capsule; 6, ureter; 7, pelvis of kidney; 8, renal artery and vein; 9, hilus; 10, calyx.
Source: From: training.seer.cancer.gov [36].

and deep to the renal cortex is the renal medulla formed from seven cone-shaped renal pyramids. The renal pyramids are aligned with their bases facing outward toward the renal cortex and their apexes point inward toward the center of the kidney. Each apex connects to a minor calyx, a small hollow tube that collects urine. The minor calyces merge to form three larger major calyces, which further merge to form the hollow renal pelvis at the center of the kidney. The renal pelvis exits the kidney at the renal hilus, where urine drains into the ureter [37].

The Nephron: This is the functional unit of the kidney and consists of a continuous tube of highly specialized heterogeneous cells, which show subspecialization along the length of nephrons and between them (Figure 12.2). The total number of nephrons varies between different species and within any one species as a function of age. The human kidney contains a total number of about 1 million nephrons and approximately 30,000 are found in the rat kidney. There are marked structural and functional differences between the nephrons arising in the cortex and those arising in the juxtamedullary regions [5,37]. The macroscopic differentiation of the kidney into distinct zones arises not only from the regional vascularity but also from the way different functional parts of the nephron are arranged within the kidney. The nephron is the primary area where urine formation takes place and where essential blood components like glucose, amino acids, and salts are reabsorbed. They can be subdivided in four major sections which differ in function, location, and cell type. There are well over 20 morphologically different cell types in the kidney based on light microscopy alone, and the diversity becomes more apparent when histochemical and immunohistochemical methods are applied to renal tissue sections. The spectrum of biochemical (and structural and functional) characteristics in these cells demonstrates the very marked heterogeneity that is the hallmark of the kidney [5].

The Glomerulus: The renal corpuscle, often simply referred to as the glomerulus, consists of two structures: the Bowman's capsule and the actual glomerulus. In the proper sense, the glomerulus describes a capillary bed or branches arising from the afferent arteriole, anastomose, and drain to the efferent arteriole. There are also communicating vessels between the branch capillaries. The glomerulus is enclosed in the Bowman's capsule, a cup-like sac, which forms the initial part of the nephron. It functions as a relatively poorly selective macromolecular exclusion filter to the hydrostatic pressure of the blood. Generally, the number of glomeruli is related to the mass of the species and the size of each depends on the environmental water balance, among other factors. Three anatomically distinct types of glomeruli have been identified: (i) those in the superficial cortex, which are part of the superficial nephrons; (ii) those arising in the midcortical area; and (iii) those of juxtamedullary origin, which continue as nephrons that loop down into the medulla. The structure of the glomerulus is complex (Figure 12.3) and has only been defined using scanning and transmission electron microscopy [40−42]. The Bowman's capsule has an outer parietal and an inner visceral layer, both of which are composed of squamous epithelium while the visceral layer is lined by podocytes which contact the capillary bed, forming filtration slits by their foot processes [37]. Three filtration barriers can, therefore, be differentiated, namely: the fenestrated capillaries (barrier for cellular components), the glomeruli basal membrane (pore size: 240−340 nm), and

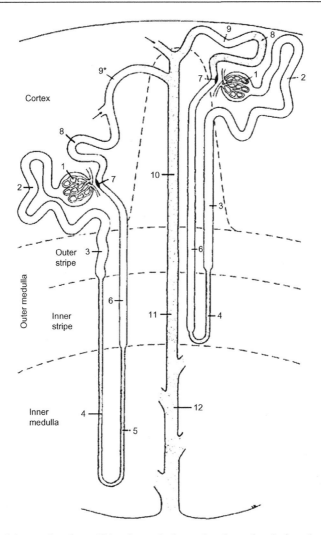

Figure 12.2 Scheme of nephron. This scheme depicts a short-looped and a long-looped nephron together with the collecting system. Within the cortex a medullary ray is delineated by a dashed line. 1, renal corpuscle including Bowman's capsule and the glomerulus (glomerular tuft); 2, proximal convoluted tubule; 3, proximal straight tubule; 4, descending thin limb; 5, ascending thin limb; 6, distal straight tubule (thick ascending limb); 7, macula densa located within the final portion of the thick ascending limb; 8, distal convoluted tubule; 9, connecting tubule of the juxtamedullary nephron that forms an arcade; 10, cortical collecting duct; 11, outer medullary collecting duct; 12, inner medullary collecting duct.
Source: From Kriz and Banir [38].

the visceral layer (filtration slits: ~25 nm). All layers together provide barrier properties of the glomerulus for molecules >70 kDa [43]. Therefore, molecules exceeding this size cannot enter the primary urine in healthy kidneys. Beside this

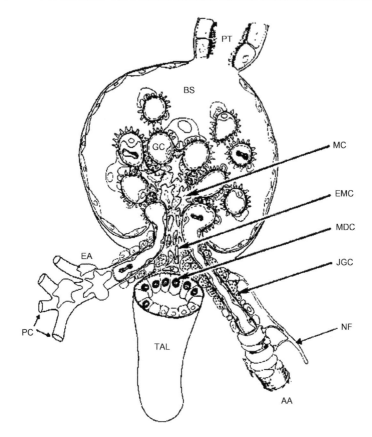

Figure 12.3 Glomerulus and juxtaglomerular complex, consisting of afferent arteriole (AA) with the granular cells (JGC) of the juxtaglomerular apparatus, the extraglomerular mesangial cells (EMC), the macula densa (MDC) segment of the ascending loop of Henle, and the efferent arteriole (EA). Also shown are the proximal tubule (PT), Bowman's space (BS), glomerular capillaries (GC), peritubular capillaries, mesangial cells (MC), and nerve fibers (NF). *Source*: From Schrier and Gottschalk [39].

molecular size exclusion, the electrical charge also plays a vital role. Uncharged or positively charged molecules are filtered more easily. The reason for this is because of the negatively charged surface of the endothelial cells. Albumins, for example, have an average molecular mass of 66 and are therefore under the barrier properties of the glomerulus. Because of the strong negative charge, the impermeability of the glomerular barriers for albumin is still at 99.97% [44].

The Tubular System: Directly behind the glomerulus is the proximal tubule which is found only in the cortex or subcortical zones of the kidney. Anatomically, each proximal tubule is divided into the convoluted portion (pars convoluta) and the shorter straight descending portion (the pars recta). The pars convoluta is confined entirely to the renal cortex while the pars recta descends into the outer medulla [45] and continues to become the descending limb of the loop of Henle. It may be subdivided, by a number of morphological and functional features, into three segments, S_1, S_2, and S_3. Segments S_1 and S_2 belong to the pars convoluta, while the S_3 segment belongs to the pars recta. Although the cellular complexity increases from S_1 to S_3 segments, the cellular structure is similar over the individual segments. The proximal tubular cells (PTCs) are characterized by a luminal surface covered with densely packed microvilli. This brush border increases the luminal surface, depending on their resorptional functionality and a putative flow sensing within the lumen [46]. In addition, the PTCs are highly packed with mitochondria for energy production and supply to various ATP-dependent transport processes.

Next to the proximal tubule is the loop of Henle which functions primarily to create a concentration gradient in the medulla by achieving a high sodium concentration near the collecting ducts and facilitating water reabsorption [47]. The loop of Henle is divided into two portions; the descending limb, which is the first part after the proximal tubule, and the ascending limb. Prominent also is their subdivision into five parts depending on their morphological structure and location within the kidney, namely:

- Thick descending limb
- Thin descending limb
- Thin ascending limb
- Thick ascending limb
- Cortical thick ascending limb

The thick and thin descending limbs are highly and freely permeable to water but their permeability to urea and ions is quite low. The thin ascending limb is characterized by its impermeability to water but not to ions. Similarly, the thick ascending limb is impervious to water and it is responsible for the reabsorption of Na^+, K^+, and Cl^- by passive transport via $Na-K-2Cl$ symporter [48]. The final area of the loop of Henle, the cortical thick ascending limb, drains urine into the distal convoluted tubule. The distal tubular cells (DTCs) are responsible for the regulation of several electrolytes like Na^+, K^+, and Ca^{2+} as well as pH by absorbing and secreting bicarbonate and protons. In contrast to the PTCs, the DTCs do not have apical brush borders and are under strong hormonal control. Aldosterone increases sodium reabsorption, although sodium and potassium levels also are controlled by secreting K^+ and absorbing Na^+. Following the distal tubule is the collecting duct which performs fine adjustment of concentration via water absorption. All the processes in the collecting duct are under the control of aldosterone and antidiuretic hormone.

General Function of the Kidney: The kidney is an essential organ which performs several important physiological functions including clearance of endogenous waste products, control of volume status, maintenance of electrolyte and acid–base balance, and endocrine function [1,2]. Other major functions include metabolism and excretion of exogenously administered therapeutic and diagnostic agents as well as environmental exposures. Furthermore, the kidney is responsible for the synthesis of erythropoietin and this underscores its direct involvement in the formation of erythrocytes [49]. Another central position of the kidney is based on the synthesis of the endopeptidase renin, a major stimulus of the renin–angiotensin–aldosterone system and an important regulator of extracellular fluid volume and blood pressure.

12.2.2   The Kidney and Xenobiotic Metabollsm

The ease with which xenobiotics enter the mammalian body and are eliminated largely depends on their water solubility. Lipophilic compounds generally are reabsorbed and often would concentrate in the body. However, they undergo series of biochemical processes (biotransformation) which convert them to more water-soluble metabolites. The metabolites formed are usually chemically distinct and more hydrophilic than the parent compound and are readily excreted in the urine. Generally, two types of enzyme reactions are involved in biotransformation: the first is phase I reactions, also known as functionalization reactions and these involve oxidation, reduction, and hydrolysis. The other, referred to as phase II reactions, are primarily conjugation reactions in which the parent compound or phase I metabolite is covalently linked to an endogenous molecule such as glucuronic acid, glutathione (GSH), amino acid, and sulfate. The addition of this moiety facilitates the excretion of these molecules from the body [50].

The involvement of the kidney in the metabolism of both endogenous and exogenous compounds did not receive much attention until about three decades ago. For a long time, most of the knowledge about drug metabolism was based on information from experimental studies on the liver which is quantitatively the most important site of biotransformation of most drugs and xenobiotics [51,52]. Evidence has since demonstrated to show that the kidney plays an active role in drug metabolism and capable of carrying out extensive phase I and phase II metabolic reactions [53,54]. Pichette and du Souich [55] demonstrated that the kidneys are responsible for 85% of furosemide total clearance, either via excretion (43%) or biotransformation (42%). In fact, the enzyme systems involved in renal drug metabolism are qualitatively similar to those found in the liver and other extrarenal tissues (Table 12.1), and certain biotransformation reactions have been shown to occur at a much faster rate in the kidney than in the liver. Glycination of benzoic acid is a typical example [56]. It is also long established that gamma-glutamyltransferase and histamine N-methyltransferase activity in mammalian tissues is highest in the kidney than in any other organ [57,58]. An important difference between the kidney and other tissues or organs involved in metabolism, however, is significant heterogeneity exhibited by the former with many of the drug metabolism enzyme systems differentially distributed among the nephron cell populations [51,59–61]. Cytochrome P450, for example, is found exclusively in renal cortex while prostaglandin synthetase, which catalyzes co-oxidation of several drugs and xenobiotics, is found in the medulla. The implication of such heterogeneity is the localization of renal drug or xenobiotic metabolism to specific cell types. While some other tissues also exhibit heterogeneity, only a few do so to the same extent as the kidneys [51].

Table 12.1 Renal Drug Metabolizing Enzymes

Phase I Enzymes	Phase II Enzymes	Ancillary Enzymes
Cytochrome P450	Esterases	GSH peroxidase
Microsomal FAD-containing monooxygenase	N-Acetyltransferase	Catalase
Alcohol and aldehyde dehydrogenases	GSH S-transferase	NADPH:Quinone oxidoreductase
Epoxide hydrolase	Thiol S-methyltransferase	NADPH-generating pathways
Prostaglanding synthetase	UDP-Glucuronosyltransferase	Superoxide dismutase
Monoamine oxidase	Sulfotransferases	GSH reductase

FAD, flavin adenine dinucleotide; UDP, uridine 5′-diphosphate; NADPH, reduced nicotinamide adenine dinucleotide phosphate.
Source: From Lash [51].

12.2.3 Renal Metabolism and Risk to Nephrotoxicity

Humans frequently encounter a wide variety of potential nephrotoxins, many of which are either prescribed or available to the general population as over-the-counter medications. Notable among therapeutic agents with known nephrotoxic potential include antimicrobial agents, chemotherapeutic agents, analgesics, and immunosuppressive agents [2,9,62−71]. Several of these agents are very potent nephrotoxins and are capable of promoting renal injury even with brief or low-level exposure. The first step in the development of nephrotoxicity requires adequate exposure to the offending agent. These toxins may injure the tubules directly, at the site of toxin transport or concentration, or by inducing renal ischemia, hemoglobinuria, or myoglobinuria [21,72]. Severity of renal injury usually increases with continuous exposure and exposure to high doses. Acute tubular or cortical necrosis and acute interstitial nephritis are the commonest of the several lesions that have been described.

Because of the kidney's large capacity for metabolism, the concept of biotransformation as a means by which reactive and toxic species are produced becomes critical in understanding the mechanism of nephrotoxicity associated with most types of xenobiotics. The mammalian kidney possesses numerous enzymatic reactions that can bioactivate chemicals. The GSH conjugation pathway is one of the major renal metabolic pathways that contribute to xenobiotic toxicity and are believed to be important factors in risk assessment. Metabolism of GSH S-conjugates of numerous halogenated hydrocarbons to cysteine S-conjugates provides a branch point in the pathway [51].

GSH-dependent pathway in xenobiotics bioactivation: Many GSH conjugates undergo further enzymatic modification by hydrolysis of the GSH S-conjugate at the γ-glutamyl bond and this reaction is catalyzed by the enzyme γ-glutamyltransferase (γ-GT). This enzyme is localized in the cell membrane of many cell types including kidney tubules which exhibits the highest activity of the enzyme in several mammals studied, including humans [52,57,73]. Although the liver has the highest levels of GSH S-transferase activity in the body and quantitatively the most important site of GSH conjugate formation, it contains minimal activities of γ-GT, the only enzyme that is capable of hydrolyzing the γ-glutamyl peptide bond of GSH and GSH S-conjugates [73]. Hence, the GSH and GSH S-conjugates formed in the liver cells are actively extruded into plasma and bile for translocation to kidney and small intestine, respectively. Both γ-GT and dipeptidases, which are present at relatively high levels in biliary epithelium, break down GSH S-conjugates secreted into bile to cysteine S-conjugates [74]. Ultimately, the toxic metabolite is generated in the renal epithelial cells owing to its high intracellular concentrations of the conjugates [75,76]. The intrarenal formation of the GSH S-conjugate of hexafluoropropene which was directly associated with nephrotoxicity corroborated this GSH-dependent mechanism of xenobiotic activation. According to this study, the GSH S-conjugate formed in the liver did not cause nephrotoxicity as it was excreted in the bile without reaching the kidneys [77].

Several alternative reactions are possible after the GSH S-conjugate is metabolized to the corresponding cysteine S-conjugate. The resulting cysteine S-conjugate may be N-acetylated to form mercapturic acid, a highly polar and chemically stable compound, which is then excreted. However, for many compounds the cysteine S-conjugate is a substrate for renal enzymes generating reactive metabolites that produce toxicity. The localization of transport systems that deliver the conjugates to kidney cells and the presence of enzymes that act on these conjugates within those cells determine the target site specificity of injury [51].

β-Lyase-dependent bioactivation of cysteine conjugate: β-Lyase is a pyridoxal phosphate-dependent enzyme found in renal cytosol and mitochondria. It cleaves the carbon–sulfur bond of several haloalkyl or haloalkenyl cysteine S-conjugates to generate reactive sulfur-containing compounds [78–80]. In addition to catalyzing a β-elimination reaction, this enzyme also has been shown to be capable of catalyzing a transamination reaction to generate the corresponding 2-keto acid [79,81,82]. Furthermore, the observation that an addition of exogenous 2-keto acids both stimulated β-lyase activity and potentiated the cytotoxicity of *S*-(1,2-dichlorovinyl)-L-cysteine in isolated rat kidney cells indicated that the keto acid/amino acid status could have a regulatory function [82]. In either the case of β-elimination or transamination, the ultimate toxic metabolite is the same, as the keto acid analogue is chemically unstable and spontaneously releases the reactive thiolate [51].

12.3 Known Nephrotoxic and Nephroprotective Plants

Recently, attention was drawn to the growing public acceptance and use of herbal medicines and issues relating to safety and adverse reactions were extensively revisited [83]. In spite of the positive perception of patients on the use of plant-derived medicines and alleged satisfaction with therapeutic outcomes coupled with their disappointment with conventional allopathic or orthodox medicines in terms of effectiveness and/or safety [84,85], the problem of safety continues to remain a major issue of concern. Documentation on folkoric use and some scientific studies, on the other hand, also have pointed to the possible benefit of plants in the treatment of nephrotoxicity and other kidney-related injuries. This section provides a brief overview of a few known nephrotoxic and nephroprotective plants in other parts of the world, although some of these plants are also found in Africa and are being investigated by scientists on the continent.

12.3.1 Known Nephrotoxic Plants

Herbal medicines and related natural products are an important and unregulated source of potentially nephrotoxic substances and a large number of them have long been recognized as capable of causing renal failure with transient or chronic requirement for dialysis [86,87]. Some medicinal plants and plant-derived compounds with nephrotoxic potential are described in this section. Generally, they have been classified into (i) agents which are directly nephrotoxic; (ii) herbal drugs which can cause electrolyte abnormalities by acting on the kidney; (iii) agents which can predispose to formation of stones (oxalate stones); (iv) agents which act as diuretics; (v) herbal drugs which contain heavy metals or other drugs; and (vi) herbal agents which can interact with other drugs, especially in the renal transplant subject [88,89].

12.3.1.1 Euphorbia paralias *(Euphorbiaceae)*

Plants of the genus Euphorbia are prolific producers of diterpenes of great biomedical interest. *Euphorbia paralias*, which belongs to this group, is a hardy perennial plant inhabiting sandy coasts and shingle beaches and it is native to the entire Mediterranean region. It is also one of the most common *Euphorbia* species growing in Egypt [90]. This plant is employed as an anti-inflammatory agent, purgative, and local anesthetic in traditional medicine [91]. Boubaker et al. reported the case of a 29-year-old male who developed renal damage following one-time ingestion of the boiled plant of *Euphorbia paralias* for treatment of edema as a part of native medical treatment. Nephrotoxicity resulting from poisoning with this plant was considered the cause of the AKI. Kidney biopsy confirmed the patient's ARF to be due to a severe tubular necrosis. Although the exact mechanism of the toxicity was not known, Euphorbiaceae plants are well known to contain irritant, cytotoxic, and tumor-promoting constituents such as ingenanes found in the latex of *Euphorbia paralias*. The underlying mechanisms responsible for renal injury resulting in ARF by this plant were believed to be toxic and immunoallergic in nature [92].

12.3.1.2 Solanum grandiflorum *(Solanaceae)*

Solanum grandiflorum, popularly known as wolf's fruit, "lobeira," or "jurubebão," is widely used in folkloric medicine, because of its relaxing, sedative, and antispasmodic properties [93]. Capsules containing fruit starch of this plant are used to control diabetes and can be found in the pharmacies around the region of south of

Minas Gerais in Brazil. Pereira et al. [94] reported that subchronic ingestion of hydroalcoholic extract of the fruits of *Solanum grandiflorum* was nephrotoxic. Their study revealed glomerular degeneration and necrosis, tubular and hypercellular lesions following treatment with the fruits of this plant in rats without any significant biochemical and hematological alterations. The high saponin content of *Solanum grandiflorum* fruits, which is known to alter permeability of the cells [95], was believed to be responsible for this renal injury.

12.3.1.3 Aristolochic Species Nephropathy

Aristolochia fangchi (Aristolochiaceae) had long been linked to the development of subacute interstitial fibrosis of the kidney (also referred to as "Chinese herbs nephropathy") in the United Kingdom and many other countries [96–99]. Consumption of herbal preparation containing the root of *Aristolochia fangchi* purchased over the internet for psoriasis led to kidney failure and death of a 75-year-old man in Australia [100]. Similar cases of nephropathy associated with chronic use of aristolochic acid-containing herbal products was reported in Taiwan and Japan [101,102]. The presence of aristolochic acids in *Aristolochia fangchi* was also linked to severe adverse reactions which led to nephrotoxic and carcinogenic events in more than 100 women using the weight-loss preparation [103]. A specific type of fibrosing interstitial nephritis was observed with slimming therapy incorporating the two Chinese herbs, *Stephania tetrandra* and *Magnolia officinalis* [104]. Aristolochic acid is a nitrophenanthrene carboxylic acid which forms DNA adducts in renal as well as other tissues after metabolic activation. The DNA adducts is known to be responsible for the characteristic renal interstitial fibrosis and extensive loss of cortical tubules [89]. Several other plants particularly from the *Asarum* and *Bragantia* genera also contain aristolochic acid and are nephrotoxic. Most patients present with renal insufficiency, moderate increase in blood urea, mild proteinuria, severe anemia, and urinary sediment. Glycosuria and sterile proteinuria which indicate tubular dysfunction are common [89].

12.3.1.4 Uncaria tomentosa *(Rubiaceae)*

Uncaria tomentosa or Cat's claw is a Peruvian herbal preparation with several traditional uses including gastritis, rheumatism, cirrhosis, gonorrhea, and cancers of the female genital tract. The use of this plant has been associated with development of acute interstitial nephritis leading to ARF [89].

12.3.1.5 Other Examples

Rhubarb (*Rheum officinale*; Polygonaceae) is a common herbal preparation with high oxalate content and can promote the formation of renal calculi. *Averrhoa carambola* (Star fruit; Oxalidaceae) ingestion has also been reported to produce acute oxalate nephropathy [105]. Juniper berries (*Juniperus communis*; Cupressaceae) contain terpine-4-ol which may cause kidney irritation and damage in excess [89].

12.3.2 Known Nephroprotective Plants

Many plants are believed to prevent kidney damage and often are used in the traditional system of medicine for the treatment of kidney failure in different parts of the world. Several studies have validated and also identified novel plants and plant-derived compounds that are capable of protecting the kidney against injury from therapeutic or environmental toxins. Some of them are discussed below and Table 12.1 also provides an additional list of other plants which have been investigated and documented as nephroprotective in experimental models.

12.3.2.1 Panax ginseng (Araliaceae)

The protective effect of ginsenoside rb-1 and quercetin, two natural antioxidants isolated from *Panax ginseng*, has been demonstrated in an experimental model of acute nephritis induced by puromycin aminonucleoside. In addition to an antioxidant property, their protective action was also attributed to their ability to suppress the formation of phosphatidylcholine hydroperoxide [106].

12.3.2.2 Sanguisorbae radix (Rosaceae)

Chen et al. [107] investigated the protective effect of *Sanguisorbae radix* extract against renal dysfunction induced by lipopolysaccharide (LPS) endotoxin in rats. Pretreatment with the plant extract significantly reduced LPS-induced increases in BUN and serum creatinine, serum nitrite/nitrate level, and iNOS activity. Results from the study indicated that extract of *Sanguisorbae radix* played an important role in preserving renal function and attenuating renal dysfunction in conditions associated with excessive generation of NO.

12.3.2.3 Garlic (Allium sativum; Alliaceae)

The nephroprotective effect of aged garlic (*Allium sativum*) extract has also been established in a rat model of gentamicin-mediated nephrotoxicity. Biochemical alteration induced by gentamicin was ameliorated following treatment with the extract and the protective effect was shown to be associated with decrease in oxidative stress and preservation of Mn-superoxide dismutase, GSH peroxidase, and GSH reductase activities in the renal cortex [108].

12.3.2.4 Nigella sativa (Ranunculaceae)

Ali [109] reported dose-dependent amelioration of the biochemical and histological alteration associated with gentamicin nephrotoxicity following treatment with *Nigella sativa* oil in rats. The study attributed the ameliorative effect to the ability of the *Nigella sativa* oil to elevate reduced GSH concentration and total antioxidant status in renal cortex.

12.3.2.5   Other Examples

Sida rhomboidea Roxb (Fam. Malvaceae), a weed found in marshy places across India, is consumed as an ethnomedicine against dysuria and other urinary disorders [110]. Thounaojam et al. [111] validated this ethnobotanical use in a study designed to assess the protective effect of the leaf extract of this plant against nephrotoxicity and renal dysfunction induced by gentamicin in rats. Results from the study revealed that leaf extract of *Sida rhomboidea* offered significant protection against gentamicin nephropathy, an effect attributed to its free radical scavenging and antioxidant property [112]. Khan et al. [113] also demonstrated that extracts of the aerial parts of *Launaea procumbens* (Asteraceae) effectively protected against CCl_4-induced renal oxidative damage in rats, through antioxidant and free radical scavenging effects of flavonoids and saponins present in the fractions. In a previous study, authors validated the folkloric use of *Digera muricata* (Amaranthaceae) in renal disorders by demonstrating the effectiveness of extracts from aerial parts of this plant in attenuating biochemical alterations associated with CCl_4-induced nephrotoxicity and restoring renal function to normal, also through an antioxidant action [114].

Furthermore, Hou et al. [115] demonstrated that iridoid glycosides isolated from Chinese herb, *Paederia scandens* (Rubiaceae), exerted protective effects against kidney damage in uric acid nephropathic rats through its uric acid-lowering, anti-inflammatory, and immunomodulatory properties. Results from this study also indicated that reduction of systolic blood pressure via up-regulation of NOS-1 expression and down regulation of TNF-α and TGF-β1 expression also contributed to the amelioration of kidney injury associated with the high uric acid-induced nephropathy. Similarly, studies have demonstrated the efficacy of silymarin, a flavonoid complex isolated from the seeds of *Silybum marianum* (milk thistle) (Asteraceae), as a powerful free radical scavenging agent [116–118]. Its protective effect against adriamycin and acetaminophen mediated nephrotoxicity has been reported [118,119]. The study by Bektur et al. [119] demonstrated its protective property against acetaminophen-induced nephrotoxicity to be mediated through antioxidant and anti-inflammatory actions. Various studies using various experimental models have also demonstrated the beneficial effects of green tea and extracts of *Moringa oleifera* (Moringaceae) in improving renal function and preventing toxic injuries to the kidneys [114,120–123].

12.4   Methods in Nephrotoxic and Nephroprotective Screenings of Medicinal Plants

There exist several established methods for assessing nephrotoxicity induced by drugs, chemicals, industrial, or environmental toxic agents. These methods also are employed commonly by researchers in the field of medicinal plant research in screening for nephrotoxic potentials of various plants. The methods also are applicable for assessing nephroprotective potential of medicinal plants, except that known nephrotoxicants are employed in the induction of renal injury. Notable among the known nephrotoxicants commonly used include aminoglycoside antibiotics (gentamicin), amphotericin B, cisplatin, cyclosporine, acetaminophen, cadmium, carbon tetrachloride, mercuric chloride, etc. Described below are available methods for assessing nephrotoxicity including the applicable methods to evaluate herbal renal toxicity in Africa. The merits and demerits of each method also are highlighted.

12.4.1 In Vivo *Animal Models*

Animal models are regarded as the "gold standard" for assessing toxicity and are still the recommended models by the European Medicines Agency (EMA) and the Food and Drug Administration (FDA) for investigating toxicity of new drugs [124,125]. Classically, acute, subacute, and chronic toxicities are assessed in rodent and non-rodent species and the major drawbacks to the use of these models are the huge expenses involved and the time-consuming nature. It is, therefore, usually not practical to use them for screening the nephrotoxicity potential of large number of samples. More important is the fact that extensive use of animals for studies involving toxicity evaluation is contrary to the current trend of applying the principle of Replacement, Reduction, and Refinement [126].

Generally, in the *in vivo* animal model method of assessing nephrotoxicity, animals (e.g., rats, mice, rabbits) are treated with the test compounds and then biomarkers of nephrotoxicity are measured either in urine or blood samples or both and kidneys are harvested and processed for histology. The classical markers of kidney injury include proteinuria, plasma creatinine, and BUN. The utilization of these traditional markers in nephrotoxicological studies, however, is limited owing to poor sensitivity. This poor sensitivity makes them very unreliable in detecting early stages of kidney damage since the kidneys possess great ability to compensate renal mass loss and to recover after acute insult. The implication of this is that plasma creatinine and BUN will most times reach appreciably high or detectable levels to indicate a reduction in renal functionality after approximately two-thirds of renal biomass has been lost [125]. Apart from low sensitivity, these markers also lack specificity. Serum creatinine concentration, a breakdown product of muscle tissue, has been shown to depend on factors such as age, gender, muscle mass, and weight. Similarly, BUN also has been shown to increase with gastrointestinal bleeding or enhanced protein catabolism and other pathologic conditions without any negative impact on the kidneys [125].

Some recent studies have described new biomarkers considered to be promising candidates in detecting acute or chronic tubulotoxicity [127,128]. These markers include calbindin, clusterin, cystatin C, GSH S-transferase-α, kidney injury molecule-1, neutrophil gelatinase-associated lipocalin/lipocalin-2, β_2-microglobulin, osteopontin, tissue inhibitor of matrix metalloproteinase-1, and vascular endothelial growth factor. The superiority of these markers over plasma creatinine and BUN in terms of sensitivity and specificity has been demonstrated in several studies and both FDA and EMA have accepted them for the detection of AKI in the context of nonclinical drug development. However, only a few studies have compared these markers with histopathological analyses and it has been suggested that biomarkers alone should not be used in the assessment of nephrotoxicity of herbal or traditional medicines [125,127,129].

The *in vivo* animal models are still the most popular and frequently used methods in Africa. Several studies conducted in different laboratories across Africa to assess nephrotoxicity as well as nephroprotective potential of various folklore herbs or medicinal plants distributed across the continent utilize experimental animals especially rats, mice, and rabbits. Most of the studies on nephroprotection employed therapeutic agents (such as cisplatin, gentamicin, cyclosporine, acetaminophen, and adriamycin) as well as environmental toxicants and toxins (such as carbon tetrachloride, cadmium, ethylene glycol, mercuric chloride, etc.) to induce renal injury [24–33].

12.4.2   Isolated Perfused Kidney

The isolated perfused kidney preparation from rat and mouse has been shown to be ideal for assessing renal vascular, tubular, and glomerular functions and has also been found useful in evaluating nephrotoxic potential of various chemicals [125,130]. The major advantage of this model is the fact that it can reproduce the *in vivo* toxicity for very acute toxicants as it is an intact organ and maintains structural integrity [131]. It is, however, a difficult technique, time consuming, and suffers from limited viability.

12.4.3   Renal Cell Culture Models

There are two types of the renal cell culture models, namely: cell lines and primary cell cultures. When compared with the primary cells, cell lines are easier to culture, possess unlimited life span, isolation procedure is not required, and they have better stability. Limited access to primary cells, particularly those from a human source, is also a major hindrance to their utilization. Cell lines are, however, more distant from *in vivo* situation than primary cells. Several renal cell lines reproducing different renal cell types are available. The major ones include:

- *Cell lines with characteristics of the proximal tubule such as normal rat kidney cells (NRK52E);*
- *Lewis lung carcinoma porcine kidney (LLC-PK1);*
- *Opossum kidney or human kidney-2 (HK-2 cells), generated by transduction of human primary PTCs with human papilloma virus;*
- *Cell lines with characteristics of the distal tubule or the collecting duct such as Madin Darby canine kidney cells or Madin Darby bovine kidney cells.*

Renal cell culture models possess the advantage of providing better means for investigating and understanding biochemical and cellular mechanisms of cytotoxicity, cheaper and less time consuming than animal experimentation. However, the nephrotoxic potency of the xenobiotic being investigated will depend on the nephron segments as well as the cell type used for study (epithelial cells from the proximal tubule are usual targets of toxic injury). Another major disadvantage associated with the renal cell culture models is inability to allow for studies related to pharmacokinetics [125].

12.4.4 Renal Slice Model

This method involves preparation of renal cortical slices from normal rat kidneys and incubation with test compound or vehicle. At the end of incubation, tissue capacity for pyruvate-stimulated gluconeogenesis and lactate dehydrogenase release are estimated as measures of cellular function and cytotoxicity [132]. This model is not suitable for chronic renal toxicity assessment and selective injury also has been observed following *in vitro* nephrotoxicant exposure to precision-cut renal slices [133,134]. It, however, has the advantage of providing interaction between various cells within the nephron and has been shown to be useful for biotransformation and toxicity studies [134,135].

12.4.5 "Omics" Methods

Advances in transcriptomics, proteomics, and metabonomics, otherwise referred to as "omics" technologies, are believed to have the potential for the development of molecular markers that may allow for detection of early changes in signal transduction, regulation, and biochemistry with high sensitivity and specificity [125].

12.5 Nephrotoxic Medicinal Plants of Africa

Renal failure is reported to be a frequent cause of morbidity and mortality in the developing world and the use of nephrotoxic herbal/traditional remedies remains a major cause [136−140]. While many of these herbal remedies have been the focus of investigation in recent times, the pathogenic mechanisms of renal injury, as well as the identities of most of the nephrotoxic compounds, are still not fully known. Although nephrotoxins are encountered in both common edible plants and medicinal herbs worldwide, the prevalence of nephropathy associated with the use of herbal remedies in many developing countries is also directly related to the lack of medical facilities, poverty, ignorance, lax legislation, and the belief in indigenous systems of medicine [23,83,141−143]. Discussed below and summarized in Table 12.2 is a list of medicinal plants/herbal remedies that have been documented as causing nephrotoxicity.

12.5.1 Alstonia boonei *(Apocynaceae)*

Alstonia boonei, a large evergreen tree, is distributed throughout the tropics and rain forests of West and Central Africa [163−165]. The stem bark is widely used

Table 12.2 African Medicinal Plants Screened for their Nephrotoxicity

Plant (Family)	Area of Plant Collection	Traditional Use	Chemical Constituents	Pharmacological Activities
Alstonia boonei (Apocynaceae)	Nigeria	Treatment of febrile illness, jaundice, painful micturition, rheumatic conditions, as an antivenom against snake bite and as an antidote against arrow poisoning [144,145]	Aqueous stem bark extract used though known to contain saponins, alkaloids, tannins, and cardiac glycosides [146]	Nephrotoxic, testicular toxicity [26,32]
Callilepis laureola (Compositae)	South Africa	Treatment of stomach problems, tape worm infestations, impotence, cough, and induction of fertility [147]	Aqueous root extract used; known to contain atractyloside, carboxyatractyloside [148–150]	Nephrotoxic, cytotoxic, hepatotoxic [148–150]
Nauclea latifolia (Rubiaceae)	Nigeria	Treatment of malaria, diabetes, high blood pressure [151,152]	Cardiac glycosides, flavonoids, saponins, terpenoids, tannin [153]	Not nephrotoxic [153]
Dioscorea quartiniana (Dioscoreaceae)	South Africa	For hunting and criminal/ritual poisoning, consumed as food [23,154]	Dioscorine, dioscine [23,154,155]	Nephrotoxic, hepatotoxic, central nervous system irritability, analeptic, antidiuretic [23,154]
Larrea tridentata (Zygophyllaceae)	South Africa	Nutritional supplement, weight-loss medication, anticancer agent, general cleansing tonic and as a remedy for arthritis [156]	Nordihydroguaiaretic acid [157]	Oxidant, nephrotoxic [158]
Catha edulis (Celastraceae)	South Africa	Leaves chewed or consumed as tea to achieve a state of euphoria and stimulation [159]	S-cathinone [160]	Nephrotoxic, hepatotoxic [161]
Aloe capensis (Liliaceae)	Zimbabwe	Laxative [154,162]	Aloin, aloinosides [23,154]	Nephrotoxic, gastrointestinal inflammation, and bleeding [136,162]

for various medicinal purposes [144,145] and has been listed in the African Pharmacopoeia as an antimalarial drug [164]. It is one of the 10 most commonly used folkloric medicinal plants in southwest Nigeria [166]. In spite of the numerous benefits, chronic consumption of *Alstonia boonei* has been shown to be capable of causing nephrotoxicity [26,32].

12.5.2 Pithecellobium lobatum *(Fabaceae)*

Djenkol bean is a pungent-smelling edible fruit of the hardwood tree, *Pithecellobium lobatum*. It is recognized as a major cause of kidney injury and ARF in the tropics [167]. It contains djenkolic acid, a sulfur-rich cysteine thioacetal of formaldehyde, which produces severe tubular necrosis with a lesser degree of glomerular cell necrosis in animals. Djenkolic acid also forms needle-like crystals, especially in concentrated acidic urine in distal tubules, leading to kidney stones and obstruction [168−170]. Djenkol bean poisoning also has been implicated in unexplained hematuria in children [171]. Similarly, a case report of djenkolism has also been reported in an adult in which renal biopsy revealed morphological change similar to those observed in animal studies [167].

12.5.3 Callilepis laureola *(Impila) (Asteraceae)*

Impila is a herb derived from the tuberous roots of *Callilepis laureola*. It is used as a traditional remedy in South Africa to treat a number of conditions some of which include inducing fertility, correcting impotence, and also as a vermicide and decongestant, among others [147]. Reports of fatal intoxication resulting from the use of this plant are well documented in literature [172−175]. It possesses marked renal toxicity and it is regarded as one of the most common causes of ARF in black population of South Africa [176]. Histopathological changes associated with impila-mediated nephrotoxic ARF include necrosis of the proximal convoluted tubule and loop of Henle and this toxic effect is believed to be mediated by atractyloside or carboxyatractyloside [148]. The mechanism of toxicity involves inhibition of ADP transport across the mitochondrial membrane and ATP synthesis, leading ultimately to cell death from energy starvation [177,178].

12.5.4 Dioscorea quartiniana *(Dioscoreaceae)*

Dioscorea species, commonly known as yam, are tuberous plants which constitute a major staple food in most parts of Africa. Many species of this plant have been consumed for years without known reports of toxicity. A few species, however, have been identified as possessing the risk of causing kidney injury, and poisoning usually results from ingestion of improperly prepared tubers, especially in cases of famine or from occasional medical use. Detoxification of the tubers involves placing them in running water for a few days, soaking in salt water, boiling for several hours and then squeezing out the juice, or roasting to detoxify them [23]. There are reports of the use of the non-detoxified plant in suicide, hunting, and for criminal and ritual poisoning [154]. Renal failure as well as CNS irritability and hepatic failure with hypoglycemia are the predominant symptoms of poisoning with *Dioscorea* species [23].

12.5.5   Larrea tridentate *(Zygophyllaceae)*

Larrea tridentate (Chaparral) enjoys use as a nutritional supplement, weight-loss medication, anticancer agent, general cleansing tonic, and as a remedy for arthritis in folk medicine [156]. It has long been associated with cystic renal disease in both rats and humans [158,179,180]. The active constituent of chaparral has been identified as nordihydroguaiaretic acid [157] with *O*-quinone as its major metabolite. *O*-quinone is believed to increase fragility of lysosomal membranes by inducing lipid peroxidation, causing autolysis, desquamation of necrotic proximal tubular epithelial cells, and accumulation of cellular debris leading to tubular obstruction [158].

12.5.6   Catha edulis *(Khat) (Celastraceae)*

Khat is very popular among the local populations of East Africa and the Arabian Peninsula where habitual chewing of the leaf is common [159]. An experimental study in rabbits demonstrated the nephrotoxic as well as hepatotoxic potential of this plant [161]. Acute tubular nephrosis and dose-related lesions with fat droplets in the upper cortical tubules were observed on histologic evaluation of the kidney in this study [161]. S-cathinone is recognized as the major active constituent of fresh khat leaf [160] and the main metabolites are norephedrine and D-norpseudoephedrine [181].

12.5.7   Aloe *Species (Aloaceae)*

Aloe capensis (Cape aloe) is a plant known to contain aloin and aloinosides and have been reported to cause parenchymatous nephritis. The estimated lethal dose for adults is between 8 and 20 g of plant material. Gold [136] reported that most cases of ARF in South Africa were partly related to herbal remedies containing aloe extract or aloin. Neuwinger [154] also reported cases of fatal poisoning from the use of aloe juice and *Aloe chabaudii*, *Aloe excelsa*, *Aloe greatheadii*, and *Aloe globuligemma* were the species implicated. Similarly, Luyckx et al. [162] reported a case of ARF following ingestion of a herbal remedy containing Cape aloes for treatment of constipation.

12.6   Nephroprotective Medicinal Plants of Africa

Medicinal plants continue to serve as important source of drugs and results from several studies show that they may serve as a vital source of potentially useful new compounds for the development of effective therapy to combat xenobiotic-induced nephrotoxicity and related conditions [182]. While the nephroprotective potential of many medicinal plants has been reported in several experimental and clinical studies, the acclaimed efficacy of several others is yet to be validated scientifically. Some African medicinal plants that have been investigated and documented as possessing nephroprotective properties are discussed in this section and summarized in Table 12.3.

Pharmacological properties exhibited by medicinal plants and reported to be responsible for the observed nephroprotection in several studies include anti-inflammatory, immunomodulatory, antioxidant, and diuretic mechanisms; reduction in proteinuria, renal interstitial fibrosis, renal ischemia/reperfusion injury, tubular and mesangial cell proliferation, blood lipid levels, blood pressure, apoptosis, renal necrosis, and calcium oxalate crystal aggregation; and stimulation of renal repair mechanisms, RNA and protein synthesis [88,234,235]. In spite of the advances in medicinal plant research and efforts in understanding their mode of action, continuous efforts are still required to identify and develop traditionally used medicinal plants in renal diseases so that more effective treatments could be made available from plants whose efficacies have been known for several hundreds of years [235].

12.6.1 Glycine max *(Soybean, Family: Fabaceae)*

Nutritional intervention studies have shown that consumption of soy-based protein is capable of reducing proteinuria and attenuating renal functional or structural damage in animals and humans with various forms of chronic renal disease [236]. In two separate studies, the effectiveness of the phenolic extract of soybean (PESB) in attenuating nephrotoxic ARF associated with gentamicin and cisplatin treatments respectively was demonstrated in rats [24,31]. PESB (obtained from soybean grown in southwestern Nigeria) significantly improved renal function and attenuated oxidative stress and tubular necrosis characterizing gentamicin and cisplatin-mediated nephropathy in both studies. Inflammatory mechanism involved in cisplatin nephrotoxicity and ARF was also significantly attenuated as evident by the marked reduction in myeloperoxidase activity and nitric oxide generation [31]. It was therefore concluded that the nephroprotection offered by the soybean extract was largely due to antioxidant and anti-inflammatory mechanisms and that these actions were related to the polyphenolic content of the plant.

12.6.2 Curcumin

Farombi and Ekor [202] also provided evidence to show that co-administration of curcumin along with gentamicin was capable of attenuating lipid peroxidative damage and restoring antioxidant status, markers of renal injury and urinary excretory indices in gentamicin nephrotoxic rats. Curcumin, a naturally occurring polyphenol and a major yellow pigment in turmeric ground rhizome of *Curcuma longa* Linn. (Zingiberaceae), is used widely as a spice and coloring agent in several foods, cosmetics and drugs [198,199]. Its anti-inflammatory and antioxidant properties are well established and it is also known to be a potent inhibitor of reactive oxygen species formation [203,204]. These attributes of curcumin might account for its efficacy in attenuating renal oxidative damage mediated by gentamicin and its

Table 12.3 African Medicinal Plants Screened for Their Nephroprotective Effects

Plant (Family)	Area of Plant Collection	Traditional Use	Chemical Constituents	Pharmacological Activities
Zingiber officinale (Zingiberaceae)	Egypt	Spice/food preservative, pain relief, promote digestion [183–185]	Volatile oil, gingerol, shogaol, resins, starch, fibers, capsaicin, paradol [30,186–188]	Anti-inflammatory, antitumor, antipyretic, gastroprotective, cardiotonic, antioxidant activities [30,185,189,190]
Moringa oleifera (Moringaceae)	Nigeria	Antispasmodic, stimulant, expectorant, diuretic, emmenagogue, abortifacient, antifungal, antibacterial [191]	Quercetin glucosides, rutin, kaempferol glycosides, and chlorogenic acids [192,193]	Antioxidant, antimicrobial, anti-inflammatory, anticancer, antifertility, antiulcer, antidiabetic, diuretic, hepatoprotective, nephroprotective [194–197]
Curcuma longa (Zingiberaceae)	Nigeria	Spice, food coloring, relieve indigestion, wound healing, liver problems, etc. [198,199]	Curcumin, termeric oil, terpenoids, curcumen (terpene), starch, albumnoids [200]	Anti-inflammatory, antioxidant, anticancer, antidiabetic, hepatoprotective, antimicrobial, antimutagenic, antistress [201–204]
Glycine max (Fabaceae)	Nigeria	Seed as source of milk and vegetable oil, animal fodder, and pasture [205,206]	Phenolic extract, isoflavones [24,31,207]	Antioxidant, anti-inflammatory, anticancer, antihypercholesterolemic [24,31,208,209]
Phyllanthus amarus (Euphorbiaceae)	Nigeria	Treatment of stomach disorders, skin diseases, cold, hepatic and urolithic diseases, diabetes, high blood pressure, pain, inflammatory conditions, and infections [210–213]	Lignin, cardiac glycosides, tannins, and saponins [214,215]	Free radical scavenging and antioxidant, antidiarrheal, antimutagenic, anticarcinogenic, anti-inflammatory [28,214,216–218]
Euclea divinorum Heirns (Ebenaceae)	Ethiopia	Treatment of scabies, inflammation of the skin, eczema, abdominal pain, gonorrhea, constipation, kidney problems [219,220]	Triterpenes, naphthoquinones [221]	Free radical scavenging and antioxidant, anticancer, nephroprotective, antimicrobial [33,222]

Species (Family)	Country	Uses	Constituents	Activities
Foeniculum vulgare (Apiaceae)	Egypt	For savory formulations, sauces, liqueurs, confectionery, consumed in salad or cooked as kitchen vegetable, natural source of antioxidants [223–225]	Estragole, *trans*-anethole limonene, fenchone, rotundifolone [226]	Anti-inflammatory [225,227]
Hyphaene thebaica (Palmae)	Egypt	For treatment of high blood pressure, serve as febrifuge and parasite expellant, also consumed as food [228,229]	Saponins, coumarins, hydroxycinnamates, essential oils, flavonoids, estrone [230,231]	Anti-inflammatory, antioxidant, antifungal, antihypertensive [227,231,232]
Phyllanthus amarus (Euphorbiaceae)	Nigeria	Treatment of stomach disorders, skin diseases, cold, hepatic and urolithic diseases, diabetes, high blood pressure, pain, inflammatory conditions, and infections [210–213]	Lignin, cardiac glycosides, tannins, and saponins [214,215]	Free radical scavenging and antioxidant, antidiarrheal, antimutagenic, anticarcinogenic, anti-inflammatory [28,214,216,218,233]

promise as a candidate for treatment of drug-induced acute renal oxidative damage and failure as reported by Farombi and Ekor [202].

12.6.3   Zingiber officinale *(Zingiberaceae)*

Recognition of the medicinal value of ginger (*Zingiber officinale* Roscoe) is growing in Africa. The crude plant material and as well as isolated constituents such as (81)-gingerol, (81)-paradol, phenolic 1,3-diketones, and zingerone have been reported to possess anti-inflammatory, antitumor, antipyretic, gastroprotective, cardiotonic, and antioxidant activities [185,189,190]. Ginger has been used extensively in treating several common ailments in folklore medicine [183] and scientific evidences are emerging in support of its beneficial properties in ameliorating different disorders [184,185]. Histological, ultrastructural, and immunohistochemical studies by Ali et al. [30] demonstrate the nephroprotective effect of ginger extract (prepared from the rhizome) in a rat model of cisplatin-induced AKI. Results from both biochemical and histological evaluation revealed significant improvement in the acute renal damage induced by cisplatin which was evident from the marked decrease in BUN and creatinine concentrations, as well as significant recovery from the glomerular and tubular injuries. The nephroprotective property of this plant extract was attributed to the individual or synergistic action of its antioxidant constituents [25,27,29].

12.6.4   Phyllanthus amarus *(Euphorbiaceae)*

Phyllanthus amarus Schum and Thonn. is another medicinal plant known for its ethnobotanical value in Africa. Different parts of the plant are employed in traditional management of diverse veterinary and human diseases. Among other uses in African traditional medicine, the plant is highly valued for its hepatoprotective, antidiabetic, antihypertensive, analgesic, anti-inflammatory, and antimicrobial properties [210]. Its effectiveness in protecting against nephrotoxicity induced by gentamicin and acetaminophen was evaluated by Adeneye and Benebo [28]. Data from the study was in support of this plant attenuating renal dysfunction and tubulonephrosis induced by these two therapeutic agents, thus validating its folkloric use in traditional African medicine. The nephroprotective action also was reported to be due to the inherent antioxidant and free radical scavenging constituent of the plant extract.

12.6.5   Euclea divinorum *(Ebenaceae)*

It is an evergreen, dioecious shrub that is widely distributed in Ethiopia, Sudan, and South Africa [237]. It is used in Ethiopian folklore medicine to treat scabies, inflammation of the skin, eczema, abdominal pain, gonorrhea, and kidney problems [220,237]. Recently, Feyissa et al. [33] assessed the protective effects of the crude leaf extract and solvent fractions of the plant against gentamicin-mediated nephrotoxicity in rats. According to this study, pre- and cotreatment with both crude leaf extract and fractions of *Euclea divinorum* reversed biochemical and morphological damage associated with gentamicin-induced nephrotoxicity. This was revealed by the observed decrease in tubular necrosis, serum, and oxidant markers as well as by an increase in antioxidant molecules.

12.7 Conclusion

The kidney remains a major organ for xenobiotic metabolism and excretion. This important physiological function exposes it to large amount of toxicants and predisposes it to nephrotoxicity, leading to acute or chronic renal injury. The incidence of xenobiotic-induced nephrotoxicity and ARF is increasingly becoming apparent, constituting a major health challenge in all subregions of Africa. Plants and plant-derived medicines are well recognized as an important source of potentially nephrotoxic substances and documented evidence from various studies continues to demonstrate the contribution and major etiologic role African medicinal plants play in the pathogenesis of kidney injury. There are also several other African medicinal plants showing promising potential in the treatment of toxicant-induced renal injury. Deliberate efforts are required toward continuous documentation of the rich African flora in order to clearly delineate between the nephrotoxic plants and those with protective and curative potential against nephrotoxic injury. Also, there is the need to standardize the methods of assessment of nephrotoxicity and nephroprotection and improve on the *in vivo* animal model commonly employed across the continent. This should include measurement of novel and promising biomarkers with superior sensitivity and specificity over plasma creatinine and BUN for assessing acute or chronic tubulotoxicity alongside detailed histopathological analyses.

References

[1] Ferguson MA, Vaidya VS, Bonventre JV. Biomarkers of nephrotoxic acute kidney injury. Toxicology 2008;245:182–93.

[2] Perazella MA. Renal vulnerability to drug toxicity. Clin J Am Soc Nephrol 2009;4:1275–83.

[3] Almeida AF. Drug-induced renal disease—prevention strategies. Med Update 2005;637–41.

[4] Karie S, Launay-Vacher V, Deray G, Isnard-Bagnis C. Drugs renal toxicity. Nephrol Ther 2010;6(1):58–74.

[5] World Health Organization. Principle and methods for the assessment of nephrotoxicity associated with exposure to chemicals. Environmental Health Criteria No. 119 (EUR 13222 EN), 1991. Available from: http://www.inchem.org/documents/ehc/ehc/ehc119.htm.

[6] Nuyts GD, Elseviers MM, De Broe ME. Health impact of renal disease due to nephrotoxicity. Toxicol Lett 1989;46:31–44.

[7] NICE National Institute for Helath and Care Excellence. Acute kidney injury: prevention, detection and management of acute kidney injury up to the point of renal replacement therapy. NICE Clin Guideline 2013;49–55:39–41.

[8] Evenepoel P. Acute toxic renal failure. Best Pract Res Clin Anaesthesiol 2004;18:37−52.

[9] Schetz M, Dasta J, Goldstein S, Golper T. Drug-induced acute kidney injury. Curr Opin Crit Care 2005;11:555−65.

[10] Choudhury D, Ahmed Z. Drug-associated renal dysfunction and injury. Nat Clin Pract Nephrol 2006;2:80−91.

[11] Kim SY, Moon A. Drug-induced nephrotoxicity and its biomarkers. Biomol Ther 2012;20(3):268−72.

[12] Liano F, Junco E, Pascual J, Madero R, Verde E. The spectrum of acute renal failure in the intensive care unit compared with that seen in other settings. The Madrid acute renal failure study group. Kidney Int 1998;66:16−24.

[13] Silvester W, Bellomo R, Cole L. Epidemiology, management, and outcome of severe acute renal failure of critical illness in Australia. Crit Care Med 2001;29:1910−5.

[14] Mehta RL, Pascual MT, Soroko S, Savage BR, Himmelfarb J, Ikizler TA, et al. Spectrum of acute renal failure in the intensive care unit: the PICARD experience. Kidney Int 2004;66:1613−21.

[15] Henrich WL. Nephrotoxicity of several newer agents. Kidney Int 2005;107−9.

[16] Singh NP, Ganguli A, Prakash A. Drug-induced kidney diseases. J Assoc Physicians India 2003;51:970−9.

[17] Schrier RW. Cancer therapy and renal injury. J Clin Invest 2002;100:743−5.

[18] Rabik CA, Dolan ME. Molecular mechanisms and toxicity associated with platinating agents. Cancer Treat Rev 2007;33(1):9−23.

[19] Naughton CA. Drug-induced nephrotoxicity. Am Fam Phys 2008;78:743−50.

[20] Kirtane AJ, Nagai J, Takano M. Molecular-targeted approaches to reduce renal accumulation of nephrotoxic drugs. Expert Opin Drug Metab Toxicol 2010;6:1125−38.

[21] Naicker S, Aboud O, Gharbi MB. Epidemiology of acute kidney injury in Africa. Semin Nephrol 2008;28:348−53.

[22] Barsoum RS. Tropical acute renal failure. Contrib Nephrol 2004;144:44−52.

[23] Steenkamp V, Stewart MJ. Nephrotoxicity associated with exposure to plant toxins, with particular reference to Africa. Ther Drug Monit 2005;27:270−7.

[24] Ekor M, Farombi EO, Emerole GO. Modulation of gentamicin-induced renal dysfunction and injury by the phenolic extract of soybean (Glycine max). Fundam Clin Pharmacol 2006;20:263−71.

[25] Ajith TA, Nivitha V, Usha S. Zingiber officinale Roscoe alone and in combination with a-tocopherol protect the kidney against cisplatin-induced acute renal failure. Food Chem Toxicol 2007;45:921−7.

[26] Oze GO, Nwanjo HU, Onyeze GO. Nephrotoxicity caused by the extract of A. boonei (De Wild) stem bark in guinea pigs. Int J Nutr Wellness 2007;3:2.

[27] Taghizadeh AA, Shirpor A, Farshid A, et al. The effect of ginger on diabetic nephropathy, plasma antioxidant capacity and lipid peroxidation in rat. Food Chem 2007;101:148−53.

[28] Adeneye AA, Benebo AS. Protective effect of the aqueous leaf and seed extract of Phyllanthus amarus on gentamicin and acetaminophen-induced nephrotoxic rats. J Ethnopharmacol 2008;118:318−23.

[29] Ajith TA, Aswathy MS, Hema U. Protective effect of Zingiber officinale Roscoe against anticancer drug doxorubicin-induced acute nephrotoxicity. Food Chem Toxicol 2008;46:3178−81.

[30] Ali DA, Abdeen AM, Ismail MF, Mostafa MA. Histological, ultrastructural and immunohistochemical studies on the protective effect of ginger extract against

cisplatin-induced nephrotoxicity in male rats Toxicol Ind Health 2013; Apr 3. [Epub ahead of print]. Available from: http://dx.doi.org/doi:10.1177/0748233713483198.

[31] Ekor M, Emerole GO, Farombi EO. Phenolic extract of soybean (*Glycine max*) attenuates cisplatin-induced nephrotoxicity in rats. Food Chem Toxicol 2010;48 (4):1005−12.

[32] Awodele O, Osunkalu VO, Akinde OR, Teixeira da Silva JA, Okunowo WO, Odogwu EC, et al. Modulatory roles of antioxidants against the aqueous stem bark extract of *Alstonia boonei* (Apocynaceae)-induced nephrotoxicity and testicular damage. Intl J Biomed Pharm Sci 2010;4(2):76−80.

[33] Feyissa T, Asres K, Engidawork E. Renoprotective effects oft the crude extract and solvent fractions of the leaves of *Euclea divinorum* Hierns against gentamicin-induced nephrotoxicity in rats. J Ethnopharmacol 2013;145:758−66.

[34] Ali BH, Al Moundhri MS. Agents ameliorating or augmenting the nephrotoxicity of cisplatin and other platinum compounds: a review of some recent research. Food Chem Toxicol 2006;44(8):1173−83.

[35] Rached E, Hoffmann D, Blumbach K, Weber K, Dekant W, Mally A. Evaluation of putative biomarkers of nephrotoxicity after exposure to ochratoxin a *in vivo* and *in vitro*. Toxicol Sci 2008;103:371−81.

[36] SEER Training Modules. Kidney and ureter cancer. Retrieved 8th January, 2014, from <http://training.seer.cancer.gov/kidney/anatomy/>.

[37] Schmidt R, Lang F, Thews G. Physiologie des Menschen. 29th ed. Heidelberg: Springer-Verlag GmbH; 2004.

[38] Kriz W, Bankir L. A standard nomenclature for structures of the kidney. Am J Physiol 1988;254:F1−8.

[39] Schrier RW, Gottschalk CW. Diseases of the kidney. Boston, MA: Little, Brown and Co; 1987.

[40] Moffat DB. New ideas on the anatomy of the kidney. J Clin Pathol 1981;34:1197−206.

[41] Moffat DB. Morphology of the kidney in relation to nephrotoxicity—portae renales. In: Bach PH, Bonner FW, Lock EA, editors. Nephrotoxicity: assessment and pathogenesis. Chichester: John Wiley; 1982.

[42] Maunsbach AN, Olsen TS, Christensen EI. Functional ultrastructure of the kidney. New York, London: Academic Press; 1980.

[43] Greaves P. Histopathology of preclinical toxicity studies. Interpretation and relevance in drug safety evaluation. London: Elsevier; 2011.

[44] Klinke R, Silbernagel S. Lehrbuch der Physiologie. 4th ed. Stuttgart, New York, NY: Georg Thieme Verlag; 2004.

[45] Boron WF, Baoulpaep EL. Medical physiology: a cellular and molecular approach. Philadelphia, PA: Saunders, Elsevier; 2008.

[46] Wang T. Flow-activated transport events along the nephron. Curr Opin Nephrol Hypertens 2006;15:530−6.

[47] Eaton DC, Pooler J. *Vander's renal physiology*. 6th ed. New York, NY: McGraw-Hill Medical; 2004.

[48] Lytle C, Xu JC, Biemesderfer D, Forbush B. Distribution and diversity of Na−K−Cl cotransport proteins: a study with monoclonal antibodies. Am J Physiol 1995;3(269): C1496−505.

[49] Sands JM, Verlander JW. Anatomy and physiology of the kidneys. In toxicology of the kidney. 3rd ed. Boca Raton, FL: CRC Press; 2005.

[50] Lock E, Reed CJ. Xenobiotic metabolizing enzymes of the kidney. Toxicol Pathol 1998;26:18−25.

[51] Lash LH. Role of renal metabolism in risk to toxic chemicals. Environ Health Persp 1994;102(11):75−9.

[52] Lohr JW, Willsky GR, Acara MA. Renal drug metabolism. Pharmacol Rev 1998;50:107−41.

[53] Anders MW. Metabolism of drugs by the kidney. Kidney Int 1980;18:636−47.

[54] Bock KW, Lipp HP, Bock-Hennig BS. Induction of drug-metabolizing enzymes by xenobiotics. Xenobiotica 1990;20:1101−11.

[55] Pichette V, du Souich P. Role of the kidneys in the metabolism of furosemide: its inhibition by probenecid. J Am Soc Nephrol 1996;7:345−9.

[56] Poon K, Pang S. Benzoic acid glycine conjugation in the isolated perfused rat kidney. Drug Metab Dispos 1995;23:255−60.

[57] Goldbarg JA, Friedman OM, Pineda EP, Smith EE, Chatterji R, Stein EJ, et al. The colorimetric determination of U-glutamyl transpeptidase with a synthetic substrate. Arch Biochem Biophys 1960;91:61−70.

[58] Bowsher RR, Verburg KM, Henry DP. Rat histamine N-methyltransferase. J Biol Chem 1983;258:12215−20.

[59] Walker LA, Valtin H. Biological importance of nephron heterogeneity. Annu Rev Physiol 1982;44:203−19.

[60] Guder WG, Ross BD. Enzyme distribution along the nephron. Kidney Int 1984;26:101−11.

[61] Mohandas J, Marshall JJ, Duggin CG, Horvath JS, Tiller DJ. Differential distribution of glutathione and glutathione-related enzymes in rabbit kidney: possible implications in analgesic nephropathy. Biochem Pharmacol 1984;33:1801−7.

[62] Elseviers MM, DeBroe ME. Analgesic nephropathy: is it caused by multi-analgesic abuse or single substance use?. Drug Saf 1999;20(1):15−24.

[63] Kintzel PE. Anticancer drug-induced kidney disorders. Drug Saf 2001;24:19−38.

[64] Gambaro G, Perazella MA. Adverse renal effects of anti-inflammatory agents: evaluation of selective and nonselective cyclooxygenase inhibitors. J Intern Med 2003;253:643−52.

[65] Markowitz GS, Fine PL, Stack JI, Kunis CL, Radhaksrishnan J, Palecki W, et al. Toxic acute tubular necrosis following treatment with zoledronate (Zometa). Kidney Int 2003;64:281−9.

[66] Rougier F, Ducher M, Maurin M, Corvaisier S, Claude D, Jeliffe R, et al. Aminoglycoside dosages and nephrotoxicity. Clin Pharmacokinet 2003;42:493−500.

[67] Alexander BD, Wingard JR. Study of renal safety in amphotericin B lipid complex-treated patients. Clin Infect Dis 2005;40(6):414−21.

[68] Izzedine H, Launay-Vacher V, Deray G. Antiviral drug-induced nephrotoxicity. Am J Kidney Dis 2005;45:804−17.

[69] Lamieire NH, Flombaum CD, Moreau D, Ronco C. Acute renal failure in cancer patients. Ann Med 2005;37:13−25.

[70] Perazella MA. Drug-induced nephropathy: an update. Expert Opin Drug Saf 2005;4:689−706.

[71] Falagas ME, Kasiakou SF. Nephrotoxicity of intravenous colistin: a prospective evaluation. Crit Care 2006;10:13−21.

[72] Nand N, Aggarwal HK, Jai D, Sharma M. Indigenous drug induced nephropathy. In: Sahay BK, editor. Medicine Update 2006;16:458−62.

[73] Hinchman CA, Ballatori N. Glutathione-degrading capacities of liver and kidney in different species. Biochem Pharmacol 1990;40:1131−5.

[74] Ballatori N, Jacob R, Barrett C, Boyer JL. Biliary catabolism of glutathione and differential reabsorption of its amino acid constituents. Am J Physiol 1988;254:1−7.

[75] Lash LH, Anders MW. Uptake of nephrotoxic S-conjugates by isolated rat renal proximal tubular ceils. J Pharmacol Exp Ther 1989;248:531−7.

[76] Zhang G, Stevens JL. Transport and activation of S-(1,2-dichlorovinyl)-L-cysteine and N-acetyl-S-(1,2-dichlorovinyl)-L-cysteine in rat kidney proximal tubules. Toxicol Appl Pharmacol 1989;100:51−61.

[77] Koob M, Dekant W. Metabolism of hexafluoropropene: evidence for bioactivation by glutathione conjugate formation in the kidney. Drug Metab Dispos 1990;18:911−6.

[78] Stevens JL. Cysteine conjugate β-lyase activities in rat kidney cortex: subcellular localization and relationship to the hepatic enzyme. Biochem Biophys Res Commun 1985;129:499−504.

[79] Elfarra AA, Jakobson I, Anders MW. Mechanism of S-(1,2-dichorovinyl)glutathione-induced nephrotoxicity. Biochem Pharmacol 1986;35:283−8.

[80] Lash LH, Elfarra AA, Anders MW. Renal cysteine conjugate β-lyase: bioactivation of nephrotoxic cysteine S-conjugates in mitochondrial outer membrane. J Biol Chem 1986;261:5930−5.

[81] Stevens JL. Isolation and characterization of a rat liver enzyme with both cysteine conjugate β-lyase and kynureninase activity. J Biol Chem 1985;260:7945−50.

[82] Elfarra AA, Lash LH, Anders MW. x-Ketoacids stimulate rat renal cysteine conjugate-lyase activity and potentiate the cytotoxicity of S-(1,2-dichlorovinyl)-L-cysteine. Mol Pharmacol 1987;31:208−12.

[83] Ekor M. The growing use of herbal medicines: issues relating to adverse reactions and challenges in monitoring safety. Front Pharmacol 2014;4(177). Available from: http://dx.doi.org/doi:10.3389/fphar.2013.00177.

[84] Huxtable RJ. The harmful potential of herbal and other plant products. Drug Saf 1990;5:126−36.

[85] Abbot NC, Ernst E. Patients' opinions about complimentary medicine. Forsch Komplementarmed 1997;4:164−8.

[86] Hecker E. New toxic, irritant and cocarcinogenic diterpenoids esters from Euphorbiaceae and from Thymelaeaceae. Pure Appl Chem 1977;49:1423−31.

[87] Sayed MD, Rizk A, Hammouda FM, El- Missiry MM, Williamson EM, Evans FJ. Constituents of Egyptian Euphorbiaceae. IX. Irritant and cytotoxic ingenane esters from Euphorbia paralias L. Experientia 1980;36:1206−7.

[88] Combest W, Newton M, Combest A, Kosier JH. Effects of herbal supplements on the kidney. Urol Nursing 2005;25(5):381−6.

[89] Singh NP, Prakash A. Nephrotoxic potential of herbal drugs. JIMSA 2011;24 (2):79−81.

[90] Tackholm V. Student flora of Egypt. 2nd ed. Beirut: Cairo University Press; 1974.

[91] Pliny the Elder. Naturalis historia 26:45. London, Cambridge, MA: William Heinemann, Harvard University Press; 1949-1954.

[92] Boubaker K, Ounissi M, Brahmi N, Goucha R, Hedri H, Abdellah TB, et al. Acute renal failure by ingestion of Euphorbia paralias. Saudi J Kidney Dis Transpl 2013;24:571−5.

[93] Guerra MO. Avaliação do potencial tóxico do fruto da lobeira (Solanum grandiflorum), administrado no período de organognese do rato. In: XIV Simpósio de Plantas Medicinais do Brasil, 144(F-245); 1996.

[94]   Pereira MC, Carvalho JCT, Lima LM, Caputo LRG, Ferreira LR, Fiorini JE, et al. Toxicity of a subchronic treatment with hydroalcoholic crude extract from *Solanum grandiflorum* (Ruiz et Pav) in rats. J Ethnopharmacol 2003;89:97−9.

[95]   Harborne JB, Baxter H. Phytochemical dictionary: a handbook of bioactive compounds from plants. London: Taylor & Francis Ltd; 1995.

[96]   Cosyns JP, Jadoul M, Squifflet JP, Wese FX, van Ypersele de Strihou C. Urothelial lesions in Chinese herb nephropathy. Am J Kidney Dis 1999;33:1011−7.

[97]   Tanaka A, Shinkai S, Kasuno K, Maeda K, Murata M, Seta K, et al. Chinese herbs nephropathy in the Kansai area: a warning report. Nihon Jinzo Gakkai Shi 1997;39:438−40.

[98]   Stengel B, Jones E. End-stage renal insufficiency associated with Chinese herbal consumption in France. Nephrologie 1998;19:15−20.

[99]   Lord GM, Tagore R, Cook T, Gower P, Pusey CD. Nephropathy caused by Chinese herbs in the UK. Lancet 1999;354:481−2.

[100] Chau W, Ross R, Li JY, Yong TY, Klebe S, Barbara JA. Nephropathy associated with use of a Chinese herbal product containing aristolochic acid. Med J Aust 2011;194:367−8.

[101] Yang SS, Chu P, Lin YF, Chen A, Lin SH. Aristolochic acid-induced Fanconi's syndrome and nephropathy presenting as hypokalemic paralysis. Am J Kidney Dis 2002;39:E14.

[102] Hong YT, Fu LS, Chung LH, Hung SC, Huang YT, Chi CS. Fanconi's syndrome, interstitial fibrosis and renal failure by aristolochic acid in Chinese herbs. Pediatr Nephrol 2006;21:577−9.

[103] Zhou S, Koh HL, Gao Y, Gong ZY, Lee EJ. Herbal bioactivation: the good, the bad and the ugly. Life Sci 2004;74:935−68.

[104] Vanherweghem JL, Depierreux M, Tielemans C, Abramowicz D, Dratwa M, Jadoul M, et al. Rapidly progressive interstitial renal fibrosis in young women: association with slimming regimen including Chinese herbs. Lancet 1993;341(8842):387−91.

[105] Chen CL, Fang HC, Chou KJ, Wang JS, Chung HM. Acute oxalate nephropathy after ingestion of star fruit. Am J Kidney Dis 2001;37(2):418−22.

[106] Lim BO, Yu BP, Oh JH, Park DK. The inhibitory effects of ginsenoside and quercetin on oxidative damage by puromycin amino nucleoside in rats. Phytother Res 1998;12 (5):375−7.

[107] Chen CP, Yokoazwa T, Kitani K. Beneficial effects of *Sanguisorbae radix* in renal dysfunction caused by endotoxin *in vivo*. Biol Pharm Bull 1999;22(12):1327−30.

[108] Maldonado PD, Barrera D, Medina-Campos ON, Hernández-Pando R, Ibarra-Rubio ME, Pedraza-Chaverrí J. Aged garlic extract attenuates gentamicin induced renal damage and oxidative stress in rats. Life Sci 2003;20(73):2543−56.

[109] Ali BH. The effect of *Nigella sativa* oil on gentamicin nephrotoxicity in rats. Am J Chin Med 2004;32(1):49−55.

[110] Pullaiah T. Encyclopaedia of world medicinal plants. New Delhi, India: Regency Publications; 2006. p. 1797.

[111] Thounaojam MC, Jadeja RN, Devkar RV, Ramachandran AV. *Sida rhomboidea.* Roxb leaf extract ameliorates gentamicin induced nephrotoxicity and renal dysfunction in rats. J Ethnopharmacol 2010;132:365−7.

[112] Thounaojam MC, Jadeja RN, Devkar RV, Ramachandran AV. Antioxidant and free radical scavenging activity of *Sida rhomboidea.* Roxb methanolic extract determined using different *in vitro* models. Boletín Latinoamericano y del Caribe de Plantas Medicinales y Aromáticas 2010;9:191−8.

[113] Khan RA, Khan MR, Sahreen S. Evaluation of *Launaea procumbens* use in renal disorders: a rat model. J Ethnopharmacol 2010;128:452−61.

[114] Khan MR, Rizvi W, Khan GN, Khan RA, Shaheen S. Carbon tetrachloride-induced nephrotoxicity in rats: protective role of *Digera muricata*. J Ethnopharmacol 2009;122:91−9.

[115] Hou SX, Zhu WJ, Pang MQ, Jeffry J, Zhou LL. Protective effect of iridoid glycosides from *Paederia scandens* (LOUR.) MERRILL (Rubiaceae) on uric acid nephropathy rats induced by yeast and potassium oxonate. Food Chem Toxicol 2014;64:57−64.

[116] De Groot H, Raven U. Tissue injury by reactive oxygen species and the protective effects of flavonoids. Fundam Clin Pharmacol 1998;12:249−55.

[117] Kren V, Walterova D. Silybin and silymarin—new effects and applications. Biomed Papers 2005;149:29−41.

[118] El-Shitanya NA, El-Haggarb S, El-Desoky K. Silymarin prevents adriamycin-induced cardiotoxicity and nephrotoxicity in rats. Food Chem Toxicol 2008;46:2422−8.

[119] Bektur NE, Sahin E, Baycu C, Unver G. Protective effects of silymarin against acetaminophen-induced hepatotoxicity and nephrotoxicity in mice. Toxicol Ind Health 2013. Available from: http://dx.doi.org/doi:10.1177/0748233713502841.

[120] Sharma V, Paliwal R, Janmeda P, Sharma S. Renoprotective effects of *Moringa oleifera* pods in 7,12-dimethylbenz [a] anthracene-exposed mice. Zhong Xi Yi Jie He Xue Bao 2012;10(10):1171−8.

[121] Shin BC, Kwon YE, Chung JH, Kim HL. The antiproteinuric effects of green tea extract on acute cyclosporine-induced nephrotoxicity in rats. Transplant Proc 2012;44 (4):1080−2.

[122] Rehman H, Krishnasamy Y, Haque K, Thurman RG, Lemasters JJ, Schnellmann RG, et al. Green tea polyphenols stimulate mitochondrial biogenesis and improve renal function after chronic cyclosporin a treatment in rats. PLoS One 2013;8(6):e65029.

[123] Ouédraogo M, Lamien-Sanou A, Ramdé N, Ouédraogo AS, Ouédraogo M, Zongo SP, et al. Protective effect of *Moringa oleifera* leaves against gentamicin-induced nephrotoxicity in rabbits. Exp Toxicol Pathol 2013;65(3):335−9.

[124] Swanepoel CR, Blockman M, Talmud J. Nephrotoxins in Africa. In: de Broe ME, Porter GA, editors. Clinical Nephrotoxins: Renal injury from Drugs and Chemicals. 3rd Edition. USA: Springer; 2008. p. 859−70.

[125] Ouedraogo M, Baudoux T, Stévigny C, Nortier J, Colet JM, Efferth T, et al. Review of current and "omics" methods for assessing the toxicity (genotoxicity, teratogenicity and nephrotoxicity) of herbal medicines and mushrooms. J Ethnopharmacol 2012;140:492−512.

[126] Van den Bulck K, Hill A, Mesens N, Diekman H, De Schaepdrijver L, Lammens L. Zebrafish developmental toxicity assay: a fishy solution to reproductive toxicity screening, or just a red herring? Reprod Toxicol 2011;32:213−9.

[127] Hoffmann D, Fuchs TC, Henzler T, Matheis KA, Herget T, Dekant W, et al. Evaluation of a urinary kidney biomarker panel in rat models of acute and subchronic nephrotoxicity. Toxicology 2010;277:49−58.

[128] Marrer E, Dieterle F. Impact of biomarker development on drug safety assessment. Toxicol App Pharmacol 2010;243:167−79.

[129] Tonomura Y, Tsuchiya N, Torii M, Uehara T. Evaluation of the usefulness of urinary biomarkers for nephrotoxicity in rats. Toxicology 2010;273:53−9.

[130] Gandolfi AJ, Brendel K. *In vitro* systems for nephrotoxicity studies. Toxicol In Vitro 1990;4:337−45.

[131] Shanley PF, Brezis M, Spokes K, Silva P, Epstein FH, Rosen S. Hypoxic injury in the proximal tubule of the isolated perfused rat kidney. Kidney Int 1986;29:1021–32.

[132] Hong SK, Anestis DK, Ball JG, Valentovic MA, Rankin GO. In vitro nephrotoxicity induced by chloronitrobenzenes in renal cortical slices from Fischer 344 rats. Toxicol Lett 2002;129:133–41.

[133] Ruegg CE. Preparation of precision-cut renal slices and renal proximal tubular fragments for evaluating segment-specific nephrotoxicity. J Pharmacol Toxicol Methods 1994;31:125–33.

[134] Gandolfi AJ, Brendel K, Fisher RL, Michaud JP. Use of tissue slices in chemical mixture toxicology and interspecies investigations. Toxicology 1995;105:285–90.

[135] Minigh JL, Valentovic MA. Characterization of myoglobin toxicity in renal cortical slices from Fischer 344 rats. Toxicology 2003;184:113–23.

[136] Gold CH. Acute renal failure from herbal and patent remedies in blacks. Clin Nephrol 1980;14:128–34.

[137] Nyazema NZ. Poisoning due to traditional remedies. Cent Afr J Med 1984;30:80–3.

[138] Ojogwu LI. Drug induced acute renal failure: a study of 35 cases. West Afr J Med 1992;11:185–9.

[139] Boonpucknavig V, Soontornniyomkij V. Pathology of renal diseases in the tropics. Semin Nephrol 2003;23:88–106.

[140] Jha V, Chugh KS. Nephropathy associated with animal, plant, and chemical toxins in the tropics. Semin Nephrol 2003;23:49–65.

[141] Zhao Z, Yuen JP, Wu J, Yu T, Huang W. A systematic study on confused species of Chinese materia medica in the Hong Kong market. Ann Acad Med Singapore 2006;35:764–9.

[142] De Smet PA. Clinical risk management of herb–drug interactions. Br J Clin Pharmacol 2007;63:258–67.

[143] Lai MN, Lai JN, Chen PC, Tseng WL, Chen YY, Hwang JS, et al. Increased risks of chronic kidney disease associated with prescribed Chinese herbal products suspected to contain aristolochic acid. Nephrology (Carlton) 2009;14:227–34.Karie S, Launay-Vacher V, Deray G, Isnard-Bagnis C. Toxicité rénale des médicaments Néphrologie Thérapeutique 2010;6:58–74.

[144] Ojewole JAO. Studies on the pharmacology of echitamine, an alkaloid from the stem bark of A. boonei L. (Apocynaceae). Int J Crude Drug Res 1984;22:121–43.

[145] Asuzu IU, Anaga AO. Pharmacological screening of the aqueous extract of A. boonei stem bark. Fitoterapia 1991;63:411–7.

[146] Fasola TR, Egunyomi A. Nigerian usage of bark in phytomedicine. Ethnobotany Res Appl 2005;3:73–7.

[147] Hutchings A, Scott AH, Lewis G, et al. Zulu medicinal plants: an inventory. Pietermaritzburg: University of Natal Press; 1996.

[148] Wainwright J, Schonland M. Toxic hepatitis in black patients in Natal. S Afr Med J 1977;51:571–3.

[149] Popat A, Shear NH, Malkiewicz I, Stewart MJ, Steenkamp V, Thomson S, et al. The toxicity of Callilepis laureola, a South African traditional herbal medicine. Clin Biochem 2001;34(3):229–36.

[150] Wainwright J, Schonland M, Candy H. Toxicity of Callilepis laureola. S Afr Med J 1977;52:313–5.

[151] Onyeyili PA, Nwosu CA, Amin JD, Jibike JI. Anti-helminthic activity of crude aqueous extract of Nauclea latifolia Stem bark against Orine nematodes. Fitoterapia 2001;72:12–21.

[152] Ajagbonna OP, Esaigun PE, Alayande MO, Akinloye OA. Anti-malaria activity and hematological effect of stem bark water extract of *Nauclea latifolia*. Biosci Res Commun 2002;14(5):481−6.
[153] Akinloye OA, Olaniyi MO. Nephrotoxicity and hepatotoxicity evaluation in Wistar Albino Rats exposed to *Nauclea latifolia* leaf extracts. Pertanika J Trop Agric Sci 2012;35(3):593−601.
[154] Neuwinger HD. African ethnobotany. Poisons and drugs. London: Chapman & Hall; 1996.
[155] Bevan CWL, Hirst J. A convulsant alkaloid of *Dioscorea dumetorum*. Chem Ind 1958;25:103.
[156] Tyler V. The honest herbal. 3rd ed. Binghamton, NY: Pharmaceutical Products Press; 1993.
[157] Walker C, Gisvold OA. Phytochemical investigation of *Larrea divaricata* Cav. J Am Pharm Assoc 1945;34:78−81.
[158] Goodman T, Grice HC, Becking GC, et al. A cystic nephropathy induced by nordihydroguaiaretic acid in the rat. Light and electron microscopic investigations. Lab Invest 1979;23:93−107.
[159] Carvalho F. The toxicological potential of khat. J Ethnopharmacol 2003;87:1−2.
[160] Al-Motarreb A, Baker K, Broadley KJ. Khat: pharmacological and medical aspects and its social use in Yemen. Phytother Res 2002;16:403−13.
[161] Al-Mamary M, Al-Habori M, Al-Aghbari AM, et al. Investigation into the toxicological effects of *Catha edulis* leaves: a short term study in animals. Phytother Res 2002;16:127−32.
[162] Luyckx VA, Ballantine R, Claeys M, et al. Herbal remedy-associated acute renal failure secondary to Cape aloes. Am J Kidney Dis 2002;39:13.
[163] Oliver-Bever B. Medicinal plants in tropical West Africa. London: Cambridge University Press; 1986.
[164] Olajide OO, Awe SO, Makinde M, Ekhelar AI, Olusola A, Morebise O, et al. Studies on the anti-inflammatory, antipyretic and analgesic properties of *A. boonei* stem bark. J Ethnopharmacol 2000;71:179−86.
[165] Abel C, Busia K. An exploratory ethnobotanical study of the practice of herbal medicine by the Akan peoples of Ghana. Altern Med Rev 2005;10:112−22.
[166] Akintonwa A, Awodele O, Afolayan G, Coker HAB. Mutagenic screening of some commonly used medicinal plants. J Ethnopharmacol 2009;125:461−70.
[167] Segasothy M, Swaminathan M, Kong NC, Bennett WM. Djenkol bean poisoning (djenkolism): an unusual cause of acute renal failure. Am J Kidney Dis 1995;25 (1):63−6.
[168] Areekul S, Kirdudom P, Chaovanapricha K. Studies on djenkol bean poisoning (djenkolism) in experimental animals. Southeast Asian J Trop Med Public Health 1976;7:551−8.
[169] H'ng PK, Nayar SK, Lau WM, Segasothy M. Acute renal failure following jering ingestion. Singapore Med J 1991;32(2):148−9.
[170] Wong JS, Ong TA, Chua HH, Tan C. Acute anuric renal failure following jering bean ingestion. Asian J Surg 2007;30(1):80−1.
[171] Vachvanichsanong P, Lebel L. Djenkol beans as a cause of hematuria in children. Nephron 1997;76:39−42.
[172] Watson AR, Coovadia HM, Bhoola KD. The clinical syndrome of Impila (*Callilepis laureola*) poisoning in children. S Afr Med J 1979;55:290−2.
[173] Seedat YK, Hitchcock P. Acute renal failure from *Callilepis laureola*. S Afr Med J 1971;45:832−3.

[174] Grobler A, Koen M, Brouwers F. Planttoksiene en lewersiektes by kinders in Pretoria en omgewing. Tijdschr Kindergeneeskd 1997;65:99–104.

[175] Bye S, Dutton M. Poisonings from the incorrect use of traditional medicaments: an introduction. In: Oliver J, editor. Proceedings of the international association of forensic toxicology. Edinburgh: Scottish Academic Press; 1992.

[176] Seedat VK. Acute renal failure among blacks and Indians in South Africa. S Afr Med J 1978;54:27–431.

[177] Heldt H, Jacobs H, Klingenberg M. Endogenous ADP of mitochondria, an early phosphate acceptor of oxidative phosphorylation as disclosed by kinetic studies with (^{14}C) labelled ADP and ATP and with atractyloside. Biochem Biophys Res Commun 1965;18:174–9.

[178] Luciani S, Carpenedo F, Tarjan E. Effects of atractyloside and carboxyatractuloside in the whole animal. In: Santi R, Luciani S, editors. Atractyloside: chemistry, biochemistry and toxicology. Padova: Piccin Medical Books; 1978. p. 109–24.

[179] Evan AP, Gardner KD. Nephron obstruction in nordihydroguaiaretic acid-induced renal cystic disease. Kidney Int 1979;15:7–19.

[180] Smith AY, Feddersen RM, Gardner KD, et al. Cystic renal cell carcinoma and acquired renal cystic disease associated with consumption of chaparral tea: a case report. J Urol 1994;152:2089–91.

[181] Brenneisen R, Geisshusler S, Schorno X. Metabolism of cathinone to (-)-norephedrine and (-)-norpseudoephedrine. J Pharm Pharmacol 1986;38:298–300.

[182] Mohana LS, Usha Kiran RT, Sandhya, Rani KS. A review on medicinal plants for nephroprotective activity. Asian J Pharm Clin Res 2012;5:8–14.

[183] Ogungbe IV, Lawal AO. The protective effect of ethanolic extracts of garlic and ascorbic acid on cadmium-induced oxidative stress. J Biol Sci 2008;8(1):181–5.

[184] Khanom F, Kayahara H, Hirota M, Tadasa K. Superoxide scavenging and tyrosinase inhibitory active compound in ginger (*Zingiber officinales* Rosc.). Pak J Biol Sci 2003;6:1996–2000.

[185] El-Sharaky AS, Newairy AA, Kamel MA, Eweda SM. Protective effect of ginger extract against bromobenzene-induced hepatotoxicity in male rats. Food Chem Toxicol 2009;47:1584–90.

[186] Evans WC. *Trease and Evans' Pharmacognosy*. 16th ed. Edinburg, UK: Saunders, Elsevier; 2009.

[187] Mustafa T, Srivastava KC, Jensen KB. Drug development report: pharmacology of ginger, *Zingiber officinale*. J Drug Dev 1993;6(24):25–89.

[188] Kiuchi F, Shibuya M, Sankawa V. Inhibitors of prostaglandin biosynthesis from ginger. Chem Pharm Bull 1993;30:754.

[189] Lantz RC, Chen GJ, Sarihan M, Slyom AM, Jolad SD, Timmermann BN. The effect of extracts from ginger rhizome on inflammatory mediator production. Phytomedicine 2007;14:123–8.

[190] Choudhury D, Das A, Bhattacharya A, Chakrabarti G. Aqueous extract of ginger shows antiproliferative activity through disruption of microtubule network of cancer cells. Food Chem Toxicol 2010;48:2872–80.

[191] Nadkarni KM. Indian Materia Medica. Bombay, India: Popular Prakashan; 2009.

[192] Ndong M, Uehara M, Katsumata S, Suzuki K. Effects of oral administration of *Moringa oleifera* Lam on glucose tolerance in Goto-Kakizaki and Wistar rats. J Clin Biochem Nutr 2007;40(3):229–33.

[193] Manguro LO, Lemmen P. Phenolics of *Moringa oleifera* leaves. Nat Prod Res 2007;21(1):56–68.

[194] Ndiaye M, Dieye AM, Mariko F, Tall A, Sall Diallo A, Faye B. Contribution to the study of the anti-inflammatory activity of *Moringa oleifera* (Moringaceae). Dakar Med 2002;47(2):210−2.

[195] Ranjan R, swarup D, Patra RC, Chandra, Vikas. *Tamarindus indica* L. and *Moringa oleifera* M. extract administration ameliorates fluoride toxicity in rabbits. Indian J Exp Biol 2009;47(11):900−5.

[196] Mishra G, Singh P, Verma R, Kumar S, Srivastav S, Jha KK, et al. Traditional uses, phytochemistry and pharmacological properties of *Moringa oleifera* plant: an overview. Der Pharmacia Lettre 2011;3(2):141−64.

[197] Iwara IA, Otu EA, Efiong EE, Igile GO, Mgbeje BAI, Ebong PE. Evaluation of the nephroprotective effects of combined extracts of *Vernonia amygdalina* and *Moringa oleifera* in diabetes induced kidney injury in albino Wistar rats. Sch J App Med Sci 2013;1(6):881−6.

[198] Okada K, Wangpoentrakul C, Tanaka T, Toyokuni S, Uchida K, Osawa T. Curcumin and especially tetrahydrocurcumin ameliorate oxidative stress-induced renal injury in mice. J Nutr 2001;131:2090−5.

[199] Joe B, Vijaykumar M, Lokesh BR. Biological properties of curcumin-cellular and molecular mechanisms of action. Crit Rev Food Sci Nutr 2004;44:97−111.

[200] Molina-Jijon E, Tapia E, Zazueta C, El Hafidi M, Zatarain-Barron ZL, Hernandez-Pando R, et al. Curcumin prevents Cr (VI)-induced renal oxidant damage by a mitochondrial pathway. Food Chem Toxicol 2011;10747:1−15.

[201] Adaramoye OA, Nwosu IO, Farombi EO. Sub-acute effect of NG-nitro-l-arginine methyl-ester (L-NAME) on biochemical indices in rats: protective effects of Kolaviron and extract of *Curcuma longa* L. Pharmacognosy Res 2012;4(3):127−33.

[202] Farombi EO, Ekor M. Curcumin attenuates gentamicin-induced oxidative renal damage in rat. Food Chem Toxicol 2006;44(9):1443−8.

[203] Biswas SK, McClure D, Jimenez LA, Megson IL, Rahman I. Curcumin induces glutathione biosynthesis and inhibits NFkappaB activation and interleukin-8 release in alveolar epithelial cells: mechanism of free radical scavenging activity. Antioxid Redox Signal 2005;7:32−41.

[204] Venkatesan N, Punithavathi D, Arumugan V. Curcumin prevents adriamycin nephrotoxicity in rats. Br J Pharmacol 2000;12:231−4.

[205] Messina M. Modern application for an ancient bean: soybeans and the prevention and treatment of chronic disease. J Nutr 1995;123(3S):567−9.

[206] Hawrylewicz EJ, Zapata JJ, Blair WH. Soy and experimental cancer: animal studies. J Nutr 1995;123(3S):709−12.

[207] Coward L, Barnes NC, Setchell KDR, Barnes S. The antitumor isoflavones, genistein and daidzein, in soybean foods of American and Asian diets. J Agric Food Chem 1993;41:1961−7.

[208] Kennedy AR. Overview: anticarcinogenic activity of protease inhibitors. In: Troll W, Kennedy AR, editors. Protease inhibitors as cancer chemopreventive agents. New York, NY: Plenum; 1993.

[209] Wilcox JN, Blumenthal BF. Thrombotic mechanisms in atherosclerosis: potential impact of soy proteins. J Nutr 1995;123(3S):631−8.

[210] Adeneye AA, Arogundade MO. Immunostimulatory and haematopoietic activities of the leaf and seed aqueous extract of *Phyllanthus amarus* in normal and cyclophosphamide-treated mice. Nig J Health Biomed Sci 2007;6(2):53−7.

[211] Kokwaro JO. Medicinal Plants of East Africa. 3rd ed. Nairobi: University of Nairobi Press; 2009.

[212] Iwu MM. Modalities of drug administration. Handbook of African Medicinal plants. Boca Raton, FL: CRC Press Inc.; 1993.

[213] Joshi H, Parle M. Pharmacological evidences for antiamnesic potentials of *Phyllanthus amarus* in mice. Afr J Biomed Res 2007;10:165−73.

[214] Kassuya CAL, Silverstre AA, Rehder V, Calixto JB. Anti-allodynic and antiedematogenic properties of the lignan from *Phyllanthus amarus* in models of persistant inflammatory and neuropathic pain. Eur J Pharmacol 2003;478:145−53.

[215] Ayoola GA, Sofidiya T, Odukoya O, Coker HAB. Phytochemical screening and free radical scavenging activity of some Nigerian medicinal plants. J Pharmaceut Sci Pharm Pract 2006;8:133−6.

[216] Odetola AA, Akojenu SM. Anti-diarrhoeal and gastro-intestinal potentials of the aqueous extract of *Phyllanthus amarus* (Euphorbiaceae). Afr J Med Med Sci 2000;29 (2):119−22.

[217] Joy KL, Kuttan R. Inhibition by *Phyllanthus amarus* of hepatocarcinogenesis induced by *N*-nitrosodiethylamine. J Biochem Nutr 1998;24:133−9.

[218] Rajeshkumar NV, Joy KL, Kuttan G, Ramsewak RS, Nair MG, Kuttan R. Antitumor and anticarcinogen activity of *Phyllanthus amarus* extract. J Ethnopharmacol 2002;81:17−22.

[219] Abate G. Etse debdabe, 159. Addis Ababa, Ethiopia: Artistic Printing Press; 1989.

[220] Wondimu T, Asfaw Z, Kelbessa E. Ethnobotanical study of medicinal plants around "Dheeraa" town, Arsi Zone, Ethiopia. J Ethnopharmacol 2007;112:152−61.

[221] Mebe PP, Cordell GA, Pezzuto JM. Pentacyclic triterpenes and naphthoquinones from *Euclea divinorum*. Phytochemistry 1998;47:311−3.

[222] Mothana RA, Lindequist U, Gruenert R, Bednarski PJ. Studies of the *in vitro* anticancer, antimicrobial and antioxidant potentials of selected Yemeni medicinal plants from the island Soqotra. BMC Complement Altern Med 2009;9:7−18.

[223] Guilled MD, Manzanons MJ. A study of several parts the plant *Foeniculum vulgare* as a source of compounds with industrial interests. Food Res Int 1996;29:86−8.

[224] Atta-Aly MA. Fennel swollen base yield and quality as affected by variety and source nitrogen fertilizer. Sci Hortic 2001;88:191−202.

[225] Barros A, Heleno SA, Carvalho AM, Ferreira ICFR. Systematic evaluation of the antioxidant potential of different parts of *Foeniculum vulgare* Mill. from Portugal. Food Chem Toxicol 2009;47:2458−64.

[226] Muckensturm B, Foechterlen D, Reduron JP, Danton P, Hildenbrand M. Phytochemical and chemotaxonomic studies of *Foeniculum vulgare*. Biochem Syst Ecol 1997;25(4):353−8.

[227] Shalby AB, Hamza AH, Ahmed HH. New insights on the anti-inflammatory effect of some Egyptian plants against renal dysfunction induced by cyclosporine. Eur Rev Med Pharmacol Sci 2012;16:455−61.

[228] Burkill HM. *The useful plants of West Tropical Africa* (Vol. 1: 61 and 2: 338.), Kew, London: Royal Botanical Gardens; 1984.

[229] Cook JA, Vanderjagt DJ, Pastuszyn A, Mounkaila G, Glew RS, Millson M, et al. Nutritional and chemical composition of 13 wild plant foods of Niger. J Food Compos Anal 2000;13:83−92.

[230] Amins ES, Paleologou AM. Estrone in *Hyphaene thebaica* kernel and pollen grains. Phytochemistry 1973;12:899−901.

[231] Hsu B, Coupar IM, Ng K. Antioxidant activity of hot water extract from the fruit of the Doum palm. Hyphaene Thebaica Food Chem Toxicol 2006;98:317−28.

[232] Irobi ON, Adedayo O. Antifungal activity of aqueous extract of dorminant fruits of *Hyphaene thebaica* (Palmea). Pharmaceut Biol 1999;37:114–7.

[233] Bibu KJ, Joy AD, Mercy AK. Therapeutic effect of ethanolic extract of *Hygrophila spinosa* T. Anders on gentamicin-induced nephrotoxicity in rats. Indian J Exp Biol 2010;48(9):911–7.

[234] Peng A, Gu Y, Lin SY. Herbal treatment for renal diseases. Ann Acad Med Singapore 2005;34:44–51.

[235] Ghayur MN, Janssen LJ. Nephroprotective drugs from traditionally used Aboriginal medicinal plants. Kidney Int 2010;77:471–2.

[236] Ranich T, Bhathena SJ, Velasquez MT. Protective effects of dietary phytoestrogens in chronic renal disease. J Ren Nutr 2001;11:183–93.

[237] Friis I, White F. In: Hedberg, Edwards S, Nemomissa S, editors. Ebenaceae., flora of Ethiopia and Eritrea, vol. 4. Addis Ababa: The National Herbarium, Addis Ababa University; 2003.

13 Cardiotoxicity and Cardioprotective Effects of African Medicinal Plants

Afolabi Clement Akinmoladun[1], Mary Tolulope Olaleye[1] and Ebenezer Olatunde Farombi[2]

[1]Phytomedicine, Drug Metabolism and Toxicology Unit, Department of Biochemistry, School of Sciences, The Federal University of Technology, Akure, Nigeria, [2]Drug Metabolism and Toxicology Unit, Department of Biochemistry, College of Medicine, University of Ibadan, Ibadan, Nigeria

13.1 Introduction

The biochemical modification of pharmaceutical substances by living organisms, usually through specialized enzymatic systems, is called drug metabolism. Drug metabolism involves a set of metabolic pathways that modify the chemical structure of drugs and other compounds which are foreign to an organism's normal biochemistry. Medicinal plants are a major source of drugs for the therapy of human and veterinary ailments. Although the liver is the principal organ responsible for the metabolism of drugs and other xenobiotics in the body, the heart is another important organ that may be concerned with drug and xenobiotic metabolism [1]. The heart is not only involved in the biotransformation of these compounds, but is also a prime target for their pharmacological effects.

Primarily, the heart functions as a pump that supplies blood and oxygen to other parts of the body. However, it also contains some enzymes used in the transformation of drugs and xenobiotics. The biotransformation of a drug can lead either to adverse outcomes to the heart (toxicity) or it can result in favorable conditions (detoxication and cardioprotection). Plant secondary metabolites also can be regarded as xenobiotics. Plant extracts contain a diverse array of these chemical constituents that can affect a wide range of organ systems, including the circulatory system. These constituents could either be medicinal (phytopharmaceuticals), toxic (phytotoxins), or both (phytopharmatoxins). Apart from its intrinsic ability to metabolize drugs, the heart is important in drug metabolism in another way; transportation of blood from one section of the body to another. The heart probably has

Toxicological Survey of African Medicinal Plants. DOI: http://dx.doi.org/10.1016/B978-0-12-800018-2.00013-3

a greater exposure to drugs and other foreign compounds than the liver and other organs *via* inhalation and other routes as it can receive drug- or toxicant-laden blood from the pulmonary and systemic circulations. This makes it susceptible to the pharmacological effects of the drugs or their metabolic products.

Different methods for the evaluation of the cardioprotective and cardiotoxic properties of medicinal plants exist. These include *in vitro*, *ex vivo*, and *in vivo* techniques. Modulation of enzyme biomarkers and the regulation of appropriate protein factors and genes as well as histoarchitectural evaluations often are employed in these evaluations. Many of the available techniques are used in Africa, but the continent still lags behind in cutting edge research and the use of some modern techniques that are easily and routinely employed in developed nations.

The cardioactivity of individual phytochemicals from medicinal plants has been extensively and elaborately discussed in the literature. Although references are made to some of these individual phytochemicals in the present work, this chapter focuses on the medicinal plants containing these bioactive entities.

13.2   The Heart and the Metabolism of Xenobiotics

13.2.1   The Heart and the Cardiovascular System

The adult human heart weighs 250–300 g and is about the size of a closed fist. It is located in the thorax at the left side of the chest between the lungs, behind the sternum, and above the diaphragm. It is surrounded by the pericardium. Walls of the heart are principally composed of cardiac muscle, called myocardium, and connective tissue. The heart muscle cells are called cardiomyocytes [2,3a]. Cardiac muscle is an involuntary striated muscle tissue found only in the heart and is responsible for the ability of the heart to pump blood. The contraction of the cardiac myocyte is myogenic. The heart pumps blood throughout the blood vessels by repeated, rhythmic contractions [3b–d].

The heart has four compartments: the right and left atria and ventricles. The two atria are smaller and are the upper chambers of the heart while the two ventricles are the larger, lower chambers of the heart. The left ventricular wall is much thicker than the right ventricular wall. This is logical as the left ventricle pumps blood to the systemic circulation, where the pressure is considerably higher than for the pulmonary circulation, which arises from right ventricular outflow. The heart has four valves. Between the right atrium and ventricle lies the tricuspid valve, and between the left atrium and ventricle is the mitral valve. The pulmonary valve lies between the right ventricle and the pulmonary artery, while the aortic valve lies in the outflow tract of the left ventricle (controlling flow to the aorta). Under normal conditions, the valves ensure that blood only flows unidirectionally in the heart [3a]. In order to pump blood through the body, the heart is connected to the vascular system of the body. This cardiovascular system is designed to transport oxygen and nutrients to the cells of the body and remove carbon dioxide and metabolic waste products from the body. The blood returns from the systemic circulation to the right atrium and from there goes through the tricuspid valve to the right ventricle. It is ejected from the right ventricle through the pulmonary valve to the lungs. Oxygenated blood returns from the lungs to the left atrium, and from there through the mitral valve to the left ventricle. Finally, blood is pumped through the aortic valve to the aorta and the systemic circulation [3a,4]. The cardiac cycle refers to the repetitive pumping process that begins with the onset of cardiac muscle contraction and ends with the beginning of the next contraction and it comprises the systole and diastole. The diastole and the systole represent the two phases of each beat or pump of the heart. During diastole, the ventricles are being filled and the atria contract while during systole, the ventricles contract while the atria are relaxed and being filled [4].

13.2.2 Energy and Drug Metabolism in the Heart

The heart is a highly metabolically active organ with the cardiac myocyte being perhaps the most physically energetic cell in the body. The mammalian heart can metabolize both fat and carbohydrate and the human heart uses several kilograms of ATP per day. This probably explains why one-third of the cardiac myocyte consists of mitochondria [5]. Myocardial energy substrate metabolism is important in cardiac health. Modulation of myocardial energy substrate metabolism by drugs could be an important intervention in some pathologies of the heart. A case in point is the success of glucose−insulin−potassium in the treatment of cardiogenic shock after hypothermic ischemic arrest of the heart. Many drugs target intermediary metabolism for the treatment of cardiac dysfunction and may contribute to the regeneration of normal cardiac myocyte function. Fatty acid metabolism dominates in the fasted state but the heart switches from fat to carbohydrate for oxidative energy production when actively stressed which may lead to loss of insulin sensitivity and metabolic flexibility (metabolic dysregulation) in the long run [5].

The myocardium has certain distinctive biochemical features, although many of its basic reaction patterns are similar to those of other tissues. Cardiac metabolism is not limited to only energy transfer but includes pleiotropic actions such as the generation of signals for cardiac growth, programmed cell death, and programmed cell survival, the formation of reactive oxygen species (ROS), and the regulation of transcription factors [6a,b]. Also, apart from being able to metabolize biomolecules such as fat and carbohydrate, cardiac cytochrome P450 (CYP) enzymes can be involved in drug metabolism within the heart and influence pharmacologic efficacy. Metabolism mediated by CYP enzymes can influence the survival of cardiomyocytes during ischemia [1]. All these present metabolic targets for the treatment of cardiac pathologies. Therefore, cardiac metabolism has an important role to play in cardioprotection and cardiotoxicity by drugs and medicinal plants.

13.2.3 Effects of Cardioactive Substances on the Heart

Cardioactive compounds can affect the integrity of cardiac myocytes or nerves controlling the heart and alter heart rate and cardiac contractile patterns, which manifest in adverse or positive outcomes depending on the specific pharmacological effects of the agents and the prevailing pathological or physiological conditions.

The heart has a network of cardiac neurons that are involved in the intrinsic electrical conduction of the heart, allowing electrical propagation by the generation of an action potential [3b–d]. Cardiomyocyte electric activation takes place by means of the same mechanism as in nerve cell, that is, from the inflow of sodium ions across the membrane. Associated with the electric activation of the cardiac muscle cell is its mechanical contraction [6c]. The structural integrity of the heart is very important, as a structurally diseased heart is likely to be electrically unstable.

Medicinal plants can show positive or negative inotropic, chronotropic, dromotropic, lusitropic, and/or bathmotropic activities as a result of changes in the electrolyte and redox homeostasis of the heart. Bioactive substances in medicinal plants can trigger antioxidant, vasoactive, inflammatory, and apoptotic cascades in the heart and can also regulate genes involved in cardiac metabolism. Overall, these effects manifest as either cardioprotection or cardiotoxicity.

13.3 Known Cardiotoxic and Cardioprotective Plants

The cardioprotective and cardiotoxic effects of medicinal plants are attributable to bioactive components with direct or indirect effect on the heart and blood vessels. The activity demonstrated on the heart by the same medicinal plant or a bioactive compound could be protective or toxic depending on the concentration employed. The cardioactive principles in these plants cut across many chemical groups. As far as the cardiovascular system is concerned, however, cardiac glycoside- and alkaloid-containing plants are very important not only in both lethal livestock and human poisonings but also in therapy of cardiovascular disorders.

13.3.1 Cardiotoxic Plants

13.3.1.1 Cardiac Glycoside-Containing Medicinal Plants

Cardiac glycosides, which are highly toxic and found in a number of plants, are usually phytochemicals consisting of an aglycone (structurally related to steroid hormones) linked to one or more sugar molecules. The aglycones of cardiac glycosides can be divided into two chemical groups, the cardenolides and bufadienolides, with the former group being the most implicated in cardioactivity. Many cardiac glycoside-containing plants have historically been known to be toxic to veterinary animals and humans. The most recognized of these plants is foxglove (*Digitalis purpurea*), found in Africa and other parts of the world. It contains the cardiac glycosides digoxin, digitoxin, and digitonin, among several others. Digoxin at therapeutic levels is used to treat congestive heart failure, but becomes toxic at high doses. In South Africa, *Acokanthera oppositifolia* (bushman poison bush, *boesmansgif*) sap contains cardenolides and has been used by the San people for applying to the tips of their hunting arrows. The cardiotoxic bufadienolides present in *Drimia sanguine* (sekanama, *slangkop*) and *Bowiea volubilis* (climbing potato, *knolklimop*) species also have been implicated in human poisoning [7]. In different parts of Africa and North America, milkweeds (*Asclepias* spp.), oleander (*Nerium oleander*), and lily of the valley (*Convallaria majalis*) are important causes of cardiac glycoside poisoning in livestock, and sometimes in other animals and humans [8]. Various toxic cardiac glycosides are present in milkweed. Oleandrin and neriine are two potent cardiac glycosides found in all parts of oleander. In lily of the valley, the cardiac glycosides convallerin and convallamarin are among at least 15 others found throughout the plant and they have similar cardiac effects to digitalis glycosides. Some exotic plants used for ornamental and other purposes, which originated from Africa and are now found in other parts of the world, e.g., *Bryophyllum* spp. and *Adonis microcarpa* also contain cardiac glycosides and pose a risk to livestock. The potent cardiac glycosides, thevitin A and B and thevetoxin are found in all parts of *Thevetia peruviana* and *Thevetia thevetioides* (yellow oleander) [8].

13.3.1.2 Alkaloid-Containing Medicinal Plants

A significant number of alkaloid-containing plants with cardioactivity have also been described in the literature. The toxicity of yews (*Taxus* species) to humans and animals has been known for many years [9]. Yews contain toxic alkaloids such as taxine A and B, collectively referred to as taxine. Members of the genus *Zigadenus* contain several steroidal alkaloids, the best known being zygacine and zygadenine [10]. The toxic species cause death in humans and some animals by hypotension. The cocoa plant (*Theobroma cacao*) contains theobromine, an alkaloid which is toxic in some animals and produces adverse effects such as hyperexcitability, sweating, and increased respiration and heart rates [11]. Medicinal plants like the *Datura* spp. (in particular *Datura stramonium*, *Datura ferox*, and *Datura innoxia*) contain the tropane alkaloids which are often extracted to be used as medicine, drug, poison, or antidote. Poisonings of humans by plants containing tropane alkaloids such as atropine and scopolamine through unintended overdoses can occur. These plants are also toxic for animals if ingested in larger amounts. These plants may cause a cholinergic syndrome and alterations in the heart rate [12]. *Atropa belladonna* (deadly nightshade) is native to North Africa and other parts of the world. The tropane alkaloids of *A. belladonna* were used as poisons, and early humans made poisonous arrows from the plant. It also has a long history of use as a medicine and cosmetic [13]. Deadly *A. belladonna* intoxication has been infrequently reported in both children and adults in the literature. Tachycardia, a dysfunction of the heart, is one of the reported symptoms [14]. Aconites are the dried rootstocks of *Aconitum* plants, e.g., *Aconitum ferox*. Despite the well-known toxicity and the low safety margin, aconite tubers and their processed products remain popular in oriental or homeopathic medicines for treating rheumatism, neuralgia, and cardiac complaints. Severe poisonings and fatalities from ingestion of aconite-containing herbal remedies have been reported from different parts of the world [15,16]. Symptoms of intoxication include rapid-onset generalized weakness, chest discomfort, hypotension, and arrhythmias. In severe poisoning, life-threatening ventricular tachycardia, if untreated, would lead to a fatality [17]. The toxicity of aconite is related to the C19-norditerpenoid ester alkaloids like aconitine, deoxyaconitine, mesaconitine, hypaconitine, and yunaconitine [18]. *Areca catechu* is a commonly used herb in India which has been introduced to some parts of Africa. It contains an alkaloid arecoline which produces ventricular arrhythmia [19]. Other commonly used medicinal plants in India, which are also found in Africa, with reported cardiotoxic/cardioprotective effects are *Cleistanthus collinus, Asparagus abscendens, Asparagus racemosus,* and *Withania somnifera. C. collinus* grows in hilly deciduous forests of Central and South India, Malaysia, and Africa [20]. All parts of the plant are potentially toxic and used for suicide and homicide in India. The toxic principles are arylnaphthalene lignan lactones—Diphyllin and its glycoside derivatives Cleistanthin A and Cleistanthin B. In addition, the lignans Cleistanthin C, Cleistanthin D, and Cleistanone are present in the plant. The toxicity of the plant has been attributed primarily to Cleistanthin A and B [20,21].

13.3.1.3   Medicinal Plants Containing Other Cardioactive Chemical Entities

Avocado (*Persea americana*) is found in different parts of Africa. Fruit, pits, leaves, and the actual plant, though generally harmless to humans, are all poisonous to some species of animals, especially birds, causing cardiac distress and ultimately heart failure. The toxic principle is persin, a fungicidal fatty acid. Several *Jatropha* species contain phorbol esters, e.g., the oil curcin. The symptoms of intoxication in humans include muscle shock, decrease of visual capacity, and an increase in heart rate [11]. *Pavetta harborii* was reported to cause gousiekte, a syndrome characterized by heart failure and sudden death, when ingested by ruminants. A number of cardiodynamic changes [22] as well as myocardial lesions that consist of a loss of myofilaments, replacement of myocytes with collagenous tissue, lengthening of sarcomeres and cardiac dilatation [23] have been reported. The cationic polyamine pavetamine has been identified as the cardiotoxin in these plants. The toxin isolated from the dahlia sea anemone *Taelia felina* (not a plant) showed negative inotropic and chronotropic effects on the heart of rat *in vitro* and substantially increased the resistance to perfusion [24a]. Atractyloside, a diterpenoid glycoside first isolated from *Atractylisn gummifera*, a plant found in Africa and other parts of the world, may cause cardiac failure in some animals [24b].

13.3.2   Cardioprotective Plants

In addition to some cardiotoxic medicinal plants with cardiotonic effects at low doses indicated above, several other medicinal plants are known mainly for cardioprotective properties. Due to the increasing cases of cardiovascular diseases and its leading cause of mortality worldwide, the list of plants with putative claims for cardioprotection has been growing. Cardioprotective medicinal plants cut across different plant families and contain varying bioactive phytoconstituents with flavonoids, cardiac glycosides, alkaloids, and other antioxidative phytochemicals as the principal bioactive principles. The criteria for the classification of a medicinal plant as cardioprotective include folkloric evidence and observed cardiotonic effects in pharmacological investigations especially amelioration of cardiovascular disease or its risk factors. Well-known plants with reported cardiotonic and cardioprotective properties include *Aloe vera* (Xanthorrhoeaceae), *Terminalia arjuna* (Combretaceae), *Allium sativum* (Liliaceae), *Curcuma longa* (Zingiberaceae), *Gingko biloba* (Ginkgoaceae), *Ocimum sanctum* (Lamiaceae), *Azadiracta indica* (Meliaceae), *Moringa oleifera* (Moringaceae), *Hibiscus sabdarifa* (Malvaceae), *W. somnifera* (Solanaceae), *Psidium guajava* (Myrtaceae), and many others. A few of these will be briefly discussed because of space constraints.

Aloe vera (*Aloe barbadensis*) (Xanthorrhoeaceae) grows in arid climates and is widely distributed in Africa, India, and other arid areas. *A. vera* has been reported to have hypolipidemic, hypoglycemic, antidiabetic, and cardioprotective effects. It was also reported to ameliorate atheromatous heart disease and myocardial ischemia. Polysaccharides, bitter principles, and antioxidative components present in the plant are credited with its bioactivity [25−27]. *A. vera* gel is rich in polysaccharides, including acemannan which has been reported as the primary active substance in the parenchyma. However, given the number of other potentially active compounds in the plant, it is possible that the biological activities of *A. vera* result from the synergistic action of a variety of compounds, rather than from a single defined component [28−30].

Terminalia arjuna (Combretaceae) is native to India but was introduced into the African continent decades ago. The cardioprotective property of the plant has been extensively discussed [31]. It has multimode cardioprotective activity and can be considered as a useful drug for coronary artery disease, hypertension, and ischemic cardiomyopathy. The plant was found to increase the force of cardiac contraction, exert negative or positive inotropic/chronotropic effects on heart, and cause a dose-dependent decrease in blood pressure (BP) and hypolipidemic effect. It improves cardiac muscle function and subsequently enhances pumping activity of the heart. It is used as a cure for congestive heart failure, coronary artery disease, myocardial necrosis, and ischemia−reperfusion injury [32,33]. Cardioactive constituents of the plant include terpenoids, glycosides, flavonoids, and tannins.

Allium sativum (garlic) (Liliaceae) is among the oldest of all cultivated plants. It has been used as a medicinal agent for thousands of years and has multiple beneficial effects including hypolipidemic, hypoglycemic, antithrombotic, and cardioprotective properties. It is a species in the onion genus, native to central Asia and has long been a staple in the Mediterranean region, as well as a frequent seasoning in Asia, Africa, and Europe [34−36]. Onion (*Allium cepa*), another popular species of the *Allium* genus, is also reported to possess remarkable cardioprotective property [33].

Curcuma longa (turmeric) (Zingiberaceae) is cultivated in all parts of India and also in southern China, Japan, as well as throughout the African continent and other parts of the world [37,38]. It is one of most essential spices all over the world with a long and distinguished human use, particularly in the Eastern civilization [39a]. Apart from its culinary uses, turmeric has been used widely in the traditional medicine system of India, Pakistan, and Bangladesh because of its several beneficial properties [39b]. The hypolipidemic, anti-ischemic, antiatherosclerotic, hypotensive, and vasorelaxant effects of turmeric have been reported [40−43]. The major bioactive principle, curcumin, was reported to demonstrate anti-ischemic, C-reactive protein lowering, cardiac repair, and amelioration of cardiac dysfunction following myocardial infarction as well as amelioration of cardiac inflammation in rats with autoimmune myocarditis among other effects [44−47].

Gingko biloba (Ginkgoaceae) is native to China but has been introduced to other parts of the world, including Africa. Terpene trilactones and flavonoid glycosides are the two major groups of bioactive constituents found in the plant. Liebgott et al. [48] investigated the complementary cardioprotective effects of flavonoid metabolites and terpenoid constituents of extract EGb 761 (extract of *G. biloba* 761) from the plant during ischemia and reperfusion and showed that part of the cardioprotection afforded by EGb 761 is due to a specific action of its terpenoid constituents and the flavonoid metabolites that are formed after *in vivo* administration of the extract, which act in a complementary manner to protect against myocardial ischemia−reperfusion injury. The cardioprotective property of *G. biloba* has also been attributed to its vasodilatory and antihypertensive properties, but one research indicated that *G. biloba* does not reduce BP or the incidence of hypertension in elderly men and women [49]. The protective effect of the plant on drug-induced cardiotoxicity has been demonstrated [50,51].

13.3.3   Mechanisms of Action of Cardioactive Substances

Cardioprotective and cardiotoxic medicinal plants exhibit their action via various mechanisms, many of which have been clearly elucidated. The primary pharmacological effect of cardiac glycosides is to inhibit the Na^+/K^+ ATPase exchanger, which increases intracellular Na^+ concentration, thus reducing the Na^+ gradient across the membrane and decreasing the amount of Ca^{2+} pumped out of the cell by the Na^+/Ca^{2+} exchanger during diastole. Consequently, the intracellular Ca^{2+} concentration rises, thereby occasioning positive inotropy. More specifically, the mechanisms of action of digoxin have been described in detail by Lelièvre and Lechat [52]. Digoxin binds to and inhibits sarcolemma-bound Na^+/K^+ ATPase. This ATPase catalyzes both an active influx of $2K^+$ ions and an efflux of $3Na^+$ ions against their respective concentration gradients, with ATP hydrolysis supplying the energy needed. Hyperkalemia is one sure sign of serious digoxin toxicity [53]. The inhibition induced by digoxin leads to an efflux of potassium from the cell and an increase in internal sodium ion concentration at the inner face of the cardiac membranes. This local accumulation of sodium causes an increase in free calcium concentrations via the Na^+-Ca^{2+} exchanger which is responsible for the inotropic action of digoxin, secondary to the release of Ca^{2+} from the sarcoplasmic reticulum. Toxic effects of digoxin (i.e., arrhythmias) occur when the cytoplasmic Ca^{2+} increases to concentrations exceeding the storage capacity of the sarcoplasmic reticulum. This internal Ca^{2+} overload necessitates several cycles of Ca^{2+} release—reuptake in order to restore the Ca^{2+} equilibrium between sarcoplasmic reticulum and cytoplasm. High internal concentrations of Ca^{2+} also activate a depolarizing current which generates delayed after-depolarizations that give rise to extrasystoles and sustained ventricular arrhythmias *in vivo* [54]. Aconite causes serious electrolyte disturbances. It has a direct effect on myocardium and adrenals, producing release of catecholamines [55]. It blocks tetrodotoxin-sensitive sodium channels and blocks inactivation leading to early and late after-depolarizations but without hyperkalemia. The mechanisms behind most of the cardiotoxic effects of aconitum alkaloids can be summarized as binding to voltage-dependent sodium channels inducing a hyperpolarized state, resulting in permanent activation of the channel; modulation of neurotransmitter release and receptors, particularly norepinephrine and acetylcholine; promotion of lipid peroxidation of the cardiac system, possibly causing cardiac arrhythmias; and induction of cellular apoptosis in the heart, liver, and other organs [56]. *C. collinus* (Phyllanthaceae) produces serious urinary potassium loss leading to hypokalemia and cardiac arrhythmia [57]. Extracts of the leaf inhibited muscle contraction by reducing excitability of nerve and muscle membranes, as well as blocking neuromuscular transmission, probably as a result of inhibition of thiol-dependent enzymes such as lactate dehydrogenase (LDH) and cholinesterase [58,59]. *C. collinus* depletes glutathione and ATPases in various tissues, including skeletal muscle and heart, in animal models. The loss of ATPase activity may be a result of oxidation of $-SH$ (thiol) groups. It appears that inhibition of the thiol/thiol enzymes by lignan lactones may be an integral mechanism in the toxicity profile of *C. collinus* [20]. Pavetamine, the active constituent of *P. harborii* (Rubiaceae), inhibits protein synthesis in the cardiac muscle of rats but has no effects on other tissues [60]. In another study where *P. harborii* extracts were administered to rats, protein synthesis in cardiac tissue was compromised to the extent that it leads to a degeneration in the myofilaments. These pathological changes in the cardiac tissue could be the underlying cause of the reduced cardiodynamic function (systolic and diastolic) associated with the disease [61].

Tropane alkaloids from *Datura* spp. have anticholinergic effect. They have the ability to bind to muscarinic acetylcholine receptors and act as competitive antagonists at these receptors [62]. The M2 subtype of muscarinic receptors occurs in the atria of the heart, at smooth muscles of the gastrointestinal tract, as well as in the central nervous system. Atropine is a nonselective antagonist of all classes of muscarinic receptors, but known to have a stimulating effect on the central nervous system, whereas scopolamine is a depressant of the central nervous system [62]. As anticholinergic compounds disturb the balance between cholinergic and adrenergic regulation of organ functions, secondary effects may occur. A prominent example is the effect of atropine on the heart rate, which commences as bradycardia at low (therapeutic) doses, progressing into tachycardia and arrhythmia at higher (toxic) doses [12]. The peripheral side effects of anticholinergics include myocardial ischemia.

Most reports on the cardioprotective mechanisms of medicinal have dwelt on the antioxidant property. Oxidative and nitrosative stresses have been implicated in the etiology of many diseases [63,64]. In particular, ROS such as superoxide radical ($O_2^{·-}$), hydroxyl radical ($OH^·$), and H_2O_2 together with reactive nitrogen species (RNS) such as peroxynitrite ($ONOO^-$), nitroxyl (NO^-), nitrosyl chloride (NOCl) and nitrogen dioxide (NO_2) are important factors in the etiology of several pathological conditions such as lipid peroxidation, protein peroxidation, DNA damage, and cellular degeneration related to cardiovascular and other diseases. Medicinal plants contain many antioxidative components which act as major defense against radical-mediated toxicity by preventing or attenuating the deleterious effects of free radicals. Although the exact mechanisms and interactions among various antioxidants are not fully understood, it is possible that one antioxidant may equilibrate with another to establish a cellular redox potential and thus all endogenous antioxidants may act in concert to protect against oxidative insult. Nonetheless, it has been suggested that antioxidants can act through several mechanisms such as: (i) scavenging ROS or their precursors, (ii) inhibiting the formation of ROS, (iii) attenuating the catalysis of ROS generation *via* binding to metals ions, (iv) enhancing endogenous antilipoprooxidant generation, and (v) reducing apoptotic cell death by upregulating the antideath gene (*Bcl-2*) [65].

This implies that the cardioprotective effect of medicinal plants and their constituents can involve mechanisms independent of direct free radical scavenging [48]. Findings by El-Boghdady [51] and Lelièvre and Lechat [52] showed that in addition to antioxidant and apoptotic mechanisms, the cardioprotective effect of medicinal plants can be mediated by the regulation of inflammatory and vasoactive mediators. They reported significant increases in cardiac tumor necrosis factor-alpha (TNF-α) and caspase-3 levels in doxorubicin-intoxicated rats which were ameliorated by *G. biloba* extract and a reversal of adriamycin-provoked increases in the levels of endothelin-1, TNF-α, and nitrite/nitrate levels by the extract.

The activation of the PI-3K, PKB/Akt pathway is one of the mechanisms that have been described for cardioprotection against anoxia/reoxygenation or ischemia–reperfusion injury [66–68]. This pathway, normally activated by various extracellular substances, has antiapoptotic effects leading to a mitigation of the negative outcomes of ischemia. Activation of PKB/Akt is also a prerequisite for glucose uptake by the heart through the two transporters Glut 1 and Glut 4. An improved ability to import and utilize glucose is cardioprotective when the heart is subjected to the absence of oxygen, as induced by ischemia. The heart then uses the energy generated by glycolysis to protect itself [69,70]. Huisamen reported that the antihypertensive and myocardial protective ability of *Prosopis glandulosa* (Fabaceae) may be linked to changes in the PI-3-kinase/PKB/Akt pathway. Hong et al. [71] showed that curcumin could improve heart function, diminish infarct size, and reverse the abnormal changes in the activities of serum LDH and creatine kinase MB (CK-MB) in rats after myocardial infarction and that of a total of 179 genes which were found to be significantly differentially expressed between sham-operated rats and coronary artery-ligated rats, cytokine—cytokine receptor interaction, ECM—receptor interaction, focal adhesions, and colorectal cancer pathway may be involved in the cardioprotective effects of curcumin. Additional molecular mechanisms of cardioprotection by curcumin and other medicinal plants abound in the literature.

13.4 Methods in Cardiotoxic and Cardioprotective Screenings of Medicinal Plants

The increasing use and growing demand for plant-based medicines and medicinal plants present the challenge of quality control and regulation. Concerns about the use and safety of herbal products and medicinal plants have led to the establishment of various regulatory frameworks by many nations but these are still lacking or ineffective compared to the regulatory standards for orthodox medicines. While the regulation of the use of medicinal plants may be difficult for many reasons, toxicity testing can greatly assist in guiding people on their safe use. Also, because a large number of plants still uninvestigated for bioactivity may be crucial in addressing many health challenges ravaging humans, there is the need for the screening of these plants so as to harness their positive potentials or avoid potentially devastating hazards. Methods used to screen medicinal plants for cardioprotective or cardiotoxic effects include *in vitro*, *ex vivo*, and *in vivo* models.

13.4.1 In Vitro *Tests*

In vitro methods to screen for the toxicity of medicinal plants in general also can be applied to cardiotoxic plants. Cytotoxicity assays using primary cultures and cell lines related to cardiovascular disease are in use for screening medicinal plants for cardiotoxicity or cardioprotection [3d]. There is increasing advancement in these techniques as well as continuous refinement of methods for more consistent and reproducible results as scientists seek for cultures and cell lines that best mimic the *in vivo* scenario. The use of human embryonic stem cell (hESC)-derived cardiomyocytes to test for drug-induced cardiotoxicity in a biosensor setup with a significantly higher technological relevance was described by Anderson et al. [72]. The hESC-derived cardiomyocytes released detectable levels of two clinically decisive cardiac biomarkers, cardiac troponin T and fatty acid binding protein 3 (FABP3), when the cardiac cells were exposed to the cardioactive compound. The release is monitored by the immuno-biosensor technique surface plasmon resonance which was particularly appropriate due to its capacity for parallel and high-throughput analysis in complex media. Apart from evaluating the modulation of biochemical markers in investigations, researchers also predicate investigations around relevant physical parameters. For example, a compound's risk for prolongation of the surface electrocardiographic QT interval and hence risk for life-threatening arrhythmias is mandated before approval of nearly all new pharmaceuticals. QT prolongation has most commonly been associated with loss of current through hERG (human ether-a-go-go related gene) potassium ion channels due to direct block of the ion channel by drugs or occasionally by inhibition of the plasma membrane expression of the channel protein. Su et al. [73] developed an efficient, reliable, and cost-effective hERG screening assay for detecting drug-mediated disruption of hERG membrane trafficking using microfluidic-based systems to improve throughput and lower cost of current. Cultured stably transfected HEK cells that overexpressed hERG (WT-hERG) were studied for their morphology, proliferation rates, hERG protein expression, and response to drug treatment. This method is suitable for drug screening applications, particularly for tests involving hydrophobic drug molecules. The hERG potassium channel is a major target of drug safety programs in cardiotoxicity. Wible et al. [74] reported the development of HERG-Lite, a comprehensive high-throughput screen for drug-induced hERG risk, which monitors the expression of hERG in mammalian cell lines. Assessment of Braam et al. [75] on their part showed that human cardiomyocytes derived from pluripotent (embryonic) stem cells (hESC) could be used as a renewable, scalable, and reproducible system on which to base cardiac safety pharmacology assays. Their analyses of extracellular field potentials in hESC-derived cardiomyocytes (hESC-CM) and generation of derivative field potential duration (FPD) values showed dose-dependent responses for 12 cardiac and noncardiac drugs. They confirmed that serum levels in patients of drugs with known effects on QT interval overlapped with prolonged FPD values derived from hESC-CM, as predicted and proposed hESC-CM FPD prolongation as a safety criterion for preclinical evaluation of new drugs in development and that assays based on hESC-CM could complement or potentially replace some of the preclinical cardiac toxicity screening tests currently used for lead optimization and further development of new drugs. Although some of these advanced techniques may not yet be feasible in many laboratories in Africa, African researchers are moving away from the traditional toxicity testing involving the use of large numbers of animals. Cell culture techniques are routinely carried out in well-equipped laboratories and biomarkers for cardiotoxicity and cardioprotection such as CK-MB, LDH, etc., as well as antioxidant and pro-oxidant indices are readily evaluated in these cell-based systems. Imaging techniques also are available in some laboratories to complement biochemical and molecular assays. Advanced techniques similar to those described here and employed in developed nations are actually feasible in some premium research laboratories in Africa.

Analytical methods to detect or monitor the level of several specific cardiotoxicants, e.g., cardiac glycosides also are available and these could be used to screen medicinal plants for potential cardiotoxic effects.

Although a common and unsophisticated method, evaluating the production of thiobarbituric acid reactive substances in animal heart homogenates is still a useful method in screening medicinal plants for cardiotoxicity and cardioprotection especially in laboratories with minimal facilities. Thiobarbituric Acid Reactive Substances (TBARS) test can be carried out in in vitro, ex vivo, or in vivo settings. This is an important method because tested drugs are in contact with tissues and cells from living systems. Evaluations for additional key biomarkers are often carried out in the same system.

13.4.2 Ex Vivo *Methods*

Ex vivo anti-ischemic studies using the Langendorff technique have remained a valuable strategy in evaluating the cardioprotection/cardiotoxicity of medicinal plants and other drugs. This method is well described in the literature and various adaptations exist. The rat heart often is used but those of other animals such as the hamster also have been employed. Perfusion can be carried out in several modes, e.g., retrogradely in the Langendorff mode or the working heart mode. Perfusion medium normally used include Krebs−Henseleit buffer and Normal HEPES (4-(2-hydroxyethyl)-1-piperazineethanesulfonic acid) Tyrode buffer. Test substances are usually introduced into the medium and ischemia is induced by regulation of the supply of oxygen from a cylinder. Several indices can be evaluated in cardiac tissue or the perfusate. Histological assessment can be carried out on the tissue. More frequently, the pumping heart is connected by a force transducer to a recording polygraph machine from which heart rates and force of cardiac contraction can be calculated. Electrocardiographic assessments also can be made from the recordings. Determination of infarct size in *ex vivo* perfused rat hearts as a measure of myocardial damage incurred by ischemia followed by reperfusion is taken as the gold standard to prove cardioprotection. Some novel approaches are being applied to this method and it is widely used in Africa and other laboratories around the globe [69,76−79]. A complex, uncommon *in vivo* surgical myocardial ischemia/reperfusion technique also exists.

13.4.3 In Vivo *Models*

A commonly used model for *in vivo* cardioprotective study in Africa is the evaluation of plant materials for protection against drug-induced cardiotoxicity. The cardiotoxicants usually employed are isoproterenol (isoprenaline) and doxorubicin (adriamycin) which are also used for *in vitro* cardioprotective studies. These methods are well described in the literature. Many cardiac parameters can be evaluated as well as histoarchitectural assessment. Protein and gene expression assays often are integrated into the model so that biochemical, histological, and molecular parameters related to the cardioprotective effect of the medicinal plant can all be evaluated [77,80−83].

Methods used to evaluate effect of medicinal plants on risk factors for cardiovascular disease could be considered as indirect cardioprotective investigations. Studies in this category routinely carried out in Africa include antihypertensive studies involving the evaluation of biochemical and hemodynamic indices, antihyperlipidemic and cholesterol lowering studies, among others [70,84a].

Important biomarkers in cardioprotective assays include LDH 1/2, cardiac troponin, C-reactive protein, angiotensin converting enzyme, myeloperoxidase, lipid profile, CK-MB, calcium/calmodulin (Ca^{2+}/CaM)-dependent protein kinase II FABP3, and myoglobin [84b,c].

Lin et al. [85] described an *in vivo* model for screening cardioactive drugs with zebrafish using pseudodynamic three-dimensional imaging. With pseudodynamic 3D imaging, individual parameters that are central to the cardiac function of zebrafish, including the ventricular stroke volume, ejection fraction, cardiac output, heart rate, diastolic filling function, and ventricular mass were derived. Both inotropic and chronotropic responses of the heart of zebrafish treated with drugs that are commonly prescribed and possess varied known cardiac activities were evaluated. It was found that the function of zebrafish exhibited a pharmacological response similar to that of humans. In view of the growing interest of using zebrafish in both fundamental and translational biomedical research, the authors envisaged that this method should be of benefit in contemporary pharmaceutical development as well as exploratory research such as gene, stem cell, or regenerative therapies targeting congenital or acquired heart diseases.

13.5 Cardiotoxic Medicinal Plants of Africa

Medicinal plants are native to specific parts of the world. However, because of transregional movements, some medicinal plants have been introduced to places beyond their native habitats. Table 13.1 itemizes cardiotoxic plants that have been investigated in Africa. Information was obtained mostly from reviews and pertained to South Africa. Cardiac glycosides, especially cardenolides, were the toxic principles in most plants and the Apocynaceae family seemed to predominate. Toxic alkaloids came a distant second. The toxic factor of *P. harborii* (Rubiaceae), a cationic polyamine is of special interest as a major deviation from the usual cardiotoxicants.

13.6 Cardioprotective Medicinal Plants of Africa

Table 13.2 shows a list of medicinal plants investigated for cardioprotection in Africa. The source of information was either original research works or reviews. Various reviews on Africa medicinal plants exist and some are excellent sources of information on cardioactive plants [86b−d,87a,90,106]. The table is far from being exhaustive, but points to some notable trends especially when compared with Table 13.1. More cardioprotective studies were carried out than cardiotoxicity studies and there is a preponderance of leading cardiovascular investigations from South Africa. Most investigators used crude herbal extracts. The studies are diverse, but those relating to the protective effect of medicinal plants or their active constituents on drug-induced toxicity are common as well as those evaluating their protective effect in ischemia/reperfusion induced myocardial dysfunction. For studies involving risk factors for cardiovascular diseases, antidiabetic and

Table 13.1 African Medicinal Plants Screened for Their Cardiotoxicity

Plant (family)	Area of Plant Collection	Traditional Use	Chemical Constituents	Pharmacological/Toxicological Activities
Nerium oleander (Apocynaceae)	South Africa [7]; Uganda	Abortifacient [86a]	Cardenolide (oleandroside and neriin) [7]	Cardiotonic; cardiotoxic [7,86a]
Digitalis purpurea (Scrophulariaceae)	South Africa [7]	Treatment of cardiac problems [7]	Cardenolide (digoxin) [7]	Positive inotropy, tachycardia [86b]
Acokanthera oppositifolia (Apocynaceae)	South Africa [7]	Arrow poison; treatment of headaches, snakebite, colds, toothache, etc. [86f]	Cardenolides, Acovenoside [86g]	Pain reliever; antisnake venom; cardiotoxic [86g]
Drimia sanguine (Hyacinthaceae/ Asparagaceae)	South Africa [7]	Blood purifier; treatment of edema, headaches, etc. [7]	Bufadienolides [7]	Cardiotoxic [7]
Bowiea volubilis (Hyacinthaceae/ Asparagaceae)	South Africa [7]	Blood purifier; treatment of edema, infertility, etc. [7]	Bufadienolides [7]	Cardiotoxic [7]
Convallaria majalis (Liliaceae)	South Africa [7]	Unconfirmed medicinal use as diuretic and cardiac tonic	Cardenolides (convallerin and convallamarin) [86h]	Cardiotoxic [86h]
Thevetia nerifolia or *Thevetia peruviana* (Apocynaceae)	West Africa [87b] South Africa [88]	Emetic, laxative, insecticide, arrow poison [87b,88]	Cardiac glycosides thevetins A and B, peruvoside, aucubine; saponins; steroids [87b,88]	Cardiotoxic, cardiotonic [87b]
Strophanthus hispidus (Apocynaceae)	West and other parts of Africa [87b,89a]	Arrow poison; treatment of parasites, malaria, dysentery, etc. [89a]	Cardenolides (strophanthins, ouabain) [89a]	Cardiotoxic, cause the heart to arrest in systole [89a]
Datura stramonium (Solanaceae)	South Africa [11,12]	For spiritual visionary purpose, analgesic, to treat asthma [89b]	Alkaloids, atropine, hyoscine, and hyoscyamine [11,12,89b]	Antiasthmatic, anticholinergic, antimicrobial, hypotensive, causes heart failure, multiple organ failure [11,12,89b]
Pavetta harborii (Rubiaceae)	South Africa [61]	Unspecified	Cationic polyamine [89c]	Cardiotoxic, can cause heart failure [61]
Pergularia daemia (Asclepiadaceae)	Nigeria [90,91]	Bronchitis and cough, rheumatic pain [90,91]	Cardenolides (e.g., uzarigenin, calotropin, pergularin) [90]	Acts on the uterus; hepatoprotective; cardiotoxic [90]
Solanum nigrum (Solanaceae)	West Africa [90,92a]	Fever, diarrhea, eye diseases, diuretic, antiepileptic, emmenagogic [90]	Alkaloids (solanine) [90,92b]	Hypotension, bradycardia [92b]

Table 13.2 African Medicinal Plants Screened for Their Cardioprotective Effects

Plant (family)	Area of Plant Collection	Traditional Use	Chemical Constituents	Pharmacological Activities
Prosopis glandulosa (Fabaceae)	South Africa [70]	Treatment of eye infections, wounds, stomach problems [92c]	Alkaloid (indolizidines) [92d]	Hypoglycemic, hypotensive, anti-ischemic, increase cardiomyocyte insulin sensitivity, cardioprotective [70,92c]
Globimetula cupulata (Loranthaceae)	Nigeria [93]	Many traditional uses including treatment of hypertension and diabetes [93]	Flavonoids, alkaloids, tannins [93]	Hypotensive, hypoglycemic, antidiabetic [93]
Curcuma longa (Zingiberaceae)	Egypt [94]	Carminative, antifungal, and as antiplatelet agent [94]	Volatile oil and curcuminoids [95b]	Protection against doxorubicin cardiotoxicity [94]
Stachys schimperi (Lamiaceae)	Egypt [96a]	Treatment of genital tumors, sclerosis of the spleen, inflammatory diseases, cough, and ulcers [96b]	Polyphenols [96a]	Protection against doxorubicin cardiotoxicity [96a]
Ginkgo biloba (Ginkgoaceae)	Egypt [97a]	Treatment of peripheral arterial disease and cerebral insufficiency in the elderly [97a]	Flavonoids, terpenoids [97a,b]	Protection against isoproterenol cardiotoxicity [97a,c,d]
Allium sativum (Alliaceae)	Egypt [98a]	Not specified	S-allyl cysteine [98a]	Protection against doxorubicin cardiotoxicity, immunomodulatory, hepatoprotective, antimutagenic, anticarcinogenic [98a,b]
Terminalia arjuna (Combretaceae)	Mauritius [31]	Aphrodisiac, expectorant, tonic, styptic, antidysenteric, purgative and laxative, cardiac ailments [31]	Flavonids, glycosides, tannins [31]	Cardiotonic, anti-ischemic, hypotensive, protection against isoproterenol cardiotoxicity [31–33]

Plant (family)	Location	Traditional use	Constituents	Pharmacological effects
Hibiscus sabdariffa (Malvaceae)	Nigeria [99a]	Treatment of hypertension, health beverage [99b]	Phenolics, anthocyanins [99a]	Attenuation of hypertension and reversal of cardiac hypertrophy [99a]
Vitis vinifera (Vitaceae)	Tunisia [95,100,101]	Treatment of hepatitis, stomach aches, etc. [95a]	Polyphenols (resveratrol) [100]	Protection against doxorubicin cardiotoxicity, antiaging, anticancer Protection against myocardial ischemia Alleviation of obesity and heart dysfunction Prevention of cardiac siderosis [100a–102]
Parkia biglobosa (Fabaceae)	Nigeria [80]	Treatment of arterial hypertension, piles, amoebiasis, bronchitis, cough, burn, zoster, and abscess [101b]	Flavonoids, cardiac glycosides, tannins, alkaloids [80]	Protection against doxorubicin cardiotoxicity [80]
Aspalathus linearis (Fabaceae)	South Africa [103a]	Health/functional beverage [103a]	Flavonoids, dihydrochalcone glucoside (aspalathin) [103b]	Protection against myocardial ischemia [103a]
Leonotis leonurus (Lamiaceae)	South Africa [104a]	Treatment of coughs, colds, influenza, bronchitis, headaches, high BP, and other cardiovascular ailments [104b]	Diterpenoid (marrubiin) [104a]	Anticoagulant, antiplatelet and anti-inflammatory effects, cardioprotective [104a]
Artemisia afra (Asteraceae)	South Africa [105a]	Treatment of heart-related diseases [105a]	Not specified	Protection against isoproterenol-induced myocardial injury [105a]
Rauvolfia vomitoria (Apocynaceae)	East Africa, West Africa [90]	Not specified	Alkaloids (raumitorine, reserpine, reserviline) [87,90,106]	Hypotensive, sedative, tranquilizing [90,105b]

(Continued)

Table 13.2 (Continued)

Plant (family)	Area of Plant Collection	Traditional Use	Chemical Constituents	Pharmacological Activities
Cryptolepis sanguinolenta (Periplocaceae)	West Africa [90]	Not specified	Alkaloids (cryptolepine) [90]	Hypotensive, causes vasodilation [90]
Morinda lucida (Rubiaceae)	West Africa [90]	Not specified	Tannins, methylanthraquinones, heteroside (morindin) [90]	Prevention of hypertension and its cerebral complications [90]
Strophantus spp. (Apocynaceae)	West and other parts of Africa [90]	Not specified	Steroid heteroside [90]	Cardiotonic [90]
Argemone mexicana (Papaveraceae)	Nigeria [90] Senegal	Diuretic, sedative and cholagogic properties [87b]	Alkaloids (berberine, protopine, coptisine, chelerythrine, etc.) [87b]	Vasodilating action [87b]
Solanum nigrum (Solanaceae)	West Africa [90,92a]	Fever, diarrhea, eye diseases, diuretic, antiepileptic, emmenagogic [90,92a]	Alkaloids, flavonoids, polysaccharides [90,92a]	Hypolipidemic, antihyperglycemic, hypotensive [90,92a]

antihypertensive studies are the most common. Some plants could be referred to as cardiovascular pharmatoxins since they are listed as both cardioprotective and cardiotoxic plants. This is due to multiple bioactive constituents. A plant can also show this behavior even though a single active compound predominates due to concentration effects.

13.7 Conclusion

Medicinal plants are the future of medicine. Phytomedicines are coming to equal prominence with orthodox medicines worldwide. However, some medicinal plants are becoming endangered species. It is interesting to note that many cardioprotective principles have been found in various plants, but these compounds are not going to clinical trials and then ultimately to pharmacy. The focal point of phytotherapy research should be drug development from medicinal plants. Phytochemicals that modulate enzymes that are targets for therapy should be characterized. Identified cardioactive phytochemicals or derived more effective analogues should be synthesized to preserve endangered species. The synergistic bioactivity of phytochemicals in plant extracts often is touted as an advantage that is difficult to replicate by single synthetic drugs. It is our suggestion that "hybrid phytochemicals" that can mimic this synergistic property could be synthesized using templates from plants.

Despite advancements in the fight against cardiovascular diseases, it has remained the number one cause of mortality worldwide. Cardioprotective medicinal plants can greatly contribute to stemming the tide of this disease. However, because only a thin line of demarcation exists between a medicinal plant being cardioprotective and its being a poison at times, proper classification and screening is necessary. This information should be available to all users of medicinal plants.

African laboratories need to be better equipped to employ newer, better, more sensitive, faster, and at times less costly and cumbersome methods of screening for cardiotoxic and cardioprotective botanicals. Governments should put a proper regulatory framework in place to ensure the safe use of cardioactive plants. Data emanating from research laboratories will be useful in this regard.

References

[1] Sato M, Yokoyama U, Fujita T, Okumura S, Ishikawa Y. The roles of cytochrome P450 in ischemic heart disease. Curr Drug Metab 2011;12:526−32.

[2] Williams PL, Warwick R. Gray's anatomy. 37th ed. Edinburgh: Churchill Livingstone; 1989. p. 1598.

[3] (a) Malmivuo J, Plonsey R. The heart. Bioelectromagnetism. Principles and applications of bioelectric and biomagnetic fields. New York, NY: Oxford University Press; 1995. p. 119−22.

(b) Barrett K, Barman S, Boitano S, Brooks H. Ganong's review of medical physiology. 23rd ed. New York, NY: McGraw-Hill; 2010.

(c) Hall JE, Guyton AC. Guyton and Hall textbook of medical physiology. 12th ed. Philadelphia, PA: Saunders/Elsevier; 2011.

(d) Parameswaran S, Kumar S, Verma RS, Sharma RK. Cardiomyocyte culture—an update on the *in vitro* cardiovascular model and future challenges. Can J Physiol Pharmacol 2013;91(12):985−98.

[4] Seeley RR, Stephens TD, Tate P. Heart Essentials of anatomy and physiology. Boston, MA: McGraw-Hill Higher Education; 2007. p. 321−52.

[5] Taegtmeyer H. Cardiac metabolism as a target for the treatment of heart failure. Circulation 2004;110:894−6.

[6] (a) Taegtmeyer H. Genetics of energetics: transcriptional responses in cardiac metabolism. Ann Biomed Eng 2000;28:871−6.

(b) Barger PM, Kelly DP. PPAR signaling in the control of cardiac energy metabolism. Trends Cardiovasc Med 2000;10:238−45.

(c) Netter FH. Heart. The ciba collection of medical illustrations. Summit, NJ: Ciba Pharmaceutical Company; 1971. p. 293.

[7] van der Bijl Jr P, van der Bijl Sr. P. Cardiotoxicity of plants in South Africa. Cardiovasc J Afr 2012;23(9):477−8.

[8] Knight APC, Walter RG. Plants affecting the cardiovascular system. A guide to plant poisoning of animals in North America. Ithaca, NY: Knight APC and Walter RG Publisher, Teton NewMedia, Jackson; 2011.

[9] Clarke ML, Harvey DG, Humphreys DJ. Veterinary toxicology. 2nd ed. London: Bailliere Tindall; 1981. p. 256−57.

[10] Majak W, McDiarmid RE, Cristofoli W. Content of zygacine in *Zygadenus venosus* at different stages of growth. Phytochemistry 1992;31:3417−8.

[11] Verstraete F. Management and regulation of certain bioactive compounds present as inherent toxins in plants intended for feed and food. In: Bernhoft A, editor. Bioactive compounds in plants—benefits and risks for man and animals. Proceedings from a symposium held at The Norwegian Academy of Science and Letters, Oslo, November 2008. Oslo: Novus forlag; 2010.

[12] European Food Safety Authority. Scientific opinion of the panel on contaminants in the food chain on a request from the European Commission on Tropane alkaloids (from *Datura* spp.) as undesirable substances in animal feed. EFSA J 2008;691:155.

[13] Roberts MF, Wink M. Alkaloids: biochemistry, ecology, and medicinal applications. New York, NY: Plenum Press; 1998.

[14] Caksen H, Odabaş D, Akbayram S, Cesur Y, Arslan S, Uner A, et al. Deadly nightshade (*Atropa belladonna*) intoxication: an analysis of 49 children. Hum Exp Toxicol 2003;22(12):665−8.

[15] Tai YT, But RR, Young K, Lau CR. Cardiotoxicity after accidental herb-induced aconite poisoning. Lancet 1992;340:1254−6.

[16] Guha S, Dawn B, Dutta F, Chakraborty T, Pain S. Bradycardia, reversible panconduction defect and syncope following self-medication with a homeopathic medicine. Cardiology 1999;91:268−71.

[17] Lin CC, Chan TY, Deng JF. Clinical features and management of herb-induced aconitine poisoning. Ann Emerg Med 2004;43:574−9.

[18] Lai CK, Poon WT, Chan YW. Hidden aconite poisoning: identification of yunaconitine and related aconitum alkaloids in urine by liquid chromatography−tandem mass spectrometry. J Anal Toxicol 2006;30(7):426−33.

[19] Dwivedi S, Aggarwal A, Sharma V. Cardiotoxicity from "safe" herbomineral formulations. Trop Doct 2011;41(2):113—5.

[20] Chrispal AJ. *Cleistanthus collinus* poisoning. J Emerg Trauma Shock 2012;5 (2):160—6.

[21] Parasuraman S, Raveendran R, Madhavrao C. GC—MS analysis of leaf extracts of *Cleistanthus collinus* Roxb. (Euphorbiaceae). Int J Ph Sci 2009;1:284—6.

[22] Pretorius PJ, Terblanche M. A preliminary study on the symptomatology and cardiodynamics of gousiekte in sheep and goats. J S Afr Vet Med Assoc 1967;38:29—53.

[23] Newsholme SJ, Coetzer JAW. Myocardial pathology of ruminants in Southern Africa. J S Afr Vet Med Assoc 1984;55:89—96.

[24] (a) Konya RS, Elliott RC. The coronary vasoconstrictor action of extract IV from the Dahlia sea anemone *Tealia felina* L. Toxicon 1996;34(2):277—82.

(b) Obatomi DK, Bach PH. Biochemistry and toxicology of the diterpenoid glycoside atractyloside. Food Chem Toxicol 1998;36(4):335—46.

[25] Ajabnoor MA. Effect of aloes on blood glucose levels in normal and alloxan diabetic mice. J Ethnopharmacol 1990;28(2):215—20.

[26] Agarwal OP. Prevention of atheromatous heart disease. Angiology 1985;36(8):485—92.

[27] Sakai T, Repko BM, Griffith BP, Waters JH, Kameneva MV. IV infusion of a drag-reducing polymer extracted from *Aloe vera* prolonged survival time in a rat model of acute myocardial ischaemia. Br J Anaesth 2007;98(1):23—8.

[28] tHart LA, Van den Berg AJ, Kuis L, van Dijk H, Labadie RP. An anti-complementary polysaccharide with immunological adjuvant activity from the leaf parenchyma gel of *Aloe vera*. Planta Med 1989;55(6):509—12.

[29] Dagne E, Bisrat D, Viljoen A, Van Wyk BE. Chemistry of *Aloe* species. Curr Org Chem 2000;4:1055—78.

[30] Hamman JH. Composition and applications of *Aloe vera* leaf gel. Molecules 2008;13:1599—616.

[31] Dwivedi S. *Terminalia arjuna*, Wight & Arn: a useful drug for cardiovascular disorders. J Ethnopharmacol 2007;114(2):114—29.

[32] Gauthaman K, Banerjee SK, Dinda AK, Ghosh CC, Maulik SK. *Terminalia arjuna* (Roxb.) protects rabbit heart against ischemic—reperfusion injury: role of antioxidant enzymes and heat shock protein. J Ethnopharmacol 2005;99(3):403—7.

[33] Tilak-Jain JA, Devasagayam TPA. Cardioprotective and other beneficial effects of some Indian medicinal plants. J Clin Biochem Nutr 2006;38:9—18.

[34] Thomson M, Ali M. Garlic (*Allium sativum*): a review of its potential use as an anti-cancer agent. Curr Cancer Drug Targets 2003;3(1):67—81.

[35] Ensminger AH. Foods & nutrition encyclopedia. Boca Raton, FL: CRC Press; 1994.

[36] Simonetti G, Schuler S. Simon & Schuster's guide to herbs and spices. New York, NY: Simon & Schuster, Inc.; 1990.

[37] Iwu MM. Handbook of African medicinal plants. Boca Raton, FL: CRC Press; 1993.

[38] Kapoor LD. Handbook of ayurvedic medicinal plants. Boca Raton, FL: CRC Press; 2000.

[39] (a) Ravindran PN. Turmeric: the golden spice of life. In: Ravindran PN, Nirmal Babu K, Sivaraman K, editors. Turmeric: the genus *Curcuma*. Boca Raton, FL: CRC Press; 2007.

(b) Chattopadhyay I, Biswas K, Bandyopadhyay U, Banerjee RK. Turmeric and curcumin: biological actions and medicinal applications. Curr Sci India 2004;87:44—53.

[40] Dixit VP, Jain P, Joshi SC. Hypolipidaemic effects of *Curcuma longa* L and *Nardostachys jatamansi* DC in triton-induced hyperlipidaemic rats. Indian J Physiol Pharmacol 1988;32(4):299—304.

[41] Mohanty I, Singh Arya D, Dinda A, Joshi S, Talwar KK, Gupta SK. Protective effects of *Curcuma longa* on ischemia-reperfusion induced myocardial injuries and their mechanisms. Life Sci 2004;75(14):1701—11.

[42] Zahid Ashraf M, Hussain ME, Fahim M. Antiatherosclerotic effects of dietary supplementations of garlic and turmeric: restoration of endothelial function in rats. Life Sci 2005;77(8):837—57.

[43] Adaramoye OA, Anjos RM, Almeida MM, Veras RC, Silvia DF, Oliveira FA, et al. Hypotensive and endothelium-independent vasorelaxant effects of methanolic extract from *Curcuma longa* L. in rats. J Ethnopharmacol 2009;124(3):457—62.

[44] Xu P, Yao Y, Guo P, Wang T, Yang B, Zhang Z. Curcumin protects rat heart mitochondria against anoxia-reoxygenation induced oxidative injury. Can J Physiol Pharmacol 2013;91(9):715—23.

[45] Sahebkar A. Are curcuminoids effective C-reactive protein-lowering agents in clinical practice? evidence from a meta-analysis. Phytother Res 2013. Available from: http://dx.doi.org/doi:10.1002/ptr.5045.

[46] Wang NP, Wang ZF, Tootle S, Philip T, Zhao ZQ. Curcumin promotes cardiac repair and ameliorates cardiac dysfunction following myocardial infarction. Br J Pharmacol 2012;167(7):1550—62.

[47] Mito S, Watanabe K, Harima M, Thandavarayan RA, Veeraveedu PT, Sukumaran V, et al. Curcumin ameliorates cardiac inflammation in rats with autoimmune myocarditis. Biol Pharm Bull 2011;34(7):974—9.

[48] Liebgott T, Miollan M, Berchadsky Y, Drieu K, Culcasi M, Pietri S. Complementary cardioprotective effects of flavonoid metabolites and terpenoid constituents of *Ginkgo biloba* extract during ischemia and reperfusion. Basic Res Cardiol 2000;95(5):368—77.

[49] Brinkley TE, Lovato JF, Arnold AM, Furberg CD, Kuller LH, Burke GL, et al. Ginkgo evaluation of memory (GEM) study investigators. Effect of *Ginkgo biloba* on blood pressure and incidence of hypertension in elderly men and women. Am J Hypertens 2010;23(5):528—33.

[50] El-Boghdady NA. Increased cardiac endothelin-1 and nitric oxide in adriamycin-induced acute cardiotoxicity: protective effect of *Ginkgo biloba* extract. Indian J Biochem Biophys 2013;50(3):202—9.

[51] El-Boghdady NA. Antioxidant and antiapoptotic effects of proanthocyanidin and *Ginkgo biloba* extract against doxorubicin-induced cardiac injury in rats. Cell Biochem Funct 2013;31(4):344—51.

[52] Lelièvre LG, Lechat P. Mechanisms, manifestations, and management of digoxin toxicity. Heart Metab 2007;35:9—11.

[53] Bauman JL, Didomenico RJ, Galanter WL. Mechanisms, manifestations, and management of digoxin toxicity in the modern era. Am J Cardiovasc Drugs 2006;6(2):77—86.

[54] Ferrier G. Digitalis arrhythmias: role of oscillatory after potentials. Prog Cardiovasc Dis 1977;19:459—74.

[55] Dwivedi S. Aconite induced cardiac arrhythmia. J Assoc Physicians India 1993;41: 618—9.

[56] Fu M, Wu M, Qiao Y, Wang Z. Toxicological mechanisms of aconitum alkaloids. Pharmazie 2006;61(9):735—41.

[57] Thomas K, Dayal AK, Narasimhan, Ganesh A, Sheshadri MS, Cherian AM, et al. Metabolic and cardiac effects of *Cleistanus collinus* poisoning. J Assoc Physc India 1991;39:312—4.

[58] Nandakumar NV, Pagala MK, Venkatachari SA, Namba T, Grob D. Effect of *Cleistanthus collinus* leaf extract on neuromuscular function of the isolated mouse phrenic nerve-diaphragm. Toxicon 1989;27(11):1219—28.

[59] Sarathchandra G, Balakrishnamurthy P. Pertubations in glutathione and adenosine triphosphatase in acute oral toxicosis of *Cleistanthus collinus*: an indigenous toxic plant. Indian J Pharmacol 1997;29:82—5.

[60] Schultz RA, Fourie N, Basson KM, Labuschagne L, Prozesky L. Effect of pavetamine on protein synthesis in rat tissue. Onderstepoort J Vet Res 2001;68:325—30.

[61] Hay L, Schultz RA, Schutte PJ. Cardiotoxic effects of pavetamine extracted from *Pavetta harborii* in the rat. Onderstepoort J Vet Res 2008;75:249—53.

[62] Brown JH, Taylor P. Muscaric receptor agonist and antagonists. In: Brunton LL, Lazo JS, Parker KL, editors. Goodman & Gilman's the pharmacological basis of therapeutics. 11th ed. New York, NY: McGraw-Hill; 2006.

[63] Tshibangu JN, Chifunder K, Kaminsky R, Wright AD, Konig GM. Screening of African medicinal plants for antimicrobial and enzyme inhibitory activity. J Ethnopharmacol 2002;80:25—35.

[64] Nicolescu AC, Reynolds JN, Barclay LRC, Gregory RJ, Thatcher GRJ. Organic nitrites and NO: inhibition of lipid peroxidation and radical reactions. Chem Res Toxicol 2004;17:185—96.

[65] Dhalla NS, Elmoselhi AB, Hata T, Naoki Makino N. Status of myocardial antioxidants in ischemia—reperfusion injury. Cardiovasc Res 2000;47:446—56.

[66] Hausenloy DJ, Tsang A, Mocanu MM, Yellon DM. Ischemic preconditioning protects by activating prosurvival kinases at reperfusion. Am J Physiol Heart Circulat Physiol 2005;288:971—6.

[67] Jonassen AK, Sack MN, Mjøs OD, Yellon DM. Myocardial protection by insulin at reperfusion requires early administration and is mediated *via* Akt and p70s6 kinase cell-survival signaling. Circ Res 2001;89:1191—8.

[68] Di R, Wu X, Chang Z, Zhao X. S6K inhibition renders cardiac protection against myocardial infarction through PDK1 phosphorylation of Akt. Biochem J 2012;441: 199—207.

[69] Huisamen B, Donthi, Lochner A. Insulin in combination with vanadate stimulates glucose transport in isolated cardiomyocytes from obese Zucker rats. Cardiovasc Drugs Ther 2001;15:445—52.

[70] Huisamen B, George C, Dietrich D, Genade S. Cardioprotective and anti-hypertensive effects of *Prosopis glandulosa* in rat models of pre-diabetes. Cardiovasc J Afr 2013;24 (2):10—6.

[71] Hong D, Zeng X, Xu W, Ma J, Tong Y, Chen Y. Altered profiles of gene expression in curcumin-treated rats with experimentally induced myocardial infarction. Pharmacol Res 2010;61(2):142—8.

[72] Andersson H, Steel D, Asp J, Dahlenborg K, Jonsson M, Jeppsson A, et al. Assaying cardiac biomarkers for toxicity testing using biosensing and cardiomyocytes derived from human embryonic stem cells. J Biotechnol 2010;150(1):175—81.

[73] Su X, Young EW, Underkofler HA, Kamp TJ, January CT, Beebe DJ. Microfluidic cell culture and its application in high-throughput drug screening: cardiotoxicity assay for hERG channels. J Biomol Screen 2011;16(1):101—11.

[74] Wible BA, Hawryluk P, Ficker E, Kuryshev YA, Kirsch G, Brown AM. HERG-Lite: a novel comprehensive high-throughput screen for drug-induced hERG risk. J Pharmacol Toxicol Methods 2005;52(1):136—45.

[75] Braam SR, Tertoolen L, van de Stolpe A, Meyer T, Passier R, Mummery CL. Prediction of drug-induced cardiotoxicity using human embryonic stem cell-derived cardiomyocytes. Stem Cell Res 2010;4(2):107−16.

[76] Skrzypiec-Spring M, Grotthus B, Szelag A, Schulz R. Isolated heart perfusion according to Langendorff—still viable in the new millennium. J Pharmacol Toxicol Methods 2007;55(2):113−26.

[77] Akinmoladun AC, Obuotor EM, Barthwal MK, Dikshit M, Farombi EO. Ramipril-like activity of *Spondias mombin* Linn against no-flow ischemia and isoproterenol-induced cardiotoxicity in rat heart. Cardiovasc Toxicol 2010;10(4):295−305.

[78] Bell RM, Mocanu MM, Yellon DM. Retrograde heart perfusion: the Langendorff technique of isolated heart perfusion. J Mol Cell Cardiol 2011;50(6):940−50.

[79] MacDougall DA, Calaghan S. A novel approach to the Langendorff technique: preparation of isolated cardiomyocytes and myocardial samples from the same rat heart. Exp Physiol 2013;98(8):1295−300.

[80] Komolafe K, Akinmoladun AC, Olaleye MT. Methanolic leaf extract of *Parkia biglobosa* protects against doxorubicin-induced cardiotoxicity in rats. Int J Appl Res Nat Prod 2013;6(3):39−47.

[81] Upaganlawar A, Balaraman R. Cardioprotective effects of *Lagenaria siceraria* fruit juice on isoproterenol-induced myocardial infarction in Wistar rats: a biochemical and histoarchitecture study. J Young Pharm 2011;3(4):297−303.

[82] Nandave M, Ojha SK, Joshi S, Kumari S, Arya DS. *Moringa oleifera* leaf extract prevents isoproterenol-induced myocardial damage in rats: evidence for an antioxidant, antiperoxidative, and cardioprotective intervention. J Med Food 2009;12(1):47−55.

[83] Li H, Xie YH, Yang Q, Wang SW, Zhang BL, Wang JB, et al. Cardioprotective effect of paeonol and danshensu combination on isoproterenol-induced myocardial injury in rats. PLoS One 2012;7(11):e48872.

[84] (a) Akinloye O, Akinmoladun AC. Modulatory effect of *Psidium guajava* Linn and *Ocimum gratissimum* Linn on lipid profile and selected biochemical indices in rabbits fed high cholesterol diet. J Complement Integr Med 2010;7:1.

(b) O'Brien PJ. Cardiac troponin is the most effective translational safety biomarker for myocardial injury in cardiotoxicity. Toxicology 2008;245(3):206−18.

(c) Erickson JR, Joiner ML, Guan X, Kutschke W, Yang J, Oddis CV, et al. A dynamic pathway for calcium-independent activation of CaMKII by methionine oxidation. Cell 2008;133(3):462−74.

[85] Lin KY, Chang WT, Lai YC, Liau I. Towards functional screening of cardioactive and cardiotoxic drugs with zebrafish *in vivo* using pseudo-dynamic three-dimensional imaging. Anal Chem 2014. Available from: http://dx.doi.org/doi:10.1021/ac403877h.

[86] (a) Adome RO, Gachihi JW, Onegi B, Tamale J, Apio SO. The cardiac effect of the crude ethanolic extract of *Nerium oleander* in the isolated guinea pig hearts. Afr Health Sci 2003;3(2):77−82.

(b) Hauptman PJ, Kelly RA. Digitalis. Circulation 1999;99(9):1265−70.

(c) Sandberg F, Cronlund A. An ethnopharmacological inventory of medicinal and toxic plants from equatorial Africa. J Ethnopharmacol 1982;5(2):187−204.

(d) Tahraoui A, El-Hilaly J, Israili ZH, Lyoussi B. Ethnopharmacological survey of plants used in the traditional treatment of hypertension and diabetes in south-eastern Morocco (Errachidia province). J Ethnopharmacol 2007;110(1):105−17.

(e) Bisset NG. Arrow and dart poisons. J Ethnopharmacol 1989;25(1):1−41.

(f) Van Wyk B, van Oudtshoorn B, Gericke N. Medicinal Plants of South Africa. Pretoria: Briza Publications; 2000.

(g) Bethwell OO. *Acokanthera oppositifolia* (Lam.) Codd. In: Schmelzer GH, Gurib-Fakim A, editors. Prota 11(1): Medicinal plants/Plantes médicinales 1 [CD-Rom]. Wageningen, the Netherlands: PROTA; 2007.

(h) Wink M. Mode of action and toxicology of plant toxins and poisonous plants. Mitt Julius Kühn-Inst 2009;421:93—112.

[87] (a) Oliver-Bever BEP. Medicinal plants in Tropical West Africa. Cambridge: University Press; 1986.

(b) Oliver-Bever B. Medicinal plants in tropical West Africa. I: Plants acting on the cardiovascular system. J Ethnopharmacol 1982;5(1):1—72.

(c) Ndhlala AR, Ncube B, Okem A, Mulaudzi RB, Van Staden J. Toxicology of some important medicinal plants in southern Africa. Food Chem Toxicol 2013;62:609—21.

[88] Burkill HM. The useful plants of West Tropical Africa. London: The Whitefrairs Press Ltd; 1985.

[89] (a) Beentje HJ. In: Schmelzer GH, Gurib-Fakim A, editors. *Strophanthus hispidus* DC. [Internet] Record from PROTA4U. Wageningen, the Netherlands: PROTA (Plant Resources of Tropical Africa/Ressources végétales de l'Afrique tropicale); 2006<http://www.prota4u.org/search.asp> [accessed 13.01.14].

(b) Soni P, Siddiqui AA, Dwivedi J, Soni V, Sahu RK. Pharmacological properties of *Datura stramonium* L. as a potential medicinal tree: an overview. Asian Pac J Trop Biomed 2012;2(12):1002—8.

(c) Fourie N, Erasmus GL, Schultz RA, Prozesky L. Isolation of toxin responsible for gousiekte, a plant induced cardiomyopathy of ruminants in southern Africa. Onderstepoort J Vet Res 1995;62:77—87.

[90] Bakhlet AO, Adam SAI. Therapeutic utility, constituents and toxicity of some medicinal plants: a review. Vet Human Toxicol 1995;37(3):255—8.

[91] Patel MS, Rowson JM. Investigation of certain Nigerian medicinal plants. Part 1: Preliminary pharmacological and phytochemical screening for cardiac activity. Planta Med 1964;12:34—42.

[92] (a) Atanu FO, Ebiloma UG, Ajayi EA. A review of the pharmacological aspects of *Solanum nigrum* Linn. Biotec Mol Biol Rev 2011;6(1):001—7.

(b) Slaughter RJ, Beasley DM, Lambie BS, Wilkins GT, Schep LJ. Poisonous plants in New Zealand: a review of those that are most commonly enquired about to the National Poisons Centre. N Z Med J 2012;125(1367):87—118.

(c) George C, Lochner A, Huisamen B. The efficacy of *Prosopis glandulosa* as antidiabetic treatment in rat models of diabetes and insulin resistance. J Ethnopharmacol 2011;137(1):298—304.

(d) Samoylenko V, Ashfaq MK, Jacob MR, Tekwani BL, Khan SI, Manly SP, et al. Indolizidine, antiinfective and antiparasitic compounds from *Prosopis glandulosa* var. glandulosa. J Nat Prod 2009;72(1):92—8.

[93] Ojewole JAO, Ojewole SO. Hypoglycaemic and hypotensive effects of *Globimetula cupilata* (DC) Van Tieghem (Loranthaceae) aqueous leaf extract in rats. Cardiovasc J South Afr 2007;18:9—15.

[94] El-Sayed EM, El-azeem AA, Afify AA, Shabane MH, Ahmed H. Cardioprotective effects of *Curcuma longa* L. extracts against doxorubicin-induced cardiotoxicity in rats. J Med Plants Res 2011;17(5):4049—58.

[95] (a) Mansour R, Haouas N, Ben Kahla-Nakbi A, Hammami S, Mighri Z, Mhenni F, et al. The effect of *Vitis vinifera* L. leaves extract on *Leishmania infantum*. Iran J Pharm Res 2013;12(3):349—55.

(b) Pothitirat W, Gritsanapan W. Variation of bioactive components in *Curcuma longa* in Thailand. Current Sci 2006;91(10):1397–400.

[96] (a) Abdel-Sattar E, El-Gayed SH, Shehata I, Ashour O, Nagy AA, Mohamadin AM. Antioxidant and cardioprotective activity of *Stachys schimperi* Vatke against doxorubicin induced Cardiotoxicity. Bull Fac Pharm, Cairo Univ 2012;50:41–7.

(b) Hartwell JL. Plants used against cancer. A survey. Massachusetts: Quarterman Publications Inc; 1982.

[97] (a) Kamel MF, Radwan DM, Amin HA. Protective effect of *Ginkgo biloba* against experimental cardiotoxicity induced by isoproterenol in adult male albino rats a histological and biochemical study. Egypt J Histol 2010;33:735–44.

(b) Lugasi A, Horvahovich P, Dworschak E. Additional information to the *in vitro* antioxidant activity of *Ginkgo biloba*. Phytother Res 1999;13:160–2.

(c) Chen X, Salwinski S, Lee JJ. Extracts of *Ginkgo biloba* and ginsenosides exert cerebral vasorelaxation via a nitric oxide pathway. Clin Exp Pharmacol Physiol 1997;24:958–9.

(d) Perry EK, Pickering AT, Wang WW, Houghton PJ, Perry NS. Medicinal plants and Alzheimer's disease: from ethnobotany to phytotherapy. J Pharm Pharmacol 1999;51(5):527–34.

[98] (a) Al-Kreathy HM, Zoheir AD, Nessar A, Mark S, Osman AM. Mechanisms of cardioprotective effect of aged garlic extract against doxorubicin-induced cardiotoxicity. Integr Canc Ther 2011;15:364–71.

(b) Agarwal K. Therapeutic actions of garlic constituents. Med Res Rev 1996;16:111–24.

(c) Ajibesin KK, Ekpo BA, Bala DN, Essien EE, Adesanya SA. Ethnobotanical survey of Akwa Ibom State of Nigeria. J Ethnopharmacol 2008;115:387–408.

(d) Ayoka AO, Akomolafe RO, Iwalewa EO, Akanmu MA, Ukponmwan OE. Sedative, antiepileptic and antipsychotic effects of *Spondias mombin* L. (Anacardiaceae) in mice and rats. J Ethnopharmacol 2006;103:166–75.

[99] (a) Odigie IP, Ettarh RR, Adigun SA. Chronic administration of aqueous extract of *Hibiscus sabdariffa* attenuates hypertension and reverses cardiac hypertrophy in 2K-1C hypertensive rats. J Ethnopharmacol 2003;86:181–5.

(b) Mozaffari-Khosravi H, Jalali-Khanabadi BA, Afkhami-Ardekani M, Fatehi F, Noori-Shadkam M. The effects of sour tea (*Hibiscus sabdariffa*) on hypertension in patients with type II diabetes. J Hum Hypertens 2009;23(1):48–54.

[100] Mokni M, Hamlaoui-Guesmi S, Amri M, Marzouki L, Limam F, Aouani E. Grape seed and skin extract protects against acute chemotherapy toxicity induced by doxorubicin in rat heart. Cardiovasc Toxicol 2012;20:158–65.

[101] (a) Mokni M, Limam F, Elkahoui S, Amri M, Aouani E. Strong cardioprotective effect of resveratrol, a red wine polyphenol, on isolated rat hearts after ischemia/reperfusion injury. Arch Biochem Biophys 2007;457(1):1–6.

(b) Olaleye MT, Komolafe K, Akindahunsi AA. Effect of methanolic leaf extract of *Parkia biglobosa* on some biochemical indices and hemodynamic parameters in rats. J Chem Pharmaceut Res 2013;5(1):213–20.

[102] Charradi K, Sebai H, Elkahoui S, Ben Hassine F, Limam F, Aouani E. Grape seed extract alleviates high-fat diet-induced obesity and heart dysfunction by preventing cardiac siderosis. Cardiovasc Toxicol 2011;11(1):28–37.

[103] (a) Pantsi WG, Marnewick JL, Esterhuyse AJ, Rautenbach F, van Rooyen J. Rooibos (*Aspalathus linearis*) offers cardiac protection against ischaemia/reperfusion in the isolated perfused rat heart. Phytomedicine 2011;18(14):1220–8.

(b) Shimamura N, Miyase T, Umehara K, Warashina T, Fujii S. Phytoestrogens from *Aspalathus linearis*. Biol Pharm Bull 2006;29:1271−4.

[104] (a) Mnonopi N, Levendal RA, Davies-Coleman MT, Frost CL. The cardioprotective effects of marrubiin, a diterpenoid found in *Leonotis leonurus* extracts. J Ethnopharmacol 2011;138(1):67−75.

(b) Bienvenu E, Amabeoku GJ, Eagles PK, Scott G, Springfield EP. Anticonvulsant activity of aqueous extract of *Leonotis leonurus*. Phytomedicine 2002;9:217−23.

[105] (a) Sunmonu TO, Afolayan AJ. Protective effect of *Artemisia afra* Jacq. on isoproterenol-induced myocardial injury in Wistar rats. Food Chem Toxicol 2010;48(7):1969−72.

(b) Kerharo J, Adam JG. La pharmacopee senegalai5e traditionelle. Paris: Vigot; 1974.

[106] Watt JM, Breyer-Brandwijk MG. Medicinal and poisonous plant" of southern and Eastern Africa. Edinburgh: E. & S. Livingstone; 1962.

14 Neurotoxicity and Neuroprotective Effects of African Medicinal Plants

Germain S. Taïwe[1] and Victor Kuete[2]

[1]Department of Zoology and Animal Physiology, University of Buea, Cameroon, [2]University of Dschang, Faculty of Science, Department of Biochemistry, Dschang, Cameroon

14.1 Introduction

Neurotoxicity refers to the ability of an agent to adversely affect the structural or functional integrity of the nervous system. Structural damage to nervous system components usually results in altered functioning, although the reverse is not always true. Alterations in nervous system function may occur through toxicant interactions with the normal signaling mechanisms of neurotransmission, resulting in little or no structural damage. Nevertheless, it is easier to identify alterations, be they structural or functional, than it is to define adversity. For example, the stimulant effect of a morning cup of coffee may be too anxiety-provoking for some individuals, but a necessity to others [1]. Many substances alter the normal activity of the nervous system. Sometimes these effects are immediate and transient, like the stimulatory effect of a cup of coffee or a headache from the fresh paint in your office. Other effects can be much more insidious, such as the movement disorders suffered by miners after years of chronic manganese intoxication. Many agents are safe or even therapeutic at lower doses, but become neurotoxic at higher levels. Trace metals and pyridoxine (vitamin B6) fall into this category of dose-dependent effects. Because these agents affirm the maxim, "the dose makes the poison," it becomes necessary to have a meaningful definition of nervous system poisoning or neurotoxicity [2]. The search for new molecules that act on the central nervous system (CNS) and that can be used for therapeutic purposes started with several studies in the nineteenth century. In fact, the first drugs used to treat pathologic conditions of the CNS were based on natural resources, specifically on plants. However, studies targeting plants with this type of bioactivity represent only a very small percentage of those investigations. In a review of the existing literature, it appears that African medicinal plants with molecules that produce this kind of activity are increasingly attractive targets for the development of new drugs.

Toxicological Survey of African Medicinal Plants. DOI: http://dx.doi.org/10.1016/B978-0-12-800018-2.00014-5

Most assessments of neurotoxic and neuroprotective activities of African medicinal plants were conducted using the *in vivo* or *in vitro* neurotoxicity studies and developmental neurotoxicity. An increasing number of herbal products, represented by *Hypericum perforatum* (Clusiaceae) known as St. John's Wort, *Panax ginseng* (Araliaceae) commonly known as ginseng, and *Ginkgo biloba* (Ginkgoaceae), have been introduced into neuropathology practice in the past decade [3]. In this chapter, a brief introduction to the CNS is presented and its functions are described. The objective of this review is to provide an overview of African medicinal plants that have been shown to have neurotoxic and/or neuroprotective effects in experimental animal models of nervous system diseases and their hypothetical mechanisms of actions.

14.2   The CNS and the Metabolism of Xenobiotics

14.2.1   Overview of the CNS

The CNS includes the brain, brain stem, and spinal cord. The brain is held in the cranial cavity of the skull and it consists of the cerebrum, cerebellum, and the brain stem. The brain and spinal cord are protected by bony structures, membranes, and fluid. The nerves involved are cranial nerves and spinal nerves [4].

14.2.2   General Functions of the CNS

The CNS represents the largest part of the nervous system, including the brain and the spinal cord. Together, with the peripheral nervous system, it has a fundamental role in the control of behavior [4]. The CNS is conceived as a system devoted to information processing, where an appropriate motor output is computed as a response to a sensory input. In addition to these basic vital functions, the nervous systems of higher organisms are responsible for feeling, thinking, and learning [5]. Many threads of research suggest that motor activity exists well before the maturation of the sensory systems and senses only influence behavior without dictating it. This has brought the conception of the CNS as an autonomous system [4].

14.2.3   Cellular Elements in the CNS

The human CNS contains about 10^{11} (100 billion) neurons. It also contains $10-50$ times this number of glial cells. The CNS is a complex organ; it has been calculated that 40% of the human genes participate, at least to a degree, in its formation. The neurons, the basic building blocks of the nervous system, have evolved from primitive neuroeffector cells that respond to various stimuli by contracting [5]. In more complex animals, contraction has become the specialized function of muscle cells, whereas integration and transmission of nerve impulses have become the specialized functions of neurons [3].

14.2.3.1 Neurons

The basic unit of the nervous system is the neuron, a type of cell that is structurally and functionally specialized to receive, integrate, conduct, and transmit information. Neurons in the mammalian CNS come in many different shapes and sizes. The cell body (soma) contains the nucleus and is the metabolic center of the neuron. Neurons have several processes called dendrites that extend outward from the cell body and arborize extensively. Particularly in the cerebral and cerebellar cortex, the dendrites have small knobby projections called dendritic spines [5]. A typical neuron also has a long fibrous axon that originates from a somewhat thickened area of the cell body, the axon hillock. The first portion of the axon is called the initial segment. The axon divides into presynaptic terminals, each ending in a number of synaptic knobs which are also called terminal buttons or boutons [4]. They contain granules or vesicles in which the synaptic transmitters secreted by the nerves are stored. Based on the number of processes that emanate from their cell body, neurons can be classified as unipolar, bipolar, and multipolar.

14.2.3.2 Glial Cells

There are two major types of glial cells in the vertebrate nervous system: microglia and macroglia. Microglia are scavenger cells that resemble tissue macrophages and remove debris resulting from injury, infection, and disease (e.g., multiple sclerosis, acquired immunodeficiency syndrome-related dementia, Parkinson disease, and Alzheimer disease). Microglia arise from macrophages outside of the nervous system and are physiologically and embryologically unrelated to other neural cell types [5].

The three types of macroglia includes oligodendrocytes, Schwann cells, and astrocytes. Oligodendrocytes and Schwann cells are involved in myelin formation around axons in the CNS and peripheral nervous system, respectively. Astrocytes, which are found throughout the brain, are of two subtypes. Fibrous astrocytes, which contain many intermediate filaments, are found primarily in white matter. Protoplasmic astrocytes are found in gray matter and have a granular cytoplasm [4,5]. Both types send processes to blood vessels, where they induce capillaries to form the tight junctions making up the blood—brain barrier. They also send processes that envelop synapses and the surface of nerve cells. Protoplasmic astrocytes have a membrane potential that varies with the external K^+ concentration but do not generate propagated potentials. They produce substances that are tropic to neurons, and they help maintain the appropriate concentration of ions and neurotransmitters by taking up K^+ and the neurotransmitters glutamate and γ-aminobutyrate [5].

14.2.4 The Blood—Brain Barrier

A highly selective barrier, known as the blood—brain barrier, exists between specific blood-borne substances and the neural tissue of the CNS. The brain is protected against the entry of many chemicals by the blood—brain barrier cells. Chemicals can enter the brain by at least two mechanisms: passive diffusion, which applies mostly to small lipophilic compounds and active transport of compounds of physiological importance. It is important to recognize that a chemical that is a potent neurotoxic agent *in vitro* is not necessarily neurotoxic *in vivo* if it does not enter the brain because of the blood−brain barrier [5].

The blood−brain barrier is not equally effective in all parts of the brain. For example, this barrier is less efficient in the so-called circumventricular organs. The capillaries of the circumventricular organs have large interendothelial pores (and exhibit active pinocytosis) instead of tight junctions. This is well demonstrated by glutamic acid, an excitatory amino acid and a putative chemical transmitter in the brain, which is neurotoxic when injected intracerebrally. When administered systemically, even at high doses, it does not accumulate in the brain although, under certain conditions, it accumulates in the nucleus arcuatus of the hypothalamus where it may cause rather selective degeneration of some ileurones [5,6]. It can be argued that some chemicals exert profound effects on the CNS without entering the brain; in fact, brain composition and/or function may change in relation to stimulation of peripheral inputs. For instance, stimulation of prolactin secretion from the hypophysis results in feedback mechanisms on dopaminergic functions in the hypothalamus. This mechanism has been suggested as an explanation for why domperidone, an antiemetic drug, increases dopamine metabolites in the striatum even though it does not cross the blood−brain barrier [7].

14.2.5   The Role of Metabolism

If a chemical is metabolized rapidly, very little may be available to the brain. Because in several cases, the metabolic products are more polar and less lipophilic than the parent compound, metabolism may represent a detoxifying mechanism with respect to the brain. However, sometimes the metabolic products are still sufficiently lipophilic to cross the blood−brain barrier; in such cases, the metabolites may contribute to the neurotoxicity of the parent compound, or even be more toxic than the parent compound [5−7]. The capacity of nervous tissue to metabolize xenobiotics is largely unknown. In some cases, metabolic activity in the brain may result in short-lived, unstable metabolites which can covalently bind to macromolecules (proteins or nucleic acids) of the brain. Alkylation and carbomylation of some nitrosoureas which accumulate in the brain is a typical example. Another case is the neurotoxin, 6-hydroxydopamine (6-OHDA), which enters catecholaminergic nerve terminals quite selectively by means of the uptake mechanism for dopamine and noradrenaline where it gives rise to unstable radicals which denature proteins [5,7]. Cytochrome P450 dependent monooxygenases act on a wide variety of endogenous substrates in physiologically significant manner. In a tissue as heterogeneous as brain, it is often difficult, if not impossible to differentiate the role this enzyme system might play in both endogenous metabolism as well as in the metabolism of foreign compounds. The cerebral effects of a xenobiotic might be interpreted as be in either due to disruption of the normal physiological function of the cytochrome P450 or by acting as a substrate of cytochrome P450, thereby leading to the formation of a reactive species. The most important role for cerebral cytochrome P450 demonstrated to date is the transformation of androgens to estrogens by aromatization and of the latter to catechol estrogens by 2-hydroxylation [6,7]. Nervous tissue has a high demand for energy, yet nerve cells can only synthesize ATP through glucose metabolism in the presence of oxygen. Critical ATP-dependent processes in the nervous system include regulation of ion gradients, release and uptake of neurotransmitters, anterograde and retrograde axonal transport, active transport of nutrients across the blood–brain barrier, P-gp function, phosphorylation reactions, assembly of mitochondria, and many others. The highest demand for energy (up to 70%) is created by the maintenance of resting potential in the form of sodium and potassium concentration gradients across the nerve cell membrane [6]. These gradients are maintained primarily by the activity of the Na^+/K^+ ATPase pump. The pump uses the energy of hydrolyzing each ATP molecule to transport three sodium ions out of the cell and two potassium ions into the cell. Maintenance of the resting potential is not the only benefit of this pump's activity, however. The gradients created by the pump also are important for maintaining osmotic balance, and for the activity of indirect pumps that make use of the sodium gradient to transport other molecules against their own concentration gradient. Neurotransmission is thus heavily dependent on the proper functioning of the Na^+/K^+ ATPase pump [8].

Another process dependent on energy metabolism is axonal transport. Axonal transport carries organelles, vesicles, viruses, and neurotrophins between the nerve nucleus and the terminal [6,8]. This distance can be quite long when one considers that the length of the sciatic nerve, for example, can be up to 1 m. Anterograde transport (from cell body to terminal) is accomplished by two mechanisms defined by their rate, fast axonal transport and slow axonal transport. Fast axonal transport proceeds at rates of approximately 400 mm/day and is mediated by the ATP-dependent motor protein kinesin. Kinesin forms cross-bridges between vesicles or organelles and microtubules, and dual projections of these cross-bridges shift back-to-front along microtubules in a coordinated, ATP-dependent manner, such that the entire molecule appears to be walking. Slow axonal transport is used to carry cytoskeletal elements such as tubulin and neurofilaments to the far ends of the axon, and it proceeds at approximately 0.2–5 mm/day. Traditionally slow transport has been regarded as passively dependent on axoplasmic flow; however, recent evidence suggests that the cytoskeletal elements actually move rather quickly but frequently stall in a stop-and-go fashion. Fast axonal transport also proceeds retrogradely, mediated by the ATP-dependent motor protein dynein. The rate of retrograde transport is about 200 mm/day. Neurons use retrograde transport for recycling membranes, vesicles, and their associated proteins. Neurotrophic factors, and some viruses and toxins (e.g., tetanus toxin) are also transported by this mechanism [8].

14.3 Known Neurotoxic and Neuroprotective Plants

14.3.1 Possible Mechanism of Action of Known Neurotoxic Medicinal Plants

The etiology of diseases of the CNS is diverse, and the medicinal plants studied for having potential effect on them operate *via* several mechanisms of action. While the correct mechanism of action responsible for neurological bioactivity for some plants is unknown, in other cases they have already been properly studied. Neurotoxicants affect the nervous system in a number of different ways. Some neurotoxicants damage the distal portions of axons without much effect on the remainder of the cell, while others produce outright cell death. Still others affect signaling processes in the nervous system, without causing structural damage. Also, some examples will be detailed to demonstrate different mechanisms of action of known neurotoxic medicinal plants.

For example, *Datura stramonium* (Solanaceae) is a wild-growing herb, known as jimson weed. It is a plant with both poisonous and medicinal properties. The neurotoxicity is attributed to the presence of tropane alkaloids which contain a methylated nitrogen atom and include the anticholinergic drugs atropine and scopolamine, as well as the narcotic cocaine. A wide range of medicinal values of this plant have contributed in scientific field and also in ethnomedicine. The phytochemical, ethnomedicinal, and toxicological works on *D. stramonium* have given a better assessment and information of this particular plant. The toxins in jimson weed are tropane belladonna alkaloids, which possess strong anticholinergic properties [9]. The tropane alkaloids, the predominant chemicals found in all species of *Datura* plants, are anticholinergic and CNS stimulants. They act by competitively and irreversibly inhibiting acetylcholine on muscarinic receptors, thereby causing both CNS and peripheral nervous system manifestation. The CNS features include restlessness, delirium, altered sensorium, and hallucination. The peripheral nervous system manifestations include hyperpyrexia, pupillary dilatation, dryness of mouth and skin, urinary retention, and reduced gastric movement. These alkaloids include: hyoscyamine (leaves, roots, seeds), hyoscine (roots); atropine and scopolamin, as well as sitosterol and proteins. It is reported that hyoscyamine is the predominant alkaloid in *D. stramonium* from the line of flowering. Thorn apple leaves contain 0.2–0.45% of total alkaloids, seeds approximately 0.2%.The tropane alkaloid contains a methylated nitrogen atom and includes the anticholinergic drugs, atropine, hyoscyamine, and scopolamine as well as the narcotic cocaine. All parts of the plant are toxic but the highest amount of the alkaloids is contained in ripe seeds. They act as competitive antagonist of acetylcholine at peripheral and central muscarinic receptor sites. Poisoning results in widespread paralysis of parasympathetic innervated organs [9–12].

Intoxication of cattle, sheep, goats, horses, and rabbits by *Nierembergia hippomanica* has been reported in Argentina, Uruguay, and South Africa. Mortalities, which may occur some hours after the mydriasis, locomotory ataxia, weakened heart action, dyspnea, excitement, and convulsions. Necropsies on acute cases reveal evidence of gastrointestinal irritation and hyperemia of the brain and meninges. Voluntary ingestion of approximately 30 g/kg fresh flowering plants by a second calf resulted in nervous signs characterized by chewing motions, protrusion of the tongue, dysphagia, hypermetria, ataxia, paresis, and lateral recumbency. Salivation, dehydration, and cardiac irregularities completed the clinical picture. Clinical chemistry changes revealed muscle damage and increased serum urea and creatinine concentrations indicative of kidney involvement. This is the first confirmed outbreak of *N. hippomanica* var. violacea intoxication of stock in South Africa [13,14].

Other members of the Solanaceae that may induce nervous signs in cattle in South Africa are *Solanum kwebense* and *Nicotiana glauca*. Ingestion of large quantities of *S. kwebense* precipitates epileptiform seizures. In this neurotoxicity, cerebellar atrophy is microscopically associated with degeneration and loss of the Purkinje cell layer [14]. *N. glauca* poisoning is characterized by salivation, irregular gait, tremors, convulsions, and dyspnea, but intoxication is extremely rare and of no practical significance [14].

Cassava (*Manihot escculenta*) is an annual tuber root crop cultivated widely in the tropics and subtropics; it serves as a major food crop of low protein but high calorie content. At the cellular level, both free cyanide and hydrogen cyanide have been found to induce degeneration via increased lysosomal activity *in vivo*. Cyanide as earlier described is a potent neurotoxic substance that can initiate a series of intracellular reactions leading to cell dysfunction and eventually cell death; it enhances N-methyl-D-aspartate receptor function. In cultured neurons, cyanide neurotoxicity is linked with N-methyl-D-aspartate receptors, which mediated a rise in calcium that in turn activates a series of biochemical reactions leading to generation of reactive oxygen species and nitric oxide, this antioxidant species then mediates peroxidation of lipids. It is concluded that oxidative stress plays an important role in cyanide-induced neurodegeneration; this phenomena can cause cell death in two ways: apoptosis and necrosis [15,16].

The pharmacology of the (noncurarizing) alkaloids occurring in African species of *Strychnos* demonstrates very clearly the regrettable paucity of our knowledge concerning their actions. Fluorocurarine, fluorocurine, mavacurine, these three alkaloids isolated from the seeds of *Strychnos icaja*, all have very weak curarizing activity. Harman and certain of its derivatives act as competitive and selective inhibitors of type A monoamine oxidase. The cytotoxic properties of the alkaloid have been established using *in vitro* cell cultures. It inhibits the synthesis of DNA by direct interaction with chromatin, probably by intercalating itself between the base pairs of the DNA. On the other hand, recent studies have shown that harman and other β-carbolines attach reversibly to the benzodiazepine receptor and block the effects of these latter compounds by displacing them from the binding sites, which also blocks the effects of GABA. This is thought to explain the convulsant and anxiogenic effects of these molecules, effects that are antagonized by benzodiazepines. Strychnine is of course, the archetypal convulsant poison. It acts on the medulla by blocking the chemical intermediate glycine; this results in paralysis of the cells of Renshaw which normally exercise an inhibitory function on the motor cells. In consequence, with toxic doses, about 50−60 mg for an adult, uncoordinated movements are propagated to all parts of the spinal cord, all the muscles begin the act, especially the extensors, i.e., the action is "tetanizing." After two or three tonic crises, contraction of the muscles of the thorax and diaphragm bring about death by asphyxia [17].

14.3.2 Possible Mechanism of Action of Neuroprotective Medicinal Plants

A search for novel pharmacotherapy from medicinal plants for neurotoxic diseases has progressed significantly in the past decade. This is reflected in the large number of herbal preparations for which neuroprotective potential has been evaluated in a variety of experimental models.

Ginkgo biloba has been claimed to be effective in treating memory problems. Indeed, a large number of studies have shown cognition-enhancing effects of *G. biloba* extracts, particularly its standard extract EGb 761, in animal models. EGb 761 significantly improved the impairment of learning and memory function induced by aging, cerebral ischemia, and abnormal cerebral glucose metabolism in rodents. EGb 761 also has the ability to enhance memory processes in normal control animals. EGb 761 possesses diverse pharmacological actions, mainly as an antioxidant, AChE inhibitor, and mediator of GABAA and h-adrenergic receptors [18,19].

Ginseng has been used in traditional Chinese medicine for centuries and is currently one of the most widely taken herbal products throughout the world. Ginseng is in fact the dried root of several species in the plant genus Panax, including *P. ginseng*, *Panax quinquefolius*, *Panax notoginseng*, and *Panax japonicus*. Among them *P. ginseng* is the most extensively investigated member. The major active constituents of ginseng are ginseng saponins (ginsenosides), of which over 30 individual examples have been characterized. The nootropic effects of a variety of *P. ginseng* extracts and ginsenosides have been shown in aged rats, scopolamine- and ethanol-treated rats, rats with medial prefrontal cortex lesions, scopolamine-treated mice, and ischemic gerbils. Male chicks exhibited a significant reduction of the number of errors during retention trials in a visual discrimination task following intraperitoneal injections of the ginseng saponin. Ginseng constituents also were effective in reducing stress and anxiety. It is well known that the hypothalamic pituitary-adrenal system plays a role in the physiological and pharmacological actions of *P. ginseng*. Ginsenosides are also capable of modulating cholinergic and GABAergic transmission, enhancing nitric oxide synthesis, and protecting neuronal cell apoptosis [20,21].

Nauclea latifolia Smith (Rubiaceae) was evaluated for its anticonvulsant, anxiolytic, and sedative activity in mice. Animal models (maximal electroshock-, pentylenetetrazol-, and strychnine-induced convulsions; N-methyl-D-aspartate induced turning behavior; elevated plus maze; stress-induced hyperthermia; open field; and diazepam-induced sleep) were used. The decoction from the bark of the roots of *N. latifolia* strongly increased the total sleep time induced by diazepam. It also protected mice against maximal electroshock-, pentylenetetrazol-, and strychnine-induced seizures. In addition, turning behavior induced by N-methyl-D-aspartate was inhibited. *N. latifolia* antagonized, in a dose-dependent manner, stress-induced hyperthermia and reduced body temperature. In the elevated plus maze, *N. latifolia* increased the number of entries into, percentage of entries into, and percentage of time in open arms, and reduced rearing, head dipping, and percentage of time in closed arms. In the open field test, *N. latifolia* increased crossing and reduced rearing and defecation [22].

While the *Hibiscus asper* Hook. f. (Malvaceae) is a traditional herb largely used in tropical region of the Africa as vegetable, potent sedative, tonic, and restorative, anti-inflammatory, and antidepressive drug, there is very little scientific data concerning the efficacy of this. Methanolic extract of *H. asper* leaves showed potent antioxidant and free radical scavenging activity. Chronic administration of methanolic extract (50 and 100 mg/kg, IP, daily, for 7 days) significantly reduced anxiety-like behavior and inhibited depression in elevated plus maze and forced swimming tests, suggesting anxiolytic and antidepressant activity. Also, spatial memory performance in Y-maze and radial arm-maze tasks was improved, suggesting positive effects on memory formation. In 6-OHDA-lesioned rats, methanolic extract of *H. asper* leaves showed potent antioxidant and antiapoptotic activities. Chronic administration of the methanolic extract (50 and 100 mg/kg, IP, daily, for 7 days) significantly increased antioxidant enzyme activities (superoxide dismutase, glutathione peroxidase, and catalase), total glutathione content, and reduced lipid peroxidation (malondialdehyde level) in rat temporal lobe homogenates, suggesting antioxidant activity. Also, DNA cleavage patterns were absent in the 6-OHDA-lesioned rats treated with methanolic extract of *H. asper* leaves, suggesting antiapoptotic activity [23].

14.4 Methods in Neurotoxic and Neuroprotective Screenings of Medicinal Plants

In most cases, the assessment of the neurotoxicity of chemicals needs to be based on data from animal experiment. Data in humans from several epidemiological studies is only infrequently available, but is particularly valuable.

14.4.1 In Vivo *Models of Neurotoxicity*

A multidisciplinary approach is required in order to adequately assess potential neurotoxic effects of compounds, due to the diverse function of the nervous system. Many effects may be measured using neurophysiological (e.g., electroencephalography, measurement of evoked potentials), neuropathological (e.g., microscopy, histochemistry, immunohistochemistry), or behavioral techniques. Due to the variety of biochemical targets and toxic effects test strategies used to evaluate, the neurotoxic potential of chemicals should be determined on a case-by-case basis. A staged approach is commonly used, consisting of three tiers aimed to identify (tier 1) and characterize (tiers 2−3) the neurotoxicity of the chemical. Tier 1 is usually a basic repeated dose toxicity of the numerous possible effects on the nervous system; batteries of tests may be necessary to ensure the detection and characterization of possible neurotoxic effects [2]. The OECD guideline 424 for the testing of chemicals was designed to either confirm or further characterize potential neurobehavioral and neuropathological effects observed in adult animals during the repeat dose toxicity study. It may be carried out as a separate study. Essentially, the neurotoxicity study comprises: detailed clinical observations; changes in skin, fur, eyes, mucous membranes; occurrence of secretions and excretions; autonomic activity (e.g., lacrimation, piloerection, pupil size, respiratory pattern, unusual urination/defecation, discolored urine); body position, activity level, coordination of movement; changes in gait, posture, reactivity to handling; clonic or tonic movements; convulsion or tremors; stereotypes (e.g., excessive grooming, repetitive circling); strange behavior (e.g., biting, walking backwards); functional tests; assessment of sensory function (e.g., sensory irritation); assessment of motor function (e.g., limb grip strength, foot splaying); assessment of cognitive function (e.g., habituation, avoidance) [2]. The functional observational battery is a standardized screening battery for assessing many aspects of behavior and neurological function in rodents and is designed to detect and quantify major overt behavioral physiological and neurological signs. The tests have been validated with many known neurotoxic chemicals and often are used in conjunction with other measures of toxicity. The functional observational battery comprises a number of tests to identify specific deficits in motor and sensory functions by measuring neuromuscular, sensory, and autonomic function [7].

14.4.2 In Vitro *Models of Neurotoxicity*

To address problems associated with increasing cost and time required for neurotoxicity testing, the large number of chemicals in commercial use that have not been investigated and animal welfare issues, considerable effort is being directed at the development of *in vitro* alternatives. These may be considered as part of a tiered system with which to identify potential neurotoxic chemicals, although such approaches have not been validated as replacements for animal studies [7]. Several *in vitro* models commonly are used in neurotoxicity evaluations, including synaptic fractions, primary cultures of rat astrocytes, rat cerebellar granule neurons, primary motor neurons of dissociated cultures of murine spinal cord, rat brain region organ cultures, and hippocampal slices. Many endpoints may be used to detect basic cytotoxicity. However, more specific endpoints are required in developing *in vitro* systems to evaluate neurotoxic potential of chemicals. Possible target sites of neurotoxicity include changes in morphological endpoints; neurite outgrowth; peripheral nerve or central nervous myelin; axonal transport, electrophysiological indices; blood–brain barrier; calcium homeostasis and neurotransmitters, and hormones [7].

14.4.3 *Developmental Neurotoxicity*

It is widely accepted that the degree of neurotoxicity varies with the developmental period and can be particularly high in the early stages of development, due to characteristics of the cellular and molecular process involved in brain development including slow maturation, immature protective systems, cell specialization, and limited capacity for regeneration [7,8]. A tiered approach to determine when plants extract or chemical should undergo developmental neurotoxicity testing, using a weight of evidence approach was proposed by the US Environmental Protection Agency (EPA). Criteria used to recommend developmental neurotoxicity testing include CNS/behavioral teratogens (and structural analogues), adult neuropathic agents, adult neuroactive agents, hormonally active compounds, and developmental toxins that do not necessarily produce CNS effects. According to this tiered scheme, chemicals not meeting any of the criteria would not be recommended for developmental neurotoxicity testing [2,8]. However, interlaboratory validations are needed to improve interpretability of data from developmental neurotoxicity studies. The International Life Science Institution neuropathology review describes the strengths and weaknesses of the various morphological approaches in developmental neurotoxicity studies that provide a number of suggestions for the appropriate use of different techniques. However, clear recommendations for the use of techniques in developmental studies are lacking [7].

14.4.4 Clinical Neurotoxicity Tests in Humans

The assessment of neurotoxicity begins with a clinical evaluation of the patient, including medical history and a standard neurological examination. The clinical evaluation of neurotoxic diseases assesses various functions, including mental status, cranial nerve function, motor function, coordination and gait, reflex, and cutaneous sensory [2]. Neurophysiological testing often is carried out during a clinical examination. The tests used are aimed at assessing a wide range of neurological functions including different aspects of verbal function, visuospatial ability, memory, attention, cognition tracking and flexibility, and psychomotor abilities [8].

14.5 Neurotoxic Medicinal Plants of Africa

A degenerative disease known as ataxic neuropathy has been prevalent in certain areas of the tropics. Chronic intoxication of dietary origin, from plant extracts, has been implicated as the cause for this disorder. A significant number of studies has been performed to find alternatives or treatments for neurotoxicity of the nervous forum by identifying structures with activity at the CNS. However, most of the screenings are usually conducted on an *ad hoc* basis and not systematically. The etiology of diseases of the CNS is diverse, and the plants studied for having potential effect on them operate via several mechanisms of action. While the correct mechanism of action responsible for neurotoxicological activity for some plants is unknown, in other cases they have already been properly studied. As the purpose of this review is not to catalogue all the mechanisms of action from the plants mentioned, but to provide a general approach, mechanisms will be summarized in Table 14.1 for the cases in which these mechanisms are understood.

In Africa, many species of higher plants and others are used as traditional medicines. These plants contain chemical substances with interesting pharmacological effects, and several of these plants are used to treat neurological and age-related disorders. In a previous study, a number of plants, including *D. stramonium* L. (Solanaceae), *N. hippomanica* var. (Violaceae), *S. kwebense* and *N. glauca* (Solanaceae), *M. esscunlenta* Crantz (Euphorbiaceae), *S. icaja* Baill (Loganiaceae), *Coscinium fenestratum* (Gaertn) Colebr (Menispermaceae), *Rauwolfia serpentina* L. (Apocynaceae), *Securidaca longipedunculata* Fresen. (Polygalaceae), and

Table 14.1 African Medicinal Plants Screened for Their Neurotoxicity

Plant (Family)	Area of Plant Collection	Traditional Use	Chemical Constituents	Pharmacological Activities
Datura stramonium L. (Solanaceae)	Tropical Africa, West Africa, South Africa	The leaves have been used to sedate hysterical and psychotic patients and to treat insomnia	Tropane alkaloids, scopolamine, atropine	Atropine and scopolamine act as CNS depressants and competitively antagonize muscarinic cholinergic receptors, convulsion [3,24,25]
Nierembergia hippomanica var. (violacea)	South Africa	No relevant additional information	Flavone glycoside	Sinus arrhythmia, tachycardia, epileptiform seizures, cerebellar atrophy [13,14]
Solanum kwebense (Solanaceae)	Zimbabwe, Kenya, Zambia, South Africa	Abdominal pain, inflammation, rheumatism	Alkaloids, solasodine, solasonine, solamargine	Cerebellar cortical degeneration with selective involvement of Purkinje neurons [14]
Nicotiana glauca (Solanaceae)	Sub-Saharan Africa, Kenya, Tanzania and Uganda	Dizziness, migraines, headaches	Piperidinepyridine alkaloid, anabasine	Convulsions and dyspnea, hallucination [3,14]
Manihot essculenta Crantz (Euphorbiaceae)	Africa	Ringworm, tumor, conjunctivitis, sores and abscesses, inflammation	Cyanide, carbohydrates	Degeneration via increased lysosomal activity, it enhances *N*-methyl-D-aspartate receptor function [14]
Strychnos icaja Baill (Loganiaceae)	From Guinea east to the Central African Republic and south to Angola	Chronic and persistent malaria, cognitive disorders	Alkaloids, flavonoids, fluorocurarine, fluorocurine, mavacurine, strychnine	It inhibits the synthesis of DNA by direct interaction with chromatin. It acts on the medulla by blocking the chemical intermediate glycine [14,25]
Coscinium fenestratum (Gaertn) Colebr (Menispermaceae)	Tropical Africa, West Africa, Congo, Cameroon	Hypoglycemic hypotensive laxative and antidiabetic activities, abdominal disorders, jaundice, fever, and general debility	Protoberberine alkaloids	Induction of stereotype behaviors, convulsion, abdominal contraction [3,24,25]

Species (Family)	Distribution	Uses	Constituents	Effects
Rauwolfia serpentina L. (Apocynaceae)	Central Africa, South Africa	Fever, pain, dementia, cognitive disorders	The alkaloid reserpine	The ability of reserpine to induce depression and deplete brain amines became one of the pillars of the monoamine theory of affective disorders [3]
Securidaca longipedunculata Fresen. (Polygalaceae)	Tropical Africa	Epilepsy and convulsions	Indole alkaloid securinine, ergot alkaloids, flavonoids	CNS depressant with motor incoordination, agitation, confusion [3]
Tephrosia capensis var. Capensis (Fabaceae)	West Africa, South Africa, Cameroon, Ghana, Nigeria, Congo, Angola	The Sotho cook the roots for palpitations and make decoctions of the plant with *Commelina africana* (Commelinaceae) for weak hearts and nervousness. It has been reported to be toxic	Flavonoids, alkaloids	Induction of stereotype behaviors, convulsion, hallucination, seizure, agitation [3,25]

Tephrosia capensis Var. *Capensis* (Fabaceae) were shown to possess the anticholinergic effects likely to produce delirium and stupor but rarely cause deep coma. Ingestion of large quantities of the extracts of these medicinal plants precipitates epileptiform seizures. In this neurotoxicity, cerebellar atrophy is microscopically associated with degeneration and loss of the Purkinje cell layer. Cyanide isolated from *M. escculenta* as earlier described is a potent neurotoxic substance that can initiate series of intracellular reactions leading to cell dysfunction and eventually cell death; it enhances *N*-methyl-D-aspartate receptor function.

14.6   Neuroprotective Medicinal Plants of Africa

In the past decade, there has been substantiated considerable interest in phytochemical bioactive constituents from herbal medicines, which can have long-term medicinal or health-promoting qualities in neurotoxic diseases. In comparison, many medicinal plants exhibit specific medicinal actions without serving a nutritional role in the human diet and may be used in response to specific health problems over short- or long-term intervals. Therefore, a scientific reexamination of these therapies in preclinical models is valuable for the development of novel neuroprotective drugs. We will review in Table 14.2 plants used in African traditional medicine used in the treatment of neurotoxic diseases.

Phytotherapy in African traditional medicine still plays an important role in the management of neurotoxic diseases, mainly among populations with very low income, and phytotherapy relies on the use of a wide variety of plant species. *Annona muricata* Linn (Annonaceae), *Annona senegalensis* Pers (Annonaceae), *Bidens pilosa* Linn (Asteraceae), *Bryophyllum pinnatum* (Lam) Oken (Crassulaceae), *Citrus sinenis* (Linn) Osbeck (Rutaceae), *Clerodendron thomsoniae* Balf (Verbenaceae), *Daniellia oliveri* (Rolfe) Hutch and Dalz (Caesalpiniaceae), *Datura stramonium* Linn (Solanaceae), *Detarium microcarpum* Guil et Perr (Caesalpiniaceae), *Euphorbia hirta* Linn (Euphorbiaceae), *Flacourtia indica* Willd (Flacourtiaceae), *Hymenocardia acida* Tul (Hymenocardiaceae), *Jatropha gossypiifolia* Linn (Euphorbiaceae), *Khaya senegalensis* A Juss (Desrousseaux) (Meliaceae), *Mentha cordifolia* Auct (Lamiaceae), *Prosopis africana* Guill and Perr (Taub) (Mimosaceae), *Ricinus communis* Linn (Euphorbiaceae), *Securidaca longepedunculata* Fres (Polygalaceae), *Senna singueana* (Delile) Lock (Caesalpiniaceae), *Terminalia glaucescens* Planch. ex Benth (Combretaceae), *Terminalia mollis* Laws (Combretaceae), *Tetrapleura tetraptera* Taub (Schum Thonn) (Sapotaceae), *Withania somnifera* L. (Dunal) (Solanaceae), *Valeriana officinalis* L. (Valerianaceae), *Rauwolfia serpentium* L. (Apocynaceae), *Ziziphus jujube* Mill (Rhamnaceae), *Hypericum perforatum* L. (Clusiaceae), *Passiflora edulis* Sims (Passifloraceae), *Ocimum sanctum* L. (Laminaceae), and *Hibiscus asper* Hook. f. (Malvaceae) are plants that are used empirically in traditional medicine in Africa to treat epilepsy and diseases related to the brain such as agitation, anxiety, convulsions, dizziness, headache, insomnia, migraine, pain, and schizophrenia according to

Table 14.2 African Medicinal Plants Screened for Their Neuroprotective Effects

Plant (Family)	Area of Plant Collection	Traditional Use	Chemical Constituents	Pharmacological Activities
Annona muricata Linn (Annonaceae)	West Africa, Tropical area	Insomnia, diabetes spasms, fever	Steroid, cardiac glycosides, flavonoids	Anticonvulsant, sedative, antinociceptive, anxiolytic [3,24,26]
Annona senegalensis Pers (Annonaceae)	Cameroon, Central Africa West Africa, South Africa	Convulsions, epilepsy sterility, diarrhea, dysentery	Flavonoids, saponin	Anxiolytic, anticonvulsant [24,26,27]
Bidens pilosa Linn (Asteraceae)	West Africa, South Africa, Tropical area	Dizziness, migraines, headaches, rheumatism	Saponin, flavonoids, sitosterol glycoside, lupeol	Anticonvulsant, anxiolytic, chemopreventive, antidiabetic [3,24,27,28]
Bryophyllum pinnatum (Lam) Oken (Crassulaceae)	Africa	Convulsions, rheumatism, arthritis, pain	Flavonoids, antraquinones	Antinociceptive, anti-inflammatory, sedative, anticonvulsant [3,24,28,29]
Citrus sinenis (Linn) Osbeck (Rutaceae)	Humid tropical areas	Epilepsy, convulsions, insomnia, agitation, headaches, malaria fever, anxiety, schizophrenia	Limonene, limonin, quercetin, hesperidine	Sedative, analgesic [24,26,28]
Clerodendron thomsoniae Balf (Verbenaceae)	West Africa, Nigeria, Cameroon, Gambi	Convulsions, headache, parasitic diseases	Flavonoids, saponin, alkaloids, coumarine, steroids, quinone	Effect on purinergic neurotransmission [24,26,29]
Daniellia oliveri (Rolfe) Hutch and Dalz (Caesalpiniaceae)	Angola, Cameroon Sudan, West Africa Central Africa	Epilepsy, migraine, headaches, epilepsy, anxiety, schizophrenia	Flavonoids, glycosides, quercetin	Antispasmodic action [24,26,28]

(Continued)

Table 14.2 (Continued)

Plant (Family)	Area of Plant Collection	Traditional Use	Chemical Constituents	Pharmacological Activities
Ginkgo biloba L. (Ginkgoaceae)	Africa	Several age-related mental health problems of the elderly, including those associated with Alzheimer's disease	Ginkgolic acid conjugate, bilobalide	Inhibition of α-adrenoreceptors, antiagonistic effects on GABA receptors, antioxidant [24,26,27]
Searsia chirindensis (Baker f.) Moffett (Anacardiaceae)	South Africa	Mental disturbances, epilepsy, pain	Flavonoids, alkaloids, glycosids	Antipsychotic, anticonvulsant, sedative GABAergic activation [24,26,28]
Euphorbia hirta Linn (Euphorbiaceae)	Africa continent	Convulsions, insomnia, epilepsy, and mental illness	Alkaloids, tannins	The leaf extracts have demonstrated GABAergic activity [3,26,28]
Flacouria indica Willd (Flacourtiaceae)	Cameroon, Tropical area	Epilepsy, headache, fever, stomachache, diarrhea, sleep disorders	Beta-sistosterol butyrolactone, steroids, flacourtine, flavonoids, coumarine, terpenoids, polyphenols	Antiplasmodial Protection against liver toxicity [3,24,27]
Datura stramonium Linn (Solanaceae)	Tropical Africa, West Africa, South Africa	Sedative, antipsychotic, neuropathic pain	Alkaloids, flavonoids, tropane alkaloids, scopolamine, atropine	Atropine and scopolamine used in ophthalmology to dilate pupils, scopolamine used to treat motion sickness, opioid activation [3,25,29]
Harpagophytum procumbens (Burch) DC (Pedaliaceae)	South Africa	Inflammation, pain, fever, agitation		Anti-inflammatory; antirheumatic, antioxidant, sedative [3,26,27]

Plant (Family)	Region	Traditional uses	Phytochemicals	Pharmacological activities
			Coumarins; phenolic glycosides, harpogosids	
Hypericum perforatum L. (Clusiaceae)	South Africa	Attention-deficit-hyperactivity disorders, anxiety, convulsion, fever, pain	Flavonoids, phenolic compounds; hyperforin	Analgesic; psychomotor disturbances; antidepressant [3,25–27]
Hibiscus asper Hook. f. (Malvaceae)		Potent sedative, tonic and restorative, anti-inflammatory and antidepressive	Flavonoids phenolic acids, polysaccharides	Antioxidant, anxiolytic, antidepressive, neuroprotective [23,26,27]
Hymenocardia acida Tul (Hymenocardiaceae)		Headaches, fever, hypotension, diabetes, sickle cells	Homoorientin, triterpenoids, alkaloids	Anti-inflammatory, antioxidant [24,26,27]
Jatropha gossypiifolia Linn (Euphorbiaceae)	Cameroon, Central Africa, West Africa	Epilepsy, schizophrenia Convulsions, fever, hypertension, convulsions	Flavonoids, saponins, triterpenoids	Antispasmolytic, antioxidant, anticonvulsant, sedative [3,24,26,29]
Khaya senegalensis A Juss (Desrousseaux) (Meliaceae)	West Africa, tropical area	Headaches, schizophrenia, cerebral malaria, abdominal pain, fever, epilepsy, infantile convulsion	Saponins, tannins, triterpenes, flavonoids, alkaloids	Anticonvulsant, sedative, anxiolytic [3,26,27]
Mentha cordifolia Auct (Lamiaceae)	West Africa, tropical area	Insomnia, muscle relaxant	Phenolic compounds, flavonoids	Antioxydant, anticonvulsant, sedative [3,26,27]
Mimosa pudica L. (Mimosaceae)	West Africa, Cameroon, Congo	Infantile convulsions, insomnia, epilepsy, hypertension, inflammation, pain	Narcisine, flavonoids	Sedative and spasmolitic activities
Ocimum sanctum L. (Laminaceae)		Epilepsy, insomnia, anxiety states, headache, migraine, agitation, fever	Flavonoids, alkaloids, triterpenoids, saponin	Antidepressant, antioxidant, CNS depressant, antinociceptive [3,24,27]

(Continued)

Table 14.2 (Continued)

Plant (Family)	Area of Plant Collection	Traditional Use	Chemical Constituents	Pharmacological Activities
Passiflora edulis Sims (Passifloraceae)		Anxiety, epilepsy	Flavonoids, triterpenoids, flavonoids, luteolin	Anticonvulsant, anxiolytic, sedative [3,25,26]
Prosopis africana Guill and Perr (Taub) (Mimosaceae)	Cameroon West Africa	Epilepsy, insomnia, anxiety states, headache, migraine, agitation, fever, vermifuge	Alkaloids, tannins, saponins, flavonoids, glycosides, and phenols	Anticonvulsant, antioxidant [25–27]
Ricinus communis Linn (Euphorbiaceae)	Central Africa, West Africa	Epilepsy, convulsions, headache, diarrhea, asthma	Flavonoids, alkaloids, ricin	Neuroleptic like properties [3,26,27]
Securidaca longepedunculata Fres (Polygalaceae)		Epilepsy, schizophrenia, pain, rheumatism	Alkaloids, flavonoids, indole alkaloid securinine, some ergot alkaloids	Anticonvulsant, anxiolytic, sedative, antinociceptive [3,26,27]
Senna singueana (Delile) Lock (Caesalpiniaceae)	Cameroon, Mali, Soudan, East and South Africa	Fever Conjunctivitis, convulsions, gonorrhea, bilharzias, stomachache, constipation, epilepsy, syphilis	Flavonoids, alkaloids, saponine, 7-methylphyscion Cassiamin A	Anticonvulsant, antioxidant, sedative [3,24,25,29]
Tetrapleura tetraptera Taub (Schum Thom) (Combretaceae)	Angola, Cameroon, Sudan, West Africa,	Epilepsy, convulsions, fevers, malaria	Saponins, tannins, flavonoids	Anticonvulsant, sedative, antioxidant [3,26,27]

Plant (Family)	Region	Traditional use	Constituents	Activity
Ziziphus jujube Mill (Rhamnaceae)	Central Africa, West Africa, Tropical area	Used in traditional medicine to nourish the heart and calm the spirit, this plant is often prescribed to aid in sleep or to calm the mind	Flavonoids, alkaloids, tanin, triterpenoids, Spinosin Jujubosides	Jujuboside A inhibits hippocampal hyperactivity [3,25,27]
Rauwolfia serpentium L. (Apocynaceae)	South Africa	Infantile convulsion, fever, inflammation	Indole alkaloids	Psychomotor disturbances; tranquilizer [3,25,27]
Valeriana officinalis L. (Valerianaceae)	South Africa	Epilepsy, insomnia, anxiety, tranquilizer	Valeranone; sesquiterpenoids	CNS depressant, anticonvulsant, GABAergic activation [3,25,27]
Withania somnifera L. (Dunal) (Solanaceae)	South Africa	Convulsion, epilepsy, pain, inflammation	Steroids; witherferin, alkaloids	Anti-inflammatory, anticonvulsant, neuroprotective, antioxidant [3,25,26]

traditional healers and the literature [3,10–12,24–27,29]. Controversially, the current review demonstrated that the whole part of *D. stramonium* has many traditional uses and toxicity to a greater extent. This plant has been used in curing different types of diseases. The alkaloids isolated from the plants administered systemically also are found to be toxic to the animals and 48 alkaloids have been determined [25,29]. Two new tropane alkaloids, 3-phenylacetoxy-6,7-epoxynortropane and 7-hydroxyapoatropine, were also identified. The seeds are the most potent part of the plant, although the whole plant has psychoactive activity. Atropine and scopolamine act as CNS depressants and competitively antagonize muscarinic cholinergic receptors. The anticholinergic syndrome results from the inhibition of central and peripheral muscarinic neurotransmission [10–12,25]. The patient presents with dry skin and mucosa, flushing, mydriasis with loss of accommodation that causes blurred vision and photophobia, altered mental status, hyperpyrexia, sinus tachycardia, urinary retention, and myoclonic jerking. Other symptoms may include ataxia, impaired short-term memory, disorientation, confusion, hallucinations, psychosis, agitated delirium, seizures, coma, respiratory failure, and cardiovascular collapse [10–12]. On the other hand, atropine and scopolamine have been used in ophthalmology to dilate pupils when they are administered locally; in anesthesia to decrease secretions and treat bradycardia, in toxicology to treat organophosphate and nerve gas poisoning, and in emergency medicine for cardiac arrest. In addition, *D. stramonium* seed extract has an analgesic effect on both acute and chronic pain which was produced by hot plate and formalin tests. It is likely that this effect can be attributed to the alkaloids which interact with opioid system. The whole plant is toxic, particularly the foliage and seeds [3,10–12,25,29].

14.7   Conclusion

The purposes of this review were to conduct a screening of African plants with neurotoxic and/or neuroprotective activities among medicinal plants that have been studied but may contain the constituents responsible for such activity, and also to find possible targets for future studies of new and alternative therapies for the treatment of neurological disorders and neurodegenerative diseases. The use of flora represents a viable resource in the search for compounds not only for activity on the CNS but also in other therapeutic applications. The traditional characteristics, which identify compounds with low molecular weight and high lipophilicity to allow activity on the CNS, are not static. We found that there should be no limit to the types of compounds in the search of possible structures with action at the CNS. More studies are needed to support existing information about these plants that have been gleaned primarily through folk sources.

References

[1] Bonita LB. Toxicology of the nervous system. In: Hodgson E, editor. A text book of modern toxicology. 3rd ed. Hoboken, NJ: John Wiley & Sons, Inc.; 2004.

[2] Miller DB. Endocrine disruption: estrogen, endrogen and nervous system. In: Tilson HA, Harry GJ, editors. Neurotoxicology. Philadelphia, PA: Taylor and Francis; 1998. p. 201−17.

[3] Zhang ZJ. Therapeutic effects of herbal extracts and constituents in animal models of psychiatric disorders. Life Sci 2004;75(14):1659−99.

[4] Kim EB, Susan MB, Scott B, Heddwen B. Ganong's review of medical physiology. 24th ed. USA: The McGraw-Hill Companies; 2010.

[5] Rang HP, Dale MM, Ritter JM. Pharmacology. 4th ed. Edinburgh: Churchill Livingstone; 1999.

[6] Harry J, Kulig B, Lotti M, Tilson H, Winneke G. Neurotoxicity risk assessment for human health. Environmental health criteria 223. Geneva: World Health Organization; 2001.

[7] Massaro E. Handbook of neurotoxicology. Totowa, NJ: Humana Press; 2002.

[8] Tilson H, Harry J. Neurotoxicology. Philadelphia, PA: Taylor and Francis; 1999.

[9] Shore PA, Giachetti A. Reserpine: basic and clinical pharmacology. In: Iversen LL, Iversen SD, Snyder SH, editors. Handbook of psychopharmacology, vol. 10. New York, NY: Plenum Press; 1978. p. 197−219.

[10] Elisabetta M, Alessandra M, Ferri S, Ida BC. Distribution of hyoscyamine and scopolamine in *Datura stramonium*. Fitoterapia 2001;72(6):644−8.

[11] Guharov SR, Barajas M. Intense stimulant effect: atropine intoxication from the ingestion and smoking of Jimson weed (*Datura stramonium*). Vet Toxicol 1991;33: 588−9.

[12] Oberndorfer S, Grisold W, Hinterholzer G, Rosner M. Coma with focal neurological signs caused by *Datura stramonium* intoxication in a young man. J Neurol Neurosurg Psychiatry 2002;73:458−9.

[13] Kellerman TS, Coetzer JAW, Naude TW. Plant poisonings and mycotoxicoses of livestock in southern Africa. Cape Town: Oxford University Press; 1988.

[14] Botha CJ, Schultz R, Vander Lugt A, Retief Elizabeth JJ, Labuschagne L. Neurotoxicity in calves induced by the plant, *Nierembergia hippomanica* Miers var. *violacea* Millan in South Africa. Onderstepoort J Vet 1999;66:237−44.

[15] Mathangi DC, Mohan V, Namasivayam A. Effect of cassava on motor coordination and neurotransmitter level in the albino rat. Food Chem Toxicol 1999;37:57−60.

[16] Okafor PN, Okoronkwo CO, Maduagwu ON. Occupational and dietary exposure of humans to cyanide from large scale cassava processing and ingestion. Food Chem Toxicol 2002;40:1001−5.

[17] Quetin-Leclercq J, Angenot L, Bisset NG. South American *strychnos* species. Ethnobotany (except curare) and alkaloid screening. J Ethnopharm 1990;28:1−52.

[18] Ponto LLB, Schultz SK. *Ginkgo biloba* extract: review of CNS effects. Ann Clin Psychiatry 2003;15:109−19.

[19] Klein J, Chatterjee SS, Loffelholz K. Phospholipid breakdown and choline release under hypoxic conditions: inhibition by bilobalide, a constituent of *Ginkgo biloba*. Brain Res 1997;755:347−50.

[20] Churchill JD, Gerson JL, Hinton KA, Mifek JL, Walter MJ, Winslow CL, et al. The nootropic properties of ginseng saponin Rb1 are linked to effects on anxiety. Integr Physiol Behav Sci 2002;37:178−87.

[21] Jaenicke B, Kim EJ, Ahn JW, Lee HS. Effect of *Panax ginseng* extract on passive avoidance retention in old rats. Arch Pharm Res 1991;14:25−9.

[22] Ngo Bum E, Taïwe GS, Moto FCO, Ngoupaye GT, Nkantchoua GN, Pelanken MM, et al. Anticonvulsant, anxiolytic and sedative properties of the roots of *Nauclea latifolia* Smith in mice. Epilepsy Behav 2009;15:434−40.

[23] Hritcu L, Foyet HS, Stefan M, Mihasan M, Asongalem AE, Kamtchouing P. Neuroprotective effect of the methanolic extract of *Hibiscus asper* leaves in 6 hydroxydopamine-lesioned rat model of Parkinson's disease. J Ethnopharmacol 2011;137:585−91.

[24] Ngo Bum E, Taïwe GS, Moto FCO, Ngoupaye GT, Vougat RRN, Sakoue VD, et al. Antiepileptic medicinal plants used in traditional. In: Afawi Z, editor. Clinical and genetic aspects of epilepsy. InTech; 2011. p. 175−92 [Chapter 8]

[25] Gomes GMN, Campos GM, Orfao MCJ, Ribeiro AFC. Plants with neurobiological activity as potential targets for drug discovery. Prog Neuropsychopharmacol Biol Psychiatry 2009;33:1372−89.

[26] Arbonnier M. Arbres, arbustes et lianes des zones sèches d'Afrique de l'Ouest trees, shrubs and lianas of West Africa dry zones. 1st ed. Mali, Ouagadougou: Centre de Coopération Internationale en Recherche Agronomique pour le développement/ Muséum national d'histoire naturelle/Union mondiale pour la nature; 2000 (CIRAD/ MNHN/UICN)

[27] Adjanohoun JE, Aboukakar N, Dramane K, Ebot ME, Ekpere JA, Enow-Orock EG, et al. Traditional medicine and pharmacopoeia. Contribution to ethnobotanical and floristic studies in Cameroon. Porto-Novo (Benin): Centre de Production de Manuels Scolaires; 1996.

[28] Dalziel JM. The useful plants of West Tropical Africa. London: The Crown Agency for the Colonies; 1937.

[29] Biholong M. Contribution à l'étude de la flore du Cameroun: les Astéracées [Thèse de doctorat]. Bordeaux, France: Université de Bordeaux III, p. 10−50; 1986.

15 Toxicity and Beneficial Effects of Some African Plants on the Reproductive System

Yakubu Musa Toyin[1], Ajiboye Taofeek Olakunle[2] and Akanji Musbau Adewunmi[1]

[1]Phytomedicine, Toxicology and Reproductive Biochemistry Research Laboratory, Department of Biochemistry, University of Ilorin, Ilorin, Nigeria, [2]Antioxidants, Free Radicals, Functional Foods and Toxicology Research Laboratory, Department of Biological Sciences, Al-Hikmah University, Ilorin, Nigeria

15.1 Introduction

Normal reproductive capacity in both males and females is a result of complex interactions of numerous mechanisms involved from all aspects of sexual behavior, sexual performance, sperm production, and ejaculation in males, and conception and parturition in females. In males, the hypothalamo-hypophysial systems regulate functioning of Sertoli cells, spermatogenic epithelium, and androgen production by Leydig cells *via* producing gonadotropin, follicle-stimulating hormone (FSH) and luteinizing hormone (LH) [1]. Similarly in females, the complex interaction between the anterior pituitary, hypothalamus, ovary, and uterus leads to 2 monthly ovarian and uterine cycles (menstrual cycle). Regulation of these monthly cycles and by extension reproductive behaviors in females is brought about by the interplay of gonadotropin-releasing hormone, FSH, LH, estrogen, and inhibin.

Human reproductive function could be injured by exposure to different xenobiotic-reproductive toxicants from the environment *via* occupational exposure, life style, or chemotherapy [2–4]. Reproductive toxicants could exert harmful effects directly or indirectly by different processes and mechanisms. Some of the mechanisms include destruction of normal processes of endocrine or paracrine regulation by hormones or antihormone-mimicking compounds, destructive action on specific cell types such as germ cells, damage to blood—organ barrier, induction or inhibition of gonads and liver enzymes system leading to increase or decrease of steroid hormone secretion or clearance, mutagenic effects on germ cells, neurotoxic

Toxicological Survey of African Medicinal Plants. DOI: http://dx.doi.org/10.1016/B978-0-12-800018-2.00015-7

actions accompanied by disorders of libido/erection/ejaculation disorders, and hormonal imbalance [1].

The side effects of these chemical compounds on the reproductive organs are as diverse as the various classes of the drugs. These include gonadal toxicity or gonadal dysfunction, ovarian failure which results in premature menopause and irregular menstruation, infertility (either primary or secondary) caused by anticancer agents, anovulation caused by sex steroids, hypospermia caused by cimetidine and phentoin, decreased desire for sex (libido disorder), and generally quality-of-life issues that may last for many years. All these may result in reproductive incompetence. Paradoxically, while some medicinal plants no doubt have adversely affected the reproductive system, there are myriad herbs that also are beneficial.

Therefore, options for managing reproductive dysfunction which includes medications, mechanical aids such as penile implants, psychological and hormonal interventions are not without their shortcomings [5]. In addition, there is also the need to explore cheaper, effective options with reduced side effects in medicinal plants. Studies abound that reported on several medicinal plants which have been scientifically proven for managing reproductive dysfunction. Therefore, the present review gives accounts of the various medicinal herbs that have been claimed and/or scientifically validated for use in the management of xenobiotic-induced reproductive dysfunction in Africa, as well as those that have some toxic effects on the reproductive system.

15.2   The Reproductive System and the Metabolism of Xenobiotics

In general, xenobiotics require metabolic activation in order to exert adverse effects *via* covalent interactions between intermediate metabolites and cellular macromolecules such as DNA or protein [6,7]. In addition, xenobiotics may gain properties of endocrine disruptors in the process of toxic bioactivation [8]. Xenobiotic-metabolizing P_{450} enzymes in the testis catalyze the biotransformation of lipophilic xenobiotic or endogenous compounds and could be responsible for testicular toxicity [9]. The greatest impact into xenobiotics metabolism, including medical drugs, belongs to isoforms of Cytochrome P_{450} mixed function oxidase system that are involved in the metabolism of xenobiotics in the body, and these include Cytochrome P_{450} 1A2 (abbreviated as CYP1A2), Cytochrome P_{450} 2C9 (abbreviated as CYP2C9), Cytochrome P_{450} 2C19 (abbreviated as CYP2C19), Cytochrome P_{450} 2D6 (abbreviated as CYP2D6), Cytochrome P_{450} 2E1 (abbreviated as CYP2E1), and Cytochrome P_{450} 3A4 (abbreviated as CYP3A4) [10]. The Cytochrome P_{450} proteins are a large and diverse group of monooxygenases that catalyze many reactions involved in drug metabolism and synthesis of cholesterol, steroids, and other lipids. At present, among this short list of isoforms, only a few have been implicated in reproductive system disorders. For example, CYP2E1 is important from toxicological point of view as it has a high inducibility and ability to massively generate oxygen reactive which results in a wide range of reproductive pathologies in the testis and uterus. Others (CYP3A4) may influence testosterone metabolism leading to antiandrogenic effects and adverse effects on the Sertoli cell function and spermatogenesis [11].

15.2.1   Male Reproductive System

The male reproductive system consists of a number of sex organs that are used during sexual intercourse and procreation. The male reproductive system consists of the main male sex organ, the penis and the testes (located outside around the pelvic region), the ducts which include the epididymis, ductus deferens, ejaculatory duct, and urethra while the accessory glands include the seminal vesicles, prostate gland, and bulbo-urethral glands. The testes and epididymes are located outside the body cavity, enclosed within the scrotum. The ductus differentia leads from the testes into the pelvis where they join the ducts of the seminal vesicle to form ampullae. The ampullae extend into the ejaculatory ducts, pass through the prostate, and empty into the urethra within the prostrate. The urethra in turn exits from the pelvis and passes through the penis to outside of the body cavity [12]. The reproductive organs of the males produce, maintain, and transport sperm (the male reproductive cells) and protective fluid (semen); discharge sperm within the female reproductive tract during sexual intercourse; and produce and secrete male sex hormones responsible for maintaining the male reproductive system.

15.2.1.1   Anatomy of Male Reproductive System

The testes are small, ovoid, or ellipsoidal organs, each about 4−5 cm long and an inch in diameter within a pouch called the scrotum [12]. The outer part of each testis is a thick, white capsule consisting mostly of fibrous connective tissues called tunica albuginea. The connective tissue enters the testis to form incomplete septa which divide each testis into 300−400 cone-shaped lobules. The substance of the testes between the septa includes seminiferous tubules, interstitial cells, or Leydig cells [12]. The seminiferous tubules empties into a set of short, straight tubules called the tubuli recti, which in turn empty into a tubular network called rete testis. The rete empties into 15−20 tubules known as the efferent ductules.

The efferent ductules become extremely convoluted into comma-shaped structures called the epididymes. Each of the epididymes consists of a head, a body, and a long tail. The head contains the convoluted efferent ductules, which empties into single convoluted tubules known as the duct of the epididymis located primarily within the body of the epididymes. The ductus deferens or vas deferens emerges from the tail, ascends along the posterior side of the testis medial to the epididymes and becomes associated with the blood vessels and nerve cells [12]. The ductus deferens, testicular artery, and venous plexus, lymphatic vessels, nerves, and fibrous remnants of the vaginalis constitute spermatic cord.

The urethra, which is about 20 cm long, extends from the urinary bladder to the distal end of the penis. The urethra is divided into prostatic urethra (connected to the bladder and passes through the urethra), membraneous urethra (shortest part of the urethra, extending from the prostate gland through the perineum), and the spongy or penile urethra (longest part and extends from the membranous urethra through the entire length of the penis) [12].

The seminal vesicles are sac-shaped glands found next to the ampulla of the ductus deferentia. Each gland which is about 5 cm long and located posterior to the urinary bladder and anterior to the rectum tapers into a short excretory duct that joins the ampulla of the ductus deferens to form the ejaculatory duct.

The prostate gland which consists of both the glandular and muscular tissues is about 4 cm long and 2 cm wide. It is dorsal to the symphysis pubis at the base of the urinary bladder. It is composed of a fibrous connective tissue capsule containing distinct muscle cells that radiate inwards toward the urethra and numerous fibrous partitions [12].

The Cowper's glands, also known as the bulbo-urethral glands, are a pair of pea-sized exocrine glands located inferior to the prostate and anterior to the anus. It is a compound mucous gland with the single duct from each bulbo-urethral gland entering the spongy urethra at the base of the penis.

The scrotum is a sac-like organ made of skin and muscles that houses the testes. It is located inferior to the penis in the pubic region. The scrotum is made up of two side-by-side fibromuscular pouches (each with a testis) divided by a median septum (raphe) which continues posteriorly to the anus and anteriorly onto the inferior surface of the penis. The outer layer of the scrotum contains the skin, superficial fascia consisting of loose connective tissues, and dartos muscle [12].

The penis is the external male sexual organ located superior to the scrotum and inferior to the umbilicus. It is roughly cylindrical in shape and contains the urethra and the external opening of the urethra. The penis contains three columns of erectile tissues; two of these erectile columns form the dorsum and sides of the penis called corpora cavernosum while the third column is called the corpus spongiosum. The corpus spongiosum forms the ventral portion of the penis and expands to form a cap called the glans penis over the distal end of the organ. The spongy urethra passes through the corpus spongiosum, penetrates the glans penis, and opens as the external urethral orifice. At the base of the penis, the corpus spongiosum expands to form the bulb of the penis. The skin of the penis is well supplied with sensory receptors. A loose fold of the skin called prepuce or foreskin covers the glans penis.

15.2.1.2 Physiology of the Male Reproductive System

The testes in adults functions as exocrine by producing and secreting sperm and endocrine by producing and secreting testosterone. It also secretes Mullerian Inhibiting Substance in the fetus to cause regression of female structures. Sperm production occurs within the seminiferous tubules from complex local events as well as distant regulatory signals. Under appropriate control of local testosterone production by the Leydig cells, the Sertoli cells within the seminiferous tubules provide an appropriate environment for the development of immature germ cells into mature spermatozoa. Other functions of Sertoli cells include support and nutrition of germ cells; release of mature germ cells into the lumen; translocation of developing germ cells in an adluminal direction; secretion of androgen-binding protein, transferrin, inhibin; cell−cell communication via gap junctions to coordinate spermatogenesis; and blood−testis barrier [13].

The epididymis functions in sperm conduit (sperm propulsion by spontaneous rhythmic contractions of the duct), fluid resorption, sperm reservoir (the cauda is a major site of sperm storage), sperm maturation (fertilizing ability and motility of sperm improves from caput to corpus to cauda) [14].

15.2.2 Female Reproductive System

The female reproductive system is made up of the ovaries, uterine tubes, uterus, vagina, and external genitalia organs. The internal reproductive organs of the females which are held in place by the ligaments are located within the pelvis between the urinary bladder and the rectum. The uterus, vagina, and ovaries are located in the midline.

15.2.2.1 Anatomy of the Female Reproductive System

The anatomy of the female reproductive system includes parts inside and outside the body. The external female reproductive structures (the genitals) function to enable sperm to enter the body and to protect the internal genital organs from infectious organisms.

The two small, oval-shaped ovaries which are attached to the posterior surface of the broad ligament by mesovarium are small organs of about 2−3.5 cm long and 1−1.5 cm wide. Two other ligaments, suspensory ligament extends from the mesovarium to the body while the ovarian ligament attaches the ovary to the superior margin of the uterus. The mesovarium is the channel through which the ovarian arteries, veins, and nerves enter the ovary [12]. The visceral peritoneum where it covers the ovary is called germinal epithelium. Immediately below this is a capsule of dense fibrous connective tissues called the tunica albuginea. The dense outer part of the ovary is called the cortex while the inner part is known as the medulla. The connective tissue of the ovary is referred to as the stroma.

The paired uterine tubes, also known as fallopian tubes or oviducts, are located on each side of the uterus and are associated with the ovary. Each tube located along the superior margin of the broad ligament opens directly into the peritoneal cavity. It expands to form the infundibulum and long thin processes called fimbriae. The part of the uterine tubes nearest to the infundibulum, called the ampulla, is the widest and longest part of the tube accounting for about 7.5−8 cm of the total length of 10 cm. The much narrower and thicker walled isthmus is the part of the uterine tube nearest the uterus. The uterine part of the tube passes through the uterine walls and ends in a very small uterine opening. The wall of the uterine tube is made up of three layers: the serosa, which is the outer layer formed by the peritoneum, the muscular middle layer of longitudinal and circular muscles, and the inner mucosa consisting mucous membrane of simple ciliated columnar epithelium.

The uterus (womb) is a hollow, pear-shaped organ that is the home to a developing fetus. It has a shape of medium-sized pear and about 7.5 cm long and 5 cm wide. It is slightly flattened anteroposteriorly and is oriented in the pelvic cavity with larger, rounded part called the fundus, directly superior and the narrower lower part, the cervix. The main part of the uterus, the body, is located between the fundus and the cervix. The isthmus marks the junction of the cervix and the body. Internally, the uterine cavity continues as the cervical canal and opens into the vagina through the ostium. Three ligaments, broad, round, and sacra holds the uterus in its position. The uterine wall is made up of three layers of peritoneum or serous (covers the uterus), myometrium (next layer, deep to the perimetrium consisting of thick layer of smooth muscle), and endometrium (the innermost simple columnar epithelium lining) [12].

The vagina or birth canal is a tube of about 10 cm long that joins the cervix (the lower part of uterus) to the outside of the body. There is the presence of longitudinal ridges called columns that extend the length of the anterior and posterior vaginal walls. The walls of the vagina are made up of outer muscular layer (allows the vagina to increase in size to accommodate the penis during sexual intercourse and during childbirth) and an inner mucous membrane (made of moist stratified squamous epithelium that protects the surface). A thin mucous membrane known as hymen covers the vaginal opening or orifice.

The external genitalia called vulva or pudendum is made up of the vestibule and its surrounding structures. The vestibule is the space to which the vagina opens in the posterior. There is also a pair of thin, longitudinal folds of skin referred to as labia minora which forms a border on each side of the vestibule. A small erectile tissue called the clitoris is found on the anterior margin of the vestibule. The clitoris, usually <2 cm in length, consists of a shaft and distal glans. The erectile tissues of the clitoris, corpora cavernosa expands at the base to form the crus of the clitoris. The two labia minora unites over the clitoris to form a fold of skin called the prepuce [12]. Laterally, the labial minora fuse to form two prominent, rounded folds of skin called the labial majora. The two labia majora unite anteriorly in an elevation over the pubis symphasis to give rise to mons pubis while the space between the two labia majora is the pupendal cleft. The labia majora, which is covered by hair after puberty, contains sweat and oil-secreting glands. The Bartholin's glands are located besides the vaginal opening and produce a fluid (mucus) secretion.

15.2.2.2   Physiology of the Female Reproductive System

Females of reproductive age experience cycles of hormonal activity that repeat at about 1-month intervals. With every cycle, a woman's body prepares for a potential pregnancy, whether or not that is the woman's intention. However, when pregnancy does not result, there will be periodic shedding of the uterine lining called menstruation. The average menstrual cycle takes about 28 days and occurs in phases: the follicular phase, the ovulatory phase (ovulation), and the luteal phase. There are four major hormones that stimulate or regulate the activity of cells or organs involved in the menstrual cycle: FSH, LH, estrogen, and progesterone.

15.2.2.2.1 Follicular Phase of the Menstrual Cycle
This phase starts on the first day of the cycle with the following events:

- The FSH and LH are released from the brain and travel in the blood to the ovaries.
- The hormones stimulate the growth of about 15–20 eggs in the ovaries, each in its own follicle.
- These hormones (FSH and LH) trigger increased production of estrogen.
- As estrogen levels rise, it turns off the production of FSH. This careful balance of hormones allows the body to limit the number of follicles that mature.
- As the follicular phase progresses, one follicle in one ovary becomes dominant and continues to mature. This dominant follicle suppresses all of the other follicles in the group. As a result, they stop growing and die. The dominant follicle continues to produce estrogen.

15.2.2.2.2 Ovulatory Phase of the Menstrual Cycle
The ovulatory phase, or ovulation, starts about 14 days after the starting of the follicular phase. The ovulatory phase is the midpoint of the menstrual cycle, with the next menstrual period starting about 2 weeks later. During this phase, the following events occur:

- The rise in estrogen from the dominant follicle triggers a surge in the amount of LH produced by the brain.
- This causes the dominant follicle to release its egg from the ovary (ovulation).
- The eggs are captured by finger-like projections on the end of the fallopian tubes (fimbriae) before being swept into the tube.
- Also during this phase, there is an increase in the amount and thickness of mucus produced by the cervix (lower part of the uterus). If sexual intercourse is permitted during this time, the thick mucus captures the sperm, nourishes it, and moves it toward the egg for fertilization.

15.2.2.2.3 Luteal Phase of the Menstrual Cycle
The luteal phase of the menstrual cycle begins immediately after ovulation and involves the following processes:

- Once it releases its egg, the empty follicle develops into a new structure called the corpus luteum.
- The corpus luteum secretes progesterone that prepares the uterus for a fertilized egg to implant.
- If fertilization had taken place, the embryo will travel through the fallopian tube to implant in the uterus for the female to be referred to as pregnant.

If the egg is not fertilized, it passes through the uterus. The lining of the uterus breaks down, sheds, and the next menstrual period begins. The vast majority of the eggs within the ovaries steadily die, until they are depleted at menopause. At birth, there are approximately 1 million eggs and by the time of puberty, only about 300,000 remain. Of these, 300–400 will be ovulated during a woman's reproductive lifetime. The eggs continue to degenerate during pregnancy, with the use of birth control pills, and in the presence or absence of regular menstrual cycles.

15.2.2.2.4   The Menstrual/Hormonal Cycle

The hormonal cycle facilitates maturation and rupture of the ovarian follicle resulting in the release of an ovum (the female reproductive or germ cell). Each month a series of changes take place that prepare the uterus for pregnancy as itemized:

- The first day of menstruation (referred to as day 1) occurs when levels of estrogen and progesterone are low. In response to these low levels, the hypothalamus secretes gonadotrophin-releasing hormone (GnRH) which triggers the anterior pituitary gland to release FSH and LH.
- FSH stimulates the development of many follicles within the ovary with one dominant follicle taking over. As it continues to grow, it produces increasing amounts of estrogen that not only stimulates the release of LH and inhibits FSH but suppresses further follicular development.
- When LH levels are elevated (LH surge), the ovarian follicle "ruptures" and releases one ovum into the fallopian tube by hair-like projections called cilia that line the fimbriae (the fringe-like end of the fallopian tube that is closest to the ovary). This process is called ovulation. Increasing estrogen levels causes the cervical mucous (vaginal secretions) to become clear and profuse and cause the orifice to dilate. These two actions may facilitate the transport of semen (containing sperm) from the vagina, through the uterus, and into the fallopian tube.
- Following ovulation, the ruptured follicle is transformed into the corpus luteum, a glandular mass that continues to produce estrogen and high levels of progesterone. The progesterone causes the endometrium to thicken, preparing it for implantation of a fertilized egg. If fertilization takes place during ovulation, hormonal levels remain high, essential for the maintenance of the pregnancy.
- If fertilization does not occur, the corpus luteum shrinks and levels of both estrogen and progesterone decrease. The withdrawal of estrogen and progesterone causes the blood vessels of the endometrial (uterine) lining to disintegrate and result in menstruation. The average menstrual cycle is 28−35 days, and menstrual flow usually continues for 3−7 days, depending on individual.
- Following menstruation, estrogen and progesterone levels are low, triggering the hypothalamus to once again release GnRH, starting the entire cycle again. If fertilization does take place, menstruation will not reoccur for the duration of the pregnancy.

15.2.2.2.5   Mechanism of Action of Contraception/Pregnancy

Most hormonal methods of birth control, including emergency contraception, work by preventing or postponing ovulation, and thickening the cervical mucus, making the environment hostile to the sperm cells. The absence of a menstrual period in a sexually active woman is presumed to indicate pregnancy.

15.2.2.2.6   Menopause

Menopause, the end of menstruation, occurs between the ages of 45 and 55 (with the average age of 51.3). An entirely normal developmental and physiological process, it can be accompanied by symptoms such as hot flashes, fatigue, moodiness, insomnia, decreased libido and sexual response, changes in memory, weight gain, and vaginal dryness.

15.3 Known Toxic and Beneficial Plants Acting on the Reproductive System

15.3.1 Mode of Action of Toxic Medicinal Plants on Reproductive System

15.3.1.1 Male Reproductive System

An ample number of medicinal plants in Africa have been reported to elicit varying modes/degree of toxicity and benefit on the male reproductive system. Depending on the dosage and time of exposure, a herb may be deleterious or beneficial to a living system. The phytoconstituents (saponins, tannins, steroids, alkaloids, glycosides, and terpenes) and elemental constituents of these plants cause dysfunction at the organ level leading to impairment in the normal functioning of the organs that make up the reproductive system. A few of these medicinal plants are highlighted.

15.3.1.1.1 Hibiscus sabdariffa

Hibiscus sabdariffa Linn. (Malvaceae) is otherwise known as Roselle or red sorrel (English) and *isapa* (Western Nigeria). It is native to Asia (India), Malaysia, and tropical Africa including Nigeria where it is indiscriminately consumed by the populace as beverages because of its aphrodisiac and other pharmacological properties [15]. Orisakwe et al. [15] reported that administration of the extract of *H. sabdariffa* calyx at the doses of 2.3, 4.6, and 1.15 g/kg body weight produced hyperplasia of testis with thickening of the basement membrane, decreased epididymal sperm counts, disintegration of sperm cells, and distortion of the tubules.

15.3.1.1.2 Chromolaena odoratum

Chromolaena odorata (L.) R.M. King & H.E. Robins (syn. *Eupatorium odoratum* L.) (Compositae), also known as Christmas bush, bitter bush, Siam weed, and baby tea (English); akintola-ta-ku (Yoruba, Western Nigeria) and ishero (Urhobo, Southwestern Nigeria) is a scrambling shrub. It is known to have originated from South and Central America but now is found in tropical rain forest and Guinea Savannah regions of Nigeria. Studies have revealed that the alkaloids significantly decreased ($p < 0.05$) the testes—body weight ratio; the concentrations of testicular total protein, glycogen, sialic acid, and cholesterol; and the activities of γ-glutamyl transferase, acid phosphatase, and alkaline phosphatase. The serum LH and FSH levels, as well as testicular and serum testosterone levels, also decreased significantly ($p < 0.05$). There were decreases in the sperm count, motility, and density, as well as morphological changes in the sperm cells. The pH and whitish gray color of the semen were not significantly affected. All of the doses of the alkaloids increased the total mean number of sperm cell abnormalities, with the secondary type predominating over the primary sperm cell abnormality [16]. The study concluded that the alterations in the levels of the hormones, secretory and synthetic

constituents of the testes, and the spermatotoxic effects by the alkaloids from
C. odorata leaves may be due to nonavailability or deprivation of testosterone to
the target organ [15,16].

15.3.1.1.3   Mondia whitei

Watcho et al. [17] reported that *Mondia whitei* (Asclepiadaceae) extract has inhibitory effects on spermatogenesis and reduces fertility in rats. Lesions of some
tubules in the caput epididymes by the extract were attributed to low testicular
androgen levels, resulting in a rarefaction of Leydig cells, or to a secondary effect
on the pituitary gland leading to the reduction of gonadotropin synthesis. The
degenerated epithelium in *M. whitei* treated rat was partly due to accumulated cholesterol. The study concluded that the mode of toxicity of *M. whitei* is shown by its
antispermatogenic activity.

15.3.1.1.4   Helianthus annus

Helianthus annus (Astereacea), commonly called sunflower, is an annual plant that
is widely distributed worldwide. Study has shown that the administration of ethanolic extract at the dose of 14 g/kg of rat weight on the histology of the testes, blood
levels of some reproductive hormones, and epididymal properties of male rats suggested the existence of some antifertility effects [18].

15.3.1.1.5   Azadirachta indica

The plant *Azadirachta indica* (Meliaceae), commonly called neem, has been used
medicinally the world over. Extracts of *A. indica* have been reported to have several antifertility effects. These include reduction in serum testosterone and LH
levels, inhibition of folliculogenesis and immunocontraceptive activity [19–22]. In
a study by Akpanatha et al. [23], the administration of aqueous leaf extract of the
plant at the doses of 200 and 400 mg/kg body weight significantly reduced the concentrations of serum LH and FSH. Histomorphologic sections of the pars anterior
revealed reduced acidophil and basophil populations, with prominent degranulated
chromophobes which were larger in the group treated with 400 mg/kg of *A. indica*
leaf extract. This group also presented hypertrophy of the basophils when compared
with the control. These results indicate that the antifertility effect of the leaf extract
may be a result of hypophysis changes that ultimately cause the reduction of LH
levels.

15.3.1.2   Female Reproductive System

15.3.1.2.1   Bambusa vulgaris

Bambusa vulgaris (L.) (Poaceae), known as bamboo (English), and by other tribes
in Nigeria as Oparun (Yoruba), Iko (Bini), and Atosi (Igbo), is found in tropical
and subtropical areas, especially in the monsoon and wet tropics. The species are
usually very big, having numerous branches at a node with one or two much larger
than the rest. Preliminary chemical screening of the aqueous extract of *B. vulgaris*
revealed the presence of alkaloids, tannins, phenolics, glycosides, saponins, flavonoids, and anthraquinones. Clinical signs of toxicity such as respiratory distress,

salivation, weight loss, dull eyes, diarrhea, changes in the appearance of fur, as well as mortality, were not observed in the rabbits at any period of the experiment. The 250 mg/kg body weight of the extract decreased ($p < 0.05$) the number of live fetuses, whereas the 500 mg/kg body weight produced no live fetus. The 250 and 500 mg/kg body weight of the extract reduced the survival rate of the fetus to 29% and 0%, whereas the same doses produced abortion at the rate of 60% and 100%, respectively. The implantation index and preimplantation loss compared well with the control. Both doses increased the resorption index and postimplantation loss. The extract also decreased the concentrations of serum progesterone, FSH and LH. While there was no effect on the weight of the uterus, uterine/body weight ratio, length of the right uterine horn, and uterine cholesterol, the alkaline phosphatase activity and glucose concentration decreased significantly. The extract also provoked vaginal opening. This study has substantiated the abortifacient potential of the aqueous extract of *B. vulgaris* leaves. The mechanism of abortion could possibly be through changes in the implantation site, altered hormone levels and, partly, estrogenicity made possible at least in part by the phytoconstituents [24].

15.3.1.2.2 Momordica charantia
Estrus cycle of rats became irregular with prolonged estrus and metestrus phases, and reduced diestrus and proestrus phases after the oral administration of petroleum ether, benzene, chloroform, and alcohol extracts of the seeds of *Momordica charantia* (Cucurbitaceae) at a dose level of 25 mg/100 g body weight. The results showed reduced ovarian weight, number of developing follicles, Graffian follicles, and *corpora lutea* and an increased number of atretic follicles in histological sections of the ovary. However, the benzene extract of *M. charantia* seeds was more effective in causing these changes as compared to other extracts [25].

15.3.1.2.3 Garcinia kola
Garcinia kola (Guttifeare) seed extract was reported to have profound effects on estrus cycle, ovulation, and fetal development in adult female Sprague-Dawley rats. In the study, estrus cycle was altered for the first 2 weeks after commencement of extract but returned to normal from the third week. This was indicated by the irregular pattern of estrus with a prolonged diestrus in the treated rats. Ovulation was partially blocked as shown by the reduced number of ova in the oviduct of the treated rats compared with control ($p < 0.05$). There was a significant decrease in the weight of fetuses of the treated rats ($p < 0.05$) while 7% of the fetuses from pregnant rats, which received treatment for the first 5 days of gestation, had malformed left upper limbs. Results suggest that *G. kola* seeds at 200 mg/kg body weight altered estrus cycles in rats partly inhibit ovulation and produced duration-dependent teratogenicity in fetal rats [26].

15.3.1.2.4 Asparagus africanus
Asparagus africanus L. (Liliaceae) known as *Saritti* (Ethiopia), *aluki, kadankabe* (Western Nigeria), and *sasarin-kura* (Northern Nigeria) are among the many plants with traditionally claimed antifertility properties in Ethiopia. It is a perennial climbing or erect shrub that can grow between 700 and 3800 m above sea level. The

aqueous extracts of the leaves and the roots showed an anti-implantation activity of 70% and 77%, respectively, while the ethanol extracts of the leaves and roots showed 48% and 61%, respectively. The antifertility activities of the aqueous and ethanol extracts were 40% (for leaves), 60% (for roots), and 20% (for leaves), 40% (for roots), respectively. All the extracts have resulted in significant ($p < 0.05$) reduction in the number of implants when compared with their respective controls. Each extract potentiated acetylcholine-induced uterine contractions in a concentration dependent manner significantly ($p < 0.05$). It was concluded that the leaves and roots of this plant possess hormonal properties that can modulate the reproductive function of the experimental rats with respect to anti-implantation and antifertility [27].

15.3.1.2.5 Rumex steudelii

Rumex steudelii Hochst (Polygonaceae), locally known as "Tult" or "Yeberemelas," is one of the traditionally used antifertility plants in Ethiopia. It is an erect, perennial herb that grows up to 1 m tall. The plant is distributed in the north and central parts of Ethiopia, at an altitudinal range of 1200–3900 m [28]. The extract reduced significantly ($p < 0.01$) the number of litters and weights of ovaries and uterus, prolonged the estrus cycle ($p < 0.05$) and the diestrus phase ($p < 0.01$) of the rats and produced dose-dependent antifertility and contraceptive effects. All these observations suggest that the extract has antifertility effect and is safe at the effective antifertility doses employed in this study [29].

15.3.1.2.6 Cnidoscolous aconitifolius

Cnidoscolous aconitifolius (Miller) I. M. Johnston (Euphorbiaceae) known as tree spinach (English), *efo iyana ipaja*, or *efo Jerusalem* (Yoruba) is commonly found growing in western parts of Nigeria. It is an ornamental, evergreen, droughtdeciduous shrub of 3–5 m tall. The large (32 cm long and 30 cm wide) palmate lobed leaves are alternately arranged. Phytochemical screening of the extract revealed the presence of alkaloids, saponins, phenolics, tannins, flavonoids, anthraquinones, phlobatannins, and triterpenes. Administration of the extract at 250, 500, and 1000 mg/kg body weight significantly ($p < 0.05$) increased ($p < 0.05$) the serum prolactin concentration whereas those of estradiol, progesterone, FSH and LH reduced significantly. The study concluded that the alterations in the female rat reproductive hormones by the extract are indications of adverse effect on the maturation and ovulation of follicles. Consequently, the extract may impair fertility and conception in female rats [30].

15.3.1.2.7 Helianthus annus

Ethanolic extract of 0.5 g/kg of *H. annus* also was investigated for fecundity in Wistar rats. The results revealed that coital frequency was not affected by the extract but pregnancy rate and average numbers of pups were significantly reduced in extract-treated animals. The study concludes that the histodegeneration of the gonads might be responsible for the reduced fecundity in the animals [31].

15.3.2 Mode of Action of Beneficial Medicinal Plants on the Reproductive System

15.3.2.1 Rauvolfia vomitoria

Rauvolfia vomitoria (Apocynaceae) is recognized for its many therapeutic effects, especially in the treatment of diarrhea, malaria, hypertension, and male infertility. Administration of the aqueous extract of *R. vomitoria* stem bark at the doses of 25, 50, 100, and 200 mg/kg body weight for 21 days, dose dependently reduced the body weight while the relative weight of testis at high doses (100 and 200 mg/kg body weight), relative weight of the prostate and seminal vesicles, daily sperm production at high doses as well as the sperm count in epididymis and vas deferens at low doses (25 and 50 mg/kg), sperm motility and transit at low doses, testicular and epididymal protein, testicular cholesterol, and testosterone significantly increased. The study concludes that treatment with the aqueous extract of *R. vomitoria* could improve the fertility of male rats [32].

15.3.2.2 Fagara tessmannii

Fagara tessmannii (*Zanthoxylum tessmannii*) is a tree from the family of Rutaceae, which grows in tropical and subtropical regions, and is locally known in the littoral region of Cameroon, Africa as "Ewoungea" or "Bongo." In a review by Lembe et al. [32], administration of ethanolic extract of *F. tessmannii*, at the doses of 0.01, 0.1, and 1 g kg/ body weight/day for 14 days, decreased the weight of epididymis and seminal vesicle at low doses (0.01 g/kg body weight), decreased prostate weight at all doses ($p < 0.05$), increased the transit of spermatozoa in cauda epididymidis at a lower dose of 0.01 g/kg body weight, increased the length of stages IX−I of the seminiferous tubule and serum testosterone level. The results suggest that *F. tessmannii*, 14 days after treatment, may improve spermatogenesis, testosterone level, and sperm transit in cauda epididymidis but negatively impair reproductive organ activities.

15.3.2.3 Cnestis ferruginea

Cnestis ferruginea Vahl ex DC (Connaraceae), also known as *Gboyin gboyin* or *Omu aja* (Yoruba, Western Nigeria), *Fura amarya* (Hausa, Northern Nigeria), *Amu nkita* (Igbo, Eastern Nigeria), *Ukpo ibieka* (Edo, Southern Nigeria), and *Usiere ebua* (Efik, Southern Nigeria), is a perennial shrub found mainly in the savannah region of tropical West Africa. The plant is about 3.0−3.6 m high with densely, rusty brown, pubescent branches. The leaves which are alternate or sometimes opposite in orientation have ovate to narrowly oblong leaflets and orange-red fruits. In a recent study to investigate the effect of *C. ferruginea* root in paroxetin-induced sexual dysfunction in male rats, Yakubu and Nurudeen [33] reported that administration of paroxetine to sexually active male rats significantly ($p < 0.05$) reduced the mount frequency (MF), intromission frequency (IF), and ejaculation frequency (EF), whereas mount latency (ML), intromission latency (IL), ejaculatory latency (EL), and postejaculatory interval (PEI) were increased. The extracts at the doses of 13, 26, and 52 progressively reversed the trends of MF, IF, EF, ML, IL, EL, and PEI in the paroxetin-treated animals toward the control values throughout the exposure period. The sexual behavior parameters compared well with the PowmaxM-treated animals, but not comparable to the distilled water administered animals. In addition, all the doses of the extract elevated ($p < 0.05$) the levels of serum LH and FSH and decreased testosterone contents. The aqueous extract of *C. ferruginea* root at the doses of 13, 26, and 52 mg/kg body weight restored sexual competence at least to a reasonable extent in sexually impaired/sluggish male rats with the highest dose producing the best efficacy.

15.3.2.4 Ananas comosus

Ananas comosus (L.) Merr (Bromeliaceae), known as pineapple, is a herbaceous, biennial, tropical plant that grows up to 1.0—1.5 m high and produces a fleshy, edible fruit whose flesh ranges from nearly white to yellow. Yakubu and Nurudeen [33] reported that the juice contained tannins, cardenolides, dienolides, cardiac glycoside, and flavonoids. The number and weights of live fetuses, number of implantation sites, corpora lutea, computed percent implantation index, resorption index, pre- and postimplantation losses were not significantly ($p > 0.05$) altered in comparison to the control. Neither fetal death nor provoked vaginal bleeding was observed in the pregnant rats. The maternal weight increased in all the experimental animals with that of the control augmenting least. The 250 and 500 mg/kg body weight increased ($p < 0.05$) the serum concentrations of progesterone and estrogen in the pregnant rats. The fruit juice of *A. comosus* did not exhibit abortifacient activity in pregnant Wistar rats but may be beneficial to reproduction as it may enhance the growth and development of the fetuses.

15.4 Methods in the Screenings of Medicinal Plants Affecting the Reproductive System

Several methods have been employed by various researchers to evaluate the effects of medicinal plants on the reproductive systems of experimental animals. These effects may be with the view of determining the positive impact like in most cases the efficacy or pharmacological effects of medicinal plants on selected or the whole reproductive organs of the animals, while those of the negative effects center on toxicity or safety evaluation. Some of these will be reviewed here to set out objective(s) of the research and are by no means exhaustive.

15.4.1 Aphrodisiac

15.4.1.1 Physical Methods

15.4.1.1.1 Mating Behavior Test in Normal Male Rats
Mating behavior tests can be carried out by the methods of Dewsbury and Davis [34] and Szechtman et al. [35]. Briefly, healthy and sexually experienced male

albino rats that show brisk sexual activity should be selected for the study. After extract administration at various concentration to various groups of the animals depending on the experimental design and objective(s) of the study, the male animals should be brought to the laboratory and exposed to dim light (in 1 w fluorescent tube in a laboratory of 14'−14') at the stipulated time of testing daily for some days (3−6 days) before the experiment. The female animals should be artificially brought into estrus (heat) as the female rats allow mating only during the estrus phase by administering either suspension of ethinyl estradiol orally at the dose of 100 μg/animal 48 h prior to the pairing and subcutaneous administration of progesterone at the dose of 1 mg/animal 6 h before the experiment or alternatively by the sequential administration of estradiol benzoate (10 μg/100 g body weight) and progesterone (0.5 mg/100 g body weight) through subcutaneous injections, 48 h and 4 h, respectively, prior to pairing [36]. The receptivity of the female animals should be confirmed before the test by exposing them to male animals, other than the control and test animals. The most receptive females should then be selected for the study. The experiment could be carried out after the commencement of the treatment of the male animals at intervals of some days (e.g., days 1, 3, and 5) or on a single day with well-defined time interval, depending on ethnobotanical information on the posology of the plant being investigated. The experiment should be conducted in the night (as most of the experimental animals are so active during this period) 20:00 hours in the same laboratory and under very dim light. The receptive female animals should be introduced into the cages of the male animals in the ratio 1:1 (female:male). The observation for mating behavior should commence immediately and continue for the first two mating series. The test should be terminated if the male fails to evince sexual interest. Any female animal that did not show receptivity should be replaced by another artificially "warmed" female. The occurrence of events and phases of mating may be called out to be recorded on audio cassette as soon as they appear. Their disappearance should also be called out and recorded. Later, the frequencies and phases can be determined from cassette transcriptions.

The parameters of male sexual behavior that should be monitored can include:

- MF is number of times the male assumes copulatory position but failed to achieve intromission, characterized by lifting the male's fore body over the hind quarter of the female and clasping her flanks with its fore hand.
- IF is number of intromissions (introduction of the male copulatory organ into the vagina) from the time of introduction of the female until ejaculation.
- EF is the number of ejaculations made during the observatory period.
- ML is time interval between the introduction of the female and the first mount by the male.
- IL is time interval from the time of introduction of the female to the first intromission by the male, usually characterized by pelvic thrusting and springing dismounts.
- EL is the time interval between the first intromission and ejaculation (the act of ejecting semen), usually characterized by longer, deeper pelvic thrusting and slow dismount followed by a period of inactivity or reduced activity.

- PEI is the time interval between ejaculation and erection of the male copulatory organ for the next phase.
- Index of Libido is defined as the ratio of number mated to number paired expressed in percentage. This can be expressed mathematically as

$$\% \text{ Index of Libido} = \frac{\text{number mated}}{\text{number paired}} \times 100$$

- Computed Male Sexual Behavior Parameters: Using the aforementioned parameters of sexual behavior, the following can thus be computed:

a. $\% \text{ Mounted} = \dfrac{\text{number mounted}}{\text{number paired}} \times 100$

b. $\% \text{ Intromitted} = \dfrac{\text{number of intromissions}}{\text{number paired}} \times 100$

c. $\text{Intromission ratio} = \dfrac{\text{number of intromission}}{\text{number of mounts} + \text{number of intromissions}}$

d. $\% \text{ Ejaculated} = \dfrac{\text{number of ejaculations}}{\text{number paired}} \times 100$

e. $\text{Copulatory efficiency} = \dfrac{\text{number of intromissions}}{\text{number of mounts}} \times 100$

f. Intercopulatory efficiency = Average time between intromissions [37].

Any medicinal plant with aphrodisiac tendencies should produce statistically significant increase in the indices of sexual vigor of MF and IF, significant decrease in ML and IL. These are indicators of stimulation of sexual arousability, motivation, and vigor [38,39]. The significant decrease in ML and IL, as well as significant increase in computed male sexual behavior parameters of percent mounted, percent intromitted, percent ejaculated, and the reduction in intercopulatory efficiency can indicate sustained increase in sexual activity and aphrodisiac property inherent in the plant extract [37].

15.4.1.1.2 Mating Behavior Test in Sexual Dysfunction in Male Rats
15.4.1.1.2.1 Induction of Sexual Dysfunction and Assessment of Mating Behavior Indices in Male Rats
Male rats would be oral administered 10 mg/kg of paroxetine hydrochloride suspension (prepared daily in Tween-80 (BDH Chemicals, Ltd., Poole, England), suspended in 0.9% saline solution) using a metal oropharyngeal cannula, once daily for 21 days to induce sexual dysfunction [40,41]. Similarly, intact, healthy female rats would be artificially brought into estrus (as female rats allow indiscriminate mating only during this period and for the data to be reliable) by a single injection of 50 μg of estradiol benzoate 36 h before testing with male rats [40]. Rats in the estrus phase would be confirmed by vaginal smear examinations as described by OECD [42]. The primed female rats that exhibit profound receptivity

would be introduced into the cages containing the male rats, after which the following sexual behavior parameters would be observed directly from the cage side and perhaps, also using camera in dim light, for 30 min: MF, IF, EF, ML, IL, EL, and PEI. The frequencies and latencies of these behavior parameters would be determined electronically from cassette transcription and would be reconciled with camera recording. Male rats that show minimum reduction of 25% in MF, IF, EF, and EL as well as minimum increase of 25% in ML, IL, and PEI would be declared as sexual dysfunction and used for the subsequent study [33].

15.4.1.1.3 Test for Libido
This test shall be carried out by adopting the procedure described by Davidson [43] and modified by Amin et al. [36]. Sexually experienced male albino rats shall be kept singly in separate cages during the experiment. The female rats shall be made receptive by hormonal treatment and all the animals shall be accustomed to the testing condition as previously presented in mating behavior test. The animals shall then be observed for the MF on the evening of specific day according to the design of the experiment (likely seventh day) at 20:00 hours. The penis shall be exposed by retracting the sheath and applying 5% xylocaine ointment 30, 15, and 5 min before starting observations. Each animal shall be placed individually in a cage with the receptive female rat in the same cage. The number of mountings shall be noted. The animals shall also be observed for intromission and ejaculation.

15.4.1.1.4 Test for Potency
The potency of the plant extract at various doses depending on the design may be studied according to the methods described by Hart [44]. The male animals shall be kept singly in separate cages during the experiment. The extracts will be administered at least 30 min to 1 h before the commencement of the experiment. On the eighth day, the test for penile reflexes shall be carried out by placing the animal on its back in a glass cylinder with partial restraint. The preputial sheath will be pushed behind the glans by means of the thumb and index finger and held in this manner for a period of 15 min. Such stimulation shall normally elicit a cluster of genital reflexes. The frequency of the following components of penile reflexes will therefore be recorded: (i) erections (E), (ii) quick flips (QF), and (iii) long flips (LF).

From the above listed components of penile erection, the total penile reflexes (TPR) can thus be computed as the sum total of each of the components of penile erection. Mathematically, it can be expressed as $E + QF + LF$. Statistically significant increase in the frequency of penile reflexes by the extract shall suggest aphrodisiac potential.

15.4.1.1.5 Penile Microcirculation Study
A Laser Doppler Flow Meter may be adopted for the determination of penile microcirculation, using the procedure described by Grotthus et al. [45]. Briefly, the animals shall be anesthetized by intravenous administration of 30 mg/kg body weight sodium pentobarbital. The central ear artery shall be cannulated for continuous monitoring of arterial blood pressure. At the beginning of the test, the penile sheath shall be retracted manually, and after 10 min of adaptation to room temperature in the

laboratory, the Laser Doppler flow detection probe shall be positioned in a holder close (2–3 mm) to the dorsal side of the male organ (penis). The result of the test will be based on the average of arbitrary flow units (flux) within 10 min of the test. The probe shall be calibrated with flux standard before each test.

15.4.1.1.6 Intracavernous Pressure (ICP) Study

Twelve hours after the administration of the last dose of the plant extract, the male animals shall be anesthetized by intraperitoneal administration of 50 mg/kg body weight of sodium pentobarbital. This shall be followed by the incision of the penile skin and degloving the prepuce to expose the corpora cavernosa. A 26-gauge needle connected to a polyethylene tube (PE-50) filled with NSS and 100 IU/ml of heparin on one side of the corpora cavernosa shall be inserted for the ICP measurement. Another 22-gauge needle shall be placed into the right carotid artery connected to a PE tube for the measurement of mean arterial pressure (MAP). Both tubes shall be connected to blood pressure transducers via transducer amplifiers to a data acquisition board. Computers can be used to see real-time display and records of pressure measurements (mmHg). Similarly, the major pelvic ganglion, pelvic and cavernous nerves can be exposed by a midline abdominal incision. The cavernous nerve can then be stimulated by using a square pulse stimulator connected to a platinum bipolar electrode positioned on the cavernous nerve using 5 V with a frequency of 50 Hz and duration of 5 min as stimulus parameters. The stimulation may be done three times and the ICP recorded. The ICP should be allowed to return to baseline before the next stimulation.

Statistically significant increase in ICP may imply their role on nitric oxide (NO) and by extension, erectile function. Medicinal plants with aphrodisiac potential should be capable of stimulating cavernous nerve which normally should lead to increase in NO and cyclic guanosine phosphate (cGMP) signaling in corpus cavernosal smooth muscle relaxation. The subsequent arteriolar dilation leading to increased arterial inflow and impaired venous return (due to engorgement of the cavernosum) builds up a pressure system within the corpora that results in penile tumescence and rigidity [46].

15.4.1.1.7 Assay for Neuronal Nitric Oxide Synthase and Androgen Receptor Protein

Nitric oxide synthase (NOS), a calcium/calmodulin-dependent enzyme, is responsible for the biosynthesis of NO from L-arginine. Because NO is responsible for the relaxation of smooth muscles of the cavernosum which eventually lead to inflow of blood into the male organ, determination of the activity of NOS in the male copulatory organ and the testes is imperative, as this will lend credence to results that will be obtained from the ICP study. The activity of NOS can be estimated by the use of Western Blot as described by Bush et al. [47]. Similarly, analysis of androgen receptor (AR) protein will further give an idea of the receptors available for the binding of the androgens notably the free or bioavailable testosterone.

15.4.2   Mating Behavior Test in Female Rats

15.4.2.1   Mating Behavior Test

For the evaluation of medicinal plants for mating behavior in females, the procedure described by Giuliano et al. [48] will be adopted. Administration of the extract shall be done an hour before pairing the female rats with the male rats. Female rats shall individually be placed in cages 5 min before their pairing with male rats and the observation shall be made for 30 min. The study shall be carried out at 20:00−24:00 hours under dim light condition at room temperature. The female sexual behavior parameters that shall be monitored on selected days of the experimental period according to the design shall include proceptive and receptive sexual behavior such as darting latency (the time taken for the female rat to initiate the short run in front of male exposing their posterior), darting frequency (the number of darting recorded for a specific period, 1 h after the administration of the extracts), hopping latency (the time taken to initiate hopping, characterized by a short jump with stiff legs followed by immobility and a presenting behavior), hopping frequency (the number of hopping recorded for a specific period, 1 h after administration of the extracts), anogenital and genital grooming (which are both orientation activities). Receptive sexual behavior shall include lordosis latency (the time taken to assume the behavior posture by the female to allow mounting by the male rats) and lordosis frequency (the number of lordosis achieved within a specified period).

All these can be complemented by determining the concentrations of female reproductive hormones such as testosterone, LH, FSH, prolactin, and estradiol in the serum of the animals. This can also be complemented with histopathological examination of the ovary.

15.4.3   Fertility Testing

15.4.3.1   Biochemical Parameters Used to Evaluate Fertility in Males

Rats will be grouped according to the experimental design and their body weights determined. The weight and size of the testis, epididymal caudal sperm functions (sperm concentration, sperm motility in percentage, sperm grade activity, forward progressive movement, abnormal sperm morphology, and sperm viability), in addition to biochemical tests of the serum of the animals which should include LH, FSH, testosterone, cholesterol, and total protein, as well as the histological examination of the testes and caudal epididymis shall be determined. Furthermore, the fertility index of the male rats shall be computed as described by Oberlander et al. [49]. In each stage, the male rat shall be caged separately with two coeval females of proven fertility in the evening for 6 days. Presence of sperms in the vaginal smears examined on the next day morning may suggest that the females had mated to the particular male and the day of mating shall be taken to be day 1 of pregnancy. Fertility test would be considered positive by the presence of implantation sites. After the evaluation of fertility by mating test, the animals shall be sacrificed by decapitation and blood collected by cardiac puncture and serum prepared. The concentration of testosterone shall be determined using standard procedures.

15.4.3.2 Androgenic Study

Matured male rats shall be assigned into groups and given their doses accordingly for a specified period of time. The animals shall be sacrificed 24 h after their last dose(s) and the following parameters of functional indices determined in the testicular homogenates/supernatants: testes—body weight ratio, testicular concentrations of total protein, glycogen, sialic acid, and total cholesterol, as well as the activities of gamma glutamyl transferase, acid and alkaline phosphatases. The concentrations of testosterone, LH, and FSH shall be assayed according to the procedures outlined in the respective manufacturer's protocols. All these may be correlated with histopathological examination of the testes.

15.4.3.3 Fertility in Females

15.4.3.3.1 Implantation Study
Adult females used in this test shall be paired overnight with vigorous sexually experienced males. Successful mating shall be confirmed by the presence of sperm in the vaginal smear the following morning (07:00—08:00) and this day shall be considered day 1 of pregnancy. Only sperm positive females will be used in the study. The rats will then be assigned into various groups according to the design of the experiment and variously administered their extracts, vehicle and any other chemical compounds for 7 days. On day 10 of pregnancy, each female shall be laparatomized under diazepam/ketamine (10/50 mg/kg, respectively) anesthesia and the number of implantation sites counted.

15.4.3.3.2 Fertility Study
The procedure described by Watcho et al. [50] shall be used to determine the long-term effect of the extract on the fertility of sexually naive female rats; 30 primipare adult animals will be used, assigned into various groups, and administered the extract and vehicle consecutively for 21 days. During the last 5 days of treatment (days 17—21), each female rats shall be allowed to mate with a vigorous male of proven fertility. The vaginal smear shall be examined daily under microscope and the sperm positive females shall be isolated and followed up until parturition. During the experimental period, the weight of the rats shall be determined. At the end of the study, the litter size and the ratio of male to female shall be recorded. The number of pups with deformity shall be observed up till day 7 after birth. The following reproductive indices shall be calculated using the expression:

- Mating index is the number of sperm positive females/number of mated females $\times$ 100
- Pregnancy index defined as number of pregnant females/number of sperm positive females $\times$ 100

- Fertility index = (number of pregnant females/number of females with successful copulation) × 100 [38]
- Gestation index defined as number of females with alive pups/No. of pregnant females × 100
- Delivery index defined as number of females delivering/number of pregnant females × 100
- Birth live index defined as number of live offspring/number of offspring delivered × 100
- Postnatal viability index defined as number of pups alive on day 4/no. of alive pups × 100
- Postimplantation loss index defined as number of implantation sites−number of live fetuses/number of implantation sites × 100
- Weaning viability index, defined as number of pups alive at day 21/no. of pups alive at day 4 × 100.

15.4.3.3.3 Uterotrophic Assay

Gonado-intact and ovariectomized immature females shall be used. Ovariectomy shall be performed as previously described [51]. The female rats shall be assigned into various groups and treated for 7 days according to the design of the study. The body weight of each animal shall be recorded daily and the animals sacrificed under ether anesthesia 24 h after the final dose. Uteri shall be excised, trimmed free of any fat and adhering nonuterine tissue. The body of the uterus shall be cut just above its junction with the cervix and at the end of the uterine horns with the ovaries. The relative weight of the uterus shall be calculated according to the expression:

$$\text{Relative weight of the uterus } (mg/100\ g) = (\text{uterus wet weight } (mg) / \text{body weight } (g)) \times 100$$

The uterine cholesterol, alkaline phosphatase, glucose, and protein shall be determined using standard methods [24].

15.4.3.3.4 Evaluation of Plant Extract for Its Effect on Female Reproductive Hormones

The female animals after appropriate treatment for specified number of days shall be sacrificed, serum prepared, and the concentrations of female reproductive hormones such as prolactin, estradiol, progesterone, LH and FSH determined according to standard procedures [52].

15.4.3.3.5 Evaluation of Medicinal Plant for Its Efficacy on Polycystic Ovarian Syndrome (PCOS)

15.4.3.3.5.1 Induction and Confirmation of PCOS Female animals shall be randomized into two groups (I and II). Group I (control) shall receive 0.5 ml of the vehicle only (1% aqueous solution of carboxymethylcellulose (CMC)) while animals in Group II shall receive same volume corresponding to 0.5 mg/kg body weight of anastrozole dissolved in 1% CMC (2 ml/kg) daily for 21 days [53]. Here anastrozole is the inducer of PCOS. The stage of estrus cyclicity in the animals shall be monitored on daily basis for 12 days for the presence of predominant cell type in the vaginal smears using a light microscope. Twenty-four hours after the last dose of anastrozole, animals shall be selected randomly from each group, sacrificed, and an aliquot of the blood sample obtained from the jugular veins. Serum shall be prepared and concentrations of serum reproductive hormones such as testosterone, progesterone, LH, and FSH shall be assayed using standard methods. Furthermore, vaginal smears shall be obtained on daily basis between 9:00 and 10:30 hours, prepared appropriately, and examined as described by Marcondes et al. [54].

Thereafter, the PCOS-positive female animals with evidence of irregularities in their estrus cycle shall be randomly assigned into groups, treated for 30 days, and the aforementioned parameters determined. Twenty-four hours after the exposure period (30 days of treatment), the animals in the estrus stage of their cycle shall be weighed individually and thereafter sacrificed under light ether anesthesia. The sera shall be prepared and used for the quantitative determination of serum testosterone, progesterone, FSH, and LH as earlier described.

15.5   Reproductive Toxicity

15.5.1   Evaluation of Medicinal Plants for Reproductive Toxicity

Antifertility effect of the extract could be assessed using the procedure described by Yakubu and Afolayan [55]. Healthy and sexually experienced male animals that show brisk sexual activity shall be selected for the study. Animals in each treatment group shall be housed individually in their cages and paired with a receptive female after their treatment doses. The precoital sexual behavior shall be observed for 60 min. The presence of sperm in vaginal smear should confirm successful mating. Sperm count shall be determined using improved Neubauer hemocytometer while the morphology of the sperm cells shall be observed microscopically at 400×. The gestation period and number of litters shall also be noted. The following reproductive parameters shall be computed:

- Percentage mating success = (number mated/number paired) × 100
- Quantal pregnancy = (number pregnant/number mated) × 100
- Fertility index = (number pregnant/number paired) × 100 [55].

15.5.2   Epididymal Semen Analysis

Analysis of the epididymal semen could be done on cauda epididymis as described by Amelar et al. [56]. Distal cauda epididymes shall be minced with scissors to release the epididymal content into a 35-mm Petri dish containing 2 ml phosphate buffer (0.1 M, pH 7.4). The samples shall be maintained at 37°C for 30 min before determining the sperm parameters as described previously for sperm count, motility, morphology, and turbidity.

15.5.3 Female Rat Fertility and Embryonic Study

The procedure described by Costa-Silva et al. [57] shall be used for the female fertility study. Medicinal plant extract at varying doses shall be administered to the animals during the organogenic period (7—14 days of pregnancy) [57]. The animals in each treatment group housed individually in their cages shall be observed for survival, changes in behavior, and signs of vaginal bleeding. The maternal weights shall also be recorded on days 1 and 20. The animals shall then be laparotomized under ether anesthesia on day 20 of pregnancy (for rats) and their uterine horns removed. The number of implants, resorptions, corpora lutea, live and dead fetuses shall be recorded. The fetuses, placentae, and ovaries shall be observed for any abnormality and their weights noted. The following can therefore be computed from these data:

- Implantation index = (total number of implantation sites/number corpora lutea) × 100
- Resorption index = (total number of resorption sites/total number of implantation sites) × 100
- Preimplantation loss = (number of corpora lutea − number of implantations/number of corpora lutea) × 100
- Postimplantation loss = (number of implantations − number of live fetuses/number of implantations) × 100.

15.5.4 Abortifacient Study

The abortifacient activity of medicinal plant extract could be evaluated using the method described by Salhab et al. [58] as modified by Yakubu et al. [52]. Briefly, female rats shall be paired overnight with the male rats in ratio 1:1 in the aluminum cages that allowed the animals free access to food and water. The day on which a vaginal plug and spermatozoa (detected with the aid of light microscope) appeared in the vaginal smear shall be considered day 0 of pregnancy. Pregnant animals shall be completely randomized into groups depending on the design of the experiment and extract and/or vehicle administered on days 10—18 of pregnancy (organogenetic period). Animals shall be lapratomized ventrally under ether anesthesia on day 19 of pregnancy (for rats, 24 h after last dose) and their fetuses, if any, shall be examined. The following parameters of abortion/implantation shall be recorded and/or computed:

- Number of live and dead fetuses
- Survival ratio = (number of live fetus/(number of live + dead fetus)) × 100
- Number of rats that aborted
- Percentage aborted = (number of rats that aborted/number of rats used) × 100
- Number of implantations
- Number of corpora lutea
- Implantation index = (total number of implantation sites/number of corpora lutea) × 100
- Resorption index = (total number of resorption sites/total number of implantation sites) × 100
- Preimplantation loss = (number of corpora lutea − number of implantations/number of corpora lutea) × 100

• Postimplantation loss = (number of implantations − number of life fetuses/number of implantations) × 100.

The initial and final weights of the rats as well as the feed and water intake shall also be computed.

15.5.5 Determination of Uterogenicity and Estrogenicity

Uterotrophic bioassay of bilaterally ovariectomized female rats shall be used to determine estrogenicity as described by Jun et al. [51]. After 8 days of ovariectomy, all the animals in various groups shall receive their daily doses for 7 days. By the 15th day, the body weights of the animals shall be determined. Thereafter, the animals shall be sacrificed, the uteri shall be removed carefully with their luminal fluid, and their weights noted. The uterine−body weight ratio shall be computed from the expression: (weight of uterus/weight of the animals) × 100, while other uterine parameters such as protein, glucose, cholesterol, and alkaline phosphatase activity shall be determined in the uterine homogenates/supernatants using standard methods. The percentage vaginal opening = (number of rats with open vagina/number of treated rats) × 100 also shall be computed. The levels of the hormones, such as progesterone, estradiol, FSH, and LH, also shall be determined using standard.

15.5.6 Contraception

Semen samples can be obtained from proven fertile animals by electrostimulation after a 3-day period of sexual abstinence. The semen samples should normally show sperm count ranging from 80 to 110×10^6/ml, motility of 60−70%, morphology of 55−60%, and viability of 70−80%. After liquefaction for 30 min at 37°C, basic semen characteristics (sperm count, motility, morphology, and viability) shall be evaluated. The different concentrations of extract shall first be tested for their spermicidal activity *in vitro* using the Sander−Cramer immobilization test [59]. The concentration of the extract that showed total immobilization of the spermatozoa in the least time shall be incorporated into a suitable gelling agent to form a vaginal gel. The vaginal gel shall then be tested for *in vitro* spermicidal activity. Dose- and time-related effects of the extract on sperm motility of the sperm cells shall be determined *in vitro* [59]. The end point shall be the total immobilization of spermatozoa after 20−30 s of incubation and shall be expressed as minimum effective concentration.

15.6 Medicinal Plants of Africa with Effects on the Reproductive System

The reproductive capacity of both male and female has been shown to be on a decline. Several studies have reported attempts to solve this problem using medicinal plants in both males and females [33,37,50,52,58,60]. Some of these medicinal plants have been summarized in Tables 15.1−15.3.

Table 15.1 African Medicinal Plants with Abortifacient, Conceptive, and Contraceptive Effects

Plant (Family)	Area of Plant Collection/ Parts Used	Traditional Uses	Chemical Constituents	Pharmacological Activities
Abortifacient				
Abrus precatorius Linn. (Fabaceae)	Native to Tropical Africa (Sudan, Kenya, Tanzania, Uganda, Rwanda, Niger, Togo, Namibia, and Madagascar/Seeds	Cures fever, stomatitis, asthma and bronchitis [61,62], and oral contraceptives [63]	Abridine [64]	The methanolic extract of *A. precatorius* caused a reversible disruption in the estrus cycle of the regularly cyclic rats and completely blocked ovulation in all the treated rats. The uterine dissection on postcoital day 12 revealed neither implantation nor resorption sites in all the animals treated with *A. precatorius*. The extract also decreased the mean body weight, mean crown-rump length, and mean tail length of fetuses of the treated rats. Abridine is claimed to be responsible for the postcoital antifertility (100% sterility) and anti-implantation in females [64]
Acalypha indica L. (Euphorbiaceae)	Tropical Africa (Nigeria, Sudan)/whole plant	Used as diuretic, purgative and anthelmintic agents; cures bronchitis, asthma, pneumonia, rheumatism scabies, emmenagogue, and contraceptive [65]	Acetates of acalyphamide aurantiamide, succinimide calypho-lactate, 2-methyl anthraquinone, tri-*o*-methylellagic acid, β-sitosterol and its β-D-glucoside (leaves); a cyanogenetic glucoside, acalyphine, two alkaloids, viz., acalyphine and triacetonamine, an essential oil *n*-octacosanol, kaempferol,	Exhibits postcoital anti-implantation, antizygotic, and blastocytotoxic due to estrogenic sterols and flavonoids [67]

(Continued)

Table 15.1 (Continued)

Plant (Family)	Area of Plant Collection/ Parts Used	Traditional Uses	Chemical Constituents	Pharmacological Activities
			quebrachitol, β-sitosterol acetate, stigmasterol [66]	
Ananas comosus Linn. Merr. (Bromeliaceae)	South and West Africa/unripe fruit juice and leaves	Abortifacient, dysuria, antidyspepsial, antidiarrheal, and antidiabetic agent [68]	β-sitosterol, ergosterol peroxide, 5α-stigmastane-3β,5,6β-triol (III) 3-monobenzoate [69]	All the compounds such as ergosterol peroxide, β-sitosterol, 5-stigmastene-3 β,7α-diol, 5-stigmastene-3 β -7 β-diol, 7-Oxo-5-stigmasten-3 β-ol, 5α-stigmastane-3 β,5, 6 β-triol from the petroleum ether and benzene extracts of the whole leaves of *A. comosus* exerted some degree of abortifacient activity when administered on day 1, but β-sitosterol and the 7-oxo derivative were devoid of activity when given on days 6–7 and ergosterol peroxide showed the maximum abortifacient effect at both stages of pregnancy, but the action was delayed (starting from days 13–16), especially when given after implantation [69]
Gloriosa superba (Liliaceae)	Africa (Botswana, Zimbabwe, Namibia, Zambia, Rwanda, South Africa)	Impotence, abortifacient, antipyretic, rheumatic fever inflammations, leprosy, piles, ulcers, intestinal worm infestations, thirst, bruises,	Colchicines and colchicoside, 2-demethylcolchicine, 3-demethylcolchicine, and N-formyl-N-deacetylcolchicine, superbine, colchicine, gloriosine,	Antispermatogenic; the 50, 100, and 200 mg/kg body weight of aqueous extract produced significant abortifacient, anti-implantation, and uterotonic activities in rats [74]

Plant	Origin/parts	Uses	Phytochemicals	Pharmacological activity
		skin problems, and snakebite [70]	gloriosol, phytosterols, and stigmasterol [71–73]	
Heliotropium indicum (Family: Boraginaceae)	West and Central Tropical Africa (Burundi, Cameroun, Central Africa Republic, Nigeria)	Skin diseases, diarrhea, malaise ulcer, fever, ophthalmic disorders, inflammation, tumor astringent, expectorant, and febrifuge [75,76]	Pyrrolizidine alkaloids, tannins, and saponins [77]	Ethanolic extract at oral doses of 200 and 400 mg/kg body weight exhibited abortifacient activity with moderate anti-implantation activity [67]
Calotropis procera (Asclepiadaceae)	Native to West Africa as far south as Angola, North and East Africa, Madagascar	Diarrhea, dysentery, eczema, leprosy, elephantiasis, convulsion, rheumatism [78]	Alkaloids, flavonoids, tannins, saponins, and cardiac glycosides [79]	The 250 mg/kg body weight of (1/4 of the LD_{50}) ethanolic extract showed strong anti-implantation activity. No antiestrogenic activity was detected [79]
Hibiscus rosa-sinensis Linn. (Malvaceae) Garden Hibiscus (English)	West Africa (Nigeria, Cameroon); North Africa (Egypt)/leaves, flower buds, stem	Abortifacient, aphrodisiac, relieves appendicitis pain, influenza, cough, asthma, tertiary syphilis [80,81]	Cyaniding, quercetin, entriacontane, calcium oxalate, thiamine, riboflavin, niacin, and ascorbic acid [82]	Anti-implantation and antispermatogenic activities [83]; the extract at a dose level of 1 g/kg body weight/day terminated pregnancy in the animals and was attributed to reduction in progesterone level and increase in uterine acid phosphatase activity
Senna alata Linn. Roxb (Leguminosae), Candle Bush (English)	Native to Mexico, but found in Africa (Nigeria)/leaves	Hepatitis, skin diseases, jaundice, gastroenteritis, ringworm, eczema, burns, wound, skin infection, diarrhea, upper respiratory tract infection, constipation, and food poisoning [85]	Alkaloids, phenolics, cardiac glycosides, saponins, flavonoids, cardenolides, and dienolides [86]	The extract at 250, 500, and 1000 mg/kg body weight significantly reduced ($p < 0.05$) the number of life foetus, weight, and survival ratio of the foetus, numbers of implantations and corpora lutea, implantation index, progesterone, prolactin, estradiol, FSH and LH whereas the number of dead fetus, number and percentage of rats that aborted, percentage vaginal opening, resorption index, pre- and postimplantation losses increased significantly. In addition, mifepristone-treated animals produced resorption index that compared well with the distilled water control. There was also no dead or live

(Continued)

Table 15.1 (Continued)

Plant (Family)	Area of Plant Collection/ Parts Used	Traditional Uses	Chemical Constituents	Pharmacological Activities
				fetus and serum progesterone concentration was increased in the animals treated with mifepristone. All cases of abortion were accompanied with vaginal bleeding. Although the final weight of the rats increased significantly, the feed and water intake were not significantly altered in all the treatment groups. The weight of the uterus, uterine–body weight ratio, length of the right uterus horn, and uterine cholesterol decreased significantly in all the treatment groups. The uterine alkaline phosphatase activity and glucose concentration increased in only the extract-treated animals whereas mifepristone decreased the uterine alkaline phosphatase activity and glucose content of the animals. Abortifacient activity via mechanisms such as changes in hormonal levels, implantation site, estrogenicity, and uterogenicity [86]
Ficus asperifolia Miq. (Moraceae)	Africa (South Africa, Senegal, Uganda, Tanzania, Madagascar, and Cameroon)	Sterility/infertility, anthelmintic and purgative	Alkaloids, saponins, sterols, and triterpenes in the fruits [87]	*F. asperifolia* at the doses of 100 and 500 mg/kg body weight did not disrupt (0%) the order of appearance of normal estrus cycle stages, namely, proestrus, estrus, metestrus, and diestrus. Short-term treatment (1 week duration) exhibited high frequency of appearance of proestrus and estrus stages while mid- (3 weeks) and long-term (6 weeks) treatments revealed constancy in the frequency of all stages irrespective to animal groups. The growth of the uterus remained unchanged

Bersama engleriana Geurke (Melianthaceae)	Africa (Senegal, Zaire, South Africa, Cameroon)	Diabetes, aphrodisiac, antitumor, antimicrobial [89]	Sterol, triterpenes, and saponins [89]	The study concludes that the plant has proimplantation, prodevelopment, and uterotrophic-like activities [88] Oral administration of aqueous and methanolic extracts of the leaves for 21 days to adult male rats increased the frequencies of erection and intromission suggesting aphrodisiac activity [90], uterotonic activity in females
Aframomum melegueta K. Schum (Zingiberaceae) Grain of Paradise (English)	Mali, Cameroon, West Africa (Togo, Nigeria)	Female sterility, gastrointestinal and respiratory infections in Mali peptic ulcer, intestinal diseases, hypertension, aphrodisiac [91]	Sterols, alkaloids, flavones, and polyterpenes such as paradol, shogoal, zingerone, and gingerol [91]	The aqueous extract of the dry seeds at 115 and 230 mg/kg body weight significantly increased the content of testosterone in serum and testis, cholesterol in testis, α-glucosidase in epididymis and fructose in seminal vesicle after 8 days of treatment of *A. melegueta*-treated rats (115 and 230 mg/kg). Results also showed that levels of cholesterol in testis, α-glucosidase in epididymis, and fructose in seminal vesicle increased by 93.34%, 83.44%, and 62.78%, respectively, after 55 days of *A. melegueta* treatment. The study concluded that the extract stimulates spermatogenesis and androgenic activity [92]
Bulbine natalensis Baker (Asphodelaceae)	South Africa/leaves and roots	Wounds, burns, rashes, itches, ringworms, cracked lips, quell vomiting, diarrhea, convulsion, venereal diseases, diabetes, and rheumatism [93]	Tannins, anthraquinones, cardiac glycosides, saponins alkaloids [94]	The aqueous extract of the plant at 25, 50, and 100 mg/kg body weight increased ($p < 0.05$) the testicular–body weight ratio as well as alkaline phosphatase activity, glycogen, sialic acid, protein, and cholesterol content of the testes except the single administration of 100 mg/kg body weight which compared well ($p > 0.05$)

(*Continued*)

Table 15.1 (Continued)

Plant (Family)	Area of Plant Collection/ Parts Used	Traditional Uses	Chemical Constituents	Pharmacological Activities
				with the controls for glycogen and cholesterol. The testicular and serum testosterone concentrations were increased except in the 100 mg/kg body weight where the effect on the tissue and serum hormone did not manifest until after the first and seven daily doses, respectively. Testicular acid phosphatase activity, serum FSH and LH concentrations also increased at all the doses except in the 100 mg/kg body weight where the effect on the enzyme and the hormone did not manifest until after 7 days. The increases were most pronounced in the 50 mg/kg body weight extract-treated animals. The study hinged all these on the anabolizing and androgenic activities of the plant [94]
Basella alba	Cameroon	Virility and fertility [95]	Proteins, fats, vitamins A, C, E, K, B9 (folic acid), riboflavin, niacin, thiamine, flavonoids, and minerals (calcium, magnesium, and iron)	The results showed that methanolic extract did not affect Leydig cell viability. At the concentration of 10 μg/ml, the extract significantly stimulated testosterone and estradiol production ($p < 0.01$ and $p < 0.03$, respectively), and enhanced aromatase mRNA level ($p < 0.04$). Fecundity was also improved by 25% in

Plant	Origin/parts used	Uses	Constituents	Comments
Dalbergia saxatilis Linn. (Fabaceae)	Nigeria/leaves, bark, and roots	Bronchial ailments, toothache, accelerate birth	Triterpenoid glycosides [97]	the extract-treated 2.5-month-old rats. The study concludes that the extract improved fecundity and stimulates testosterone production in rats [95,96] The triterpenoid glycoside, DSS, isolated from the root of *D. saxatilis* at the dose of 200 mg/kg body weight inhibited conception by 71.4% in female Wistar rats. It did not significantly alter the fertility of rats at the first and second trimesters of pregnancy but significantly decreased ($p < 0.05$) the mean day 20 fetal crown-rump length and mean maternal body weights [97]
Spondias mombin (Anacardiaceae) Hog Plum (English)	Nigeria	Induce delivery, postpartum infections of the uterus [78]	Alkaloids, resin, tannins, saponins [78]	The effects of intraperitoneal administration of 70% aqueous ethanol leaf extract of *S. mombin* revealed that only two out of the five animals (40%) administered the leaf extract for 4 consecutive days after mating were found to be pregnant when autopsied at the 20th day of pregnancy. All the rats (100%) were found with fetuses attached to the endometrium. The body weight of the animals increased generally as the experiment progressed but was lower in animals treated with the extract at the beginning of gestation. All the animals that received the extract at the beginning of the second trimester had live fetuses in

(Continued)

Table 15.1 (Continued)

Plant (Family)	Area of Plant Collection/Parts Used	Traditional Uses	Chemical Constituents	Pharmacological Activities
				their uterine horns. The study concluded that the extract has significant anticonceptive activity attributed to a direct action of the extract on the uterus [98]
Asparagus africanus (Lam) (The African Asparagus, English)	Egypt, Eritrea, Ethiopia, Somalia, Sudan, Kenya, Tanzania, Uganda, Burkina Faso, Gambia, Ghana, Nigeria, Senegal, Malawi, Mozambique, Zimbabwe, and South Africa	Venereal diseases, facilitate childbirth, diuretic, laxative, sedative, aphrodisiac [98]	Steroidal saponins [99]	The aqueous extracts of the leaves and the roots showed an anti-implantation activity of 70% and 77%, respectively, while the ethanol extracts of the leaves and roots showed 48% and 61%, respectively. The antifertility activities of the aqueous and ethanol extracts were 40% (for leaves), 60% (for roots), and 20% (for leaves), 40% (for roots), respectively. All the extracts have resulted in significant ($p < 0.05$) reduction in the number of implants as compared with their respective controls. Each extract potentiated acetylcholine-induced uterine contractions in a concentration dependent manner significantly ($p < 0.05$). The study concluded that the leaves and roots of this

Plant name (family)	Region/part	Chemical constituents	Ethnomedicinal uses	Reproductive/pharmacological activity
Rumex steudelii Hochst Ex. A. Rich (Polygonaceae)	Africa (Ethiopia)/root	Polyphenols, phytosterols, *O*-anthraquinones glycoside, tannins, hydrolyzable tannins, and saponins [101]	Rectal proplase, hemorrhoid, wounds, eczema, swelling, leprosy, tonsillitis, abdominal colic and *Tinea nigra*, hemostatic and oxytocic agents [100]	plant may possess hormonal properties that can modulate the reproductive function (uterine contractile and antifertility) of the experimental rats [27]. Anti-implantation, antifertility by causing atrophic changes in the uterus and disrupting ovarian folliculogenesis; inhibiting further development of the recruited ovarian follicles [101]
Achyranthes aspera L. (Amaranthaceae)	Africa (Kenya, Tanzania, Uganda, Ethiopia)/leaves, roots/whole plant	Alkaloids, potassium salt	Fertility control, in placental retention, and in postpartum bleeding hasten delayed labor, causes abortion [78]	The extract showed significant ($p < 0.05$) abortifacient activity and increased pituitary and uterine wet weights in ovarectimized rats. The extract, however, did not significantly influence serum concentration of the ovarian hormones and various lipids except lowering high-density lipoprotein at doses tested. Furthermore, the chloroform and ethanol extracts exhibited 100% anti-implantation activity when given orally at 200 mg/kg body weight. Both the extracts at the dose of 200 mg/kg body weight also exhibited estrogenic activity. The study concludes on antifertility, abortifacient [102], estrogenic, and anti-implantation activities [103]

Table 15.2 African Medicinal Plants Affecting Spermatogenesis and Ovulation

Plant (Family)	Area of Plant Collection	Traditional Use	Chemical Constituents	Pharmacological Activities
Spermatogenesis				
Dalbergia sissoo Roxb. (Fabaceae)	Nigeria	Astringent, abortifacient, expectorant, anthelmintic, skin diseases (leprosy, eczema, scabies, leukoderma) and stomach troubles, gonorrhea, leucorrhea, and wound healing	Steroids, saponins, flavonoids, triterpenoids, glycosides	Antispermatogenic activity; *in vitro* study of the effect of ethanol extract of stem bark of *D. sissoo* Roxb. on 15 healthy fertile men aged 25–35 years produced dose- and time-dependent effects on sperm motility and sperm viability. Ethanol extract at a concentration of 20 mg/ml caused complete immobilization within 3 min. Sperm viability and hypo-osmotic swelling was significantly reduced at this concentration. The *in vivo* studies of the extract at 200 mg/kg body weight significantly decreased ($p < 0.001$) the weight of the testis and epididymis as well as sperm motility and sperm count in the epididymis of mice [104]
Gloriosa superba (Liliaceae)	Zambia	Impotence and abortifacient antipyretic, rheumatic fever, inflammations, leprosy, piles, ulcers, intestinal worm infestations, thirst, bruises, skin problems, and snakebite [78]	Colchicines and colchicoside, 2-demethylcolchicine, 3-demethylcolchicine, and *N*-formyl-*N*-deacetylcolchicine. superbine, colchicine, gloriosine, gloriosol, phytosterols, and stigmasterol [71–73]	Leads to shrinkage of the seminiferous tubules and significantly reduced sperm count and motility. All these supports its antispermatogenic activity [105]

Plant	Country	Uses	Constituents	Activity
Mucuna urens	Nigeria	Anthelmintic, rubrefacient [78]	Cardiac glycosides, saponins, flavonoids, anthranoids, polyphenols, alkaloids, anthraquinones, and hydroxyl methyl anthraquinones [106]	The ethanolic extract of the plant at 70, 140, and 210 mg/kg body weight decreased sperm count and motility. Sperm abnormalities included unusual head with large acrosome, looped tailpiece, midpiece with distal droplet, pin head, pyriform head, and long hook [106]. All these suggest antispermatogenesis activity
Fagara tessmannii (Rutaceae)	Cameroon	Treatment of infertility, uterine leiomioma, and sexual weakness	Triterpenes [107]	*F. tessmannii*, 14 days after treatment, may improve spermatogenesis, testosterone level, and sperm transit in cauda epididymidis but negatively impair reproductive organ activities [107]
Morinda lucida Benth (Rubiaceae)	West Africa (Nigeria)	Purgative, emetic, diuretic, diabetics [78]	Alkaloids, anthraquinones, anthraquinols, oruwalol, and oruwal	Possessed antispermatogenic activity by reducing sperm motility and viability, and the epididymal sperm counts of rats as well as increased sperm morphological abnormalities [108]
Ovulation				
Garcinia kola Heckel (Guttiferae)	Nigeria	Liver disorders, hepatitis, diarrhea, laryngitis, and bronchitis [78]	Garcinia bioflavonoids GB1 and GB2, garcinal and garcinoic acid [26]	Alters estrus cycle in rats, partly inhibits ovulation and may produce duration-dependent teratogenicity in fetal rats [26]

(*Continued*)

Table 15.2 (Continued)

Plant (Family)	Area of Plant Collection	Traditional Use	Chemical Constituents	Pharmacological Activities
Aspilia africana Pers (Asteraceae)	Nigeria, South Africa	Remedy for lumbago and sciatica neuralgia [63,109]	Saponins, tannins, and flavonoids, Germacrene D, alpha-pinene, and limonene [110]	Methanolic extract reduced wet weights of the ovaries and derangement of granulosa cells, degeneration and reduction of follicles and poor vascularity of ovarian stroma [111]
Momordica charantia (Cucurbitaceae)	Africa (Nigeria)/ seeds	Menstrual irregularities, antifertility, antiovulatory and abortifacient activities, anticancer, antidiabetic [78]	Steroids, triterpinoids, reducing sugars, sugars, alkaloids, phenolic compounds, flavonoids, and tannins [112]	Irregular pattern of estrus cyclicity and significantly increased the length of the estrus cycle in rats [112]
Curcuma longa (Zingiberaceae) (Turmeric/Haldi)	South Africa, Kenya, Nigeria	Antibacterial, antiulcer, anticancer, antidiabetes	Flavonoids, amino acids, and alkaloids	Suppressed estrus phase and of ovulation due to the antiestrogenic property of phytochemicals which either block the estrogen receptors or diminished estrogen synthesis due to diminished cholesterol metabolism or both [113]
Jatropha gossypifolia (Euphorbiaceae)	Tropical Africa (Nigeria) and South Africa	Used as thermogenic, anodyne, astringent, vermifuge, odonalgia and in treatment of strangury, paralysis, arthralgia, and melalgia [78]	Saponin, tannin, flavonoid, reducing sugars, cardiac glycosides, terpenoids, triterpenoids, steroids, xanthoprotein, and starch [114]	Prolonged the estrus cycle with increase in duration of diestrus stage altered the release of LH, FSH and prolactin, and estradiol secretion [114]

Table 15.3 African Medicinal Plants with Effects on Sexual Behavior of Animals

Plant (Family)	Area of Plant Collection	Traditional Use	Chemical Constituents	Pharmacological Activities
Tulbaghia violacea Harv. (Alliaceae) (Society Garlic-English)	South Africa	Respiratory diseases (tuberculosis and asthma), esophageal cancer, gastrointestinal problems, as well as colds and fever, aphrodisiac medicine [115]	Dithiapentane, p-xylene, chloromethylmethyl sulfide, O-xylene, thiodiglycol, and p-xylol	The extract significantly increased ($p < 0.05$) LH-induced testosterone production suggesting androgenic effects [116]
Mondia whitei (Periplocacea)	Malawi, Cameroon/ roots	Used as spices, aphrodisiac, and for the treatment of urinary tract infection, jaundice, and headache [117]	Glucosides, alkaloids, and 2-hydroxy-4-methoxybenzaldehyde [118]	The aqueous and hexane extracts at 100 and 500 mg/kg body weight significantly ($p < 0.001$) reduced the ML and the hexane extract was found to be more efficient than the aqueous extract. The treatment had no significant effect ($p > 0.05$) on intromission, ejaculation, and erection. The study concluded that *M. whitei* had sexual enhancement of the sexually inexperienced male rats [117]

(Continued)

Table 15.3 (Continued)

Plant (Family)	Area of Plant Collection	Traditional Use	Chemical Constituents	Pharmacological Activities
Fadogia agrestis Schweinf (Rubiaceae) Black aphrodisiac (English)	Nigeria	Aphrodisiac, blood tonic	Saponins, alkaloids, anthraquinones, and flavonoids [39]	The aqueous extract of *F. agrestis* stem at 18, 50, and 100 mg/kg body weight significantly increased the MF and IF; significantly prolonged the EL ($p < 0.05$) and reduced ML and IL ($p < 0.05$). There was also a significant increase in serum testosterone concentrations in all the groups in a manner suggestive of dose dependence ($p < 0.05$). Increase in the blood testosterone concentrations and this may be the mechanism responsible for its aphrodisiac effects and various masculine behaviors [39]
Asparagus racemosus Willd. (Liliaceae)	Tropical Africa (Republic of Congo), Nigeria, and South Africa	Used as a uterine tonic, galatogogue, hyperacidity, and as a general health tonic	Asparagamine A, steroidal saponins, shatavaroside A and shatavaroside B, and filiasparoside C [119]	Hydroalcoholic extract at higher concentration (400 mg/kg body weight) showed significant aphrodisiac activity on male Wistar albino rats as evidenced by an increase in number of mounts and mating performance [119]. On the other hand, hydroalcoholic extract at lower dose (200 mg/kg body weight) and aqueous extract (400 mg/kg body weight) showed moderate aphrodisiac activity.

Massularia acuminata (G. Don) Bullock ex Hoyl.	Sierria Leone, Nigeria, and Democratic Republic of Congo	The juice from the fruit is used as antibiotics for the treatment of eye infections, stems as chewing stick for oral hygiene, aphrodisiac, and anticarcinogen [121]	*M. acuminata* stem contains alkaloids (0.22%), saponins (1.18%), anthraquinones (0.048%), flavonoids (0.032%), tannins (0.75%), and phenolics (0.066%) [120]	The extract at 500 and 1000 mg/kg body weight significantly ($p < 0.05$) increased the frequencies of mount and intromission. In addition, the EL was significantly prolonged ($p < 0.05$). The latencies of mount and intromission were reduced significantly whereas EF increased. The extract also significantly ($p < 0.05$) increased the serum testosterone content of the animals. The improved sexual appetitive behavior in male rats at the doses of 500 and 1000 mg/kg body weight of *M. acuminata* stem may be attributed, at least in part, to the alkaloids, saponins, and/or flavonoids as these phytochemicals have engorgement, androgen enhancing, and antioxidant properties [120]
Myristica fragrans Houtt. (nutmeg-English)	Whole of Africa	Used as spice, aphrodisiac, stomachic, carminative, tonic nervous stimulant, aromatic, narcotic, astringent, hypolipidemic, antithrombotic, antifungal, antidysenteric, and anti-inflammatory properties [121]	Contains volatile oil, a fixed oil, proteins, fats, starch, mucilage, myristin and myristic acids, pinene, sabinene, camphene, myristicin, elemicin, isoelemicin, eugenol, isoeugenol, methoxyeugenol, safrole, dimeric phenylpropanoids, lignanes, and neolignanes [122,123]	It significantly increased the MF, IF, IL, and caused significant reduction in the ML and PEI. It also significantly increased MF with penile anesthetization as well as Erection, Quick Flip, Long Flip, and the aggregate of penile reflexes with penile stimulation. It possesses aphrodisiac activity, increasing both libido and potency, which might be attributed to its nervous stimulating property [124]

(Continued)

Table 15.3 (Continued)

Plant (Family)	Area of Plant Collection	Traditional Use	Chemical Constituents	Pharmacological Activities
Lecaniodiscus cupanioides Planch. Ex. Bth	Nigeria, Senegal, South Africa	Wound, sores, abdominal swelling, fever, measles, aphrodisiac	Alkaloids, anthraquinones, phenolics, saponins, and tannins	MF, IF, EF, and testosterone, FSH, and LH concentrations were reduced significantly ($p < 0.05$) in paroxetine-treated rats. Administration of 25, 50, and 100 mg/kg body weight of the aqueous root extract of *L. cupanioides* significantly ($p < 0.05$) reversed the paroxetine-mediated alterations in MF, IF, EF, ML, IL, EL, PEI, and testosterone, FSH, and LH concentrations dose dependently [125]

15.6.1 Medicinal Plants of Africa with Abortifacient, Conceptive, and Contraceptive Activities

Medicinal plants can exhibit diverse effects on the reproductive system. Some of these include abortifacient (by altering hormonal levels, creating an environment that is not conducive for the growth and development of the fetus, altering uterine milieu), conceptive (by enhancing spermatogenesis, stimulating ovulation, and creating conducive environment in the females for the viability, motility, and durability of the sperm cells when deposited in the vagina), and contraception (by destroying the sperm cells and incapacitating it from fertilizing and disrupting ovulation). All these are summarized in Table 15.1.

15.6.2 Medicinal Plants of Africa Affecting Spermatogenesis and Ovulation

The effects of medicinal plants on the spermatogenesis are evidenced from reduction in the viability, count, motility, and morphologies of sperm cells whereas the adverse effects on ovulation may manifest as prolongation in the menstrual cycle as a consequence of disturbance in the normal hormonal profile of the females, as presented in Table 15.2.

15.6.3 Medicinal Plants of Africa with Affecting Sexual Behavior and Reproductive Hormones

Medicinal plants of Africa origin that enhance sexual behavior in animals stimulate the synthesis and secretion of testosterone in male animals whereas in females, it elevates the concentrations of estrogen and reduces the prolactin, as summarized in Table 15.3.

15.7 Conclusion

The reproductive system of animals is vulnerable to the effects of medicinal plants of African origin. Although some of these plants have been scientifically proven to be beneficial to the reproductive system by enhancing the normal functioning of the organs, some when consumed at specific doses and exposure period can be deleterious and hamper the reproductive process, leading to dysfunction and finally infertility. Therefore, extreme care should be exercised when consuming these herbs. Of particular interest is the large number of medicinal plants that have been validated for beneficial effects on the reproductive system. This level of interest should be sustained and efforts should be made at isolating the active ingredient with a view to formulating them into useful drugs.

References

[1] Shayakhmetova GM, Bondarenko LB. Cytochrome P4502E1 and other xenobiotics metabolizing isoforms in pathogenesis of male reproductive disorders. J Physiobiochem Metab 2013;2:1−7.

[2] Homan GF, Davies M, Norman R. The impact of lifestyle factors on reproductive performance in the general population and those undergoing infertility treatment: a review. Hum Reprod Update 2007;13:209−23.

[3] Younglai EV, Wu YJ, Foster WG. Reproductive toxicology of environmental toxicants: emerging issues and concerns. Curr Pharm Des 2007;13:3005−19.

[4] Ten J, Mendiola J, Torres-Cantero AM, Moreno-Crau JM, Moreno-Crau S, Roca M, et al. Occupational and lifestyle exposures on male infertility: a mini review. Open Reprod Sci J 2008;1:16−21.

[5] Yakubu MT, Oladiji AT, Akanji MA. Evaluation of biochemical indices of male rat reproductive function and testicular histology in Wistar rats following chronic administration of aqueous extract of *Fadogia agrestis* (Schweinf. Ex Heirn) stem. Afr J Biochem 2007;1:156−63.

[6] Liebler DC. Protein damage by reactive electrophiles: targets and consequences. Chem Res Toxicol 2008;21:117−28.

[7] Henkler F, Stolpmann K, Luch A. Exposure to polycyclic aromatic hydrocarbons: bulky DNA adducts and cellular responses. EXS 2012;101:107−31.

[8] De Coster S, van Larebeke N. Endocrine-disrupting chemicals: associated disorders and mechanisms of action. J Environ Public Health 2012;713696.

[9] Schuppe HC, Wieneke P, Donat S, Fritsche E, Köhn FM, Abel J. Xenobiotic metabolism, genetic polymorphisms and male infertility. Andrologia 2000;32:255−62.

[10] Wienkers LC, Heath TG. Predicting *in vivo* drug interactions from *in vitro* drug discovery data. Nat Rev Drug Discov 2005;4:825−33.

[11] Bonde JP. Occupational risk to male reproduction. G Ital Med Lav Ergon 2002;24:112−7.

[12] Seeley R, Stephens TD, Tate P. Reproductive system. Anatomy and physiology. 8th ed. New York, NY: McGraw Hill; 2008International Edition; p. 1031−75.

[13] Lipshultz LI, Howards SS, editors. Infertility in the male. New York, NY: Churchill Livingstone; 1983. p. 217−48.

[14] Fawcett DW. The cell biology of gametogenesis in the male. Perspect Biol Med 1979;22:S56−73.

[15] Orisakwe OE, Husaini DC, Afonne OJ. Testicular effects of sub-chronic administration of *Hibiscus sabdariffa* calyx aqueous extract in rats. Reprod Toxicol 2004;18:295−8.

[16] Yakubu MT. Effect of a 60-day oral gavage of a crude alkaloid extracts from *Chromolaena odoratum* leaves on hormonal and spermatogenic indices of male rats. J Androl 2012;33(6):1199−207.

[17] Watcho P, Kamtchouing P, Sokeng S, Moundipa PF, Tantchou J, Essame JL, et al. Reversible antispermatogenic and antifertility activities of *Mondia whitei* L. in male albino rat. Phytother Res 2001;15:26−9.

[18] Ejebe DE, Siminialayi IM, Nwadito C, Emudainowho JOT, Akonghrere R, Onyesom I, et al. Effects of ethanol extract of leaves of *Helianthus annuus* (Common sunflower) on the reproductive system of male Wistar rats: testicular histology, epididymal sperm properties and blood levels of reproductive hormones. Biomed Pharmacol J 2008; 1(1):65−78.

[19] Awasthy KS. Genotoxicity of a crude leaf extract of Neem in male germ cells of mice. Cytobios 2001;106(Suppl. 2):151−64.

[20] Raji Y, Udoh US, Mewoyeka OO, Ononye FC, Bolarinwa AF. Implication of reproductive endocrine malfunction in male anti-fertility efficacy of *Azadirachta indica* extract in rats. Afr J Med Sci 2003;32:152−65.

[21] Roop JK, Dhaliwal PK, Garaya SS. Extracts of *Azadirachta indica* and *Melia azedarach* seeds inhibit folliculogenesis in albino rats. Braz J Med Biol Res 2005; 38(6):943−7.

[22] Akpantah AO, Ekong MB, Uruakpa KC, Akpaso M, Eluwa MA, Ekanem TB. Gonadal histomorphologies and serum hormonal milieu in female rats treated with *Azadirachta indica* leaf extract. Iranian J Reprod Med 2010;8(4):185−90.

[23] Akpantah AO, Ekong MB, Obeten KE, Akpaso MI, Ekanem TB. Hormonal and histomorphologic effects of *Azadirachta indica* leaf extract on the Pars Anterior of Wistar rats. Int J Morphol 2011;29(2):441−5.

[24] Yakubu MT, Bukoye BB. Abortifacient potentials of the aqueous extract of *Bambusa vulgaris* leaves in pregnant Dutch rabbits. Contraception 2009;80(3):308−13.

[25] Shivalingappa H, Satyanarayan ND, Purohit MG, Sharanabasappa A, Patil SB. Effect of ethanol extract of *Rivea hypocrateriformis* on the estrous cycle of the rat. J Ethnopharmacol 2002;82:11−7.

[26] Akpantah AO, Oremosu AA, Noronhna CC, Ekanem JB, Okanlawon AO. Effect of *Garcinia kola* seed extracts on ovulation, oestrous cycle and foetal development in cyclic female Sprague Dawley rat. Nig J Physiol Sci 2005;20(1−2):58−62.

[27] Tafesse G, Mekonnen Y, Makonnen E. Antifertility effect of aqueous and ethanol extracts of the leaves and roots of *Asparagus africanus* in rats. Afr Health Sci 2006;6:81−5.

[28] Edwards S, Tadesse M, Demissew S, Hedberg I. Flora of Ethiopia and Eritrea: Magnoliaceae to Flacourtiaceae. Volume 2, Part 1. Addis Ababa, Ethiopia/Uppsala, Sweden: The National Herbarium; 2000. p. 336−47.

[29] Gebriea E, Makonnena E, Debellab A, Zerihunc L. Phytochemical screening and pharmacological evaluations for the antifertility effect of the methanolic root extract of *Rumex steudelii*. J Ethnopharmacol 2005;96:139−43.

[30] Yakubu MT, Akanji MA, Oladiji AT, Olatinwo AWO, Adesokan AA, Yakubu MO, et al. Effect of *Cnidoscolous aconitifolius* (Miller) I.M. Johnston leaf extract on reproductive hormones of female rats. Iranian J Reprod Med 2008;6:149−55.

[31] Ejebe DE, Siminialayi IM, Emudainohwo JOT, Kagbo HD, Amadi P. Effects of ethanol extract of leaves of *Helianthus annus* on the fecundity of Wistar rats. Asian Pac J Trop Med 2010;435−8.

[32] Lembe DM, Koloko BL, Bend EF, Domkam J, Oundoum PC, Njila MN, et al. Fertility enhancing effects of aqueous extract of *Rauvolfia vomitoria* on reproductive functions of male rats. J Exp Integr Med 2014;4(1):43−9.

[33] Yakubu MT, Nurudeen QO. Effects of aqueous extract of *Cnestis ferruginea* (Vahl ex De Cantolle) root on paroxetine-induced sexual dysfunction in male rats. Asian Pac J Reprod 2012;1(2):111−6.

[34] Dewsbury DA, Davis Jr. HN. Effect of reserprine on the copulatory behaviour of male rats. Physiol Behav 1970;5:1331−3.

[35] Szechtman H, Hershkowitz M, Simantov R. Sexual behavior decreases pain sensitivity and stimulates endogenous opioids in male rats. Eur J Pharmacol 1981;70:279−85.

[36] Amin KMY, Khan MN, Zillur-Rehman S, Khan NA. Sexual function improving effect of *Mucuna pruriens* in sexually normal male rats. Fitoterapia 1996;67:53—8.

[37] Yakubu MT. Aphrodisiac potentials and toxicological evaluation of aqueous extract of *Fadogia agrestis* stem in male rats [Ph.D. thesis]. Ilorin, Nigeria: Department of Biochemistry, University of Ilorin; 2006.

[38] Ratnasooriya WD, Dharmasiri MG. Effects of *Terminalia catappa* seeds on sexual behaviour and fertility of male rats. Asian J Androl 2000;2:213—9.

[39] Yakubu MT, Akanji MA, Oladiji AT. Aphrodisiac potentials of the aqueous extract of *Fadogia agrestis* (Schweinf. Ex Hiern) stem in male albino rats. Asian J Androl 2005;7:399—404.

[40] Chan JSW, Waldingger MD, Oliver B, Oosting RS. Drug-induced sexual dysfunction in rats. Curr Protocol Neurosci 2010;9:9—34.

[41] Malviya N, Jain S, Gupta VB, Vyas S. Effect of garlic bulb on paroxetine-induced sexual dysfunction in male rats. Asian J Pharm Biol Res 2011;1(2):218—21.

[42] OECD. Guidance or genetic document histological evaluation of endocrine and reproductive tests in rodents. Preparation, reading and reporting of vaginal smears; 2009.

[43] Davidson JM. Sexology, sexual biology, behaviour and therapy: selected papers of Fifth World Congress of SexologyIn: Zewi H, editor. Holland, Amsterdam: Excerpta Medica, Princeton-Oxford; 1981. p. 42—7.

[44] Hart BL. Activation of sexual reflexes of male rats by dihydrotestosterone but not estrogen. Physiol Behav 1979;23:107—9.

[45] Grotthus B, Piasecki T, Pieśniewska M, Marszalik P, Kwiatkowska J, Skrzypiec-Spring M, et al. The influence of prolonged beta-blockers treatment on male rabbit's sexual behavior and penile microcirculation. Int J Impot Res 2006;19:49—54.

[46] Traish AM, Park K, Dhir V, Kim NN, Moreland RB, Goldstein I. Effects of castration and androgen replacement on erectile function in a rabbit model. Endocrinol 1999;140:1861—8.

[47] Bush PA, Gonzalez NE, Griscavage JM, Ignarro LJ. Nitric oxide synthase from cerebellum catalyzes the formation of equimolar quantities of nitric oxide and citrulline from L-arginine. Biochem Biophys Res Commun 1992;185:960—6.

[48] Giuliano F, Rampin O, McKenna K. In: Carson C, Kirbey R, Goldstein I, editors. Textbook of erectile dysfunction. Oxford: ISIS Medical Media; 1999. p. 43—50.

[49] Oberlander G, Yeung CH, Cooper TG. Induction of reversible infertility in male rats by oral ornidazole and its effects on sperm motility and epididymal secretions. J Reprod Fertil 1994;100(2):551—9.

[50] Watcho P, Zelefack F, Nguelefack TB, Ngouela S, Telefo PB, Kamtchouing P, et al. Effects of the aqueous and hexane extracts of *Mondia whitei* on the sexual behaviour and some fertility parameter of sexually inexperienced male rats. Afr J Tradit Complem Altern Med 2007;4(1):37—46.

[51] Jun K, Onyon L, Haseman J. The OECD program to validate the rat uterotrophic bioassay to screen compounds for *in vivo* estrogenic responses. Environ Health Perspect 2001;209(1):8—13.

[52] Yakubu MT, Adeshina AO, Oladiji AT, Akanji MA, Oloyede OB, Jimoh GA, et al. Abortifacient potential of aqueous extract of *Senna alata* leaves in rats. J Reprod Contraception 2010;21(3):163—77.

[53] Kafali H, Iriadam M, Ozardali I, Demir N. Letrozole-induced polycystic ovaries in the rat: a new model for cystic ovarian disease. Arch Med Res 2004;35:103—8.

[54] Marcondes FK, Blanchi FJ, Tanno AP. Determination of the estrous cycle phases of rats: some helpful considerations. Braz J Biol 2002;62(4A):609—14.

[55] Yakubu MT, Afolayan AJ. Reproductive toxicologic evaluations of *Bulbine natalensis* baker stem extract in albino rats. Theriogenol 2009;72:322−32.

[56] Amelar RD, Dubin L, Schoenfeld C. Semen analysis. Urol 1973;2:605−11.

[57] Costa-Silva JH, Lyra MMA, Lima CR, Arruda VM, Araujo AV, Ribeiro E, et al. A toxicological evaluation of the effect of *Carapa guianensis* Aublet on pregnancy in Wistar rats. J Ethnopharmacol 2007;112:122−6.

[58] Salhab AS, Al-Tamimi SO, Gharaibehand MMN, Shomaf S. The abortifacient effects of castor bean extract and ricin-A chain in rabbits. Contraception 1998;58:193−7.

[59] Sander EV, Cramer SD. A practical method for testing the spermicidal action of chemical contraceptives. Hum Fertil 1941;6:134−7.

[60] Yakubu MT, Olawepo OJ, Fasoranti GA. *Ananas comosus*: is the unripe fruit juice an abortifacient in pregnant Wistar rats? Eur J Contraception Reprod Health Care 2011;16 (5):397−402.

[61] Kirtikar KR, Basu BD. Indian medicinal plants. 2nd ed. Dehradun, India: International Book Distributor; 1987. p. 86−7.

[62] Kirtikar KR, Basu BD. In: Blatter E, Caius JF, editors. Indian medicinal plants. 2nd ed. Allahabad: Lalit Mohan Basu; 1975.

[63] Watt JM, Breyer-Brandwijk MG. Medicinal and poisonous plants of Southern and Eastern Africa. 2nd ed. Edinburgh: E. & S. Livingstone; 1962.

[64] Zia-ul-Haque A, Qazi MH, Hamdard ME. Studies on the antifertility properties of active components from seeds of *Abrus precatorius* Linn. Pak J Zool 1983;15:141−6.

[65] Bourdy G, Walter A. Maternity and medicinal plants in Vanuatu. I: The cycle of reproduction. J Ethnopharmacol 1992;37:179−96.

[66] Raj J, Singh KP. *Acalypha indica*. CCRH Q Bull 2000;22(1−2):1−6.

[67] Kumar D, Kumar A, Prakash O. Potential antifertility agents from plants: a comprehensive review. J Ethnopharmacol 2012;140:1−32.

[68] Xie W, Wang W, Su H, Xing D, Pan Y, Du L. Effect of ethanolic extracts of *Ananas comosus* L. leaves on insulin sensitivity in rats and HepG2. Comp Biochem Physiol C Toxicol Pharmacol 2006;143:429−35.

[69] Pakrashi A, Basak B. Abortifacient effect of steroids from *Ananas comosus* and their analogs on mice. J Reprod Fertil 1976;46:461−2.

[70] Burkill HM, Dalziel JM. The useful plants of West Tropical Africa: families M-R. Royal Botanic Gardens; 1997.

[71] Shanmugam H, Rathinam R, Chimathamba A, Venkatesan T. Antimicrobial and mutagenic properties of the root tubers of *Gloriosa superba* Linn. (Kalihari). Pak J Bot 2009;41(1):293−9.

[72] Joshi CS, Priya ES, Mathela CS. Isolation and anti-inflammatory activity of colchicinoids from *Gloriosa superba* seeds. Pharm Biol 2010;48:206−9.

[73] Kayode J, Kayode GM. Ethnomedicinal survey of botanicals used in treating sexually transmitted diseases in Ekiti state, Nigeria. Ethnobot Leaflet 2008;12:44−55.

[74] Malpani AA, Aswar UM, Kushwaha SK, Zambare GN, Bodhankar SL. Effect of the aqueous extract of *Gloriosa superba* Linn. (Langli) roots on reproductive system and cardiovascular parameters in female rats. Trop J Pharmaceut Res 2011;10:169−76.

[75] Adelaja AA, Ayoola MD, Otulana JO, Akinola OB, Olayiwola A, Ejiwunmi AB. Evaluation of the histo-gastroprotective and antimicrobial activities of *Heliotropium indicum* Linn. (boraginaceae). Malaysia J Med Sci 2008;15:22−30.

[76] Catalfamo JI, Martin WB, Birecka H. Accumulation of alkaloids and their necines in *Heliotropium curassavicum*, *H. spathulatum* and *H. indicum*. Phytochemistry 1982;21:2669−75.

[77] Oluwatoyin SM, Illeogbulam NG, Joseph A. Phytochemical and antimicrobial studies on the aerial parts of *Heliotropium indicum* Linn. Annal Biol Res 2011;2:129—36.
[78] Gill LS. Ethnomedical uses of plants in Nigeria. Benin, Nigeria: Uniben Press; 1992. p. 1—276.
[79] Ranab AC, Kamath JV. Preliminary study on antifertility activity of *Calotropis procera* roots in female rats. Fitoterapia 2002;73(1):111—5.
[80] Noumi E, Djeumen C. Abortifacient plants of the Buea region, their participation in the sexuality of adolescent girls. Indian J Tradit Knowl 2007;6:502—7.
[81] Olagbende-Dada SO, Ezeobika EN, Duru FI. Anabolic effect of *Hibiscus rosasinensis* Linn. leaf extracts in immature albino male rats. Nig Q J Hosp Med 2007;17:5—7.
[82] Nair R, Kalariya T, Chanda S. Antibacterial activity of some selected Indian medicinal flora. Turkish J Biol 2005;29:41—7.
[83] Murthy DR, Reddy CM, Patil SB. Effect of benzene extract of *Hibiscus rosasinensis* on the estrous cycle and ovarian activity in albino mice. Biol Pharm Bull 1977;20:756—8.
[84] Pakrashi A, Bhattacharya K, Kabir SN, Pal AK. Flowers of *Hibiscus rosasinensis*, a potential source of contragestative agent. III: Interceptive effect of benzene extract in mouse. Contraception 1986;34:523—36.
[85] El-Mahmood AM, Doughari JH, Chanji FJ. *In vitro* antibacterial activities of crude extract of *Nauclea latifolia* and *Daniella oliveri*. Sci Res Essays 2003;31(3):102—5.
[86] Yakubu MT, Musa IF. Effect of post-coital administtration of the alkaloid from *Senna alata* Linn. (Roxb) leaves on some fetal and maternal outcomes of pregnant rats. J Reprod Infertil 2012;13(4):211—7.
[87] Ankush RA, Singh A, Sharma A, Singh N, Kumar P, Bhatia V. Antifertility activity of medicinal plants on reproductive system of female rat. Int J Bio-Engr Sci Technol 2011;2(3):44—50.
[88] Ngadjui E, Watch P, Nquelefack TB, Kamanyi A. Effects of *Ficus asperifolia* on normal rats estrus cyclicity. Asian Pac J Tropical Biomed 2013;3(1):53—7.
[89] Kuete V, Mbaveng AT, Tsaffack M, Beng VP, Etoa FX, Nkengfack AE, et al. Antitumour, antioxidant and antimicrobial activities of *Bersama engleriana* (Melianthaceae). J Ethnopharmacol 2008;115(3):494—501.
[90] Watcho P, Mekemdjio A, Nguelefack TB, Kamanyi A. Sexual stimulant effects of the aqueous and methanolic extracts from the leaves of *Bersama engleriana* in adult male rats. Pharmacologyonline 2007;1:464—74.
[91] Umukoro S, Ashorobi BR. Further pharmacological studies on aqueous seed extract of *Aframomum melegueta* in rats. J Ethnopharmacol 2008;115:489—93.
[92] Mbongue GYF, Kamtchouing P, Dimo T. Effects of the aqueous extract of dry seeds of *Aframomum melegueta* on some parameters of the reproductive function of mature male rats. Andrologia 2012;44:53—8.
[93] Pujol J. Naturafrica—the Herbalist Handbook: African Flora, medicinal plants. Durban, South Africa: Jean Pujol Natural Healers' Foundation; 1990. p. 25—8.
[94] Yakubu MT, Afolayan AJ. Anabolic and androgenic activities of *Bulbine natalensis* Baker stem in adult male Wistar rats. Pharml Biol 2010;48(5):568—76.
[95] Nantia EA, Manfo PF, Beboy NE, Travert C, Carreau S, Monsees TK, et al. Effect of methanol extract of *Basella alba* L. (Basellaceae) on the fecundity and testosterone level in male rats exposed to flutamide *in utero*. Androl 2012;44:38—45.
[96] Moundipa PF, Beboy NS, Zelefack F, Ngouela S, Tsamo E, Schill WB, et al. Effects of *Basella alba* and *Hibiscus macranthus* extracts on testosterone production of adult rat and bull Leydig cells. Asian J Androl 2005;7:411—7.

[97] Uchendu CN, Kamalu TN, Asuzu IU. A preliminary evaluation of antifertility activity of a triterpenoid glycoside (DSS) from *Dalbergia saxatilis* in female Wistar rats. Pharmacol Res 2000;41:521−5.

[98] Uchendu CN, Isek T. Antifertility activity of aqueous ethanolic leaf extract of *Spondias mombin* (Anacardiaceae) in rats. Afr Health Sci 2008;8:163−7.

[99] Debella A, Haslinger E, Kunert O, Michl G, Abebe D. Steroidal saponins from *Asparagus africanus*. Phytochemistry 1999;51:1069−75.

[100] Abebe D, Ayehu A. Medicinal plants and enigmatic health practices of northern Ethiopia. BSPE; 1993.

[101] Gebrie E, Makonnen E, Debella A, Zerihun L. Phytochemical screening and pharmacological evaluations for the antifertility effect of the methanolic root extract of *Rumex steudelii*. J Ethnopharmacol 2005;96:139−43.

[102] Shibeshi W, Makonnen E, Zerihun L, Debella A. Effect of *Achyranthes aspera* L. on fetal abortion, uterine and pituitary weights, serum lipids and hormones. Afr Health Sci 2006;6:108−12.

[103] Vasudeva N, Sharma SK. Estrogenic and pregnancy interceptory effect of *Achyranthas aspera* Linn. root. Afr J Trad Compl Alt Med 2007;4(1):7−11.

[104] Vasudeva N, Vats M. Anti-spermatogenic activity of ethanol extract of *Dalbergia sissoo* Roxb. stem bark. J Acupunct Meridian Study 2011;4(2):116−22.

[105] Dixit VP, Joshi S, Kumar A. Possible antispermatogenic activity of *Gloriosa superba* (ETOH extract) in male gerbils (*Meriones hurrianae* Jerdon): a preliminary study. Comp Physiol Ecol 1983;81:17−22.

[106] Etta HE, Bassey UP, Eneobong EE, Okon OB. Anti-spermatogenic effects of ethanol extract of *Mucuna urens*. J Reprod Contraception 2009;20:161−8.

[107] Lembe DM, Gasco M, Rubio J, Yucra S, Ngo-Sock E, Gonzales GF. Effect of the ethanolic extract from *Fagara tessmannii* on testicular function, sex reproductive organs and hormone level in adult male rats. Androl 2011;43:139−44.

[108] Raji Y, Akinsomisoye OS, Salman TM. Anti spermatogenic activity of *Morinda lucida* extract in male rats. Asian J Androl 2005;7(4):405−10.

[109] Okoli CO, Akah PA, Nwafor SV, Anisiobi AI, Ibegbunam IN, Erojikwe O. Anti-inflammatory activity of hexane leaf extract of *Aspilia africana* C.D. Adams. J Ethnopharmacol 2007;109:219−25.

[110] Kuiate JR, Amvam ZPH, Lamaty G, Menut C, Bessière JM. Composition of the essential oils from the leaves of two varieties of *Aspilia africana* (Pers.) C.D. Adams from Cameroon. Flavour Fragr J 1999;14:167−9.

[111] Okwuonu C, Oluyemi K, Grillo B, Adesanya O, Ukwenya V, Odion B, et al. Effects of methanolic extract of *Aspilia africana* leaf on the ovarian tissues and weights of Wistar rats. Int J Altern Med 2006;5:1.

[112] Ifeanyi AC, Eboetse YO, Ikechukwu DF, Adewale OA, Carmel NC, Olugbenga OA. Effect of *Momordica charantia* on estrous cycle of Sprague-Dawley rats. Pac J Med Sci 2011;8(1):37−48.

[113] Ghosh AK, Das AK, Patra KK. Studies on antifertility effect of rhizome of *Curcuma longa* Linn. Asian J Pharm Life Sci 2011;1(4):349−53.

[114] Jain S, Choudhary GP, Jain DK. Pharmacological evaluation and anti-fertility activity of ethanolic extract of *Jatropha gossypifolia* leaf in rats. Biomed Res Int 2013; [Article ID 125980], 5 pages.

[115] Dyson A. Discover indigenous healing plants of the herb and fragrance gardens at Kirstenbosch Botanical Garden (Afrikaans Edition). National Botanical Institute; 1998.

[116] Ebrahim M, Pool EJ. The effect of *Tulbaghia violacea* extracts on testosterone secretion by testicular cell cultures. J Ethnopharmacol 2010;132(1):359−61.

[117] Watcho P, Zelefack F, Nguelefack TB, Ngouela S, Telefo PB, Kamtchouing P, et al. Effects of the aqueous and hexane extracts of *Mondia whitei* on the sexual behaviour and some fertility parameters of sexually inexperienced male rats. Afr J Tradit Complemt Altern Med 2006;4(1):37−46.

[118] Kubo I, Kinst-Hori I. 2-Hydroxy-4-methoxybenzaldehyde: a potent tyrosinase from African medicinal plants. Planta Medica 1999;65:19−22.

[119] Javeed AW, Rajeshwara N, Achur RKN. Phytochemical screening and aphrodisiac property of *Asparagus racemosus*. Int J Pharm Drug Res 2011;3(2):112−5.

[120] Yakubu MT, Akanji MA. Effect of aqueous extract of *Massularia acuminata* stem on sexual behaviour of male rats. Evid Based Complem Altern Med 2011; [Article ID 738103], 10 pages.

[121] Nadkarni AK. 3rd ed. Nadkarni's Indian Materia Medica, vol. 1. Mumbai: Popular Prakashan Pvt. Ltd; 2002. p. 830−34.

[122] Isogai A, Suzuki A, Tamura S. Structure of dimeric phenoxypropanoids from *Myristica fragrans*. Agar Biol Chem 1973;37:193−4.

[123] Janssen J, Laeckman GM. Nutmeg oil: identification and quantification of its most active of platelet constituents as inhibitors aggregation. J Ethnopharmacol 1990;29:179−88.

[124] Tajuddin A, Shamshad A, Abdul L, Iqual AQ, Kunwar MYA. An experimental of sexual function improving effect of *Myristica fragrans* Houtt. (nutmeg). BMC Compl Altern Med 2005;5:16−22.

[125] Ajiboye TO, Nurudeen QO, Yakubu MT. Aphrodisiac effect of aquoeus root extract of *Lecaniodiscus cupanioides* in sexually impaired rats. J Basic Clin Physiol Pharmacol 2013; [Epub ahead of print].

16 African Plants with Dermatological and Ocular Relevance

Danielle Twilley and Namrita Lall

Department of Plant Sciences, University of Pretoria, Pretoria, South Africa

16.1 Introduction

The human body is constantly exposed to potentially dangerous viruses, bacteria, harmful pathogens, and environmental factors which can cause harm to the human body, both internally and externally. The human body is therefore equipped with two major kinds of defense mechanisms which assist in getting rid of these threats: innate immunity and acquired immunity (Figure 16.1). The innate immunity consists of external barriers, such as the skin, mucous membranes, and tears, as well as internal cellular (macrophages, phagocytes, and dendritic cells) and chemical defenses (antimicrobial proteins) which combat pathogens that penetrate the external defense barriers. The acquired immunity is developed only once the body is exposed to foreign substances or pathogens and is highly specific. The acquired immunity consists of the humoral response, where B lymphocytes secrete antibodies which attach to the antigens of foreign substance and therefore act as markers for it to be destroyed, and the cell-mediated response where cytotoxic lymphocytes directly destroy the foreign substance [1]. In this chapter, the skin and eyes as external defense barriers will be discussed in more detail.

The skin is the largest bodily organ and the primary defense mechanism against the external environment. It is able to defend against various factors such as UV radiation, various microbes such as viruses, bacteria, and other harmful pathogens, as well as potentially harmful substances in the environment. Not only is the skin important as a defense mechanism, but it is also the most important thermal regulator of body temperature [2]. The first layer of the skin, the epidermis, is made up of specialized cells such as keratinocytes (Figure 16.2). These keratinocytes play an important role in regulating the immune response as it is a major producer and secretor of inflammatory mediators such as histamines, cytokines, prostaglandins, leukotrienes, and eicosanoids in response to various stimuli [3]. Furthermore, the

Toxicological Survey of African Medicinal Plants. DOI: http://dx.doi.org/10.1016/B978-0-12-800018-2.00016-9

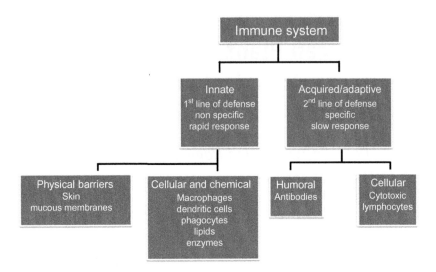

Figure 16.1 Overview of innate and adaptive immunity.

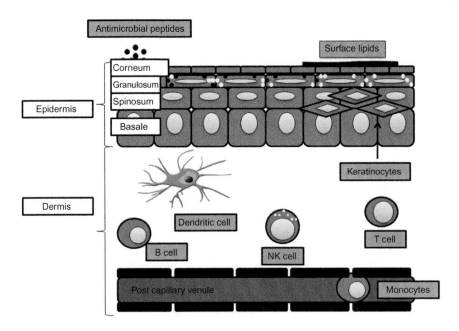

Figure 16.2 Defense mechanism in the epidermis and dermis layer of the skin.

keratinocytes are important as they are able to regulate the body's temperature due to the expression of heat shock proteins and are able to protect against UV radiation due to the presence of UV-absorbing molecules such as melanin, vitamin D, and trans-urocanic acid. The surface also consists of antimicrobial peptides, iron-binding proteins, and lipids which add to the antimicrobial activity of the skin. The antimicrobial peptides act by directly inhibiting the growth of bacteria, viruses, and fungi or indirectly by activating the cellular and acquired immunity. The antimicrobial peptides furthermore increase inflammation which increases the proliferation of cells and therefore, influences wound healing. Last, the skin barrier plays a major role in the permeability of substances and molecules through to the internal system which also adds to the defensive function of the body which is regulated by the stratum corneum, the outermost part of the skin [2]. One of the major regulators of the permeability barrier is the pH of the skin. The slight acidity of skin, which ranges between pH 4.5 and 5.5, is important in maintaining the integrity of the stratum corneum, which in turn is important in establishing a good permeability and antimicrobial barrier [4].

The eye contains important defense mechanisms which protect it from infections caused by viruses, fungi, parasites, and bacteria [5]. The eye, like the skin, contains many specific and nonspecific mechanisms which provides defense against pathogens. The eye possesses a first line of defense which consists of the eyelids and the eyelashes. Naturally, when the eyelids are shut it protects the tissue from exposure to any potential pathogens and allergens. The blinking reflex of the eye is triggered when something comes into contact with the eyelids and eyelashes, and this in turn serves to wash the eye of any foreign pathogens or particles. Tears are also a highly important function in the nonspecific defense of the eye. Apart from the washing which causes the removal of pathogens, tears also contain an enzyme known as lysozyme which hydrolyzes the peptidoglycan in bacterial cell walls. Furthermore, tears contain lactofferin and β-lysin which respectively interfere with pathogen metabolism and disrupt the integrity of the bacteria membranes [5]. Tears also contain other defense mechanism mentioned in Table 16.1.

Table 16.1 Summary of More Ocular Defense Mechanism Carried out by the Tear Film

Component	Function
Mucin and immunoglobulins	Prevents pathogens from binding to ocular surface
Cytokines	Recruitment of phagocytic cells
Phospholipase A_2	Enzymatically breaks down phospholipids in pathogen membranes
Defensins	Assist in inhibiting the growth of pathogens
Lactoferrin	Binds iron on ocular surface reducing its availability to pathogens

The conjunctiva is a part of the eye tissue that contains the most immunological functions on the surface of the eye. It consists of two layers, namely the epithelium and substantia propria. The epithelium layer of the eye provides protection through six layers of epithelial cells, which also consists of melanocytes, inflammatory cells, and Langerhans cells, as well as tight junctions between the cells, which provides a physical barrier to incoming pathogens, and a basement membrane where mast cells and eosinophils reside beneath. The substantia propria, however, provides protection more through an inflammatory response using various cells such as mast cells, plasma cells, lymphocytes, and neutrophils [6].

There are, however, many drugs that can cause side effects in both the skin and eyes, when taken to treat other diseases of the body. Naturally, there are many factors to take into consideration when prescribing a drug to a patient, and these factors include: the nature of the drug, the amount of the drug, the route of administration, the health of the patient, any inherited abnormalities, and whether any drugs are being taken simultaneously [7].

There are many types of drugs that can cause unwanted effects in the skin. A group of drugs that can cause notable effects on the skin include anticonvulsant and antiepileptic drugs. These drugs are commonly taken for the treatment of seizures and epilepsy. Examples of such drugs include carbamazepine and phenobarbital, which can cause urticaria and photosensitivity [8]. Examples of drugs causing possible ocular side effects include antispasmodics such as atropine and antihypertensives such as hexamethonium, which are used for peptic ulcerations and hypertension, respectively, but can both possibly cause cycloplegia, which is difficulty with near vision. Another good example is digitalis, which is a cardiac glycoside used for congestive heart failure, but could cause snowy vision and conjunctivitis. There are also types of antidepressents such as imipramine and amphetamines which could cause cycloplegia, glaucoma, and diplopia, which is commonly known as double vision [7].

In Africa, there are many reports of medicinal plants used for the treatment of various skin and eye disorders. Reports on treatment of skin and eye disorders are mainly found from southern African traditional healers where decoctions and infusions are the most used method of medicinal plant preparation. The aim of this chapter is to discuss the protective barriers of the skin and eyes as well as traditional African medicinal plants used in the treatment of skin and eye diseases.

16.2   The Skin and the Metabolism of Xenobiotics

The skin, as the largest organ of the body, not only plays an important role in the body's first line of defense, but it also plays an important role in the metabolism and elimination of xenobiotics from the body. Xenobiotics can be defined as any foreign substances or exogenous chemicals which the body does not recognize such

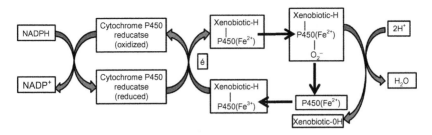

Figure 16.3 Phase I enzyme reactions mediated by cytochrome P450.

as drugs, pollutants, as well as some food additives and cosmetics. Xenobiotics are transformed and broken down by two systems, namely the phase I and phase II enzyme systems [9].

Phase I enzyme systems are based on transforming the xenobiotic into a more polar inactive metabolite by forming functional groups such as $-OH$, $-NH_2$, and $-SH$ (Figure 16.3). Reactions responsible for these transformations include N- and O-dealkylation (N-, O-, S-) as well as N- and O-oxidation, deamination, aliphatic, and aromatic hydroxylation and dehalogenation reactions with cytochrome P450 as the main enzyme responsible [10]. These reactions are known as oxidative reactions which are mediated by a heme protein (contained within the cytochrome P450 enzyme), NADPH, the reductase enzyme, phosphotidylcholine, and molecular oxygen.

Phase II enzymes are responsible for transforming metabolites into more easily excretable forms as well as inactivation of active metabolites. These reactions are known as conjugation reactions which include glucoronidation, sulfation, methylation, acetylation, as well as glutathione and amino acid conjugation. This is generally known as the detoxification phase of xenobiotics [11].

As mentioned in the introduction, the skin acts as a barrier between the external environment and the body to defend from potentially harmful substances. The mechanical barrier physically inhibits the penetration of foreign substances into the body, whereas the chemical barrier acts through the xenobiotic transformation system as well as the scavenging of reactive oxygen species and excretion of xenobiotics. The skin contains all the enzymes present in the phase I and phase II systems such as cytochrome P450, alcohol dehydrogenase, monoamine oxidase, aldehyde dehydrogenase, flavin-dependent monooxygenase, glutathione S-transferase, NADP(H): quinone oxidoreductase, and catechol-O-methyltransferase. These phase I and phase II enzymes are only induced once a xenobiotic has entered the body [12,13]. The skin also contains an antioxidant system which is able to scavenge reactive oxygen species which are generated through xenobiotic metabolism. These antioxidant enzymes include catalase, glutathione peroxidase, and superoxide dismutase. The reactive oxygen species are the contributors of oxidative stress which lead to various diseases and disorders such as cardiovascular disease, cancer, aging, and various neurodegenerative diseases [14]. Not only is the skin able to metabolize xenobiotics and scavenge the reactive oxygen species generated by their metabolism, but it can also excrete xenobiotics or toxic products generated by their metabolism. The skin contains up to 4 million sweat glands, where water-soluble substances such as drugs, steroids, cytokines, and metals are eliminated, which is increased during an increase in body temperature [9]. The excretion of fat-soluble substances such as lipids and cholesterol is largely mediated through the sebaceous glands found in skin.

16.3   The Eye and the Metabolism of Xenobiotics

The eye is not often considered an organ which functions in drug metabolism, however, evidence shows that phase 1 and phase 11 enzymes are present and therefore suggests metabolic activities within the eye [15].

The eye has a rich blood supply which could bring along unwanted metabolic byproducts and toxins which the eye then needs to remove [16]. The eye also has a blood−ocular barrier which is important in maintaining homeostasis by restricting the movement of unwanted compounds into the eye. The phase I metabolic activities, more specifically the cytochrome p450-dependent activities, are higher in ocular sights which are closest to the uveal blood flow, such as the capillary body, iris, retina, and lens. During the phase II metabolic activity studies in rabbits, it was found that there was a presence of metabolic enzymes such as glutathione S-transferase in the iris and capillary body as well as N-acetyl transferase, sulfo-transferase and Uridine 5'-diphospho-glucuronosyl transferase (UDP)-glucuronosyl transferase in the cornea. Further in this study it was noted that the glutathione S-transferase activity in the cornea was 70% of the activity in the intestine, which is likely due to the high exposure of the cornea to foreign compounds and substances [17].

16.4   African Plants Affecting the Skin

An example of a plant native to South Africa that causes skin irritation is *Plumbago auriculata*, commonly known as blue plumbago. All parts of *P. auriculata* contain 2-methyl juglone, which has been reported to cause blistering of the skin [18]. There are, however, many plants that are used in cosmetics to help improve the health of skin. A very well-known plant which is native to South Africa is known as Rooibos (*Aspalathus linearis*). It is a well-known antioxidant, which is rich in polypehnols, flavonoids, phenolic acids, oligosaccharides, and polysaccharides. It was first used 300 years ago by the Khoi-Khoi tribe in the Cederberg, Western Cape as a tea. However, it has also been used for people who suffer from eczema [19]. There is also ongoing research to determine whether plant extracts possessed any antityrosinase activity. A high activity of the tyrosinase enzyme leads to various dermatological disorders such as melanoma, age spots, and actinic damage. In a study conducted by Mapunya et al. [20], it was found that an ethanolic bark extract from *Harpephyllum caffrum* resulted in good inhibition of the tyrosinase enzyme and overall good melanin inhibition *in vitro*, which makes this plant a potential substitute for other available skin-lightening products.

16.5 African Traditional Medicines Used for the Treatment of Dermatological Disorders

The World Health Organization (WHO) has estimated a value of 65−80% of the world's population in developing countries that depend primarily on plants for their primary healthcare needs. This is mainly due to the inaccessibility of poor income countries to access modern Western medicines [21]. Furthermore, it is estimated that 35,000−70,000 plant species around the world have been used for their medicinal value.

In African traditional medicine (ATM), there is a high demand for medicinal plants which are used for the treatment of skin ailments such as wounds and treatment of burns as well as other skin conditions such as infection and ulcers [22]. In southern Africa alone, there have been over 100 plants identified as having medicinal value for various skin ailments. Most of these medicinal plants are used for wound healing and various skin diseases, which includes bacterial and viral infections [23]. Table 16.2 is a summary of medicinal plants reported to have properties which are able to rid certain skin diseases [23−39].

The most common mode of preparation and possibly the most simple way of preparing medicinal plants for administration is by preparing decoctions, where plants are boiled in water or other solvents to extract the active ingredients, or infusions, where the plant is immersed in hot or cold water for a certain period of time and then topically applied to the affected area. Other methods include preparing powders of the plant parts as well as pastes, juice, ointments, and poultices, and some unspecified methods. Pastes are generally the next most popular choice of administration. The ointments and poultices normally are used as a dressing where the plant material is heated and then applied either heated or cooled onto the area. Most of the plant parts used for the treatment of skin diseases include leaves, roots, bark, the whole plant, as well as fruits, rhizomes, bulbs, and flowers (Figure 16.4). The most common plant part used is leaves, which are possibly due to the easy accessibility and sustainability of the plant part [23].

16.6 African Plants Affecting the Eyes

A plant family known to cause great toxicity and inflammation of the skin and eyes is the *Euphorbiaceae* family. Ocular toxicity caused by various species of *Euphorbia*

Table 16.2 Traditionally Used Southern African Plants in the Treatment of Various Skin Ailments and Diseases

Scientific Name/Family	Common Name	Plant Part Used	Preparation/ Administration	Traditional Usage	Area of Plant Collection	References
Abrus precatorius L. (Fabaceae)	Crab's eye	Leaves	Juice applied topically	Skin spots	Nigeria	[24]
Acacia ataxacantha DC. (Fabaceae)	Flamethorn	Leaves	Juice applied topically	Burns and sores	Nigeria	[24]
Agathosma betulina (P.J. Bergius) Pillans. (Rutaceae)	Buchu	Buchu vinegar with leaf	Infusion/tincture	Bruises	Southern Africa	[25]
Embelia ruminata (E.Mey. ex A.Dc.) Mez. (Myrsinaceae)	Vidanga	Leaves	Paste	Wound healing	Southern Africa	[26]
Aloe ferox Mill. (Xanthorrhoeaceae)	Bitter aloe	Leaf sap, leaves and roots	Sap applied topically	Bruises and burns	Southern Africa	[25]
Arctopus echinatus L. (Apiaceae)	Bear's foot	Roots	Infusion applied topically	Skin irritation	Southern Africa	[27–29]
Annona stenophylla Engl. & Diels (Annonaceae)	Dwarf-Custard apple	Roots	Topically applied	Boils	Zimbabwe	[30]
Flueggea virosa (Roxb. ex Wiild.) Voigt (Euphorbiaceae)	Common bushweed	Roots	Powder applied topically	Wound healing	Zimbabwe	[30]
Erythrina abyssinica Lam. Ex DC. (Fabaceae)	Flame tree	Roots	Topically	Wound healing	Zimbabwe	[30]
Kirkia acuminate Oliv. (Kirkiaceae)	White seringa	Fruit	Applied directly to affected area	Wound healing	Zimbabwe	[30]
Carpobrotus edulis (L.) L. Bolus. (Aizoeceae)	Sour fig	Leaf juice and pulp	Juice applied to affected area	Wound healing and eczema/burns	Southern Africa	[23]

Species (Family)	Common name	Part used	Preparation	Use	Region	Reference
Cassine transvaalensis (Celastraceae)	Saffronwood	Bark	Infusion	Skin inflammation and rash	Southern Africa	[27,31]
Centaurea benedicta (L.) L. (Asteraceae)	Holy thistle	Whole plant	Topically	Wound healing/ulcers	Southern Africa	[27]
Centella asiatica (L.) Urb. (Apiaceae)	Pennywort	Leaves	Tinctures	Wound healing/leprosy and acne	Southern Africa	[23]
Cinnamomum camphora (L.) Presl. (Lauraceae)	Camphor tree	Essential oil	Topically	Skin inflammation/antiseptic	Southern Africa	[27]
Cissampelos capensis Thunb. (Menispermaceae)	Davidjies	Rhizomes, roots, and leaves	Paste	Wound healing/sores and ulcers/boils	Southern Africa	[27,32]
Strychnos spinosa Lam. (Loganiaceae)	Spiny Monkey Orange	Fruits	Extract orally ingested	Gonorrhea and genital warts	Zimbabwe	[30]
Cnicus benedictus L. (Asteraceae)	Holy thistle	Whole plant	Paste	Wound healing/ulcers	Southern Africa	[23]
Crinum macrowanii Baker. (Amaryllidaceae)	River lily	Bulbs and leaves	topically	Sores/boils and acne	Southern Africa	[23]
Datura sramonium L. (Solanaceae)	Jimson weed	Leaves	Skin patch	Wound healing/boils and abscesses	Southern Africa	[23]
Dioscorea dregeana T. Durand & Scinz. (Dioscoreaceae)	Wild yam	Fresh tubers	Decoctions applied topically	Sores and cuts	Southern Africa	[23]
Dioscorea sylvatica Eckl. (Dioscroreaceae)	Elephants foot	Fresh rhizomes	Topically	Skin problems	Zimbabwe	[33]
Distemonanthus benthamianus Baill. (Fabaceae)	African satinwood	Root bark	Decoction ingested	Skin spots	Nigeria	[24]
Dodonaea angustifolia L.f. (Sapindaceae)	Sand olive	Leaves and twig tips	Decoction applies topically	Antipruritic/boils	Southern Africa	[23]

(Continued)

Table 16.2 (Continued)

Scientific Name/Family	Common Name	Plant Part Used	Preparation/Administration	Traditional Usage	Area of Plant Collection	References
Entandrophragma caudatum (Sprague) Sprague (Meliaceae)	Wooden banana	Fruit	Burnt fruit peels applied directly	Genital warts	Zimbabwe	[30]
Erythrina lysistemon Hutch. (Fabaceae)	Common coral tree	Bark	Poultice or powdered burnt bark	Wound healing/sores and ulcers/abscesses	Southern Africa	[23]
Ximenia caffra Sond. (Olacaceae)	Large sourplum	Roots	Powder topically applied	Wound healing	Zimbabwe	[30]
Harpagophytum procumbens DC. ex Meisn. (Pedaliaceae)	Devils claw	Roots	Paste or ointment	Sores/boils	Southern Africa	[27,28]
Harpephyllum caffrum Bernh. ex Krauss. (Anacardiaceae)	Wild plum	Bark	Applied topically	Eczema and acne	Southern Africa	[27]
Jatropha curcas L. (Euphorbaceae)	Purging nut	Rhizomes	Topically	Wound healing and boils	Southern Africa	[27,34]
Kigelia Africana (Lam.) Benth. (Bignoniaceae)	Sausage tree	Fruit	Topically	Ulcers and sores/skin inflammation/abscesses	Southern Africa	[23]
Brillantaisia cicatricosa Lindau (Acanthaceae)	Giant salvia	Leaves	Unknown	Leprosy and eczema	Rwanda	[35]
Lobostemon fruticosus (L.) H. Buek. (Boraginaceae)	Pajama bush	Leaves and twigs	Ointment	Wound healing	Southern Africa	[23]
Mentha longifolia L. (Lamiaceae)	Wild mint	Leaves	Topically	Wound healing	Southern Africa	[27,36]

Species (Family)	Common name	Part used	Preparation	Use	Region	Reference
Merwilla natalensis (Planchon) Speta (Hyacinthaceae)	Blue squill	Bulb	Boiled water extract applied topically	Boils and veld sores	South Africa	[25]
Myrothamnus flabellifolius Welw. (Myrothammaceae)	Resurrection plant	Leaves and twigs	Powdered leaves applied topically	Wound healing and burns	Southern Africa	[27]
Osmitopsis asteriscoides Cass. (Asteraceae)	Mountain daisy	Leaves	Topically	Cuts/skin inflammation	Southern Africa	[27,37]
Chenopodium ugandae (Aellen) Aellen (Chenopodiaceae)	Unknown	Leaves	Unknown	Skin disease and leprosy	Rwanda	[35]
Dolichopentas longiflora Oliv. (Rubiaceae)	Unknown	Leaves and roots	Unknown	Skin disease, scabies, and skin mycosis	Rwanda	[35]
Psidium guajava L. (Myrtaceae)	Guava	Leaves	Infusion applied topically	Wound healing/ulcers and boils	Southern Africa	[23]
Ricinus communis L. (Euphorbaceae)	Caster bean tree	Leaf and pulverized or burnt seeds and bark	Poultice	Wound healing/sores and boils	Southern Africa	[27,28]
Protea madiensis Oliv. (Proteaceae)	Tall wooden sugarbush	Roots and bark	Unknown	Skin diseases and hyperpigmentation	Rwanda	[35]
Scabiosa columbaria L. (Dipsacaceae)	Wild scabious	Leaves and roots	Ointment	Wound healing	Southern Africa	[23]
Scadoxus puniceus (L.) Friis & Nordal. (Amaryllidaceae)	Red paintbrush	Bulbs and roots	Decoction applied topically	Wound healing/ulcers and sores/skin allergies	Southern Africa	[27,28]

(Continued)

Table 16.2 (Continued)

Scientific Name/Family	Common Name	Plant Part Used	Preparation/ Administration	Traditional Usage	Area of Plant Collection	References
Sclerocarya birrea (A. Rich.) Hochst. subsp. caffra (Anacardiaceae)	Marula	Leaves	Essence from leaves applied topically	Abscesses, burns and spider bites	South Africa	[25]
Sesamum angolense Welw. (Pedaliaceae)	Unknown	Leaves	Unknown	Hyperpigmentation	Rwanda	[35]
Securidaca longepedunculata Fresen. (Polygalaceae)	Violet tree	Leaves and bark	Ointment	Wound healing/sores	Southern Africa	[23]
Terminalia sericea Burch, ex DC. (Combretaceae)	Silver cluster leaf	Root sap or bark	Applied topically	Wound healing and leprosy	Southern Africa	[23]
Urginea altissima (Linn. f.) Bak (Hyacinthaeae)	Tall squill	Fresh bulb	Topically	Skin problems	Zimbabwe	[33]
Vitex doniana Sweet (Lamiaceae)	Black plum	Root	Poultice applied topically	Leprosy	Nigeria	[24]
Withania somnifera (L.) Dunal. (Solanaceae)	Poison gooseberry	Leaves and berries	ointment	Wound healing/skin inflammation and abscesses	Southern Africa	[23]
Xysmalobium undulatum R. Br. (Apocynaceae)	Milk bush	Roots	Powder applied topically	Wound healing/sores and abscesses	Southern Africa	[23]
Zantedeschia aethiopica (L.) Spreng. (Araceae)	Arum lily	Leaves	Leaf applied directly	Wound healing/ Sores and boils	Southern Africa	[27,28]
Ziziphus mucronata Willd. (Rhamnaceae)	Buffalo-thorn	Leaves, roots and bark	Decoction applied topically	Sores/skin inflammation and boils	Southern Africa	[23]

Plant part

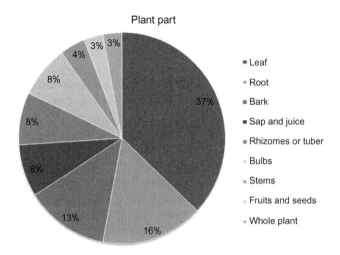

Figure 16.4 Plant parts used to treat various skin infections and disorders.

ranges from mild conjunctivitis to severe kerato-uveitis and even permanent blindness in some cases [40]. *Euphorbia* species also have been reported to cause corneal scarring, photo dermatitis, and corneal ulceration [41]. An example of a species found in western Africa is *Euphorbia trigona*, commonly known as the African milk tree. The sap of this species was able to cause loss of corneal epithelium 16 h after direct exposure to the sap in a 60-year-old male. Other species around the world known to have caused corneal edema and conjunctival hyperemia include *Euphorbia neriifolia* (Indian Spurge tree) and *Euphorbia milii* (Crown-of-thorns houseplant) [42]. The main phytochemicals that have been found in the latex of *Euphorbia* are di-and tri-terpene esters, 1-inositol, pyrrogallic and catechuic tannins, and a xanthoramnine alkaloid [43]. Another well-known example of a *Euphorbia* species that has a damaging effect on the eyes is known as *Euphorbia tirucalli* (Pencil tree); it also has been known to cause blindness. This plant is found in tropical areas of Africa where it has traditionally been used to treat cancerous tumors and as a fish poison [42]. There also have been reports on traditionally used medicines in Oman that have toxicity in the eye. In a case study reported by Shenoy et al., a patient with a chalazion tried *Caloptropis procera* plant extract as part of the treatment and developed a decrease in vision in the eye [44].

Although there are many plants in Africa and around the world that cause toxicity toward the eye, there is also ongoing research which suggests that there are plant extracts which could potentially be used as treatments for various disorders and diseases of the eye as well as have protective effects on the eye. An example is a polyherbal eye drop formulation, which contains *Berberis aristata*, *Cassia absus*, *Coptis teeta*, *Symplocos racemosa*, *Azadirachta indica*, and *Rosa damascene*. It is generally used as an anti-inflammatory for inflammation of the conjunctiva [45].

16.7   Ocular Effects of Medicinal Plants of Africa

The use of traditional medicine to treat eye diseases and disorders has increased worldwide in the last 20 years. This is possibly due to the ever-increasing problem of poor income communities not being able to access primary eye care facilities, as well as cultural beliefs which lead patients to seek medical advice from a traditional healer rather than from a hospital. The use of traditional eye medicines normally consists of crude or processed plant or animal products and can also include the use of chemical substances [46].

Traditional African plant medicines are used for a range of eye disorders which includes chalazion, entropion, trichiasis, conjunctivitis, corneal leucoma, cataracts, various eye injuries, tumors, and refractive errors. The most common type of preparations for medicinal plants which are used for the treatment of various eye disorders include decoctions, infusions, or plant sap as eye drops (Figure 16.5). For these various concoctions which are made, the water is normally taken from taps, and wells as well as river water and in some cases, even the saliva of the traditional healer is used [47].

The use of traditional eye medicine in Africa is still an ongoing practice with reports from Kenya, Malawi, Tanzania, Nigeria, as well as Zimbabwe [30,47−51]. Table 16.3 is an overview of plants which are used for various eye disorders in Kenya, Zimbabwe, and Tanzania. Plant parts used for the treatment of eye disorders using African medicinal plants are also depicted in Figure 16.6.

Although treatment of eye disorders through ATM is a common practice, it is not always the safest practice. In a study conducted in Malawi, it was found that 26% of blindness in blind school children was attributed to the use of traditional eye medicine [52].

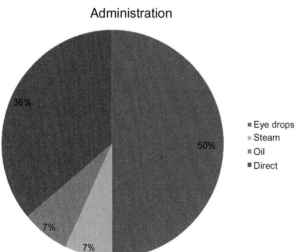

Administration

Figure 16.5 Modes of administration for medicinal plants used in Africa to treat disorders of the eye.

36%

50%

7%

7%

■ Eye drops
▩ Steam
▨ Oil
■ Direct

Table 16.3 Medicinal Plants Used to Treat Various Ocular Disorders in Africa

Scientific Name/Family	Common Name	Plant Part Used	Administration	Traditional Usage	Area of Plant Collection	Reference
Abrus precatorius L. subsp. *africanus* Verdc. (Fabaceae)	Bead vine	Seeds or roots	Seeds taken orally or sap of root applied directly	Cataracts, refractive errors, and ocular pains	Kenya	[47]
Ageratum conyzoides Linn. (Asteraceae)	Goat weed	Leaves	Sap from leaves used as eye drops	Eye problems	Nigeria	[50]
Boscia coriacea Pax. (Capparaceae)	Unknown	Leaves	Leaves chewed and mixed with saliva and applied to conjunctival sac	Trachoma	Kenya	[47]
Cordia sinensis Lam. (Boraginaceae)	Grey leaved saucer berry	Leaves	Crushed leaves rubbed directly on tarsal conjunctiva until it bleeds	Trachoma	Kenya	[47]
Dichrocephala integrifolia L. f. Kuntze. (Asteraceae)	Veronia	Leaves	Juice of crushed leaves applied directly as eye drops	Eye infections	Tanzania	[51]
Hoslundia opposite Vahl. (Lamiaceae)	Orange bird berry	Leaves	Extract drops	Cataracts	Zimbabwe	[30]
Lannea discolor (Sond.) (Anacardiaceae)	Live long	Roots	Eye drops	Sore eyes	Zimbabwe	[30]
Manihot esculenta Crantz. (Euphorbiaceae)	Cassava	Roots	Eye drops	Eye problems	Nigeria	[50]
Panicum deustum Thunb. (Poaceae)	Reed panicum	Leaves	Chewed leaves applied to wounds on eye lid	Trachoma	Kenya	[47]
Peltophorum africanum Sond. (Fabaceae)	African wattle	Root	Eye drops	Sore eyes	Zimbabwe	[30]

(Continued)

Table 16.3 (Continued)

Scientific Name/Family	Common Name	Plant Part Used	Administration	Traditional Usage	Area of Plant Collection	Reference
Pterocarpus angolesis DC. (Fabaceae)	Wild teak	Sap	Applied directly	Sore eyes	Zimbabwe	[30]
Ricinus communis L. (Euphobiaceae)	Castor oil plant	Seeds	Oil	Sore eyes	Zimbabwe	[30]
Sclerocarya birrea (A. Rich.) Hochst. (Anacardiaceae)	Marula	Roots	Steam	Sore eyes	Zimbabwe	[30]
Strychnos madagascariensis Poir. (Strychnaceae)	Black monkey orange	Roots	Extract drops	Sore eyes	Zimbabwe	[30]

Plant part

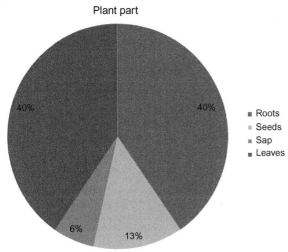

40%

40%

6%

13%

■ Roots
▨ Seeds
■ Sap
■ Leaves

Figure 16.6 Plant parts used for the treatment of eye disorders using African medicinal plants.

16.8 Conclusion

The use of traditional African medicinal plants for the treatment of various skin and eye disorders is an ongoing practice. The reasons for the use of traditional medicine are the inaccessibility of some regions to Western medicine, as well as the cost effectiveness of using traditional medicine as opposed to conventional medicine [53]. However, the main reason for the use of herbal remedies is their importance in the culture and tradition of African people [54]. Although ATMs are still widely used, there are still various side effects that can be caused by their use.

References

[1] Campbell NA, Reece JB. Biology. 7th ed. San Francisco, CA: Benjamin Cummings; 2005.
[2] Lee SH, Jeong SK, Ahn SK. An update of the defensive barrier function of skin. Yonsei Med J 2006;47(3):293−306.
[3] Albanesi C, Scarponi C, Giustizieri ML, Girolomoni G. Keratinocytes in inflammatory diseases. Curr Drug Targets Inflamm Allergy 2005;4:329−34.
[4] Parra JL, Paye M. EEMCO Group. EEMCO guidance for the *in vivo* assessment of skin surface pH. Skin Pharmacol Appl Skin Physiol 2003;16:188−202.
[5] Jett BD. Foundation: host defense against ocular infection, vol. 2, Chapter 45 Duane's opthalmology. Available from: <http://www.oculist.net/downaton502/prof/ebook/duanes/pages/v8/v8c045.html>; 2014.
[6] Irek M, Bozkurt B. Epithelial cells in ocular allergy. Curr Allergy Asthma Rep 2003;3 (4):352−7.
[7] Willetts GS. Ocular side-effects of drugs. Br J Ophthalmol 1969;53:252−62.

[8] Herbert AA, Ralston JP. Cutaneous reactions to anticonvulsant medications. J Clin Psychiatry 2001;62(14):22−6.

[9] Zhou SS, Li D, Zhou YM, Cao JM. The skin function: a factor of anti-metabolic syndrome. Diabetol Metabol Syndr 2012;4:15.

[10] Anzenbacher P, Anzenbacherova E. Cytochrome p450 and metabolism xenobiotics. Cell Mol Life Sci 2001;58(5−6):737−47.

[11] Janocovaa P, Anzenbacherb P, Anzenbacherovaa E. Phase II drug metabolizing enzymes. Biomed Pap Med Fac Univ Palacký Olomouc Czech Repub 2010;154 (2):103−11.

[12] Block KW, Lipp HP, Block-Hennig BS. Induction of drug-metabolizing enzymes by xenobiotics. Xenobiotica 1990;20:1101−11.

[13] Ahmad N, Mukhar H. Cytochrome p450: a target for drug development for skin disease. J Invest Dermatol 2004;123:417−25.

[14] Ames B. Micronutrients prevent cancer and delay aging. Toxicol Lett 1998; 102−103:5−18.

[15] Al-Gananeem AM, Crooks PA. Phase I and phase II ocular metabolic activities and the role of metabolism in ophthalmic pro-drug and co-drug design and delivery. Molecules 2007;12:373−88.

[16] Moorthy R, Valluri S. Ocular toxicity associated with systemic drug therapy. Curr Opin Ophthalmol 1999;10:438−46.

[17] Watkins J, Wirthwein DP, Sanders RA. Comparative study of phase II biotransformation in rabbit ocular tissue. Drug Metab Dispos 1991;19:708−13.

[18] de Ruijter A. *Plumbago auriculata* Lam. In: Schmelzer GH, Gurib-Fakim A, editors. Prota 11(1): medicinal plants/Plantes médicinales 1. CD-Rom. Wageningen, the Netherlands: PROTA; 2006.

[19] Van Niekerk C, Viljoen A. Indigenous South African medicinal plants: Part 11: *Aspalathus linearis* (Rooibos). SA Pharma J 2008;41−2.

[20] Mapunya MB, Nikolova RV, Lall N. Melanogenesis and antityrosinase activity of selected South African plants. Evid Based Complement Altern Med 2012;2012: 374017.

[21] Tag H, Kalita P, Dwivedi P, Das AK, Namsa ND. Herbal medicines used in the treatment of diabetes mellitus in Arunachal Himalaya, northeast, India. J Ethnopharmacol 2012;141:786−95.

[22] Naidoo KK, Coopoosamy RM. A comparative analysis of two medicinal plants used to treat common skin conditions in South Africa. Afr J Pharm Pharmacol 2011;5:393−7.

[23] Mabona U, Van Vuuren SF. Southern African medicinal plants used to treat skin disorders. S Afr J Bot 2013;87:175−93.

[24] Ajibesin KK. Ethnobotanical survey of plants used for skin diseases and related ailments in Akwa Ibom State, Nigeria. Ethnobot Res Appl 2012;10:463−522.

[25] Street RA, Prinsloo G. Commercially important Medicinal plants of South Africa: a review. J Chem 2013;2013:205048.

[26] Kumara Swamy HM, Krishna V, Shankarmurthy K, Abdul Rahiman B, Mankani KL, Mahadevan KM, et al. Wound healing activity of embelin iaolted from the ethanol extract of leaves of *Embelia ribes* Burn. J Ethnopharmacol 2007;109:529−34.

[27] Van Wyk BE, Van Oudtshoorn B, Gericke N. Medicinal plants of Southern Africa. 2nd ed. South Africa: Briza; 2000.

[28] Watt JM, Breyer-Brandwijk MG. The medicinal and poisonous plants of Southern and Eastern Africa. 2nd ed. London: Livingstone; 1962.

[29] Magee AR, Van Wyk BE, Van Vuuren SF. Ethnobotany and antimicrobial activity of sieketroos (*Arctopus* species). S Afri J Bot 2007;73:159−62.

[30] Maroyi A. Traditional use of medicinal plants in south-central Zimbabwe: review and perspective. J. Ethnobiol Ethnomed 2013;9:31−48.

[31] Steenkamp V, Fernandes AC, Van Rensburg CEJ. Screening of Venda medicinal plants for antifungal activity against *Candida albicans*. S Afr J Bot 2007;73:256−8.

[32] Babajide OJ, Mabusela WT, Greem IR, Ameer F, Weitz F, Iwuoha EI. Phytochemical screening and biological activity studies of five South African indigenous medicinal plants. J Med Plant Res 2010;2:1924−32.

[33] Cogne AL, Marston A, Mari S, Hostettmann K. Study of two plants used in traditional medicine in Zimbabwe for skin problems and rheumatism: *Dioscorea sylvatica* and *Urginea altissima*. J Ethnopharmacol 2001;75(1):51−3.

[34] Perumal Samy RP, Ignacimuthu S, Sen A. Screening of 34 Indian medicinal plants for antibacterial properties. J Ethnopharmacol 1998;62:173−82.

[35] Kamagaju L, Morandini R, Bizuru E, Nyetera P, Nduwayeza JB, Stévigny C, et al. Tyrosinase modulation by five Rwandese herbal medicines traditionally used for skin treatment. J Ethnopharmacol 2013;146(3):824−34.

[36] Gulluce M, Sahin F, Sokeman M, Ozer H, Daferera D, Sokeman A, et al. Antimicrobial and antioxidant properties of the essential oils and methanol extract from *Mentha longifolia* L. ssp. *longifolia*. Food Chem 2007;103:1449−56.

[37] Viljoen AM, Van Vuuren S, Ernst E, Klepser M, Demirci B, Baser H, et al. *Osmitopsis asteriscoides* (Asteraceae) the antimicrobial activity and essential oil composition of Cape-Dutch remedy. J Ethnopharmacol 2003;88:137−43.

[38] Van Wyk BE. A broad view of commercially important southern African medicinal plants. J Ethnopharmacol 2008;119:342−55.

[39] Van Wyk BE. A review of Khoi-San and Cape Dutch medical ethnobotany. J Ethnopharmacol 2008;119:331−41.

[40] Sofat BK, Sood GC, Chandel RD, Mehrotra SK. *Euphorbia royaleana* latex keratitis. Am J Ophthalmol 1972;74:634−7.

[41] Hsueh KF, Lin PY, Lee SM, Hsieh CF. Ocular Injuries from plant Sap of Genera *Euphorbia* and *Dieffenbachia*. J Chin Med Assoc 2004;67:93−8.

[42] Basak SK, Bakshi PK, Basu S, Basak S. Keratouveitis casued by *Euphorbia* plant sap. Indian J Ophthalmol 2009;57(4):311−3.

[43] Karini I, Yousefi J, Ghashghai A. Ocular toxicity caused by *Euphorbia* sap: a case report. Iran J Pharmacol Ther 2010;9(1):37−9.

[44] Shenoy R, Bialasiewicz A, Khandekar R, Barwani BA, Belushi HA. Tradiational medicine in Oman: its role pin ophthamology. Middle East Afri J Ophthalmol 2009;16(2):92−6.

[45] Abdul L, Abdul R, Sukul RR, Nazish S. Anti-inflammatory and antihistaminic study of a Unani eye drop formulation. J Ophthalmol Eye Dis 2010;2:17−22.

[46] Exe BI, Chuka-Okosa CM, Uche JN. Traditional eye medicine use by newly presenting ophthalmic patients to a teaching hospital in south-eastern Nigeria: socio-demographic and clinical correlates. BMC Complement Altern Med 2009;9:40−6.

[47] Klauss V, Adala HS. Traditional herbal eye medicine in Kenya. World Health Forum 1994;15:138−43.

[48] Courtright P, Lewallan S, Kanjaloti S, Divala DJ. Traditional eye medicine use among patients with corneal disease in rural Malawi. Br J Ophthalmol 1994;78:810−2.

[49] Mselle J. Visual impact of using traditional medicine on the injured eye in Africa. Acta Trop 1998;70:185−92.

[50] Osahon AI. Consequences of traditional eye medication in U.B.T.H Benin city. Niger J
      Ophthalmol 1995;3:51—4.
[51] Moshi MJ, Otieno DF, Mbabazi PK, Weisheit A. Ethnomedicine of the Kagera
      Region, north western Tanzania. Part 2: The medicinal plants used in Katoro Ward,
      Bukoba district. J Ethnobiol Ethnomed 2010;6:19—23.
[52] Lewallen S, Coutright P. Peripheral corneal ulcers associated with the use of African
      traditional eye medicines. Br J Opthalmol 1995;79:343—6.
[53] Light ME, Sparg SG, Stafford GI, Van Staden J. Riding the wave: South Africa's
      contribution to ethnopharmacological research over the last 25 years. J Ethnopharmacol
      2005;100:127—30.
[54] Fennell CW, Lindsey KL, McGaw LJ, Sparg SG, Stafford HI, Elgorashi EE, et al.
      Assessing African medicinal plants for efficacy and safety: pharmacological screening
      and toxicology. J Ethnopharmacol 2004;94:205—17.

17 Toxicity and Protective Effects of African Medicinal Plants on the Spleen and Lung

Armel J. Seukep, Doriane E. Djeussi, Francesco K. Touani, Aimé G. Fankam, Igor K. Voukeng, Jaurès A.K. Noumedem, Simplice B. Tankeo, Alfred Ekpo Itor, Ngueguim K. Glawdys and Victor Kuete

Department of Biochemistry, Faculty of Science, University of Dschang, Dschang, Cameroon

17.1 Introduction

The spleen and lungs are two organs of the upper abdominal cage. The human lungs are a pair of large, spongy organs optimized for gaseous exchange between blood and the air. The body requires oxygen to survive. The lungs provide vital oxygen and remove carbon dioxide before it can reach hazardous levels [1]. The spleen is a secondary lymphoid organ of vertebrates. Its structure is similar to the lymph node; it acts primarily as a blood filter and plays important roles in regard to red blood cells and the immune system. It removes old red blood cells and holds a reserve of blood, which can be valuable in case of hemorrhagic shock, and it also recycles irons. As a part of the mononuclear phagocyte system, it metabolizes hemoglobin removed from senescent erythrocytes [2,3]. Plant-based drugs are widely used worldwide and particularly in Africa for their diverse curative activities, including organ protective effects [4]. Their toxicity toward the spleen and lungs is usually neglected. After administration, xenobiotics reach the systemic circulation and can act on lungs and spleen. This chapter reviews available data related to the toxic and protective effects of African medicinal plants on the lung and spleen. It also provides updated literature on the anatomy and physiology of the two organs, as well as their roles in the metabolism of xenobiotics.

Toxicological Survey of African Medicinal Plants. DOI: http://dx.doi.org/10.1016/B978-0-12-800018-2.00017-0

17.2 The Anatomy and Physiology of the Spleen and Lung

17.2.1 Anatomy and Physiology of the Spleen

The spleen is a large vascular lymphatic organ lying in the upper part of the abdominal cavity on the left side between the stomach and diaphragm, composed of white and red pulps; the white consists of lymphatic nodules and diffused lymphatic tissue; the red consists of venous sinusoids between which are splenic cords. The stroma of both red and white pulps are reticular fibers and cells. A framework of fibro-elastic trabeculae extending from the capsule subdivides the structure into poorly defined lobules. In healthy adult humans, the spleen is approximately 7−14 cm in length with a weight ranging between 150 [5] and 200 g [6]. The spleen develops in the cephalic part of dorsal mesogastrium (from its left layer, during the sixth week of intrauterine life) into a number of nodules that fuse and form a lobulated spleen. Notching of the superior border of the adult spleen is evidence of its multiple origins [7]. Its possess two ends of which the anterior is expanded and is more like a border directed forward and downward to reach the midaxillary line. The posterior end is rounded and is directed upward and backward; the spleen's three borders are the superior, inferior, and intermediate. The superior border is notched by the anterior, while the inferior border is rounded and the intermediate border directs toward the right.

The two surfaces of the spleen are the diaphragmatic and visceral. The diaphragmatic surface is smooth and convex while the visceral surface is irregular, concave with impressions. The gastric impression is the largest and most concave, for the fundus of the stomach. The renal impression is for the left kidney and lies between the inferior and intermediate borders. The colic impression is for the splenic flexure of the colon. Its lower part is related to the phrenicocolic ligament. The pancreatic impression for the tail of the pancreas lies between the hilum and colic impression. The splenic circulation is adapted for the separation and storage of the red blood cells, as this organ has superior and inferior vascular segments based on the blood supply. The vascular system traverses the spleen and permeates it [8]. Approximately 350 L of blood pass through the spleen per day. The arterial supply of the spleen comes from the splenic artery which reaches the spleen's hilum by passing through the splenorenal ligament. It divides into multiple branches at the hilum into straight vessels called penicillins, ellipsoids, and arterial capillaries in the spleen. Its terminal branches besides the splenic artery also gives off branches to the pancreas, five to seven short gastric branches, and the left gastro-omental (gastroepiploic) artery. Sympathetic fibers are derived from the celiac plexus [9]. The splenic vein is formed by the union of five or more veins that emerges from the hilum. It runs the lienorenal ligament, to the right and across the front of the left kidney, the left diaphragmatic crus and the aorta, lying in groove in the back of pancreas. It usually receives the inferior mesenteric vein and ends behind the neck of the pancreas by joining the superior mesenteric vein. There is no evidence of lymphatic vessels within the parenchyma of the spleen. However, some arise from the capsule and trabeculae and drain to the pancreaticosplenic lymph nodes.

17.2.2 Anatomy and Physiology of the Lungs

The anatomy and physiology of the respiratory tract is quite complex (Figure 17.1). Each anatomic segment performs in concert with the others and is accountable for a wide variety of physiological responsibilities. The lungs consist of right and left sides. The right lung has three lobes namely upper lobe, middle lobe, and lower lobe, while the left lung has two lobes known as the upper lobe and lower lobes. On its medial side, the left lung has a concavity, the cardiac notch which is a concave impression molded to accommodate the shape of the heart which sits in the mid-chest extending into the left side [11]. Each lobe is surrounded by a pleural cavity, which helps to lubricate the lungs, as well as providing surface tension to keep the lung surface in contact with the rib cage. The lung parenchyma denotes alveolar tissue with respiratory bronchioles, alveolar ducts, and terminal bronchioles, and often includes any form of lung tissue. This also includes bronchioles, bronchi, blood vessels, and lung interstitium. Two large tubes called bronchi start from the trachea, separate, and distribute air to the left and right sides of the lungs. The tubes gradually form more generations, like branches of a tree which become smaller and

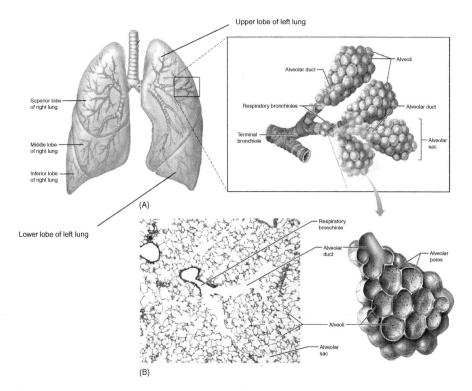

Figure 17.1 Respiratory zone structures. (A) Diagrammatic view of respiratory bronchioles, alveolar ducts, alveolar sacs, and alveoli. (B) Photomicrograph of a section of human lung, showing the respiratory structures that form the final divisions of the bronchial tree [10].

smaller, as they spread to the ends of the lungs they eventually form a grape-like structure known as the alveoli. The lungs contain approximately 2400 km of airways and 300−500 million alveoli, having a total surface area of about 70−100 m^2 in adults [12]. The diaphragm is a large dome-shaped muscle that contracts and relaxes during breathing. It also separates the chest and abdominal cavity. Muscles near the ribs also help expand the chest for breathing. Air reaches the alveoli (air sacs) where oxygen then moves from the air sacs into the capillaries through their thin walls. Capillaries are tiny blood vessels that carry oxygenated blood to the bloodstream that supplies the body. Carbon dioxide is released from the blood to the lungs and exhaled, then moves from capillaries into the alveoli.

Pulmonary arteries from the right side of the heart supply blood to the lungs. There are about 1 billion capillaries in the lungs surrounding alveoli. The blood in the capillaries is separated from the air in the alveoli only by the extremely thin alveolar and capillary walls which allow gases to be exchanged between the blood and the lungs in a process called respiration. Inhaled oxygen enters the blood from the alveoli. Carbon dioxide leaves the blood and enters the alveoli to be exhaled. The lungs are traversed by another important system of vessels called the lymphatics [13]. Pulmonary veins form post alveoli to carry oxygenated blood back to the heart.

Each lung contains a network of lymphatic vessels that carry a mixture of fluid and proteins called lymph. Lymph is carried from the lung tissues through a series of lymph nodes to filter the fluid before it is returned to the bloodstream. This network of lymph vessels and lymph nodes is an important part of the body's immune system [13].

17.3   The Roles of the Spleen and Lung in the Metabolism of Xenobiotics

17.3.1   The Role of the Spleen in the Metabolism Xenobiotics

The process of pharmacokinetic or the kinetics of drugs is gathered in four phases: absorption, distribution, metabolism, and excretion (ADME concept). Drug metabolism, also known as xenobiotic metabolism, is the biochemical modification of pharmaceutical substances or xenobiotics respectively by living organisms, usually through specialized enzymatic systems [14]. The metabolic transformations which most xenobiotics undergo in the body are numerous and diverse, but they may be conveniently classified into two main phases. In phase I, compounds may undergo either oxidation, reduction, or hydrolysis reactions generally resulting in the addition or exposure of functional groups, such as −OH (hydroxyl group) or COOH (carboxylic group), which may then undergo phase II or synthetic (conjugation) reactions. The net result of such phase I and II metabolism is often the production of more polar, less lipid soluble, more readily excretable, and less biologically active compounds [15]. These reactions often act to detoxify poisonous compounds. However, in some cases, the intermediates in xenobiotic metabolism can themselves be the cause of toxic effects. The rate of metabolism determines the duration and intensity of a drug's pharmacological action. Quantitatively, the smooth endoplasmic reticulum of the liver cell is the principal organ of drug metabolism, although every biological tissue such as spleen has some ability to metabolize drugs. Similar in structure to a large lymph node, the spleen acts primarily as a blood filter [2].

For intravenously administered drugs, the spleen can be considered the draining lymphoid tissue. Following opsonization, the macromolecules undergo phagocytosis by macrophages and Kupffer cells. The uptake by reticuloendothelial cells takes place within seconds after opsonization [16], and it is responsible for the fast clearance of the drug delivery systems from the blood and their accumulation in the spleen, and to a lesser degree in the lung, kidney, and lymph nodes. After phagocytosis, the nanomaterials have different metabolism and excretion pathways according to their specific characteristics, such as biodegradability, size, and other properties. The metabolism of the polymer and its elimination from the body are two very important features of a conjugate design, yet they are often overlooked. A successful conjugate will not only deliver the drug to its target and release it there, but it will also be eliminated from the body, preventing an iatrogenic illness, emerging from the accumulation of the polymer in the tissue or in the clearance organs of the spleen or others organs (kidneys, liver) [17]. The spleen contains the enzyme UDP-glucuronosylltransferase, a phase II enzyme. It plays a role in glucuronidation of products of the mixed-function oxidase reactions. This family of enzyme catalyzes the transfer of glucuronic acid to a multitude of xenobiotics compounds, including drugs [15].

17.3.2   The Role of the Lung in the Metabolism of Drugs

The lung is the major portal of entry into the body of all inhaled compounds such as those contained in polluted city air, tobacco smoke, and medicinal and household aerosols. As the entire cardiac output passes through the lungs, all chemicals such as drugs or xenobiotics present in the circulation also must traverse the lungs [18]. Delivering therapeutic agents to the lungs requires a deep understanding of the kinetics and dynamics of drugs in this biologically and physiologically complex system. The role of the lungs in the metabolism of drugs (xenobiotics) is discussed in this chapter. The lung is exposed to a wide variety of exogenous chemicals (xenobiotics) that may exert profound pharmacological, therapeutic, or toxicological actions. These chemicals may be metabolized by the lung, possibly increasing or decreasing their toxicity or pharmacological action [19]. The enzymes responsible for the pulmonary metabolism of these xenobiotics include both mixed-function oxidases and enzymes catalyzing conjugation reactions [18]. Pulmonary drug metabolism depends on an interplay between factors occurring prior to target interaction (deposition, dissolution, and mucociliary transport), and events competing with target interaction (biotransformation, receptor interaction, and nonspecific retention), and following target interaction (systemic absorption). An inhaled drug substance may be eliminated from the lung by metabolism in the mucus or lung tissue [20]. This mechanism may act in parallel with others such as mucociliary or cough clearance to the gastrointestinal tract, passive or active absorption into the capillary blood network and have an important role in determining the drug's duration of action in the lungs and the airway selectivity of the inhaled drug [21].

The body's primary detoxification enzymes, the cytochrome P450 (CYP) families, show the highest expression levels in hepatocytes and enterocytes but also are expressed in the lungs and other organs, providing a line of defense against ingested or inhaled xenobiotics. The human drug-metabolizing enzymes belong to the families CYP1, CYP2, and CYP3. These enzymes catalyze an oxidation reaction in which a functional group is introduced to the molecule, serving as a site of conjugation, increasing hydrophilicity and thereby facilitating elimination [22]. In the lungs, several CYP isoforms are expressed as well as other biotransformation enzymes such as sulfotransferases, UDP glucuronosyl transferases, glutathione S-transferases, esterases, peptidases, cyclo-oxygenases, flavine mono-oxygenases [23,24]. The wide range of biotransformation enzymes enables metabolism of a broad spectrum of chemically different substrates. Local metabolism of several inhaled compounds, pharmaceutical drugs as well as tobacco smoke components, pollutants, and toxicants have been demonstrated in lung tissue [25,26].

The lung is a very complex organ composed of at least 40 different cell types with different expression patterns of metabolizing enzymes [20]. The balance of activating and deactivating enzymes and their cofactors in a particular cell type in the lung may in part determine the cellular specificity of the xenobiotic. It was shown that much of the total lung content of cytochrome P450, the mixed-function oxidase responsible for the activation of many chemicals to reactive metabolites, was localized in the nonciliated bronchiolar epithelial cells (also called Clara cells). This provided a metabolic basis for the susceptibility of this cell type to many pulmonary toxins including 4-ipomeanol, 3-methylfuran, and naphthalene [18].

The drug-metabolizing capacity of the lungs is in general substantially lower than that of the liver, and there is little evidence of a major contribution to systemic clearance. In most cases, *in vitro* enzymatic activities of liver are significantly higher than those of lung, but these differences may be offset *in vivo* by other factors such as blood flow and distribution. This can be observed with compounds such as certain amines that are concentrated in lung tissue [27]. The contribution of the lung to the overall metabolism of the compound may be far greater than can be determined from *in vitro* enzymatic activities [18].

Many small molecules have near complete bioavailability via lung absorption as a consequence of low metabolic activity and relatively rapid absorption [28−30]. Several drugs such as budesonide, ciclesonide, salmeterol, fluticasone propionate, and theophylline are substrates to enzymes present in the lungs [31−35]. Because formation of local metabolites cannot be excluded, and are sometimes anticipated, studies are conducted during drug development to screen for lung metabolism and assess the risks for metabolic interactions and toxicity [20]. However, local metabolism in the lungs can be taken advantage of, as several of the drugs listed above demonstrate. One example is the use of prodrugs where the administered form of the drug is activated *in situ* by metabolic enzymes. The reversible pulmonary biotransformation (e.g., ciclesonide) is metabolized into its active form by esterases in the lungs and further reversibly conjugated to fatty acids [33]; for budesonide, the conjugation to fatty acids in the lungs results in the formation of a compound that has a substantially lower elimination rate [35] contributes to the prolonged lung retention and duration of effect observed with a compound [35,36]. Most pulmonary drug-metabolizing enzymes are inducible as a response to increased exposure to their substrates. An organism adapts to the new situation with an increased metabolic capacity in order to handle the "threat" [37−39]. By contrast, suppression of enzyme levels can also occur as a response to xenobiotics or disease [40,41]. An important example of enzyme induction that changes the metabolizing capacity of the lungs is the effects of tobacco smoke. The induction of drug-metabolizing enzymes may result in increased metabolism of drugs, which could result in impaired therapeutic effect. Clinically relevant induction of CYP1A1 activity has been observed in patients who smoke during treatment with theophylline (alkaloid extracts from the sheets of tea and using in the composition of drugs against asthma) compared with nonsmokers [42]. Several compounds have the ability to inhibit drug-metabolizing enzymes and thereby cause clinically adverse interactions [43]. Drug interactions resulting from inhibition are however not likely to have a major significance in the lungs because of the limited contribution of lung metabolism to systemic clearance, although the local metabolism pattern could be affected by such interactions [20]. Because of the pioneering work of the Millers and many other groups, it is now recognized that many xenobiotics require metabolic activation before exerting their pharmacological effect or toxicity [19]. This activation may take place in the lung. The possible contribution of extra-pulmonary metabolic activation of drugs to metabolites that are toxic to the lung will depend in part on the relative stability of such metabolites and the inherent susceptibility to toxicity of the lung compared to other tissues. Alternatively a stable, proximate, toxic metabolite may be formed in an extra-pulmonary site, transported to the lung, and then metabolized to the ultimate toxic form. Finally, the lung may be responsible for the formation of the toxic metabolite from the parent xenobiotic [18,19].

17.4 Overview of the African Plants Known to Affect the Spleen and Lung

17.4.1 Medicinal Plants Screened Out of Africa with Hurtful Effect on the Spleen

Medicinal plants extract and derived products offer unlimited opportunities for the discovery of new drugs. Most of the natural products used in folk remedy have solid scientific evidence in regard to their biological activities. However, there is little information or evidence available concerning the possible toxicity that medicinal plants may cause to the consumers [44]. Some plants with toxic effect include *Digitalis purpurea* (Plantaginaceae), *Ricinus communis* (Euphorbiaceae), and *Allium ascalonicum* (Liliaceae).

Digitalis purpurea is an attractive biennial plant that is cultivated in Canada and is naturalized in several provinces. Upon ingestion, this plant can cause toxic reactions that lead to severe sickness and death in animals and in humans. Postmortem examination of pigs showed gastrointestinal inflammation, punctiform necrosis of the border of the spleen, and fatty degeneration of some nerve fibers in the heart. The presence of digitoxin (an important pharmaceutical drug derived from *Digitalis purpurea*) in the body tissues confirms *Digitalis purpurea* poisoning in animals [45]. Many studies have demonstrated the toxicity of ricin on the spleen. It is a naturally occurring lectin (a carbohydrate-binding protein) produced in the seeds of the *Ricinus communis*. When administered [46], it was observed that ricin triggers complete relocation of T and B lymphocyte populations, with destruction of B cells. In addition, a study on geese found that onion (*Allium ascalonicum*) killed them after they have eaten it, with an autopsy revealed swelling of the liver and spleen [47].

17.4.2 Medicinal Plants Screened Out of Africa with Protective Effect on the Spleen

Generally, spleen disorders are usually of a deficient nature, requiring the tonification of central "*qi*" and the stimulation of the ascending movement of clear "*yang qi*." Herbs that eliminate pathological dampness are common spleen therapies. Some of them include *Curcuma longa* (Zingiberaceae), *Astragalus* spp. (Fabaceae), *Dioscorea oppositifolia* (Dioscoreaceae), *Radix glycyrrhizae* (Apiaceae), *Ziziphus jujuba* (Rhamnaceae) and *Biota orientalis* (Cupressaceae), *Atractylodes macrocephala* (Asteraceae), *Pinellia ternata, Citrus reticulatus* (Rutaceae), and *Glycyrrhizae uralensis* (Fabaceae).

Turmeric is a product of *Curcuma longa*. It is a rhizomatous herbaceous perennial plant native to tropical South Asia. Turmeric showed antioxidant potential by lowering oxidative stress in animals. A study showed that a diet containing 0.1% turmeric fed for 3 weeks to retinol-deficient rats lowered lipid peroxidation rates by 18.0% in the spleen [48].

Astragalus (*Astragalus* spp.) is one of the most famous tonic herbs from China. In traditional Chinese medicine (TCM), it is said to tonify the blood and spleen and aid the defensive "*qi*." *Astragalus* is said to protect and enhance the functioning of distressed organs [49]. Numerous studies showed that the herb enhances immune function by increasing natural killer cell activity, [50] increasing T-cell activity [51]. The tuber of *Dioscorea oppositifolia* is used as herbal tonic which stimulates the stomach and spleen [52].

Radix glycyrrhizae properties are sweet, neutral; it tonifies the spleen and strengthens the "*qi*." While *Ziziphus jujuba* and *Biota orientalis* are used to support several channels such as the spleen, kidney, and large intestine. Also, *Angelica sinensis* is used to support the spleen and liver. It stimulates appetite and the immune system. Rhizoma *Atractylodes macrocephalae* properties include nontoxic, sweet, and warm. It supplements spleen and "*qi*" and is used in the treatment of indigestion and stomach disorders [53].

On the other hand, combination of some herbs in herbal therapy is also used to maintain the spleen in good health. Such combination include *Pinellia ternata, Citri reticulatus, Glycyrrhizae uralensis* and it is used to dry Damp, regulate "*qi*," and harmonize the middle warmer (stomach and spleen) [54].

17.4.3   Medicinal Plants Screened Out of Africa with Hurtful Effect on the Lung

Several phytochemicals were found to display hurtful effect on the lungs. *Jatropha curcas* (Euphorbiaceae) seeds at certain concentrations of diet induces pathological changes in the lungs, small intestines, liver, and heart [55]. *Verbesina encelioides* (Asteraceae) compromised respiratory function characterized by rapid respiration and frothly exudate from the nacres [56]. The pulmonary toxicity of sweet potatoes (*Ipomoea batatas*; Convolvulaceae) originated from India has been ascribed to a lung edema factor which is produced in the roots in response to microbial infection [57]. Pyrrolizidine alkaloids (PA) detected in some *Crotalaria* species are toxic in the lung and liver. An example of PA with such effects includes monocrotaline, a pneumotoxin, producing pulmonary arterial hypertension and right ventricular hypertrophy [58].

17.4.4   Medicinal Plants Screened Out of Africa with Beneficial Effect on the Lung

Although certain plants have toxic effects on the lungs, there are those which have beneficial effects. TCM contains numerous ingredients that they may be considered to have anti-lung cancer activity as well as serving as immunomodulators [59]. For examples, acetone extract of *Angelica sinensis* (Apiaceae) induced the activation of caspase 9/3 mediated by the suppression of Bcl-2 and cdk 4 expression in the A549, HT29 and J5DBTRG-05MG lung cancer cells lines. The ethanolic extract of *Scutellaria barbata* (Lamiaceae) activates caspase 3/7 [59]. Some Thai medicinal plants used as the ingredients of a Southern Thai traditional formula for cancer treatment were selected to test cytotoxicity activity against large cell lung carcinoma [60]. Other plants such as *Bridelia ovata* (Phyllanthaceae), *Curcuma zedoaria* (Zingiberaceae), *Derris scandens* (Fabaceae), *Dioscorea membranacea* (Dioscoreaceae), *Nardostachys jatamansi* (Valerianaceae), and *Rhinacanthus nasutus* (Acanthaceae) which displayed cytotoxic activity toward lung carcinoma [60]. *Pulmonaria officinalis* (Poraginaceae), usually called lungwort and found in Europe, is used to treat lung diseases (tuberculosis, bronchopneumonia) [61]. Some Native Americans made roots tea of *Helianthus annuus* (sun flower; Asteraceae) cultivated in North America to treat various lung problems [61]; *Angelica sinensis* called angelica was used in Europe and America to treat respiratory diseases, as well as a variety of other hollow-stemmed plants. Eucalyptus oil (*Eucalyptus globulus*; Myrtaceae) extracted from the leaves of this plant found in Australia helps open up respiratory passages in colds, bronchitis, and other respiratory conditions [61]. *Scutellaria baicalensis* (Lamiaceae) also known as Chinese skull-cap or *Huang-qin* is used to treat a wide variety of conditions, including respiratory infection conditions [61]. It also is used to treat diseases, including improving brain function, lowering blood pressure and cholesterol, diuresis, and improving digestion. Alantolactone compound contents in *Inula helenium's* (Asteraceae) roots shows some action against the bacillus that causes tuberculosis, according to German research in 1999, and it had been used in the past as a treatment for tuberculosis [61]. *Adenophora stricta* (Campanulaceae; found in Europe and Asia) is a stimulant for the respiratory and cardiac systems and has been used to treat tuberculosis, chronic bronchitis, and dry cough. A mucilaginous substance in the fresh leaves of *Tussilago farfara* (Asteraceae) was used for cough remedies. Ancient and modern herbalists have recommended smoking the dried leaves of this plant to treat cough and symptoms of asthma and chronic bronchitis. The herb is still used in Europe for acute respiratory conditions. Chronic usage may result in liver damage. The rhizomes of *Belamcanda chinensis* (Iriddaceae) are used in the TCM for asthma, croup, swelling, and pain in the throat, and several other conditions [61].

17.5   African Plants Affecting the Spleen

17.5.1   Harmful Effects of African Plants on the Spleen

The harmful effects of medicinal plants have been demonstrated in animal models [62,63]. Several African plants were found to affect the body organs like liver, lung, heart, brain, kidney, and spleen [63]. The harmful effects on the concerned organ are generally identified through histological observation and mainly the changes in their relative weight [64,65], increase of the total number of nucleated cells [66], hemorrhage [63], etc. Table 17.1 summarizes the harmful effects of African plants on the spleen.

17.5.2   Protective Effects of African Plants on the Spleen

Despite the scarcity of published data, many African medicinal plants are traditionally used to protect the spleen [80,81]. Table 17.2 provides the evidence-based data on two plants having protective effects on the spleen.

17.6   African Plants Affecting the Lung

17.6.1   Harmful Effects of African Plants on the Lung

Several African plants were reported to have hurtful effects on the lungs. Some of them include *Ocimum gratissimum*, *Crotalaria dura*, *Lasiospermum bipinnatum*, *Azadirachta indica*, *Turraeanthus africanus*, *Sacoglottis gabonensis*, and *Azadirachta indica* (Table 17.3).

Table 17.1 African Medicinal Plants with Harmful Effect on the Spleen

Plant (Family)	Traditional Use	Area of Plant Collection	Observed Effects
Hydnora johannis (Hydnoraceae)	Diarrhea, cholera, dysentery, and swelling tonsillitis [67].	Sudan	Root dry powder and ethanolic extract showed toxic effect probably by dose dependent and/or frequency of administration on Wistar rats [68].
Cylicodiscus gabunensis (Mimosaceae)	Chewing stick, diarrhea, gastrointestinal disorder, rheumatism, filariasis, and headache [69].	Cameroon	The ethyl acetate bark extract affects the spleen, judging by the significant change in its relative weight suggesting possible immunotoxic effect [62].
Khaya ivorensis (Meliaceae)	Malaria [70].	Nigeria	The ethanol stem bark extract was demonstrated to exhibit a severe effect on the spleen, when compared with the weight of the controlled rats [70].
Ficus exasperate (Moraceae)	Management of cardiac arrhythmias, asthma, bronchitis, tuberculosis, emphysema, chest pain, eye troubles and stomach pains, bleeding arrest [71].	Nigeria	Toxicity studies in rats at doses of $0.1-1$ g/kg bwt showed that the aqueous crude extracts of *Ficus exasperate* decreased spleen weight [72].
Berlina grandiflora (Leguminosae)	Gastrointestinal disorders, chewing stick, constipation; purgative [64].	Nigeria	The aqueous methanolic stem bark extract affected spleen judging by the significant changes in its relative weight at 250 and 500 mg/kg [64].
Azadirachta indica (Maliaceae)	Trypanosomosis [73]	Nigeria	The ethanolic extract of *Azadirachta indica* stem bark induced enlargement of the spleen in the rats infected by *Trypanosoma brucei* [73].

(Continued)

Table 17.1 (Continued)

Plant (Family)	Traditional Use	Area of Plant Collection	Observed Effects
Dichapetalum barteri (Dichapetaceae)	–	Nigeria	The spleen of treated animals by ether extract showed vasculitis and perivasculitis with round mononuclear cells [74].
Murraya koenigii (Rutaceae)	Tonic, febrifuge, stomachic, vomiting, dysentery, diarrhea, skin eruptions, bites of venomous animals, stimulant, diabetes [75].	Nigeria	The possible toxicity of the plant methanolic extract was indicated by changes in spleen weight through chemical injury induced by the administered extract to the cells and tissues [75].
Azadirachta indica (Meliaceae)	–	Nigeria	Atrophy of the spleen due to the reduction of their absolute and the computed organ-body weight ratio by the ethanolic extract of *Azadirachta indica* stem bark [65].
Terminalia avicennioides (Combretaceae)	Astringent, dysentery laxative and diuretic [63].	Zaria, Nigeria	The butanol extract of *Terminalia avicennioides* revealed lesions (mainly necrosis, organ enlargement and hemorrhage) were observed in the liver, spleen and kidney at 13 g/kg bwt [63].
Morinda lucida (Rubiaceae)	Malaria, febrifuge, analgesic, laxative, and infections [76]; diabetes, hypertension, cerebral congestion, dysentery, stomachache, ulcers, leprosy, and gonorrhea [77]; diarrhea, febrifuge, analgesic and laxative; irregular menstruation, insomnia, and jaundice [66].		After treating by bark aqueous extracts of *Morinda lucida*, the total number of nucleated cells in the spleen of malaria-infected mice increased enormously before the animals died [66].

(Continued)

Table 17.1 (Continued)

Plant (Family)	Traditional Use	Area of Plant Collection	Observed Effects
Hibiscus sabdariffa (Malvaceae)	Diuretic, stomachic, laxative, aphrodisiac, antiseptic, astringent, cholagogue, sedative, hypertension, and cardiac diseases [78].	Ibadan, Nigeria	A significant reduction in the weight of the spleen of the animals administered with ethanol and water extracts was observed [79].

bwt, body weight.

Table 17.2 African Medicinal Plants with Protective Effect on the Spleen

Plant (Family)	Traditional Use	Area of Plant Collection	Observed Effects
Carica papaya (Caricaceae)	Warts, cancer, tumors, and injurations of the skin [82,83]; nervous pains, elephantoid growth, and asthma [84].	Nigeria	The ethanolic extract of unripe pulp of *Carica papaya* may be protective (spleen and brain were well protected at 500 mg/kg bwt of the extract) against potassium bromate induced tissue damage in Wistar rats [80].
Vernonia amygdalina (Asteraceae)	Malaria, purgative, parasites, treatment of eczema, maintaining healthy blood glucose levels [85].	Nigeria	The ethanolic extract of *Vernonia amygdalina* has a protective potential against tissue damage induced by potassium bromated by significantly lowering body weight ratio of the organ, total protein level, malondialdehyde concentration, and amino acid level [81].

bwt, body weight.

Table 17.3 African Medicinal Plants with Harmful Effect on the Lung

Plant (Family)	Traditional Use	Area of Plant Collection	Observed Effects
Ocimum gratissimum (Lamiaceae)	Malaria, convulsion, stomach pain, and catarrh [86].	Nigeria	Aqueous extract of has shown adverse effect on the lungs of the albino rat when administered at high doses [86].
Crotalaria dura (Fabaceae)	–	South Africa	Associated with chronic pulmonary disease in horses and mules [87].
Lasiospermum bipinnatum (Asteraceae)	–	South Africa	It causes bronchiolar dilatation, and interstitial pneumonia characterized by proliferation of nonciliated epithelial cells to cattle and sheep [88].
Azadirachta indica (Meliaceae)	–	Nigeria	Ethanolic extract of stem bark causes hypertrophy of rat's lungs compared to the control when administered at 50, 100, 200, and 300 mg/kg [65].
Turraeanthus africanus (Meliaceae)	Infertility, sexual weakness, fibroma, and cancer [89].	Cameroon	Histology of lungs presented hemorrhage, alveoli necrosis, and cell infiltration after treatment with water extract [89].
Sacoglottis gabonensis (Humiriaceae)	–	Nigeria	Water extract of stem bark causes congestion and edema of lungs, bronchi, and bronchioles in rats when administered by intraperitoneal way [90].
Azadirachta indica (Maliaceae)	Tuberculosis and skin infections, pests, malaria [73]	Nigeria	Intraperitoneal administration of ethanol extract causes congestion and edema of the trachea, bronchi, bronchioles, lungs [73].

17.6.2   Protective Effects of African Plants on the Lung

A number of African plants were reported for their health benefit action on the lungs. In Table 17.4, a synopsis of some of these plants is provided.

17.7   Conclusion

The lung and spleen are two minor organs involved in the metabolisms of xenobiotics. Nonetheless, damage induced by drug could alter their function and induce

Table 17.4 African Medicinal Plants with Protective Effect on the Lung

Plant (Family)	Traditional Use	Area of Plant Collection	Observed Effects
Entada africana (Mimosaceae)	Bronchitis, cough, whooping cough dysentery, fever, wound [91].	Nigeria	Water extract of the roots reduces the number of cough induced by citrus; histamine-induced bronchoconstriction was also reduced [91].
Pelargonium sidoides (Geraniaceae)	Treatment of respiratory tract infections [92]; tuberculosis, coughs, colds, sore throat.	South Africa	A modern aqueous-ethanolic formulation of the roots (Umckaloabo®) is effective due to antibacterial activities and/or stimulation of the nonspecific immune system [92]; clinical trials shown that aqueous-ethanolic modern formulation of *Pelargonium sidoides* have shown efficacious results during the treatment of bronchitis and sinusitis [87].
Trichilia emetica (Emiliaceae)	Mixed with milk or honey, used against asthma, cough, bronchial trouble, fever, and vomiting [93].	Mali	Polysaccharides extract when given by peroral administration reduces the number of cough efforts [93].
Artemisia afra (Asteraceae)	Coughs, colds, whooping cough, bronchitis, asthma, fever, mumps swelling, pneumonia, pimples, and skin rashes [94].	South Africa	Water, methanol and dichloromethane extract of leaves possess moderate activity against *Klebsiella pneumonia* and *Mycobacterium aurum* [94].
Crossopteryx febrifuga (Rubiaceae)	Fever, respiratory tract diseases, bronchitis, cough, bronchopneumonia, cough, inflammation of oropharynx, and hiccough [95].	Mali	Water extract inhibited citric acid-induced cough [95].

(Continued)

Table 17.4 (Continued)

Plant (Family)	Traditional Use	Area of Plant Collection	Observed Effects
Prunus africana (Rosaceae)	Chest pain, diarrhea, genito-urinary complaints, allergies, inflammation, kidney diseases, malaria, fever, stomachache [96].	Kenya, South Africa	Compounds contained in *Prunus africana* extract are believed to block leukotrienes, reduce edema and inflammation, as well as diminished histamine and therefore ease asthmatic symptoms [97].
Warburgia ugandensis (Canellaceae)	Expectorant, fever, malaria, stomachache, constipation, and diarrhea [98].	Kenya	The water extract induces the serum IgE level and BALF eosinophil percentage [97].
Tulbaghia violacea (Alliaceae)	Colds, fever, asthma, tuberculosis, stomachache [94].	South Africa	Dichloromethane extract has moderate activity on *Klebsiella pneumoniae* and *Mycobacterium aurum* [94].
Abrus precatorius (Fabaceae)	Asthma, aphrodisiac, diabetes, cough, fever, sexually transmitted disease, schistosomiasis, stomach troubles [99,100].	Tanzania, Ghana	Methanolic extract avoid contractions induced by histamine and acetylcholine and possess bronchodilator activity [99].

severe side effects in the whole body. In this chapter, we compile data related to the protective roles as well as the harmful effects of plants used in the traditional African medicine to treat a variety of ailments. The reported data brings awareness to persons using herbal treatments to avoid those plants with damaging effects on the spleen and lungs, which can affect the entire body system.

References

[1] Taylor T. Lungs and respiratory system of the chest: lungs. In: Anatomy systems, <http://www.innerbody.com/anatomy/respiratory/lungs>; 1999 [accessed on 06.03.14].
[2] Mebius RE, Kraal G. Structure and function of the spleen. Nat Rev Immunol 2005;5:605−16.
[3] Steiniger B. Spleen. Encyclopedia of life sciences. Chichester: John Willey & Sons; 2005 <www.els.net>.

[4]   Amri E. The role of selected plant families with dietary ethnomedicinal species used as anticancer. J Med Plants Stud 2014;2(1):28−39.

[5]   Draper DJ, Sacher RA, Dessypris EN, Kaplan LJ. eMedicine, Splenomegaly. Updated: October 4, 2009, <http://www.emedicine.com/med/topic2156.htm> [accessed on 07.01.2014].

[6]   Spielmann AL, DeLong DM, Kliewer MA. Sonographic evaluation of spleen size in tall healthy athletes. Am J Roentgenol 2005;184:45−9.

[7]   Sadler TW. Digestive system. Langman's medical embryology. 11th ed. Philadelphia, PA: Lippincott Williams & Wilkins; 2009 [chapter 14] p. 215−16.

[8]   Snell RS. 5-The abdomen. In part II. The abdominal cavity. Clinical anatomy by regions. 8th ed. Baltimore, MD: Lippincott Williams & Wilkins; 2007. p. 259−60.

[9]   Lee McA, Decker GAG, Du Plessis DJ. 8-The spleen. Lee McGregor's synopsis of surgical anatomy. 12th ed. Oxford, UK: Butterworth−Heinemann; 1986. p. 106−13.

[10] Marieb EN, Hoehn K. 22-The respiratory system human. In anatomy & physiology. 9th ed. Redwood City, CA: Pearson; 2012.

[11] Tomco R. Lungs and mechanics of breathing. In: AnatomyOne. Amirsys, Inc. Retrieved 2012-09-28, <https://app.anatomyone.com/login?destination=%2Fsystemic %2Frespiratory-system%2Flungs-and-mechanics-of-breathing>.

[12] Notter RH. Lung surfactants: basic science and clinical applications. New York, NY: Marcel Dekker; 2000. p. 120.

[13] NMA (Nucleus Medical Art). The lungs and respiratory system, <www.nucleusinc. com>; 2005 [accessed on 4.03.14].

[14] Mizuno N, Niwa T, Yotsumoto Y, Sugiyama Y. Impact of drug transporter studies on drug discovery and development. Pharmacol Rev 2003;55(3):425−61.

[15] Parke DV. The biochemistry of foreign compounds. Oxford: Pergamon Press; 1968.

[16] Panagi Z, Beletsi A, Evangelatos G, Livaniou E, Ithakissios DS, Avgoustakis K. Effect of dose on the biodistribution and pharmacokinetics of PLGA and PLGA—mPEG nanoparticles. Int J Pharm 2001;221:143−52.

[17] Markovsky E, Baabur-Cohen H, Eldar-Boock A, Omer L, Tiram G, Ferbe S, et al. Administration, distribution, metabolism and elimination of polymer therapeutics. J Control Release 2012;161:446−60.

[18] Cohen GM. Pulmonary metabolism of Foreign compounds: its role in metabolic activation. Envir Health Pers 1990;85:31−41.

[19] Boyd MR. Biochemical mechanisms in chemical-induced lung injury: roles of metabolic activation. Crit Rev Toxicol 1980;7:103−76.

[20] Olsson B, Bondesson E, Borgström L, Edsbäcker S, Eirefelt S, Ekelund K, et al. 2-Pulmonary drug metabolism, clearance, and absorption. In: Smyth HDC, Hickey AJ, editors. Controlled pulmonary drug delivery. Advances in delivery science and technology. London: Springer; p. 21−49.

[21] Edsbäcker S, Wollmer P, Selroos O, Borgström L, Olsson B, Ingelf J. Do airway clearance mechanisms influence the local and systemic effects of inhaled corticosteroids? Pulm Pharmacol Ther 2008;21:247−58.

[22] Parkinson A. Biotransformation of xenobiotics. In: Curtis Klaassen D, editor. Casarett & Doull's toxicology—the basic science of poisons. Ontario, Canada: McGraw-Hill; 1996.

[23] Hukkanen J, Pelkonen O, Hakkola J, Raunio H. Expression and regulation of xenobiotic-metabolizing cytochrome P450 (CYP) enzymes in human lung. Crit Rev Toxicol 2002;32:391−411.

[24] Patton JS, Fishburn CS, Weers JG. The lungs as a portal of entry for systemic drug delivery. Proc Am Thorac Soc 2004;1:338−44.

[25] Kelly JD, Eaton DL, Guengerich FP, Coulombe RA. Aflatoxin B1 activation in human lung. Toxicol Appl Pharmacol 1997;144:88−95.

[26] Nebert DW, Dalton TP, Okey AB, Gonzalez FJ. Role of aryl hydrocarbon receptor-mediated induction of the CYP1 enzymes in environmental toxicity and cancer. J Biol Chem 2004;279:23847−50.

[27] Bend JR, Serabjit-Singh CJ, Philpot RM. The pulmonary uptake, accumulation and metabolism of xenobiotics. Ann Rev Pharmacol 1985;25:97−125.

[28] Brown RA, Schanker LS. Absorption of aerosolized drugs from the rat lung. Drug Metab Dispos 1983;11:355−60.

[29] Schanker LS, Mitchell EW, Brown RA. Species comparison of drug absorption from the lung after aerosol inhalation or intratracheal injection. Drug Metab Dispos 1986;14:79−88.

[30] Tronde A, Nordén B, Marchner H, Wendel A-K, Lennernäs H, Bengtsson UH. Pulmonary absorption rate and bioavailability of drugs in vivo in rats: structure−absorption relationships and physicochemical profiling of inhaled drugs. J Pharm Sci 2003;92:1216−33.

[31] Cazzola M, Testi R, Matera MG. Clinical pharmacokinetics of salmeterol. Clin Pharmacokinet 2002;41:19−30.

[32] Ha HR, Chen J, Freiburghaus AU, Follath F. Metabolism of theophylline by cDNA expressed human cytochromes P-450. Br J Clin Pharmacol 1995;39:321−6.

[33] Nave R, Fisher R, Zech K. In vitro metabolism of ciclesonide in human lung and liver precision-cut tissue slices. Biopharm Drug Dispos 2006;27:197−207.

[34] Pearce RE, Leeder JS, Kearns GL. Biotransformation of fluticasone: in vitro characterization. Drug Metab Dispos 2006;34:1035−40.

[35] Tunek A, Sjödin K, Hallström G. Reversible formation of fatty acid esters of budesonide, an antiasthma glucocorticoid, in human lung and liver microsomes. Drug Metab Dispos 1997;25:1311−7.

[36] Van DBK, Boorsma M, Staal-Van DB, Edsbäcker S, Wouters EF, Thorsson L. Evidence of the in vivo esterification of budesonide in human airways. Br J Clin Pharmacol 2008;66:27−35.

[37] Hahn ME. Aryl hydrocarbon receptors: diversity and evolution. Chem Biol Interact 2002;141:131−60.

[38] Honkakoski P, Negishi M. Regulation of cytochrome P450 (CYP) genes by nuclear receptors. Biochem J 2000;347:321−37.

[39] Pavek P, Dvorak Z. Xenobiotic-induced transcriptional regulation of xenobiotic metabolizing enzymes of the cytochrome P450 superfamily in human extrahepatic tissues. Curr Drug Metab 2008;9:129−43.

[40] Morgan ET. Regulation of cytochrome P450 by inflammatory mediators: why and how? Drug Metab Dispos 2001;29:207−12.

[41] Zhang K, Kuroha M, Shibata Y, Kokue E, Shimoda M. Effect of oral administration of clinically relevant doses of dexamethasone on regulation of cytochrome P450 subfamilies in hepatic microsomes from dogs and rats. Am J Vet Res 2006;67:329−34.

[42] Kroon LA. Drug interactions with smoking. Am J Health-System Pharm 2007;64:1917−21.

[43] Guengerich FP. Role of cytochrome P450 enzymes in drug−drug interactions. Adv Pharmacol 1997;43:7−35.

[44] Dias FDL, Takahashi CS. Cytogenetic evaluation of aqueous extracts of the medicinal plants Alpinia mutans rose (Zingerberaceae) and Pogostemum hyneanus benth

(Labitae) on Wistar rats and *Allium cepa* (Liliaceae) root tip cells. Braz J Genet 1994;17(2):175–80.

[45] Cooper MR, Johnson AW. Poisonous plants in Britain and their effects on animals and man. London, England: Her Majesty's Stationery Office; 1984.

[46] Leek MD, Griffiths GD, Green MA. Pathological aspects of ricin toxicity in mammalian lymph node and spleen. Med Sci Law 1990;30(2):141–8.

[47] Crespo R, Chin RP. Effect of feeding green onions (*Allium ascalonicum*) to White Chinese geese (*Threskiornis spinicollis*). J Vet Diagn Invest 2004;16(4):321–5.

[48] Kaul S, Krishnakantha TP. Influence of retinol deficiency and curcumin/turmeric feeding on tissue microsomal membrane lipid peroxidation and fatty acids in rats. Mol Cell Biochem 1997;175:43–8.

[49] Dharmananda S. Astragalus. Bestways 1988;30–32:66–7.

[50] Yang YZ, Jin PY, Gúo Q, Wang QD, Li ZS, Ye YC, et al. Effect of *Astragalus membranaceus* on natural killer cell activity and induction of alpha- and gamma-interferon in patients with Coxsackie B viral myocarditis. Chin Med J (Engl) 1990;103(4):304–7.

[51] Zhao KS, Mancini C, Doria G. Enhancement of the immune response in mice by *Astragalus membranaceus* extracts. Immunopharmacol 1990;20(3):225–33.

[52] Das S, Choudhury MD, Mazumder PB. *In vitro* propagation of genus dioscorea—a critical review. Asian J Pharm Clin Res 2013;6(3):26–30.

[53] Woo TM, Wynne AL. Phamacotherapeutics for nurse practitioner prescribers. 3rd ed; Philadelphia: F.A. Davis Company; 2011.

[54] Tierra M. Integrating the traditional Chinese theory and treatment of the lungs with that of western physiology, <http://www.planetherbs.com/theory/integrating-the-traditional-chinese-theory-and-treatment-of-the-lungs-with-that-of-western-physiology.html>; 2013 [accessed on 2.03.14].

[55] Adam SEI, Magzoub M. Toxicity of *Jatropha curcas* for greats. Toxicology 1975;4 (3):388–9.

[56] Keeler RF, Baker DC, Panter KE. Concentration of *Verbesina encelioides* and *Galega officinalis* and the toxic and pathogenic effect induced by the plants. J Environ Pathol Toxicol Oncol 1992;11(2):11–7.

[57] Boyd MR, Burka LT, Hairs TM, Wilson BJ. Lung toxic furanoterpernoid produced by sweet potatoes (*Ipomoea batatas*) following microbial infection. Biochim Biophys Acta 1974;337(2):184–95.

[58] Asres K, Sporer F, Wink M. Patterns of pyrrolizidine alkaloids in 12 Ethiopian *Crotalaria* species. Biochem Sys Ecol 2004;32:915–30.

[59] Zhou Y, Gao W, Li K. Chinese herbal medicine in the treatment of lung cancer review. Asian J Trad Med 2008;3(1):1–11.

[60] Saetung A, Itharat A, Dechsukum C, Wattanapiromsakul C, Keawpradub N, Ratanasuwan P. Cytotoxic activity of Thai medicinal plants for cancer treatment. J Sci Technol 2005;27:469–78.

[61] Hull K. Indiana medical history museum: guide to the medicinal plant garden. Indianapolis, IN: Indiana Medical History Museum; 2010. p. 58.

[62] Kouitcheu MLB, Penlap Beng V, Kouamb J, Essame O, Etoa FX. Toxicological evaluation of ethyl acetate extract of *Cylicodiscus gabunensis* stems barks (Mimosaceae). J Ethnopharmacol 2007;111:598–606.

[63] Sulaiman MM, Mamman M, Aliu YO, Ibrahim NDG. Acute toxicity studies of the butanol extract of the root of *Terminalia avicennioides* Guill. & Perr in rats. Nig J Pharm Sci 2013;12(1):1–6.

[64] Aniagu SO, Nwinyi FC, Olanubi B, Akumka DD, Ajoku GA, Izebe KS, et al. Is *Berlina grandiflora* (Leguminosae) toxic in rats? Phytomedicine 2004;11:352−60.

[65] Ashafa AOT, Orekoya LO, Yakubu MT. Toxicity profile of ethanolic extract of *Azadirachta indica* stem bark in male Wistar rats. Asian Pac J Trop Biomed 2012;811−7.

[66] Lawal HO, Etatuvie SO, Fawehinmi AB. Ethnomedicinal and pharmacological properties of *Morinda lucida*. J Nat Prod 2012;5:93−9.

[67] El Ghazali GE. The promising medicinal plants of the Sudan. Sudan: National Council for Research, Khartoum University Press; 1997.

[68] Yagi S, Yagi AI, Abdel Gadird EH, Henrya M, Chapleura Y, Laurain-Mattar D. Toxicity of *Hydnora johannis* Becca. dried roots and ethanol extract in rats. J Ethnopharmacol 2011;137:796−801.

[69] Adjanohoun JE, Aboubakar N, Dramane K, Ebot ME, Ekpere JA, Enow-Orock EG, et al. Contribution to ethno-botanical and floristic studies in cameroon: traditional medicine and pharmacopoeia. Lagos: Technical and Research Commission of Organisation of African Unity (OAU/STRC); 1996. p. 240−41.

[70] Agbedahunsi JM, Fakoyab FA, Adesanyac SA. Studies on the anti-inflammatory and toxic effects of the stem bark of *Khaya ivorensis* (Meliaceae) on rats. Phytomedicine 2004;11:504−8.

[71] Bafora EE, Igbinuwen O. Acute toxicity studies of the leaf extract of *Ficus exasperata* on haematological parameters, body weight and body temperature. J Ethnopharmacol 2009;123:302−7.

[72] Irene II, Chukwunonso CAA. Body and organ weight changes following administration of aqueous extracts of *Ficus exasperata* Vahl on white albino rats. J Anim Vet Adv 2006;5:277−9.

[73] Mbaya AW, Ibrahimb UI, Goda OT, Ladi S. Toxicity and potential anti-trypanosomal activity of ethanolic extract of *Azadirachta indica* (Maliacea) stem bark: an *in vivo* and *in vitro* approach using *Trypanosoma brucei*. J Ethnopharmacol 2010;128:495−500.

[74] Nwude N, Parsons LE, Adaudi AO. Acute toxicity of the leaves and extracts of *Dichapetalum barteri* (Engl.) in mice, rabbits and goats. Toxicology 1977;7:23−9.

[75] Adebajoa AC, Ayoolab OF, Iwalewac EO, Akindahunsi AA, Omisore NOA, Adewunmi CO, et al. Anti-trichomonal, biochemical and toxicological activities of methanolic extract and some carbazole alkaloids isolated from the leaves of *Murraya koenigii* growing in Nigeria. Phytomedicine 2006;13:246−54.

[76] Makinde JM, Obih PO. Screening of *Morinda lucida* leaf extract for antimalaria action on *Plasmodium berghei* in mice. Afr J Med Sci 1985;14:59−63.

[77] Adesida GA, Adesogan EK. Oruwal, a novel dihydroanthraquinone pigment from *Morinda lucida* Benth. J Chem Soc 1972;1:405−6.

[78] Olaleye MT. Cytotoxicity and antibacterial activity of methanolic extract of *Hibiscus sabdariffa*. J Med Plants Res 2007;1:09−13.

[79] Fakeye TO, Pal A, Bawankule DU, Yadav NP, Khanuja SPS. Toxic effects of oral administration of extracts of dried calyx of *Hibiscus sabdariffa* Linn. (Malvaceae). Phytother Res 2009;23(3):412−6.

[80] Josiah SJ, Nwangwu SCO, Akintola AA, Allu T, Usunobun U, Helen N, et al. The protective effect of ethanolic extract of unripe pulp of *Carica papaya* (Pawpaw) against potassium bromate induced tissue damage in Wistar rats. Curr Res J Biol Sci 2011;3(6):597−600.

[81] Josiah SJ, Nwangwu SCO, Akintola AA, Usunobun U, Oyefule FS, Ajeigbe OK, et al. Protective role of ethanolic extract of *Vernonia amygdalina* against potassium bromate induced tissue damage in Wistar rats. Pak J Nutr 2012;11(1):54−7.

[82] Huxtable RJ. The myth of beneficent nature the risk of herbal preparations. Ann Inter Med 1992;177(2):165−6.

[83] Youdim K, McDonald J, Kalt W, Joseph J. Potential role of dietary flavonoids in reducing micro-vascular endothelium volubility to oxidative and inflammatory. J Nut Biochem 2002;13(5):282–8.

[84] Iwu MM. Handbook of African medicinal plants. Boca Raton, FL: CRC Press, Inc; 1993 PPO. 1

[85] Nwanjo HU, Nwokoro EA. Antidiabetic and biochemical effects of aqueous extract of *Vernonia amygdalina* leaf in normoglycaemic and diabetic rats. J Innov Life Sci 2004;7:6–10.

[86] Efiri EC. The effect of *Ocimum gratissimum* on the histology of the lungs of albino rat, <http://christopher-efiri.blogspot.com/2009/02/effect-of-ocimum-gratissimum-on. html>; 2012 [accessed on 04.03.14].

[87] Botha CJ, Penrith M-L. Poisonous plants of veterinary and human importance in southern Africa. J Ethnopharmacol 2008;119:549–58.

[88] Penrith ML, Van Vollenhoven E. Pulmonary and hepatic lesions associated with suspected ganskweek (*Lasiospermum bipinnatum*) poisoning of cattle. J S Afr Vet Assoc 1994;65:122–4.

[89] Lembe MD, Sonfack A, Gouado I, Dimo T, Dongmo A, Demasse MFA, et al. Evaluations of toxicity of *Turraeanthus africanus* (Meliaceae) in mice. Andrologia 2009;41(6):341–7.

[90] Nwosu CO, Eneme TA, Onyeyili PA, Ogugbuaja OV. Toxicity and anthelmintic efficacy of crude aqueous of extract of the bark of *Sacoglottis gabonensis*. Fitoterapia 2008;79:101–5.

[91] Mann A, Amupitan JO, Oyewale AO, Okogun JI, Ibrahim K, Oladosu P, et al. Evaluation of *in vitro* antimycobacterial activity of Nigerian plants used for treatment of respiratory diseases. Afr J Biotechnol 2008;7(11):1630–6.

[92] Kolodziej H. Aqueous ethanolic extract of the roots of *Pelargonium sidoides*—new scientific evidence for an old anti-infective phytopharmaceutica. Planta Med 2008;74: 661–6.

[93] Sutovska M, Franova S, Priseznakova L, Nosalova G, Togola A, Diallo D, et al. Antitussive activity of polysaccharides isolated from the Malian medicinal plants. Int J Biol Macromol 2009;44:236–9.

[94] Buwa LV, Afolayan AJ. Antimicrobial activity of some medicinal plants used for the treatment of tuberculosis in the Eastern Cape Province, South Africa. Afr J Biotechnol 2009;8(23):6683–7.

[95] Occhiuto F, Sanogo R, Germano MP, Keita A, D'angelo V, De Pasquale R. Effects of some Malian medicinal plants on the respiratory tract of guinea-pigs. J Pharm Pharmacol 1999;51(11):1299–303.

[96] Eldeen IMS, Elgorashi EE, van Staden J. Antibacterial, anti-inflammatory, anticholinesterase and mutagenic effects of extracts obtained from some trees used in South African traditional medicine. J Ethnopharmacol 2005;102(3):457–64.

[97] Karani LW, Tolo FM, Karanja SM, Khayeka CW. Safety and efficacy of *Prunus africana* and *Warburgia ugandensis* against induced asthma in BALB/c mice. Eur J Med Plants 2013;3(3):345–68.

[98] El Kamali H, El Kijalifa KE. Treatment of malaria through herbal drug in the central Sudan. Fitoterapia 1997;6:527–8.

[99] Mensah AY, Bonsu AS, Fleischer TC. Investigation of the bronchodilator activity of *Abrus precatorius*. Int J Pharm Sci Rev Res 2011;6(2):9–13.

[100] Hedberg I, Hedberg O, Madati PJ, Mshigeni KE, Mshiu EN, Samuelsson G. Inventory of plants used in traditional medicine in Tanzania. Part III. Plants of the families Papilionaceae–Vitaceae. J Ethnopharmacol 1983;9(2/3):237–60.

18 Safe African Medicinal Plants for Clinical Studies

Theophine Chinwuba Okoye[1], Phillip F. Uzor[2], Collins A. Onyeto[1] and Emeka K. Okereke[1]

[1]Department of Pharmacology and Toxicology, Faculty of Pharmaceutical Sciences, University of Nigeria, Nsukka 410001, Enugu State, Nigeria, [2]Department of Pharmaceutical and Medicinal Chemistry; Faculty of Pharmaceutical Sciences, University of Nigeria, Nsukka 410001, Enugu State, Nigeria

18.1 Introduction

The use of plants for medicinal purposes is as old as human civilization. Medicinal plants have been the basis of traditional medicine (TM) used for treatment of various diseases in diverse cultures of the world. Herbs and herbal-derived medicines have played crucial role in health and disease management for many centuries. Many ancient civilizations have shown documented evidence for the use of herbal extracts, concoctions, and various forms of plants preparations for the treatment of different kinds of diseases and ailments [1]. In Africa, as in other developing nations, about 80% of the human population still depends on the plant-based TM for their health-care needs [2,3]. This could be attributable to their easy availability, affordability, accessibility, and promising efficacy of the herbal preparations within the locality. Also during the last decade, use of TM has expanded globally and has gained popularity. It has not only continued to be used for primary health care of the poor rural dweller in developing countries, but has also been used in countries where conventional medicine is predominant in the national health-care system [4]. Therefore, with the tremendous expansion in the use of TM, worldwide safety and efficacy as well as quality control of herbal medicines have become an important concern for both health-care practitioners and the public who are the end users.

Most of these plants in addition to their medicinal values have obvious economic, cosmetic, and social applications. These plants contain diverse plants secondary metabolites or constituents which are responsible for their pharmacological and toxicological effects to both humans and animals. These secondary metabolites are also potential sources of new drugs or lead compounds in the development of new drug molecules. Despite the growing market demand for herbal medicines,

Toxicological Survey of African Medicinal Plants. DOI: http://dx.doi.org/10.1016/B978-0-12-800018-2.00018-2

there are still concerns associated with not only their use but also their safety. Reports state that less than 10% of herbal products in the world market are truly standardized to known active components [1].

Toxicological studies of medicinal plants are of paramount importance to ascertain the level of safety in human and animal use. Much of these studies have not been adequately carried out and reported on the diverse African medicinal plants, or AMPs. Long historical use of many of the AMPs, including experience passed on from generation to generation, has indeed demonstrated the safety and efficacy of these AMPs in most cases. However, scientific research is needed to provide additional evidence of safety through certain toxicological studies and tests. The results of such empirical studies are often *sine qua non* for the safety and approval of such medicinal plants by the regulatory authorities, although the World Health Organization (WHO) [4] has indicated that in conducting research and evaluating TM, knowledge and experience obtained through a long history of established practices should be respected. The toxicological studies of AMPs are to determine the level of safety in both acute and chronic use as well as the exposure level where the effect occurred. Also, important biochemical and hematological studies are often performed to determine the level of safety with respect to tissues, organs, and systems. Prior to the subjection of herbal medicines to clinical studies, it is the intended pharmacological practice that the results of the preclinical toxicological tests are significantly safe, without adverse reactions or organ toxicities. Such tests could be performed in lower animals or rodents (with ethical approval) and also by using cell-based cytotoxicity tests *via* the use of cultured cells [1]. Only medicinal plants proved to possess good pharmacological activity without obvious significant adverse effect, cytotoxicity, organ toxicity, and pathological effects could qualify as candidates for clinical studies in humans. Acute toxicity tests or acute lethality (LD$_{50}$), subacute, subchronic, and chronic toxicological tests are among the required tests. Such studies or tests are appropriate in order to ascertain the relative safety in humans based on the animal studies. It has been reported that some AMPs are very toxic to humans and animal and as such could not see the light of the day in clinical studies. Although many plants have been screened with success for several pharmacological activities, some plants have proven very toxic [3]. Therefore, AMPs with potent pharmacological effects and with significantly nontoxic effects would be discussed, as these are possible candidates for clinical studies. Clinical trials begin when toxicity studies on experimental animals indicate that the administration of a potentially therapeutic dose of the agent (e.g., new drugs, plant constituent, plant extract) to humans may not produce any harmful or serious adverse effects.

18.2   Some Medicinal Plants with Good Toxicological Profiles

18.2.1   Alstonia boonei *De Wild (Apocynaceae)*

Alstonia boonei is a medicinal plant used widely (stem bark) in Nigeria for the treatment of malaria and other ailments such as fever, insomnia, chronic diarrhea, and rheumatic pains [5,6]. Several pharmacological effects of the stem bark of the plant have been reported such as antiinflammatory, antipyretic, analgesic, and antimalarial properties. The results of acute toxicity study in mice showed that there was no mortality at all the doses tested. The LD_{50} was estimated to be greater than 5000 mg/kg [6]. Although nephrotoxicity caused by the stem bark extract has been reported [7], clinical trials are still recommended.

18.2.2   Andrographis paniculata *(Burm. F) Nees (Acanthaceae)*

This is a herbaceous plant commonly known as the "King of Bitters." The extracts of *Andrographis paniculata* and its major bioactive constituent, andrographolide, have been found to possess the following pharmacological activities: immunostimulatory, antidiabetic, hypolidemic, antiviral, antibacterial, antitumor, antidiabetic, antimalaria, hepatoprotective effects [8,9], and antiscorpion venom activity [10]. Subacute and chronic toxicity studies on laboratory animals showed that andrographolide exhibited no pathological changes with high safety profile [11], thus presenting the plant as a recommended candidate for clinical trials.

18.2.3   Annona senegalensis *Pers. (Annonaceae)*

This is widely known as African custard apple or wild sour sop (Figure 18.1) with widely reported pharmacological activities (leaves and root bark) such as analgesic, antiinflammation, anticonvulsant, antimalarial, trypanocidal, antioxidant,, and antimicrobial [12]. A diterpenoid, kaurenoic acid, isolated from the root bark of *Annona senegalensis* has been reported to possess anticonvulsant and antimicrobial properties with an LD_{50} of 3800 mg/kg in mice [13,14]. Acute and subacute toxicity study on the root bark extract of *Annona senegalensis* in rodents revealed that

Figure 18.1 Leaves of *Annona senegalensis* (Annonaceae).

the plant is relatively safe though with caution in doses greater than 400 mg/kg [12]. It should therefore be subjected to clinical trials.

18.2.4 Annona squamosa *L. (Annonaceae)*

This is commonly known as the custard apple tree with mainly the leaves in use apart from the root and stem. It has been reported to exhibit the following pharmacological activities: antibacterial, antihyperlipidemic, analgesic, antiinflammatory, antidiabetic, hepatoprotective, and antiulcer effects [15−17]. Aqueous extract of the leaves have shown to be safe in acute and subacute toxicities studies. There were no significant changes in biochemical, hematological, and histopathological parameters [18] and hence portrays *Annona squamosa* (Figure 18.2) as a candidate for clinical trials.

18.2.5 Anthocleista grandiflora *Gil. (Gentianaceae)*

Anthocleista grandiflora is commonly known as the forest fever tree, used in African traditional medicine (ATM) for treatment of malaria, anthelmintic, antidiarrheal, diabetes, high blood pressure, venereal diseases, and epilepsy [19,20]. The acute toxicity study of the stem bark indicated that extract did not cause mortality of mice up to 1000 mg/kg body weight (bw), which is more than three times the minimum effective dose (MED), while gross physical and behavioral observations of the experimental mice also revealed no visible signs of acute toxicity [20]. Generally, if the LD_{50} of the test substance is three times more than the MED, the substance is considered a good candidate for further studies [21]. The safety of the

Figure 18.2 *Annona squamosa* L. (Annonaceae).

plant from this report together with the fact that the plant has been used by the local people in TM for treatment of malaria (especially in the Choba area in southern Nigeria) makes the plant a good candidate for clinical trials.

18.2.6 Artemisia afra *(Asteraceae)*

Artemisia afra is perennial shrub commonly found in most areas of South Africa and is used to treat disease such as chest conditions, coughs, colds, heartburns, hemorrhoids, fevers, malaria, asthma, diabetes mellitus, and sore throat [22,23]. Scientific reports have indicated that the plant has bronchodilator, antiinflammatory, antihistaminic, and narcotic analgesic properties [23,24]. Acute toxicity studies in mice showed an LD_{50} of 2450 and 8960 mg/kg for the intraperitoneal and oral administration respectively. Also the chronic toxicity study (92-day treatment with the extract) exhibited no significant changes in the hematological and biochemical parameters, except for a transient decrease in aspartate aminotransferase activity [23]. This high level of safety in animal studies and good biological activities present this plant as a good candidate for clinical trials.

18.2.7 Azadirachta indica *Juss (Meliaceae)*

This is commonly known as neem plant (Figure 18.3), indigenous to India but widely cultivated in most tropical and subtropical regions of Africa. The leaves of the plant have been used since World War II for in the treatment of malaria infection [25]. Moreover, other pharmacological activities reported include antimicrobial, antiinflammatory, anticancer, antiviral, and antidiabetic [26−28]. Toxicological effects indicated that the extracts of *Azadirachta indica* altered biochemical parameters with no significant effect on hematological parameters, body weight, and

Figure 18.3 *Azadirachta indica* Juss (Meliaceae).

organ weight. However, the adverse effects of biochemical parameters are dose dependent with lower doses being safe in laboratory animals [29,30]. Because neem has been used for a long time in ATM to treat many disorders and that the reported toxic effects are dose dependent in laboratory animals, it should be recommended for clinical trials.

18.2.8   Berlina grandiflora *Hutch and Dalz (Leguminasae)*

Berlina grandiflora is used in ethnomedicine for the treatment of gastrointestinal disorders [31]. The stem is used as chewing stick and in the preparation of enemas against constipation while the bark and fruits are used to stupefy fish [31], and its anthelminthic activity has been reported [32,33]. The acute toxicity study showed no lethality with an LD_{50} above 5 g/kg bw in mice [34]. Hence, *Berlina grandiflora* can be a candidate for clinical trials.

18.2.9   Buchholzia coriacea *(Capparidaceae)*

Various parts (leaves, seeds, bark) of *Buchholzia coriacea* known as "Wander Kola" (called "okpokolo" in Igbo) are used in the traditional treatment of various diseases such as rheumatism, kidney pain, headache, sinusitis, and nasal congestion, smallpox, skin itching, microbial infections, and in the treatment of diabetes [35]. Also reported are antiplasmodial, anthelminthic, antispasmodic, and antidiarrhea activities as well as synergistic effects with standard antidiabetic agents [35,36]. Acute toxicity studies have shown an LD_{50} greater than 5000 mg/kg bw in mice [36]. The above indicated that the plant is quite safe and a good candidate for trials in humans, especially as it is used for many diseases in traditional settings.

18.2.10   Caesalpinia bonduc *(L.) Roxb (Caesalpiniaceae)*

Caesalpinia bonduc, commonly known as Gray Nicker, is distributed in the tropical regions of all major continents including Africa. The aqueous solution of the outer shell of the seeds of the plant traditionally is used for the relief of the symptoms of diabetes mellitus [37]. Antibacterial, antifungal, and antidiabetic activities of the seeds and a diterpene, bondenlide, from the plant have been reported [37,38]. On histopathological examination in alloxan-induced diabetic rats, body weights of the animals were increased and the pancreas, kidney, and liver histology indicate significant recovery with the seed extract administration [37]. The efficacy of the seed extract in reducing blood glucose level, improving body weight, and rejuvenating the damaged pancreas of alloxan-induced diabetic rats make this plant a candidate for clinical studies.

18.2.11   Cassia occidentalis *L. (Caesalpiniaceae)*

This is also known as *Coffee Senna* and has its various parts (seeds, roots, leaves, and stems) been used in TM for a long time. Pharmacological activities such as analgesic, laxative, diuretic, hepatoprotective, tuberculosis, gonorrhea, dysmenor-rheal, and antibacterial effects have been reported with *Cassia occidentalis* [39]. The plant's toxicological profile has shown a good safety profile with the acute, subacute, and subchronic toxicity studies in rats [39,40], hence qualifying it for clinical studies.

18.2.12 Chrysophyllum albidum *L. (Sapotaceae)*

Chrysophyllum albidum is popularly known as African Star Apple. Various parts of the plant (roots, leaves, and bark) are used for ethnomedicinal purposes such as yel-low fever, malaria, skin eruptions, and stomach ache [41,42]. Also eleagnine, an alkaloid isolated from *Chrysophyllum albidum* seed cotyledon, has been reported to have antinociceptive, antiinflammatory, and antioxidant activities [43]. The acute toxicity study in male albino mice showed an LD_{50} of 1850 mg/kg with no obvious acute toxicity, while the organs such as hearts, brains, lungs, kidneys, livers, spleens, and stomach showed no gross and histological changes [44]. Thus African Star Apple could be a good candidate for clinical trials.

18.2.13 Cissus cornifolia *(Vitaceae)*

Cissus cornifolia is commonly called splanch and *riigarbirri* (rope of the monkey) in Hausa [45]. In ATM, the plant is used for the treatment of diabetes, gonorrhea, sedative malaria, septic tonsils, and pharyngitis [46]. Results of the oral acute tox-icity study shows that the LD_{50} of the extract was above 5000 mg/kg bw (rats). Also, the histopathological examination of the diabetic rats treated with the leaf extract showed restoration of the pancreatic damage induced by alloxan [46]. This plant is quite safe and possesses good antidiabetic activity; it is therefore a good candidate for clinical trials.

18.2.14 Enantia chlorantha *Oliv (Annonaceae)*

Enantia chlorantha is locally known as Awogba, Oso pupa or Dokita igbo (Yoruba), Osomolu (Ikale), Kakerim (Boki), and Erenba-vbogo (Bini) and is widely distributed along the coasts of West and Central Africa [47]. The root and stem bark are used traditional in various diseases including management of malaria and jaundice, leprosy rickettsia, fever, typhoid fever, infective hepatitis, as well as hemostatic agent and uterine stimulant [47,48]. The antimicrobial, antimalarial, and antipyretic properties of the stem bark have been reported [47]. Acute toxicological study of the extracts of the plant in mice shows no mortality at an oral dose of 20 g/kg with LD_{50} estimated as 43.65 g/kg. In subchronic study for 5 weeks, no fatality was recorded and no significant damage to the body organs was observed [49]. The relative safety of the plant, its wide usage in TM, and its potent pharma-cological activities have justified its trials in humans.

18.2.15   Garcinia kola *Heckel (Guttiferae)*

Garcinia kola, often known as Bitter kola, is a flowering plant found mostly in the tropical rain forest region of Central and West Africa. In folkloric medicine, every part such as the seeds, stem, and leaves has medicinal value. The seeds are edible and are consumed as an adjuvant to the true kola (*Cola nitida*) and for medicinal purposes [50]. In ethnomedicine, *Garcinia kola* has been used as a purgative, antiparasitic, and antimicrobial agent, throat infection, diarrhea, bronchitis, and as an aphrodisiac [50,51]. Reported pharmacological effects of the seeds include antidiabetic, antiinflammatory, antipyretic, immunomodulatory, antiatherogenic, antimicrobial, and liver disorders [51,52]. Toxicological studies on the seeds of *Garcinia kola* has shown an LD_{50} above 5000 mg/kg in both mice and rats with no obvious signs of acute intoxication after a 48-h observation period [50−52]. The good toxicological profile and the long-time usage of *Garcinia kola* in ATM have indicated its suitability for clinical trials.

18.2.16   Grewia crenata *L. (Malvaceae)*

In northwestern Nigeria, *Grewia crenata* leaves popularly known as "kamomowa" are used locally in the treatment of fractured bones, wound healing, and inflammatory conditions [53]. The acute and subchronic oral toxicity studies of extracts of *Grewia crenata* leaves were evaluated in rats with an LD_{50} greater than 5000 mg/kg, while showing no obvious sign of toxicity or mortality during 14 days of treatment. In the subchronic study (28-day oral toxicity study), administration of 900, 1800, 2700, and 3600 mg/kg of body weight *Grewia crenata* leaves extracts revealed no significant difference ($p < 0.05$) in the hematological parameters, serum liver enzymes, electrolytes, and creatinine in the extract-treated groups compared to control groups [53]. The plant is therefore suitable for clinical trials because of its safety.

18.2.17   Hybanthus enneaspermus *(L.) F. Muell (Violaceae)*

Hybanthus enneaspermus is an herb or shrub (commonly known as *abiwere* in Yoruba land, Nigeria) distributed in tropical and subtropical regions of the world. It is used as an aphrodisiac, demulcent, tonic, anticonvulsant, antimalarial, antidiabetes, antiinflammatory, male sterility, urethra discharge, and in traditional child birth [54−56]. Reported pharmacological activities include antiplasmodial, antiinflammatory, antitussive, anticonvulsant, antidiabetic, free-radical scavenging, uterotonic, and antinociceptive activities [54,56]. Toxicological studies in Swiss albino mice indicated that the acute toxicity of the whole plant extract shows no lethality with an LD_{50} greater than 5000 mg/kg [56]. Thus the plant possesses sterling qualities for clinical trial.

18.2.18   Hyptis suaveolens *Poit. (Lamiaceae)*

Hyptis suaveolens, commonly called curry leaf, is widely distributed in West Africa, especially in Northern Nigeria [57,58]. Traditionally, the plant is used as a local seasoning and also used in ethnomedicine for the treatment of diabetes mellitus, fever, eczema, flatulence, cancers, and headache [58]. Acute toxicity studies of the leaf extract showed an LD_{50} of 2154.1 mg/kg bw in rats. The safety of the plant, potent pharmacological activities, together with the fact that the plant has been in use for long by the rural populace makes the plant a candidate for clinical studies.

18.2.19   Lithocarpus dealbata *(Miq.) Rehder (Fagaceae)*

The seed extract (and also the bark) of *Lithocarpus dealbata* was reported to have antidiarrheal, dysentery, and antihemorrhagic activities [59]. The toxicity study showed no mortality or any visible signs acute toxicity while the serum biochemical tests did not reveal any noticeable changes in aspartate transaminase (AST) or aspartate aminotransferase, alanine transaminase (ALT) or alanine aminotransferase, cholesterol, and total protein levels in animals [59,60]. The plant could be subjected to clinical trials to ascertain the toxic profile in humans.

18.2.20   Monochoria vaginalis *(Burm. F.) C. Presl ex Kunth (Pontederiaceae)*

This is commonly referred to as heartleaf, false pickerelweed, or oval-leafed pondweed. Reported pharmacological activities include analgesic, cough, antiasthmatic, antioxidant, hepatoprotective, antidiabetic, and hypolipidemic effects [61−63]. Toxicological studies show that the LD_{50} was found to be 1000 mg/kg while the histopathological changes induced by carbon tetrachloride (CCl_4) in rats were significantly ($p \leq 0.05$) reduced by the root extract treatment [62]. The plant is a good candidate for clinical trials because the leaves and roots are quite safe, considering the results of toxicity studies.

18.2.21   Moringa oleifera *Lam. (Moringaceae)*

Moringa oleifera is a tree that is sometimes called the "Tree of Life" or "Miracle Tree." It is an economically important tree and vegetable, which has been variously used in ATM for treatment of different diseases. The leaves of the plant have been reported to possess many pharmacological activities which include analgesic, antiinflammatory, antiasthmatic, antiulcer, antispasmodic, antibacterial, antihyperglycemia, antioxidant, anticancer, and larvicidal activities [64−67]. Toxicological studies on *Moringa oleifera* have shown absence of severe hepatotoxicity and organ damage except in very high doses. The acute lethality (LD_{50}) test of *Moringa oleifera* has been found to be relatively safe with the subchronic toxicity studies, eliciting no significant difference in sperm quality, hematological and biochemical parameters in the treated rats compared to the control [68−70]. *Moringa oleifera* has recently been reported to possess potent anticancer effects and exhibited synergistic effects with cisplatin [71,72]. These and its longtime use in ethnomedicine have shown *Moringa oleifera* to be a good candidate for clinical trial studies.

18.2.22   Mormodica charantia L. (Cucurbitaceae)

Mormodica charantia, also called Bitter melon or Balsam pear, is tropical vegetable which has been used in ATM for the treatment of diabetes [73]. The plant also has the following reported pharmacological activities: antibacterial, antiulcer, antiviral, immunostimulant, anthelmintic, hypotensive, and antioxidant effects [73]. The LD_{50} has been reported to be 91.9 and 362.34 mg/100 g bw, for the juice and alcohol extracts respectively of subcutaneously injected mice with no significant toxicity on biochemical parameters tested [74]. Therefore, *Mormodica charantia* could be a good candidate for clinical trials.

18.2.23   Newbouldia laevis *(Bignoniaceae)*

This is a tropical plant variously found in Nigeria and Ghana. Reported pharmacological effects of the stem, leaves, and roots include treatment of breast cancer, antibacterial, hepatoprotective, and antioxidant [45,75,76]. Histopathological examination shows that the CCl_4-induced hepatic damage in rats was ameliorated by treatment with the leaf extracts of *Newbouldia laevis* [76]. In acute toxicity study, the lethal dose (LD_{50}) of the plant was found to be greater than 3000 mg/kg bw (rats), thus good for clinical trials.

18.2.24   Nigella sativa L. (Ranunculaceae)

The seeds of *Nigella sativa* (commonly known as black seed or black cumin) are used in folk medicine for the treatment and prevention of a number of diseases which include asthma, diarrhea, and dyslipidemia. Much of the biological activity of the seeds has been shown to be due to thymoquinone. Thymoquinone has been reported to possess anticonvulsant as well as anticancer effects [77]. Reported pharmacological actions of the seeds and some of its active constituents include protection against nephrotoxicity and hepatotoxicity, as well as antiinflammatory, analgesic, antipyretic, antimicrobial, antineoplastic, immunological, antihypertensive, respiratory stimulating, hematinic, and trypanocidal activities [78,79]. The seeds extract has exhibited a low degree of toxicity and has been shown not to induce significant adverse effects on liver or kidney functions. It would appear that the beneficial effects of the use of the seeds and thymoquinone might be related to their cytoprotective and antioxidant actions, and to their effect on some mediators of inflammation [78].

18.2.25   Oldenlandia corymbosa Lam. (Rubiaceae)

Oldenlandia corymbosa is used traditionally in the treatment of various diseases such as hepatitis in Africa and other parts of the world [80]. Pharmacological activities such as antioxidant and anticancer activity of the plant have been reported [81]. The LD_{50} of the leaf extract in Swiss albino mice was calculated to be 14.14 and 10.56 g/kg bw using Thompson and Finney methods respectively [82]. The plant is safe with potential anticancer effect and therefore good for clinical studies.

18.2.26 Parkia biglobosa *(Jacq.) (Fabaceae)*

Parkia biglobosa, popularly known as the "African locust bean tree," has been used in the Nigerian and other West African rural communities to treat a variety of diseases such as malaria, diabetes mellitus, infections, and inflammatory diseases [83–85]. The efficacy of the various preparations (seeds, leaves, and stem barks) of *Parkia biglobosa* is widely acclaimed for the treatment of malaria, diabetes mellitus, and painful conditions [83,84,86]. The acute toxicity study in mice showed no lethality and LD_{50} is greater than 5 g/kg bw, hence the plant can be a candidate for clinical trial studies.

18.2.27 Parquetina nigrescens *(Afzel.) Bullock (Asclepiadaceae)*

Different parts of *Parquetina nigrescens* (known as *Ewe Ogbo* in Yoruba) are used in Nigeria and other parts of Africa for the treatment of gastrointestinal diseases, rickets, diarrhea, gonorrhea, and insanity [87,88]. Scientific investigations have shown its antimicrobial, gastrointestinal-protective, hematinic, cardiotonic, and catecholamine-like, analgesic, antiinflammatory, and antipyretic properties [89,90]. The LD_{50} of the leaf extract in Swiss albino mice was calculated to be 12.60 and 13.10 g/kg bw using Thompson and Finney methods respectively [82]. The plant is therefore quite safe and has a good pharmacological profile, thus making it a good candidate for clinical studies.

18.2.28 Phyllanthus muellerianus *(Kuntze) Excell (Euphorbiaceae)*

The leaves and stem of *Phyllanthus muellerianus* are widely used in West Africa to treat several diseases such as intestinal troubles, severe dysentery, constipation, stomach ache, jaundice, urethral discharges, venereal diseases, toothache, and also for wound dressing [91,92]. The acute toxicity of the stem bark was found to be greater than 4 g/kg. At the dose of 4 g/kg, kidney and liver function tests of rats were not affected [93]. The safety and potential antimicrobial properties of this plant make it a good candidate for clinical studies.

18.2.29 Piptadeniastum africana *Pip (Hook f.) Brenan. (Mimosaceae)*

Decoctions of *Piptadeniastum africana* are said to be used in Gabonese TM in the treatment of hemorrhoids, paralysis, and ulcerative incurable wounds [94]. The antimicrobial and anticancer activities of leaves and root extracts of the plant have been reported [93,94]. The *in vivo* acute toxicity study carried out on the methanolic extract of the leaf indicated that the plant was not toxic. At the dose of 4 g/kg bw (rats), kidney and liver function tests indicated that plant induced no adverse effect on these organs [93]. The safety of the plant and its anticancer potentials make the plant a good candidate for clinical studies.

18.2.30   Pittosporum viridiflorum *Sims (Pittosporaceae)*

Pittosporum viridiflorum is a South African plant traditionally used for the management of opportunistic fungal infections in human immune deficiency virus/acquired immune deficiency syndrome (HIV/AIDS) patients. It is also used to treat fever, malaria, inflammation, stomach ache, and as an antidote for insect bites [95]. The leaves have been reported to possess antimicrobial properties [96]. Based on Meyer's toxicity index in the brine shrimp assay method, extracts of the plant with LC_{50} values greater than 1 mg/ml were considered nontoxic and may be further explored for the development of plant-based pharmaceuticals [95]. The wide usage in TM and its good toxicological profile justifies its suitability for clinical trials.

18.2.31   Psidium guajava *L. (Myrtaceae)*

In ATM, *Psidium guajava* (guava) (Figure 18.4) leaf has been used in the treatment of various diseases. It serves as an economic, vegetable, and medicinal plant. Reported pharmacological activities include diabetes mellitus, diarrhea, cough, painful menstruation, hypertension, antimicrobial, and sore throat, as well as disinfectant for wounds [97,98]. Also antioxidant, antihyperlipidemic, and anticancer effects have variously been reported [98−100]. Major phytochemical constituents isolated from *Psidium guajava* is quercetin, a flavonoid [101]. The toxicological profile indicated that even the level of toxic metals in guava is significantly low and incapable of causing harm [97], thus making it a good candidate for clinical studies.

18.2.32   Solanum aethiopicum *L. (Solanaceae)*

Solanum aethiopicum, popularly known as African garden egg, is mostly found in Asia and tropical Africa. The fruits are edible and have a long history of consumption in West Africa, especially for their nutritive effects. The plant has been

Figure 18.4 *Psidium guajava* L. (Myrtaceae).

reported to exhibit purgative, sedative, antidiabetic, antiulcerogenic, and antiin-flammatory properties [102,103]. Toxicological profile indicated an LD_{50} in mice greater than 5000 mg/kg with no mortality or any significant behavioral changes in the animals [103]. The age-long human consumption, the safety profile, and the various potent pharmacological activities suggest suitability of the plant for further studies in humans.

18.2.33   Sorghum bicolor *(L.) Pers. (Gramineae, Poaceae)*

Sorghum bicolor is an annual plant having its different parts widely used in TM. Ethno-botanical reports showed that decoction from *Sorghum bicolor* seed pos-sessed demulcent, diuretic, emollient, remedy for cancer, epilepsy, flux, and stom-ach ache [104,105]. *Sorghum bicolor* leaves are one of the four herbal components of the sickle cell drug (NIPRISAN®) developed by National Institute for Pharmaceutical Research and Development (NIPRD), Abuja, Nigeria [104] and is also one of the three components of Jubi Formular®, a commercial herbal hema-tinic manufactured by Health Forever Products Ltd., Lagos, Nigeria [105]. Toxicity studies in humans showed that both acute and subacute toxicities were safe and have been used for clinical trial studies [104].

18.2.34   Stachytarpheta cayennensis *C. Rich (Val) (Verbenaceae)*

Stachytarpheta cayennensis is used in Africa for the treatment of inflammatory and digestive disorders. Reported pharmacological activities include antiallergy, antiin-flammatory, analgesic, antispasmodic, and antimicrobial, as well as a liver protec-tive [106]. In some South American countries, the leaves are taken as a tonic [107]. The toxicological profile indicated an LD_{50} greater than 5000 mg/kg in mice with no obvious signs of toxicity, which is an indication of high level of safety for use in humans [106]. Hence, *Stachytarpheta cayennensis* portrays a good candidate for clinical studies.

18.2.35   Strychnos henningsii *Gil. (Loganiaceae)*

Strychnos henningsii is a small evergreen tree or shrub with leathery leaves, mostly distributed in East and South Africa, used in the treatment of various ailments in ATM such as gynecological complaints, abdominal pain, snake bite, gastrointesti-nal pain, rheumatism, malaria, and diabetes mellitus [108,109]. The acute and sub-acute toxicity studies showed no signs of toxicity and or organ change or damage with an LD_{50} is above 5 g/kg bw (rodents) [109], hence its high level of safety; it can be recommended for clinical trial studies.

18.2.36   Tropaeolum majus *L. (Tropaeolaceae)*

Tropaeolum majus is used traditionally for chest colds, formation of new blood cells, urinary tract infections, antiseptic, and expectorant qualities. Also it has been reported to exhibit pronounced diuretic and antihypertensive effects [110]. Subchronic toxicity study of the hydroethanolic extract obtained from leaves of *Tropaeolum majus* in rats showed no significant alterations in the animal's body weight gain, relative organs weight and serum biochemical, hematological parameters of the liver, kidneys, and spleen [111]. The plant's good toxicological profile and potential source of antihypertensive agent make it a candidate for clinical studies.

18.2.37   Urena lobata L. (Malvaceae)

The seeds of *Urena lobata* were reported in ethnomedicine for the treatment of diarrhea, colic, skin disease, boils, pneumonia, rheumatism, cough, and diabetes [59,112,113]. The extract of roots exhibited antibacterial activity against Gram-positive and Gram-negative microorganisms [112], while antioxidant activity and inhibitory actions against nitric oxide release from macrophages has been reported, too [114]. The acute toxicity test showed no mortality nor visible signs of toxicity or differences in food and water uptake up to 3200 mg/kg bw doses in the animals. Thus, the plant could be evaluated in clinical trials.

18.2.38   Ximenia americana (Olacaceae)

Ximania americana is found mainly in tropical regions, especially Africa and Brazil. The bark and seed of the plants are used in TM in menstruation, stomach ulcers, and as a purgative [115]. Extracts and constituents of the plant have shown several biological activities such as antimicrobial, antifungal, anticancer, antineoplastic, moluscicide, pesticidal, antitrypanosomal, antirheumatic, antioxidant, analgesic, antidiabetic, hepatic, and hematological effects [115,116]. Acute toxicity study of the leaf and stem bark extracts revealed an oral LD_{50} in rodents greater than 5000 mg/kg while hematological and histopathological examination of the animals administered with the stem bark extract did not show any significant ($p < 0.05$) toxicities or weight changes [115,116]. The results, therefore, suggest that *Ximania americana* is safe and has good pharmacological activities, hence suitable for clinical studies.

References

[1] Obidike I, Salawu O. Screening of herbal medicines for potential toxicities. In: Gowder S, editor. New insights into toxicity and drug testing. Rijeka, Croatia: InTech Online Publishers; 2013.
[2] World Health Organization. WHO traditional medicine strategy 2002–2005. Geneva: World Health Organization; 2002 < http://apps.who.int/medicinedocs/en/d/Js2297e/ > [accessed 14.04.14.]

[3] Tchacondo T, Karou SD, Agban A, Bako M, Batawila K, Bawa ML, et al. Medicinal plants use in central Togo (Africa) with emphasis on the timing. Pharmacognosy Res 2012;4:92−103.

[4] World Health Organization. WHO General Guidelines for Methodologies on Research and Evaluation of Traditional Medicine, Geneva. Online: < http://whqlibdoc.who.int/ hq/2000/WHO_EDM_TRM_2000.1.pdf > ; 2000.

[5] Ojewole JAO. Studies on the pharmacology of echitamine, an alkaloid—from the stem bark of *Alstonia boonei* L. (Apocynaceae). Int J Crude Drug Res 1984;22:121−43.

[6] Iyiola OA, Tijani AY, Lateef KM. Antimalarial activity of ethanolic stem bark extract of *Alstonia boonei* in Mice. Asian J Biol Sci 2011;4:235−43.

[7] Oze G, Nwanjo H, Onyeze G. Nephrotoxicity caused by the extract of *Alstonia boonei* (De Wild) stem bark in Guinea pigs. Int J Nutr Wellness 2007;3:2.

[8] Chowdhury A, Biswas SK, Raihan SZ. Pharmacological potentials of *Andrographis paniculata*—an overview. Int J Pharmacol 2012;8(1):6−9.

[9] Jayakumar T, Hsieh CY, Lee JJ, Sheu JR. Experimental and clinical pharmacology of *Andrographis paniculata* and its major bioactive phytoconstituent, Andrographolide. Evid Based Complement Alternat Med 2013;2013:846740.

[10] Brahmane RI, Pathak SS, Wanmali VV, Salwe KJ, Premendran SJ, Shinde BB. Partial *in vitro* and *in vivo* red scorpion venom neutralization activity of *Andrographis paniculata*. Pharmacognosy Res 2011;3(1):44−8.

[11] Dey YN, Kumari S, Ota S, Srikanth N. Phytopharmacological review of *Andrographis paniculata* (Burm F.) Wall. Int J Nutr Pharmacol Neurol Dis 2013;3(1):3−10.

[12] Okoye TC, Akah PA, Ezike AC, Okoye MO, Ndukwu F, Ohaegbulam E, et al. Evaluation of the acute and sub acute toxicity of *Annona sennegalensis* root bark extracts. Asian Pac J Trop Med 2012;5(4):277−82.

[13] Okoye TC, Akah PA, Okoli CO, Ezike AC, Omeje EO, Odoh UE. Antimicrobial effects of a lipophilic fraction and kaurenoic acid isolated from the root bark extracts of *Annona senegalensis*. Evid Based Complement Alternat Med 2012;2012:831327.

[14] Okoye TC, Akah PA, Omeje EO, Okoye FBC, Nworu CS. Anticonvulsant effect of kaurenoic acid isolated from the root bark of *Annona senegalensis*. Pharmacol Biochem Behav 2013;109:38−43.

[15] Alluri R, Anil KS, Pasala PK. Evaluation of gastric antiulcer and antioxidant activities of aqueous extracts of *Annona squamosa* and *Achyranthes aspera* in rats. Int J Phytopharmacol 2011;2(2):66−9.

[16] Aher PS, Shinde YS, Chavan PP. *In vitro* evaluation of antibacterial potentials of *Annona squamosa* L. against pathogenic bacteria. Int J Pharm Sci Res 2012;3(5): 1457−60.

[17] Rajkumar P, Surendra J, Shailly M, Azhar A. Pharmacological review on leaves of *Annona squamosa* in GI tract ulcer in albino rats. World J Pharm Sci 2012;1 (2):499−524.

[18] Darwin RC, Vijaya C, Sujith K, Sadhavil K, Sushma L. Acute and sub-acute toxicological evaluation of aqueous and ethanol fractions of *Annona squamosa* root; a traditional medicinal herb. Pharmacology Online 2011;2:34−6.

[19] Palmer E, Pitman N. Trees of Southern Africa covering all known indigenous species in the Republic of South Africa, South-West Africa, Botswana, Lesotho & Swaziland, vol. 3. Cape Town, SA: AA Balkema Publishers; 1972. p. 1845−7

[20] Odeghe OB, Uwakwe AA, Monago CC. Antiplasmodial activity of methanolic stem bark extract of *Anthocleista grandiflora* in mice. Int J Appl Sci Technol 2012;2 (4):142−8.

[21] Carol A. Acute, sub chronic and chronic toxicology. In: CRC , editor. Handbook of toxicology. Boca Raton, FL: CRC Press Inc.; 1995. p. 51−104.

[22] Dyson A. Discovering indigenous healing plants of the herb and fragrance gardens at Kirstenbosch National Botanical Garden. Cape Town: National Botanical Institute, the Printing Press; 1998. p. 9−10.

[23] Mukinda JT. Acute and chronic toxicity of the flavonoid-containing plant, *Artemisia afra* in rodents. M.Sc Thesis. Department of Pharmacology, University of the Western Cape; 2005.

[24] Harris L. An evaluation of the bronchodilator properties of *Mentha longifolia* and *Artemisia afra*, traditional medicinal plants used in the Western Cape. M. Thesis, Discipline of pharmacology. Bellville: School of Pharmacy, University of the Western Cape; 2002.

[25] Anyaehie UB. Medicinal properties of fractionated acetone/water Neem (*Azadirachta indica*) leaf extract from Nigeria, a review. Niger J Physiol Sci 2009;24(2):157−9.

[26] Ebong PE, Atangwho IJ, Eyong EU, Egbung GE. The antidiabetic efficacy of combined extracts from two continental plants, *Azadirachta indica* (A. Juss) (Neem) and *Vernonia amygdalina* (Del.) (African Bitter Leaf). Am J Biochem Biotechnol 2008;4(3):239−44.

[27] Bisht S, Sisodia SS. Anti-hyperglycemic and antidyslipidemic potential of *Azadirachta indica* leaf extract in STZ-induced diabetes mellitus. J Pharm Sci Res 2010;2 (10):622−7.

[28] Awah FM, Uzoegwu PN, Ifeonu P. *In vitro* and anti-HIV and immunomodulatory potentials of *Azadirachta indica* (Meliaceae) leaf extract. Afr J Pharm Pharmacol 2011;5(11):1353−9.

[29] Asahafa AOT, Orekoya LO, Yakubu MT. Toxicity profile of ethanolic extract of *Azadirachta indica* stem bark in male Wistar rats. Asian Pac J Trop Biomed 2012;2 (10):811−7.

[30] Dafalla MB, Konozy EH, Saad HA. Biochemical and histological studies on the effects of *Azadirachta indica* seeds kernel extract on albino rats. Int J Med Plants Res 2012;1 (6):082−92.

[31] Asuzu IU, Nwelle OC, Anaga AO. The pharmacological activities of the methanolic bark of *Berlina grandiflora*. Fitoterapia 1993;64:529−34.

[32] Enwerem NM, Okogun JL, Wambebe CO, Okorie DA, Akah PA. Antihelminthic activity of the stem bark extracts of *Berlina grandiflora* and one of its active principles, Betulinic acid. Phytomedicine 2001;8:112−4.

[33] Enwerem NM, Wambebe CO, Okogun JL, Akah PA, Gamaniel KS. Antihelminthic screening of the stem bark of *Berlina grandiflora*. J Nat Remedies 2001;1(1):17−20.

[34] Akuodor GC, Idris-Usman M, Ugwu TC, Akpan JL, Ghasi SI, Osunkwo UA. *In vivo* schizonticidal activity of ethanolic leaf extract of *Gongronema latifolium* on *plasmodium berghei berghei* in mice. Afr J Biotech 2010;9(5):2316−21.

[35] Anowi CF, Ike C, Ezeokafor E, Ebere C. The phytochemical, antispasmodic and anti-diarrhoea properties of the methanol extract of the leaves of *Buchholzia Coriacea* family Capparaceae. Int J Curr Pharm Res 2012;4(3):52−5.

[36] Okoye TC, Akah PA, Ilogu CL, Ezike AC, Onyeto CA. Anti-diabetic effects of methanol extract of the seeds of *Buchholzia coriacea* and its synergistic effects with metformin. Asian J Biomed Pharm Sci 2012;2(12):32−6.

[37] Aswar PB, Kuchekar BS. Assessment of hypoglycemic and antidiabetic effects of *Caesalpinia bonduc* (L.) Roxb. seeds in alloxan induced diabetic rat and its phytochemical, microscopic, biochemical and histopathological evaluation. Asian J Plant Sci Res 2011;1(3):91−102.

[38] Simin K, Khaliq-uz-Zaman SM, Ahmad VU. Antimicrobial activity of seed extracts and bondenolide from *Caesalpinia bonduc* (L.) Roxb. Phytother Res 2001;15(5): 437—40.

[39] Silva MGB, Aragoo TP, Vasconcelos CFB, Ferreira PA, Andrade BA, Costa IMA, et al. Acute and sub-acute toxicity of *Cassia occidentalis* L. stem and leaf in Wistar rats. J Ethnopharmacol 2011;136:341—6.

[40] Essa'a VJ, Medoua GN. Subchronic toxicity of the beverage made from *Cassia occidentalis* seeds in mice. Int J Nutr Food Sci 2013;2(5):237—42.

[41] Adebayo AH, Abolaji AO, Opata TK, Adegbenro IK. Effects of ethanolic leaf extract of *Chrysophyllum albidum* G. on biochemical and haematological parameters of albino Wistar rats. Afr J Biotechnol 2010;9(14):2145—50.

[42] Oyebade BA, Ekeke BA, Adeyemo FC. Fruits categorization and diagnostic analysis of *Chrysophylum albidum* (G. Don) in Nigeria. Adv Appl Sci Res 2011;2(1):7—15.

[43] Idowu TO, Iwalewa EO, Aderogba MA, Akinpelu BA, Ogundaini AO. Antinociceptive, anti-inflammatory and antioxidant activities of eleagnine, an alkaloid isolated from seed cotyledon of *C. albidum*. J Biol Sci 2006;6(6):1029—34.

[44] Adewoye EO, Salami AT, Taiwo VO. Anti-plasmodial and toxicological effects of methanolic bark extract of *Chrysophyllum albidum*. J Physiol Pathophysiol 2010;1(1):1—9.

[45] Burkhill HM. The useful plants of West Tropical Africa. 2nd ed. Kew: Royal Botanical Garden; 2000. p. 293—4.

[46] Jimoh A, Tanko Y, Mohammed A. Anti-diabetic effect of methanolic leaf extract of *Cissus cornifolia* on alloxan-induced hyperglycemic in Wistar rats. Ann Biol Res 2013;4(3):46—54.

[47] Adesokan AA, Yakubu MT, Owoyele BV, Akanji MA, Soladoye AO, Lawal OK. Effect of administration of aqueous and ethanolic extracts of *Enantia chlorantha* stem bark on brewer's yeast-induced pyresis in rats. Afr J Biochem Res 2008;2(7):165—9.

[48] Gill LS. Ethnomedical uses of plants in Nigeria. Benin, Nigeria: Uniben Press; 1992. p. 143.

[49] Agbaje EO, Onabanjo AO. Toxicological study of the extracts of anti-malarial medicinal plant *Enantia chlorantha*. Cent Afr J Med 1994;40(3):71—3.

[50] Udenze ECC, Braide VB, Okwesilieze CN, Akuodor GC. Pharmacological effects of *Garcinia kola* seed powder on blood sugar, lipid profile and atherogenic index of alloxan-induced diabetes in rats. Pharmacologia 2012;3(12):693—9.

[51] Falang KD, Uguru MO, Nnamonu NL. Anti-pyretic activity of *Garcinia kola* seed extract. Eur J Med Plants 2014;4(5):511—21.

[52] Nworu CS, Akah PA, Okoli CO, Esimone CO, Okoye FBC. The effects of methanol seed extract of *Garcinia kola* on some specific and non-specific immune responses in mice. Int J Pharmacol 2007;3(4):347—51.

[53] Ukwuani AN, Abubakar MG, Hassan SW, Agaie BM. Toxicological studies of hydromethanolic leaves extract of *Grewia crenata*. Int J Pharm Sci Drug Res 2012;4(4):245—9.

[54] Afolabi AO, Oluwakanmi ET, Salahdeen HM, Oyekunle AO, Alagbonsi IA. Antinociceptive effect of ethanolic extract of *Hybanthus enneaspermus* leaf in rats. Br J Med Med Res 2014;4(1):322—30.

[55] Hemalatha S, Wahi AK, Singh PN, Chansouria JPN. Anticonvulsant and free radical scavenging activity of *Hybanthus enneaspermus*, a preliminary screening. Ind J Tradit Knowl 2003;2(4):383—8.

[56] Patel DK, Kumar R, Prasad SK, Sairam K, Hemalatha S. Antidiabetic and *in vitro* antioxidant potential of *Hybanthus enneaspermus* (Linn) F. Muell in streptozotocin-induced diabetic rats. Asian Pac J Trop Biomed 2011;1(4):S316—22.

[57] Abdullahi M, Muhammed G, Abdulkadir NU. Medicinal and economic plants of Nupe land. Bida, Nigeria: Jube-Evans Books & Publications; 2003. p. 179.

[58] Danmalam UH, Abdullahi LM, Agunu A, Musa KY. Acute toxicity studies and hypo-glycemic activity of the methanol extract of the leaves of *Hyptis suaveolens* Poit. (Lamiaceae). Niger J Pharm Sci 2009;8(2):87−92.

[59] Arun KY, Tangpu V. Antidiarrheal activity of *Lithocarpus dealbata* and *Urena lobata* extracts, therapeutic implications. Pharm Biol 2006;45:223−9.

[60] Kala CP. Ethnomedicinal botany of the Apatani in the Eastern Himalayan region of India. J Ethnobiol Ethnomed 2005;1:11.

[61] Chinna RR, Periyasamy M, Muthukumar A, Anand G. Antidiabetic and hypolipidemic activity of *Monochoria vaginalis* Presl. on alloxan induced diabetic rats. Intl J Pharm Res Scholars 2013;2(1−3):98−102.

[62] Latha B, Latha MS. Antioxidant and curative effect of *Monochoria vaginalis* methano-lic extract against carbon tetrachloride induced acute liver injury in rats. Der Pharma Chem 2013;5(2):294−300.

[63] Kumar GM, Mukesh Y, Sushil S, Ravi D, Shyam KR. Analgesic activity of roots of *Monochoria vaginalis* Presl. Int J Res Pharm Sci 2012;2(1):109−15.

[64] Jaiswal D, Rai PK, Kumar A, Mehta S, Watal G. Effects of *Moringa oleifera* Lam. leaves aqueous extract therapy on hyperglycemic rats. J Ethnopharmacol 2009;123: 392−6.

[65] Dewangan G, Koley KM, Vadlamudi VP, Mishra A, Podder A, Hirpurkar SD. Antibacterial activity of *Moringa oleifera* (drumstick) root bark. J Chem Pharm Res 2010;2(6):424−8.

[66] Pandey A, Pandey RD, Tripathi P, Gupta PP, Haider J. *Moringa oleifera* Lam. (sahijan)—a plant with a plethora of diverse therapeutic benefits, an updated retrospec-tion. Med Aromat Plants 2012;1(6):325−7.

[67] Wadhwa S, Panwar MP, Saini N, Rawat S, Singhal S. A review of the commercial, tra-ditional uses, phytoconstituents and pharmacological activity of *Moringa oleifera*. Global J Trad Med Sys 2013;2(1):1−13.

[68] Awodele O, Oreagba IA, Odoma S, da Silva JA, Osunkalu VO. Toxicological evaluation of the aqueous leaf extract of *Moringa oleifera* Lam. (Moringaceae). J Ethnophamacol 2012;139(2):330−6.

[69] Isitua CC, Ibeh IN. Toxicological assessment of aqueous extract of *Moringa oleifera* and *Caulis bambusae* leaves in rabbits. J Clin Toxicol 2003;S12:003.

[70] Ugwu OPC, Nwodo OFC, Joshua PE, Bawa A, Ossai EC, Odo CE. Phytochemical and acute toxicity studies of *Moringa oleifera* ethanol leaf extract. Int J Life Sci Biotechnol Pharm Res 2013;2(2):66−71.

[71] Sreelatha S, Jeyachitra A, Padma PR. Antiproliferation and induction of apoptosis by *Moringa oleifera* leaf extract on human cancer cells. Food Chem Toxicol 2011;49: 1270−5.

[72] Berkovich L, Earon G, Ron I, Rimmon A, Vexler A, Lev-Ari S. *Moringa oleifera* aqueous leaf extract down-regulates nuclear factor-Kappa B and increases cytotoxic effect of chemotherapy in pancreatic cancer cells. BMC Complement Altern Med 2013;13:212.

[73] Kumar DS, Sharathnath KV, Yogeswaran P, Harani A, Sudhakar K, Sudha P, et al. A medicinal potency of *Momordica charantia*. Int J Pharm Sci Rev Res 2010;1:95.

[74] Abd EI, Sattar EI, Batran S, EI-Gengaihi SE, EI Shabrawy OA. Some toxicological studies of *Momordica charantia* L on albino rats in normal and alloxan diabetic rats. J Ethnopharmacol 2006;108(2):236−42.

[75] Azuine MA, Ibrahim K, Enweren M, Gamaniel K, Wambebe C. Cytotoxic and anticancer activity of *Newbouldia laevis* in mice. West Afr J Pharmacol Drug Res 1995;11:72−9.

[76] Hassan SW, Salawu K, Ladan MJ, Hassan LG, Umar RA, Fatihu MY. Hepatoprotective, antioxidant and phytochemical properties of leaf extracts of *Newbouldia laevis*. Int J PharmTech Res 2010;2(1):573−84.

[77] Gurung RL, Lim SN, Khaw AK, Soon JFF, Shenoy K, Ali SM, et al. Thymoquinone induces telomere shortening, DNA damage and apoptosis in human glioblastoma cells. Plos One 2010;5(8):e12124. Available from: http://dx.doi.org/doi:10.1371/journal.pone.0012124.

[78] Ali BH, Blunden G. Pharmacological and toxicological properties of *Nigella sativa*. Phytother Res 2003;17(4):299−305.

[79] Ekanem JT, Yusuf OK. Some biochemical and haematological effects of black seed (*Nigella sativa*) oil on T. *brucei*-infected rats. Afr J Biomed Res 2008;11:79−85.

[80] Ahmad R, Mahbob ENM, Noor ZM, Ismail NH, Lajis NH, Shaari K. Evaluation of antioxidant potential of medicinal plants from Malaysian Rubiaceae (subfamily Rubioideae). Afr J Biotech 2010;9(46):7948−54.

[81] Pandey K, Sharma PK, Dudhe R. Anticancer activity of *Parthenium hysterophorus* Linn and *Oldenlandia corymbosa* Lam by Srb Method. Sci Rep 2012;1(6):325.

[82] Awolabi FO, Omorodion-Osagie E, Olatunji-Bello II, Adegoke OA, Adeleke TI. Acute oral toxicity test and phytochemistry of some West African medicinal plants. Nig Q J Hosp Med 2009;19(1):53−8.

[83] Asase A, Alfred A, Yeboah O, Odamtten GT, Simmonds SJ. Ethno botanical study of some Ghanaian antimalarial plant. J Ethnopharmacol 2005;99:273−9.

[84] Gronhaug TE, Glaeserud S, Skogsrud M, Ballo N, Bah S, Diallo D, et al. Ethno pharmacological survey of six medicinal plants from Mali, West-Africa. J Ethnopharmacol 2008;4:4−26.

[85] Abo KA, Fred-Jayesimi AEA. Ethno botanical studies of medicinal plants used in the management of diabetes mellitus in southern Nigeria. J Ethnopharmacol 2008;115:67−71.

[86] Tijani AY, Okhale SE, Salawu TA, Onigbanjo HO, Obianodo LA, Akingbasote JA, et al. Anti-diarrheal and antibacterial properties of crude aqueous stem bark extract and fractions of *P. biglobosa (Jacq)* R.Br Ex G. Don. Afr J Pharm Pharmacol 2009;7:347−53.

[87] Sofowora A. Medicinal plants and traditional medicines in Africa. Ibadan, Nigeria: Spectrum Book Ltd; 1993. p. 52.

[88] Iwu MM. A handbook of African medicinal plants. Florida: C.R.C. Inc; 1993. p. 2−25.

[89] Owoyele BV, Nafiu AB, Oyewole IA, Oyewole LA, Soladoye AO. Studies on the analgesic, anti-inflammatory and antipyretic effects of *Parquetina nigrescens* leaf extract. J Ethnopharmacol 2009;122:86−90.

[90] Odetola AA, Oluwole FS, Adeniyi BA, Olatiregun AM, Ikupolowo OR, Labode O, et al. Antimicrobial and gastrointestinal protective properties of *Parquetina nigrescens* (Afzel.) Bullock. J Biol Sci 2006;6(4):701−5.

[91] Ofokansi KC, Attama AA, Uzor PF, Ovri MO. Antibacterial activities of the combined leaf extract of *Phyllanthus muellerianus* and Ciprofloxacin against urogenital isolates of *Staphylococcus aureus*. Clin Pharmacol Biopharm 2012;1(4):1−5.

[92] Katsayal UA, Lamal RS. Preliminary phytochemical and antibacterial screening of the ethanolic stem bark extract of *Phyllanthus muellerianus*. Niger J Pharm Sci 2009;8: 121−5.

[93] Assob JCN, Kamga HLF, Nsagha DS, Njunda AL, Nde PF, Asongalem EA, et al. Antimicrobial and toxicological activities of five medicinal plant species from Cameroon Traditional Medicine. BMC Complement Altern Med 2011;11:70.

[94] Mengome LE, Feuya-Tchouya GR, Eba F, Nsi-Emvo E. Antiproliferative effect of alcoholic extracts of some Gabonese medicinal plants on human colonic cancer cells. Afr J Tradit Complement Altern Med 2009;6(2):112−7.

[95] Otang WM, Grierson DS, Ndip RN. Assessment of potential toxicity of three South African medicinal plants using the brine shrimp (*Artemia salina*) assay. Afr J Pharm Pharmacol 2013;7(20):1272−9.

[96] Seo Y, Berger JM, Hoch J, Neddermann KM, Bursuker I, Mamber SW, et al. A new triterpene saponin from *Pittosporum viridiflorum* from Madagascar rainforest. J Nat Prod 2002;65:65−8.

[97] Dhiman A, Nanda A, Ahmad S. Metal analysis in *Citrus sinensis* fruit peel and *Psidium guajava* leaf. Toxicol Int 2011;18(2):163−7.

[98] Ryu NH, Park KR, Kim SM, Yun HM, Nam D, Lee SG, et al. A hexane fraction of guava leaves (*Psidium guajava* L.) induces anticancer activity by suppressing AKT/mammalian target of Rapamycin/Ribosomal p70 S6 kinase in human prostrate cancer cells. J Med Food 2012;15(3):231−41.

[99] Deguchi Y, Miyazaki K. Antihyperglycemic and anti-hyperlipidemic effects of guava leaf extract. Nutr Metabol 2010;7:9.

[100] Ling LT, Palanisamy UD. Review: potential antioxidants from tropical plants. In: Valdez B, editor. Food industrial processes—methods and equipment. Rijeka, Croatia: InTech; 2012. Available Online: <www.Intechopen/books/food-industrial-processes-methods-and-equipment/review-potential-antioxidants-from-tropical-plants>

[101] Metwally AM, Omar AA, Harraz FM, El Sohafy SM. Phytochemical investigation and antimicrobial activity of *Psidium guajava* L. leaves. Pharmacogn Mag 2010;6 (23):212−8.

[102] Ezeugwu CO, Okonta JM, Nwodo NJ. Antidiabetic properties of ethanolic fruit extract of *Solanum aethiopicum* L. Res J Pharm Allied Sci 2004;2(2):251−4.

[103] Anosike CA, Obidoa O, Ezeanyika LUS. Membrane stabilization as a mechanism of the anti-inflammatory activity of methanol extract of garden egg (*Solanum aethiopicum*). Daru 2012;20:76−83.

[104] Wambebe C, Khamofu H, Momoh JA, Ekpeyong M, Audu BS, Njoku OS, et al. Double blind, placebo controlled, randomized cross-over clinical trial of NIPRISAN in patients with sickle cell disorder. Phytomedicine 2001;8(4):252−61.

[105] Erah PO, Asonye CC, Okhamafe AO. Response of *Trypanosoma brucei brucei* induced anaemia to a commercial herbal preparation. Afr J Biotechnol 2003;2 (9):307−11.

[106] Okoye TC, Aguwa CN, Okoli CO, Akah PA, Nworu CS. Evaluation of the anticonvulsant and sedative effects of leaf extracts of *Stachytarpheta cayennensis*. J Trop Med Plants 2008;9(1):17−22.

[107] Mesia-vela S. Pharmacological study of *Stachytarpheta cayennensis* Val in Rodents. Phytomedicine 2004;11(7):616−24.

[108] Leeuwenberg AJ. The Loganiaceae of Africa VIII. Strychnos III. Mededdelingen Landbouwhogeschool Wageningen 1969;69:1−316.

[109] Oyedemi SO, Bradley G, Afolayan AJ. Ethno botanical survey of medicinal plants used for the treatment of diabetes mellitus in the Nkonkobe Municipality of South Africa. J Med Plants Res 2009;3:1040−4.

[110] Gasparotto Jr A, Gasparotto FM, Lourenço EL, Crestani S, Stefanello ME, Salvador MJ, et al. Antihypertensive effects of isoquercitrin and extracts from *Tropaeolum majus* L, evidence for the inhibition of angiotensin converting enzyme. J Ethnopharmacol 2011;134(2):363−72.

[111] Gomes C, Lourenço EL, Liuti ÉB, Duque AO, Nihi F, Lourenço AC, et al. Evaluation of subchronic toxicity of the hydroethanolic extract of *Tropaeolum majus* in Wistar rats. J Ethnopharmacol 2012;142(2):481−7.

[112] Mazumder UK, Gupta M, Manikandan L, Bhattacharya S. Antibacterial activity of *Urena lobata* roots. Fitoterapia 2001;72:927−9.

[113] Pinto MM, Sousa ME, Nascimento MS. Xanthone derivatives, new insights in biological activities. Curr Med Chem 2005;12:2517−38.

[114] Choi EM, Hwang JK. Screening of Indonesian medicinal plants for inhibitor activity on nitric oxide production of RAW264.7 cells and antioxidant activity. Fitoterapia 2005;76:194−203.

[115] Monte FJQ, de Lemos TLG, de Araújo MRS, de Sousa Gomes E. *Ximenia americana*, chemistry, pharmacology and biological properties, a review. In: Rao V, editor. Phytochemicals—a global perspective of their role in nutrition and health. Europe: InTech; 2012. Available from <http://www.intechopen.com/books/phytochemicals-a-global-perspective-of-their-role-in-nutrition-and-health/ximenia-americana-chemistry-pharmacology-and-biological-properties-a-review>.

[116] Siddaiah M, Jayaveera KN, Souris K, Yashodha Krishna JP, Kumar PV. Phytochemical screening and antidiabetic activity of methanolic extract of leaves of *Ximenia americana* in rats. Int J Innov Pharm Res 2011;2(1):78−83.

19 Harmful and Protective Effects of Terpenoids from African Medicinal Plants

Armelle T. Mbaveng[1], Rebecca Hamm[2] and Victor Kuete[1]

[1]Department of Biochemistry, Faculty of Science, University of Dschang, Dschang, Cameroon, [2]Department of Pharmaceutical Biology, Institute of Pharmacy and Biochemistry, University of Mainz, Mainz, Germany

19.1 Introduction

Terpenoids or isoprenoids represent large and diverse classes of naturally occurring organic compounds that are derived from isoprene units (C5) and assembled and modified in thousands of ways. They are extraordinarily diverse in nature, but they all originate from the condensation of the universal phosphorylated derivatives of hemiterpene, isopentenyl diphosphate, and dimethylallyl diphosphate giving geranyl pyrophosphate [1]. They represent the most widespread group of natural products and can be found in all classes of living things. Terpenoids include monoterpenes (C10; e.g., carvone, geraniol, D-limonene, and perillyl alcohol) and sesquiterpenes (C15; e.g., farnesol) which are the main constituents of the essential oils while diterpenes (C20; e.g., retinol and trans-retinoic acid)-, Sesterterpenes (C25)-, triterpenes (C30; e.g., betulinic acid, lupeol, oleanic acid, and ursolic acid), tetraterpenes (C40; e.g., α-carotene, β-carotene, lutein, and lycopene) and other terpenes are constituents of balsams, resins, waxes, and rubber [1–5]. Terpenoids are biosynthesized via the mevalonic acid pathway and the 2-*C*-methyl-D-erythritol 4-phosphate/1-deoxy-D-xylulose 5-phosphate pathway (MEP/DOXP pathway), also known as (non-mevalonate pathway) or mevalonic acid-independent pathway, taking place in the plastids of plants and apicomplexan of protozoa, and in many bacteria [1–4]. Many defensive compounds (phytoalexins) include sesquiterpenoids and diterpenoids from angiosperm species [6]. For example, oleoresin is roughly equal to a mixture of turpentine (85% monoterpenes and 15% sesquiterpenes) and

Toxicological Survey of African Medicinal Plants. DOI: http://dx.doi.org/10.1016/B978-0-12-800018-2.00019-4

rosin (diterpene) that acts in several conifer species to seal wounds and is toxic to both invading insects and their pathogenic fungi [7]. Some common terpenoids include citral, menthol, camphor, salvinorin identified in *Salvia divinorum*, the cannabinoids found in *Cannabis*, the ginkgolide and bilobalide (*Ginkgo biloba*), and the curcuminoids (turmeric and mustard seeds). The steroids and sterols (C27) in animals are produced biologically from terpenoid precursors [4]. Many terpenes have biological activities (against cancer, malaria, inflammation, and a variety of infectious diseases such as viral and bacterial) and are used for the treatment of human diseases. The worldwide sales of terpene-based pharmaceuticals in 2002 were approximately US $12 billion [8]. Among these pharmaceuticals, the anticancer drug Taxol® and the antimalarial drug Artemisinin were two of the most renowned terpene-based drugs [8]. Nevertheless, some terpenoids are known to induce toxic effects in humans and animals. Some terpenoids have local irritant effects and, thus, are capable of causing gastrointestinal disorders [9]. Central nervous system (CNS) manifestations may range from an altered mental status to seizures and coma [9]. The popular *Salvia officinalis* tree oil constituent, thujone, has hallucinogenic properties and has become an abused drug [10]. According to the 2009 Annual Report of the *American Association of Poison Control National Poison Data System*, two deaths over 3362 single exposures to disinfectants containing pine oil, one death over 10,714 single exposures to camphor, and no deaths over 422 single exposures to turpentine were reported in the United States [11]. On the other side, bioactive terpenoids were also identified in African medicine with some of them having organ-protective effects while few are known for their nonbeneficial properties in humans. In this chapter, we will discuss both the harmful and protective effects of the most common terpenoids occurring in African medicinal plants.

19.2   Plant Terpenoids and Their Medicinal Importance

Terpenoids constitute one of the largest group of natural products accounting for more than 40,000 individual compounds, with several new compounds being discovered every year [12]. They have been found to be useful in the prevention and therapy of several diseases, including cancer, and also possess antimicrobial, antifungal, antiparasitic, antiviral, antiallergenic, antispasmodic, antihyperglycemic, anti-inflammatory, and immunomodulatory properties [12]. Monoterpenes are useful in the prevention and therapy of several cancers, including mammary, skin, lung, stomach, colon, pancreatic, and prostate carcinomas [12]. Many sesquiterpenes have antimicrobial and antiplasmodial activities while some have anticancer, anti-feedant, anti-mycotoxigenic, antioxidant, vascular mycorelaxing, mollucidal, trypanocidal, anti-inflammatory, anti-protozoal, herbicidal, hepatoprotective, neuroproliferative, anti-diabetic, analgesic, and cytotoxic activities. Sesquiterpenes such as costunolide, parthenolide and its derivative 1,10-epoxyparthenolide were documented for their antimicrobial properties against *Helminthosporium* spp. Warburganal and muzigadial also had significant antifungal activities against *Saccharomyces cerevisiae* (MIC values of 3.13 and 1.56 μg/mL), *Candida utilis*, and *Sclerotinia libertiana* (MIC of 3.13 μg/mL) [1]. The diterpenoids aframolins A and B, labdan-8(17),12(*E*)-diene-15,16-dial, and aframodial isolated from the seeds of *Aframomum longifolius* were tested for their antimicrobial activities, but only aframodial was active against *Cryptococcus neoformans*, *Staphylococccus aureus*, and methycillin-resistant *Staphylococccus aureus* with similar MIC value of 20 μg/mL [3]. The anticancer potential of diterpenoids has also been documented for caseanigrescens A (IC_{50} of 1.4 μM), B (IC_{50} of 0.83 μM), C (IC_{50} of 1.0 μM), and D (IC_{50} of 1.0 μM) against the A2780 human ovarian cancer cell line [3]. Aulacocarpinolide (IC_{50} of 12.5 μg/mL) and aulacocarpin B (IC_{50} of 25 μg/mL) were also cytotoxic toward murine leukemia L1210 cells [3]. Maleastrones A and B are two limonoids isolated from *Malleastum* sp. collected from the Madagascar rain forest. They displayed significant cytotoxic activities against a panel of cancer cell lines including A2780, MDA-MB-435, HT-29, H522-T1, U937 with IC_{50} values varying between 0.19 and 0.63 μM for the first compound and 0.20 and 0.49 μM for the second compound [3]. The most successful terpenoids in pharmaceuticals are the anticancer drug Taxol and the antimalarial drug Artemisinin [8]. However, considering the number of new biologically active terpenoids isolated worldwide every year, it is evident that others will also be used in the future. This explains why it is necessary to draw the attention of the scientific community not only to the importance of the chemicals belonging to terpenoids, but also to their possible side effects as already reported, with emphasis on African medicinal plants.

19.3 Toxic Terpenoids and Their Mode of Action

Chemically, plant poisons are not only peptides but also low molecular weight compounds, belonging to alkaloids, terpenoids, phenolics, or other secondary metabolites. Plants produce a wide variety of secondary metabolites which can interfere with the biochemistry and physiology of herbivores on the one hand and some with bacteria, fungi, viruses, and even competing plants on the other hand [13]. Plants produce and accumulate not only single entities but mixtures of secondary metabolites that mostly belong to several classes [13]. The plants' defenses against herbivores include repellence, deterrence, toxicity, and growth inhibition as well as toxicity against microorganisms [13].

Toxins and poisons are classified in four categories according to their oral toxicity determined in rat experiments:

1. Class Ia: extremely hazardous (5 mg or less per kilogram body weight).
2. Class Ib: highly hazardous (5−50 mg/kg body weight).
3. Class II: moderately hazardous (50−500 mg/kg body weight).
4. Class III: slightly hazardous (500 mg and more per kilogram body weight) [13].

In this classification, toxins, which fall into the classes Ia, Ib, and II, interfere with central functions in animals. The most poisonous substances are neurotoxins which affect the nervous system, followed by cytotoxins and metabolic poisons that disturb liver, heart, kidneys, respiration, muscles, and reproduction. It has been shown that the morbidity and mortality associated with exposure to toxic terpenes is largely related to the degree of CNS depression when the compound is aspired [14]. However, it should be noted that morbidity induced by terpenoids is extremely low.

19.3.1 Neurotoxicity

A seizure is an episode of neurologic dysfunction caused by abnormal neuronal activity that results in a sudden change in behavior, sensory perception, or motor activity. The so-called "epilepsy" is the recurrent, unprovoked seizure from known or unknown causes, while "ictus" describes the period in which the seizure occurs, and "postictal" refers to the period after the seizure has ended, but before the patient has returned to his or her baseline mental status [9]. It has been demonstrated that neurotoxins can affect important ion channels of neuronal cells, such as Na^+, K^+, and Ca^{2+} channels, either by activating or inactivating them permanently [13]. Both actions will stop neuronal signal transduction and thus block the activity of the CNS but also neuromuscular signaling [13], which eventually leads to paralysis of both striated and smooth muscles of heart, lungs, and skeleton. A special case is the Na^+, K^+-ATPase, which is the most important ion pump in neuronal and other cells to maintain an ion gradient important for action potentials and transport mechanisms [13]. Cardiac glycosides, the secondary metabolites used in the treatment of congestive heart failure and cardiac arrhythmia, occurring in several plant families and even in toad skins (genus *Bufo*), are strong inhibitors of this pump. Due to the fact that this pump is extremely important, cardiac glycosides are considered to be toxins of class Ia. It should be noted that drugs such as ouabain (**1**) and digoxin (**2**) are cardiac glycosides (Figure 19.1). However, compound **1** from the foxglove plant is used clinically, whereas compound **2** is used only experimentally due to its extremely high potency. The sesquiterpenes 13-*O*-acetylsolstitialin A (**3**) and cynaropicrin (**4**) (Figure 19.1), isolated from *Centaurea solstitialis*, were found to be responsible for the ability of the plant to cause neurodegenerative changes in the brain of horses [15].

19.3.2 Inhibition of Cellular Respiration

In cellular respiration, taking place in mitochondria, ATP is generated. This energy is essential for all cellular and organic functions making respiration a vulnerable target in animals. Plant metabolites can attack this target with HCN, which binds to iron ions of the terminal cytochrome oxidase in the mitochondrial respiratory chain [13]. HCN does not occur in a free form, but is stored as cyanogenic glycosides in plant vacuoles [13]. A plant's cytosolic enzymes, such as β-glucosidase and nitrilase, hydrolyze the cyanogenic glycosides and the extremely toxic HCN is released [13]. The diterpene atractyloside (**5**) (Figure 19.1) is a potent inhibitor of the mitochondrial ADP/ATP transporter and thus inhibits the ATP supply of a cell [13]. Atractyloside is an inhibitor of the adenine nucleotide translocator that inhibits

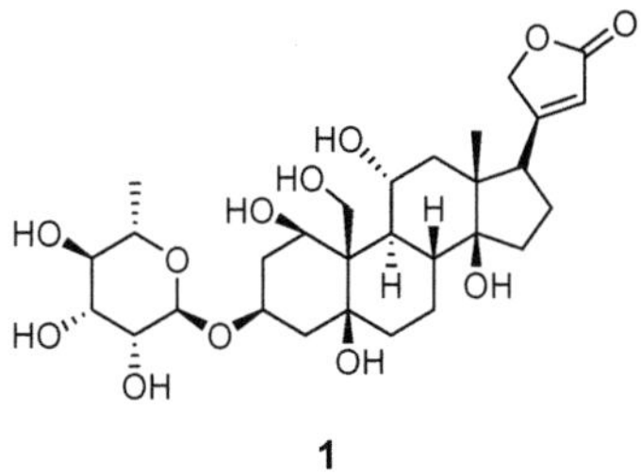

Figure 19.1 Chemical structures of some known toxic terpenoids. Ouabain (**1**), digoxin (**2**), 13-*O*-acetylsolstitialin A (**3**), cynaropicrin (**4**), atractyloside (**5**), ptaquiloside (**6**).

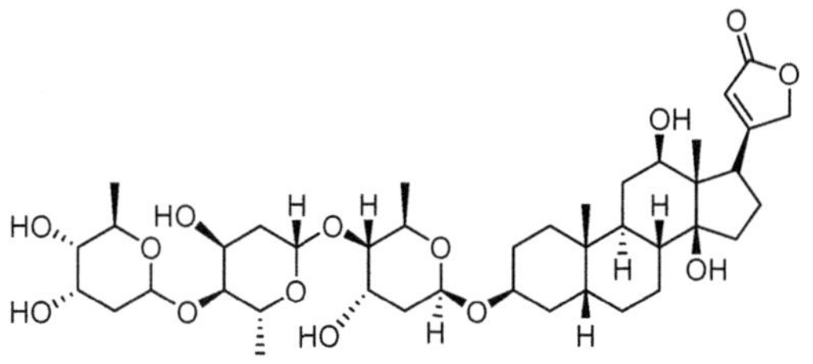

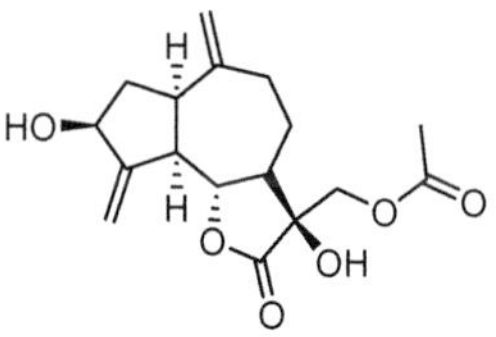

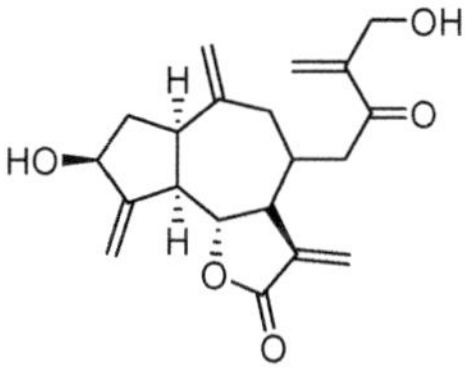

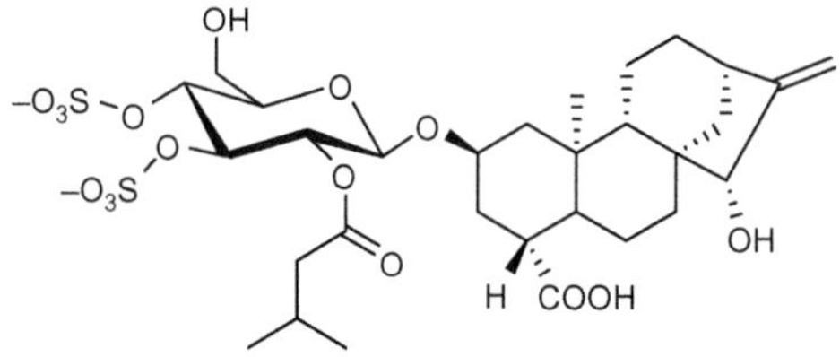

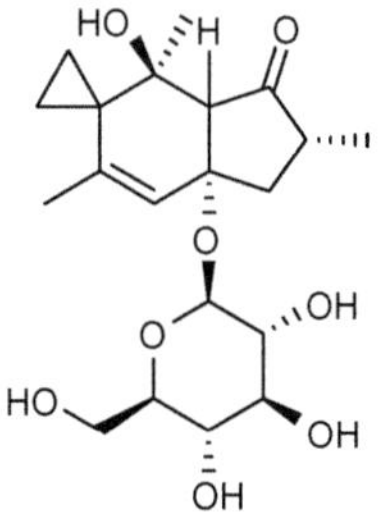

oxidative phosphorylation by blocking the transfer of adenosine nucleotides through the mitochondrial membrane [16].

19.3.3 Cytotoxins

A cytotoxin is any substance that has a toxic effect on an important cellular function, such as venom or a chemical agent. An important target in this context are biomembranes, which regulate the import and export of metabolites and ions [13]. Membrane fluidity and integrity can be severely disturbed by both steroidal and triterpenoid saponins. Saponins are usually stored as inactive bidesmosidic saponins in plant vacuoles; upon wounding and decompartmentation, they are converted into the membrane-active monodesmosidic saponins, which are amphiphilic holding detergent activities [13]. Within cells, other important targets include several enzymes and proteins but also DNA/RNA.

19.3.4 Alkylating and Intercalating DNA Toxins

An unstable glycoside ptaquiloside (**6**) (Figure 19.1), containing a reactive cyclo-propane ring, has been isolated from ferns and is a potent carcinogenic agent [17]. Many plant alkaloids were also found to act as alkylating agents (see Chapter 21).

19.3.5 Toxins of Skin and Mucosal Tissues

Phytochemicals also can affect the skin and mucosal tissue of animals. The diterpenes known as phorbol esters from the Euphorbiaceae and Thymelaeaceae, which resemble the endogenous signal compound diacylglycerol, are activators of the key enzyme protein kinase C [13]. When in contact with skin, mucosal tissue, or the eyes, they cause severe and painful inflammation, with ulcers and blister formation [13].

19.4 Protective Terpenoids from Plants

Terpenoids are produced by a variety of plants, particularly conifers and some insects, having strong smells and thus may have a protective function. Terpenoids have been found useful in the prevention and therapy of several diseases, such as cancer, and also displayed antimicrobial, antifungal, antiparasitic, antiviral, antial-lergenic, antispasmodic, antihyperglycemic, anti-inflammatory, and immunomodu-latory properties [2−4,12,18,19]. They can also be used as protective substances in storing agriculture products as they are known to have insecticidal properties [12]. The diterpenoids (Figure 19.2) 4-epi-abietol (**7**) and sugiol (**8**) from Pakistan plant

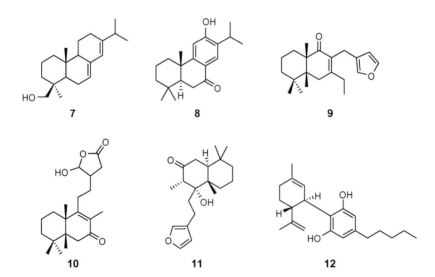

Figure 19.2 Chemical structures of some known protective terpenoids. 4-epi-abietol (**7**), sugiol (**8**), labdane diterpenes T1 (**9**) and T2 (**10**), hispanolone (**11**), cannabidiol (**12**).

Juniperus procera were found to have hepatoprotective effects *via* the reduction of elevated liver enzymes [20]. Furthermore, it has been demonstrated that immune-stimulating complexes formulated with *Quillaja* saponin preparations induced specific cytotoxic T-lymphocyte responses and have been reported to induce antibody responses and/or protective immunity in guinea-pig, turkey, cat, rabbit, dog, seal, sheep, pig, cow, horse, and monkeys [21]. Labdane diterpenes T1 (**7**) and T2 (**8**) derived from the hispanolone (**9**) exert significant cardio protection against anoxia/ reperfusion injury through inhibition of apoptosis and activation of survival pathways [22]. A cannabinoid found in the genus *Cannabis*, Cannabidiol (**10**) showed strong antioxidant properties and have been shown to play a role in the compound's neuroprotective and anti-ischemic effects [23]. The strong antioxidant effect of compound **12** has been proven to be effective in treating an often drug-induced set of neurological movement disorders known as dystonia [24].

19.5 Harmful Effects of Terpenoids from African Medicinal Plants

Some terpenoids identified in African medicinal plants showed various side effects (Table 19.1). However, indirect correlation between the presence of the terpenoids and plants was discovered. For example, furanosesquiterpenoids, identified in four plants of the family Asteraceae, namely *Lasiospermum bipinnatum, Athanasia minuta, Athanasia trifurcata,* and *Nidorella foetida,* induced hepatotoxicity and photosensitivity [28,37]. It was also reported that a syndrome known as "vermeersiekte" (*Geigeria ornativa* O. Hoffm. poisoning) was observed in sheep, induced by the consumption of plants of the genus *Geigeria* (Asteraceae) in South Africa [38]. The toxic principles are suggested to be sesquiterpene lactones, and their main target appears to be striated muscle, in which marked pathological changes occur [39]. Tetranortriterpenes were also identified as the toxins of *Melia azedarach* (Meliaceae) causing tremors, dyspnea, vomiting, and diarrhea in humans and animals [28]. Furthermore, a mixture of six saponins identified as quinovic acid glycosides (saponins A, D, E, G, J, K) from the Gabonese medicinal plant *Nauclea diderrichii* (Rubiaceae), induced synergistic *in vitro* DNA damage and chromosome mutations in mammalian cells [40]. This genotoxic activity was suggested to be due to the capacity of *Nauclea* saponins to reduce cell defense against oxidative stress through the inhibition of glutathione-*S*-transferase activity [40]. In this section, the synopsis of the four most common toxic terpenoids identified in African plants is presented.

19.5.1 Atractyloside

Atractyloside (**5**) is a diterpenoid glycoside produced by several plants (Table 19.1) used in ethnomedicines in Europe, Africa, South America, Asia, and the far East [26]. Throughout Africa and the Mediterranean, many of the plants

Table 19.1 Toxic Terpenoids from African Medicinal Plants

Toxins	Plant Source	Toxic Syndrome	Toxic Effects in Organ or Cells
Atractyloside (5) (diterpene glycoside)	*Atractylis gummifera, Atractylis carduus, Atractylodes lancea, Iphiona aucheri, Callilepsis laureola, Xanthium strumarium, Wedelia glauca* and *Wedelia asperrima* (Asteraceae) [25]	Gastrointestinal hemorrhage, hepatic necrosis [25]	Liver and kidney injuries [26]
Cicutoxin (11) (polyacetylenes)	*Cicuta virosa* (Apiaceae) [27]	Salivation, dilated pupils, abdominal pain, vomiting, muscular spasms, and convulsions [27]	Respiratory failure, CNS attack [27]
Daphnetoxin (14) (diterpenoid)	*Gnidia polycephala, Gnidia burchellii* (Thymeleaeceae) [28], *Daphne gnidium* L., *Daphne giraldi* L., (Thymeleaeceae) [29]	Diarrhea, emphysema [28] (ruminants and humans)	Kidneys, circulatory system and CNS injuries
Digoxin (2) (cardenolide)	*Digitalis lanata* [30] and *Digitalis purpurea* [28]	Arrhythmia, nausea and vomiting and diarrhea [31]	—
Gibberellic acid (15) (diterpenoid)	*Capsicum annuum* L. (Solanaceae) [32]	Toxic for mammalians [33,34]	Breast, lung [33], kidney and liver injuries [34], carcinogenic effects [33], neurotoxicity [35,36]

(−), not available.

used in ethnomedicines, herbal and alternative medications, contain **5** and it was found to cause acute proximal tubule necrosis leading to renal failure and death in humans [26]. This compound is sometimes responsible for human intoxication (gastrointestinal hemorrhage, hepatic necrosis) [25]. Compound **5** was reported to competitively inhibit the adenine nucleoside carrier in isolated mitochondria and thus blocking oxidative phosphorylation and this has been suggested to explain changes in carbohydrate metabolism and the toxic effects in liver and kidney [26]. It should be noted that, despite significant human exposure and many reported deaths, there is no rational approach to limit or prevent the compound

5-induced toxicity [26]. However, precaution should be taken when using plants with diterpenoid 5 in ethno-medicine.

19.5.2 Cicutoxin

Cicutoxin (13) (Figure 19.3), a major violent toxin of the *Cicuta* species, is a chemical in the class of C(17)-polyacetylenes bearing a long pi-bond conjugation system, a terminal hydroxyl, and an allylic hydroxyl in its structure, and a variety of its analogues have been isolated from the plant [41]. *Cicuta virosa* is a plant that occurs in South Africa [42]. This compound is not really a terpenoid but we preferred to discuss its side effects in this chapter, due to the relative similarity of it structure to linear terpenoids. Compound 13 acts as a GABA-receptor antagonist, and its action in the CNS is considered responsible for the highly dramatic and often fatal intoxication following ingestion [27]. The clinical signs may appear within an hour and include salivation, dilated pupils, abdominal pain, vomiting, muscular spasms, and convulsions [27]. Death from respiratory failure may occur within a few hours. Due to the acute nature of the poisoning, animals are often found dead not far from the habitat of the plant, where the plant's roots may have become exposed following a previous drop in the water level [27]. There are no characteristic pathological findings postmortem. Incidents, mainly involving cattle, occur sporadically in the Nordic countries [27].

19.5.3 Daphnetoxin

The daphnane type orthoester diterpene, found exclusively in plants of the family Thymelaeaceae, daphnetoxin (14) (Figure 19.3) identified in the South African *Gnidia* spp. (*Gnidia polycephala, Gnidia burchellii*), turned out to be toxic to ruminants, causing diarrhea and emphysema [28]. It was also found that the presence of compound 12 and the chemically related compound mezerein was responsible for the high toxicity of the Thymelaeaceae plant *Daphne mezereum* and that the poison is absorbed by the skin in humans. Compound 14 induces strong inflammation at the junction leading to serious damage of the kidneys, the circulatory system and CNS. It was also demonstrated that daphnetoxin-induced toxicity includes increase of proton leakage in the inner mitochondrial membrane, induction of the mitochondrial permeability transition pore, inhibition of ATP synthase and inhibition of the

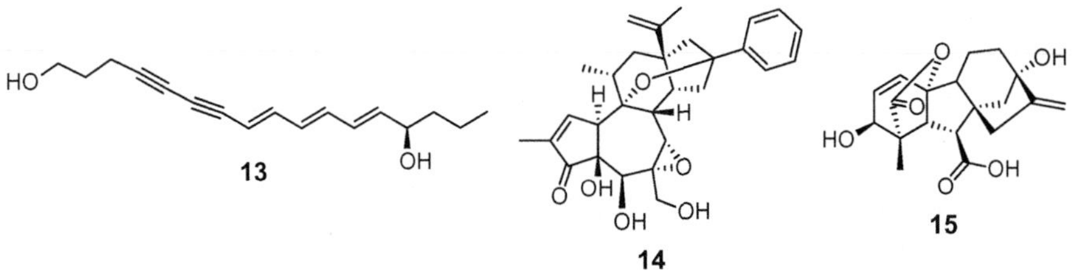

Figure 19.3 Chemical structures of cicutoxin (13), daphnetoxin (14), and gibberellic acid (15).

mitochondrial respiratory chain [29]. Peixoto et al. [43] also demonstrated that **14** can interact with mitochondrial membranes, disrupting the coupling between oxidation and phosphorylation, suggesting that the compound can promote large bioenergetic deficits leading to the loss of several functions that are important for the survival of the cell and the organism. However, it was also been shown that compound **14** has a potential cholesterol-lowering activity [44].

19.5.4 Digoxin

Digoxin (**2**) is a cardiac glycoside extracted from *Digitalis lanata* [30] and was also reported in the South African medicinal plant *Digitalis purpurea*, used traditionally for the treatment of congestive heart failure [28]. Its corresponding aglycone is known as digoxigenin, and its acetyl derivative as acetyldigoxin. This compound is widely used in the treatment of various heart conditions, namely atrial fibrillation, atrial flutter, and sometimes heart failure that cannot be controlled by other medication [45]. Despite its wide therapeutic use, compound **2** has a low therapeutic index and toxicity is common. It was found that toxicity can occur within the therapeutic range with old patients being especially at risk of developing digoxin toxicity [31]. Possible side effects include arrhythmia such as ventricular extrasystoles, ventricular bigeminy/trigeminy and atrial tachycardia, nausea and vomiting and occasionally, diarrhea, confusion especially in the elderly, yellow vision (xanthopsia), blurred vision, and photophobia [31].

19.5.5 Gibberellic Acid

Gibberellic acid (**15**) (Figure 19.3), a pentacyclic diterpene acid also known as gibberellin A3, GA, and GA3 is an endogenous plant growth regulator used worldwide in agriculture. Gibberellic acid occurs in some vegetables such as pepper (*Capsicum annuum*) [32] and olive [46]. Compound **15** plays an important role in many cellular processes, where it promotes stem elongation, overcomes dormancy in seeds and buds involved in parthenocarpic fruit development, flowering, and the mobilization of food reserves in grass seed germination [47].

Compound **15** has also been shown to cause alarming toxicity to mammalian systems, particularly in the breast, lung [33], kidney, and liver [34] of adult mice. It was reported that GA3 induces carcinogenic effects in adult Swiss albino mice [33]. Ozmen et al. [48] reported that compound **15** affects sexual differentiation and some physical parameters in laboratory mice. Celik and Tuluce [35] and Troudi et al. [36] also demonstrated that **15** can induce oxidative stress, leading to the generation of free radicals and causing cell damage in many organs, including heart, kidney, stomach, spleen, and liver of adult rats. Compound **15** was reported to have a neurotoxic effect on the cerebrum and cerebellum of suckling rats [49]. The neurotoxicity of gibberellic acid consisted of a significant increase in the malondialdehyde level and a decrease in the antioxidant enzyme activities of catalase, superoxide dismutase, glutathione peroxidase in the cerebrum and cerebellum of suckling pups, a decline of glutathione content and vitamin C and abnormal development of the external granular layer and a loss of Purkinje cells in the cerebellum of rats [49].

19.6 Organ-Protective Effect of Terpenoids from African Medicinal Plants

Some pharmacologically active terpenoids (Table 19.2) identified in African medicinal plants also showed protective effects on human and animal organs. This section is not aimed at reviewing the pharmacological activities of the reported terpenoids, but to discuss their protective effects at cellular or organ levels.

19.6.1 Betulinic Acid

Betulinic acid (16) (Figure 19.4) is a pentacyclic triterpenoid with antiretroviral, antimalarial, and anti-inflammatory and anticancer properties [56]. Its anticancer actions involve the inhibition of topoisomerase [56]. Compound 13 has been isolated in several African medicinal plants among which were *Diospyros caniculata* (Ebenaceae) [50], and *Irvingia gabonensis* (Ixonanthaceae) [52] as well as in many other plants worldwide. Compound 16 showed vascular protective effects via inhibition of reactive oxygen species and NF-kappaB activation in human umbilical vein endothelial cells [57]. Ekşioğlu-Demiralp et al. [58] demonstrated the renal protective effects of 16 as the compound attenuates renal ischemia/reperfusion-induced oxidant responses, improved microscopic damage and renal function by regulating the apoptotic function of leukocytes and inhibiting neutrophil infiltration.

19.6.2 Lupeol

Lupeol (17) (Figure 19.4) is a pharmacologically active pentacyclic triterpenoid found in several medicinal plants worldwide. Some African medicinal plants in which compound 17 occurs include *Ficus cordata* Thunb (Moraceae) [59], *Albizia adianthifolia* (Fabaceae) [60], *Mimosa invisa* (Mimosaceae) [61], *Klainedoxa gabonensis* Pierre ex Engl. (Irvingiaceae) [62], and *Turraeanthus africanus* (Meliaceae) [63]. Compound 17 showed hepatoprotective effects against Aflotoxin B1-induced damage in rats [72]. Donfack et al. [59] also demonstrated that 17 has hepatoprotective effects against CCl$_4$ intoxication. The protective effect of compound 17 in ameliorating the renal injury associated with hypercholesterolemia was also reported [73]. Prasad et al. [84] showed that 17 has protective effects against Benzo-[a]-pyrene-induced clastogenic changes in Swiss albino mice. It was also shown that this triterpene possesses cardioprotective effects which can be beneficial in hypercholesterolemic condition as it minimized the lipid abnormalities and abnormal biochemical changes induced by cholesterol and cholic acid fed rats [70]. It was also demonstrated that 17 reverses the alterations of electrolytes in serum

Table 19.2 Protective Terpenoids from African Medicinal Plants

Compounds	Plant Source	Pharmacological Activities	Protective Effects in Organ or Cells
Betulinic acid (**16**) (triterpenoid)	*Diospyros canaculata* and *Diospyros leucomelas* (Ebenaceae) [50,51], *Irvingia gabonensis* (Ixonanthaceae) [52], *Betula pubescens* (Betulaceae) [53], *Ziziphus mauritiana* (Rhamnaceae), *Prunella vulgaris* (Lamiaceae), *Triphyophyllum peltatum* (Dioncophyllaceae), *Ancistrocladus heyneanus* (Ancistrocladaceae), *Tetracera boiviniana* (Dilleniaceae), *Syzygium formosanum* (Myrtaceae) [51], *Chaenomeles sinensis* (Rosaceae) [54], *Pulsatilla chinensis* (Ranunculaceae) [55]	Antiretroviral, antimalarial, and anti-inflammatory, anticancer [56], antimicrobial [52] activities	Vascular [57] and renal [58] protections
Lupeol (**17**) (triterpenoid)	*Ficus cordata* Thunb (Moraceae) [59], *Albizia adianthifolia* (Fabaceae) [60], *Mimosa invisa* (Mimosaceae) [61], *Klainedoxa gabonensis* Pierre ex Engl. (Irvingiaceae) [62], *Turraeanthus africanus* (Meliaceae) [63], *Allanblackia monticola* Staner L.C. and *Allanblackia monticola* (Guttiferae) [64,65], *Fagara tessmannii* Engl. (Rutaceae) [66], *Harungana madagascariensis* (Guttiferae) [67], *Manilkara zapota* (L.) Van Royen (Sapotaceae) [68]	Antiprotozoal, antimicrobial, anti-inflammatory, antitumor, and chemopreventive properties [69]	Cardioprotective [70,71], hepatoprotective [59,72], renal protection [73]
Oleanolic acid (**18**) (triterpenoid)	*Aralia chihensis* L. var nude Nakai (Araliaceae), *Beta vulgaris* L. var. cicla L. (Chenopooliaceae), *Calendula officinalis* L. (Compositae), *Eugenia jaumbolana* Lam. (Myrtaceae), *Ganoderma lucidum* Karst, *Glechoma hederacea* L. (Labiatae), *Ligstrum lucidum* Ait. (Oleaceae), *Luffa cyclindrica* Roem. (Cucurbitaceae), *Oieandra nerii/blia* L. Panar ginseng C.A. Meyer (Araliaceae), *Sapindus mukorossi* Gaertn (Sapindaceae), *Swertia mileensis* He et Shi (Gentianaceae), *Swertia japonica* Makino (Gentianaceae), *Tetrapanax papyriferum* L. (Araliaceae), *Tinospora sagittata* G. (Menispermaceae) [74], *Irvingia gabonensis* (Ixonanthaceae) [52], *Syzygium aromaticum* (Myrtaceae) [75], *Gardenia volkensii* (Rubiaceae) [76]	Weak anti-HIV [77] and weak anti-HCV, antitumor and antiviral [74], antimicrobial [52] activities	Cardioprotective [78], hepatoprotective [74]

| Ursolic acid (19) (triterpenoid) | *Stereospermum acuminatissimum* K. Schum. (Bignoniaceae) [79], *Ficus mucuso* (Moraceae) [80], *Diospyros crassiflora* (Hiern) (Ebenaceae) [81], *Cussonia bancoensis* (Araliaceae) [82], *Mitragyna stipulosa* (Rubiaceae) [83], *Calluna vulgaris* (Ericaceae), *Eribotrya japonica* Lindl. (Rosaceae), *Eucalyptus hybrid* (Myrtaceae), *Glechoma heakracea* L. (Labiatae), *Melaleuca leucadendron* L. (Myrtaceae), *Ocimum sanctum* L. (Labiatae), *Rosmarinus oficinalis* L. (Labiatae), *Pyrola rotundifolia* (Pyrolaceae), *Psychotria serpens* L. (Rubiaceae), *Sambucus chinesis* Lindl (Caprifoliaceae), *Solanwn incanum* L. (Solanaceae), *Tripierospermum taiwanense* (Gentianaceae) [74] | — | Hepatoprotective [74] |

HCV, hepatitis C virus; HIV, human immunodeficiency virus; (−), not reported.

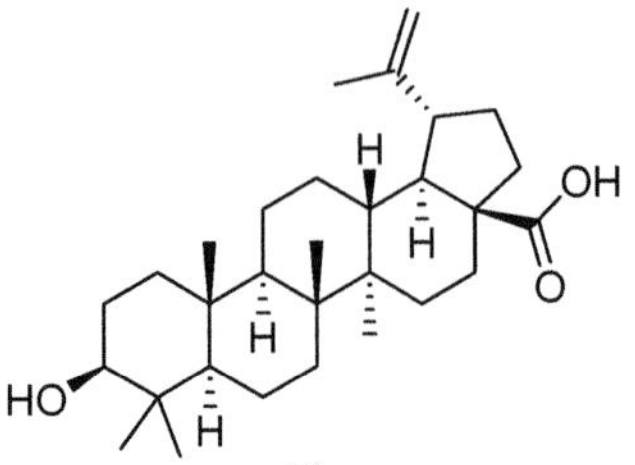

Figure 19.4 Chemical structures of four organ-protective terpenoids identified in African medicinal plants. Betulinic acid (**16**), lupeol (**17**), oleanolic acid (**18**), ursolic acid (**19**).

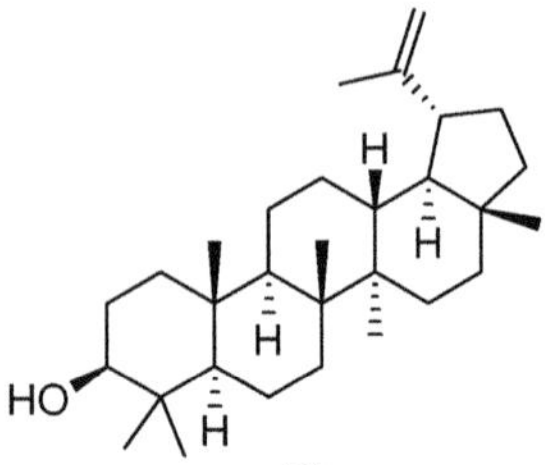

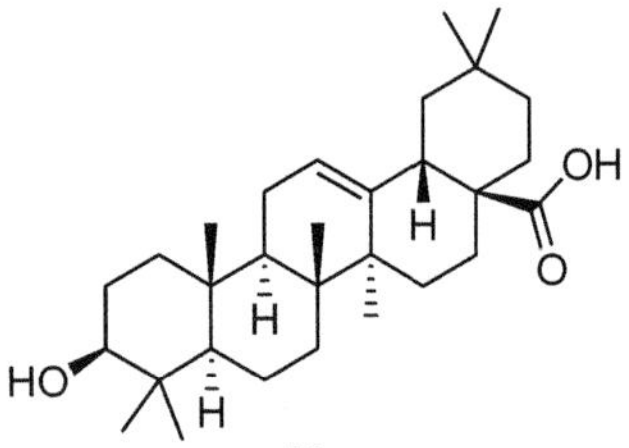

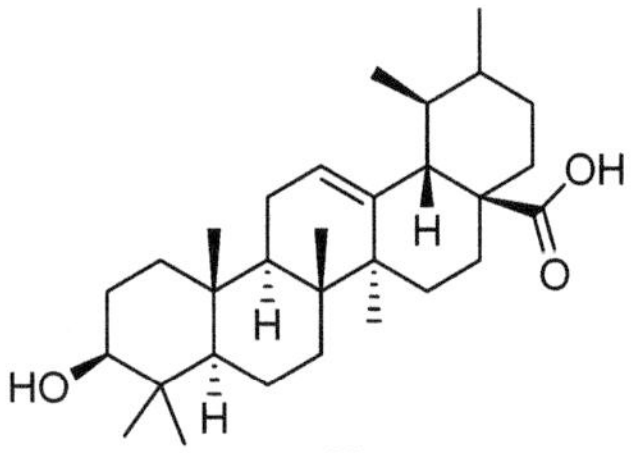

and cardiac tissue induced by cyclophosphamide. Furthermore, compound **17** could preserve membrane permeability, highlighting its protective effect against cyclophosphamide-induced cardiotoxicity [71].

19.6.3 Oleanolic Acid

Oleanolic acid (**18**) (Figure 19.4) was reported in several African medicinal plants such as *Irvingia gabonensis* [52], *Syzygium aromaticum* [75], *Gardenia volkensii* [76], and many other plants worldwide (Table 19.1). Compound **18** was found to decrease oxidative stress, apoptosis, hexosamine biosynthetic pathway flux (HBP) and proteasomal activity following ischemia-reperfusion [78]; Compound **18** attenuates acute and chronic hyperglycemia-mediated pathophysiologic molecular events (oxidative stress, apoptosis, HBP, ubiquitin–proteasome system) and thereby improves contractile function in response to ischemia-reperfusion, explaining its cardioprotective effects [78]. The mechanism of hepatoprotection by **18** and its isomer, ursolic acid (**19**) involve the inhibition of toxicant activation and the enhancement of the body defense systems [74].

19.6.4 Ursolic Acid

Ursolic acid (**19**) (Figure 19.4) is a pentacyclic triterpene acid also known as urson, prunol, or malol. Compound **19** was isolated from many African medicinal plants among which were *Stereospermum acuminatissimum* K. Schum. (Bignoniaceae) [79],

Ficus mucuso (Moraceae) [80], *Diospyros crassiflora* (Hiern) (Ebenaceae) [81], *Cussonia bancoensis* (Araliaceae) [82], and *Mitragyna stipulosa* (Rubiaceae) [83]. Compound **19** is used in cosmetics and is also capable of inhibiting various cancer cell types by blocking the STAT3 (signal transducer and activator of transcription 3) activation pathway [85,86]. It was found that many medicinal plants containing **19** such as *Eucalyptus hybrid*, *Sambucus chinensis* Lindl (Caprifoliaceae), *Solanum incanum* L. (Solanaceae), and *Tripierospermum taiwanense* (Gentianaceae) have hepatoprotective effects [74]. As reported above, the mechanism of hepatoprotection by **19** is similar to that of **18** [74]. Compound **19** has a wide range of pharmacological activity in humans, displaying antiprotozoal, antimicrobial, anti-inflammatory, antitumor, and chemopreventive properties [69].

19.7 Conclusion

In this chapter, we call attention to both harmful and protective effects of some terpenoids and mostly those identified in African medicinal plants. Although we focused only on five toxic terpenoids, namely cicutoxin, atractyloside, daphnetoxin, digoxin, and gibberellic acid, it is worth to see that they occur in several medicinal plants among which some were cited in here. Therefore, medicinal consumption of plant bearing these compounds is not totally safe and caution should be taken. In the other part of this chapter, we also discussed four most commonly occurring plant terpenoids, betulinic acid, lupeol, oleanolic acid, and ursolic acid, as well as their occurrence in several medicinal plants. The information pooled in this review provides a useful tool when selecting medicinal plants for the screening of their toxicological studies, as well as for their protective effects.

References

[1] Awouafack MD, Tane P, Kuete V, Eloff JN. 2—Sesquiterpenes from the medicinal plants of Africa. In: Kuete V, editor. Medicinal plant research in Africa. Oxford: Elsevier; 2013. p. 33–103.

[2] Tchimene MK, Okunji CO, Iwu MM, Kuete V. 1—Monoterpenes and related compounds from the cedicinal plants of Africa. In: Kuete V, editor. Medicinal plant research in Africa. Oxford: Elsevier; 2013. p. 1–32.

[3] Sandjo LP, Kuete V. 3—Diterpenoids from the medicinal plants of Africa. In: Kuete V, editor. Medicinal plant research in Africa. Oxford: Elsevier; 2013. p. 105–33.

[4] Sandjo LP, Kuete V. 4—Triterpenes and steroids from the medicinal plants of Africa. In: Kuete V, editor. Medicinal plant research in Africa. Oxford: Elsevier; 2013. p. 135–202.

[5] Lawal OA, Ogunwande IA. 5—Essential oils from the medicinal plants of Africa. In: Kuete V, editor. Medicinal plant research in Africa. Oxford: Elsevier; 2013. p. 203–24.

[6] Stoessl A, Stothers JB, Ward EWB. Sesquiterpenoid stress compounds of the solanaceae. Phytochemistry 1976;15(6):855–72.

[7] Steele CL, Katoh S, Bohlmann J, Croteau R. Regulation of oleoresinosis in grand fir (*Abies grandis*). Differential transcriptional control of monoterpene, sesquiterpene, and diterpene synthase genes in response to wounding. Plant Physiol 1998;116(4):1497—504.

[8] Wang G, Tang W, Bidigare R. Terpenoids as therapeutic drugs and pharmaceutical agents. In: Zhang L, Demain A, editors. Natural products. Totowa, NJ: Humana Press; 2005. p. 197—227.

[9] Kashani JS. Terpene toxicity. <http://emedicine.medscape.com/article/818675-overview#a0104>; 2011 [accessed 16.07.13].

[10] Bucheler R, Gleiter CH, Schwoerer P, Gaertner I. Use of nonprohibited hallucinogenic plants: increasing relevance for public health? A case report and literature review on the consumption of *Salvia divinorum* (Diviner's Sage). Pharmacopsychiatry 2005;38 (1):1—5.

[11] Bronstein AC, Spyker DA, Cantilena Jr LR, Green JL, Rumack BH, Giffin SL. Annual Report of the American Association of Poison Control Centers' National Poison Data System (NPDS): 27th Annual Report. Clin Toxicol (Phila) 2010;48(10): 979—1178.

[12] Thoppil RJ, Bishayee A. Terpenoids as potential chemopreventive and therapeutic agents in liver cancer. World J Hepatol 2011;3(9):228—49.

[13] Wink M. Mode of action and toxicology of plant toxins and poisonous plants. Wirbeltierforschung in der Kulturlandschaft 2009;421:93—112.

[14] Soo Hoo GW, Hinds RL, Dinovo E, Renner SW. Fatal large-volume mouthwash ingestion in an adult: a review and the possible role of phenolic compound toxicity. J Intensive Care Med 2003;18(3):150—5.

[15] Wang Y, Hamburger M, Chen C, Costall B, Naylor R, Jenner P, et al. Neurotoxic sesquiterpenoids from the yellow star thistle *Centaurea solstitialis* L. (Asteraceae). Helv Chim Acta 1991;74:117—23.

[16] Stewart MJ, Steenkamp V. The biochemistry and toxicity of atractyloside: a review. Ther Drug Monit 2000;22(6):641—9.

[17] Potter DM, Baird MS. Carcinogenic effects of ptaquiloside in bracken fern and related compounds. Br J Cancer 2000;83(7):914—20.

[18] Kuete V, Viertel K, Efferth T. 18—Antiproliferative potential of African medicinal plants. In: Kuete V, editor. Medicinal plant research in Africa. Oxford: Elsevier; 2013. p. 711—24.

[19] Ndhlala AR, Amoo SO, Ncube B, Moyo M, Nair JJ, Van Staden J. 16—Antibacterial, antifungal, and antiviral activities of African medicinal plants. In: Kuete V, editor. Medicinal plant research in Africa. Oxford: Elsevier; 2013. p. 621—59.

[20] Alqasoumi SI, Abdel-Kader MS. Terpenoids from *Juniperus procera* with hepatoprotective activity. Pak J Pharm Sci 2012;25(2):315—22.

[21] Francis G, Kerem Z, Makkar HP, Becker K. The biological action of saponins in animal systems: a review. Br J Nutr 2002;88(6):587—605.

[22] Cuadrado I, Fernandez-Velasco M, Bosca L, de Las Heras B. Labdane diterpenes protect against anoxia/reperfusion injury in cardiomyocytes: involvement of AKT activation. Cell Death Dis 2011;2:e229.

[23] Fouad AA, Al-Mulhim AS, Jresat I. Cannabidiol treatment ameliorates ischemia/reperfusion renal injury in rats. Life Sci 2012;91(7—8):284—92.

[24] Yoo KY, Park SY. Terpenoids as potential anti-Alzheimer's disease therapeutics. Molecules 2012;17(3):3524—38.

[25] Sanchez JF, Kauffmann B, Grelard A, Sanchez C, Trezeguet V, Huc I, et al. Unambiguous structure of atractyloside and carboxyatractyloside. Bioorg Med Chem Lett 2012;22(8):2973—5.

[26] Obatomi DK, Bach PH. Biochemistry and toxicology of the diterpenoid glycoside atractyloside. Food Chem Toxicol 1998;36(4):335—46.

[27] Bernhoft A. Bioactive compounds in plants: benefits and risks for man and animals. Proceedings from a symposium held at The Norwegian Academy of Science and Letters, Oslo, 13—14 November 2008. Oslo: Novus forlag; 2008 2010.

[28] Botha CJ, Penrith ML. Poisonous plants of veterinary and human importance in southern Africa. J Ethnopharmacol 2008;119(3):549—58.

[29] Diogo CV, Félix L, Vilela S, Burgeiro A, Barbosa IA, Carvalho MJM, et al. Mitochondrial toxicity of the phyotochemicals daphnetoxin and daphnoretin—relevance for possible anti-cancer application. Toxicol In Vitro 2009;23(5):772—9.

[30] Hollman A. Digoxin comes from *Digitalis lanata*. Br Med J 1996;312:912.

[31] Digoxin toxicity. GP notebook. <http://www.gpnotebook.co.uk/simplepage.cfm?ID=-1321926632>; 2001. [accessed 24.07.13].

[32] Arous S, Boussaid M, Marrakchi M. Plant regeneration from zygotic embryo hypocotyls of Tunisian chilli (*Capsicum annuum* L.). J Appl Hortic 2011;3:17—22.

[33] El-Mofty MM, Sakr SA, Rizk AM, Moussa EA. Carcinogenic effect of gibberellin A3 in Swiss albino mice. Nutr Cancer 1994;21(2):183—90.

[34] Ustun H, Tecimer T, Ozmen M, Topcuoglu S, Bozcuk F. Effects of gibberellic acid and benzoprenin on mice: histopathologic review. Ank Patol Bült 1992;9:36—40.

[35] Celik I, Turker M, Tuluce Y. Abcisic acid and gibberellic acid cause increased lipid peroxidation and fluctuated antioxidant defense systems of various tissues in rats. J Hazard Mater 2007;148(3):623—9.

[36] Troudi A, Mahjoubi Samet A, Zeghal N. Hepatotoxicity induced by gibberellic acid in adult rats and their progeny. Exp Toxicol Pathol 2010;62(6):637—42.

[37] Kellerman T, Coetzer J, Naudé T, Botha C, Town C. Plant poisonings and mycotoxicoses of livestock in Southern Africa. 2nd ed. Cape Town: Oxford University Press; 2005.

[38] Kellerman TS, Naude TW, Fourie N. The distribution, diagnoses and estimated economic impact of plant poisonings and mycotoxicoses in South Africa. Onderstepoort J Vet Res 1996;63(2):65—90.

[39] van der Lugt JJ, van Heerden J. Experimental vermeersiekte (*Geigeria ornativa* O. Hoffm. poisoning) in sheep. II: histological and ultrastructural lesions. J S Afr Vet Assoc 1993;64(2):82—8.

[40] Liu W, Di Giorgio C, Lamidi M, Elias R, Ollivier E, De Meo MP. Genotoxic and clastogenic activity of saponins extracted from *Nauclea* bark as assessed by the micronucleus and the comet assays in Chinese Hamster Ovary cells. J Ethnopharmacol 2011; 137(1):76—83.

[41] Uwai K, Ohashi K, Takaya Y, Ohta T, Tadano T, Kisara K, et al. Exploring the structural basis of neurotoxicity in C(17)-polyacetylenes isolated from water hemlock. J Med Chem 2000;43(23):4508—15.

[42] Chapter 2. Material and methods. <https://ujdigispace.uj.ac.za/bitstream/handle/10210/239/Chapter2.pdf?> [accessed 03.08.13].

[43] Peixoto F, Carvalho MJ, Almeida J, Matos PA. Daphnetoxin interacts with mitochondrial oxidative phosphorylation and induces membrane permeability transition in rat liver. Planta Med 2004;70(11):1064—108.

[44] Zhang Y, Zhang H, Hua S, Ma L, Chen C, Liu X, et al. Identification of two herbal compounds with potential cholesterol-lowering activity. Biochem Pharmacol 2007;274 (6):940−7.

[45] Sticherling C, Oral H, Horrocks J, Chough SP, Baker RL, Kim MH, et al. Effects of digoxin on acute, atrial fibrillation-induced changes in atrial refractoriness. Circulation 2000;102(20):2503−8.

[46] Chaari-Rkhis A, Maalej M, Ouled Messaoud S, Drira N. *In vitro* vegetative growth and flowering of olive tree in response to GA3 treatment. Afr J Biotechnol 2006;5: 297−302.

[47] Salisbury F, Ross C. Plant physiology. Belmont, CA: Wadsworth; 1992.

[48] Ozmen M, Topcuogolu S, Bozcuk S, Bozcuk N. Effects of abscisic acid and gibberellic acid on sexual differentiation and some physiological parameters of laboratory mice. Trends J Biol 1995;19:357−64.

[49] Troudi A, Bouaziz H, Soudani N, Ben Amara I, Boudawara T, Touzani H, et al. Neurotoxicity and oxidative stress induced by gibberellic acid in rats during late pregnancy and early postnatal periods: biochemical and histological changes. Exp Toxicol Pathol 2012;64(6):583−90.

[50] Tangmouo J, Lontsi D, Ngounou F, Kuete V, Meli A, Manfouo R, et al. Diospyrone, a new coumarinylbinaphthoquinone from *Diospyros canaliculata* (Ebenaceae): structure and antimicrobial activity. Bull Chem Soc Ethiop 2005;19:81−8.

[51] Zuco V, Supino R, Righetti SC, Cleris L, Marchesi E, Gambacorti-Passerini C, et al. Selective cytotoxicity of betulinic acid on tumor cell lines, but not on normal cells. Cancer Lett 2002;175(1):17−25.

[52] Kuete V, Wabo GF, Ngameni B, Mbaveng AT, Metuno R, Etoa FX, et al. Antimicrobial activity of the methanolic extract, fractions and compounds from the stem bark of *Irvingia gabonensis* (Ixonanthaceae). J Ethnopharmacol 2007;114(1): 54−60.

[53] Tan Y, Yu R, Pezzuto JM. Betulinic acid-induced programmed cell death in human melanoma cells involves mitogen-activated protein kinase activation. Clin Cancer Res 2003;9(7):2866−75.

[54] Gao H, Wu L, Kuroyanagi M, Harada K, Kawahara N, Nakane T, et al. Antitumor-promoting constituents from *Chaenomeles sinensis* Koehne and their activities in JB6 mouse epidermal cells. Chem Pharm Bull (Tokyo) 2003;51(11):1318−21.

[55] Ji ZN, Ye WC, Liu GG, Hsiao WL. 23-Hydroxybetulinic acid-mediated apoptosis is accompanied by decreases in bcl-2 expression and telomerase activity in HL-60 cells. Life Sci 2002;72(1):1−9.

[56] Chowdhury AR, Mandal S, Mittra B, Sharma S, Mukhopadhyay S, Majumder HK. Betulinic acid, a potent inhibitor of eukaryotic topoisomerase I: identification of the inhibitory step, the major functional group responsible and development of more potent derivatives. Med Sci Monit 2002;8(7):BR254−65.

[57] Yoon JJ, Lee YJ, Kim JS, Kang DG, Lee HS. Protective role of betulinic acid on TNF-alpha-induced cell adhesion molecules in vascular endothelial cells. Biochem Biophys Res Commun 2010;391(1):96−101.

[58] Eksioglu-Demiralp E, Kardas ER, Ozgul S, Yagci T, Bilgin H, Sehirli O, et al. Betulinic acid protects against ischemia/reperfusion-induced renal damage and inhibits leukocyte apoptosis. Phytother Res 2010;24(3):325−32.

[59] Donfack J, Kengap R, Ngameni B, Chuisseu P, Tchana A, Buonocore D, et al. *Ficus cordata* Thunb (Moraceae) is a potential source of some hepatoprotective and antioxidant compounds. Pharmacologia 2011;2(5):137−45.

[60] Tamokou Jde D, Simo Mpetga DJ, Keilah Lunga P, Tene M, Tane P, Kuiate JR. Antioxidant and antimicrobial activities of ethyl acetate extract, fractions and compounds from stem bark of *Albizia adianthifolia* (Mimosoideae). BMC Complement Altern Med 2012;12:99.

[61] Nana F, Sandjo LP, Keumedjio F, Kuete V, Ngadjui BT. A new fatty aldol ester from the aerial part of *Mimosa invisa* (Mimosaceae). Nat Prod Res 2012;26(19):1831−6.

[62] Wansi JD, Chiozem DD, Tcho AT, Toze FA, Devkota KP, Ndjakou BL, et al. Antimicrobial and antioxidant effects of phenolic constituents from *Klainedoxa gabonensis*. Pharm Biol 2010;48(10):1124−9.

[63] Djemgou PC, Gatsing D, Hegazy ME, El-Hamd Mohamed AH, Ngandeu F, Tane P, et al. Turrealabdane, turreanone and an antisalmonellal agent from *Turraeanthus africanus*. Planta Med 2010;76(2):165−71.

[64] Azebaze AG, Menasria F, Noumi LG, Nguemfo EL, Tchamfo MF, Nkengfack AE, et al. Xanthones from the seeds of *Allanblackia monticola* and their apoptotic and antiproliferative activities. Planta Med 2009;75(3):243−8.

[65] Nguemfo EL, Dimo T, Dongmo AB, Azebaze AG, Alaoui K, Asongalem AE, et al. Anti-oxidative and anti-inflammatory activities of some isolated constituents from the stem bark of *Allanblackia monticola* Staner L.C (Guttiferae). Inflammopharmacology 2009;17(1):37−41.

[66] Mbaze LM, Poumale HM, Wansi JD, Lado JA, Khan SN, Iqbal MC, et al. Alpha-glucosidase inhibitory pentacyclic triterpenes from the stem bark of *Fagara tessmannii* (Rutaceae). Phytochemistry 2007;68(5):591−5.

[67] Kouam SF, Ngadjui BT, Krohn K, Wafo P, Ajaz A, Choudhary MI. Prenylated anthronoid antioxidants from the stem bark of *Harungana madagascariensis*. Phytochemistry 2005;66(10):1174−9.

[68] Fayek NM, Monem AR, Mossa MY, Meselhy MR. New triterpenoid acyl derivatives and biological study of *Manilkara zapota* (L.) Van Royen fruits. Pharmacognosy Res 2013;5(2):55−9.

[69] Gallo M, Sarachine M. Biological activities of lupeol. Int J Biomed Pharm Sci 2009; 3(1):46−66.

[70] Sudhahar V, Kumar SA, Sudharsan PT, Varalakshmi P. Protective effect of lupeol and its ester on cardiac abnormalities in experimental hypercholesterolemia. Vascul Pharmacol 2007;46(6):412−8.

[71] Sudharsan PT, Mythili Y, Selvakumar E, Varalakshmi P. Lupeol and its ester inhibit alteration of myocardial permeability in cyclophosphamide administered rats. Mol Cell Biochem 2006;292(1−2):39−44.

[72] Preetha SP, Kanniappan M, Selvakumar E, Nagaraj M, Varalakshmi P. Lupeol ameliorates aflatoxin B1-induced peroxidative hepatic damage in rats. Comp Biochem Physiol C Toxicol Pharmacol 2006;143(3):333−9.

[73] Sudhahar V, Ashok Kumar S, Varalakshmi P, Sujatha V. Protective effect of lupeol and lupeol linoleate in hypercholesterolemia associated renal damage. Mol Cell Biochem 2008;317(1−2):11−20.

[74] Liu J. Pharmacology of oleanolic acid and ursolic acid. J Ethnopharmacol 1995;49(2): 57−68.

[75] Khathi A, Masola B, Musabayane CT. Effects of *Syzygium aromaticum*-derived oleanolic acid on glucose transport and glycogen synthesis in the rat small intestine. J Diabetes 2013;5(1):80−7.

[76] Kinuthia EW, Langat MK, Mwangi EM, Cheplogoi PK. Constituents of Kenyan *Gardenia volkensii*. Nat Prod Commun 2012;7(1):13−4.

[77] Mengoni F, Lichtner M, Battinelli L, Marzi M, Mastroianni CM, Vullo V, et al. *In vitro* anti-HIV activity of oleanolic acid on infected human mononuclear cells. Planta Med 2002;68(2):111—4.

[78] Mapanga RF, Rajamani U, Dlamini N, Zungu-Edmondson M, Kelly-Laubscher R, Shafiullah M, et al. Oleanolic acid: a novel cardioprotective agent that blunts hyperglycemia-induced contractile dysfunction. PLoS One 2012;7(10):e47322.

[79] Ramsay KS, Wafo P, Ali Z, Khan A, Oluyemisi OO, Marasini BP, et al. Chemical constituents of *Stereospermum acuminatissimum* and their urease and alpha-chymotrypsin inhibitions. Fitoterapia 2012;83(1):204—8.

[80] Bankeu JJ, Mustafa SA, Gojayev AS, Lenta BD, Tchamo Noungoue D, Ngouela SA, et al. Ceramide and Cerebroside from the stem bark of *Ficus mucuso* (Moraceae). Chem Pharm Bull (Tokyo) 2010;58(12):1661—5.

[81] Akak CM, Djama CM, Nkengfack AE, Tu PF, Lei LD. New coumarin glycosides from the leaves of *Diospyros crassiflora* (Hiern). Fitoterapia 2010;81(7):873—7.

[82] Tapondjou LA, Lontsi D, Sondengam BL, Choi J, Lee KT, Jung HJ, et al. *In vivo* antinociceptive and anti-inflammatory effect of the two triterpenes, ursolic acid and 23-hydroxyursolic acid, from *Cussonia bancoensis*. Arch Pharm Res 2003;26(2):143—6.

[83] Tapondjou LA, Lontsi D, Sondengam BL, Choudhary MI, Park HJ, Choi J, et al. Structure—activity relationship of triterpenoids isolated from *Mitragyna stipulosa* on cytotoxicity. Arch Pharm Res 2002;25(3):270—4.

[84] Prasad S, Kumar Yadav V, Srivastava S, Shukla Y. Protective effects of lupeol against benzo[a]pyrene induced clastogenicity in mouse bone marrow cells. Mol Nutr Food Res 2008;52(10):1117—20.

[85] Shishodia S, Majumdar S, Banerjee S, Aggarwal BB. Ursolic acid inhibits nuclear factor-kappaB activation induced by carcinogenic agents through suppression of IkappaBalpha kinase and p65 phosphorylation: correlation with down-regulation of cyclooxygenase 2, matrix metalloproteinase 9, and cyclin D1. Cancer Res 2003;63 (15):4375—83.

[86] Pathak AK, Bhutani M, Nair AS, Ahn KS, Chakraborty A, Kadara H, et al. Ursolic acid inhibits STAT3 activation pathway leading to suppression of proliferation and chemosensitization of human multiple myeloma cells. Mol Cancer Res 2007;5(9): 943—55.

20 Harmful and Protective Effects of Phenolic Compounds from African Medicinal Plants

Armelle T. Mbaveng[1], Qiaoli Zhao[2] and Victor Kuete[1]

[1]Faculty of Science, Department of Biochemistry, University of Dschang, Dschang, Cameroon, [2]Department of Pharmaceutical Biology, Institute of Pharmacy and Biochemistry, University of Mainz, Mainz, Germany

20.1 Introduction

Phenolic compounds are classes of chemicals consisting of hydroxyl group (-OH) bonded directly to an aromatic hydrocarbon group. They comprise flavonoids, phenolic acids, tannins, lignans, coumarins, quinones, xanthones, among others [1−8]. Flavonoids constitute the largest group of plant phenolics, accounting for over half of the 8000 naturally occurring phenolic compounds [9]. Variations in substitution patterns to ring C in the structure of these compounds result in the major flavonoid classes such as flavonols, flavones, flavanones, flavonols, isoflavones, and anthocyanidins [4,10]. Phenolic acids as well as flavonoids also constitute an important class of phenolic compounds with bioactive functions, usually found in plants and food products [10]. Phenolic acids comprise two subgroups, namely the hydroxybenzoic and the hydroxycinnamic acids [1,10]. The most commonly found hydroxybenzoic acids include gallic, *p*-hydroxybenzoic, protocatechuic, vanillic, and syringic acids, while commonly occurring hydroxycinnamic acids are caffeic, ferulic, *p*-coumaric, and sinapic acids [10,11]. However, quinones, lignans, xanthones, coumarins, and other groups also exist in considerable numbers as well as many simple monocyclic phenolics [12]. In plants, flavonoids play a role in flower and seed pigmentation, in plant fertility and reproduction, and in various defense reactions to protect against abiotic stresses such as UV light or biotic stresses such as predator and pathogen attacks [13]. Several bioactive polyphenolic compounds promote human health as they have a wide range of effects *in vivo* and *in vitro*, including chelation of metals such as iron, it is prudent to test whether the regular consumption of bioactive polyphenolic components impair the utilization of dietary iron [14]. Phenolic compounds

Toxicological Survey of African Medicinal Plants. DOI: http://dx.doi.org/10.1016/B978-0-12-800018-2.00020-0

have been intensively studied for their possible toxic effects on the environment and threats to human health [15]. Some phenolics are extremely toxic for aquatic organisms such as fish and shellfish at very low concentrations [15]. Phenolic compounds such as catechol (**1**) and hydroquinone (**2**) (Figure 20.1) are broadly produced in plants. Catechol is a common intermediate in the aerobic degradation of aromatic compounds by microorganisms [15]. It has been shown that catechol can be cell transformed and is genotoxic to cultured microorganisms and mammalian cells [15]. Hydroquinone also showed positive response in mammalian cell gene mutation assays *in vitro*, suggesting its mutagenic potency [15,16]. In African plants, the polyphenolic compound gossypol (**3**), a pigment present in cotton seed (*Gossypium* spp.), has been found highly toxic to pigs causing dyspnea, anorexia, unthriftiness, and diarrhea [15]. Nonetheless, the state of the art in phenolic-induced threats from African medicinal plants has not been reviewed. This chapter therefore identifies and pools together the phenolic compounds with potential toxic effects detected in African medicinal plants. We also discussed compounds with beneficial effects on human heath such as their cardio-, hepato-, and nephroprotective effects, as well as their occurrence in African medicinal plants.

20.2   Medicinal Importance of Plant Phenolics

When discussing the importance of this group of secondary metabolites, it should be noted that it comprises a variety of classes with different functional groups, and therefore we cannot generalize the beneficial effects of phenolic compounds. However, it was suggested that polyphenols could be serious candidates to explain the protective effects of plant-derived foods and beverages [17]. Natural phenolics have been reported to have antimicrobial, antioxidant, antiviral, anti-inflammatory, cytotoxic, and vasodilatory effects [1,4,18−21]. Resveratrol (**4**) activates human SIRT1 (gene encoding Sirtuin 1), extends the lifespan of budding yeast, *Saccharomyces cerevisiae*, and may be a sirtuin-activating compound [22,23]. Other sirtuin-activating phenolics include butein (**5**), piceatannol (**6**), isoliquiritigenin (**7**), fisetin (**8**), and quercetin (**9**) [22]. Resveratrol has been shown to increase longevity of *Caenorhabditis elegans* and *Drosophila melanogaster* [24]. It has been demonstrated that the longevity increase may be due to a caloric restriction effect [25]. Based on the up-to-date reports, it is agreed that the specific intake of foods and

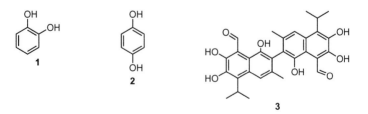

Figure 20.1 Chemical structures of some catechol (**1**), hydroquinone (**2**) and gossypol (**3**).

beverages containing relatively high concentrations of flavonoids plays an important role in reducing cardiovascular disease (CVD) risk as they improve the vascular function and modulate inflammation [17]. This is very interesting as CVD remains the worldwide leading cause of mortality and a major cause of morbidity and disability [26]. The mechanisms by which polyphenols express their beneficial properties involve their interaction with molecular signaling pathways and related machinery that regulate cellular processes such as inflammation [17,27]. These include **4**, rutin (**10**) and quercitrin (**11**), glycosides of quercetin, curcumin (**12**), piceatannol, genistein (**13**), morin (**14**), hesperidin (**15**), and the hemisynthetic diosmin (**16**) [17] (Figure 20.1). Several flavonoids and related polyphenols are active in rheumatoid arthritis models such as collagen- and adjuvant-induced arthritis, including **4**, **9**, **10**, **15**, nobiletin (**17**), alpha-glucosylhesperidin, catechin (**18**), the chalcone derivative 1-(2,4-dichlorophenyl)-3-(3-(6,7-dimethoxy-2-chloroquinolinyl))-2-propen-1-one, epigallocatechin-3-gallate or EGCG (**19**), genistein (**13**), and others [17]. Polyphenols also combat tissue damage originated by ischemia reperfusion episodes such as resveratrol, theaflavin, 3-methoxypuerarin, catechin, and EGCG [17]. Tea catechins protect against postischemic myocardial remodeling *via* anti-inflammatory actions [17]. Doxorubicin, an anthracycline antibiotic, closely related to the natural product daunomycin, is clinically used in the treatment of cancer.

20.3 Toxic Phenolics and Mode of Action

Natural phenolics were found to have both anticarcinogenic effect and a carcinogenic, DNA-, as well as mutagenic potential. However, it also has been shown that mature individuals seem to rapidly metabolize most of phenolic compounds, suggesting that toxic and mutagenic effects could be minimized to low concentrations. Though EGCG (**19**) for example rapidly induces detoxifying nuclear factor (erythroid-derived 2)-like 2 (Nrf2) transcription factor activity (antioxidant effect), it also leads to rapid degradation of the phenolic molecules. The basic moiety of phenolic compounds, phenol itself is a strong neurotoxin, and its presence in the bloodstream can lead to instant death as it shuts down the neural transmission system. Phenol and its vapors have corrosive effects on eyes, skin, and respiratory tract, while hydroquinone (**2**) is also toxic [28,29]. It was also reported that high intake of flavonoids during pregnancy could increase risk of neonatal leukemia [30,31]. Naturally occurring polyphenols bind with non-heme iron *in vitro* in model systems [32], possibly reducing its absorption especially in vegetarian nutrition, a situation that can lead to anemia. It has been shown that rotenoids produced by some legumes can inhibit the mitochondrial respiratory chain [33]. They appear therefore to be inhibitors of cellular respiration. Phenolic compounds such as cucurbitacins block cell division and vesicle transport along microtubules [33] therefore acting as cytotoxins. Several furanocoumarins are mutagenic and carcinogenic [33] and react through alkylating and DNA intercalating. Furanocoumarins from Apiaceae can penetrate skin and intercalate dermal cells. When the skin is exposed to sun light, the furanocoumarins alkylate DNA, which kills the cells and induces strong blister formation and necrosis [33]. Rodents metabolize coumarin largely to 3,4-coumarin epoxide, a toxic unstable compound that on further differential metabolism may explain coumarin's ability to cause liver cancer in rats and lung tumors in mice [34,35]. Coumarin also was found to be hepatotoxic in rats (but less so in mice). Its detoxification involves the metabolism to its hydroxylated derivative, 7-hydroxycoumarin, having lower toxicity. The "tolerable daily intake" (TDI) of 0.1 mg coumarin per kg body weight (bw) has been established by the German Federal Institute for Risk Assessment [35]. However, the US Occupational Safety and Health Administration (OSHA) regards coumarin to be not classifiable as carcinogenic to humans [36].

20.4 Harmful Effects of Phenolics from African Medicinal Plants

In Africa, apart from hemodynamic causes and infections, herbal remedies contribute to both morbidity and mortality, although these causes often go unrecognized [37]. Many phenolic compounds found in African medicinal plants showed various side effects (Table 13.1). Below we discuss four of the naturally occurring toxic phenolic compounds. Their chemical structures are depicted in Figures 20.2 and 20.3.

20.4.1 Chamuvaritin

Chamuvaritin (20) is a dihydrobenzylchalcone isolated from *Uvaria chamae*, an African medicinal plant used as purgative and febrifugal [38]. This phytochemical was mutagenic in *Salmonella typhimurium* strains TA98 and TA100 *via* the activation by the hepatic microsomal enzyme preparation. It was found that metabolic activation was necessary for the expression of the mutagenic activity of 20 in *S. typhimurium* and that it can induce a dose-dependent mutation [38]. The heterocyclic and the seemingly hydrophobic nature of 20 were suggested as properties which give cause for suspicion of its carcinogenic (or mutagenic) properties [38]. Uwaifo and coworkers [38] also demonstrated that 20 induced a specific frameshift mutation in TA98 strain, strongly suggesting that it may also induce tumor formation in animal systems [38]. Frameshift mutations elicit error prone DNA-repair response which enhances the possibility of mutagenicity resulting in carcinogenicity [38].

20.4.2 Gossypol

Gossypol (3) is a yellow polyphenolic pigment, present in cotton seeds (*Gossypium* spp.) that permeates cells and acts as an inhibitor for protein kinase C (PKC) [39] and several dehydrogenase enzymes. It is a male antifertility agent which has been reported to have antiviral activity and has been tested as a male oral contraceptive in China [39,40]. Compound 3 has also long been known to possess antimalarial properties and was reported to have anticancer and anti-HIV (human immunodeficiency virus) properties [39,40]. Compound 3 was found to be toxic to many

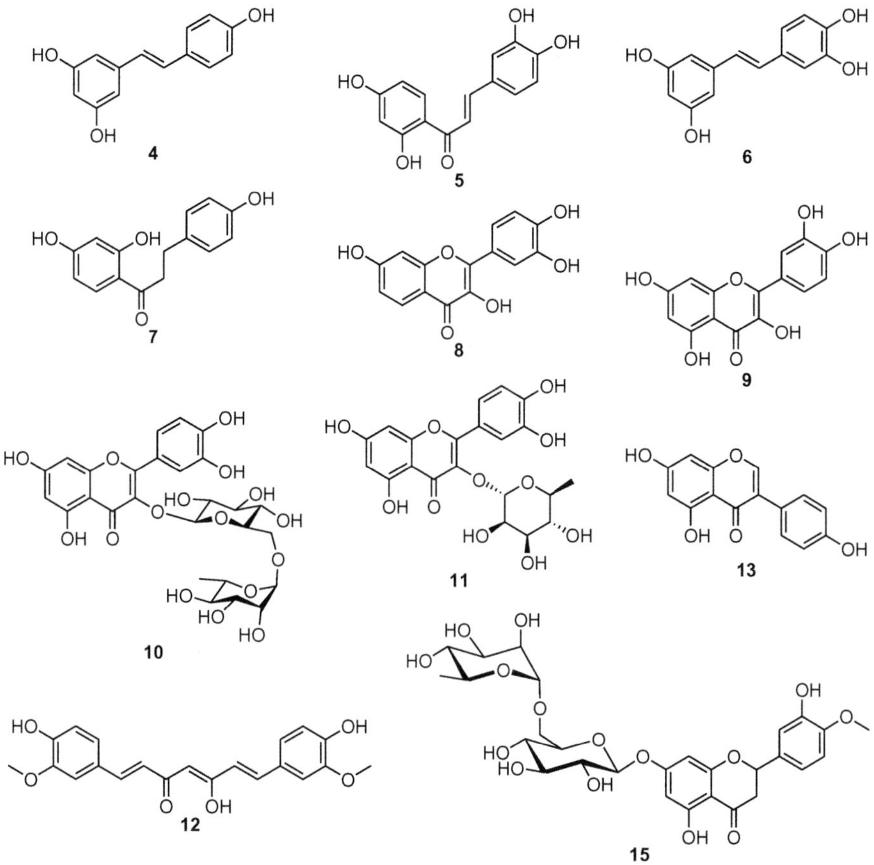

Figure 20.2 Chemical structures of some flavonoids with beneficial effects in human. Resveratrol (**4**); butein (**5**); piceatannol (**6**); isoliquiritigenin (**7**); fisetin (**8**); quercetin (**9**); rutin (**10**); quercitrin (**11**); curcumin (**12**); genistein (**13**); morin (**14**); hesperidin (**15**), diosmin (**16**); nobiletin (**17**); catechin (**18**); epigallocatechin-3-gallate or EGCG (**19**).

animal species. It is inactivated by binding to proteins, which occurs in the rumen, and adult ruminants can therefore tolerate higher amounts of **3** in the diet than monogastric animals, including young ruminants [41]. Pigs are severely affected [41]. Clinical signs that may include dyspnea, anorexia, unthriftiness, and diarrhea appear 1−3 months after gossypol has been included in the feed [41]. On the other hand, pigs may die suddenly without warning during transport to the abattoir, without premonitory signs [41]. Postmortem examination reveals severe cardiomyopathy as well as severe hepatosis with centrilobular necrosis and hemorrhage that may affect the whole lobule [41]. The hepatic lesions may be more pronounced than the lesions in the heart. Gossypol poisoning has also been associated with various manifestations of infertility in monogastric species including humans [42].

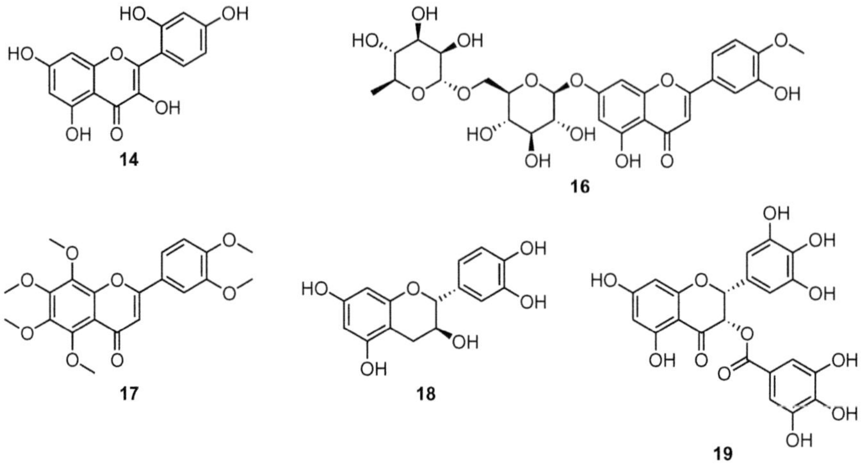

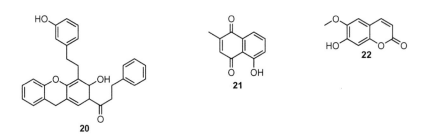

Figure 20.2 (Continued)

Figure 20.3 Chemical structures of chamuvaritin (**20**), plumbagin (**21**) and scopoletin (**22**).

Compound **3** is also toxic to human erythrocytes *in vitro* by stimulating cell death contributing to the side effect of hemolytic anemia [40].

20.4.3 Plumbagin

Plumbagin (**21**), a naphthoquinone originally found in the genus *Plumbago*, has been reported to have many beneficial properties but also has been reported to have many side effects [43−46]. Compound **21** is also commonly found in the carnivorous plant genera *Drosera* and *Nepenthes* [47−49]. Phytochemical **21** was isolated from many African medicinal plants such as *Diospyros crassiflora* and *Diospyros canaliculata* used to treat infectious diseases, including gonorrhea and tuberculosis (*D. crassiflora*) as well as leprosis, bacterial infections, and as antipoison (*D. canaliculata*) [50,51]. Plumbagin showed antimicrobial [45,52], antimalarial [47], anti-inflammatory [53], anticarcinogenic [44,54,55], cardiotonic [56], immunosuppressive [57], antifertility action [58], neuroprotective [59], antiatherosclerosis effects [60]. Nonetheless, **21** is not only potentially beneficial for the treatment of various diseases; Compound **21** was together with juglone considered as a toxin [61] and was also found genotoxic [62] and mutagenic [63]. It was reported that the compound has many side effects such as diarrhea, skin rashes, increase in white blood cell counts, increase in serum phosphate and acid phosphate level, hepatic toxicity, and reproductive system toxicity in male and female rats [62]. Without affecting the lymphoma L5178Y cell viability, **21** was found to induce significant DNA damage at concentrations as low as 0.25 ng/ml [62].

20.4.4 Scopoletin

Scopoletin (**22**) is a coumarin phytoalexin biosynthetized by several plants among which are the *Scopolia carniolica or Scopolia japonica, Artemisia scoparia*, and *Viburnum prunifolium*. This phytochemical also is found in vinegar [64] and some whiskies. Compound **22** also occurs in several African medicinal plants among which are *Convolvulus tricolor* (Convolvulaceae) [65] from Algeria, *Garcinia brevipedicellata* (Clusiaceae) [66], *Macaranga barteri* (Euphorbiaceae) [67], *Scaphopetalum thonneri* (Sterculiaceae) [68] from Cameroon, *Torilis radiata* Moench (Apiaceae) [69] from Egypt, *Pentas longiflora* (Rubiaceae) [70] from Kenya, *Tachiadenus longiflorus* Grisebach (Gentianaceae) [71] from Madagascar, and *Artemisia afra* Jacq. ex Willd. (Asteraceae) [72] from South Africa. Experimental evidence strongly suggested a pharmacological and/or toxicological role for **22**. In fact, treatment with **22** significantly reduced the fructose and citric acid contents of the male reproductive organs of the guinea pigs, therefore suggesting that this compound can induce testicular failure at the level of sperm maintenance [73]. The recent isolation [74] of **22** and other coumarin compounds from cassava and its products raises the possibility that some of the diseases (e.g., endemic goiter, cretinism, mental retardation, slowly developing tropical neuropathy characterized by optic atrophy, nerve deafness, ataxia, scrotal dermatitis, stomatitis, and glossitis) associated with the consumption of cassava diet staples such as *gari* and which have been attributed to the cyanogenic glycosides may in fact not be entirely due to the cyanogens [75−77]. This is based on the fact that a greater proportion of the cyanide content of cassava is lost during processing, which is not the case with **22** whose level is not significantly altered during the processing of cassava into flour or *gari* [77].

20.5 Protective Effects of Phenolics from African Medicinal Plants

Many phenolic compounds having protective effects in organs and other systems in humans or animals were identified in African medicinal plants. In this section, we will not emphasize the pharmacological activities of the reported phytochemicals, but their protective effects at cellular or organ levels (Figure 20.4).

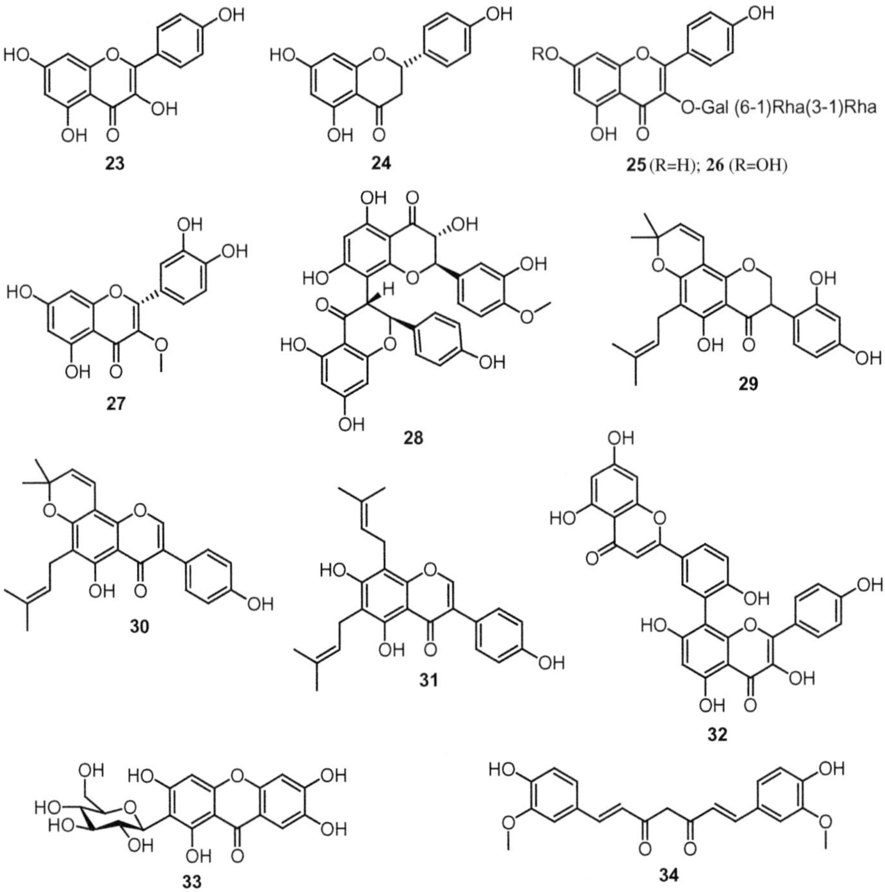

Figure 20.4 Chemical structures of some protective phenolics identified in African medicinal plants. Kaempferol (**23**); naringenin (**24**); 3-O-β-Isorhamninoside (**25**); rhamnocitrin 3-O-β-isorhamninoside (**26**); 3-methoxyquercetin (**27**); kolaviron (**28**); 2, 3-dihydro-2'-hydroxyosajin (**29**), osajin (**30**) and 6, 8-diprenylgenistein (**31**); amentoflavone (**32**); mangiferin (**33**); curcumin (**34**).

20.5.1   Catechin

Catechin (**18**) is a naturally occurring flavonol and a well-demonstrated antioxidant phytochemical originally derived from catechu, which is the tannic juice or boiled extract of *Acacia catechu* L.f (Fabaceae) [78]. Catechin is a constituent of several medicinal plants found throughout Africa such as *Ficus mucuso* [79], *Ficus gnaphalocarpa* [80], *Ficus cordata* (Moraceae) [81] and *Khaya grandifoliola* C.D.C. (Meliaceae) [82] harvested in Cameroon, *Holothuria atra* (Holothuriidae) [83], *Thalassodendron ciliatum* (Cymodoceaceae) [84] from Egypt, *Psidium guajava*

(Myrtaceae) [85], *Acalypha wilkesiana* "Godseffiana" Muell Arg (Euphorbiaceae) [86] from Nigeria, *Terminalia sericea* (Combretaceae) [87], *Peltophorum africanum* (Fabaceae) [88] and *Guibourtia coleosperma* (Fabaceae) [89] from South Africa. Compound **18** is a pharmacologically active compound found to induce longevity in the nematode worm *Caenorhabditis elegans* [90], to inhibit intestinal tumor formation in mice [91], to inhibit the oxidation of low-density lipoprotein [92], to suppress expression of Kruppel-like factor 7 [93], to show an enhancement of the antifungal effect of amphotericin B against *Candida albicans* [94] and to prevent human plasma oxidation [95]. Transcriptomic studies also indicated that **18** could reduce atherosclerotic lesion development in apo E-deficient mice [96]. It was demonstrated that catechin-like compounds were strong therapeutic candidates for protection against the cognitive decline caused by the HIV. In fact, epicatechin, EGCG (**19**) and other catechin flavonoids may protect against neurotoxic oxidative stress caused by the HIV-Tat protein [97]. Epicatechin was found able to cross the blood−brain barrier and activate brain-derived neurotrophic factor (BDNF) pathways [97], suggesting its neuroprotective effects. Compound **18** was reported as histidine decarboxylase inhibitor, inhibiting the conversion of histidine to histamine suggesting to have beneficial effects through reduction of potentially damaging, histamine-related local immune response [98]. This phytochemical is also a selective monoamine oxidase inhibitor (MAOI) of type MAO-B, showing its ability to reduce the symptoms of Parkinson and Alzheimer diseases [99].

20.5.2 Epigallocatechin-3-gallate

Epigallocatechin-3-gallate or epigallocatechin gallate or EGCG (**19**), an ester of epigallocatechin and gallic acid, is a type of catechin and the most abundant catechin in tea. Compound **19** has antioxidant properties that may have therapeutic applications in the treatment of cancer and several other disorders [100−103]. In addition to the presence of **19** in tea and other beverage in Africa, it is also a constituent of several medicinal plants of the continent, including *Limoniastrum feei* (Girard) Batt. (Plumbaginaceae) [104] harvested in Algeria, *Rumex vesicarius* L. (Polygonaceae) [105] from Egypt, *Aspalathus linearis* (Fabaceae) [106] and *Sideroxylon inerme* L. (Sapotaceae) [107] found in South Africa. Mahler and coworkers [108] reported that an important mechanism that underlay the major health benefits of consuming **19**, such as its anticancer and anti-inflammatory properties, was its suppressive effect on the growth of different cell types. In fact, T cells are highly active cells and an effective T cell-mediated immune response depends on rapid T cell expansion. Physiological levels (0.5−10 μM) of **19** inhibited the proliferation of primary T cells from mice [108,109]. This effect was mediated by inhibition of cell cycle progression and cell division [109]. Therefore, **19** may exert an effect on autoimmune and inflammatory diseases that involve excessive T cell activation, such as multiple sclerosis (MS) [109]. It was found in animal models of human MS with experimental autoimmune encephalomyelitis (EAE) [110−113], that oral application of **19** prevented and reversed the disability and inhibited myelin-specific inflammatory responses. This effective protection against relapsing central nervous system (CNS) autoimmune disease resulted in a favorable long-term clinical outcome [108]. Furthermore, **19** reduced axonal damage and neuronal cell death by directly targeting reactive oxygen species (ROS) formation [108]. Decreased proliferation of human CD4 + T cells incubated with **19** could be linked to its interference with cell cycle, NF-κB activation and protein degradation pathway (proteasome) [108,114]. Neuronal loss in Alzheimer disease is accompanied by the deposition of amyloid-beta protein (Aβ) in senile plaques. ROS-induced lipid peroxidation has been suggested to play an important role in Aβ- mediated neurotoxicity. In cultured hippocampal neuronal cells, **19** prevented Aβ-induced cell death and lipid peroxidation through its antioxidant property, suggesting that this phytochemical can be useful in the treatment of Alzheimer disease [108]. Also, it was found that **19** promoted the generation of the non-amyloidogenic soluble form of amyloid precursor protein (APP) *via* a PKC-dependent activation of α-secretase [115−117] and in Alzheimer transgenic mice; it reduced cerebral amyloidosis through promotion of alpha-secretase activity [118]. Besides, **19** markedly reduced secreted Aβ levels in the conditioned medium of Chinese hamster ovarian cells, overexpressing the "Swedish" mutated APP (CHO/ΔNL) [119] and in primary neuronal cells derived from transgenic mice bearing the APP Swedish mutation [118]. Chemicals altering proteolytic cleavage of APP (by enhancing α-, or inhibiting β- and γ-secretase activities) thereby reducing Aβ peptides are interesting therapeutic options for Alzheimer disease treatment [120]. It was demonstrated that administration of **19** *via* drinking water recovered Aβ-induced memory dysfunction in mice [121].

20.5.3 Genistein

Genistein (**13**) is a phytoestrogen and an isoflavone originally isolated from *Genista tinctoria* (Fabaceae). It occurs in several African medicinal plants including *Ficus chlamydocarpa* (Moraceae) [81] and *Erythrina indica* (Fabaceae) [122] from Cameroon, *Pongamia pinnata* (Fabaceae) [123] from Egypt and *Cajanus cajan* (Fabaceae) [124] from Ghana. Compound **13** was found to be a potent agent in both prophylaxis and treatment of cancer as well as other chronic diseases [125−127]. At the molecular level, **13** acts *via* an estrogen receptor-mediated mechanism and also inhibits the activity of Adenosine triphosphate (ATP) utilizing enzymes such as: tyrosine-specific protein kinases, topoisomerase II and enzymes involved in phosphatidylinositol turnover [125]. Compound **13** was found to induce apoptosis and differentiation in cancer cells, inhibit cell proliferation, modulate cell cycling, exert antioxidant effects, inhibit angiogenesis, and suppress osteoclast and lymphocyte functions [125]. Besides, **13** health beneficial effects have been shown in osteoporosis, CVDs, and menopause [128]. In fact, it was demonstrated that **13** could protect against pro-inflammatory factor-induced vascular endothelial barrier dysfunction and inhibit leukocyte-endothelium interaction, thereby modulating vascular inflammation, a major event in the pathogenesis of atherosclerosis [128]. It has been successfully used as immunosuppressive agent both *in vitro* and *in vivo* [125]. It was stated that the main advantage of **13** as a potential drug was its multidirectional action in the living cells and its very low toxicity [125]. Though it was found *in vitro* that **13** could induce apoptosis of testicular cells at certain levels, raising concerns about effects it could have on male fertility [129], it was later demonstrated that isoflavones had no observable effect on endocrine measurements, testicular volume, or semen parameters in healthy males over 2 months [130].

20.5.4 Kaempferol

Kaempferol (**23**) is a naturally occurring flavonol (flavonoid) that has been isolated from edible plants such as the tea *Camellia sinensis* (Theaceae), the broccoli *Brassica oleracea* (Brassicaceae), the grapefruit *Citrus* × *paradisi* (Rutaceae), beans, endive, tomato, strawberries, the apples *Malus domestica* (Rosaceae), as well as in plants or botanical products commonly used in traditional medicine (e.g., *Ginkgo biloba, Tilia spp, Equisetum spp, Moringa oleifera, Sophora japonica*, and propolis) [131]. Compound **23** is also a constituent of several African medicinal plants, including *Hedysarum coronarium* [132], *Tylosema esculentum* (Fabaceae) [133], *Vahlia capensis* (Vahliaceae) [134] from Botswana, *Bryophyllum pinnatum* (Lank.) Oken (Crassulaceae) [135], *Vismia laurentii* De Wild (Guttiferae) [136] from Cameroon, *Acalypha wilkesiana* "Godseffiana" Muell Arg (Euphorbiaceae) [86] from Nigeria, *Helichrysum simillimum* (Asteraceae) [137] and *Combretum erythrophyllum* (Combretaceae) [138] from South Africa. Some epidemiological studies have found a positive association between the consumption of foods containing **23** and a reduced risk of developing several disorders such as cancer and CVDs [131]. Numerous preclinical studies have shown that **23** and some glycosides of kaempferol have a wide range of pharmacological activities, including antioxidant, anti-inflammatory, antimicrobial, anticancer, cardioprotective, neuroprotective, antidiabetic, antiosteoporotic, estrogenic/antiestrogenic, anxiolytic, analgesic, and antiallergic activities [131]. *In vitro* and *in vivo* investigations have shown plausible mechanisms by which flavonoids may confer cancer and cardiovascular protection [139]. It was reported that the consumption of kampferol-containing foods was associated with a reduced mortality from coronary heart disease [131,140], with a weak risk reduction for coronary heart disease death [141], with a reduced incidence of myocardial infarction [131,140,142], and with a lower incidence of cerebrovascular disease [143]. The protective role of **23** in CVDs may be mediated by different mechanisms. Because oxidative stress [144] and inflammation [145] are known cardiovascular risk factors, it makes sense to think that the known antioxidant and anti-inflammatory properties of **23** may play a critical role in this protective effect. The ability of **23** to prevent atherosclerosis may be mediated, for instance, by its capacity to prevent the oxidation of the low-density lipoproteins [146]. Compound **23** also has been shown to inhibit angiotensin-converting enzyme (which converts angiotensin I to angiotensin II and causes an elevation of blood pressure) [131,147−149], to induce vasodilator effects [149,150], and to cause antiplatelet and antithrombotic effects [151,152]. Compound **23** and some glycosides of kaempferol also may decrease triglycerides levels, cholesterol levels, and/or reduce body weight [131,153−155].

20.5.5 Morin

Morin (**14**) is a phytochemical of the class flavonol found in several medicinal plants including *Maclura pomifera*, *Maclura tinctoria* (Moraceae), and *Psidium guajava* (Myrtaceae) [156]. Compound **14** was shown to have bacteriostatic [156], antihypertensive [157], anti-inflammatory, and antioxidant effects [158]. Compound **14** displayed the protective effect against diabetic induced osteopenia due to its anti-inflammatory and antioxidant properties [158]. This compound also shows a cardioprotection against myocardial injury in rats [159,160]. Oral administration of **14** (100 mg/kg bw) to alcohol-intoxicated rats for 30 days showed significant decreases in lipid peroxidation and restoration of antioxidant, hepatic, and renal markers to normal suggesting its hepato- and nephroprotective properties [161]. Besides, it was found that **14** might be beneficial to reduce the nephrotoxicity of imipenem *via* the inhibition of human organic anion transporter 3 (OAT3)-mediated renal excretion of imipenem [162]. Lee and collaborators [163] also showed that **14** had hepatoprotective and antifibrogenic effects against dimethylnitrosamine -induced hepatic injury. Compound **14** was found to protect erythrocytes from lytic attack by peroxyl radicals generated with 2,2′-azo-bis (2-amidinopropane) dihydrochloride [164].

20.5.6 Naringenin

Naringenin (**24**) is a naturally occurring flavanone (flavonoid) known to have a bioactive effect on human health [165]. This phytochemical is the predominant flavanone in citrus fruits such as grapefruits, oranges, and tomatoes [166,167]. Phytochemical **24** is found in several African medicinal plants such as *Catharanthus roseus* (Apocynaceae), the Madagascar periwinkle [168], *Acalypha wilkesiana* "Godseffiana" Muell Arg (Euphorbiaceae) [86] from Nigeria, and *Elaeodendron croceum* (Celastraceae) from South Africa. Compound **24** possesses various biological activities such as antidiabetic [169−173], antiatherogenic [166,174], antidepressant [175,176], immunomodulatory [166], antitumor [177,178], anti-inflammatory [179], DNA protective [180], hypolipidaemic [172,173], antioxidant [181,182], peroxisome proliferator-activated receptors (PPARs) activator [174], and memory improving [165]. Compound **24** was found to decrease oxidative stress in rats by depleting elevated lipid peroxide and nitric oxide and elevating reduced glutathione levels [165]. Cholinergic function was improved by **24** through the inhibition of elevated cholinesterase (CHE) activity, suggesting that it acted as an antioxidant and CHE inhibitor against type-2 diabetes-induced memory dysfunction [165]. Besides, it was shown that when administered to mice at 4.5 mg/kg bw, **24** significantly ameliorated scopolamine-induced amnesia strengthen its role as chemopreventive agent against Alzheimer disease [182]. It was also found that rats pretreated with **24** showed a clear protection of the number of tyrosine hydroxylase (TH)-positive cells in the substantia nigra and dopamine levels in the striata suggesting the ability of this compound to exhibit neuroprotection in the 6-hydroxydopamine model of Parkinson disease may be *via* its antioxidant capabilities and capability to penetrate into the brain [175]. In the study conducted by Badary and collaborators [183], **24** reduced the extent of cisplatin-induced nephrotoxicity, as evidenced by significant reduction in serum urea and creatinine concentrations, decreased polyuria, reduction in body weight loss, marked reduction in urinary fractional sodium excretion and glutathione *S*-transferase activity, and increased creatinine clearance. Compound **24** also has a protective role in the abatement of doxorubicin-induced cardiac toxicity that resides, at least in part, on its antiradical effects and regulatory role on NO production [184].

20.5.7 Quercetin

Quercetin (**9**) is a naturally occurring flavonol (flavonoid) found in many fruits, vegetables, leaves, and grains. Higher concentrations of **9** occur in red onions [185] and in tomatoes [186] while it is also found in honey [187]. Compound **9** also occurs in several medicinal plants in all regions of Africa, including *Citrullus colocynthis* (Cucurbitaceae) [188], *Pituranthos scoparius* (Coss. and Dur.) Benth. and Hook. (Apiaceae) [189] from Algeria, *Vahlia capensis* (Vahliaceae) [134] from Botswana, *Ficus gnaphalocarpa* (Moraceae) [80], *Psorospermum androsaemifolium* BAKER (Clusiaceae) [190] from Cameroon, *Leptadenia reticulata* (Retz.) Wt. et Arn) (Asclepiadaceae) [191] from Ghana, *Euphorbia stenoclada* [192] (Euphorbiaceae) from Madagascar, *Musa paradisiaca* (Musaceae) [193] from Nigeria, *Buddleja salviifolia* (L.) Lam. (Scrophulariaceae) [194], *Curtisia dentata* (Burm.f) C.A (Cornaceae) [195], *Helichrysum melanacme* (Asteraceae) [196] and *Onobrychis viciifolia* (Fabaceae) [197] from South Africa, etc. Compound **9** has been shown to exhibit antioxidative [198,199], anticarcinogenic [200–202], anti-inflammatory [203], anti-aggregatory [204], and vasodilating [205] effects. It was found that **9** could induce insulin secretion by activation of L-type calcium channels in the pancreatic β-cells [206]. Liu and coworkers [207] also demonstrated that **9** could protect mouse brain against lead-induced neurotoxicity. It was also reported that **9** could protect cardiomyocytes from anoxia/reoxygenation injury by increasing the expression of PKC epsilon type (PKCε) therefore enhancing the activity of its downstream pathway [208]. The cardioprotective effects of **9** in cardiomyocyte under ischemia/reperfusion injury also was confirmed by Chen et al. [209]. This compound also potentially augmented the cardioprotective effect of losartan against chronic doxorubicin cardiotoxicity *via* its antioxidant and anti-inflammatory properties [210]. It was also showed that exposure doses of letrozole that were equal to the daily recommended human dose had toxic effects on the spermatogenic lineage in rats, while simultaneous treatment of **9** and letrozole could prevent the deleterious effects on testicular tissue caused by letrozole administration [211]. Moreover, **9** showed protective on chloroquine-induced oxidative stress and hepatotoxicity in mice [212]. Gao et al. [213] also reported that **9** could attenuate the progression of monocrotaline-induced pulmonary hypertension in rats. The protective effects of **9** on cadmium-induced cytotoxicity in primary cultures of rat proximal tubular cells were also documented [214]. Despite its numerous protective effects, compound **9** has neither been confirmed scientifically as a specific therapeutic for any condition nor approved by any regulatory agency until now.

20.5.8   Resveratrol

Resveratrol (**4**) is a stilbenoid and a phytoalexin naturally occurring in red wine, grapes [215], peanuts, certain berries as well as in several other plants, especially the roots of the *Fallopia japonica* (Polygonaceae), from which it is extracted commercially. Compound **4** is found in several African medicinal plants including *Senna italica* (Fabaceae) [216] harvested in South Africa, *Terminalia sericeae* (Combretaceae) from Tanzania [217], *Elephantorrhiza goetzei* (Fabaceae) from Botswana [218], etc. Many reports carried out with **4** have shown that it possesses strong antiplatelet and vasodilating properties, as well as the ability to protect renal function [219−223]. Compound **4** has been shown to be a potent activator of sirtuins, which act in the regulation of apoptosis and cells differentiation [22] and therefore it seems to protect neurons [224]. Compound **4** also showed anticancer activities and was found to inhibit angiogenesis [215,222]. This compound is also known to have anti-inflammatory activity, which is due to prostaglandin regulation by inhibition of both cyclooxygenases (COX-1 and COX-2) [222,225−227]. The effects of **4** on the lifespan of many model organisms remain controversial, with uncertain effects in fruit flies, nematode worms [24], and short-lived fish. In humans, however, while reported effects are generally positive, it is also suggested that **4** may have lesser benefits [228]. In one positive human trial, extremely high doses up to 3−5 g of **4**, in a proprietary formulation designed to enhance its bioavailability, significantly lowered blood sugar [229]. Compound **4** was shown to have cardioprotective effects [230]. Compound **4** also showed beneficial effects in cell culture systems against the accumulation of the amyloid-beta peptide, a main culprit in Alzheimer disease [231] suggesting its neuroprotective effects.

20.5.9   Rutin

Rutin (**10**), also known as rutoside, quercetin-3-*O*-rutinoside and sophorin, is a common dietary flavonoid glycoside that is consumed in fruits, vegetables, and plant-derived beverages [232]. Compound **10** is a citrus flavonoid glycoside found in many plants such as *Fagopyrum esculentum* an especially in *Fagopyrum tataricum* Gaertn (Polygonaceae) [233]. Rutin is also found in the fruit of *Dimorphandra mollis* (from Brazil), the fruits and flowers of the *Styphnolobium spp.*, the fruits and fruit rinds especially of the citrus fruits orange, grapefruit, lemon, lime, and apple; berries such as mulberry, ash tree fruits, and cranberries [233]. It was identified in several African medicinal plants such as *Cussonia barteri* (Araliaceae) [234] from Cameroon, *Moringa oleifera* Lam. (Moringaceae) [235], *Garcinia kola* (Heckel) (Guttiferae) [236], *Alchornea laxiflora* (benth) pax and hoffman. (Euphorbiaceae) [237] from Nigeria, *Cussonia bojeri* SEEM [238], *Cussonia racemosa* (Araliaceae) [239] from Madagascar, *Convolvulus hystrix* Vahl. (Convolvulaceae) [240a], *Pongamia pinnata* (Linn.) Pierre (Leguminosae) [123] from Egypt, among others. Compound **10** was reported to have several pharmacological properties including antioxidant, anticarcinogenic, cytoprotective, antiplatelet, anti-thrombic, vasoprotective, and cardioprotective activities [240b]. It has also been suggested that compound **10**'s uptake can significantly decrease the weights of body, liver organ, and adipose tissue as well as the levels of hepatic triglycerides and cholesterol levels in high-fat diet (HFD) rats [241]. Wu and collaborators also demonstrated the hypolipidemic effects of **10** [232]. Olaleye and Akinmoladun [242] demonstrated the anti-ulcerogenic potentials of low doses of rutin in ethanol-, acetic acid-, and stress-induced ulcers in rats, suggesting its gastro-protective effects. It was lately shown that **10** could help to prevent blood clots, so it could be used to treat patients at risk of heart attacks and strokes [243] while evidence also showed that this phytochemical could be used to treat hemorrhoids, varicosis, and microangiopathy [244]. Compound **10** also showed nephroprotective effects, reducing attenuated hexachlorobutadiene-induced nephrotoxicity [245]. Abarikwu and coworkers [246] lately proved that **10** could reverse cadmium-induced alterations of antioxidant defense system and can significantly prevent cadmium-induced testes damage as well as depletion of plasma and testicular selenium levels in rats. Another study strengthened the hypothesis that **10** worked as an antioxidant *in vivo* by scavenging ROS suggesting that this could serve to prevent oxidative potassium bromate-induced nephrotoxicity in rats [247].

20.5.10   Other Protective Phenolics from African Plants

20.5.10.1   Antimutagenic and Genoprotective Effects

Two flavonoid glycosides, 3-O-β-Isorhamninoside (**25**) and rhamnocitrin 3-O-β-isorhamninoside (**26**) isolated from the leaves of Tunisian plant *Rhamnus alaternus* L. (Rhamnaceae), traditionally used as a digestive, diuretic, laxative, hypotensive, and for the treatment of hepatic and dermatological complications [248], reduced strongly both aflatoxine B1 and nitrofurantoine mutagenicity [249]. Compounds **25** and **26** also exhibited a preventive effect against H_2O_2-induced DNA damages in human lymphoblastoid TK6 cells [250].

20.5.10.2   Hepatoprotective Effects

Hubert et al. [80] demonstrated that 3-methoxyquercetin (**27**) isolated from the Cameroonian medicinal plant *Ficus gnaphalocarpa* (Moraceae) had significant antioxidant and hepatoprotective activities as indicated by its ability to prevent liver cell death and lactate dehydrogenase leakage during CCl_4 intoxication. The biflavonoid kolaviron (**28**), isolated from the seeds of *Garcinia kola* (Guttiferae) harvested in Nigeria exerts a protective action against carcinogen-induced liver damage in rats, acting as an *in vivo* natural antioxidant and enhancing drug-detoxifying enzymes such as microsomal aniline hydroxylase, aminopyrine *N*-demethylase, ethoxyresorufin *O*-demethylase, and *p*-nitroanisole *O*-demethylase [251]. Kolaviron also occurs in another Nigerian medicinal plant, *Curcuma longa* (Zingiberaceae) [252]. The diprenylated isoflavonoids 2, 3-dihydro-2'-hydroxyosajin (**29**), osajin (**30**) and 6, 8-diprenylgenistein (**31**) isolated from the Cameroonian medicinal plant *Erythrina senegalensis* (Fabaceae) were also found to have *in vitro* hepatoprotective activities against CCl_4-induced hepatitis in rat liver slices [253].

20.5.10.3   Neuroprotective Effects

It was demonstrated that compound **28** could protect against gamma-radiation-induced oxidative stress in the brain of exposed rats *via* the reduction the cellular level of malondialdehyde, urea, alanine and aspartate aminotransferases [254]. The flavonoid amentoflavone (**32**) identified in the Nigerian medicinal plant *Cnestis ferruginea* (CF) Vahl DC (Connaraceae) used in the management of psychiatric disorders produces antidepressant effect through interaction with 5-hydroxytryptamine (5-HT2) receptor and α1-, and α2-adrenoceptors as well as anxiolytic effect *via* the ionotropic gamma-Aminobutyric acid (GABA) receptor [255]. Compound **32** also occurs in other African medicinal plants such as *Calophyllum flavoramulum* (Guttiferae) [256] from Algeria, *Allanblackia monticola* (Guttiferae) [257], *Dorstenia barteri* (Moraceae) [258] and *Ouratea sulcata* (Ochnaceae) [259] from Cameroon, *Cupressus sempervirens* L. (Cupressaceae) [260] from Egypt, *Cnestis ferruginea* Vahl ex DC (Connaraceae) [261] from Nigeria and *Garcinia livingstonei* (Guttiferae) [262] from South Africa. While analyzing the molecular mechanisms underlying neuroprotection by mangiferin (**33**) in an *in vitro* model of excitotoxic neuronal death involving N-methyl-D-aspartate (NMDA) receptor overactivation, Campos-Esparza et al. [263] demonstrated that **33** displayed excellent antioxidant and antiapoptotic properties, supporting their clinical application as trial neuroprotectors in pathologies involving excitotoxic neuronal death. Compound **33** is a xanthone glycoside identified in African plants among which are *Bersama engleriana* Engl. (Melianthaceae) [264] from Cameroon, *Bombax malabaricum* (Malvaceae) [265] from Egypt, *Aphloia theiformis* (Aphloiaceae) [266] from Madagascar, and *Mangifera indica* Linn. (Anacardiaceae) [267] from South Africa.

20.5.10.4   Reproduction Organ Protective Effects

Adaramoye et al. [268] demonstrated that **28** could protect against gamma-irradiation testicular toxicity *via* the enhancement of antioxidant defense system in rats such as the decrease of the activities of superoxide dismutase, catalase and glutathione S-transferase as well as glutathione level in the serum and testes. The phenolic compound curcumin (**34**) isolated from the rhizome of *Curcuma longa* L. (Zingiberaceae) harvested in Nigeria showed protective effects di-*n*-butylphthalate (DBP)-induced testicular damage in rats [269]. The chemoprotective effects of **34** was suggested to be due to its intrinsic antioxidant properties and as such proved that it could be useful in combating phthalate-induced reproductive toxicity [269].

20.5.10.5   Miscellaneous

The xanthone glycoside **33** afforded gastroprotection against gastric injury induced by ethanol and indomethacin most possibly through the anti-secretory and antioxidant mechanisms of action [270]. Compound **33** was also reported to have protective effects on postmyocardial infarction ventricular remodeling and improve cardiac function in rats [271].

20.6 Conclusion

Taken together, the information available in this chapter shows that plant phenolics do not only display advantageous pharmacological roles to human or animals, but some can also show unlike properties. Herein, it can be noted that we describe more protective phenolics than those having side effects, obviously due to the fact that most of the medicinal plants can also contain normally more compounds with positive biological activities. However, the most prominent achievement of this review is the fact that we highlight the toxic potential of phenolics available in medicinal plants used in African traditional medicine, suggesting that plants with such phytochemicals should be consumed with caution. Indications should be given by herbalists to patients when using plants containing compounds such as chamuvaritin, gossypol, plumbagin, and scopoletin. On the other hand, privilege should be made to the used of medicinal plants containing catechin, EGCG, genistein, kaempferol, morin, naringenin, quercetin, resveratrol, rutin, due to their beneficial pharmacological activities, as well as their protective roles in various biological systems.

References

[1] Kougan GB, Tabopda T, Kuete V, Verpoorte R. Simple phenols, phenolic acids, and related esters from the medicinal plants of Africa. In: Kuete V, editor. Medicinal plant research in Africa. Oxford: Elsevier; 2013. p. 225−49.

[2] Kuete V. Phenylpropanoids and related compounds from the medicinal plants of Africa. In: Kuete V, editor. Medicinal plant research in Africa, 2013. Oxford: Elsevier; 2013. p. 251−60.

[3] Poumale HMP, Hamm R, Zang Y, Shiono Y, Kuete V. Coumarins and related compounds from the medicinal plants of Africa. In: Kuete V, editor. Medicinal plant research in Africa. Oxford: Elsevier; 2013. p. 261−300.

[4] Ngameni B, Fotso GW, Kamga J, Ambassa P, Abdou T, Fankam AG, et al. Flavonoids and related compounds from the medicinal plants of Africa. In: Kuete V, editor. Medicinal plant research in Africa. Oxford: Elsevier; 2013. p. 301−50.

[5] Eyong KO, Kuete V, Efferth T. Quinones and benzophenones from the medicinal plants of Africa. In: Kuete V, editor. Medicinal plant research in Africa. Oxford: Elsevier; 2013. p. 351−91.

[6] Mazimba O, Nana F, Kuete V, Singh GS. Xanthones and anthranoids from the medicinal plants of Africa. In: Kuete V, editor. Medicinal plant research in Africa. Oxford: Elsevier; 2013. p. 393−434.

[7] Tsopmo A, Awah FM, Kuete V. Lignans and stilbenes from African medicinal plants. In: Kuete V, editor. Medicinal plant research in Africa, 2013. Oxford: Elsevier; 2013. p. 435−78.

[8]   Shahat AA, Marzouk MS. Tannins and related compounds from medicinal plants of
      Africa. In: Kuete V, editor. Medicinal plant research in Africa. Oxford: Elsevier; 2013.
      p. 479–555.
[9]   Harborne J, Baxter H, Moss G. Phytochemical dictionary: handbook of bioactive
      compounds from plants. London: Taylor & Francis; 1999.
[10] Martins S, Mussatto SI, Martínez-Avila G, Montañez-Saenz J, Aguilar CN, Teixeira
      JA. Bioactive phenolic compounds: production and extraction by solid-state fermenta-
      tion. A review. Biotechnol Adv 2011;29(3):365–73.
[11] Bravo L. Polyphenols: chemistry, dietary sources, metabolism, and nutritional signifi-
      cance. Nutr Rev 1998;56(11):317–533.
[12] Ruiz JM, Romero L. Bioactivity of the phenolic compounds in higher plants. In: Atta ur R,
      editor. Studies in natural products chemistry. Amsterdam: Elsevier; 2001. p. 651–81.
[13] Rispail N, Morris P, Webb J. Phenolic compounds: extraction and analysis. Dordrecht:
      Springer; 2005. [chapter 7.5].
[14] Kim EY, Ham SK, Shigenaga MK, Han O. Bioactive dietary polyphenolic compounds
      reduce nonheme iron transport across human intestinal cell monolayers. J Nutr
      2008;138(9):1647–51.
[15] Chen H, Yao J, Wang F, Zhou Y, Chen K, Zhuang R, et al. Toxicity of three phenolic
      compounds and their mixtures on the gram-positive bacteria *Bacillus subtilis* in the
      aquatic environment. Sci Total Environ 2010;408(5):1043–9.
[16] Roza L, de Vogel N, van Delft JH. Lack of clastogenic effects in cultured human lym-
      phocytes treated with hydroquinone. Food Chem Toxicol 2003;41(10):1299–305.
[17] Habauzit V, Morand C. Evidence for a protective effect of polyphenols-containing
      foods on cardiovascular health: an update for clinicians. Ther Adv Chronic Dis 2012;3
      (2):87–106.
[18] Kuete V, Eichhorn T, Wiench B, Krusche B, Efferth T. Cytotoxicity, anti-angiogenic,
      apoptotic effects and transcript profiling of a naturally occurring naphthyl butenone,
      guieranone A. Cell Div 2012;7(1):16.
[19] Kuete V, Ngameni B, Mbaveng AT, Ngadjui B, Meyer JJM, Lall N. Evaluation of
      flavonoids from *Dorstenia barteri* for their antimycobacterial, antigonorrheal and anti-
      reverse transcriptase activities. Acta Trop 2010;116(1):100–4.
[20] Kuete V, Efferth T. Cameroonian medicinal plants: pharmacology and derived natural
      products. Front Pharmacol 2010;1:123.
[21] Kuete V. Medicinal plant research in Africa. Oxford: Elsevier; 2013.
[22] Howitz KT, Bitterman KJ, Cohen HY, Lamming DW, Lavu S, Wood JG, et al. Small
      molecule activators of sirtuins extend *Saccharomyces cerevisiae* lifespan. Nature
      2003;425(6954):191–6.
[23] Kaeberlein M, McDonagh T, Heltweg B, Hixon J, Westman EA, Caldwell SD, et al.
      Substrate-specific activation of sirtuins by resveratrol. J Biol Chem 2005;280
      (17):17038–45.
[24] Bass TM, Weinkove D, Houthoofd K, Gems D, Partridge L. Effects of resveratrol on
      lifespan in *Drosophila melanogaster* and *Caenorhabditis elegans*. Mech Ageing Dev
      2007;128(10):546–52.
[25] Quideau S. Plant "polyphenolic" small molecules can induce a calorie restriction-
      mimetic life-span extension by activating sirtuins: will "polyphenols" someday be used
      as chemotherapeutic drugs in western medicine?. Chembiochem 2004;5(4):427–30.
[26] World Health Organization (WHO). The global burden of disease. Available
      at: <http://whqlibdoc.who.int/publications/2008/9789241563710_eng.pdf>; 2004
      [accesed on 29.7.13].

[27] Gonzalez R, Ballester I, Lopez-Posadas R, Suarez MD, Zarzuelo A, Martínez-Augustin O, et al. Effects of flavonoids and other polyphenols on inflammation. Crit Rev Food Sci Nutr 2011;51(4):331−62.

[28] Budavari S. The Merck Index: an encyclopedia of chemical, drugs, and biologicals. Whitehouse Station, NJ: Merck; 1996.

[29] Greenlee WF, Sun JD, Bus JS. A proposed mechanism of benzene toxicity: formation of reactive intermediates from polyphenol metabolites. Toxicol Appl Pharmacol 1981;59(2):187−95.

[30] Strick R, Strissel PL, Borgers S, Smith SL, Rowley JD. Dietary bioflavonoids induce cleavage in the MLL gene and may contribute to infant leukemia. Proc Natl Acad Sci USA 2000;97(9):4790−5.

[31] Ross JA. Dietary flavonoids and the MLL gene: a pathway to infant leukemia? Proc Natl Acad Sci USA 2000;97(9):4411−3.

[32] Matuschek E, Svanberg U. Oxidation of polyphenols and the effect on *in vitro* iron accessibility in a model food system. J Food Sci 2002;67(1):420−4.

[33] Wink M. Mode of action and toxicology of plant toxins and poisonous plants. Wirbeltierforschung in der Kulturlandschaft 2009;421:93−112.

[34] Born SL, Api AM, Ford RA, Lefever FR, Hawkins DR. Comparative metabolism and kinetics of coumarin in mice and rats. Food Chem Toxicol 2003;41(2):247−58.

[35] The German Federal Institute for Risk Assessment. Frequently asked questions about coumarin in cinnamon and other foods, <http://www.bfr.bund.de/en/frequently_asked_questions_about_coumarin_in_cinnamon_and_other_foods-8487.html>; 2006 [accessed 20.07.13].

[36] <https://www.osha.gov/dts/chemicalsampling/data/CH_229620.html> Coumarin [accessed 20.07.13].

[37] Steenkamp V, Stewart MJ. Nephrotoxicity associated with exposure to plant toxins, with particular reference to Africa. Ther Drug Monit 2005;27(3):270−7.

[38] Uwaifo AO, Okorie DA, Bababunmi EA. Mutagenicity of chamuvaritin: a benzyldihydrochalcone isolated from a medicinal plant. Cancer Lett 1979;8(1):87−92.

[39] Polsky B, Segal SJ, Baron PA, Gold JW, Ueno H, Armstrong D. Inactivation of human immunodeficiency virus *in vitro* by gossypol. Contraception 1989;39(6):579−87.

[40] Zbidah M, Lupescu A, Shaik N, Lang F. Gossypol-induced suicidal erythrocyte death. Toxicology 2012;302(2−3):101−5.

[41] Botha CJ, Penrith ML. Poisonous plants of veterinary and human importance in southern Africa. J Ethnopharmacol 2008;119(3):549−58.

[42] Nicholson S. Cotton seed toxicity. New York, NY: Academic Press; 2007.

[43] van der Vijver LM. Distribution of plumbagin in the mplumbaginaceae. Phytochemistry 1972;11(11):3247−8.

[44] Parimala R, Sachdanandam P. Effect of plumbagin on some glucose metabolising enzymes studied in rats in experimental hepatoma. Mol Cell Biochem 1993;125(1):59−63.

[45] de Paiva SR, Figueiredo MR, Aragao TV, Kaplan MA. Antimicrobial activity *in vitro* of plumbagin isolated from *Plumbago* species. Mem Inst Oswaldo Cruz 2003;98(7):959−61.

[46] Demma J, Hallberg K, Hellman B. Genotoxicity of plumbagin and its effects on catechol and NQNO-induced DNA damage in mouse lymphoma cells. Toxicol In Vitro 2009;23(2):266−71.

[47] Likhitwitayawuid K, Kaewamatawong R, Ruangrungsi N, Krungkrai J. Antimalarial naphthoquinones from *Nepenthes thorelii*. Planta Med 1998;64(3):237−41.

[48] Wang W, Luo X, Li H. Terahertz and infrared spectra of plumbagin, juglone, and menadione. Carniv Pl Newslett 2010;39(3):82−8.

[49] Rischer H, Hamm A, Bringmann G. Nepenthes insignis uses a C2-portion of the carbon skeleton of L-alanine acquired via its carnivorous organs, to build up the allelochemical plumbagin. Phytochemistry 2002;59(6):603−9.

[50] Kuete V, Tangmouo JG, Marion Meyer JJ, Lall N. Diospyrone, crassiflorone and plumbagin: three antimycobacterial and antigonorrhoeal naphthoquinones from two Diospyros spp. Int J Antimicrob Ag 2009;34(4):322−5.

[51] Kuete V, Efferth T. Pharmacogenomics of Cameroonian traditional herbal medicine for cancer therapy. J Ethnopharmacol 2011;137(1):752−66.

[52] Didry N, Dubreuil L, Pinkas M. Activity of anthraquinonic and naphthoquinonic compounds on oral bacteria. Pharmazie 1994;49(9):681−3.

[53] Checker R, Sharma D, Sandur SK, Subrahmanyam G, Krishnan S, Poduval TB, et al. Plumbagin inhibits proliferative and inflammatory responses of T cells independent of ROS generation but by modulating intracellular thiols. J Cell Biochem 2010;110(5):1082−93.

[54] Hsu YL, Cho CY, Kuo PL, Huang YT, Lin CC. Plumbagin (5-hydroxy-2-methyl-1,4-naphthoquinone) induces apoptosis and cell cycle arrest in A549 cells through p53 accumulation via c-Jun NH2-terminal kinase-mediated phosphorylation at serine 15 in vitro and in vivo. J Pharmacol Exp Ther 2006;318(2):484−94.

[55] Subramaniya BR, Srinivasan G, Sadullah SS, Davis N, Subhadara LB, Halagowder D, et al. Apoptosis inducing effect of plumbagin on colonic cancer cells depends on expression of COX-2. PLoS One 2011;6(4):e18695.

[56] Itoigawa M, Takeya K, Furukawa H. Cardiotonic action of plumbagin on guinea-pig papillary muscle. Planta Med 1991;57(4):317−9.

[57] McKallip RJ, Lombard C, Sun J, Ramakrishnan R. Plumbagin-induced apoptosis in lymphocytes is mediated through increased reactive oxygen species production, upregulation of fas, and activation of the caspase cascade. Toxicol Appl Pharmacol 2010;247(1):41−52.

[58] Bhargava SK. Effects of plumbagin on reproductive function of male dog. Indian J Exp Biol 1984;22(3):153−6.

[59] Son TG, Camandola S, Arumugam TV, Cutler RG, Telljohann RS, Mughal MR, et al. Plumbagin, a novel Nrf2/ARE activator, protects against cerebral ischemia. J Neurochem 2010;112(5):1316−26.

[60] Ding Y, Chen ZJ, Liu S, Che D, Vetter M, Chang CH. Inhibition of Nox-4 activity by plumbagin, a plant-derived bioactive naphthoquinone. J Pharm Pharmacol 2005;57 (1):111−6.

[61] <drugs.com>. Black Walnut. <http://wwwdrugscom/npp/black-walnuthtml; 2009 [accessed 27.07.13].

[62] Demma J, Hallberg K. Genotoxicity of plumbagin and its effects on catechol and NQNO-induced DNA damage in mouse lymphoma cells. Björn Hellman 2009;23(2):266−71.

[63] Farr S, Natvig D, Kogoma T. Toxicity and mutagenicity of plumbagin and the induction of a possible new DNA repair pathway in Escherichia coli. J Bacteriol 1985;164 (3):1309−16.

[64] Gálvez M, Barroso C, Pérez-Bustamante J. Analysis of polyphenolic compounds of different vinegar samples. Z Lebensm Unters Forsch 1994;199(1):29−31.

[65] Kacem N, Hay AE, Marston A, Zellagui A, Rhouati S, Hostettmann K. Antioxidant compounds from Algerian Convolvulus tricolor (Convolvulaceae) seed husks. Nat Prod Commun 2012;7(7):873−4.

[66] Ngoupayo J, Tabopda TK, Ali MS, Tsamo E. Alpha-glucosidase inhibitors from *Garcinia brevipedicellata* (Clusiaceae). Chem Pharm Bull (Tokyo) 2008;56(10):1466−9.

[67] Ngoumfo RM, Ngounou GE, Tchamadeu CV, Qadir MI, Mbazoa CD, Begum A, et al. Inhibitory effect of macabarterin, a polyoxygenated ellagitannin from *Macaranga barteri*, on human neutrophil respiratory burst activity. J Nat Prod 2008;71(11): 1906−10.

[68] Vardamides JC, Azebaze AG, Nkengfack AE, Van Heerden FR, Fomum ZT, Ngando TM, et al. Scaphopetalone and scaphopetalumate, a lignan and a triterpene ester from *Scaphopetalum thonneri*. Phytochemistry 2003;62(4):647−50.

[69] Ezzat SM, Abdallah HM, Fawzy GA, El-Maraghy SA. Hepatoprotective constituents of *Torilis radiata* Moench (Apiaceae). Nat Prod Res 2012;26(3):282−5.

[70] El-Hady S, Bukuru J, Kesteleyn B, Van Puyvelde L, Van TN, De Kimpe N. New pyranonaphthoquinone and pyranonaphthohydroquinone from the roots of *Pentas longiflora*. J Nat Prod 2002;65(9):1377−9.

[71] Randrianarivelojosia M, Langlois A, Mulholland DA. Investigations of the Malagasy species *Tachiadenus longiflorus* Grisebach (Gentianaceae): linking chemical finding and traditional usage. J Ethnopharmacol 2006;105(3):456−8.

[72] More G, Lall N, Hussein A, Tshikalange TE. Antimicrobial Constituents of *Artemisia afra* Jacq. ex Willd. against periodontal pathogens. Evid Based Complement Alternat Med 2012;2012:252758.

[73] Obidoa O, Ezeanyika L, Okoli A. Effect of scopoletin on male guinea pig reproductive organs. I. Levels of citric acid and fructose. Nutr Res 1999;19(3):443−8.

[74] Richard J. Physiological deterioration of cassava roots. J Sci Food Agric 1985;36:167−76.

[75] Ekpechi OL, Dimitriadou A, Fraser R. Goitrogenic activity of cassava (a staple Nigerian food). Nature 1996;210(5041):1137−8.

[76] Osuntokun BO. Cassava diet, chronic cyanide intoxication and neuropathy in the Nigerian Africans. World Rev Nutr Diet 1981;36:141−73.

[77] Obidoa O, Obasi SC. Coumarin compounds in cassava diets: 2 health implications of scopoletin in gari. Plant Foods Hum Nutr 1991;41(3):283−9.

[78] Zheng LT, Ryu GM, Kwon BM, Lee WH, Suk K. Anti-inflammatory effects of catechols in lipopolysaccharide-stimulated microglia cells: inhibition of microglial neurotoxicity. Eur J Pharmacol 2008;588(1):106−13.

[79] Bankeu JJ, Mustafa SA, Gojayev AS, Lenta BD, Tchamo Noungoue D, Ngouela SA, et al. Ceramide and Cerebroside from the stem bark of *Ficus mucuso* (Moraceae). Chem Pharm Bull (Tokyo) 2010;58(12):1661−5.

[80] Hubert DJ, Dawe A, Florence NT, Gilbert KD, Angele TN, Buonocore D, et al. *In vitro* hepatoprotective and antioxidant activities of crude extract and isolated compounds from *Ficus gnaphalocarpa*. Inflammopharmacology 2011;19(1):35−43.

[81] Kuete V, Ngameni B, Simo CC, Tankeu RK, Ngadjui BT, Meyer JJ, et al. Antimicrobial activity of the crude extracts and compounds from *Ficus chlamydocarpa* and *Ficus cordata* (Moraceae). J Ethnopharmacol 2008;120(1):17−24.

[82] Bickii J, Njifutie N, Foyere JA, Basco LK, Ringwald P. *In vitro* antimalarial activity of limonoids from *Khaya grandifoliola* C.D.C. (Meliaceae). J Ethnopharmacol 2000;69 (1):27−33.

[83] Esmat AY, Said MM, Soliman AA, El-Masry KS, Badiea EA. Bioactive compounds, antioxidant potential, and hepatoprotective activity of sea cucumber (*Holothuria atra*) against thioacetamide intoxication in rats. Nutrition 2013;29(1):258−67.

[84] Hamdy AH, Mettwally WS, El Fotouh MA, Rodriguez B, El-Dewany AI, El-Toumy SA, et al. Bioactive phenolic compounds from the Egyptian Red Sea seagrass *Thalassodendron ciliatum*. Z Naturforsch C 2012;67(5−6):291−6.

[85] Adesida A, Farombi EO. Free radical scavenging activities of guava extract *in vitro*. Afr J Med Med Sci 2012;41(Suppl.):81−90.

[86] Ikewuchi JC, Onyeike EN, Uwakwe AA, Ikewuchi CC. Effect of aqueous extract of the leaves of *Acalypha wilkesiana* 'Godseffiana' Muell Arg (Euphorbiaceae) on the hematology, plasma biochemistry and ocular indices of oxidative stress in alloxan induced diabetic rats. J Ethnopharmacol 2011;137(3):1415−24.

[87] Nkobole N, Houghton PJ, Hussein A, Lall N. Antidiabetic activity of *Terminalia sericea* constituents. Nat Prod Commun 2011;6(11):1585−8.

[88] Theo A, Masebe T, Suzuki Y, Kikuchi H, Wada S, Obi CL, et al. *Peltophorum africanum*, a traditional South African medicinal plant, contains an anti HIV-1 constituent, betulinic acid. Tohoku J Exp Med 2009;217(2):93−9.

[89] Bekker M, Bekker R, Brandt VE. Two flavonoid glycosides and a miscellaneous flavan from the bark of *Guibourtia coleosperma*. Phytochemistry 2006;67(8):818−23.

[90] Saul N, Pietsch K, Menzel R, Sturzenbaum SR, Steinberg CE. Catechin induced longevity in *C. elegans*: from key regulator genes to disposable soma. Mech Ageing Dev 2009;130(8):477−86.

[91] Weyant MJ, Carothers AM, Dannenberg AJ, Bertagnolli MM. (+)-Catechin inhibits intestinal tumor formation and suppresses focal adhesion kinase activation in the min/+ mouse. Cancer Res 2001;61(1):118−25.

[92] Mangiapane H, Thomson J, Salter A, Brown S, Bell GD, White DA. The inhibition of the oxidation of low density lipoprotein by (+)-catechin, a naturally occurring flavonoid. Biochem Pharmacol 1992;43(3):445−50.

[93] Cho SY, Park PJ, Shin HJ, Kim YK, Shin DW, Shin ES, et al. (-)-Catechin suppresses expression of Kruppel-like factor 7 and increases expression and secretion of adiponectin protein in 3T3-L1 cells. Am J Physiol Endocrinol Metab 2007;292(4): E1166−72.

[94] Hirasawa M, Takada K. Multiple effects of green tea catechin on the antifungal activity of antimycotics against *Candida albicans*. J Antimicrob Chemother 2004;53 (2):225−9.

[95] Lotito SB, Fraga CG. (+)-Catechin prevents human plasma oxidation. Free Radic Biol Med 1998;24(3):435−41.

[96] Auclair S, Milenkovic D, Besson C, Chauvet S, Gueux E, Morand C, et al. Catechin reduces atherosclerotic lesion development in apo E-deficient mice: a transcriptomic study. Atherosclerosis 2009;204(2):e21−7.

[97] Nath S, Bachani M, Harshavardhana D, Steiner JP. Catechins protect neurons against mitochondrial toxins and HIV proteins *via* activation of the BDNF pathway. J Neurovirol 2012;18(6):445−55.

[98] Reimann HJ, Lorenz W, Fischer M, Frolich R, Meyer HJ, Schmal A. Histamine and acute haemorrhagic lesions in rat gastric mucosa: prevention of stress ulcer formation by (+)-catechin, an inhibitor of specific histidine decarboxylase *in vitro*. Agents Actions 1997;7(1):69−73.

[99] Hou WC, Lin RD, Chen CT, Lee MH. Monoamine oxidase B (MAO-B) inhibition by active principles from *Uncaria rhynchophylla*. J Ethnopharmacol 2005;100 (1−2):216−20.

[100] Zhou L, Elias RJ. Antioxidant and pro-oxidant activity of (-)-epigallocatechin-3-gallate in food emulsions: influence of pH and phenolic concentration. Food Chem 2013;138(2−3):1503−9.

[101] Hu B, Ting Y, Zeng X, Huang Q. Bioactive peptides/chitosan nanoparticles enhance cellular antioxidant activity of (-)-epigallocatechin-3-gallate. J Agric Food Chem 2013;61(4):875−81.

[102] Du GJ, Zhang Z, Wen XD, Yu C, Calway T, Yuan CS, et al. Epigallocatechin gallate (EGCG) is the most effective cancer chemopreventive polyphenol in green tea. Nutrients 2012;4(11):1679−91.

[103] Wang H, Bian S, Yang CS. Green tea polyphenol EGCG suppresses lung cancer cell growth through upregulating miR-210 expression caused by stabilizing HIF-1alpha. Carcinogenesis 2011;32(12):1881−9.

[104] Chaabi M, Beghidja N, Benayache S, Lobstein A. Activity-guided isolation of antioxidant principles from *Limoniastrum feei* (Girard) Batt. Z Naturforsch C 2008;63 (11−12):801−7.

[105] El-Hawary SA, Sokkar NM, Ali ZY, Yehia MM. A profile of bioactive compounds of *Rumex vesicarius* L. J Food Sci 2011;76(8):C1195−202.

[106] Snijman PW, Swanevelder S, Joubert E, Green IR, Gelderblom WC. The antimutagenic activity of the major flavonoids of rooibos (*Aspalathus linearis*): some dose-response effects on mutagen activation-flavonoid interactions. Mutat Res 2007;631 (2):111−23.

[107] Momtaz S, Mapunya BM, Houghton PJ, Edgerly C, Hussein A, Naidoo S, et al. Tyrosinase inhibition by extracts and constituents of *Sideroxylon inerme* L. stem bark, used in South Africa for skin lightening. J Ethnopharmacol 2008;119 (3):507−12.

[108] Mahler A, Mandel S, Lorenz M, Ruegg U, Wanker EE, Boschmann M, et al. Epigallocatechin-3-gallate: a useful, effective and safe clinical approach for targeted prevention and individualised treatment of neurological diseases? EPMA J 2013;4 (1):5.

[109] Wu D, Guo Z, Ren Z, Guo W, Meydani SN. Green tea EGCG suppresses T cell proliferation through impairment of IL-2/IL-2 receptor signaling. Free Radic Biol Med 2009;47(5):636−43.

[110] Steinman L, Zamvil SS. How to successfully apply animal studies in experimental allergic encephalomyelitis to research on multiple sclerosis. Ann Neurol 2006;60 (1):12−21.

[111] Mix E, Meyer-Rienecker H, Zettl UK. Animal models of multiple sclerosis for the development and validation of novel therapies: potential and limitations. J Neurol 2008;255(Suppl. 6):7−14.

[112] Vesterinen HM, Sena ES, ffrench-Constant C, Williams A, Chandran S, Macleod MR. Improving the translational hit of experimental treatments in multiple sclerosis. Mult Scler 2010;16(9):1044−55.

[113] Baker D, Gerritsen W, Rundle J, Amor S. Critical appraisal of animal models of multiple sclerosis. Mult Scler 2011;17(6):647−57.

[114] Aktas O, Prozorovski T, Smorodchenko A, Savaskan NE, Lauster R, Kloetzel PM, et al. Green tea epigallocatechin-3-gallate mediates T cellular NF-kappa B inhibition and exerts neuroprotection in autoimmune encephalomyelitis. J Immunol 2004;173 (9):5794−800.

[115] Choi YT, Jung CH, Lee SR, Bae JH, Baek WK, Suh MH, et al. The green tea polyphenol (-)-epigallocatechin gallate attenuates beta-amyloid-induced neurotoxicity in cultured hippocampal neurons. Life Sci 2001;70(5):603−14.

[116] Walton NM, Shin R, Tajinda K, Heusner CL, Kogan JH, Miyake S, et al. Adult neurogenesis transiently generates oxidative stress. PLoS One 2012;7(4):e35264.

[117] Levites Y, Amit T, Mandel S, Youdim MB. Neuroprotection and neurorescue against abeta toxicity and PKC-dependent release of nonamyloidogenic soluble precursor protein by green tea polyphenol (-)-epigallocatechin-3-gallate. FASEB J 2003;17 (8):952−4.

[118] Rezai-Zadeh K, Shytle D, Sun N, Mori T, Hou H, Jeanniton D, et al. Green tea epigallocatechin-3-gallate (EGCG) modulates amyloid precursor protein cleavage and reduces cerebral amyloidosis in Alzheimer transgenic mice. J Neurosci 2005;25 (38):8807−14.

[119] Reznichenko L, Amit T, Zheng H, Avramovich-Tirosh Y, Youdim MB, Weinreb O, et al. Reduction of iron-regulated amyloid precursor protein and beta-amyloid peptide by (-)-epigallocatechin-3-gallate in cell cultures: implications for iron chelation in Alzheimer's disease. J Neurochem 2006;97(2):527−36.

[120] Citron M. Beta-secretase inhibition for the treatment of Alzheimer's disease—promise and challenge. Trends Pharmacol Sci 2004;25(2):92−7.

[121] Lee JW, Lee YK, Ban JO, Ha TY, Yun YP, Han SB, et al. Green tea (-)-epigallocatechin-3-gallate inhibits beta-amyloid-induced cognitive dysfunction through modification of secretase activity via inhibition of ERK and NF-kappaB pathways in mice. J Nutr 2009;139(10):1987−93.

[122] Nkengfack AE, Azebaze AG, Waffo AK, Fomum ZT, Meyer M, van Heerden FR. Cytotoxic isoflavones from Erythrina indica. Phytochemistry 2001;58(7):1113−20.

[123] Marzouk MS, Ibrahim MT, El-Gindi OR, Abou Bakr MS. Isoflavonoid glycosides and rotenoids from Pongamia pinnata leaves. Z Naturforsch C 2008;63 (1−2):1−7.

[124] Duker-Eshun G, Jaroszewski JW, Asomaning WA, Oppong-Boachie F, Brogger Christensen S. Antiplasmodial constituents of Cajanus cajan. Phytother Res 2004;18 (2):128−30.

[125] Polkowski K, Mazurek AP. Biological properties of genistein. A review of in vitro and in vivo data. Acta Pol Pharm 2000;57(2):135−55.

[126] Steele VE, Pereira MA, Sigman CC, Kelloff GJ. Cancer chemoprevention agent development strategies for genistein. J Nutr 1995;125(Suppl. 3):713S−6S.

[127] Peterson G. Evaluation of the biochemical targets of genistein in tumor cells. J Nutr 1995;125(Suppl. 3):784S−9S.

[128] Si H, Liu D. Phytochemical genistein in the regulation of vascular function: new insights. Curr Med Chem 2007;14(24):2581−9.

[129] Kumi-Diaka J, Rodriguez R, Goudaze G. Influence of genistein (4′,5,7-trihydroxyiso-flavone) on the growth and proliferation of testicular cell lines. Biol Cell 1998;90 (4):349−54.

[130] Mitchell JH, Cawood E, Kinniburgh D, Provan A, Collins AR, Irvine DS. Effect of a phytoestrogen food supplement on reproductive health in normal males. Clin Sci (Lond) 2001;100(6):613−8.

[131] Calderon-Montano JM, Burgos-Moron E, Perez-Guerrero C, Lopez-Lazaro M. A review on the dietary flavonoid kaempferol. Mini Rev Med Chem 2011;11 (4):298−344.

[132] Tibe O, Meagher LP, Fraser K, Harding DR. Condensed tannins and flavonoids from the forage legume sulla (*Hedysarum coronarium*). J Agric Food Chem 2011;59 (17):9402–9.

[133] Mazimba O, Majinda RR, Modibedi C, Masesane IB, Cencic A, Chingwaru W. *Tylosema esculentum* extractives and their bioactivity. Bioorg Med Chem 2011;19 (17):5225–30.

[134] Majinda RR, Motswaledi M, Waigh RD, Waterman PG. Phenolic and antibacterial constituents of *Vahlia capensis*. Planta Med 1997;63(3):268–70.

[135] Tatsimo SJ, Tamokou JdeD, Havyarimana L, Csupor D, Forgo P, Hohmann J, et al. Antimicrobial and antioxidant activity of kaempferol rhamnoside derivatives from *Bryophyllum pinnatum*. BMC Res Notes 2012;5:158.

[136] Nguemeving JR, Azebaze AG, Kuete V, Eric Carly NN, Beng VP, Meyer M, et al. Laurentixanthones A and B, antimicrobial xanthones from *Vismia laurentii*. Phytochemistry 2006;67(13):1341–6.

[137] Elgorashi E, van Heerden F, van Staden J. Kaempferol, a mutagenic flavonol from *Helichrysum simillimum*. Hum Exp Toxicol 2008;27(11):845–9.

[138] Martini ND, Katerere DR, Eloff JN. Biological activity of five antibacterial flavonoids from *Combretum erythrophyllum* (Combretaceae). J Ethnopharmacol 2004;93 (2–3):207–12.

[139] Middleton Jr. E, Kandaswami C, Theoharides TC. The effects of plant flavonoids on mammalian cells: implications for inflammation, heart disease, and cancer. Pharmacol Rev 2000;52(4):673–751.

[140] Hertog MG, Feskens EJ, Hollman PC, Katan MB, Kromhout D. Dietary antioxidant flavonoids and risk of coronary heart disease: the Zutphen Elderly Study. Lancet 1993;342(8878):1007–11.

[141] Lin J, Rexrode KM, Hu F, Albert CM, Chae CU, Rimm EB, et al. Dietary intakes of flavonols and flavones and coronary heart disease in US women. Am J Epidemiol 2007;2165(11):1305–13.

[142] Geleijnse JM, Launer LJ, Van der Kuip DA, Hofman A, Witteman JC. Inverse association of tea and flavonoid intakes with incident myocardial infarction: the Rotterdam study. Am J Clin Nutr 2002;75(5):880–6.

[143] Knekt P, Kumpulainen J, Jarvinen R, Rissanen H, Heliovaara M, Reunanen A, et al. Flavonoid intake and risk of chronic diseases. Am J Clin Nutr 2002;76(3):560–8.

[144] Dhalla NS, Temsah RM, Netticadan T. Role of oxidative stress in cardiovascular diseases. J Hypertens 2000;18(6):655–73.

[145] Willerson JT, Ridker PM. Inflammation as a cardiovascular risk factor. Circulation 2004;109(21 Suppl. 1):II2–10.

[146] Tu YC, Lian TW, Yen JH, Chen ZT, Wu MJ. Antiatherogenic effects of kaempferol and rhamnocitrin. J Agric Food Chem 2007;55(24):9969–76.

[147] Oh H, Kang DG, Kwon JW, Kwon TO, Lee SY, Lee DB, et al. Isolation of angiotensin converting enzyme (ACE) inhibitory flavonoids from *Sedum sarmentosum*. Biol Pharm Bull 2004;27(12):2035–7.

[148] Olszanecki R, Bujak-Gizycka B, Madej J, Suski M, Wolkow PP, Jawien J, et al. Kaempferol, but not resveratrol inhibits angiotensin converting enzyme. J Physiol Pharmacol 2008;59(2):387–92.

[149] Xu YC, Yeung DK, Man RY, Leung SW. Kaempferol enhances endothelium-independent and dependent relaxation in the porcine coronary artery. Mol Cell Biochem 2006;287(1–2):61–7.

[150] Xu YC, Leung GP, Wong PY, Vanhoutte PM, Man RY. Kaempferol stimulates large conductance Ca2 + -activated K + (BKCa) channels in human umbilical vein endothelial cells via a cAMP/PKA-dependent pathway. Br J Pharmacol 2008;154 (6):1247−53.

[151] Chung MI, Gan KH, Lin CN, Ko FN, Teng CM. Antiplatelet effects and vasorelaxing action of some constituents of Formosan plants. J Nat Prod 1993;56(6):929−34.

[152] Hannum SM. Potential impact of strawberries on human health: a review of the science. Crit Rev Food Sci Nutr 2004;44(1):1−17.

[153] Belguith-Hadriche O, Bouaziz M, Jamoussi K, El Feki A, Sayadi S, Makni-Ayedi F. Lipid-lowering and antioxidant effects of an ethyl acetate extract of fenugreek seeds in high-cholesterol-fed rats. J Agric Food Chem 2010;58(4):2116−222.

[154] Zern TL, West KL, Fernandez ML. Grape polyphenols decrease plasma triglycerides and cholesterol accumulation in the aorta of ovariectomized guinea pigs. J Nutr 2003;133(7):2268−72.

[155] Yu SF, Shun CT, Chen TM, Chen YH. 3-O-beta-D-glucosyl-(1− > 6)-beta-D-glucosyl-kaempferol isolated from Sauropus androgenus reduces body weight gain in Wistar rats. Biol Pharm Bull 2006;29(12):2510−3.

[156] Rattanachaikunsopon P, Phumkhachorn P. Bacteriostatic effect of flavonoids isolated from leaves of Psidium guajava on fish pathogens. Fitoterapia 2007;78(6):434−6.

[157] Prahalathan P, Kumar S, Raja B. Effect of morin, a flavonoid against DOCA-salt hypertensive rats: a dose dependent study. Asian Pac J Trop Biomed 2012;2 (6):443−8.

[158] Abuohashish HM, Al-Rejaie SS, Al-Hosaini KA, Parmar MY, Ahmed MM. Alleviating effects of morin against experimentally-induced diabetic osteopenia. Diabetol Metab Syndr 2013;5(1):5.

[159] Al Numair KS, Chandramohan G, Alsaif MA, Baskar AA. Protective effect of morin on cardiac mitochondrial function during isoproterenol-induced myocardial infarction in male Wistar rats. Redox Rep 2012;17(1):14−21.

[160] Pogula BK, Maharajan MK, Oddepalli DR, Boini L, Arella M, Sabarimuthu DQ. Morin protects heart from beta-adrenergic-stimulated myocardial infarction: an electrocardiographic, biochemical, and histological study in rats. J Physiol Biochem 2012;68(3):433−46.

[161] Shankari SG, Karthikesan K, Jalaludeen AM, Ashokkumar N. Hepatoprotective effect of morin on ethanol-induced hepatotoxicity in rats. J Basic Clin Physiol Pharmacol 2010;21(4):277−94.

[162] Lim SC, Im YB, Bae CS, Han SI, Kim SE, Han HK. Protective effect of morin on the imipenem-induced nephrotoxicity in rabbits. Arch Pharm Res 2008;31(8):1060−5.

[163] Lee HS, Jung KH, Park IS, Kwon SW, Lee DH, Hong SS. Protective effect of morin on dimethylnitrosamine-induced hepatic fibrosis in rats. Dig Dis Sci 2009;54 (4):782−8.

[164] Wu TW, Zeng LH, Wu J, Fung KP. Morin: a wood pigment that protects three types of human cells in the cardiovascular system against oxyradical damage. Biochem Pharmacol 1994;47(6):1099−103.

[165] Rahigude A, Bhutada P, Kaulaskar S, Aswar M, Otari K. Participation of antioxidant and cholinergic system in protective effect of naringenin against type-2 diabetes-induced memory dysfunction in rats. Neuroscience 2012;226(0):62−72.

[166] Wilcox L, Borradaile N, Huff M. Antiatherogenic properties of naringenin, a citrus flavonoid. Cardiovasc Drug Rev 1999;17:160−78.

[167] Felgines C, Texier O, Morand C, Manach C, Scalbert A, Regerat F, et al. Bioavailability of the flavanone naringenin and its glycosides in rats. Am J Physiol Gastrointest Liver Physiol 2000;279(6):G1148−54.

[168] Schroder G, Wehinger E, Lukacin R, Wellmann F, Seefelder W, Schwab W, et al. Flavonoid methylation: a novel 4′-O-methyltransferase from *Catharanthus roseus*, and evidence that partially methylated flavanones are substrates of four different flavonoid dioxygenases. Phytochemistry 2004;65(8):1085−94.

[169] Borradaile NM, de Dreu LE, Barrett PH, Huff MW. Inhibition of hepatocyte apoB secretion by naringenin: enhanced rapid intracellular degradation independent of reduced microsomal cholesteryl esters. J Lipid Res 2002;43(9):1544−54.

[170] Borradaile NM, de Dreu LE, Barrett PH, Behrsin CD, Huff MW. Hepatocyte apoB-containing lipoprotein secretion is decreased by the grapefruit flavonoid, naringenin, *via* inhibition of MTP-mediated microsomal triglyceride accumulation. Biochemistry 2003;42(5):1283−91.

[171] Ortiz-Andrade RR, Sanchez-Salgado JC, Navarrete-Vazquez G, Webster SP, Binnie M, Garcia-Jimenez S, et al. Antidiabetic and toxicological evaluations of naringenin in normoglycaemic and NIDDM rat models and its implications on extra-pancreatic glucose regulation. Diabetes Obes Metab 2008;10(11):1097−104.

[172] Mulvihill EE, Allister EM, Sutherland BG, Telford DE, Sawyez CG, Edwards JY, et al. Naringenin prevents dyslipidemia, apolipoprotein B overproduction, and hyperinsulinemia in LDL receptor-null mice with diet-induced insulin resistance. Diabetes 2009; 58(10):2198−210.

[173] Rayidi S. Effect of naringenin on carbohydrate metabolism in STZ-nicotinamide induced diabetic rats. Biomirror 2011;2:1−9.

[174] Goldwasser J, Cohen PY, Yang E, Balaguer P, Yarmush ML, Nahmias Y. Transcriptional regulation of human and rat hepatic lipid metabolism by the grapefruit flavonoid naringenin: role of PPARalpha, PPARgamma and LXRalpha. PLoS One 2010;5(8):e12399.

[175] Zbarsky V, Datla KP, Parkar S, Rai DK, Aruoma OI, Dexter DT. Neuroprotective properties of the natural phenolic antioxidants curcumin and naringenin but not quercetin and fisetin in a 6-OHDA model of Parkinson's disease. Free Radic Res 2005;39 (10):1119−25.

[176] Olsen HT, Stafford GI, van Staden J, Christensen SB, Jager AK. Isolation of the MAO-inhibitor naringenin from *Mentha aquatica* L. J Ethnopharmacol 2008;117(3):500−2.

[177] Park J, Lee J, Paik H, Cho S, Nah S, Park Y, et al. Cytotoxic effects of 7-O-butyl naringenin on human breast cancer MCF-7 cells. Food Sci Biotechnol 2010;19:717−24.

[178] Park HR, Park M, Choi J, Park KY, Chung HY, Lee J. A high-fat diet impairs neurogenesis: involvement of lipid peroxidation and brain-derived neurotrophic factor. Neurosci Lett 2010;482(3):235−9.

[179] Ribeiro I, Rocha J, Sepodes B, Mota-Filipe Y, Ribeiro M. Effect of naringin enzymatic hydrolysis towards naringenin on the anti-inflammatory activity of both compounds. J Mol Catal B Enzym 2008;52:13−8.

[180] Orsolic N, Gajski G, Garaj-Vrhovac V, Dikic D, Prskalo ZS, Sirovina D. DNA-protective effects of quercetin or naringenin in alloxan-induced diabetic mice. Eur J Pharmacol 2011;656(1−3):110−8.

[181] Heo HJ, Kim DO, Shin SC, Kim MJ, Kim BG, Shin DH. Effect of antioxidant flavanone, naringenin, from *Citrus junoson* neuroprotection. J Agric Food Chem 2004;52 (6):1520−5.

[182] Heo HJ, Kim MJ, Lee JM, Choi SJ, Cho HY, Hong B, et al. Naringenin from *Citrus junos* has an inhibitory effect on acetylcholinesterase and a mitigating effect on amnesia. Dement Geriatr Cogn Disord 2004;17(3):151−7.

[183] Badary OA, Abdel-Maksoud S, Ahmed WA, Owieda GH. Naringenin attenuates cisplatin nephrotoxicity in rats. Life Sci 2005;76(18):2125−35.

[184] Arafa HM, Abd-Ellah MF, Hafez HF. Abatement by naringenin of doxorubicin-induced cardiac toxicity in rats. J Egypt Natl Canc Inst 2005;17(4):291−300.

[185] Smith C, Lombard K, Peffley E, Liu W. Genetic analysis of quercetin in onion (*Allium cepa* L.) Lady Raider. Texas J Agric Natl Res 2003;16:24−8.

[186] Mitchell AE, Hong YJ, Koh E, Barrett DM, Bryant DE, Denison RF, et al. Ten-year comparison of the influence of organic and conventional crop management practices on the content of flavonoids in tomatoes. J Agric Food Chem 2007;55(15):6154−9.

[187] Jaganathan SK. Can flavonoids from honey alter multidrug resistance? Med Hypotheses 2011;76(4):535−7.

[188] Benariba N, Djaziri R, Bellakhdar W, Belkacem N, Kadiata M, Malaisse WJ, et al. Phytochemical screening and free radical scavenging activity of *Citrullus colocynthis* seeds extracts. Asian Pac J Trop Biomed 2013;23(1):35−40.

[189] Dahia M, Siracusa L, Laouer H, Ruberto G. Constituents of the polar extracts from Algerian *Pituranthos scoparius*. Nat Prod Commun 2009;4(12):1691−2.

[190] Poumale HM, Randrianasolo R, Rakotoarimanga JV, Raharisololalao A, Krebs HC, Tchouankeu JC, et al. Flavonoid glycosides and other constituents of *Psorospermum androsaemifolium* BAKER (Clusiaceae). Chem Pharm Bull (Tokyo) 2008;56 (10):1428−30.

[191] Pal A, Sharma PP, Pandya TN, Acharya R, Patel BR, Shukla VJ, et al. Phytochemical evaluation of dried aqueous extract of Jivanti [*Leptadenia reticulata* (Retz.) Wt. et Arn]. Ayu 2012;33(4):557−60.

[192] Chaabi M, Freund-Michel V, Frossard N, Randriantsoa A, Andriantsitohaina R, Lobstein A. Anti-proliferative effect of *Euphorbia stenoclada* in human airway smooth muscle cells in culture. J Ethnopharmacol 2007;109(1):134−9.

[193] Shodehinde SA, Oboh G, Faoziyat SA. Antioxidant properties of aqueous extracts of unripe *Musa paradisiaca* on sodium nitroprusside induced lipid peroxidation in rat pancreas *in vitro*. Asian Pac J Trop Biomed 2013;3(6):449−57.

[194] Pendota SC, Aderogba MA, Ndhlala AR, Van Staden J. Antimicrobial and acetylcholinesterase inhibitory activities of *Buddleja salviifolia* (L.) Lam. leaf extracts and isolated compounds. J Ethnopharmacol 2013;148(2):515−20.

[195] Oyedemi SO, Oyedemi BO, Arowosegbe S, Afolayan AJ. Phytochemicals analysis and medicinal potentials of hydroalcoholic extract from *Curtisia dentata* (Burm.f) C. A. Sm stem bark. Int J Mol Sci 2012;13(5):6189−203.

[196] Lall N, Hussein AA, Meyer JJ. Antiviral and antituberculous activity of *Helichrysum melanacme* constituents. Fitoterapia 2006;77(3):230−2.

[197] Marais JP, Mueller-Harvey I, Brandt EV, Ferreira D. Polyphenols, condensed tannins, and other natural products in *Onobrychis viciifolia* (Sainfoin). J Agric Food Chem 2000;48(8):3440−7.

[198] Hayek T, Fuhrman B, Vaya J, Rosenblat M, Belinky P, Coleman R, et al. Reduced progression of atherosclerosis in apolipoprotein E-deficient mice following consumption of red wine, or its polyphenols quercetin or catechin, is associated with reduced susceptibility of LDL to oxidation and aggregation. Arterioscler Thromb Vasc Biol 1997;17(11):2744−52.

[199] Chopra M, Fitzsimons PE, Strain JJ, Thurnham DI, Howard AN. Nonalcoholic red wine extract and quercetin inhibit LDL oxidation without affecting plasma antioxidant vitamin and carotenoid concentrations. Clin Chem 2000;46:1162—70.

[200] Verma AK, Johnson JA, Gould MN, Tanner MA. Inhibition of 7,12-dimethylbenz(a) anthracene- and N-nitrosomethylurea-induced rat mammary cancer by dietary flavonol quercetin. Cancer Res 1988;48(20):5754—8.

[201] Deschner EE, Ruperto J, Wong G, Newmark HL. Quercetin and rutin as inhibitors of azoxymethanol-induced colonic neoplasia. Carcinogenesis 1991;12(7):1193—6.

[202] Pereira MA, Grubbs CJ, Barnes LH, Li H, Olson GR, Eto I, et al. Effects of the phytochemicals, curcumin and quercetin, upon azoxymethane-induced colon cancer and 7,12-dimethylbenz[a]anthracene-induced mammary cancer in rats. Carcinogenesis 1996;17(6):1305—11.

[203] Ferry DR, Smith A, Malkhandi J, Fyfe DW, deTakats PG, Anderson D, et al. Phase I clinical trial of the flavonoid quercetin: pharmacokinetics and evidence for *in vivo* tyrosine kinase inhibition. Clin Cancer Res 1996;2(4):659—68.

[204] Pignatelli P, Pulcinelli FM, Celestini A, Lenti L, Ghiselli A, Gazzaniga PP, et al. The flavonoids quercetin and catechin synergistically inhibit platelet function by antagonizing the intracellular production of hydrogen peroxide. Am J Clin Nutr 2000;72 (5):1150—5.

[205] Perez-Vizcaino F, Ibarra M, Cogolludo AL, Duarte J, Zaragoza-Arnaez F, Moreno L, et al. Endothelium-independent vasodilator effects of the flavonoid quercetin and its methylated metabolites in rat conductance and resistance arteries. J Pharmacol Exp Ther 2002;302(1):66—72.

[206] Bardy G, Virsolvy A, Quignard JF, Ravier MA, Bertrand G, Dalle S, et al. Quercetin induces insulin secretion by direct activation of L-type calcium channels in pancreatic beta cells. Br J Pharmacol 2013;169(5):1102—13.

[207] Liu CM, Zheng GH, Cheng C, Sun JM. Quercetin protects mouse brain against lead-induced neurotoxicity. J Agric Food Chem 2013;61(31):7630—5.

[208] Tang L, Peng Y, Xu T, Yi X, Liu Y, Luo Y, et al. The effects of quercetin protect cardiomyocytes from A/R injury is related to its capability to increasing expression and activity of PKCepsilon protein. Mol Cell Biochem 2013;382(1—2):145—52.

[209] Chen YW, Chou HC, Lin ST, Chen YH, Chang YJ, Chen L, et al. Cardioprotective effects of quercetin in cardiomyocyte under ischemia/reperfusion Injury. Evid Based Complement Alternat Med 2013;2013:364519.

[210] Matouk AI, Taye A, Heeba GH, El-Moselhy MA. Quercetin augments the protective effect of losartan against chronic doxorubicin cardiotoxicity in rats. Environ Toxicol Pharmacol 2013;36(2):443—50.

[211] Selim ME, Aleisa NA, Daghestani MH. Evaluation of the possible protective role of quercetin on letrozole-induced testicular injury in male albino rats. Ultrastruct Pathol 2013;37(3):204—17.

[212] Kumar Mishra S, Singh P, Rath SK. Protective effect of quercetin on chloroquine-induced oxidative stress and hepatotoxicity in mice. Malar Res Treat 2013;2013:141734.

[213] Gao H, Chen C, Huang S, Li B. Quercetin attenuates the progression of monocrotaline-induced pulmonary hypertension in rats. J Biomed Res 2012;26(2):98—102.

[214] Wang L, Lin SQ, He YL, Liu G, Wang ZY. Protective effects of quercetin on cadmium-induced cytotoxicity in primary cultures of rat proximal tubular cells. Biomed Environ Sci 2013;26(4):258—67.

[215] Cao Y, Fu ZD, Wang F, Liu HY, Han R. Anti-angiogenic activity of resveratrol, a natural compound from medicinal plants. J Asian Nat Prod Res 2005;7(3):205−13.

[216] Mokgotho MP, Gololo SS, Masoko P, Mdee LK, Mbazima V, Shai LJ, et al. Isolation and chemical structural characterisation of a compound with antioxidant activity from the roots of *Senna italica*. Evid Based Complement Alternat Med 2013;2013:519174.

[217] Joseph CC, Moshi MJ, Innocent E, Nkunya MH. Isolation of a stilbene glycoside and other constituents of *Terminalia sericeae*. Afr J Tradit Complement Altern Med 2007;4(4):383−6.

[218] Wanjala CC, Majinda RR. A new stilbene glycoside from *Elephantorrhiza goetzei*. Fitoterapia 2001;72(6):649−55.

[219] Bertelli AA, Migliori M, Panichi V, Origlia N, Filippi C, Das DK, et al. Resveratrol, a component of wine and grapes, in the prevention of kidney disease. Ann N Y Acad Sci 2002;957:230−8.

[220] Chander V, Tirkey N, Chopra K. Resveratrol, a polyphenolic phytoalexin protects against cyclosporine-induced nephrotoxicity through nitric oxide dependent mechanism. Toxicology 2005;210(1):55−64.

[221] Delmas D, Jannin B, Latruffe N. Resveratrol: preventing properties against vascular alterations and ageing. Mol Nutr Food Res 2005;49(5):377−95.

[222] Olas B, Wachowicz B. Resveratrol, a phenolic antioxidant with effects on blood platelet functions. Platelets 2005;16(5):251−60.

[223] Falchi M, Bertelli A, Galazzo R, Vigano P, Dib B. Central antalgic activity of resveratrol. Arch Ital Biol 2010;148(4):389−96.

[224] Parker JA, Arango M, Abderrahmane S, Lambert E, Tourette C, Catoire H, et al. Resveratrol rescues mutant polyglutamine cytotoxicity in nematode and mammalian neurons. Nat Genet 2005;37(4):349−50.

[225] Bertelli A, Falchi M, Dib B, Pini E, Mukherjee S, Das DK. Analgesic resveratrol? Antioxid Redox Sign 2008;10:403−4.

[226] Szewczuk LM, Forti L, Stivala LA, Penning TM. Resveratrol is a peroxidase-mediated inactivator of COX-1 but not COX-2: a mechanistic approach to the design of COX-1 selective agents. J Biol Chem 2004;279(21):22727−37.

[227] Torres-Lopez JE, Ortiz MI, Castaneda-Hernandez G, Alonso-Lopez R, Asomoza-Espinosa R, Granados-Soto V. Comparison of the antinociceptive effect of celecoxib, diclofenac and resveratrol in the formalin test. Life Sci 2002;70(14):1669−76.

[228] <http://lpioregonstateedu/infocenter/phytochemicals/resveratrol> Resveratrol [accessed 28.7.13].

[229] Elliott PJ, Jirousek M. Sirtuins: novel targets for metabolic disease. Curr Opin Investig Drugs 2008;9(4):371−8.

[230] Das DK, Maulik N. Resveratrol in cardioprotection: a therapeutic promise of alternative medicine. Mol Interv 2006;6(1):36−47.

[231] Marambaud P, Zhao H, Davies P. Resveratrol promotes clearance of Alzheimer's disease amyloid-beta peptides. J Biol Chem 2005;280(45):37377−82.

[232] Wu CH, Lin MC, Wang HC, Yang MY, Jou MJ, Wang CJ. Rutin inhibits oleic acid induced lipid accumulation *via* reducing lipogenesis and oxidative stress in hepatocarcinoma cells. J Food Sci 2011;76(2):T65−72.

[233] Kreft S, Knapp M, Kreft I. Extraction of rutin from buckwheat (*Fagopyrum esculentum* Moench) seeds and determination by capillary electrophoresis. J Agric Food Chem 1999;47(11):4649−52.

[234] Papajewski S, Vogler B, Conrad J, Klaiber I, Roos G, Walter CU, et al. Isolation from Cussonia barteri of 1'-O-chlorogenoylchlorogenic acid and 1'-O-chlorogenoyl-neochlorogenic acid, a new type of quinic acid esters. Planta Med 2011;67(8):732−6.

[235] Atawodi SE, Atawodi JC, Idakwo GA, Pfundstein B, Haubner R, Wurtele G, et al. Evaluation of the polyphenol content and antioxidant properties of methanol extracts of the leaves, stem, and root barks of *Moringa oleifera* Lam. J Med Food 2010;13 (3):710−6.

[236] Onunkwo GC, Egeonu HC, Adikwu MU, Ojile JE, Olowosulu AK. Some physical properties of tabletted seed of *Garcinia kola* (Heckel). Chem Pharm Bull (Tokyo) 2004;52(6):649−53.

[237] Ogundipe OO, Moody JO, Houghton PJ, Odelola HA. Bioactive chemical constituents from *Alchornea laxiflora* (benth) pax and hoffman. J Ethnopharmacol 2001;74 (3):275−80.

[238] Harinantenaina L, Kasai R, Yamasaki K. A new ent-kaurane diterpenoid glycoside from the leaves of *Cussonia bojeri*, a Malagasy endemic plant. Chem Pharm Bull (Tokyo) 2002;50(8):1122−3.

[239] Harinantenaina Liva RR, Kasai R, Yamasaki K. Clerodane and labdane diterpene gly-cosides from a Malagasy endemic plant, *Cussonia racemosa*. Phytochemistry 2002;60 (4):339−43.

[240a] Donia AEl Raheim Mohammed, Alqasoumi SI, Awaad AS, Cracker L. Antioxidant activity of *Convolvulus hystrix* Vahl and its chemical constituents. Pak J Pharm Sci 2011;24(2):143−7.

[240b] Chua LS. A review on plant-based rutin extraction methods and its pharmacological activities. J Ethnopharmacol 2013;150(3):805−17.

[241] Hsu CL, Wu CH, Huang SL, Yen GC. Phenolic compounds rutin and o-coumaric acid ameliorate obesity induced by high-fat diet in rats. J Agric Food Chem 2009;57 (2):425−31.

[242] Olaleye MT, Akinmoladun AC. Comparative gastroprotective effect of post-treatment with low doses of rutin and cimetidine in rats. Fundam Clin Pharmacol 2013;27 (2):138−45.

[243] Reporter DM. Chemical found in apples, onions and green tea can help beat blood clots, <http://www.dailymail.co.uk/health/article-2141602/Chemical-apples-onions-green-tea-help-beat-blood-clots.html>; 2009 [accessed 28.07.13].

[244] <http://wwwnaturalstandardcom/index-abstractasp?create-abstract=/monographs/ herbssupplements/patient-rutinasp> Rutin [accessed 29.07.13].

[245] Sadeghnia HR, Yousefsani BS, Rashidfar M, Boroushaki MT, Asadpour E, Ghorbani A. Protective effect of rutin on hexachlorobutadiene-induced nephrotoxicity. Ren Fail 2013;35(8):1151−5.

[246] Abarikwu SO, Iserhienrhien BO, Badejo TA. Rutin- and selenium-attenuated cadmium-induced testicular pathophysiology in rats. Hum Exp Toxicol 2013;32(4):395−406.

[247] Khan RA, Khan MR, Sahreen S. Protective effects of rutin against potassium bromate induced nephrotoxicity in rats. BMC Complement Altern Med 2012;12:204.

[248] Boukef K. Rhamnus alaternus. Essaydali 2001;81:34−5.

[249] Bhouri W, Sghaier MB, Kilani S, Bouhlel I, Dijoux-Franca MG, Ghedira K, et al. Evaluation of antioxidant and antigenotoxic activity of two flavonoids from *Rhamnus alaternus* L. (Rhamnaceae): kaempferol 3-O-beta-isorhamninoside and rhamnocitrin 3-O-beta-isorhamninoside. Food Chem Toxicol 2011;49(5):1167−73.

[250] Bhouri W, Boubaker J, Kilani S, Ghedira K, Chekir-Ghedira L. Flavonoids from *Rhamnus alaternus* L. (Rhamnaceae): Kaempferol 3-O-β-isorhamninoside and rhamnocitrin 3-O-β-isorhamninoside protect against DNA damage in human lymphoblastoid cell and enhance antioxidant activity. S Afr J Bot 2012;80(0):57−62.

[251] Farombi EO. Mechanisms for the hepatoprotective action of kolaviron: studies on hepatic enzymes, microsomal lipids and lipid peroxidation in carbontetrachloride-treated rats. Pharmacol Res 2000;42(1):75−80.

[252] Adaramoye OA, Nwosu IO, Farombi EO. Sub-acute effect of N(G)-nitro-L-arginine methyl-ester (L-NAME) on biochemical indices in rats: Protective effects of Kolaviron and extract of *Curcuma longa* L. Pharmacognosy Res 2012;4(3):127−33.

[253] Donfack J, Nico F, Ngameni B, Tchana A, Chuisseu P, Finzi P, et al. *In vitro* hepatoprotective and antioxidant activities of diprenylated isoflavonoids from *Erythrina senegalensis* (Fabaceae). Asian J Tradit Med 2008;3(5):172−8.

[254] Adaramoye OA. Protective effect of kolaviron, a biflavonoid from *Garcinia kola* seeds, in brain of Wistar albino rats exposed to gamma-radiation. Biol Pharm Bull 2010;33(2):260−6.

[255] Ishola IO, Chatterjee M, Tota S, Tadigopulla N, Adeyemi OO, Palit G, et al. Antidepressant and anxiolytic effects of amentoflavone isolated from *Cnestis ferruginea* in mice. Pharmacol Biochem Behav 2010;103(2):322−31.

[256] Ferchichi L, Derbre S, Mahmood K, Toure K, Guilet D, Litaudon M, et al. Bioguided fractionation and isolation of natural inhibitors of advanced glycation end-products (AGEs) from *Calophyllum flavoramulum*. Phytochemistry 2012;78:98−106.

[257] Azebaze AG, Dongmo AB, Meyer M, Ouahouo BM, Valentin A, Laure Nguemfo E, et al. Antimalarial and vasorelaxant constituents of the leaves of *Allanblackia monticola* (Guttiferae). Ann Trop Med Parasitol 2007;101(1):23−30.

[258] Mbaveng AT, Ngameni B, Kuete V, Simo IK, Ambassa P, Roy R, et al. Antimicrobial activity of the crude extracts and five flavonoids from the twigs of *Dorstenia barteri* (Moraceae). J Ethnopharmacol 2008;116(3):483−9.

[259] Pegnyemb DE, Mbing JN, de Theodore Atchade A, Tih RG, Sondengam BL, Blond A, et al. Antimicrobial biflavonoids from the aerial parts of *Ouratea sulcata*. Phytochemistry 2005;66(16):1922−6.

[260] Ibrahim NA, El-Seedi HR, Mohammed MM. Phytochemical investigation and hepatoprotective activity of *Cupressus sempervirens* L. leaves growing in Egypt. Nat Prod Res 2007;21(10):857−66.

[261] Ishola IO, Agbaje OE, Narender T, Adeyemi OO, Shukla R. Bioactivity guided isolation of analgesic and anti-inflammatory constituents of *Cnestis ferruginea* Vahl ex DC (Connaraceae) root. J Ethnopharmacol 2012;142(2):383−9.

[262] Kaikabo AA, Eloff JN. Antibacterial activity of two biflavonoids from *Garcinia livingstonei* leaves against *Mycobacterium smegmatis*. J Ethnopharmacol 2011;138 (1):253−5.

[263] Campos-Esparza MR, Sanchez-Gomez MV, Matute C. Molecular mechanisms of neuroprotection by two natural antioxidant polyphenols. Cell Calcium 2009;45 (4):358−68.

[264] Djemgou PC, Hussien TA, Hegazy ME, Ngandeu F, Neguim G, Tane P, et al. C-Glucoside xanthone from the stem bark extract of *Bersama engleriana*. Pharmacognosy Res 2010;2(4):229−32.

[265] Shahat AA, Hassan RA, Nazif NM, Van Miert S, Pieters L, Hammuda FM, et al. Isolation of mangiferin from *Bombax malabaricum* and structure revision of shamimin. Planta Med 2003;69(11):1068−70.

[266] Danthu P, Lubrano C, Flavet L, Rahajanirina V, Behra O, Fromageot C, et al. Biological factors influencing production of xanthones in *Aphloia theiformis*. Chem Biodivers 2010;7(1):140—50.

[267] Ojewole JA. Antiinflammatory, analgesic and hypoglycemic effects of *Mangifera indica* Linn. (Anacardiaceae) stem-bark aqueous extract. Methods Find Exp Clin Pharmacol 2005;27(8):547—54.

[268] Adaramoye OA, Adedara IA, Farombi EO. Possible ameliorative effects of kolaviron against reproductive toxicity in sub-lethally whole body gamma-irradiated rats. Exp Toxicol Pathol 2012;64(4):379—85.

[269] Farombi EO, Abarikwu SO, Adedara IA, Oyeyemi MO. Curcumin and kolaviron ameliorate di-*n*-butylphthalate-induced testicular damage in rats. Basic Clin Pharmacol Toxicol 2007;100(1):43—8.

[270] Carvalho AC, Guedes MM, de Souza AL, Trevisan MT, Lima AF, Santos FA, et al. Gastroprotective effect of mangiferin, a xanthonoid from *Mangifera indica*, against gastric injury induced by ethanol and indomethacin in rodents. Planta Med 2007; 73(13):1372—6.

[271] Zheng D, Hou J, Xiao Y, Zhao Z, Chen L. Cardioprotective effect of mangiferin on left ventricular remodeling in rats. Pharmacology 2012;90(1—2):78—87.

21 Health Effects of Alkaloids from African Medicinal Plants

Victor Kuete

Faculty of Science, Department of Biochemistry, University of Dschang, Dschang, Cameroon

21.1 Introduction

Alkaloids are naturally occurring, nitrogen-containing organic compounds with the exception of amino acids, peptides, purines and derivatives, amino sugars, and antibiotics [1]. The majority of alkaloids are *true alkaloids* which are derived from alpha-amino acid precursors. Other alkaloids derived from terpenes and steroids are named pseudo alkaloids, because the relatively late amination process occurs in a transamination reaction by donating a nitrogen atom of an amino acid source [1]. Alkaloids are widely distributed in higher plants of the families Apocyanaceae, Ranunculaceae, Papaveraceae, Solanaceae, and Rutaceae and are used as phytomedicine or as weapons toxins [1]. They have also been reported in lower plants, insects, marine organisms, and microorganisms. Alkaloids are known to display arrays of pharmacological effects and are used as medications, as recreational drugs, or in entheogenic rituals but many of them are toxic to other organisms [1−3]. They are local anesthetics and stimulants (cocaine), psychedelics (psilocin), stimulants (caffeine, nicotine), analgesics (morphine), antibacterials (berberine, kokusaginine, nkolbisine), anticancer drugs (vinblastine, vincristine), antihypertensive agents (reserpine), cholinomimerics (galantamine), spasmolysis agents (atropine), vasodilators (vincamine), antiarrhythmia (quinidine), antiasthma therapeutics (ephedrine), antimalarials (quinine) [1−5]. Pyrrolizidine alkaloids (PAs) possessing a 1,2-double bond in their base moiety (necine) were found to be hepatotoxic, carcinogenic, genotoxic, teratogenic, and sometimes pneumotoxic [6]. For a long time it has also been well established that humans can be affected by toxic PA and many reports demonstrated that PA-containing plants are hazardous for livestock [6]. Also plants containing tropane alkaloids have been used as hallucinogenic agents and have also been connected with sorcery, witchcraft, native medicine, and magico-religious rites [7]. Vinca alkaloids such as vinblastine, vincristine, vindesine, and vinorelbine from *Catharanthus roseus* are clinically used in the treatment of various type of cancers. However, some side effects also were reported as the result of their chemotherapeutic use. In Africa, several types of alkaloids were

Toxicological Survey of African Medicinal Plants. DOI: http://dx.doi.org/10.1016/B978-0-12-800018-2.00021-2

identified in medicinal plants [1]. In this chapter, we will discuss their hurtful and protective effects, as well as their occurrence in African medicinal plants.

21.2   Toxic Alkaloids and Their Mode of Action

The toxicity of alkaloids has been observed in animals and humans; it was found that neuroreceptors are targets for many alkaloids, which structurally resemble the endogenous neurotransmitters, such as acetylcholine, dopamine, noradrenaline, serotonin, adrenaline, gamma-aminobutyric acid (GABA), or glutamate [8]. Neuroactive alkaloids were reported to either function as agonists, which overstimulate a neuroreceptor or as antagonists, which would block a certain neuroreceptor [8]. Some alkaloids inhibit the enzymes that break down neurotransmitters, such as cholinesterase (AChE) and monoamine oxidase (MAO) [8]. Higher doses lead to death by either cardiac or respiratory arrest [8]. Cytotoxins (see Chapter 19) can also target the protein biosynthesis in ribosomes. Many plant toxins, including the alkaloids emetine from *Psychotria ipecacuanha*, amanitins from *Amanita phalloides*, or the lectins, inhibit ribosomal protein biosynthesis [8]. Plant toxic alkaloids known as microtubule poisons, such as colchicine, podophyllotoxin, vinblastine, chelidonine, and noscapine [8] block the cell division but also vesicle transport along microtubules. Phytochemicals also are known to attack DNA and RNA, by either intercalation or alkylation. Intercalating alkaloids including β-carboline alkaloids, emetine, sanguinarine, athraquinones, or furanocoumarins stabilize DNA and thus inhibit DNA replication [8]. PAs of Boraginaceae and Asteraceae, aristolochic acids (from *Aristolochia*), cycasine (from cycads) and several furanocoumarins or secondary metabolites with epoxide or aldehyde groups (which are common in Apiaceae) can cause mutations and even cancers, abortion, or malformation of the fetus in herbivores [8].

21.2.1   Toxicity of PAs

According to the International Program on Chemical Safety (IPCS), "consumption of contaminated grain or the use of PA-containing plants as herbal medicine, beverages, or food by man, or grazing on contaminated pastures by animals, may cause acute or chronic disease" [9]. It was found that there is a toxic risk related to the consumption of phytomedicine containing PA [6]. It was demonstrated that the so-called "bush-teas" containing PA induced several liver diseases reported in Jamaica, West Indies, as well as in several African countries [6]. PAs themselves display more or less low acute toxicity but it was reported that, *in vivo*, they undergo a metabolic toxication process in the liver showing why this organ is their first toxicity target [6].

After absorption, the hydroxyl group is introduced in the necine (**1** and **2**) at positions 3 or 8 by the cytochrome P450 monoxogenase enzyme complex in the liver (Figure 21.1; compounds **3** and **4**). The instability of hydroxy-PAs (OHPAs)

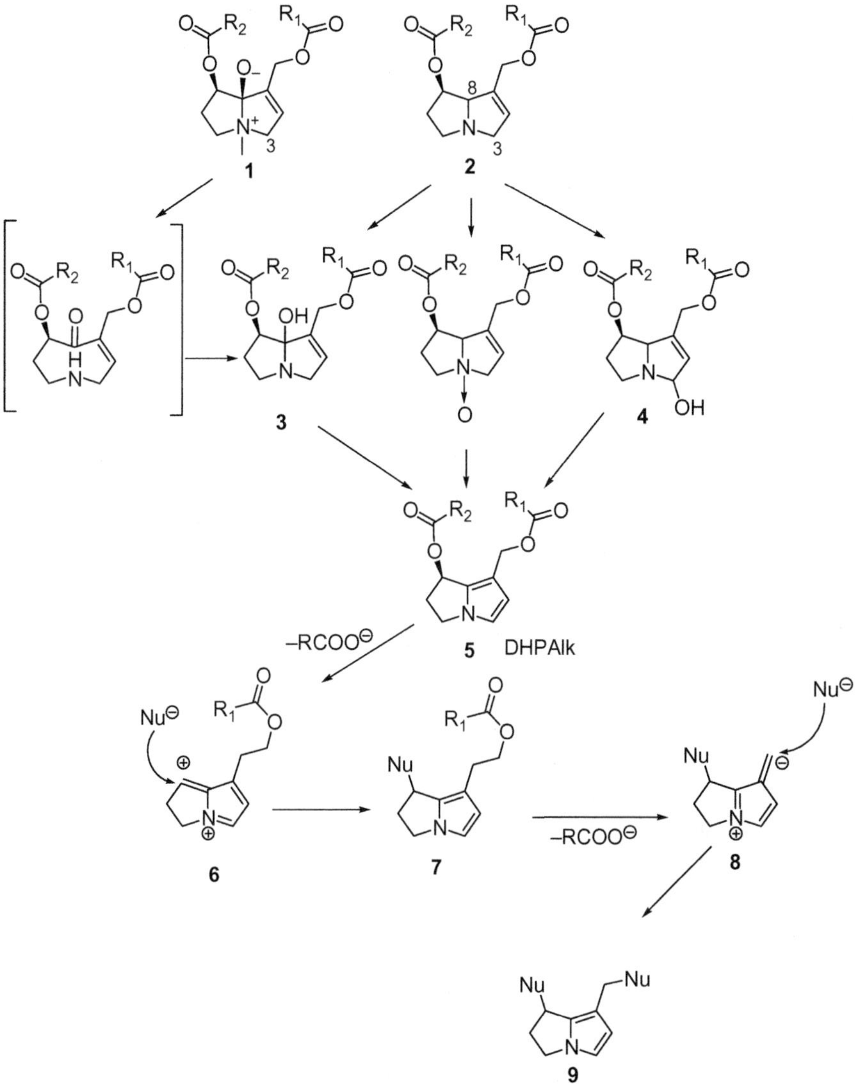

Figure 21.1 Enzymatic hydroxylation of pyrrolizidine and carbanium ion building from dehydropyrrolizidine products.
Source: Adapted from Wiedenfeld [6].

facilitates their rapid dehydration to the didehydropyrrolizidine alkaloids (DHPAlk; Figure 21.1, **5**) resulting in a second double bond in the necine followed by sponta- neous rearrangement to an aromatic pyrrole-system **5**. Otonecine-type PAs (Figure 21.1, **1**) with a methyl function at the nitrogen and a quasi keto function at the bridgecarbon 8 are metabolized to the OHPAs [6]. After hydroxylation of the

N-methyl group, it is lost as formaldehyde leaving an NH-function which undergoes condensation with the C8 keto group to produce product **3** (Figure 21.1) which spontaneously dehydrates to the DHPAlk III [6]. The metabolites **5** are able to generate stabilized carbonium ions (Figure 21.1, **6** and **8**) by loss of hydroxy groups or ester functions as hydroxyl or acid anions. These carbonium ions can react rapidly with nucleophiles (Figure 21.1, **9**) [6]. *In vivo* metabolites **6** and **8** react rapidly with nucleophilic mercapto, hydroxyl and amino groups on proteins and the amino groups of purine and pyrimidine bases in nucleosides such as DNA and RNA [6]. It was found that these alkylated products show abnormal functions and in the case of DNA mutations are possible [6]. Acute to chronic PA-poisoning was reported in humans and symptoms include hemorrhagic necrosis, hepatomegaly, and ascites; endothelial proliferation and medial hypertrophy leading to an occlusion of hepatic veins [6]. The veno-occlusive diseases cause centrilobular congestion, necrosis, fibrosis, and liver cirrhosis while PA-induced death is caused by liver failure due to necrosis and liver dysfunctions [6]. PA from *Senecio* species derived mainly from the necines, retronecine (**10**), and otonecine (**11**) were found to be carcinogenic, mutagenic, genotoxic, fetotoxic, and teratogenic [6].

21.2.2   Toxicity of Tropane Alkaloids

Classic anticholinergic poisoning is the main sign of toxicity from plants containing tropane alkaloids, and symptoms usually appear 30−60 min after ingestion; these symptoms could continue for 24−48 h due to their ability to delay the gastric emptying and absorption [7]. They occur mostly in the family Solanaceae. Plants containing tropane alkaloids atropine (**12**), scopolamine (**13**), and hyoscyamine (**14**) (Figure 21.2) such as *Datura* species, *Hyoscyamus niger, Atropa belladonna* and *Mandragora officinarum* (Solanaceae) have been used as hallucinogenic agents and

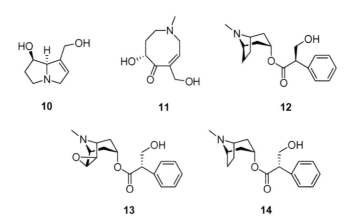

10   **11**   **12**

13   **14**

Figure 21.2 Chemical structures of some known alkaloids found in toxic plants. Retronecine (**10**); otonecine (**11**); atropine (**12**); scopolamine (**13**); hyoscyamine (**14**).

also have been connected with sorcery, witchcraft, native medicine, and magico-religious rites [7]. However, it should be noted that Chinese herbal medicines containing tropane alkaloids have been used to treat asthma, chronic bronchitis, pain, and flu symptoms [7]. Also, it was reported that smoke leaves from *Datura* is used in Africa to relieve asthma and pulmonary problems; meanwhile, other plants with tropane alkaloids (particularly *Datura* species) are added to alcoholic beverages to increase intoxication [7]. Side effects of *Datura* used as a recreational hallucinogen in the United States was reported in sporadic cases of anticholinergic poisoning and death [7]. The causes of *Datura* intoxication were determined as medication overdose, misuse of edible vegetables, deliberate abuse as hallucinogens, homicide or robbery, and accidental intoxication from contaminated food [10]. The reported symptoms of toxicity include dizziness, dry mouth, flushed skin, palpitation, nausea, drowsiness, tachycardia, blurred vision, mydriasis, hyperthermia, disorientation, vomiting, agitation, delirium, urine retention, hypertension, and coma [10]. Also, anticholinergic poisoning from *Belladonna* alkaloid contaminants was reported in foods, including commercially purchased Paraguay tea, hamburger, honey [11], stiff porridge made from contaminated millet, and homemade "moon flower" wine [7]. Other reported tropane alkloids-related intoxications include the large epidemic in New York and the eastern United States that resulted from heroin contaminated with compound **13** [10,12]. Also, toxic absorption after mucosal application was reported in 24 h of atropinism sustained by a woman who used a toothpaste mixed with the leaves and flowers of *Datura* sp., table salt, vinegar, and an alcoholic beverage [13]. Compound **13**, also known as levoduboisine or hyoscine, is a tropane alkaloid drug with muscarinic antagonist effects available as a drug sold as "*Scopoderm.*" In the psychopharmacology of learning and memory, scopolamine has taken a prominent position as an amnesia-inducing drug. It is noteworthy that treatment with **13** as a premedication for anesthesia frequently led to amnesia and has stimulated research on the effects of cholinergic blockade on human memory [14,15]. Compound **13** exerts its effects by acting as a competitive antagonist at muscarinic acetylcholine receptors; it is thus classified as an anticholinergic, antimuscarinic drug. Compound **13** occurs as secondary metabolite of plants from the family Solanaceae, such as *Hyoscyamus niger*, *Datura stramonium*, *Brugmansia* spp., and *Duboisia* spp. [16].

21.2.3 Toxicity of Vinca Alkaloids

Vinca alkaloids include vinblastine, vincristine, vindesine, and vinorelbine originally derived from *Catharanthus roseus* (Apocynaceae). They are well-known clinical cytotoxic drugs inhibiting the ability of cancer cells to divide [17]. However, some side effects were reported as results of the therapeutic use of *Vinca* alkaloids. The principal toxicity symptom of vincristine is neurotoxicity while neutropenia is the main toxicity sign of the other *Vinca* alkaloids. Other side effects include constipation, alopecia, mucositis, and vomiting [18]. It was reported that accidental intrathecal vincristine administration was found to be fatal [18].

21.3 Alkaloids Involved in Organ Protection

Many alkaloids (Figure 21.3) were documented as protective compounds from medicinal plants. The alkaloid extract from *Leonurus heterophyllus* (Lamiaceae) showed neuroprotective effects on cerebral ischemic injury [19]. In fact, stroke is the third most common cause of death worldwide and the leading cause of disability in adult [20]. The neuronal damage following cerebral ischemia is a serious risk to stroke patients. It was also demonstrated that oxyindole alkaloids such as isorhynchophylline (**15**), isocorynoxeine (**16**) and rhynchophylline (**17**) as well as indole alkaloids such as hirsuteine (**18**) and hirsutine (**19**) from *Uncaria sinensis* (Rubiaceae) can protect against glutamate-induced neuronal death in cultured cerebellar granule cells by inhibition of Ca^{2+} influx [21]. Morphinane alkaloids *N*-demethylsinomenine (**20**), 7,8-didehydro-4-hydroxy-3,7-dimethoxymorphinan-6-ol (**21**) and sinoacutine (**22**)

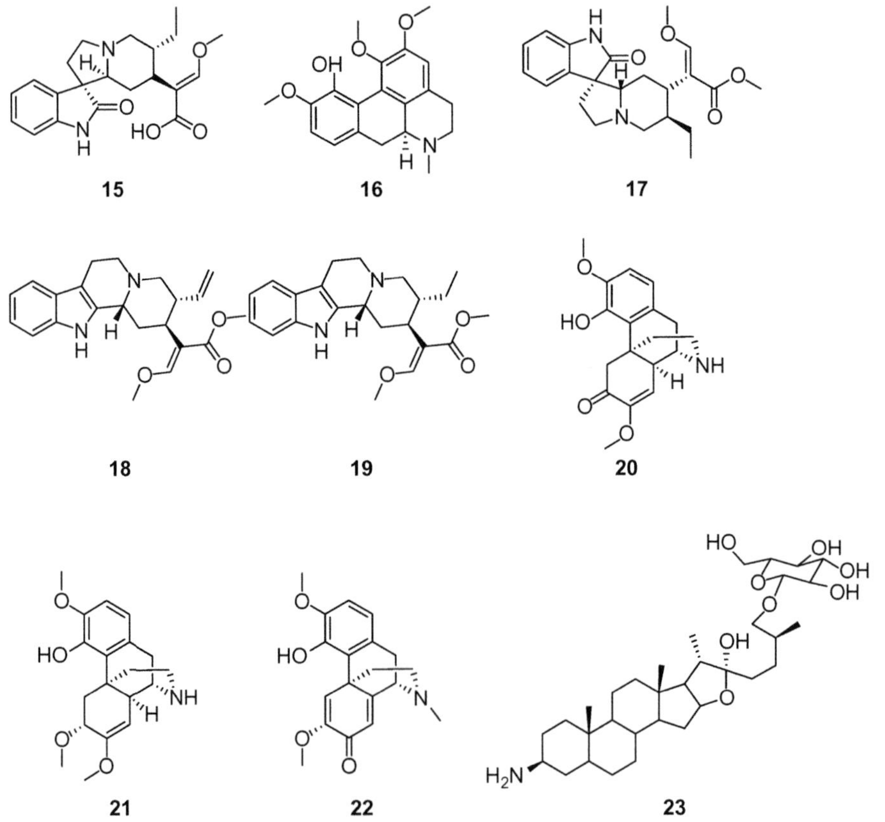

Figure 21.3 Chemical structures of some health benefit alkaloids. Isorhynchophylline (**15**); isocorynoxeine (**16**); rhynchophylline (**17**); hirsuteine (**18**); hirsutine (**19**); *N*-demethylsinomenine (**20**); 7,8-didehydro-4-hydroxy-3,7-dimethoxymorphinan-6-ol (**21**); sinoacutine (**22**); jurubine (**23**).

isolated from from *Sinomenium acutum* (Menispermaceae) have protective effects against hydrogen peroxide-induced cell injury [22]. Vieira and co-workers [23] demonstrated that steroidal alkaloids jurubine (**23**) isolated from *Solanum paniculatum* (Solanaceae) strongly protected cells against mitomycin C aneugenic and/or clastogenic activities as well as modulated mitomycin C cytotoxic action.

21.4 Harmful Effects of Alkaloids from African Medicinal Plants

Several African medicinal or food plants contain unwholesome nitrogenous compounds for animals and humans. For example, the consumption of species of South African *Senecio* spp. such as *Senecio latifolius* and *Senecio retrorsus* (Asteraceae) that contain PAs resulted in hepatotoxicosis [24]. It was shown that exposure to PA through the use of herbal remedies also may be a contributing factor to the high rates of liver cancer and cirrhosis seen in Africa [24,25]. Another PA-containing plant that is often used as a herbal medicine and also has been associated with poisoning is *Symphyttum officinale* (Boraginaceae) [26]. Also the toxicity of *Argemone* spp. (Papaveraceae) was suggested to be due to the presence of isoquinoline alkaloids among which were protopine, sanguinarine, and dihydrosanguinarine [24]. The alkaloids present in the seeds of *Argemone* spp. caused vasodilation and increased vascular permeability [24,27]. Many species of the family Solanaceae, including food plants such as tomatoes and potatoes, contain glycoalkaloids that may reach toxic levels under particular conditions (e.g., unripe tomatoes, potato tubers that have turned green after exposure to light) [24]. However, it was shown that poisoning by Solanaceae is relatively rare, possibly due to a combination of unpalatability of the unripe fruits and the fact that solanine (**29**) is rapidly hydrolyzed to the less toxic aglycone in the gastrointestinal tract. Besides, **29** is also poorly absorbed from the gastrointestinal tract [24,28]. *Datura stramonium* and *Datura ferox* (Solanaceae) are cosmopolitan weeds that contain parasympatholytic alkaloids such as atropine and scopolamine (**13**) [24]. *Nicotiana glauca* (wild tobacco) (Solanaceae) is another plant of which the young seedlings may be mistakenly collected, which contains a pyridine alkaloid, anabasine (**24**), which is very similar to nicotine (**27**) [24]. Ingestion may result in nausea, vomiting, gait abnormalities, tremors, confusion, and convulsive seizures [24,29]. It was also shown that *Boophane disticha* (Amaryllidaceae) contains various alkaloids such as buphanidrine, buphanisine, and buphanamine. Poisoning usually occurs in humans that utilize the bulb for medicinal purposes. Acute poisoning induces vomiting, weakness, coma, and mortality [24,30].

21.4.1 Anabasine

Anabasine (**24**), a structural isomer of **27** is a pyridine and piperidine alkaloid found in the Solanaceae plant *Nicotiana glauca* that is a close relative to *Nicotiana tabacum*, historically used as an insecticide. Compound **24** (Figure 21.4) is present

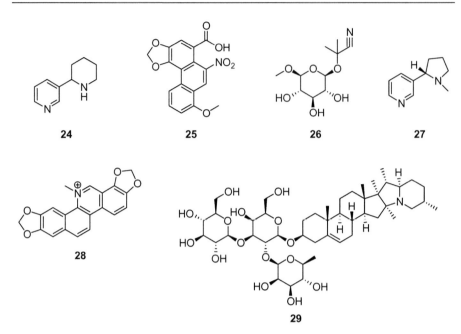

Figure 21.4 Chemical structures of some toxic alkaloids identified in African medicinal plants. Anabasine (**24**); aristolochic acid I (**25**); linamarin (**26**); nicotine (**27**); sanguinarine (**28**); solanine (**29**).

in trace amounts in tobacco smoke where it can be used to assess tobacco consumption in persons undergoing nicotine replacement therapy [31]. This phytochemical was reported in the Egyptian medicinal plant *Haloxylon salicornicum* (Moq.) Bunge ex Boiss. (Chenopodiaceae) traditionally used to treat diabetes and also used as an antiseptic and anti-inflammatory [32]. Compound **24** is a nicotinic acetylcholine receptor agonist. In high doses, it produces a depolarizing block of nerve transmission, which can cause symptoms similar to those of nicotine poisoning and, ultimately, death by asystole [33]. In larger amounts it is thought to be teratogenic in swine [34]. The intravenous LD_{50} of the $(+)$-R-anabasine rich fraction of *N. glauca* was determined as 11 mg/kg and that of the (-)-S-anabasine-rich fraction was found to be 16 mg/kg [35].

21.4.2 Aristolochic Acids

Aristolochic acids are a family of carcinogenic, mutagenic, and nephrotoxic compounds commonly found in Aristolochiaceae plants, including *Aristolochia* and *Asarum* (wild ginger), which are commonly used in Chinese herbal medicine [36]. Aristolochic acid I (**25**) (Figure 21.4) is the most abundant of the aristolochic acids and is found in almost all *Aristolochia* species [37]. Several *Aristolochia* species are used as medicinal plants in Africa. Some of them include *Aristolochia ringens*

(Vahl.) (used in Nigeria and several African countries for the management of snakebite venom, gastrointestinal disturbances, rheumatoid arthritis and insomnia, among others) [38], *Aristolochia bracteolata* Lam. (used to treat convulsions, malaria, abdominal pain, scorpion stings, flu, vomiting and pneumonia, polymeorrhea, edema) [39], *Aristolochia albida* Duch. (believed to have a "magical sense" in South Africa, to treat inflammation and various cancers in Nigeria; used as antimicrobial, stomach tonic, to treat malarial fevers and parasites in Mali) [39], *Aristolochia clematitis* L. (used as diaphoretic, an emmenagogue, and as a stimulant for chilbird in Egypt) [39], *Aristolochia elegans* Mast (used to cure snakebite, malaria, abdominal complaints including pain and swelling in Uganda, in rheumatism, gout, and as an emmenagogue in Libya) [39], *Aristolochia fimbriata* Cham and Schltdl (used to manage colds, chills, fevers, and in the treatment of asthma in South Africa) [39], *Aristolochia gadakura* (used for snakebites, scorpion stings, and as anthelmintic in Nigeria) [39], *Aristolochia heppii* Merxm (used in case of diarrhea in Zimbabwe) [39], *Aristolochia paucinervis* Pomel (used as anti-bacterial, to cure skin infections, abdominal pain, and upper respiratory tract infections in Morocco) [39] and *Aristolochia rotunda* L. (used in the treatment of snakebites) [39]. Compound **25** also occurs in the Cameroonian medicinal and threatened species *Thecacoris annobonae* Pax and K. Hoffm (Euphorbiaceae) [40]. Aristolochic acids are suggested to be the causative agents in Balkan endemic nephropathy, a chronic tubulo-interstitial nephritis seen primarily in countries in the Balkan Peninsula [41] and the so-called *Chinese herb nephropathy* (chronic tubulo-interstitial renal diseases associated with urothelial carcinoma) [42]. Exposure to aristolochic acid is associated with a high incidence of uro-epithelial tumorigenesis and was linked to aristolochic acid-associated urothelial cancer in a 2012 Taiwanese study [43].

21.4.3 Linamarin

Linamarin (**26**) is a cyanogenic glucoside found in the leaves and roots of plants such as cassava *Manihot esculenta;* Euphorbiaceae), lima beans, and flax. It is a glucoside of acetone cyanohydrin as is known to be the toxic compound of cassava, *Manihot esculenta* (Euphorbiaceae) [44]. It was shown that the roots and leaves contain the highest amount of **26** [44]. Compound **26** produces the toxic compound hydrogen cyanide (HCN) which can be hazardous to the consumer as the toxicity caused by the resulted free cyanide (CN^-) has already been reported [44]. Compound **26** is the most representative glucoside accounting for about 80% of the total cassava glucoside [44]. The toxicity of **26** in higher animals results from the combination of cyanide with Fe^{2+} which accounts for the formation of cyanohemo-globin [44]. The toxicity of cassava is discussed in Chapter 22.

21.4.4 Nicotine

Nicotine (**27**) is a potent parasympathomimetic alkaloid mostly found in Solanaceae and a stimulant acting as nicotinic acetylcholine receptor agonist. It is found in high concentration (approximately 0.6−3.0% of the dry weight) in *Nicotiana spp.* (tobacco) [45] and is present in the range of 2−7 µg/kg of various edible plants of the family (Solanaceae) such as tomatoes, potatoes, aubergines, and peppers [46]. Conflicting and controversial data concerning **27** concentrations in tea were reported in the literature, although nicotine biosynthesis in teas has not been demonstrated [46,47]. Contamination from the use of **27** as an insecticide has been proposed as a source of nicotine in tea [46−48]. Compound **27** is considered in this chapter due to it occurrence in above food plants which are also used medicinally in Africa as well as in *Nicotiana* species found throughout the continent. Compound **27** (Figure 21.4) is an antiherbivore chemical and was also widely used as an insecticide in the past [49,50]. In smaller concentration (about 1 mg), **27** acts as a stimulant in mammals, while high doses (30−60 mg) can be fatal [51,52]. It is well known that addiction to **27** is one of the most difficult to break, while the pharmacological and behavioral characteristics that determine tobacco addiction are similar to those that determine addiction to heroin and cocaine. However, it was demonstrated that **27** inhibits chromatin-modifying enzymes (class I and II histone deacetylases) which increases the ability of cocaine to cause an addiction [53]. The LD_{50} of **27** is 50 mg/kg for rats and 3 mg/kg for mice; 30−60 mg (0.5−1.0 mg/kg) can be a lethal dosage for adult humans [51,54]. It was found that spilling a high concentration of **27** onto the skin can cause intoxication or even death, as **27** readily passes into the bloodstream following dermal contact [55]. Compound **27** was found to alter the expression of a number of endothelial genes whose products play major roles in regulating the vascular tone and thrombogenicity, explaining the effects of cigarette smoking on the development of coronary atherosclerosis [56]. Besides, **27** showed acute vasoactive and mitogenic effects on vascular tissues and was also suggested to alter the function of the blood-brain barrier and disrupt normal endothelial cell function [57].

21.4.5 Sanguinarine

Sanguinarine (**28**), a benzophenanthridine from the group of benzylisoquinoline alkaloids have been extracted from many plants, including *Argemone mexicana*, *Chelidonium majus*, *Macleaya cordata*, bloodroot *Sanguinaria canadensis*, *Bocconia frutescens* and *Bocconia frutescens* (Papaveraceae) and *Poppy fumaria* (Fumariaceae), [58,59]. *Chelidonium majus* is found in Egypt [60] and in South Africa [61]. There are several controversial data regarding the pharmacological and the toxicological effects of this compound. However, in this chapter, **28** (Figure 21.4) will be considered for its hurtful properties as some sanguinarine-containing plants were more or less toxic. This alkaloid was shown to possess antimicrobial, antioxidant, anti-inflammatory, and antitumor properties [62] and is widely used in toothpaste and mouthwash for the prevention/treatment of gingivitis and other inflammatory conditions [63−65]. It has been shown that **28** imparts cell growth inhibitory responses in human squamous carcinoma A431 cells *via* an induction of apoptosis, while no apoptosis of the normal human epidermal keratinocytes was observed [66]. Adhami and co-workers [67] demonstrated that **28** can induce growth inhibition and antiproliferative effects in human prostate carcinoma cells, irrespective of their androgen status. The involvement of cyclin kinase inhibitor-cyclin-cyclin-dependent kinase machinery during cell cycle arrest and apoptosis of prostate cancer cells by **28** was also demonstrated, suggesting that it may be developed as an agent for the management of prostate cancer [67]. Nonetheless, no product containing **28** is currently approved by the agencies of the US Department of Health and Human Services, The Food and Drug Administration (FDA or USFDA) for the treatment of cancer. The FDA warns that unapproved bloodroot *Sanguinaria canadensis*, (a plant containing **28**) preparations are ineffective and dangerous [68]. Besides, **27** is regarded as a toxin that kills animal cells through its action on the Na^+-K^+-ATPase transmembrane protein [69]. Epidemic dropsy (a form of edema of extremities due to intoxication) is a disease that results from ingesting **28** [70]. The application of **28** to the skin kills cells and may destroy tissue. Besides, the bleeding wound may produce a massive scab called "eschar" explaining why this compound is considered as escharotic [71]. However, the acute toxicity studies showed a low toxicity of **28** in rats administrated orally and intravenously and the LD_{50} values were 1658 and 29 mg/kg, respectively [59,72]. No toxic effects occurred in rats fed up to 150 ppm of **28** in the diet for 14 days or in rats receiving **28**, at doses up to 0.6 mg/kg/day for 30 days by gavage. The acute intradermal LD_{50} in rabbits was found to be greater than 200 mg/kg [59,72]. Dalvi [73] demonstrated that the hepatotoxic effects at a single intraperitoneal dose of 10 mg/kg of **28** caused liver damage in rats, increasing the activity of alanine amino-transaminase and aspartate amino-transaminase and caused significant loss of microsomal cytochrome P450 and benzphetamine *N*-demethylase activities; nevertheless, similar effects were not seen by Kosina and co-workers [74] who attributed Dalvi's findings to the route of administration and high-acute dose. In another study, similar to that of Dalvi [73], **28**, at 10 mg /kg IP caused significant decreases in liver glutathione and P450 enzymes activities and increased serum sorbitol dehydrogenase and alanine aminotransferase levels in male mice, suggesting liver damage [75]. Several studies support the mutagenic effect of **28** and attribute it to its DNA intercalating activity [76]. In regards to the important pharmacological profile of **28**, this compound should subjected to rigorous toxicity testing before any claims for its therapeutic usefulness and/or safety can be made.

21.4.6 Solanine

Solanine (**29**) is a glycoalkaloid poison first extracted from the berries *Solanum nigrum* and mostly found in all parts of species of the family Solanaceae, such as *Solanum melongena, Solanum tuberosum*, and *Solanum lycopersicum*. *Solanum nigrum* naturally occurs in Africa and is used as food as well as medicinal plant in Cameroon to treat pneumonia, aching teeth, stomach ache, tonsillitis, wing worms, pain, inflammation and fever, tumor, inflammation, and also as hepaprotective, diuretic, antipyretic [77]. *Solanum melongena* L.Var inerme D.C Hiern. is also used as a food and medicinal plant in Cameroon [78]. Compound **29** (Figure 21.4) has fungicidal and pesticidal properties, and it is one of the plant's natural defenses.

It was found that the potato (*Solanum tuberosum*) alkaloids are toxic to the nervous system by interfering with the body's ability to regulate acetylcholine [79]. Compound **29** poisoning is primarily displayed by gastrointestinal and neurological disorders. Symptoms include nausea, diarrhea, vomiting, stomach cramps, burning of the throat, cardiac dysrhythmia, nightmare, headache, and dizziness [79] as well as hallucinations, loss of sensation, paralysis, fever, jaundice, dilated pupils, hypothermia, and death in more severe cases. It was suggested that doses of 2−5 mg/kg of body weight can cause toxic symptoms, and doses of 3−6 mg/kg of body weight can be fatal [79]. It was reported that *Solanum* glycoalkaloids can inhibit cholinesterase, disrupt cell membranes, and can be teratogenic [80].

21.5   Protective Effect of Alkaloids from African Medicinal Plants

The pharmacological effects of many alkaloids extracted in African medicinal are well known [1]. The protective effects of some them will be discussed below, as well as their occurrence in African plants.

21.5.1   Atropine

Atropine (**12**) is a naturally occurring tropane alkaloid extracted in several plants of the Solanaceae including *Atropa belladonna, Datura inoxia, Datura metel, Datura stramonium, Brugmansia* spp. and *Hyoscyamus* spp. Plants such as *Datura metel L., Datura stramonium* naturally occur in tropical West Africa [81]. In tropical West Africa, *Datura spp.* are used in palm wine to add a narcotic and stupefying effects [81]. *Datura stramonium* also occurs in South Africa where it used frequently as an antiasthmatic treatment [82]. Compound **12** is an anticholinergic drug acting as a competitive antagonist for the muscarinic acetylcholine receptor types M1, M2, M3, M4, and M5 [83]. Topical application of **12** is used as a cycloplegic, to temporarily paralyze the accommodation reflex, and as a mydriatic, to dilate the pupils. Due to its slow degradation, compound **12** is generally used as a therapeutic mydriatic.

21.5.2   Berberine

Berberine (**30**), (Figure 21.5) a protoberberine is an isoquinoline alkaloid extracted from several plants of the genus *Berberis* such as *Berberis aquifolium, Berberis vulgaris, Berberis aristata* (Berberidaceae) and other genera including *Coptis chinensis, Hydrastis canadensis, Xanthorhiza simplicissima* (Ranunculaceae), *Phellodendron amurense* (Rutaceae), *Tinospora cordifolia* (Menispermaceae), *Argemone mexicana* and *Eschscholzia californica* (Papaveraceae) [84]. Compound **30** has also been extracted from several African medicinal plants among which are *Enantia chlorantha* (Annonaceae) [85] from Cameroon and *Berberis aristata* from Egypt [86].

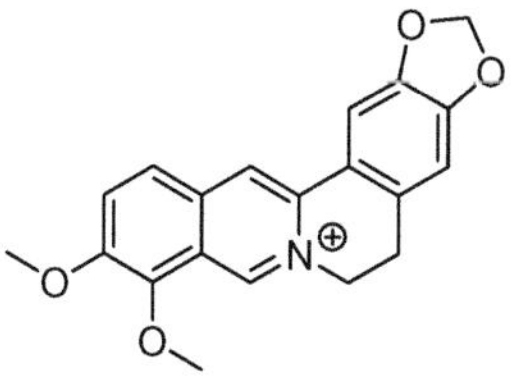

30 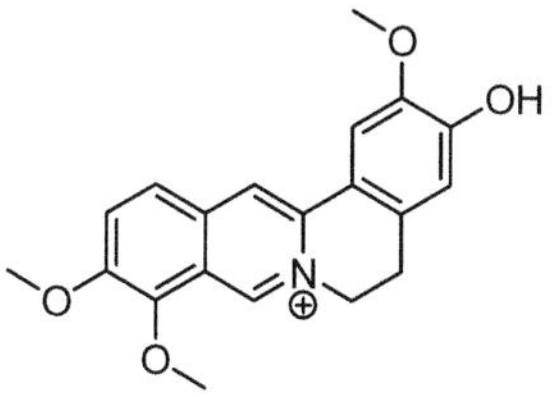							**31** 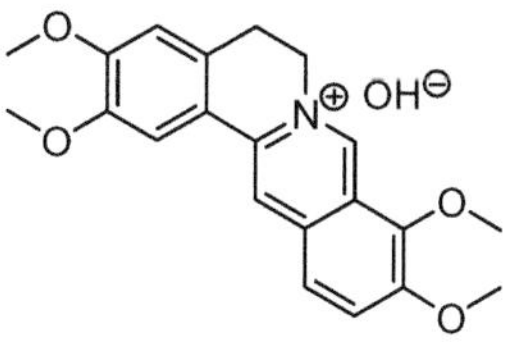							**32**

Figure 21.5 Chemical structures of berberine (**30**), jatrorrhizine (**31**), and palmatine (**32**).

This compound has several pharmacological properties including the antibacterial, antiamoebic, antifungal, antihelminthic, leishmanicidal and tuberculostatic, and anti-inflammatory activities, the inhibition effects on the formation bacterial enterotoxin, the inhibition effects on intestinal fluid accumulation and ion secretion, the inhibition effects on the smooth muscle contraction, the inhibition of platelet aggregation, the stimulation of bile and bilirubin secretion [86−89]. Compound **30** also prevented ischemia-induced ventricular tachyarrhythmia, stimulated cardiac contractility, and lowered peripheral vascular resistance and blood pressure *via* the suppression of delayed after-depolarization in the ventricular muscle in rats [90−92]. It was also suggested upon animal study that **30** may have a vasodilatory/ hypotensive effect attributable to its acetylcholine potentiating properties in rats [90]. Other health beneficial effects assigned to **30** include immunostimulation *via* increased blood flow to the spleen, macrophage activation, elevation of platelet counts in cases of primary and secondary thrombocytopenia, and increased excretion of conjugated bilirubin in experimental hyperbilirubinemia [89]. Besides, **30** may possess antitumor promoting properties as evidenced by inhibition of the enzyme cyclooxygenase-2 (COX-2) transcription and *N*-acetyltransferase activity in colon and bladder cancer cell lines [93−95].

21.5.3 Hirsuteine and Hirsutine

Hirsuteine (**18**) and hirsutine (**19**) are the major indole alkaloids of the *Uncaria* species (Rubiaceae) such *Uncaria rhynchophylla*, *Uncaria sinensis* and *Uncaria macrophylla* used in traditional Chinese herbal medicine as spasmolytic, analgesic, and sedative treatments for many symptoms associated with hypertension and cerebrovascular disorders [96]. *Uncaria rhynchophylla* as well as several plants of the genus *Uncaria* also are found in Africa [81,97]. Compounds **18** and **19** displayed a central depressive effect in mice, a weak noncompetitive antispasmodic action in the mouse intestine, and a hypotensive effect in rats [98]. It was also shown that **19** has a preventive effect on the development of gastric erosions in mice and antiarrhythmic effects on both aconitine-induced arrhythmias in mice and ouabain-induced arrhythmias in guinea pigs [98]. Compound **18** and **19** isolated from *Uncaria rhynchophylla* Miq. showed vasodilative effects in the hindlimb artery of anesthetized dogs after intra-arterial administration [99]. However, it was found that the vasodilative potency of **19** is stronger than that of **18** [99]. Shimada et al. [21] also showed that **18** and **19** isolated from the hooks and stems of *Uncaria sinensis* have protective effects against glutamate-induced neuronal death in cultured cerebellar granule cells by inhibition of Ca^{2+} influx. Besides, Jung and collaborators [100] showed that **19** reduces the production of various neurotoxic factors in activated microglial cells and possesses neuroprotective activity in a model of inflammation-induced neurotoxicity. In fact, **19** was found to block lipopolysaccharide (LPS)-related hippocampal cell death and production of nitric oxide (NO), prostaglandin (PG) E2, and interleukin-1β, to effectively inhibit LPS-induced NO release from cultured rat brain microglia, to reduce the LPS-stimulated production of prostaglandin E2 (PGE2) and intracellular reactive oxygen species and finally to decrease LPS-induced phosphorylation of the mitogen-activated protein kinases and Akt (protein kinase B) signaling proteins [100]. It was also suggested that the mechanism of action of **19** in hypoxic neonatal rat cardiomyocytes may be related to its antioxidant and antiapoptotic properties as treatment of rats increased Bcl-2 protein level and decreased Bax protein level [101].

21.5.4   Isorhynchophylline and Isocorynoxeine

Isorhynchophylline (**15**) and isocorynoxeine (**16**), two isorhynchophylline-related alkaloid are oxindole alkaloids found in *Uncaria spp.* (Rubiaceae) [102,103]. They were identified in *Uncaria rhynchophylla* (Miq.) Jacks [103], *Uncaria tomentosa* [104], *Uncariae ramulus* [105], *Uncaria sinensis* (Oliv.) Havil. [21]. Compound **16** was extracted from the Egyptian plant *Papaver rhoeas* (Papaveraceae) [106]. Compound **15** was also identified in the African medicinal plant, *Mitragyna inermis* (Willd.) O. Kuntze (Rubiaceae) commonly used in traditional medicine to treat malaria [107]. Compound **15** was reported mainly acting on cardiovascular and central nervous system diseases including hypertension, brachycardia, arrhythmia, and sedation, vascular dementia, and amnesia. Compound **15** also had effects on anticoagulation, inhibited vascular smooth muscle cell apoptosis and proliferation, also displayed endotoxemic and antispasmodic activities [102]. Compounds **15** and **16** showed inhibitory effects, similar to that of verapamil, on contractile response to high concentration of potassium ion (rats), $CaCl_2$ (rats), norepinephrine in normal and Ca^{2+}-free medium (rats and rabbits) and $^{45}Ca^{2+}$-uptake (rats) in thoracic aorta with activities two fold lesser than that of verapamil [105]. Compounds **15** and **16** also demonstrated a protective effects against glutamate-induced neuronal death in cultured cerebellar granule cells by inhibition of Ca^{2+} influx [21].

21.5.5   Jatrorrhizine

Jatrorrhizine (**31**) (Figure 21.5) also known as jateorrhizine or neprotin or jatrochizine or jatrorhizine or yatrorizine is a protoberberine alkaloid isolated from *Enantia chlorantha* (Annonaceae), *Mahonia aquifolium* (Berberidaceae) [108] and many other plant species. In Cameroon, the aqueous extract of *Enantia chlorantha* Oliver stem bark is widely used for the traditional treatment of gastritis and stomach problems [109]. This compound displayed antibacterial and antifungal [110] activities and showed anti-inflammatory effects and improved blood flow and mitotic activity in thioacetamide-traumatized rat livers [108,111−113]. Compound 31 interferes with multidrug resistance by cancer cells *in vitro* when exposed to a chemotherapeutic agent [114]. Compound 31 has blocking actions on both alpha 1- and alpha 2-adenoreceptors, which may be relevant to its hypotensive and antiarrhythmic actions [115]. Yan et al. have reported the hypoglycemic activity of jatrorrhizine [116].

21.5.6 Palmatine

Palmatine (32) (Figure 21.5) is a protoberberine alkaloid found in several plants including *Phellodendron amurense* (Rutaceae), *Rhizoma coptidis* (Ranunculaceae), *Corydalis yanhusuo* (Papaveraceae) and was found to be the major component of the protoberberine extract from *Enantia chlorantha* (Annonaceae) [117,118]. Compound 32 is a close structural analog of berberine (30), and is found in all plant families where 30 occurs, but is present to a much lower extent [119]. This compound has been extracted in African medicinal plants among which were *Enantia chlorantha* from Cameroon [120] and Nigeria [113] and *Heptacyclum zenkeri* Engl. (Menispermaceae) collected in Ghana [121]. It has been studied for its potential use in the treatment of jaundice, dysentery, hypertension, inflammation, and liver-related diseases [119,122,123]. The 8- and 13-alkyl derivatives of 32 were found to have antimicrobial and antimalarial activities, like 30 [124]. This compound also has weak *in vitro* activity against flavivirus (a virus that may cause encephalitis) [125]. In fact, it was demonstrated that 32 could inhibit West Nile virus NS2B-NS3 protease activity, suppress dengue virus and yellow fever virus in a dose-dependent manner [125]. Therefore, it was suggested that 32 could potentially be developed for the treatment of flavivirus infections [125]. Compound 32 blocked the delayed rectifier potassium current and had inhibition effect on $I_{Ca,L}$ (L-type calcium current) in guinea pig ventricular myocytes [122,123,126−130]. It was also shown that 32 blocked K^+ channel and decreased the extracellular K^+ to regulate the metabolic processes in the liver [131]. It has also been demonstrated that 32 could inhibit ICRAC (Ca^{2+} release-activated Ca^{2+} current) effectively and protected hepatocytes from calcium overload *via* the inhibition of ICRAC [132,133]. Compound 32 is known to inhibit both Ca^{2+} and cAMP-activated Cl^- secretion in isolated rat distal colon [119,134]. Compound 32 as well as 30 were found to bind easily to DNA and induced guanine-specific DNA photooxidation *via* singlet oxygen ($^1O_2$) generation [135,136]. Compound 32 is also useful as a sedative because it inhibits the dopamine biosynthesis [137], but 32 compared to 30 is less phototoxic to DNA damage and cell death in keratocytes [138]. This pytochemical has significant antitumor activity against HL-60 leukemia cells [139]. Inhibition of reverse transcriptase and topoisomerase I and II are suggested to be the probable reasons for its antitumor activity [119,140].

21.6 Conclusion

In this chapter, we discussed the hurtful and positive health effects of alkaloids occurring in African medicinal plants. The most prominent achievement of this section is the fact that we highlighted the toxic potential of alkaloids identified in medicinal plants used in African traditional medicine, clearly indicating that caution should be taken when following a therapy with samples containing such phytochemicals. Careful attention should be made when consuming medicinal plants or food plants containing compounds such as anabasine, aristolochic acid I, nicotine, sanguinarine, solanine, as well as other toxic alkaloids reported herein, even if the toxicity of the plant was not documented. In parallel, phytomedicine containing alkaloids with protective effects as described in this chapter could be explored more for their *in vivo* safety and for further clinical studies to finally produce improved herbal remedies.

References

[1] Wansi JD, Devkota KP, Tshikalange E, Kuete V. Alkaloids from the medicinal plants of Africa. In: Kuete V, editor. Medicinal plant research in Africa. Oxford: Elsevier; 2013. p. 557−605.

[2] Kuete V, Efferth T. Cameroonian medicinal plants: pharmacology and derived natural products. Front Pharmacol 2010;1:123.

[3] Kuete V, Wansi JD, Mbaveng AT, Kana Sop MM, Tadjong AT, Beng VP, et al. Antimicrobial activity of the methanolic extract and compounds from *Teclea afzelii* (Rutaceae). S Afr J Bot 2008;74(4):572−6.

[4] Alkaloid, <http://science.howstuffworks.com/alkaloid-info.htm> [accessed 21.07.13].

[5] Zofou D, Kuete V, Titanji VPK. Antimalarial and other antiprotozoal products from African medicinal plants. In: Kuete V, editor. Medicinal plant research in Africa. Oxford: Elsevier; 2013. p. 661−709.

[6] Wiedenfeld H. Plants containing pyrrolizidine alkaloids: toxicity and problems. Food Addit Contam Part A Chem Anal Control Expo Risk Assess 2011;228(3):282−92.

[7] Medscape. Tropane alkaloid poisoning, <http://emedicine.medscape.com/article/816657-overview>; 2011 [accessed 21.07.13].

[8] Wink M. Mode of action and toxicology of plant toxins and poisonous plants. Wirbeltierforschung in der Kulturlandschaft 2009;421:93−112.

[9] International Programme on Chemical Safety (IPCS). Pyrrolizidine alkaloids health and safety guide. Health and Safety Guide No. 26. Geneva, Switzerland: WHO; 1989.

[10] Chang SS, Wu ML, Deng JF, Lee CC, Chin TF, Liao SJ. Poisoning by *Datura* leaves used as edible wild vegetables. Vet Hum Toxicol 1999;41(4):242−5.

[11] Ramirez M, Rivera E, Ereu C. Fifteen cases of atropine poisoning after honey ingestion. Vet Hum Toxicol 1999;41(1):19−20.

[12] Hamilton RJ, Perrone J, Hoffman R, Henretig FM, Karkevandian EH, Marcus S, et al. A descriptive study of an epidemic of poisoning caused by heroin adulterated with scopolamine. J Toxicol Clin Toxicol 2000;38(6):597−608.

[13] Pereira CA, Nishioka Sde D. Poisoning by the use of *Datura* leaves in a homemade toothpaste. J Toxicol Clin Toxicol 1994;32(3):329—31.

[14] Kopelman M. The cholinergic neurotransmitter system in human learning and memory: a review. Q J Exp Psychol 1986;38A:535—73.

[15] Hardy TK, Wakely D. The amnesic properties of hyoscine and atropine in pre-anaesthetic medication. Anaesthesia 1962;17:331—6.

[16] Muranaka T, Ohkawa H, Yamada Y. Continuous production of scopolamine by a culture of *Duboisia leichhardtii* hairy root clone in a bioreactor system. Appl Microbiol Biotechnol 1993;40(2—3):219—23.

[17] Takimoto C, Calvo E. Chapter 3: Principles of oncologic pharmacotherapy. New York, NY: CMP United Business Media; 2008.

[18] Dorr RT, Alberts DS. Vinca alkaloid skin toxicity: antidote and drug disposition studies in the mouse. J Natl Cancer Inst 1985;74(1):113—20.

[19] Liang H, Liu P, Wang Y, Song S, Ji A. Protective effects of alkaloid extract from *Leonurus heterophyllus* on cerebral ischemia reperfusion injury by middle cerebral ischemic injury (MCAO) in rats. Phytomedicine 2011;18(10):811—8.

[20] Wong KS, Chen C, Ng PW, Tsoi TH, Li HL, Fong WC, et al. FISS-tris Study Investigators , Low-molecular-weight heparin compared with aspirin for the treatment of acute ischaemic stroke in Asian patients with large artery occlusive disease: a randomised study. Lancet Neurol 2007;6(5):407—13.

[21] Shimada Y, Goto H, Itoh T, Sakakibara I, Kubo M, Sasaki H, et al. Evaluation of the protective effects of alkaloids isolated from the hooks and stems of *Uncaria sinensis* on glutamate-induced neuronal death in cultured cerebellar granule cells from rats. J Pharm Pharmacol 1999;51(6):715—22.

[22] Bao GH, Qin GW, Wang R, Tang XC. Morphinane alkaloids with cell protective effects from *Sinomenium acutum*. J Nat Prod 2005;68(7):1128—30.

[23] Viera P, Marinho L, Ferri S, Chen-Chen L. Protective effects of steroidal alkaloids isolated from *Solanum paniculatum* L. against mitomycin cytotoxic and genotoxic actions. An Acad Bras Cienc 2013;85(2):553—60.

[24] Botha CJ, Penrith ML. Poisonous plants of veterinary and human importance in southern Africa. J Ethnopharmacol 2008;119(3):549—58.

[25] Steenkamp V, Stewart MJ, Zuckerman M. Clinical and analytical aspects of pyrrolizidine poisoning caused by South African traditional medicines. Ther Drug Monit 2000;22(3):302—6.

[26] Betz J, Page S. Perspectives on plant toxicology and public health. Wallingford: CAB International; 1998.

[27] Kingsbury J. Poisonous plants of the United States and Canada. Englewood Cliffs, NJ: Prentice Hall Inc.; 1964.

[28] Steyn D. The toxicology of plants in South Africa. Johannesburg, South Africa: Central News Agency Limited; 1934.

[29] Steenkamp PA, van Heerden FR, van Wyk BE. Accidental fatal poisoning by *Nicotiana glauca*: identification of anabasine by high performance liquid chromatography/photodiode array/mass spectrometry. Forensic Sci Int 2002;127(3):208—17.

[30] Steenkamp P. Chemical analysis of medicinal and poisonous plants of forensic importance in South Africa. Johannesburg: University of Johannesburg; 2005.

[31] Jacob P, Yu L, Shulgin AT, Benowitz NL. Minor tobacco alkaloids as biomarkers for tobacco use: comparison of users of cigarettes, smokeless tobacco, cigars, and pipes. Am J Public Health 1999;89(5):731—6.

[32] Qasheesh M. Phytochemical study of *Haloxylon salicornicum* (fam. Chenopodiaceae) growing in Saudi Arabia. Riyadh: King Saud University; 2004.

[33] Mizrachi N, Levy S, Goren ZQ. Fatal poisoning from *Nicotiana glauca* leaves: identification of anabasine by gas-chromatography/mass spectrometry. J Forensic Sci 2000;45 (3):736−41.

[34] Bush L, Crowe M. Nicotiana alkaloids. Boca Raton, FL: CRC Press Inc.; 1989.

[35] Lee ST, Wildeboer K, Panter KE, Kem WR, Gardner DR, Molyneux RJ, et al. Relative toxicities and neuromuscular nicotinic receptor agonistic potencies of anabasine enantiomers and anabaseine. Neurotoxicol Teratol 2006;28(2):220−8.

[36] Heinrich M, Chan J, Wanke S, Neinhuis C, Simmonds MS. Local uses of *Aristolochia* species and content of nephrotoxic aristolochic acid 1 and 2—a global assessment based on bibliographic sources. J Ethnopharmacol 2009;125(1):108−44.

[37] Wu T, Damu A, Su C, Kuo PC. Chemical constituents and pharmacology of *Aristolochia* species. Houston, TX: Gulf Professional Publishing; 2005.

[38] Adeyemi OO, Aigbe FR, Badru OA. The antidiarrhoeal activity of the aqueous root extract of *Aristolochia ringens* (Vahl.) Aristolochiaceae. Nig Q J Hosp Med 2012;22 (1):29−33.

[39] Heinrich M, Chan J, Wanke S, Neinhuis C, Simmonds MSJ. Local uses of *Aristolochia* species and content of nephrotoxic aristolochic acid 1 and 2—A global assessment based on bibliographic sources. J Ethnopharmacol 2009;125(1):108−44.

[40] Kuete V, Poumale HMP, Guedem AN, Shiono Y, Randrianasolo R, Ngadjui BT. Antimycobacterial, antibacterial and antifungal activities of the methanol extract and compounds from *Thecacoris annobonae* (Euphorbiaceae). S Afr J Bot 2010;76 (3):536−42.

[41] Gluhovschi G, Margineanu F, Velciov S, Gluhovschi C, Bob F, Petrica L, et al. Fifty years of Balkan endemic nephropathy in Romania: some aspects of the endemic focus in the Mehedinti county. Clin Nephrol 2011;75(1):34−48.

[42] De Broe ME. Chinese herbs nephropathy and Balkan endemic nephropathy: toward a single entity, aristolochic acid nephropathy. Kidney Int 2012;81(6):513−5.

[43] Chen CH, Dickman KG, Moriya M, Zavadil J, Sidorenko VS, Edwards KL, et al. Aristolochic acid-associated urothelial cancer in Taiwan. Proc Natl Acad Sci USA 2012;109(21):8241−6.

[44] Cereda MP, Mattos MCY. Linamarin: the toxic compound of cassava. J Venom Anim Toxins 1996;2:06−12.

[45] Hoffmann D, Hoffmann I. Chapter 3. Chemistry and toxicology, <http://cancercontrol. cancer.gov/brp/tcrb/monographs/9/>; 2012 [accessed 03.08.13].

[46] Siegmund B, Leitner E, Pfannhauser W. Determination of the nicotine content of various edible nightshades (Solanaceae) and their products and estimation of the associated dietary nicotine intake. J Agric Food Chem 1999;47(8):3113−20.

[47] Davis RA, Stiles MF, deBethizy JD, Reynolds JH. Dietary nicotine: a source of urinary cotinine. Food Chem Toxicol 1991;29(12):821−7.

[48] Sheen SJ. Detection of nicotine in foods and plant materials. J Food Sci 1988;53 (5):1572−3.

[49] Perfetti R, Thomas A. The chemical components of tobacco and tobacco smoke. Boca Raton, FL: CRC Press Inc.; 2009.

[50] Ujváry I. Nicotine and other insecticidal alkaloids. Tokyo: Springer-Verlag; 1999. p. 29−69.

[51] <http://www.inchem.org/documents/pims/chemical/nicotine.htm#PartTitle:7.%20TOXI COLOGY> [accessed 03.08.13].

[52] How drugs can kill, <http://learngeneticsutahedu/content/addiction/drugs/overdosehtml> [accessed 03.08.13].

[53] Volkow ND. Epigenetics of nicotine: another nail in the coughing. Sci Transl Med 2011;3(107): 107ps43.

[54] Okamoto M, Kita T, Okuda H, Tanaka T, Nakashima T. Effects of aging on acute toxicity of nicotine in rats. Pharmacol Toxicol 1994;75(1):1−6.

[55] Lockhart L. Nicotine poisoning. Br Med J 1993;1(246.4):246−7.

[56] Zhang S, Day I, Ye S. Nicotine induced changes in gene expression by human coronary artery endothelial cells. Atherosclerosis 2001;154(2):277−83.

[57] Hawkins BT, Brown RC, Davis TP. Smoking and ischemic stroke: a role for nicotine? Trends Pharmacol Sci 2002;23(2):78−82.

[58] Santos A, Adkilen P. The alkaloids of *Argemone Mexicana*. J Am Chem Soc 1932;54 (7):2923−4.

[59] Mackraj I, Govender T, Gathiram P. Sanguinarine. Cardiovasc Ther 2008;26(1):75−83.

[60] Khayyal MT, el-Ghazaly MA, Kenawy SA, Seif-el-Nasr M, Mahran LG, Kafafi YA, et al. Antiulcerogenic effect of some gastrointestinally acting plant extracts and their combination. Arzneimittelforschung 2001;51(7):545−53.

[61] Panzer A, Joubert AM, Bianchi PC, Seegers JC. The antimitotic effects of Ukrain, a *Chelidonium majus* alkaloid derivative, are reversible *in vitro*. Cancer Lett 2000;150 (1):85−92.

[62] Weerasinghe P, Hallock S, Brown RE, Loose DS, Buja LM. A model for cardiomyocyte cell death: insights into mechanisms of oncosis. Exp Mol Pathol 2013;94(1):289−300.

[63] Adhami VM, Aziz MH, Mukhtar H, Ahmad N. Activation of prodeath Bcl-2 family proteins and mitochondrial apoptosis pathway by sanguinarine in immortalized human HaCaT keratinocytes. Clin Cancer Res 2003;9(8):3176−82.

[64] Sun M, Lou W, Chun JY, Cho DS, Nadiminty N, Evans CP, et al. Sanguinarine suppresses prostate tumor growth and inhibits survivin expression. Genes Cancer 2010;1 (3):283−92.

[65] Holy J, Lamont G, Perkins E. Disruption of nucleocytoplasmic trafficking of cyclin D1 and topoisomerase II by sanguinarine. BMC Cell Biol 2006;7:13.

[66] Malikova J, Zdarilova A, Hlobilkova A. Effects of sanguinarine and chelerytherthrine on the cell cycle and apoptosis. Biomed Pap Med Fac Univ Palacky Olomouc Czech Repub 2006;150(1):5−12.

[67] Adhami VM, Aziz MH, Reagan-Shaw SR, Nihal M, Mukhtar H, Ahmad N. Sanguinarine causes cell cycle blockade and apoptosis of human prostate carcinoma cells via modulation of cyclin kinase inhibitor-cyclin-cyclin-dependent kinase machinery. Mol Cancer Ther 2004;3(8):933−40.

[68] Administration USFaD. Drugs. New Hampshire, MD: <http://www.fda.gov/Drugs/ GuidanceComplianceRegulatoryInformation/EnforcementActivitiesbyFDA/ ucm171057.htm>; 2009 [accessed 04.08.13].

[69] Pitts B, Meyerson L. Inhibition of Na,K-ATPase activity and ouabain binding by sanguinarine. Drug Develop Res 1981;1(1):43−9.

[70] Das M, Khanna SK. Clinicoepidemiological, toxicological, and safety evaluation studies on argemone oil. Crit Rev Toxicol 1997;27(3):273−97.

[71] Cienki JJ, Zaret L. An Internet misadventure: bloodroot salve toxicity. J Altern Complement Med 2010;16(10):1125−7.

[72] Becci PJ, Schwartz H, Barnes HH, Southard GL. Short-term toxicity studies of sanguinarine and of two alkaloid extracts of *Sanguinaria canadensis* L. J Toxicol Environ Health 1987;20(1−2):199−208.

[73] Dalvi RR. Sanguinarine: its potential as a liver toxic alkaloid present in the seeds of *Argemone mexicana*. Experientia 1985;41(1):77—8.

[74] Kosina P, Walterova D, Ulrichova J, Lichnovsky V, Stiborova M, Rydlova H, et al. Sanguinarine and chelerythrine: assessment of safety on pigs in ninety days feeding experiment. Food Chem Toxicol 2004;42(1):85—91.

[75] Williams MK, Dalvi S, Dalvi RR. Influence of 3-methylcholanthrene pretreatment on sanguinarine toxicity in mice. Vet Hum Toxicol 2000;42(4):196—8.

[76] Stiborova M, Simanek V, Frei E, Hobza P, Ulrichova J. DNA adduct formation from quaternary benzo[c]phenanthridine alkaloids sanguinarine and chelerythrine as revealed by the 32P-postlabeling technique. Chem Biol Interact 2002;140(3):231—42.

[77] Noumedem JA, Mihasan M, Lacmata ST, Stefan M, Kuiate JR, Kuete V. Antibacterial activities of the methanol extracts of ten Cameroonian vegetables against Gram-negative multidrug-resistant bacteria. BMC Complement Altern Med 2013;13:26.

[78] Fankam AG, Kuete V, Voukeng IK, Kuiate JR, Pages JM. Antibacterial activities of selected Cameroonian spices and their synergistic effects with antibiotics against multidrug-resistant phenotypes. BMC Complement Altern Med 2011;11:104.

[79] Cantwell M. A review of important facts about potato glycoalkaloids. Perishables Handling Newslett 1996;87:26—7.

[80] Friedman M, McDonald GM. Postharvest changes in glycoalkaloid content of potatoes. Adv Exp Med Biol 1999;459:121—43.

[81] Bep O. Medicinal plants in tropical West Africa. Cambridge: University Press; 1986.

[82] Pretorius E, Marx J. *Datura stramonium* in asthma treatment and possible effects on prenatal development. Environ Toxicol Pharmacol 2006;21(3):331—7.

[83] Rang H, Dale M, Ritter J, Moore P. Pharmacology. 5th ed. Churchill Livingstone, Edinburgh: Elsevier; 2003.

[84] Zhang Q, Cai L, Zhong G, Luo W. Simultaneous determination of jatrorrhizine, palmatine, berberine, and obacunone in *Phellodendri Amurensis* Cortex by RP-HPLC. Zhongguo Zhong Yao Za Zhi 2010;35(16):2061—4.

[85] Virtanen P, Lassila V, Njimi T, Ekotto Mengata D. Regeneration of D-galactosamine-traumatized rat liver with natural protoberberine alkaloids from *Enantia chlorantha*. Acta Anat (Basel) 1988;132(2):159—63.

[86] Soffar SA, Metwali DM, Abdel-Aziz SS, el-Wakil HS, Saad GA. Evaluation of the effect of a plant alkaloid (berberine derived from *Berberis aristata*) on *Trichomonas vaginalis in vitro*. J Egypt Soc Parasitol 2001;31(3):893—904.

[87] Akhter M, Sabir M, Bhide N. Possible mechanism of antidiarrhoel effect of berberine. Indian J Med Res 1979;70:233—41.

[88] Birdsall T, Kelly G. Berberine: therapeutic potential of an alkaloid found in several medicinal plants. Altern Med Rev 1997;2:94—103.

[89] Gibbs P, Seddon K. Berberine. Altern Med Rev 2000;5(2):175—7.

[90] Chun YT, Yip TT, Lau KL, Kong YC, Sankawa U. A biochemical study on the hypotensive effect of berberine in rats. Gen Pharmacol 1979;10(3):177—82.

[91] Marin-Neto JA, Maciel BC, Secches AL, Gallo Junior L. Cardiovascular effects of berberine in patients with severe congestive heart failure. Clin Cardiol 1988;11(4):253—60.

[92] Wang YX, Yao XJ, Tan YH. Effects of berberine on delayed afterdepolarizations in ventricular muscles *in vitro* and *in vivo*. J Cardiovasc Pharmacol 1994;23(5):716—22.

[93] Lin JG, Chung JG, Wu LT, Chen GW, Chang HL, Wang TF. Effects of berberine on arylamine *N*-acetyltransferase activity in human colon tumor cells. Am J Chin Med 1999;27(2):265—75.

[94] Fukuda K, Hibiya Y, Mutoh M, Koshiji M, Akao S, Fujiwara H. Inhibition by berberine of cyclooxygenase-2 transcriptional activity in human colon cancer cells. J Ethnopharmacol 1999;66(2):227—33.

[95] Creasey WA. Biochemical effects of berberine. Biochem Pharmacol 1979;28 (7):1081—4.

[96] Nakazawa T, Banba K, Hata K, Nihei Y, Hoshikawa A, Ohsawa K. Metabolites of hirsuteine and hirsutine, the major indole alkaloids of Uncaria rhynchophylla, in rats. Biol Pharm Bull 2006;29(8):1671—7.

[97] Sun G, Zhang X, Xu X, Yang J, Zhong M, Yuan J. A new triterpene from the plant of Uncaria macrophylla. Molecules 2012;17(1):504—10.

[98] Ozaki Y. Pharmacological studies of indole alkaloids obtained from domestic plants, Uncaria rhynchophylla Miq. and Amsonia elliptica Roem. et Schult. Nihon Yakurigaku Zasshi 1989;94(1):17—26.

[99] Ozaki Y. Vasodilative effects of indole alkaloids obtained from domestic plants, Uncaria rhynchophylla Miq. and Amsonia elliptica Roem. et Schult. Nihon Yakurigaku Zasshi 1990;95(2):47—54.

[100] Jung HY, Nam KN, Woo BC, Kim KP, Kim SO, Lee EH. Hirsutine, an indole alkaloid of Uncaria rhynchophylla, inhibits inflammation-mediated neurotoxicity and microglial activation. Mol Med Rep 2012;7(1):154—8.

[101] Wu LX, Gu XF, Zhu YC, Zhu YZ. Protective effects of novel single compound, Hirsutine on hypoxic neonatal rat cardiomyocytes. Eur J Pharmacol 2011;650 (1):290—7.

[102] Zhou JY, Zhou SW. Isorhynchophylline: a plant alkaloid with therapeutic potential for cardiovascular and central nervous system diseases. Fitoterapia 2012;83(4): 617—26.

[103] Ndagijimana A, Wang X, Pan G, Zhang F, Feng H, Olaleye O. A review on indole alkaloids isolated from Uncaria rhynchophylla and their pharmacological studies. Fitoterapia 2013;86:35—47.

[104] Mohamed AF, Matsumoto K, Tabata K, Takayama H, Kitajima M, Watanabe H. Effects of Uncaria tomentosa total alkaloid and its components on experimental amnesia in mice: elucidation using the passive avoidance test. J Pharm Pharmacol 2000;52(12):1553—61.

[105] Yamahara J, Miki S, Matsuda H, Kobayashi G, Fujimura H. Screening test for calcium antagonist in natural products. The active principles of Uncariae ramulus et uncus. Nihon Yakurigaku Zasshi 1987;90(3):133—40.

[106] El-Masry S, El-Ghazooly MG, Omar AA, Khafagy SM, Phillipson JD. Alkaloids from Egyptian Papaver rhoeas. Planta Med 1981;41(1):61—4.

[107] Sinou V, Fiot J, Taudon N, Mosnier J, Martelloni M, Bun SS, et al. High-performance liquid chromatographic method for the quantification of Mitragyna inermis alkaloids in order to perform pharmacokinetic studies. J Sep Sci 2010;33 (12):1863—9.

[108] Vollekova A, Kost'alova D, Kettmann V, Toth J. Antifungal activity of Mahonia aquifolium extract and its major protoberberine alkaloids. Phytother Res 2003;17 (7):834—7.

[109] Tan PV, Boda M, Etoa FX. In vitro and in vivo anti-Helicobacter/Campylobacter activity of the aqueous extract of Enantia chlorantha. Pharm Biol 2010;48(3):349—56.

[110] Cancer in Africa, <http://www.cancer.org/acs/groups/content/@epidemiologysurveilance/documents/document/acspc-031574.pdf1>; 2012 [accessed 13.08.13].

[111] Arens H, Fischer H, Leyck S, Romer A, Ulbrich B. Antiinflammatory compounds from *Plagiorhegma dubium* cell culture1. Planta Med 1985;51(1):52−6.

[112] Virtanen P, Lassila V, Njimi T, Mengata DE. Natural protoberberine alkaloids from *Enantia chlorantha*, palmatine, columbamine and jatrorrhizine for thioacetamide-traumatized rat liver. Acta Anat (Basel) 1988;131(2):166−70.

[113] Moody JO, Bloomfield SF, Hylands PJ. *In-vitro* evaluation of the antimicrobial activities of *Enantia chlorantha* Oliv. extractives. Afr J Med Med Sci 1995;24(3):269−73.

[114] Zhang H, Yang L, Liu S, Ren L. Study on active constituents of traditional Chinese medicine reversing multidrug resistance of tumor cells *in vitro*. Zhong Yao Cai 2001;24(9):655−7.

[115] Han H, Fang DC. The blocking and partial agonistic actions of jatrorrhizine on alpha-adrenoceptors. Zhongguo Yao Li Xue Bao 1989;10(5):385−9.

[116] Yan F, Benrong H, Qiang T, Qin F, Jizhou X. Hypoglycemic activity of jatrorrhizine. J Huazhong Univ Sci Tech Med Sci 2005;25:491−3.

[117] Wang YM, Zhao LB, Lin SL, Dong SS, An DK. Determination of berberine and palmatine in cortex phellodendron and Chinese patent medicines by HPLC. Yao Xue Xue Bao 1989;24(4):275−9.

[118] Virtanen P, Njimi T, Ekotto Mengata D. Clinical trials of hepatitis cure with protoberberine alkaloids of *Enantia Chlorantha* (abstract). Eur J Clin Pharmacol 1989;36: A123.

[119] Bhadra K, Kumar GS. Therapeutic potential of nucleic acid-binding isoquinoline alkaloids: binding aspects and implications for drug design. Med Res Rev 2011;31 (6):821−62.

[120] Nkwengoua ET, Ngantchou I, Nyasse B, Denier C, Blonski C, Schneider B. *In vitro* inhibitory effects of palmatine from *Enantia chlorantha* on *Trypanosoma cruzi* and *Leishmania infantum*. Nat Prod Res 2009;23(12):1144−50.

[121] Duah FK, Owusu PD, Knapp JE, Slatkin DJ, Schiff Jr. PL. Constituents of West African medicinal plants. Planta Med 1981;42(7):275−8.

[122] Niu XW, Zeng T, Qu AL, Kang HG, Dai SP, Yao WX, et al. Effects of 7-bromoethoxybenzene-tetrahydropalmatine on voltage-dependent currents in guinea pig ventricular myocytes. Zhongguo Yao Li Xue Bao. 1996;17(3):227−9.

[123] Chang YL, Usami S, Hsieh MT, Jiang MJ. Effects of palmatine on isometric force and intracellular calcium levels of arterial smooth muscle. Life Sci 1999;64(8): 597−606.

[124] Vennerstrom JL, Klayman DL. Protoberberine alkaloids as antimalarials. J Med Chem 1988;31(6):1084−7.

[125] Jia F, Zou G, Fan J, Yuan Z. Identification of palmatine as an inhibitor of West Nile virus. Arch Virol 2010;155(8):1325−9.

[126] Pereira GC, Branco AF, Matos JA, Pereira SL, Parke D, Perkins EL, et al. Mitochondrially targeted effects of berberine [Natural Yellow 18, 5,6-dihydro-9,10-dimethoxybenzo(g)-1,3-benzodioxolo(5,6-a) quinolizinium] on K1735-M2 mouse melanoma cells: comparison with direct effects on isolated mitochondrial fractions. J Pharmacol Exp Ther 2007;323(2):636−49.

[127] Yang BF, Zong XG, Wang G, Yao WX, Jiang MX. The mechanism of antiarrhythmic action of 7-bromoethoxybenzene-tetrahydropalmatine. Yao Xue Xue Bao 1990;25 (7):481−4.

[128] Xu C, Sun MZ, Li YR, Yang BF, Wang LJ, Li JM. Inhibitory effect of tetrahydropalmatine on calcium current in isolated cardiomyocyte of guinea pig. Zhongguo Yao Li Xue Bao. 1996;17(4):329−31.

[129] Lau CW, Yao XQ, Chen ZY, Ko WH, Huang Y. Cardiovascular actions of berberine. Cardiovasc Drug Rev 2001;19(3):234–344.

[130] Li Y, Fu LY, Yao WX, Xia GJ, Jiang MX. Effects of benzyltetrahydropalmatine on potassium currents in guinea pig and rat ventricular myocytes. Acta Pharmacol Sin 2002;23(7):612–6.

[131] Wang F, Zhou HY, Cheng L, Zhao G, Zhou J, Fu LY, et al. Effects of palmatine on potassium and calcium currents in isolated rat hepatocytes. World J Gastroenterol 2003;9(2):329–33.

[132] Takanashi H, Sawanobori T, Kamisaka K, Maezawa H, Hiraoka M. Ca^{2+}-activated K^+ channel is present in guinea-pig but lacking in rat hepatocytes. Jpn J Physiol 1992;42(3):415–30.

[133] Rychkov G, Brereton HM, Harland ML, Barritt GJ. Plasma membrane Ca^{2+} release-activated Ca^{2+} channels with a high selectivity for Ca^{2+} identified by patch-clamp recording in rat liver cells. Hepatology 2001;33(4):938–47.

[134] Wu DZ, Yuan JY, Shi HL, Hu ZB. Palmatine, a protoberberine alkaloid, inhibits both Ca(2+)- and cAMP-activated Cl(-) secretion in isolated rat distal colon. Br J Pharmacol 2008;153(6):1203–13.

[135] Hirakawa K, Kawanishi S, Hirano T. The mechanism of guanine specific photooxidation in the presence of berberine and palmatine: activation of photosensitized singlet oxygen generation through DNA-binding interaction. Chem Res Toxicol 2005;18 (10):1545–52.

[136] Hirakawa K, Hirano T. The microenvironment of DNA switches the activity of singlet oxygen generation photosensitized by berberine and palmatine. Photochem Photobiol 2008;84(1):202–8.

[137] Shin JS, Kim EI, Kai M, Lee MK. Inhibition of dopamine biosynthesis by protoberberine alkaloids in PC12 cells. Neurochem Res 2000;25(3):363–8.

[138] Inbaraj JJ, Kukielczak BM, Bilski P, He YY, Sik RH, Chignell CF. Photochemistry and photocytotoxicity of alkaloids from Goldenseal (Hydrastis canadensis L.). 2. Palmatine, hydrastine, canadine, and hydrastinine. Chem Res Toxicol 2006;19 (6):739–44.

[139] Kuo CL, Chou CC, Yung BY. Berberine complexes with DNA in the berberine-induced apoptosis in human leukemic HL-60 cells. Cancer Lett 1995;93(2):193–200.

[140] Sethi ML. Enzyme inhibition VI: inhibition of reverse transcriptase activity by protoberberine alkaloids and structure-activity relationships. J Pharm Sci 1983;72(5):538–41.

22 Physical, Hematological, and Histopathological Signs of Toxicity Induced by African Medicinal Plants

Victor Kuete

Department of Biochemistry, Faculty of Science,
University of Dschang, Dschang, Cameroon

22.1 Introduction

Toxicology is an applied science that incorporates several fields such as biology, chemistry, physiology, pathology, physics, statistics, and sometimes immunology or ecology to help solve problems in forensic medicine, clinical treatments, pharmacy and pharmacology, public health, industrial hygiene, veterinary science, agriculture, and more, as well as giving basic insight into how an organism functions [1]. Toxicology studies the adverse effects of any agent capable of producing a deleterious response in a biological system. Such substances known as toxicants or xenobiotics can induce changes from an organism's normal state that is irreversible, at least for a period of time. Producing an adverse effect depends on the concentration of the active compound at the target site. Hence, by convention, toxicology also includes the study of harmful effects caused by physical phenomena, such as radiation of various kinds and noise [2]. In practice, however, many complications exist beyond these simple definitions, both in bringing more precise meaning to what constitutes a poison and to the measurement of toxic effects [2]. Broader definitions of toxicology, such as "the study of the detection, occurrence, properties, effects, and regulation of toxic substances," although more descriptive, do not resolve the difficulties [2]. Medicinal plants are xenobiotics and are fully involved in several physical, behavioral, biochemical, or physiological changes in biological systems. Behavioral toxicology deals with the effects of toxicants on animal and human behavior, which is the final integrated expression of nervous function in the intact animal. This involves both the peripheral and central nervous systems, as well as effects mediated by other organ systems, such as the endocrine glands [2]. In this chapter we will focus on physical changes induced by African medicinal plants, including behavioral as well as hematological and histological modifications arising upon absorption of herbal drugs.

Toxicological Survey of African Medicinal Plants. DOI: http://dx.doi.org/10.1016/B978-0-12-800018-2.00022-4

22.2 Methods and Interpretation of Data of Important Physical Parameters Altered in Medicinal Plant Toxicity Studies

Several physical aspects are taken into account when investigating the toxicological profiles of medicinal plants. These include the mobility of animals, the sensitivity to environmental factors such as touch or noise, the stools appearance, the mortality and hence the dose causing the death of 50% of the experimental animals known as LD_{50} or medium lethal dose. The mobility as well as the ability to communicate of the experimental animals is indicative of the adverse effect of a toxicant. This is made experimentally by simple observation of treated animals as compared to controls. Xenobotics with adverse effects at high doses significantly reduce the ability of an animal to move or to interact together. These are characterized by depression, drowsiness and unsteady gait, paralysis of the hind limbs, and dyspnea, which could lead to coma and death [3]. Animals can communicate with one another through various types of signals and this was demonstrated experimentally with rats. It was shown that stress as induced by experimental treatment may give rise to the production of signals that affect nontreated animals housed nearby [4]. Such communication between test and control animals may cause biased results and disturbed welfare of the latter [4]. In toxicological studies, communication of stress may be prevented by separate housing of control and test animals, but this could introduce another source of bias [4]. Gastrointestinal complaints such as constipation and diarrhea are among the common complaints and often reported as adverse drug reactions [5] and should be considered when exploring the toxicological profiles of a substance or medicinal plant. Histopathological studies are necessary to access the physical effects of xenobiotics at tissue level. This should be done for dead or killed animals at the end of the study after necropsy. The dose-response relationship should be established clearly to confirm the effects of the toxicant on animal of the two sexes. The pathological lesions generally noted at necropsy are mainly congestion and edema of the organs and hepatomegaly with focal necrosis of liver cells [3]. Hematological indices as determined in blood samples collected from the vein of treated and control animals into a heparin containing tube also should be made for red blood cells, white blood cells as well as platelets, hemoglobin concentration and hematocrit estimation [6]. All significant changes as indicated by increase of decrease of these parameters for treated animals in comparison with untreated controls are indication of possible adverse effects of the tested substances [6].

22.3 Experimental and Plant-Induced Gastrointestinal Disorders

Side effects of medications frequently include constipation and diarrhea. Constipation is the most common gastrointestinal complaint that leads to physician visits, diagnostic tests, and medications for treatment [7,8]. Medication-induced diarrhea accounts for about 7% of all adverse drug effects and there are more than 700 drugs that have been implicated in causing diarrhea [9]. Chemotherapy involving medicinal plant also leads to gastrointestinal disorders. Animal models commonly used experimentally to induce diarrhea and to study mechanisms of action of plants and their active principles are rodents. *In vivo*, castor oil, Prostaglandin E2 (PG-E2), and heat-labile enterotoxin are commonly used agents to induce diarrhea in animals [10]. The diarrheal effect of castor oil is mediated through ricinoleic acid which causes irritation and inflammation of intestinal mucosa, leading to the stimulation of intestinal motility and increased secretion of fluid and electrolytes [10]. This model is used to assess the anti-secretory and anti-motility potential of medicinal plants. Prostaglandin E2 causes entero-pooling by stimulating fluid secretion and increasing propulsive activity in the colon [10]. Heat-labile enterotoxin (LT) is the virulent factor of *Escherichia coli* and induces diarrhea by accumulation of salt and water in the intestinal lumen [10,11]. Therefore, the LT-induced diarrheal model is suitable to study inhibitory effects of plant extracts on bacterial toxins [10]. Also, the charcoal meal test and charcoal-gum acacia-induced hyperperistalsis in animals are helpful to identity the effect of potential medicinal plants on intestinal motility [10]. It is well known that several medicinal plants are used worldwide to treat diarrhea experimentally and in humans. Some of them include *Ficus bengalensis, Ficus racemosa* (Moraceae), *Eugenia jambolana* (Myrtaceae), *Leucas lavandulaefolia* (Labiatae) and *Thespesia populnea* (Malvaceae) (used folk medicine in Khatra region of West Bengal, India), *Geranium mexicanum* (Geraniaceae) (used in traditional Mexican Medicine), *Galla chinensis* (Anacardiaceae) and *Chaenomeles speciosa* (Rosaceae) (used traditionally in China to treat gastrointestinal disorders), and *Mitragyna speciosa* (Rubiaceae) (used in Thailand) [10,12−14]. The extracts from the leaves of *Psidium guajava* (Myrtaceae) for example was found to decrease spasms associated with induced diarrhea in rodents. Reduced defecation, severity of diarrhea, and intestinal fluid secretion reductions have also been reported [15]. This activity was found to be associated with the ability of quercetin and its derivatives to affect smooth muscle fibers *via* calcium antagonism, to inhibit intestinal movement, and to reduce capillary permeability in the abdominal cavity [15]. In contrast, other medicinal plants used in the treatment of ailments induced diarrhea as adverse effects. For example, it was demonstrated that a prolonged usage of *Aloe vera* (Xanthorrhoeaceae) juice as a laxative can increase the risk of constipation. Moreover, the intake of the latex of *Aloe vera* was induced a depletion of the potassium from the cells in the intestinal lining, explaining why its laxatives effects were banned by the Food and Drug Administration (FDA) in 2002 [16]. Besides, the presence of anthraquinones in the juice of *Aloe vera* was found to increase the risk of diarrhea, especially when consumed in large quantities [16]. Diarrhea caused by laxatives in this juice is often severe, and accompanied by pain, stomach cramps, and dehydration [16].

22.4   Photosensitivity, Skin Irritation, and Sleepiness Induced by Herbal Medicine

Medicinal plants used in treatment of human ailments as well as any substance ingested into the body can cause side effects among which are some physical manifestations such as photosensitivity, skin irritation, and sleepiness. Side effects can also occur from the interaction of herbal medicines with pharmaceutical drugs [17]. Plants such as *Hypericum perforatum* (Hypericaceae) widely known as St. John's wort is a herbal medicine to treat depression or anxiety and was reported to cause skin sensitivity to the sun [17]. Individuals submitted to treatment with St. John's wort may have their skin burn more easily [17]. It was observed that typical cases of photosensitivity occur when people take very high doses of St. John's wort, or take it over a long period of time [17]. Therefore it is sound advice that when taking St. John's wort, the patient should avoid too much sun exposure [17]. Some medicinal plant were also found to cause skin irritation; in fact, topical herbal antifungal and antibacterial agents such as tea tree oil and lavender may cause rashes or skin irritation, especially if used at full strength [17]. Tea tree oil is taken from the leaves of the *Melaleuca alternifolia* (Myrtaceae), which is native to Southeast Queensland and the northeast coast of New South Wales, Australia. Tea tree oil should not be confused with tea oil, the sweet seasoning and cooking oil from pressed seeds of the tea plant *Camellia sinensis* (beverage tea) or the tea oil plant *Camellia oleifera* (Theaceae). It therefore recommended that before using any topical herbal product, individuals are encouraged to try a skin patch test by placing a small amount of the product on the inside of the elbow on one arm only, then wait a few days and observe any occurring change [17]. Sleepiness has also been reported as a consequence of the treatment with herbal drugs. It was found that medicinal plants used to treat anxiety, depression, and insomnia can cause excessive daytime sleepiness in certain individuals. Some of them include Chamomile or camomile (daisy-like plants of the family Asteraceae such as *Matricaria chamomilla, Chamaemelum nobile, Anthemis arvensis, Anthemis cotula, Cladanthus multicaulis, Cota tinctoria, Eriocephalus punctulatus, Matricaria discoidea, Tripleurospermum inodorum*), the valarian *Valeriana officinalis* (Valerianaceae) and the kava, *Piper methysticum* (Piperaceae) [17]. Valerian and kava were reported as the most likely culprits [17].

22.5 Effects of African Medicinal Plants on Physical, Hematological, and Histopathological Parameters in Toxicological Studies

Several African medicinal plants were screened for their safety and their toxicological profiles were established. Numbers of these plants altered several biochemical, physical, hematological, histological parameters in experimental animals. In a review of plant poisonings from herbal medication of persons admitted to a Tunisian toxicological intensive care unit from 1983 to 1998, Hamouda and collaborators [18] revealed that several plants were involved, mostly *Atractylis gummifera* (32%), *Datura stramonium* (25%), *Ricinus communis* (9%), *Nerium oleander* (7%) and *Peganum harmala* (7%). Reported poisonings involved neurological (91%), gastrointestinal (73%) and cardiovascular systems (18%) [18]. Hamouda et al. [18] highlighted that the lethal cases of liver failure involved *Atractylis gummifera* poisonings. The chemical structures of some potentially toxic compound are summarized in Figure 22.1. In this section, a synopsis of the hurtful effects of the above-mentioned as well as other African medicinal plants altering the physical parameters or involved in toxic symptoms in humans will be provided. Plants affecting the biochemical indices such proteins, enzymes, urea, bilirubin are discussed in Chapter 23.

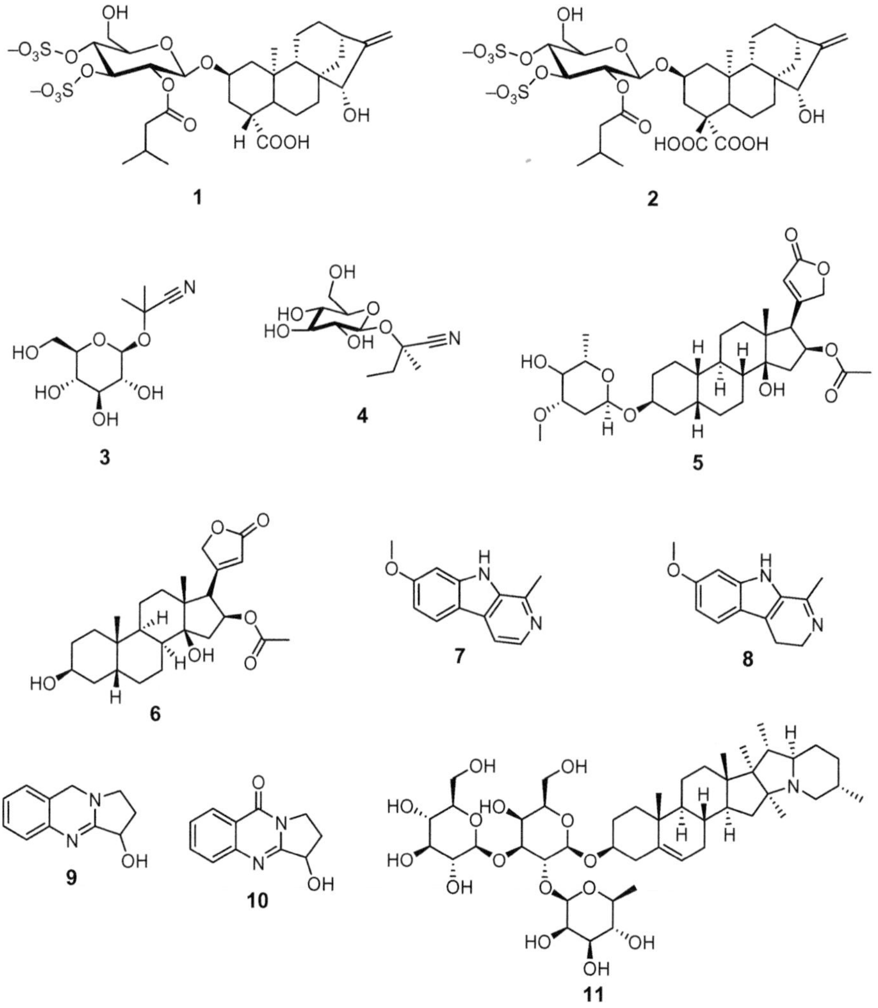

Figure 22.1 Chemical structure of some compounds from toxic African medicinal plants. Atractyloside (**1**); carboxyatractyloside (**2**); linamarin (**3**); lotaustralin (**4**); oleandrin (**5**); oleandrigenin (**6**); harmine (**7**); harmaline (**8**); vasicine (**9**); vasicinone (**10**); solanine (**11**).

22.5.1 Abrus precatorius

Abrus precatorius L. (Fabaceae), commonly known as "Rosary Pea" is an ornamental, twining, woody vine which grows to a height of 10−20 ft when supported
by other plants [19]. Leaves are alternate, compound, feather-like, with small
oblong leaflets while the flowers are numerous and appear in the leaf axils along
the stems [19]. The flowers are also small and occur in clusters 1−3 inches long,
usually red to purple, or occasionally white [19]. The fruit is a legume about 3 cm
long containing hard ovoid seeds about 1 cm long [19]. The leaves and roots of this
plant are used in South Africa to cure tuberculosis, bronchitis, whooping cough,
chest complaints, and asthma [20]. The leaves of the plants also are used as tea by
Tanzanian traditional healers to treat epilepsy [21]. In Zimbabwe, the plant is popularly used against schistosomiasis [22]. The seeds of *Abrus pretorius* are used in
Nigeria to treat diarrhea [23]. However, *Abrus precatorius* (mostly the seeds) was
reported as actually one of the world's most dangerous plants, after the castor oil
plant *Ricinus communis* L. (Euphorbiaceae) and spurge laurel, *Daphne laureola*
[24]. The toxicity of the plant was found to be due to the presence of a lectin poison called abrin, so toxic that, if swallowed or chewed, it will result in almost
immediate death [24]. Abrin is one of the most lethal known poison, inducing
severe vomiting, high fever, drooling, highly elevated levels of nervous tension,
liver failure, bladder failure, bleeding from the eyes, and convulsive seizures [24].
Abrin is a toxalbumen very similar to ricin found in castor seeds [19]. It is a lectin
composed of two polypeptide chains (A and B) connected by a disulfide bridge
[19]. This basic structure of two peptide chains linked by a single disulfide chain is
similar to that of botulinum toxin, tetanus toxin, cholera toxin, diphtheria toxin,
and insulin [19,25]. The usual fatal dose of *Ricinus communis* is reported to be just
2−3 seeds for an average adult [19]. Abrus poisoning generally causes severe
vomiting and abdominal pain, bloody diarrhea, convulsions, and alteration of sensorium with depression of central nervous system [19]. Toxicological analysis was
reported not helpful in cases of abrus poisoning while thin layer chromatography
using the seed extract and patient's serum could be helpful [19]. The decontamina-
tion (by stomach wash) was suggested as the major mode of treatment of abrus poisoning as no antidote is available yet. However, as the cause of death in most
reported cases appears to be renal failure, hemodialysis is indicated in severe poisoning with associated renal compromise [19].

22.5.2 Atractylis gummifera

Atractylis gummifera L. (Asteraceae) is a thistle distributed worldwide but especially abundant in the Mediterranean regions, in Northern Africa (Algeria,
Morocco, and Tunisia) and in Southern Europe (Italy, Greece, Spain, and Portugal)
[26]. The thistle grows commonly in dry areas, and the juice of the rhizome is poisonous [27]. This plant has a long rhizome that can reach a length of 30−40 cm
with a diameter of 7−8 cm, and the leaves are deeply divided into prickly lobes
and grouped into rosettes [26]. The plant has pink flowers grouped into capitulum

surrounded by bracts covered with spikes [26]. After the fruit is ripe, a yellowish white latex exudes from the base of the bracts [26,28]. Both therapeutic and toxic properties of this plant have been reported. In folk medicine, *Atractylis gummifera* has been used to treat several conditions including intestinal parasites, ulcers, snakebite poisoning, hydropsy, and drowsiness [26]. In traditional Arabic medicine, it was used to cauterize abscesses [26]. *Atractylis gummifera* was also known for its antipyretic, diuretic, purgative, and emetic properties [26,29]. In the popular medicine of Northern African, it is still used to treat syphilitic ulcers, induce abortion and bleach the teeth [26,30,31]. The plant also is used against parasites in folk veterinary medicine [32]. The dry rhizome is also usually burned in Arabic countries as incense to ward off bad fate [26,27]. Though the toxicity of *Atractylis gummifera* is well known, its ingestion continues to be a common cause of poisoning. The toxicity of the plant is due to the presence of two diterpenoid glucosides, atractyloside (**1**) (see Chapter 19) and carboxyatractyloside (**2**), which are able to inhibit mitochondrial oxidative phosphorylation through its permeabilization [26]. It was found that tissues of high metabolic activity are the main target organs [26]. *Atractylis gummifera* glucosides cause a severe hepatitis with fatal liver failure common. The plant's poisonous glucosides interact with detoxication and/or transformation systems in the liver even at doses not likely to induce cytolysis by blocking ADP-ATP conversion through inhibition of P450 cytochrome [26]. Poisoning symptoms generally include nausea, vomiting, epigastric and abdominal pain, diarrhea, anxiety, headache, and convulsions, often followed by coma [26]. Clinical manifestations are related to an induced hypoglycemia and neurovegetative disorders or subsequent renal failure. Liver transplantation or immunotherapy may improve the often fatal prognosis [26]. Georgiou and collaborators [31] reported the intoxication by *Atractylis gummifera* in a 7-year-old boy who drank an extract made from the plant's root as traditional medicine. The reported physical symptoms included coma, with epigastric pain, vomiting, and general anxiety [31]. In this case, severe hepatocellular damage and acute renal failure were observed [31] while a postmortem histopathological study of the liver confirmed the hepatic necrosis and allowed the differential diagnosis of the intoxication from Reye syndrome [31]. No specific pharmacological treatment for *Atractylis gummifera* intoxication is yet available and all the current therapeutic approaches are only symptomatic [26]. However, it was shown *in vitro* that some compounds such as verapamil or dithiothreitol could protect against the toxic effects of atractyloside, but only if administered before atractyloside exposure [26].

22.5.3 Butyrospermum paradoxum

Butyrospermum paradoxum (Sapotaceae), a small deciduous tree usually reaching a height of 7−15 m but may reach 25 m and a trunk diameter up to 2 m. The bark is corky, leaves oblong and fruits are ellipsoidal, 4−5 cm long, with fleshy pulp and usually one seed. The oil is used in soap and candle making, cosmetics, and as an ingredient in the fillings used for chocolate cream. The plant has been found useful in the traditional treatment of several human and animal diseases. The decoction of

Figure 22.2 *Butyrospermum paradoxum.* *Source*: V. Kuete.

this plant is widely used by the natives of north-eastern Nigeria in the traditional treatment of both animal and human trypanosomosis [33]. The water extract of this plant was found toxic to rats with an intraperitoneal LD_{50} of 240 mg/kg body weight (bw) [33]. Following intraperitoneal administration, the extract induced behavioral changes, morbidity, and mortality in the rats [33]. The symptoms observed included anorexia, dehydration, depression, prostration, coma, and death. However, the symptoms of toxicity were noted at high doses (>800 mg/kg) only [33]. At necropsy, the reported pathological lesions were congestion and edema of the lungs, bronchi, bronchioles and the kidney, hepatomegaly, and focal necrosis of the liver cells [33] (Figure 22.2).

22.5.4   Chenopodium ambrosioides

Chenopodium ambrosioides L. (Chenopodiaceae), commonly known as Mexican tea, is a polymorphic annual, and perennial herb growing to a height of over 1 m and covered with aromatic glandular hair. It is widely distributed in West Africa especially in Nigeria, Senegal, Ghana, and Cameroon [34]. This herb is used in folk medicine in the form of teas, poultices, and infusions for inflammatory problems, contusions, and lung infections, and as purgative, analgesic, as a vermifuge to expel round-worms and hook-worms, and as an antifungal [34] [35,36]. Chronic toxicity studies in albino rats with high doses of the plant extract ranging from 12.31 to 31.89 g/kg for 42 days revealed pathological features such as congestion of the lungs, metaplastic changes in the mucosal surface of the stomach, and necrosis of the kidney tubules were noticed [36]. This suggests that caution should be taken when using this plant in any chronic treatment (Figure 22.3).

22.5.5   Datura stramonium

Datura stramonium L., (also known by some as the "Devils Trumpet" or "Thorn" or "Apple" or "Locoweed" or "Jimson weed") is a wild-growing plant of the

Figure 22.3 Young (A) and flowering (B) *Chenopodium ambrosioides*.
Source: V. Kuete.

Solanaceae family, widely distributed and easily accessible with a large and coarse shrub of about 3−4 ft in height and 6 ft on rich soil [24,37]. The root is large, whitish in color, with a taproot system giving off many fibers while the stem is green or purple, hairless, cylindrical, erect and leafy, smooth, branching repeatedly in a forked manner [37]. Leaves and a single, erect flower arise through the forks of the branches [37]. The fruits are initially green but become brown with maturity and can be divided into four segments to release the seeds. The seeds are dull, irregular, and dark-colored; their surface may be pitted or slightly reticulated [37]. The plant is used in Eastern medicine, especially in Ayurvedic medicine, to treat various human ailments, such as ulcers, wounds, inflammation, rheumatism and gout, sciatica, bruises and swellings, fever, asthma and bronchitis, and toothache [37]. In Nigeria, the seeds are mixed with palm oil and applied to severe cases of insect bites and stings [38]. Ethno-medicinally, the frequent recreational abuse of this plant has resulted in toxic syndromes. All parts of the plants contain dangerous levels of the tropane alkaloids atropine, hyoscyamine, and scopolamine (alkaloids are discussed in Chapter 21) [24] which are classified as deliriants, or anticholinergics. There is a high risk of fatal overdose among uninformed users and it was reported that many hospitalizations occur among recreational users who ingest the plant for its psychoactive effects [39]. It was reported that intoxication due to *Datura stramonium* typically produces delirium, hyperthermia, tachycardia, bizarre behavior, severe mydriasis with resultant painful photophobia that can last several days, and pronounced amnesia [40]. The symptoms of poisoning generally occur approximately 30 min to an hour after smoking the plant and generally last from 24 to 48 h but have been reported in some cases to last as long as 2 weeks [41]. Adegoke and Alo [42] reported the case of two children admitted at the Hospital in Ekiti State (Nigeria) following ingestion of extract of *Datura stramonium*, and described symptoms related to neurotoxicity such as confusion, agitation, mydriasis, and hallucination. As with other cases of anticholinergic poisoning, intravenous physostigmine administration can serve as an antidote [40].

22.5.6 Manihot esculenta

Manihot esculenta Crantz. (Euphorbiaceae) also called Cassava, manioc, yuca, balinghoy, mogo, mandioca, kamoteng kahoy, tapioca is a perennial woody shrub in the Euphorbiaceae (spurge family) native to South America but now grown in tropical and subtropical areas worldwide for the edible starchy roots (tubers), which are a major food source in the developing world, in equatorial regions including Africa, South America, and Oceania. Also known as yuca, the dried root is the source of tapioca. The cassava shrub may grow to 2.75 m tall, with leaves deeply divided into 3−7 lobes. This food plant is also medicinally used to treat hypertension, headache, and other pains, irritable bowel syndrome and fever [43]. The bitter variety leaves of *Manihot esculenta* are also used to treat hypertension, headache, and pain. As *Manihot esculenta* is a gluten-free, natural starch, it is use in Western cuisine as a wheat alternative for patients with celiac disease. The cassava plant contains two different cyanogenic glucosides, linamarin (**3**), and lotaustralin (**4**) [44]. Fresh roots and leaves contain linamarin and hydrocyanic acid at levels that may be toxic, but if properly treated (in a labor-intensive process that may include roasting, soaking, or fermentation), the cyanide content is negligible. The average lethal dose of cyanide for higher animals was experimentally obtained as 1 mg/kg of live weight [44]. The amount of cyanide in the root allow the classification of cassava roots into toxic or nontoxic food [44]. Linamarase, a naturally occurring enzyme in *Manihot esculenta* catabolizes **3** and **4**, liberating hydrogen cyanide (HCN) [44]. It was demonstrated that a dose of 25 mg of pure cyanogenic glucoside from *Manihot esculenta*, which contains 2.5 mg of cyanide, is sufficient to kill a rat [45]. Excess cyanide residue from improper preparation is known to cause acute cyanide intoxication, and goiters, and has been linked to ataxia (a neurological disorder affecting the ability to walk, also known as konzo), as well as to the tropical calcific pancreatitis in humans, leading to chronic pancreatitis [46]. Symptoms of acute cyanide intoxication appear 4 or more hours after ingesting raw or poorly processed cassava and include vertigo, vomiting, and collapse. In some cases, death may result within 1 or 2 h. It can be treated easily with an injection of thiosulfate (which makes sulfur available for the patient's body to detoxify by converting the poisonous cyanide into thiocyanate) [47]. It was also found that chronic low-level cyanide exposure can cause goiter and with tropical ataxic neuropathy, a nerve-damaging disorder that renders a person unsteady and uncoordinated. Severe cyanide poisoning, particularly during famines, has been found to be associated with outbreaks of a debilitating, irreversible paralytic disorder called konzo and, in some cases, death [47]. Taking into account the presence of to toxic compound in this plant, the roots must be cooked properly to detoxify it before it is eaten. Medicinal use of any part of the plant also should include boiling to reduce it toxic potential. It was reported that a simple boiling of fresh root pieces is not always reliable because the cyanide may be only partially liberated, and only part of the **3** may be extracted in the cooking water [47]. The reduction of cyanides depends on whether the product is placed in cold water (27°C) or directly into boiling water at 100°C [47]. After 30 min cooking, the remaining cyanides are, in the first case, 8% of the initial value, and in the second case about 30% [47] (Figure 22.4).

Figure 22.4 *Manihot esculenta.* *Source*: V. Kuete.

22.5.7 Nerium oleander

Nerium oleander L. (Apocynaceae), the only species currently classified in the genus *Nerium*, is an evergreen shrub or small with all parts of the plant being toxic [24]. It is cultivated worldwide and is one of the most poisonous of commonly grown garden plants. In traditional medicine, the inhalation of the vapors arising from a heated decoction of the roots is used to treat headaches and colds [48]. Decoctions of the leaves are used for skin diseases and against paralysis and pain in extremities [48]. In the Sinai desert (Egypt), the plant is widely used in traditional Bedouin medicine to treat cancers [48]. *Nerium oleander* contains several lethal toxins, neroside, and oleandroside, the cardiac glycosides oleandrin (**5**) and oleandrigenin (**6**) [24]. It was reported that the toxins of this plant, upon ingestion, shut down the cardiovascular and nervous systems, after causing severe nausea and projectile vomiting [24]. Toxicity studies of animals administered *Nerium oleander* extract showed that rodents and birds were relatively insensitive to the cardiac glycosides of this plant [49]. *Nerium oleander* leaves have been found toxic to cattle, causing death within 36 h [49]. Toxic symptoms involved arrhythmia and auriculo-ventricular block, subendocardial and abomasal hemorrhages [49]. It was demonstrated that other mammals including dogs and humans, are relatively sensitive to the effects of cardiac glycosides and the clinical manifestations of glycoside intoxication [50,51]. The sap of this plant can cause skin irritations, severe eye inflammation and irritation, and allergic reactions characterized by dermatitis [52]. In general, the reactions to ingestion of this plant can include both gastrointestinal and cardiac effects. The reported gastrointestinal effects were nausea and vomiting, excess salivation, abdominal pain, and diarrhea [53] while cardiac reactions consist of irregular heart rate. It was also reported that poisoning and reactions to oleander plants are evident quickly, requiring immediate medical care [52] that can include induced vomiting and gastric lavage to reduce absorption of the toxic compounds. Charcoal has also been suggested to be administered to help absorb any remaining toxins [53].

22.5.8 Peganum harmala

Peganum harmala L. (*Pgh*) (Zygophyllaceae) commonly known as "Harmal" is native from the eastern Iranian region west to India, and grows spontaneously in semiarid and pre-desertic regions of southeast Morocco and distributed in North Africa and the Middle East [54,55]. In Moroccan traditional medicine, the seeds of this plant were used as powder, decoction, maceration, or infusion for fever, diarrhea, abortion, and subcutaneous tumors and is widely used as a remedy for various health conditions such as rheumatic pain, painful joint and intestinal pain [55,56]. The plant is also used for treatment of asthma, jaundice, lumbago, and many other human ailments [55,57]. The plant is rich in β-carbolines alkaloids such as harmine, harmaline, harmalol and harman and quinazoline derivatives such as vasicine and vasicinone [58]. The alkaloid extract of *Peganum harmala* showed opioid receptors mediated antinociceptive effects [55]. All parts of the plant were reported to be toxic and severe intoxication occurs in domestic animals [58]. Digestive and nervous syndromes have been reported in animals after consumption of a sublethal amount of the plant [58]. Physical symptoms of poisoning of animals appeared in a narcotic state interrupted by occasional short periods of excitement [58]. Abortion is frequent in animals that digest this plant [58] clearly warning that its therapeutic use to pregnant women should be prohibited. The signs of *Peganum harmala* overdose include hallucinations and neurosensorial syndromes, bradycardia, nausea, and vomiting [58]. Also, the toxicity of this plant traditionally used in the Middle East has been reported in humans [58] with poisoning symptoms similar to what has been reported for domestic animals [58]. Traditional medicinal use of the plant in Morocco as sedative and emmenagogue resulted in overdose and many case of poisoning were reported [59]. In fact, 200 cases of poisoning collected in poison control and the pharmacovigilance center of Morocco highlighted the toxicity of this plant used primarily for therapeutic purposes [59]. The physical aspect of poisoning included neurological, gastrointestinal, and cardiovascular signs with seven deaths being deplored [59]. The toxicity of the plant is due to some of its alkaloids. It was demonstrated that intravenous injection of harmine (**7**) and harmaline (**8**) (9 mg/kg) into cattle have toxic effects such as accelerated breathing and chronic muscular spasms [60]. It is believed that quinazoline alkaloids such as vasicine (**9**) and vasicinone (**10**) are responsible for the abortifacient activity of the plant [61]. They were found to have a uterine stimulatory effect, apparently through the release of prostaglandins [62,63].

22.5.9 Phaseolus vulgaris

Phaseolus vulgaris L. (Fabaceae) also known as the common bean or French bean is a herbaceous annual plant, grown worldwide for its edible beans, used both as the dry seed and as unripe fruits. The leaf is occasionally used as a vegetable, and the straw is used for fodder. This herb has been used as carminative, diuretic, emollient, and also in the treatment of diabetes, diarrhea, dysentery, and kidney problems [64]. This plant contains the toxic compound phytohemagglutinin (PHA), a lectin that is present in many common bean varieties. PHA consists of two closely related proteins, leucoagglutinin (PHA-L, proteins agglutinating erythrocytes) and PHA-E (proteins agglutinating leukocytes). PHA has carbohydrate-binding specificity for a complex oligosaccharide containing galactose, *N*-acetylglucosamine, and mannose. As a toxin, it can cause poisoning in monogastric animals, such as humans, through the consumption of raw or improperly prepared kidney beans. The primary symptoms of the poisoning induced by this lectin include nausea, vomiting, and diarrhea. Onset is from 1 to 3 h after consumption of improperly prepared beans, and symptoms typically resolve within a few hours [65]. Outbreaks of poisoning have been associated with cooking kidney beans in slow cookers [65]. However, it is important to note that PHA can be deactivated by boiling beans for 10 min at 100°C and that the US FDA recommends an initial soak of at least 5 h in water which should then be discarded [65]. If the beans are cooked at a temperature below boiling (without a preliminary boil), as in a slow cooker, the toxic effect of hemagglutinin is increased. It was therefore found that beans cooked at 80°C are five times as toxic as raw beans [65] (Figure 22.5).

22.5.10 Ricinus communis

Ricinus communis L. (Euphorbiaceae), commonly known as castor oil plant, is a soft wooden small tree developed throughout tropics and warm temperature regions [66]. This plant is indigenous to the southeastern Mediterranean Basin, Eastern Africa, and India but is widespread throughout tropical regions and is widely used as an ornamental plant [66,67]. The plant is known to display antimicrobial activity and has been used to treat several ailments [66]. Its leaf, root, and seed oil are used in inflammation treatment, liver disorders, hypoglycemic, and as a laxative [68,69] [66]. In Tunisia, the plant is used as a contraceptive [66]. The plant is also used in African folk medicine in the treatment of warts, cold tumors, and indurations of mammary glands, corns, and moles [70−72]. The anti-inflammatory, antioxidant, antimicrobial, and cytotoxic activities of the plant was demonstrated [66,73]. However, *Ricinus communis* is classified as the most poisonous plant on earth for humans [24]. The toxicity of raw castor beans is due to the presence of ricin [24], a naturally occurring lectin (a carbohydrate-binding protein). Ricin is a globular, glycosylated heterodimer of approximately 60−65 kDa [74]. Ricin toxin A chain (RTA) and ricin toxin B chain (RTB) are of similar molecular weights, approximately 32 and 34 kDa, respectively. RTA is an *N*-glycoside hydrolase composed of 267 amino acids [75]. It has three structural domains with approximately 50% of

Figure 22.5 *Phaseolus vulgaris.*
Source: V. Kuete.

the polypeptide arranged into alpha-helices and beta-sheets [76]. RTB is a lectin composed of 262 amino acids that is able to bind terminal galactose residues on cell surfaces [77]. Ricin is so toxic that the amount contained in a single bean would kill a human in just a few minutes, and there is no antidote [24]. The median lethal dose (LD_{50}) of ricin is around 22 μg/kg in humans if exposure is from injection or inhalation while oral exposure to this lectin is less toxic (though still toxic) and a lethal dose can be up to 20−30 mg/kg [78]. However, ingestion of seeds of the castor oil plant is rare [74]. The 2007 edition of Guinness World Records also ranks *Ricinus communis* as the most poisonous plant in the world. *Ricinus communis* acute poisoning episodes in humans are the result of oral ingestion of castor beans, five to twenty of which could prove fatal to an adult. Physical symptoms include nausea, diarrhea, tachycardia, hypotension, and seizures persisting for up to a week with blood, plasma, or urine concentration of ricin been detectable and measured to confirm diagnosis [79,80].

22.5.11 Sacoglottis gabonensis

Sacoglottis gabonensis Baill. (Humiriaceae) is a traditional rainforest plant that grows in widespread areas of the lowland swampy forests of southern Nigeria [3].

The plant grows as a tree of about 25 m high with a reddish brown and shaggy bark that easily peels off. Almost all parts of the plant are claimed to have medicinal value against various human ailments [81,82]. The stem bark, which is locally used as a preservative to improve the inebriating effect of raffia/palm wine, is also claimed to have medicinal value in the traditional control of gastrointestinal helminthiasis [3]. It was demonstrated that the intraperitoneal administration of doses ranging from 400 to 3200 mg/kg of the aqueous stem bark extract produced varying degrees of toxicity including depression, drowsiness and unsteady gait, paralysis of the hind limbs, dyspnea, coma, and death in rats [3]. The pathological lesions noted at necropsy were mainly identified as congestion and edema of the lungs, bronchi and bronchioles and hepatomegaly with focal necrosis of liver cells [3].

22.5.12 *Solanum nigrum*

Solanum nigrum (Solanaceae) commonly known as Makoi or black nightshade, usually grows as a weed in moist habitats in different kinds of soils, including dry, stony, shallow, or deep soils, and can be cultivated in tropical and subtropical agro climatic regions by sowing the seeds during April—May in well-fertilized nursery beds; it can be used for reclaiming the degraded land as well [83]. It is medicinally used in the management of several ailments, such as pneumonia aching teeth, stomache ache, tonsilitis, wing worms, pain, inflammation, fever, tumor, as tonic, as antioxidant, as anti-inflammatory, as hepaprotective, as diuretic, and as antipyretic [43,84]. The plant is frequently used as an elemental ingredient for clinical traditional Chinese medicine cancer therapy [85]. Parts of this plant can be highly toxic to livestock and humans. All parts of the plant except the ripe fruit contain the toxic glycoalkaloid solanine [11] and consumption of unripe fruits can result in toxicity with solanine-like poising symptoms (see Chapter 21). It has been stipulated that the cooked ripe fruit of black *Solanum nigrum* is safe to eat; however, it was found that detoxification cannot be attributed to normal cooking temperatures because the decomposition temperature of solanine is much higher at about 243°C [86]. *Solanum nigrum* leaves are recommended to be boiled as a vegetable with the cooking water being discarded and replaced several times to remove toxins (Figure 22.6).

22.5.13 *Solanum tuberosum*

Solanum tuberosum L. (Solanaceae) known as potatoes is a globally important crop plant producing high yields of nutritionally valuable food in the form of tubers [87]. Ingestion of green potatoes or potato sprouts can cause various symptoms which can be severe in some cases. Potato is mainly used as a staple food but also has a number of medicinal values. Moderate consumption of the juice from the tubers is used in the treatment of peptic ulcers, bringing relief from pain and acidity [88]. However, excessive doses of potato juice can be toxic and it is advised not to drink the juice of more than one large potato per day [88]. The poultice from boiling potatoes in water can be borne to rheumatic joints, swellings, skin rashes and hemorrhoids [88]. Uncooked potatoes have also been applied cold as a soothing

Figure 22.6 *Solanum nigrum.*
Source: V. Kuete.

plaster to burns and scalds [88]. Potato skins are used in India to treat swollen gums and to heal burns [88]. Potatoes contain toxic compounds known as glycoalkaloids, of which the most prevalent are solanine and chaconine. The toxicity of solanine is discussed in Chapter 21. Solanine is also found in other members of the Solanaceae plant family, which includes *Atropa belladonna* and *Hyoscyamus niger*. The concentration of glycoalkaloid in wild potatoes suffices to produce toxic effects in humans. It was also demonstrated that the exposure to light, physical damage, and age increase glycoalkaloid content within the tuber [89], the highest concentrations occurring just underneath the skin. Poisoning from cultivated potatoes occurs very rarely however, as the toxic compounds in the potato plant are, in general, concentrated in the green portions of the plant and in the fruits and cultivated potato varieties contain lower toxin levels [89]. Furthermore, it was found that at high temperatures (over 170°C) or during cooking, the toxin is partly destroyed (Figure 22.7).

22.5.14 Syzigium aromaticum

Syzigium aromaticum L. (Myrtaceae), formerly known as *Eugenia caryophyllata*, and commonly known as "clove," is one of the most widely used African medicinal plant [90]. It is an evergreen tree in the Myrtle family and grows in warm climates but is cultivated commercially in Tanzania, Sumatra, the Maluku (Molucca) Islands, and South America [91]. The tree grows up to 20 m and has leathery leaves. The clove spice is the dried flower bud measuring between 12 and 22 mm in length, with four projecting calyx lobes folded to form a hood, which hides numerous stamens [90,92]. Clove oil had long been recognized as safe and used in foods, beverages and tooth pastes, but toxicities which could be life-threatening were reported much later [90,93–95]. The systemic use of the essential oil has been restricted to three drops per day for an adult as excessive use can cause severe kidney damage [96]. The decoction of the dried bud of this plant is widely use in

Figure 22.7 *Solanum tuberosum.*
Source: V. Kuete.

the treatment of several ailments, including gastrointestinal disorders [90]. It was reported that boiled aqueous extract, as employed in Nigeria for treating gastrointestinal tract diseases and also used as food spices severely altered the hematological parameters in adult albino rats of both sexes [90]. In effect, long-term administration for 90 days revealed significant effects on hemoglobin, red blood cells, platelets, and granulocytes as well as an insignificant reduction of total white blood cells [90]. The histopathological study even at moderate doses demonstrated a significantly alteration in body organs such as liver, kidney, lungs, stomach, spleen, heart and brain, their enzymes as well as the various functions [90]. The LD_{50} for both intraperitoneal and oral routes of the extract (263 and 2500 mg/kg, respectively) were the indications of its toxic potential [90], suggesting that its prolonged usage must be avoided (Figure 22.8).

22.5.15   *Tabernaemontana crassa*

Tabernaemontana crassa Benth. (Apocynaceae) is a native species of the West African forest. It is frequently used in traditional medicine of several African countries [6]. Extremely caustic, the sap is an ingredient of an arrow-poison and may cause blindness; it is commonly applied as a healing dressing to sores, abscesses, furuncles and to anthrax pustules, and to dermal infections such as filaria, ringworm and other fungal troubles [34]. The latex is used to treat ringworm, wound infections, coryza, headaches, abscesses, boils, and carbuncles [92,97]. A paste of the fruit is applied topically in treating contusions and sprained backs [97]. The

Figure 22.8 *Syzigium aromaticum.*
Source: V. Kuete.

decoction of the leaves is taken in West Africa as a tonic [98]. The stem bark is used to treat several ailments including coryza sinusitis, stomach disorders, hematuria, gonorrhea, wound infection [97−99]. A crude ethanolic extract of this plant was also reported to have decreased motor activity and muscle relaxation effect [100]. The histopathological studies revealed that there was a dose-related effect in liver, lungs, and kidneys in rats receiving 6 weeks' daily treatment of 0.5 g/kg bw [6]. In the subacute toxicity studies, it was demonstrated that significant changes in the bw were observed for the animal taking 6 weeks daily and it was suggested that consumption of higher doses up to 6 g/kg bw should be avoided as it may result at organ injuries [6]. The toxicity of the plant was suggested to be due to the presence of indole alkaloids [6].

22.6 Conclusion

This review highlights the physical, hematological, and histopathological signs observed with some medicinal plants used in Africa. The value of this chapter lies not only in serving as awareness of the side effects of healing and food plants but also an invitation to scientist to focus more on the toxicological survey of plants used throughout the continent to treat several ailments. In the chapter, we highlighted the potential risk associated with improper use of 15 African medicinal plants, namely *Abrus precatorius* L. (Fabaceae), *Atractylis gummifera* L. (Asteraceae), *Butyrospermum paradoxum* (Sapotaceae), *Chenopodium ambrosioides* L. (Chenopodiaceae), *Datura stramonium* L., (Solanaceae), *Manihot esculenta* Crantz. (Euphorbiaceae), *Nerium oleander* L. (Apocynaceae), *Peganum harmala* L. (*Pgh*) (Zygophyllaceae), *Phaseolus vulgaris* L. (Fabaceae), *Ricinus communis* L. (Euphorbiaceae), *Sacoglottis gabonensis* Baill (Humiriaceae), *Solanum nigrum* and *Solanum tuberosum* L. (Solanaceae), *Syzigium aromaticum* L. (Myrtaceae) and *Tabernaemontana crassa* Benth. (Apocynaceae). The toxic constituent of these plants were also highlighted in this chapter and should help to identify other medicinal plant with poisoning risks. The toxicity of most of these compounds is also discussed in Chapters 19—21 and clearly highlight the toxic symptoms for animals and patients after inappropriate consumption. In addition, we discussed the toxicity associated to other compounds such as PHA from *Phaseolus vulgaris* and ricin from *Ricinus communis*.

References

[1] Casarez E. Basic principles of toxicology, <http://wwwgooglecom/url?sa = t&rct = j &q = &esrc = s&source = web&cd = 5&ved = 0CEUQFjAE&url = http%3A%2F%2F coeppharmacyarizonaedu%2Fcurriculum%2Ftox_basics%2Ftoxhandoutsdoc&ei = 1x MTUtXLJoWr4ASAyYD4Dw&usg = AFQjCNFkHAaX8w-yI4Mzbgv3m6RaPUYpsA>; 2001 [accessed 20.08.13].

[2] Hodgson E. Introduction to toxicology. In: Hodgson E, editor. A textbook of modern toxicology. 3rd ed. Hoboken, NJ: John Wiley & Sons, Inc.; 2004. p. 1—12.

[3] Nwosu CO, Eneme TA, Onyeyili PA, Ogugbuaja VO. Toxicity and anthelmintic efficacy of crude aqueous of extract of the bark of *Sacoglottis gabonensis*. Fitoterapia 2008;79(2):101—5.

[4] Beynen AC. Communication between rats of experiment-induced stress and its impact on experimental results. Anim Welf 1992;1(3):153—9.

[5] Fosnes GS, Lydersen S, Farup PG. Constipation and diarrhoea—common adverse drug reactions? A cross sectional study in the general population. BMC Clin Pharmacol 2011;11:2.

[6] Kuete V, Manfouo RN, Beng VP. Toxicological evaluation of the hydroethanol extract of *Tabernaemontana crassa* (Apocynaceae) stem bark. J Ethnopharmacol 2010;130 (3):470—6.

[7] Sonnenberg A, Koch TR. Physician visits in the United States for constipation: 1958 to 1986. Dig Dis Sci 1989;34(4):606—11.

[8] Locke 3rd GR, Pemberton JH, Phillips SF. AGA technical review on constipation. American gastroenterological association. Gastroenterology 2000;119(6):1766—78.

[9] Chassany O, Michaux A, Bergmann JF. Drug-induced diarrhoea. Drug Saf 2000;22 (1):53—72.

[10] Prasad Kota B, Teoh A, Roufogalis B. Pharmacology of traditional herbal medicines and their active principles used in the treatment of peptic ulcer, diarrhoea and inflammatory Bowel disease. In: Brzozowski T, editor. New advances in the basic and clinical gastroenterology. Rijeka, Croatia/Shanghai, China: InTech; 2012. p. 297—310.

[11] Spangler BD. Structure and function of cholera toxin and the related *Escherichia coli* heat-labile enterotoxin. Microbiol Rev 1992;56(4):622—47.

[12] Chen JC, Ho TY, Chang YS, Wu SL, Hsiang CY. Anti-diarrheal effect of *Galla Chinensis* on the *Escherichia coli* heat-labile enterotoxin and ganglioside interaction. J Ethnopharmacol 2006;103(3):385—91.

[13] Chen JC, Chang YS, Wu SL, Chao DC, Chang CS, Li CC, et al. Inhibition of *Escherichia coli* heat-labile enterotoxin-induced diarrhea by *Chaenomeles speciosa*. J Ethnopharmacol 2007;113(2):233—9.

[14] Chittrakarn S, Sawangjaroen K, Prasettho S, Janchawee B, Keawpradub N. Inhibitory effects of kratom leaf extract (*Mitragyna speciosa* Korth.) on the rat gastrointestinal tract. J Ethnopharmacol 2008;116(1):173—8.

[15] Gutierrez RM, Mitchell S, Solis RV. *Psidium guajava*: a review of its traditional uses, phytochemistry and pharmacology. J Ethnopharmacol 2008;117(1):1−27.

[16] Solanki P. *Aloe vera* juice side effects, <http://wwwbuzzlecom/articles/aloe-vera-juice-side-effectshtml>; 2013 [accessed 21.08.13].

[17] Grunert J. Side effects of herbal medicine, <http://herbs.lovetoknow.com/Side_Effects_of_Herbal_Medicine>; 2013 [accessed 22.08.13].

[18] Hamouda C, Amamou M, Thabet H, Yacoub M, Hedhili A, Bescharnia F, et al. Plant poisonings from herbal medication admitted to a Tunisian toxicologic intensive care unit, 1983−1998. Vet Hum Toxicol 2000;42(3):137−41.

[19] Pillay VV, Bhagyanathan PV, Krishnaprasad R, Rajesh RR, Vishnupriya N. Poisoning due to white seed variety of *Abrus precatorius*. J Assoc Physicians India 2005;53:317−9.

[20] Madikizela B, Ndhlala AR, Finnie JF, Staden JV. *In vitro* antimicrobial activity of extracts from plants used traditionally in South Africa to treat tuberculosis and related symptoms. Evid Based Complement Alternat Med 2013;2013:840719.

[21] Moshi MJ, Kagashe GA, Mbwambo ZH. Plants used to treat epilepsy by Tanzanian traditional healers. J Ethnopharmacol 2005;97(2):327−36.

[22] Molgaard P, Nielsen SB, Rasmussen DE, Drummond RB, Makaza N, Andreassen J. Anthelmintic screening of Zimbabwean plants traditionally used against schistosomiasis. J Ethnopharmacol 2001;74(3):257−64.

[23] Nwodo OF, Alumanah EO. Studies on *Abrus precatorius* seeds. II: antidiarrhoeal activity. J Ethnopharmacol 1991;31(3):395−8.

[24] Buzzz G. Nine most toxic plants for humans, <http://greenbuzzznet/environment/nine-most-toxic-plants-for-humans/>; 2011 [accessed 03.09.13].

[25] Balint GA. Ricin: the toxic protein of castor oil seeds. Toxicology 1974;2(1):77−102.

[26] Daniele C, Dahamna S, Firuzi O, Sekfali N, Saso L, Mazzanti G. *Atractylis gummifera* L. poisoning: an ethnopharmacological review. J Ethnopharmacol 2005;97(2):175−81.

[27] Hamouda C, Hedhili A, Ben Salah N, Zhioua M, Amamou M. A review of acute poisoning from *Atractylis gummifera* L. Vet Hum Toxicol 2004;46(3):144−6.

[28] Bruneton J. Toxic plants. Dangerous to humans and animals. Paris: Tec & Doc; 1999.

[29] Larrey D, Pageaux GP. Hepatotoxicity of herbal remedies and mushrooms. Semin Liver Dis 1995;15(3):183−8.

[30] Carpenedo F, Luciani S, Scaravilli F, Palatini P, Santi R. Nephrotoxic effect of atractyloside in rats. Arch Toxicol 1974;32(3):169−80.

[31] Georgiou M, Sianidou L, Hatzis T, Papadatos J, Koutselinis A. Hepatotoxicity due to *Atractylis gummifera* L. J Toxicol Clin Toxicol 1988;26(7):487−93.

[32] Viegi L, Pieroni A, Guarrera PM, Vangelisti R. A review of plants used in folk veterinary medicine in Italy as basis for a databank. J Ethnopharmacol 2003;89(2−3):221−44.

[33] Mbaya AW, Nwosu CO, Onyeyili PA. Toxicity and anti-trypanosomal effects of ethanolic extract of *Butyrospermum paradoxum* (Sapotaceae) stem bark in rats infected with *Trypanosoma brucei* and *Trypanosoma congolense*. J Ethnopharmacol 2007;111 (3):526−30.

[34] Burkill H. The useful plants of West Tropical Africa. London: Royal Botanic Garden Kew; 1985.

[35] Trivellato Grassi L, Malheiros A, Meyre-Silva C, Buss Zda S, Monguilhott ED, Frode TS, et al. From popular use to pharmacological validation: a study of the anti-inflammatory, anti-nociceptive and healing effects of *Chenopodium ambrosioides* extract. J Ethnopharmacol 2013;145(1):127−38.

[36] Amole O, Izegbu M. Chronic toxicity of *Chenopodium ambrosioides* in rats. Biomed Res 2005;16(2):111−3.

[37] Gaire BP, Subedi L. A review on the pharmacological and toxicological aspects of *Datura stramonium* L. J Integr Med 2013;11(2):73−9.

[38] Egharevba R, Ikhatua M. Ethno-medical uses of plants in the treatment of various skin diseases in Ovia North East, Edo State, Nigeria. Res J Agric Biol Sci 2008;4(1):58−64.

[39] Giannini A. Drugs of Abuse—second edition. Los Angeles, CA: Practice Management Information Corporation; 1997.

[40] Goldfrank L, Flommenbaum N. Goldfrank's toxicologic emergencies. New York, NY: McGraw-Hill Professional; 2006.

[41] Pennacchio M, Jefferson L, Havens K. Uses and abuses of plant-derived smoke: its ethnobotany as hallucinogen, perfume, incense, and medicine. Oxford, UK: Oxford University Press; 2010.

[42] Adegoke SA, Alo LA. *Datura stramonium* poisoning in children. Niger J Clin Pract 2013;16(1):116−8.

[43] Noumedem JA, Mihasan M, Lacmata ST, Stefan M, Kuiate JR, Kuete V. Antibacterial activities of the methanol extracts of ten Cameroonian vegetables against Gram-negative multidrug-resistant bacteria. BMC Complement Altern Med 2013;13:26.

[44] Cereda MP, Mattos MCY. Linamarin: the toxic compound of cassava. J Venom Anim Toxins 1996;2:06−12.

[45] EFSA. Opinion of the scientific panel on food additives, flavourings, processing aids and materials in contact with food (AFC) on hydrocyanic acid in flavourings and other food ingredients with flavouring properties. EFSA J 2004;105:1−28.

[46] Bhatia E, Choudhuri G, Sikora SS, Landt O, Kage A, Becker M, et al. Tropical calcific pancreatitis: strong association with SPINK1 trypsin inhibitor mutations. Gastroenterology 2002;123(4):1020−5.

[47] Food and Agriculture Organization of the United Nations. Roots, tubers, plantains and bananas in human nutrition", Rome, 1990, Ch. 7 "Toxic substances and antinutritional factors", under sub-heading "Acute cyanide intoxication, <http://wwwfaoorg/docrep/t0207e/T0207E08htm#Cassava%20toxicity>; 1990 [accessed 27.08.13].

[48] El-Seedi HR, Burman R, Mansour A, Turki Z, Boulos L, Gullbo J, et al. The traditional medical uses and cytotoxic activities of sixty-one Egyptian plants: discovery of an active cardiac glycoside from *Urginea maritima*. J Ethnopharmacol 2013;145(3):746−57.

[49] Mahin L, Marzou A, Huart A. A case report of *Nerium oleander* poisoning in cattle. Vet Hum Toxicol 1984;26(4):303−4.

[50] Szabuniewicz M, Schwartz W, McCrady J, Russell L, Camp B. Treatment of experimentally induced oleander poisoning. Arch Int Pharmacodyn Ther 1971;189:12−21.

[51] Hougen T, Lloyd B, Smith T. Effects of inotropic and arrhythmogenic digoxin doses and of digoxin-specific antibody on myocardial monovalent cation transport in the dog. Circ Res 1979;44:23−31.

[52] Goetz RJ. Oleander. Indiana plants poisonous to livestock and pets. South Bend, IN: Cooperative Extension Service, Purdue University; 1998.

[53] International Programme on Chemical Safety (INCHEM). *Nerium oleander* L, <http://wwwinchemorg/documents/pims/plant/pim366htm>; 2005 [accessed 07.09.13].

[54] el Bahri L, Chemli R. *Peganum harmala* L: a poisonous plant of North Africa. Vet Hum Toxicol 1991;33(3):276−7.

[55] Farouk L, Laroubi A, Aboufatima R, Benharref A, Chait A. Evaluation of the analgesic effect of alkaloid extract of *Peganum harmala* L.: possible mechanisms involved. J Ethnopharmacol 2008;115(3):449−54.

[56] Bellakhdar J. La pharmacopée marocaine traditionnelle. Medecine arabe ancienne et savoirs populaires. Paris: Ibis Press; 1997.

[57] Nadikarni K. Indian materia medica, vol. 1. Bombay: Popular Pakistan Limited; 1976.

[58] Mahmoudian M, Jalilpour H, Salehian P. Toxicity of *Peganum harmala*: review and a case report. Iran J Pharmacol Ther 2002;1(1):1−4.

[59] Achour S, Rhalem N, Khattabi A, Lofti H, Mokhtari A, Soulaymani A, et al. *Peganum harmala* L. poisoning in Morocco: about 200 cases. Therapie 2012;67(1):53−8.

[60] Glasby J. Encyclopedia of the alkaloids. London: Plenum Press; 1978.

[61] Shapira Z, Terkel J, Egozi Y, Nyska A, Friedman J. Abortifacient potential for the epigeal parts of *Peganum harmala*. J Ethnopharmacol 1989;27(3):319−25.

[62] Gupta OP, Anand KK, Ghatak BJ, Atal CK. Vasicine, alkaloid of *Adhatoda vasica*, a promising uterotonic abortifacient. Indian J Exp Biol 1978;16(10):1075−7.

[63] Zutshi U, Rao PG, Soni A, Gupta OP, Atal CK. Absorption and distribution of vasicine a novel uterotonic. Planta Med 1980;40(4):373−7.

[64] Borkar A, More AD. Induced flower colour mutations in *Phaseolus vulgaris* Linn through physical and chemical mutagens. Adv Biores 2010;1(1):22−8.

[65] U.S. Food and Drug Administration. Phytohaemagglutinin, <http://wwwfdagov/Food/FoodborneIllnessContaminants/CausesOfIllnessBadBugBook/ucm071092htm>; 2013 [accessed 26.08.13].

[66] Zarai Z, Ben Chobba I, Ben Mansour R, Bekir A, Gharsallah N, Kadri A. Essential oil of the leaves of *Ricinus communis* L.: *in vitro* cytotoxicity and antimicrobial properties. Lipids Health Dis 2012;11:102.

[67] Roger P, Rix M. Annuals and biennials. London: Macmillan; 1999.

[68] Dhar M, Dhar M, Dhawan B, Mehrotra B, Ray C. Screening of Indian plants for biological activity. Part I. Indian J Exp Biol 1968;6:632−47.

[69] Capasso F, Mascolo N, Izzo AA, Gaginella TS. Dissociation of castor oil-induced diarrhoea and intestinal mucosal injury in rat: effect of NG-nitro-L-arginine methyl ester. Br J Pharmacol 1994;113(4):1127−30.

[70] Huguet-Termes T. New World materia medica in Spanish renaissance medicine: from scholarly reception to practical impact. Med Hist 2001;45(3):359−76.

[71] Gibbs S, Harvey I, Sterling J, Stark R. Local treatments for cutaneous warts systemic review. BMJ 2002;352:461−4.

[72] Wilcox M, Bodeker G. Traditional herbal medicines for malaria. BMJ 2004;329:1156−9.

[73] <http://www.gpnotebook.co.uk/simplepage.cfm?ID = -1321926632> digoxin toxicity GP notebook; 2001 [accessed 24.07.13].

[74] Aplin PJ, Eliseo T. Ingestion of castor oil plant seeds. Med J Aust 1997;167(5):260−1.

[75] Olsnes S, Pihl A. Different biological properties of the two constituent peptide chains of ricin, a toxic protein inhibiting protein synthesis. Biochemistry 1973;12(16):3121−6.

[76] Weston SA, Tucker AD, Thatcher DR, Derbyshire DJ, Pauptit RA. X-ray structure of recombinant ricin A-chain at 1.8 a resolution. J Mol Biol 1994;244(4):410−22.

[77] Wales R, Richardson PT, Roberts LM, Woodland HR, Lord JM. Mutational analysis of the galactose binding ability of recombinant ricin B chain. J Biol Chem 1991;266 (29):19172−9.

[78] European Food Safety Authority (EFSA). Ricin (from *Ricinus communis*) as undesirable substances in animal feed: Scientific Opinion of the Panel on Contaminants in the Food Chain. EFSA J 726:1−38.

[79] Krieger R. Hayes' handbook of pesticide toxicology. 3rd ed. Amsterdam: Elsevier; 2010.

[80] Baselt R. Disposition of toxic drugs and chemicals in man. 8th ed. Foster City, CA: Biomedical Pubilcations; 2008.

[81] Maduka H, Okoye Z. The effect of *Sacoglottis gabonensis* and its isolate bergenin on doxorubicin-ferric ions (Fe^{3+})-induced degradation of deoxyribose. J Med Sci 2001;1:316−9.

[82] Okolo CO, Johnson PB, Abdurahman EM, Abdu-Aguye I, Hussaini IM. Analgesic effect of *Irvingia gabonensis* stem bark extract. J Ethnopharmacol 1995;45(2):125−9.

[83] Kiran, Kudesia R, Rani M, Reclaiming P. Reclaiming degraded land in India through the cultivation of medicinal plants. Bot Res Int 2009;2:174−81.

[84] Jain R, Sharma A, Gupta S, Sarethy IP, Gabrani R. *Solanum nigrum*: current perspectives on therapeutic properties. Altern Med Rev 2011;16(1):78−85.

[85] An L, Tang JT, Liu XM, Gao NN. Review about mechanisms of anti-cancer of *Solanum nigrum*. Zhongguo Zhong Yao Za Zhi 2006;31(15):1225−6.

[86] Tull D. Edible and useful plants of texas and the Southwest: a practical guide. Austin, TX: University of Texas Press; 1999.

[87] Millam S. Potato (*Solanum tuberosum* L.). Methods Mol Biol 2006;344:25−36.

[88] MDidea Extracts Professional. Potato papa or *Solanum tuberosum*, what is the fame of the Potato except Potato famines and more... <http://wwwmdideacom/products/new/new07012html>; 2012 [accessed 26.08.13].

[89] Friedman M, Roitman JN, Kozukue N. Glycoalkaloid and calystegine contents of eight potato cultivars. J Agric Food Chem 2003;51(10):2964−73.

[90] Agbaje EO, Adeneye AA, Daramola AO. Biochemical and toxicological studies of aqueous extract of *Syzigium aromaticum* (L.) Merr. & Perry (Myrtaceae) in rodents. Afr J Tradit Complement Altern Med 2009;6(3):241−54.

[91] Bisset N. Herbal drugs and phytopharmaceuticals. Stuttgart, Germany: CRC Press; 1994.

[92] Dalziel J. The useful plants of West Tropical Africa. London: Crown Agents for the Colonies; 1937.

[93] Lane BW, Ellenhorn MJ, Hulbert TV, McCarron M. Clove oil ingestion in an infant. Hum Exp Toxicol 1991;10(4):291−4.

[94] Brown SA, Biggerstaff J, Savidge GF. Disseminated intravascular coagulation and hepatocellular necrosis due to clove oil. Blood Coagul Fibrinolysis 1992;3(5):665−8.

[95] Prashar A, Locke IC, Evans CS. Cytotoxicity of clove (*Syzygium aromaticum*) oil and its major components to human skin cells. Cell Prolif 2006;39(4):241−8.

[96] Bensky D, Clavey S, Stoger E. Chinese herbal medicine: materia medica. Seattle, WA: Eastland Press; 2004.

[97] Sandberg F. Etude sur les plantes medicinales et toxiques d'Afrique équatoriale. Paris: Cahiers de la Maboké; 19653.

[98] Kerharo J, Bouquet A. Plantes médicinales de la Côte d'Ivoire et Haute Volta. Paris: Vigot; 1950.

[99] Baumer M. Compendium des plantes medicinales des Cornores des Mascareignes et des Seychelles Paris: Agence de Cooperation Culturelle et Technique; 1979.

[100] Sandberg F, Cronlund A. An ethnopharmacological inventory of medicinal and toxic plants from equatorial Africa. J Ethnopharmacol 1982;5(2):187−204.

23 Biochemical Parameters in Toxicological Studies in Africa: Significance, Principle of Methods, Data Interpretation, and Use in Plant Screenings

Jean P. Dzoyem[1,2], Victor Kuete[1] and Jacobus N. Eloff[2]

[1]Department of Biochemistry, Faculty of science, University of Dschang, Cameroon, [2]Phytomedicine Programme, Department of Paraclinical Sciences, Faculty of Veterinary Science, University of Pretoria, South Africa

23.1 Introduction

Tropical and subtropical Africa are endowed with approximately 45,000 species of plant with developmental potentials, out of which 5000 species are used medicinally [1]. Medicinal plants are integral part of the culture and traditions of African population; meanwhile, up to 80% of the world's population use herbal substances medicinally [2]. Although it is widely alleged that medicinal plants are safe because they are natural, many are potentially toxic. There is a growing safety concern, and the toxicity of plants used in traditional medicines is becoming more widely recognized [3]. As the use of medicinal plants increases, screening of their toxicity is crucial to guarantee the safety of the users. Furthermore, the assessment of the toxicological status of these plants is required to establish the scientific evidence of their safety for consumption. It has to be kept in mind that in determining toxicity, dosage plays a critical role. *In vitro* and *in vivo* methods and guidelines have been established to determine the toxicological profile of chemicals. A range of biochemical markers to be monitored has been recommended to examine the adverse effects of a chemical substance including plant extracts [4]. It is important to determine toxicological parameters for a herbal medicine once it enters the market, as it can cause serious adverse reaction on consumers. The measurement of numerous enzymes levels is important in the evaluation of the toxicological potential of therapeutic agents [5].

Toxicological Survey of African Medicinal Plants. DOI: http://dx.doi.org/10.1016/B978-0-12-800018-2.00023-6

This is because the variation of these enzymes in the serum is indication of tissue or cellular damage, leading to an abnormal release of intracellular components into the blood. Changes in biochemical parameters provide information about the mechanism of toxicological action, the functional status of major organ systems and the identification of target tissues (liver, kidney, hematopoietic and immune systems, etc.). In this regard, several biochemical indices can be measured to determine the safety of medically used medicinal plants. In the two last decades, efforts have been made in the screening of the toxicological profile of African medicinal plants with emphasis on the analysis of induced biochemical changes in animal models [6−8]. Following the World Health Organization (WHO) and the Organization for Economic Cooperation and Development (OECD) guidelines; acute, subacute, and chronic toxicities as well as genotoxicity and mutagenicity of a number of African medicinal plants have so far been investigated in animal experimental models, with rats and mice being the most frequently used animals.

The most commonly evaluated biochemical indices include enzymes such as aspartate aminotransferase (AST), alanine aminotransferase (ALT), lactate dehydrogenase (LDH), alkaline phosphatase (ALP), and creatine phosphokinase (CPK). Nonenzymatic parameters mostly investigated in the toxicological survey of African medicinal plants include measuring bilirubin (BIL), creatinine (CRE), blood urea nitrogen (BUN), malondialdehyde (MDA), glucose (GLU), total protein (TP), albumin (ALB), globulin (GLO), cholesterol (CHO), and triglyceride (TG) [9,10]. It is noteworthy that common electrolytes and minerals such as sodium, potassium, phosphate, magnesium, calcium, and chloride were not broadly investigated. Standard biochemical methods, automated clinical chemistry analyzer techniques, as well as commercial kits have been used to evaluate these parameters [11−13]. This chapter aims to overview the major biochemical markers screened in the toxicity study of African medicinal plants with emphasis on enzymes as well as nonenzymatic indices of toxicological relevance. The state of the art on the safety profile of African medicinal plants investigated is also discussed in regard to the literature retrieved in scientific websites such as Pubmed, Scholar Google, Scopus, Scirus, Sciencedirect, and Web-of-knowledge.

23.2   Important Biochemical Indices in Toxicity Studies

The measurement of biochemical parameters provides information about the functional status of major organ systems such as the liver, kidney, and hematopoietic and immune systems. Enzymes are widely used in toxicological studies as markers of detection and evaluation of cell damage. ALT and AST are among the first enzymes to become established as biomarkers of toxicity [14]. In most species dehydrogenases are effective markers of liver damage and are widely explored. Two of the urea cycle enzymes (ornithine carbamoyltransferase (OCT), arginase) also have emerged as useful indicators of liver and kidney damage. Changes in plasma lipids and proteins can be indicators of the synthetic capacity in many

Table 23.1 Biochemical Parameters of Toxicological Relevance

Enzymes	Nonenzymatic Parameters	Electrolytes, Minerals
Alanine aminotransferase (ALT)	Bilirubin (BIL)	Sodium
Aspartate aminotransferase (AST)	Malondialdehyde (MDA)	Potassium
γ-glutamyltransferase (GGT)	Creatinine (CRE)	Chloride
Ornithine carbamoyltransferase (OCT)	Blood urea nitrogen (BUN)	Calcium
Glutathione S-transferase(GST)	Glutathione (GSH)	Magnesium
Malate dehydrogenase (MDH)	Cholesterol (CHO)	Phosphorus
Sorbitol dehydrogenase (SDH)	Total Protein (TP), Albumin (ALB), Globulin (GLO)	Iron
Lactate dehydrogenase (LDH)	Glucose (GLU)	
Glutamate dehydrogenase (GDH)	Triglycerides (TG)	
Alkaline phosphatase (ALP)	Troponin I (cTnI)	
Paraoxonase (PON)	Total serum bile acids (TSBA)	
N-Acetyl-β-D-glucosaminidase (NAG)	Bile acids (BA)	
5′-nucleotidase (5′NT)		
Ornithine decarboxylase (ODC)		
Creatine kinase(CK)		
Choline Esterase (ChE)		
Hydroxyphenylpyruvate dioxygenase (HPD)		
Arginase		
Lipase		
Amylase		

organs, but these are normally preceded by more obvious changes in other diagnostic tests and only become significant when the dysfunction has developed to a serious extent. Measurements of other nonenzymatic markers have been used in toxicity studies as the diagnosis of tissue damage. Standardized lists of clinical biochemistry parameters are monitored in toxicity screenings. Changes in these parameters may provide important information regarding the mechanism of toxic responses observed in animals and the extrapolation to humans. Minimal recommendations for the panel of biochemical parameters to be measured as part of guidelines for safety assessment have been established [3]. The core parameters (enzymatic and nonenzymatics, electrolytes) usually measured during the toxicological profile assessment of drugs, including African medicinal plants, are summarized in Table 23.1. It should be noted that the list provided in Table 23.1 can be modified to include parameters that are optimized to the study of interest or to specific patterns of an organ/tissue injury.

23.3 Enzymes of Toxicological Relevance

Enzymes are highly specialized proteins that facilitate biochemical reactions that otherwise would proceed at a much lower rate. The enzymes are classified under six main group headings, which relate to the chemical reactions involved. Each class is then subdivided in order to classify the other features of the reactions. The International Union of Biochemistry uses a four-digit Enzyme Commission number (EC) to define the enzyme. The main headings include:

- Oxidoreductases (E.C.1): transfer of electrons
- Transferases (E.C.2): transfer a functional group (e.g., a methyl or phosphate group)
- Hydrolases (E.C.3): hydrolysis reactions (transfer of functional groups to water)
- Lyases (E.C.4): cleave various bonds by means other than hydrolysis and oxidation
- Isomerases (E.C.5): catalyze isomerization changes within a single molecule
- Ligases (E.C.6): join two molecules with covalent bonds

Enzyme assessment is of great importance in an early detection of metabolic change, in detection of organ-specific effects, in the determination of toxic mechanism, and in the establishment of *"No Effect"* level of a drug. As with a drug, absorption of a medicinal plant extract can exert biochemical and/or physiological changes at the cell, tissue, organ, or organism levels. The basic principle of using the enzyme levels as marker of cellular or tissue damage is based on the comparison of the changes in activity in the serum or plasma enzymes which are mainly present in the intracellular environment and are released into the serum in very low concentrations. The details about the major enzymes of toxicological significances are discussed below. Particular enzymes are investigated to ascertain the toxic effect of drug on both liver injury and function. These are grouped into liver enzymes and subsequently subdivided into leakage (ALT, AST, sorbitol dehydrogenase (SDH), glutamate dehydrogenase (GDH), and LDH) and cholestatic enzymes (ALP and γ-glutamyl transferase (GGT)). Some are much more appropriate to muscle (AST, creatine kinase (CK), and LDH), while others tend to be involved in many tissues.

23.3.1   Hepatocellular Leakage Enzymes: Transaminases

Transaminases (aminotransferases) are a group of enzymes that catalyze the interconversion of amino acids and oxoacids by transfer of amino groups. Pyridoxal-5'-phosphate (P5'P) functions as coenzyme in the amino transfer reactions. In all amino transfer reactions, 2-oxoglutarate and L-glutamate serve as one amino group acceptor and donor pair respectively. The two most important aminotransferases are AST, formerly known as glutamate oxaloacetate transaminase (GOT), and ALT, formerly called glutamate pyruvate transaminase (GPT). They are widely distributed in human tissues and are particularly active in liver, heart muscle, skeletal muscle, and kidney. These two enzymes are of greatest clinical significance. Assay of the serum activity of transaminases has become the most useful tool in the assessment of liver toxicity [15]. Therefore, the standard test for liver damage is to measure blood concentrations of AST and ALT [16].

23.3.1.1   Alanine Aminotransferase (GPT)

ALT (E.C.2.6.1.2) is an important enzyme in the intermediary metabolism of glucose and protein enzyme involved in the transfer of an amino group from the amino acid, alanine, to α-ketoglutaric acid to produce glutamate and pyruvate (Figure 23.1). Early biochemical and cytogenetic studies have suggested the existence of two isoforms of ALT (products of different genes) in mice, rats, pigs, and humans [17]. Humans express GPT1 (ALT1) and GPT2 (ALT2) and these two forms of the enzyme are approximately 70% identical [18]. GPT2 mRNA is highly expressed in muscle, fat, brain, and kidney tissues, whereas GPT1 mRNA is mainly expressed in kidney, liver, fat, and heart tissues [18]. GPT2 mRNA is the predominant form in muscle and fat tissues [18]. Independent reports show that high GPT2 mRNA levels were detected from human heart and skeletal muscle tissue [19].

The enzyme ALT is located primarily in liver and kidney, with lower amounts in heart and skeletal muscle. In case of increased liver activity, it can be found in the serum and in the cerebrospinal fluid (CSF), but not in urine [20]. Because ALT is located only in the cytoplasm, serum levels tend to be relatively high as a result of hepatocyte damage. Therefore it has been used as a marker for liver injury in preclinical toxicity studies. Serum ALT activity is significantly elevated in a variety of liver conditions, including drug toxicity [21].

23.3.1.2 Aspartate Aminotransferase (GOT)

During amino acid catabolism, AST (E.C.2.6.1.1) transfers amino groups from glutamate to oxaloacetate, to form α-ketoglutarate and aspartate respectively (Figure 23.2); aspartate is then used as a source of nitrogen in the urea cycle.

In humans, distinguishing isozymes include GOT1 which is the cytosolic isoenzyme derived mainly from red blood cells and heart and GOT2, the mitochondrial isoenzyme is present predominantly in the liver. These isoenzymes are thought to have evolved from a common ancestral AST *via* gene duplication, and they share a sequence homology of approximately 45% [22]. AST has also been found in a number of microorganisms, including *Escherichia coli, Haloferax mediterranei,* and *Thermus thermophilus* [23,24]. AST is found in high concentrations in liver. However, AST is not liver specific and is also found in red blood cells, cardiac muscle, skeletal muscle, and kidney [25]. Any injury in these organs may raise the

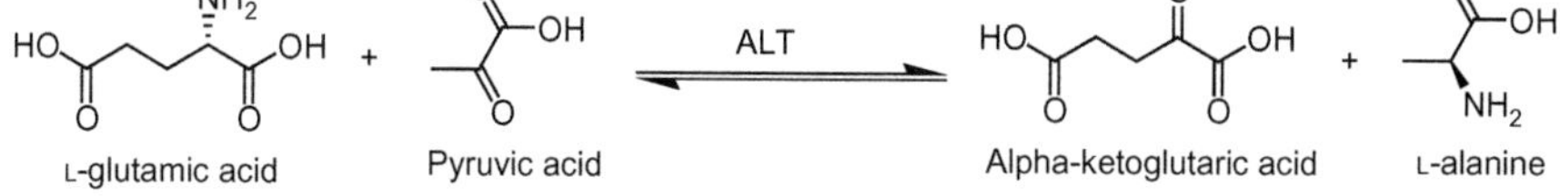

Figure 23.1 Biochemical reaction catalyzed by ALT.

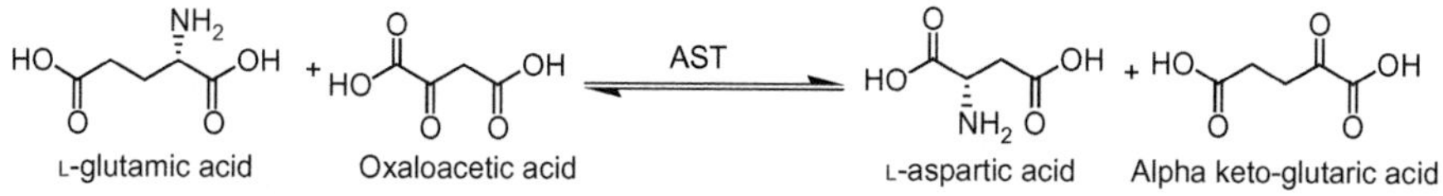

Figure 23.2 Biochemical reaction catalyzed by AST.

AST level significantly. AST is present in both cytoplasm and mitochondria of cells [26]. Other less important sources of AST that do not raise the AST level significantly in the blood are brain, pancreas, lungs, leukocytes, and erythrocytes in decreasing order of concentration. AST has been used as a biochemical marker in many toxicity studies [27].

AST analysis measures the amount of this enzyme in the blood. Low levels of AST are normally found in the blood. When body tissues or organs such as the heart or liver are diseased or damaged, additional AST is released into the blood. The concentration of AST in the blood is directly related to the extent of the tissue damage. The AST test may be done at the same time as a test for ALT. The ratio of AST to ALT may help determine whether the liver or another organ has been damaged. However, both ALT and AST levels can indicate liver damage. AST is similar to ALT in that both enzymes are associated with liver parenchymal cells. The difference is that ALT is found predominantly in the liver, with clinically negligible quantities found in the kidneys, heart, and skeletal muscle, while AST is found in the liver, heart (cardiac muscle), skeletal muscle, kidneys, brain, and red blood cells. As a result, ALT is a more specific indicator of liver inflammation than AST, as AST may be elevated also in diseases affecting other organs [28]. Because aminotransaminases are ubiquitous in their cellular distribution, serum elevation may occur with a variety of non hepatobiliary disorders. However, elevations exceeding 10 to 20 times the reference value are uncommon in the absence of hepatic cell injury. The activities of both enzymes are high in tissues especially liver, heart, and muscles. Any damage or injury to the cells of these tissues may cause release of these enzymes into the blood leading to increase of their activities in the serum. The importance of both AST and ALT elevations in serum is generally related to the number of hepatocytes affected [29]. However, the level cannot be used to predict either the type of lesion, or whether cell damage is reversible (leakage) or irreversible (frank necrosis). The ratio of AST to ALT is sometimes useful in differentiating whether the liver or another organ has been damaged [30].

23.3.1.3 Glutamate Dehydrogenase

GDH (EC1.4.1.2) is a mitochrondrial enzyme that catalyzes the conversion of glutamate to α-ketoglutarate. The reaction (Figure 23.3) occurred in two steps: the first step involves a Schiff base intermediate being formed between ammonia and α-ketoglutarate. This Schiff base intermediate is crucial because it establishes the alpha carbon atom in glutamate's stereochemistry. The second step involves the Schiff base intermediate being protonated, which is done by the transfer of a hydride ion from Nicotinamide adenine dinucleotide phosphate (reduced form)

Figure 23.3 The general reaction catalyzed by GDH.

(NADPH) producing L-glutamate. GDH is unique because it is able to utilize both Nicotinamide adenine dinucleotide (NAD$^+$) and Nicotinamide adenine dinucleotide phosphate (oxidized form) (NADP$^+$) [31]. NADP$^+$ is utilized in the forward reaction of α-ketogluterate and free ammonia, which are converted to L-glutamate via a hydride transfer from NADPH to glutamate [31]. NAD$^+$ is utilized in the reverse reaction, which involves L-glutamate being converted to α-ketoglutarate and free ammonia via an oxidative deamination reaction [32].

The analysis of the role of GDH is complicated in the way that different isoenzymes function in different directions *in vivo*. Moreover, as the isoenzyme pattern is sensitive to a wide range of abiotic factors, the possibility arises of GDH functioning in different ways under different conditions [33]. There are three isoforms of GDH based on which cofactor is used:

- GDH EC1.4.1.2 (GDH2) catalyzes essentially the formation of 2-oxoglutarate using NAD(P)$^+$ [34]
- GDH EC1.4.1.3 (GDH3) catalyzes both the formation of 2-oxoglutarate and the reverse reaction, thus having a dual coenzyme specificity [NAD(P)$^+$/NAD(P)H] [35]
- GDH EC1.4.1.4 (GDH4) catalyzes the formation of glutamate using NAD(P)H [36,37]

GDH is a sensitive and specific marker of liver disease in all animals, including nonmammalian species [38]. It is essentially a mitochondrial enzyme occurring in many tissues in the body including hepatocytes, kidney, intestine, muscle, and salivary gland. GDH rises significantly with hepatic necrosis. This enzyme is highly concentrated in liver tissue and is located in cell mitochondria and, therefore, complete cell disruption is required before it is released in large quantities. Hence any significant rise in serum GDH is indicative of hepatic necrosis.

23.3.1.4 Sorbitol Dehydrogenase

SDH (EC 1.1.1.14), also known as L-iditol 2-dehydrogenase, is a 357 amino acid member of the zinc-containing alcohol dehydrogenase family, catalyzing the zinc-dependent interconversion of polyols, such as such as sorbitol and xylitol, to their respective ketoses. Together with aldose reductase, it provides a way for the body to produce fructose from glucose without using adenosine triphosphate (ATP). SDH uses NAD + as a cofactor (Figure 23.4)

SDH has been identified in several human and animal tissues. It is widely distributed in tissues throughout the body, though it is found primarily in the cytoplasm and mitochondria of liver, kidney, and seminal vesicles [39]. SDH is a sensitive enzyme marker for liver necrosis but should be combined with measurements of ALT or other enzymes. Khayrollah et al. [40] also reported SDH as a specific

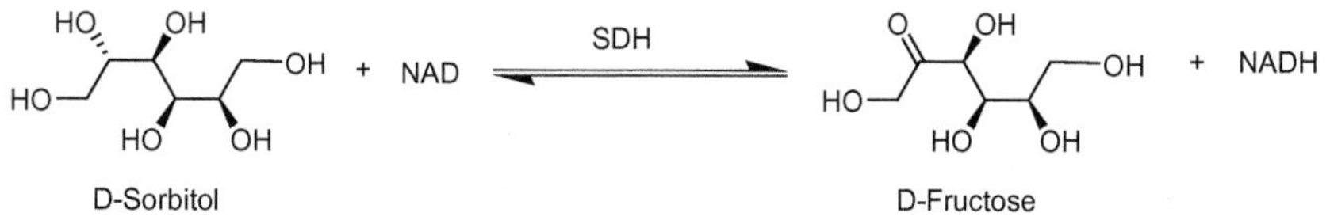

D-Sorbitol   D-Fructose

Figure 23.4 Reaction catalyzed by SDH.

indicator of liver cell damage and parenchymal hepatic diseases. SDH activity in serum is usually low but increases during acute episodes of liver damage [41].

23.3.2 Cholestatic Enzymes

The main importance of ALP and γ-GGT remains in their use in the diagnosis of cholestasis that is the impairment of bile flow, which can be caused by physical or functional obstruction of the biliary tract within or outside of the liver. Significant cholestasis leads to an increase in bilirubin in the blood. The key advantages of ALP and GGT is their greater sensitivity to determine this abnormality as compared to serum bilirubin levels alone. Nonetheless, due to the fact the activity of ALP is influenced by many other factors, it is not entirely specific for this purpose. Generally, GGT appears to be more specific, and more sensitive in some instances.

23.3.2.1 Alkaline Phosphatase

The ALPs (orthophosphoric monoester phosphohydrolase, EC 3.1.3.1) are a group of relatively nonspecific enzymes that hydrolyzes a variety of ester orthophosphates under alkaline conditions, and exhibits optimal activity between pH 9 and 10 [42]. The chemical reaction catalyzed by ALP is shown in Figure 23.5. The ALP is a nonspecific phosphatase; unlike most enzymes it recognizes a wide variety of molecules as substrates. There are two isoenzymes, but several isoforms (e.g., liver, bone, intestinal, kidney, and placental forms) have been identified in humans and preclinical species. The isoenzymes are produced from intestinal and tissue nonspecific ALP genes and differ in amino acid sequence. The major isoforms that can be measured in animals are liver-ALP (L-ALP), corticosteroid-ALP (C-ALP, only in dogs), bone-ALP (B-ALP) and intestinal-ALP (I-ALP). In humans, ALP is present in all tissues throughout the entire body, but is particularly concentrated on the border membranes of the bile canaliculi and on sinusoidal surfaces of the liver, the intestinal mucosa, the osteoblasts of bone, the renal proximal tubules, the placenta, and the mammary glands. However, only a few organs actually contribute to the circulating enzyme level. ALP is anchored to cell membranes by glycophosphatidylinositol (GPI) proteins. Cleavage of these proteins by bile acids, phospholipase D, and proteases releases ALP from membranes, resulting in increased ALP levels in serum/plasma [43].

Increased ALP levels have been associated with drug-induced cholestasis [27].

It is primarily a marker of hepatobiliary effects and cholestasis (moderate to marked elevations) [44]. It may be elevated if bile excretion is inhibited by liver

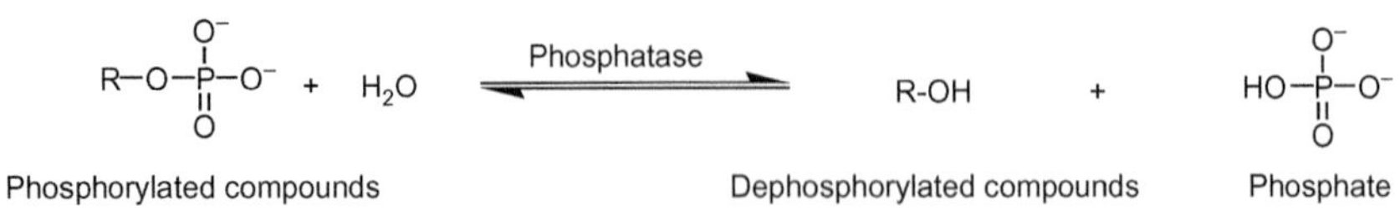

Phosphorylated compounds   Dephosphorylated compounds   Phosphate

Figure 23.5 The general reaction catalyzed by phosphatase enzymes.

damage (cholestasis) [45]. Hepatotoxicity leads to elevation of the normal values due to the body's inability to excrete it through bile as result of the obstruction of the biliary duct.

23.3.2.2 γ-Glutamyl Transferase

GGT gamma-glutamyl transpeptidase (EC 2.3.2.2) comes under the peptidase group of enzymes which specifically catalyzes the transfer of γ-glutamyl group from peptides and other compounds that contain it to the substrate itself, some amino acid or peptide, or even water, in which case a simple hydrolysis takes place [46]. It cleaves glutathione (GSH) to facilitate the recapture of cysteine for synthesis of intracellular GSH. GGT is involved in the transfer of amino acids across the cellular membrane and leukotriene metabolism [47]. It is important in glutathione metabolism, amino acid absorption, and protection against oxidant injury. An example of *in vitro* reaction catalyzed by GGT is shown in Figure 23.6.

Although GGT is present in the cell membranes of many tissues, including the kidneys, bile duct, pancreas, gallbladder, spleen, heart, brain, and seminal vesicles, the main source is the liver (primarily biliary epithelium). GGT is found primarily in the membrane and in microsomes (from SER) as aggregates. A small portion ($<5\%$) is found in the cytoplasm. Disaggregation (solubilization) and increased synthesis result in increased activity in serum [48]. GGT is a specific biomarker of hepatobiliary injury, especially cholestasis and biliary effects [49]. It was reported as a specific indicator of bile duct lesions in the rat liver [50].

23.3.3 Muscle Leakage Enzymes

Although there are many other enzymes that can be used to assess the muscle function, only markers of muscle injury (muscle leakage enzymes) such as AST, CK and LDH are of toxicological relevance. It is noteworthy that AST is also a marker of liver injury as discussed above.

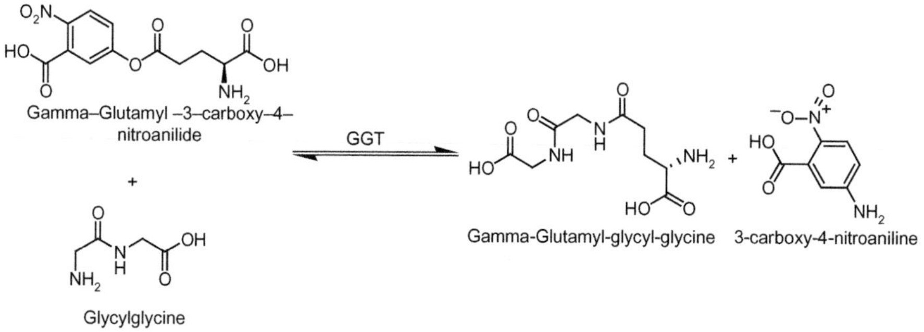

Figure 23.6 *In vitro* reaction catalyzed by gamma-glutamyl transpeptidase.

23.3.3.1 Creatine Kinase

CK, formerly known as CPK or phospho-CK (EC 2.7.3.2), is an intracellular enzyme present in the greatest amounts in skeletal muscle, myocardium, and brain; smaller amounts occur in other visceral tissues. During muscle contraction, ATP is consumed to form adenosine diphosphate (ADP), and this ADP is again rephosphorylated to ATP by enzyme CK using phosphocreatine as a phosphate donor (Figure 23.7). Phosphocreatine is the major phosphorylated compound present in muscles, eight times more than that of ATP. Enzyme activity is inhibited by excess ADP, urate, cysteine, and metal ions such as Mn^{2+}, Ca^{2+}, Zn^{2+} and Cu^{2+}. Mg^{2+} is required for the activity of CK but excess of it inhibits the CK activity [51]. CK catalyzes the conversion of creatine and consumes ATP to create phosphocreatine (PCr) and ADP. This CK enzyme reaction is reversible and thus ATP can be generated from PCr and ADP [52].

CK is a dimeric molecule composed of two subunits designated M and B. Combinations of these subunits form the isoenzymes. There are four principle isoenzymes of CK: CK-1: BB isoenzyme, found mostly in brain, prostate, gut, lungs, bladder, uterus, placenta and thyroid. Injury to brain tissue may increase CK-1 activity in CSF, but rarely results in raised total serum CK activity. CK-2: MB isoenzyme is present in varying degree in heart muscle (25−46% of CK activity) and some in skeletal muscle. CK-3: MM isoenzyme predominates in skeletal muscle with lesser amounts heart muscle. This isoenzyme accounts for increased activity in most cases with raised total CK activity in animals. CK-Mt isoenzyme is found in the mitochondria between the inner and outer membranes. It is different from other forms both immunologically and in electrophoretic mobility and comprises 15% of total cardiac CK activity [53]. More recently it has been recognized that these isoenzymes are glycoproteins [54]. Disruption of cell membranes due to hypoxia or other injury releases CK from the cellular cytosol into the systemic circulation. Serum CK activity is elevated in tissue damages involving skeletal muscle, heart muscle, and brain injury. Elevation of particular CK isoenzymes activity in serum is of toxicological significance; therefore detection of increased activity in serum is useful as an indicator of muscle injury [55].

23.3.3.2 Lactate Dehydrogenase

LDH (L-lactate: NAD oxidoreductase (LDH); EC 1.1.1.27) is a tetrameric protein catalyzing the reversible conversion of pyruvate to lactate (Figure 23.8) [56]. The enzyme uses NAD/Nicotinamide adenine dinucleotide (reduced form) (NADH) as

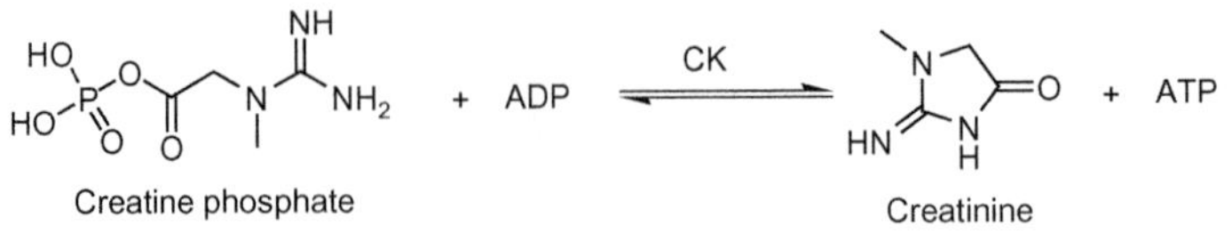

Creatine phosphate   + ADP ⇌ Creatinine + ATP

Figure 23.7 Biochemical reaction catalyzed by CK.

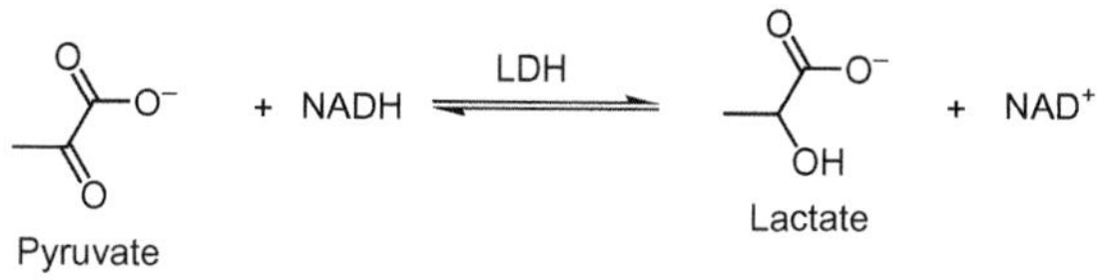

Figure 23.8 The general reaction catalyzed by LDH.

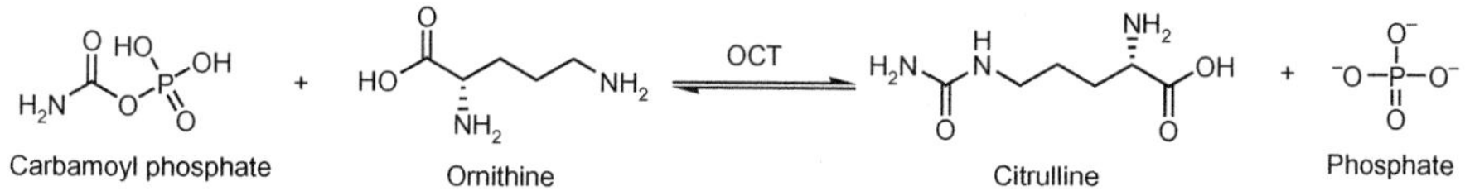

Figure 23.9 Biochemical reaction catalyzed by OCT.

cofactor and exists in six isoforms, five depending on the combination of the two subunits M (muscle) and H (heart) giving: LDH-1 (4H) in the heart; LDH-2 (3H1M) in the reticuloendothelial system, LDH-3 (2H2M) in the lungs, LDH-4 (1H3M) in the kidneys and LDH-5 (4M) in the liver and striated muscle. The sixth isoform is a homotetrameric LDH-C in the testis [57].

LDH is an intracellular enzyme which is widely distributed throughout the body and is found at high levels in tissues that utilize glucose for energy; it is therefore not organ specific. Isoenzymes are predominantly distributed in the tissue specific manner. LDH-1 and LDH-2 are predominantly present in cardiac muscles, kidney, and erythrocytes. Enough care should be taken while taking samples for the LDH analysis for myocardial infarction as sample hemolysis may give erroneous results. LDH-4 and LDH-5 isoenzymes are predominant in liver and skeletal muscle. LDH2-4 is found in many other tissues such as the spleen, lungs, endocrine glands, and platelets [58]. As a result of this distribution, an increase in LDH can reflect damage to a number of different tissues (skeletal or cardiac muscle, kidney, liver). LDH levels may be increased whenever there is cell necrosis or when neoplastic proliferation of cells causes an increase LDH production.

23.3.4 Other Enzymes of Toxicological Relevance

Several other enzymes in addition to those reported above are also used as markers to detect the damage induced by chemicals or plants. Some of them are discussed below.

23.3.4.1 Ornithine Carbamoyltransferase

Ornithine transcarbamoylase (OTC) or OCT (EC 2.1.3.3) is an enzyme that catalyzes the reaction of citrulline formation from L-ornithine and carbamoyl phosphate (Figure 23.9). In mammals it is almost exclusively located in the mitochondria of hepatocytes and is part of the urea cycle. Activity increases in both acute and chronic liver disease. The high concentration of OCT in hepatic tissue relative to other tissues leads to it being regarded as a "liver-specific" enzyme; however, its usefulness is sometimes limited by the technical requirements for the assay [59].

This enzyme measurement should not be confused with ornithine decarboxylase (ODC) which is of interest in studies of polyamine metabolism [60].

23.3.4.2 Glutathione S-transferase

Glutathione S-transferase (GST; EC 2.5.1.18) represents a major group of detoxification enzymes. GST isoenzymes are ubiquitously distributed in nature, being found in organisms as diverse as microbes, insects, plants, fish, birds, and mammals [61]. The transferases displayed various activities and participate in several different types of reaction. Most of these enzymes can catalyze the conjugation of reduced glutathione (GSH) with compounds that contain an electrophilic center through the formation of a thioether bond between the sulfur atom of GSH and the substrate (Figure 23.10). In addition to conjugation reactions, a number of GST isoenzymes exhibit other GSH-dependent catalytic activities including the reduction of organic hydroperoxides and isomerisation of various unsaturated compounds [62]. These enzymes also have several noncatalytic functions that relate to the sequestering of carcinogens, intracellular transport of a wide spectrum of hydrophobic ligands, and modulation of signal transduction pathways [62]. GSTs represent a complex grouping of proteins. Two entirely distinct super families of enzyme with transferase activity have evolved [63]. The first enzymes to be characterized were the cytosolic, or soluble, GSTs [64]. To date, at least 16 members of this superfamily have been identified in humans [63,65]. On the basis of their degree of sequence identity, the soluble mammalian enzymes have been assigned to eight families, or classes, designated Alpha (α), Mu (μ), Pi (π), Sigma (σ), Theta (θ), Zeta (ζ), Omega (ω), and Kappa (κ) [62]. Four additional classes of this superfamily, called Beta (β), Delta (δ), Phi (φ), and Tau (τ) are represented in bacteria, insects, and plants [66].

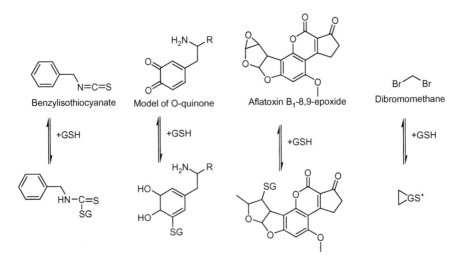

Figure 23.10 Examples of reaction catalyzed by GST.

The primary role of GSTs is to detoxify xenobiotics by catalyzing the nucleophilic attack by GSH on electrophilic carbon, sulfur, or nitrogen atoms of said nonpolar xenobiotic substrates, thereby preventing their interaction with crucial cellular proteins and nucleic acids [67]. Glutathione transferases (GST) that are essential in many detoxification processes, have until recently not been used widely due to the low levels in plasma, poor enzymatic stability and inhibition by bilirubin and bile acids in the assay [43]. The toxicological importance of GST alpha to assess acute hepatotoxicity has been compared to serum ALT and AST activities in studies on both humans and rats. Unlike ALT and AST, GST is found in high concentration in centrolobular cells, and therefore is more sensitive to injury in this metabolic zone of the liver.

23.3.4.3 Malate Dehydrogenase

Malate dehydrogenase (MDH, L-malate: NAD oxidoreductase, EC 1.1.1.37) catalyzes the NAD/NADH-dependent interconversion of the substrates malate and oxaloacetate (Figure 23.11). This reaction plays a key part in the malate/aspartate shuttle across the mitochondrial membrane, and in the tricarboxylic acid cycle within the mitochondrial matrix [68].

Several isoforms of MDH have been identified, differing in their subcellular localization and their specificity to the coenzyme NAD or NADP. In eukaryotic cells, at least two forms of the enzyme can be found. One isoform is a principal enzyme of the citric acid cycle operating within mitochondria [68]. The other is found in the cytosol where it participates in the malate/aspartate shuttle. This shuttle exchanges reducing equivalent across the mitochondrial membranes in the form of malate/oxaloacetate rather than as NAD/NADH. A third isoenzyme was found in the glyoxysomes of yeast, where it converts malate produced from glyoxylate [69]. MDH has been shown to be localized in two cellular compartments, the mitochondria and extramitochondria compartment, 10% and 90%, respectively [70]. Generally, the enzyme activity found in the serum is the extra-mitochondrial form, but in severe cellular damage, the mitochondrial form also can be detected in serum. The absolute activity in the cytoplasm is greatest in liver, followed by heart, skeletal muscle, and brain [71]. MDH is also a periportal enzyme that is released into the serum indicating tissue damage. Serum MDH activity is correlated with both liver and heart injury.

23.3.4.4 Paraoxonase 1

Paraoxonases (PONs) are a group of enzymes involved in the hydrolysis of organophosphates [72]. There are three known genotypic forms of PONs. They are coded

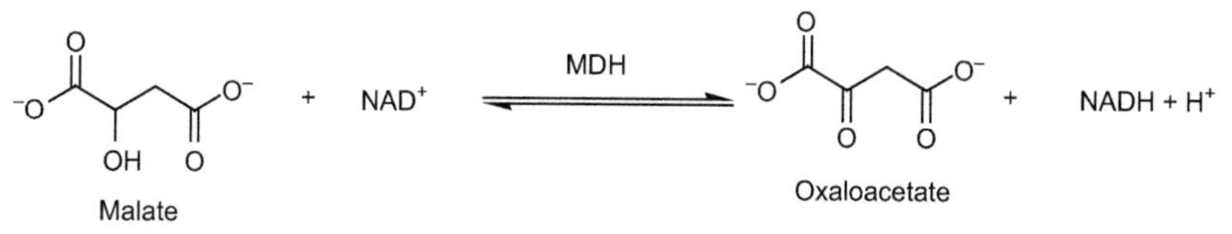

Malate + NAD⁺ ⇌ (MDH) Oxaloacetate + NADH + H⁺

Figure 23.11 Biochemical reaction catalyzed by MDH.

for by the PON set of genes—PON1, PON2 and PON3—located on the long arm of chromosome 7 [73,74]. The differences between them lie in their location and activity. PON1 is synthesized in the liver and transported along with high density lipoprotein (HDL) in the plasma. It functions as an antioxidant; it prevents the oxidation of low density lipoprotein (LDL). Its serum concentration is influenced by inflammatory changes and the levels of serum oxidized-LDL. PON2 is a ubiquitously expressed intracellular protein that can protect cells against oxidative damage [75]. PON3 is similar to PON1 in activity but differs from it in substrate specificity. Additionally, it is not regulated by inflammation and levels of oxidized lipids [76]. PON/arylesterase of human serum (PON1; EC 3.1.1.8) also known as aromatic esterase 1 or serum aryldialkylphosphatase, and of several other mammalian species catalyzes the hydrolysis of organophosphates, aromatic carboxylic acids, and possibly carbamates (Figure 23.12). Human serum PON1 is a calcium-dependent esterase and is closely associated with the HDL complex [77]. There is very little information available on PON2. However, PON2 might be able to lower the intracellular oxidative stress of a cell and prevent the cell-mediated oxidation of LDL. Cells overexpressing PON2 are less able to oxidize LDL and show considerably less intracellular oxidative stress when exposed to either H_2O_2 or oxidized phospholipid. Because PON2 is ubiquitously expressed not just in cells of the artery environment but in tissue throughout the body, it is likely that PON2 plays a role in reducing intracellular or local oxidative stress [74,75]. PON1 is widely distributed among tissues such as liver, kidney, intestine, and also serum, where it is associated with HDL, while PON2 is ubiquitously expressed in nearly every human tissue, with the highest expression in liver, lung, placenta, testis, and heart [74,78]. Although PON1 activity does not appear to show high specificity for liver damage, it has been recently identified as a potential biomarker of hepatic injury utilizing proteomic analyses by [79].

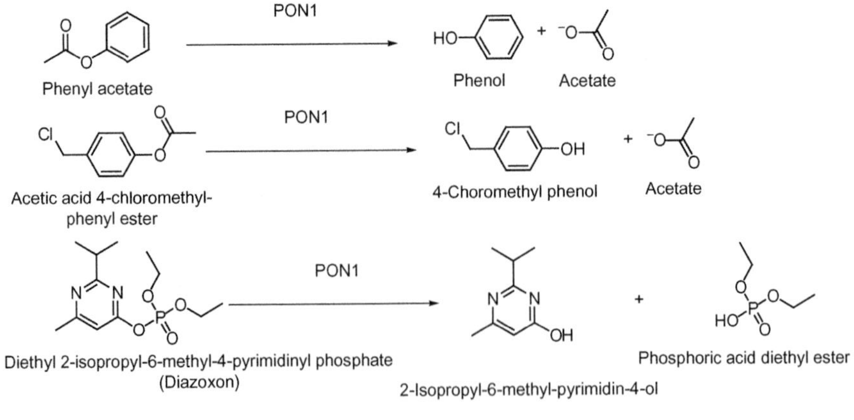

Figure 23.12 Example of biochemical reaction catalyzed by PON 1.

23.3.4.5   N-Acetyl-β-ᴅ-Glucosaminidase

N-acetyl-β-ᴅ-glucosaminidase (NAG, EC 3.2.1.30) is a high molecular weight lysosomal enzyme found in many tissues of the body damage [80]. Two types of NAG have been described, the tissue A-form present in liver, kidney and spleen that can be distinguished from the As-form present in serum on the basis of its behavior on Diethylaminoethyl (DEAE)-cellulose chromatography and gel electrophoresis. The serum As-form was also found to be present in CSF and urine from a number of patients with renal disease [81]. NAG cannot pass into glomerular ultrafiltrate due to its high molecular weight. Thus, urinary NAG is of renal origin. However, this enzyme shows high activity in renal proximal tubular cells, and leaks into the tubular fluid as the ultrafiltrate passes through proximal tubules. When proximal tubular cells are injured due to any disease process, its urine level increases and thus is used as a reflection of proximal tubular cell necrosis [82]. The assessment of urinary NAG is considered as a useful marker of renal tubular impairment in various disease states. It is extensively used both in routine practice as well as for research purposes, when it comes to the evaluation of tubular function. NAG has proven to be sensitive, persistent, and robust biomarkers of acute kidney injury [83]. Other urinary enzymes such as ALT, AST, and ALP are also sensitive indicators of kidney parenchymal damage compared to functional measurements. However, urinary NAG remains the most widely used marker of renal tubular impairment [84].

23.3.4.6   5'-Nucleotidase

The enzyme 5'-nucleotidase (5'NT, EC 3.1.3.5) is an alkaline phosphomonoesterase involved in release of inorganic phosphate only from the nucleoside 5'-phosphate, for example, catalyzes the hydrolysis of nucleoside 5'monophosphates, for example, adenosine-5'-monophosphate and inosine-5'-monophosphate [43].

This enzyme catalyzes the chemical reaction showed in Figure 23.13. The enzyme is a dimer of two identical 70 KDa subunits bound by a glycosyl phosphatidyl inositol linkage to the external face of the plasma membrane of various cells, including subsets of human lymphocytes; it has been extensively studied in rat liver and heart, and in murine and human placenta [85].

5'-NT is an intrinsic membrane glycoprotein present as an ectoenzyme in a wide variety of mammalian cells. It appears to be distributed widely in the body tissues,

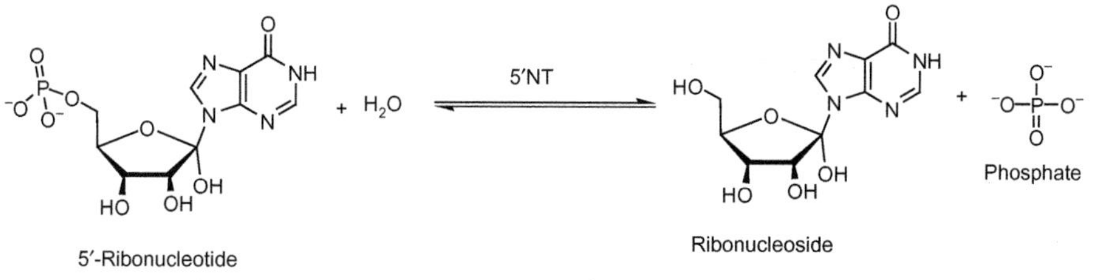

Figure 23.13 Reaction catalyzed by 5'-nucleotidase.

and to be mainly a membrane-bound enzyme, which is useful in studies of hepato-biliary injury [86]. In human lymphocytes, two soluble 5′-nucleotidases are present: the first form reveals a preference for Adenosine monophosphate (AMP) as sub-strate and is known as e-Ns; the second form, called c-N-11, has a preference for Inosine monophosphate (IMP) as substrate and controls intracellular levels of nucleoside 5′ [87].

23.3.4.7 Ornithine Decarboxylase

The enzyme L-ODC (EC 4.1.1.17) catalyzes the pyridoxal-P-dependent decarboxyl-ation of ornithine (a product of the urea cycle) to form putrescine (Figure 23.14) [88].

This reaction is the key regulator of the synthesis of the naturally occurring polyamines, which regulate proliferation and which are closely linked to neoplastic growth [89]. ODC activity has been measured in liver, prostate, skin, kidney, neo-plasma, other tissue, and in cells in culture [60].

23.3.4.8 Arginase

The arginase (EC 3.5.3.1, arginine amidinase, canavanase, L-arginase, arginine transamidinase) is a hydrolase manganese-containing enzyme that catalyzes the catabolism of L-arginine to urea and L-ornithine (Figure 23.15) [90].

Arginase is a primordial enzyme, widely distributed in the biosphere and repre-sented in all primary kingdoms, it occurs in numerous organisms and tissues where there is no functioning urea cycle [91]. In most mammals, two isozymes of this enzyme exist; the first, Arginase I, functions in the urea cycle, and is located pri-marily in the cytoplasm of the liver. The second isozyme, Arginase II, has been implicated in the regulation of the arginine/ornithine concentrations in the cell. It is located in mitochondria of several tissues in the body, with most abundance in the kidney and prostate. It may be found at lower levels in macrophages, lactating mammary glands, and brain [92]. The second isozyme may be found in the absence of other urea cycle enzymes [93]. Arginase I is highly liver specific, making it a

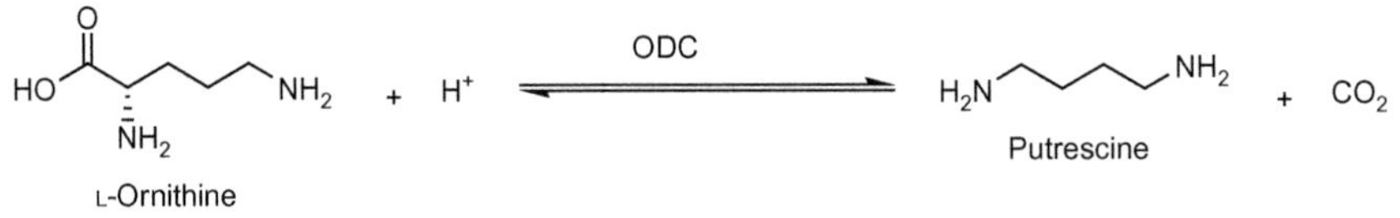

Figure 23.14 Biochemical reaction catalyzed by ODC.

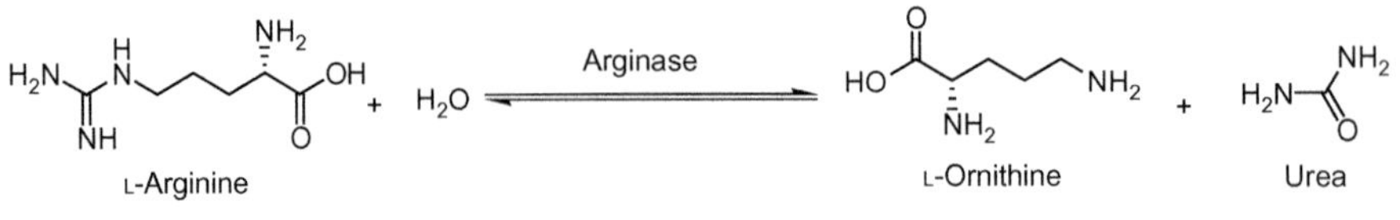

Figure 23.15 Reaction catalyzed by arginase.

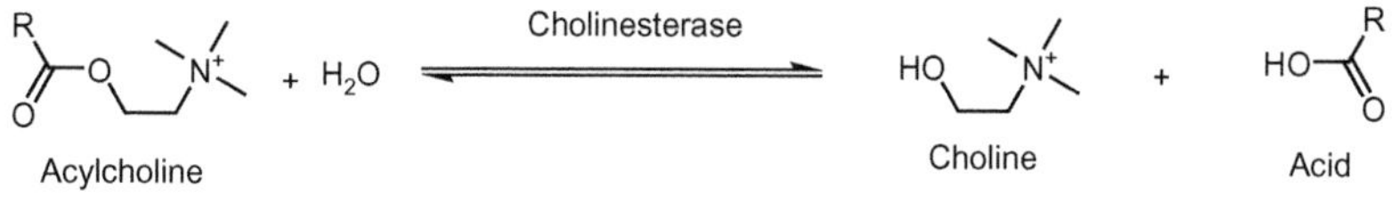

Figure 23.16 Reaction catalyzed by cholinesterase.

candidate biomarker of liver toxicity that shows higher specificity compared to the liver enzymes [94].

23.3.4.9 Cholinesterase

Cholinesterase is a family of enzymes present in the central nervous system, particularly in nervous tissue, muscle and red cells, which catalyze the hydrolysis of the neurotransmitter acetylcholine into choline and acetic acid (Figure 23.16), a reaction necessary to allow a cholinergic neuron to return to its resting state after activation [95].

It is one of many important enzymes needed for the proper functioning of the nervous systems of humans. There are two types: acetylcholinesterase (AChE, EC 3.1.1.7) and pseudocholinesterase (BChE, EC 3.1.1.8). AChE was found primarily in the blood on red blood cell membranes, in neuromuscular junctions, and in neural synapses while it is produced in the liver and found primarily in plasma [96]. AChE hydrolyzes acetylcholine more quickly while BChE hydrolyzes butyrylcholine more quickly, making the difference between the two enzymes. BChE levels may be reduced in patients with advanced liver disease and decrease must be greater than 75% before significant prolongation of neuromuscular blockade occurs with succinylcholine [97]. Elevation of plasma pseudocholinesterase was observed in 90.5% cases of acute myocardial infarction and this can be used as marker of substance toxicities [98]. AChE is primarily found in the blood on red blood cell membranes, in neuromuscular junctions, and in neural synapses, while BChE is produced in the liver and found primarily in plasma [96]. The difference between the two types of cholinesterase is their relative preferences for substrates: AChE hydrolyzes acetylcholine faster while BChE hydrolyzes butyrylcholine faster.

23.4 Nonenzymatic Markers in Toxicity Studies

In toxicological studies, a variety of biochemical measurements can be used to evaluate a broad range of physiological and metabolic functions, identifying possible target organs, measuring impaired organ function, assessing the persistence and severity of tissue injury. In addition to enzymes, several nonenzymatic parameters are of particular relevance in toxicity studies, for example, urea and creatinine for glomerular function and bilirubin for hepatotoxicity. Both enzymatic and nonenzymatic parameters are interrelated and when combined, these analyses can provide better information. Several nonenzymatic parameters have been assessed during the evaluation of African medicinal plants safety. Some of the most frequently used are discussed below.

23.4.1   Bilirubin

Bilirubin, a brownish yellow pigment of bile is the breakdown product of normal heme catabolism. It is secreted by the liver in vertebrates, which gives to solid waste products (feces) their characteristic color [99]. The production of bilirubin from heme occurs mainly in the spleen (macrophages) and liver (Kupffer cells), but also all over the body by macrophages, and in renal tubular cells. The cells that perform this job are known collectively as the reticuloendothelial system. Heme is catabolized to biliverdin by heme oxygenase enzymes. Biliverdin reductase then is converted biliverdin to unconjugated bilirubin (UBL). This form is often referred to as indirect reacting, free or prehepatic. UBL is a relatively insoluble, nonpolar molecule, it is bound to ALB in the plasma and transported as suchto the liver where it is conjugated with glucuronic acid in the hepatocytes. This conjugation is catalyzed by glucuronyl transferase. The process of conjugation makes the bilirubin water soluble, and thus easier to excrete. Conjugated bilirubin (CBL) is secreted into the bile and enters the duodenum. In the intestine, it is converted to urobilinogen by the action of bacterial flora [100]. Some of the urobilinogen gets reabsorbed and undergoes enterohepatic circulation as well as excretion by the kidneys. Total bilirubin (TBL) is a composite of indirect (nonhepatic) and direct (hepatic) bilirubin.

In liver disease however, such as obstruction in the biliary duct system, bilirubin excretion usually becomes impaired, causing a rise in the plasma bilirubin concentration. Increases in conjugated bilirubin indicate cholestasis but are not a specific marker. Increases can occur secondary to hemolysis, by parenchymal liver disease that interferes with biliary excretion, by affecting bile transporters or physically impeding bile flow, sepsis, bile sludging, and physical obstructions to bile flow in the biliary system. Therefore, bilirubin is a marker of hepatobiliary injury, especially cholestasis and biliary effects [101]. In acute human hepatic injury, TBL can be a better indicator of disease severity compared to ALT [102]. Hepatotoxicity may lead to an increase in the urobilinogen in urine. Increased urobilinogen has been observed during alcoholic liver damage and hemolysis [103].

23.4.2   Creatinine and Urea

Creatinine is a chemical waste molecule that is generated from muscle metabolism. Phosphocreatine, a molecule of major importance for energy production in muscles, undergoes spontaneous cyclization to form creatine and inorganic phosphorous; during the reaction, creatine and phosphocreatine are catalyzed by CK, and a spontaneous conversion to creatinine may occur [104]. This spontaneous degradation of creatine to creatinine occurs at a rather constant and uniform daily rate. Creatinine is freely filtered by the glomerulus and clearance of creatinine from the plasma to the urine can be used to provide an approximation of the glomerular filtration rate. A small amount of creatinine is secreted by proximal tubules in the kidney but, in contrast to urea, none is resorbed by the tubules [105]. The kidneys maintain the blood creatinine in a normal range. As the kidneys become impaired for any reason, the creatinine level in the blood will rise due to poor clearance of creatinine by the kidneys. Creatinine has been found to be a fairly reliable indicator of kidney function. Elevated creatinine level signifies impaired kidney function or kidney disease [106].

Urea is the major end product of nitrogen metabolism in humans and mammals. It is synthesized by hepatocytes from ammonia generated by catabolism of amino acids derived either from digestion of proteins in the intestines or from endogenous tissue proteins [107]. Urea is excreted by the kidneys, intestine, saliva, and sweat. In ruminants, urea is excreted into the gastrointestinal system where it is converted to amino acids and ammonia which are then used for protein production. The kidney is the most important route of urea excretion and as a result, urea has long been used as a barometer of renal function. Urea appears in the glomerular filtrate in the same concentration as is found in the blood [108]. Decreased glomerular filtration increases urea. Urea concentration is measured as urea nitrogen, and the test is usually called BUN or serum urea nitrogen (SUN). BUN levels may be elevated (a condition called uremia) in both acute and chronic renal (kidney) failure. Various diseases damage the kidney and cause faulty urine formation and excretion. Congestive heart failure leads to a low blood pressure and consequent reduced filtration rates through the kidneys, therefore, BUN may be elevated. Urinary tract obstructions also can lead to an increased BUN. In severe cases, hemodialysis is used to remove the soluble urea and other waste products from the blood [109].

Urea and creatinine are used in conjunction to assess renal function. Azotemia is defined as an increase in urea nitrogen (UN) and creatinine and can result from a variety of disorders including, but not limited to, renal failure [110]. Urea and creatinine are relatively insensitive to small amounts of damage to the kidney. Serum creatinine concentrations tend to parallel changes in urea with renal toxicity. Creatinine diffuses throughout body fluid at a slower rate than urea, and takes about 4 hours to equilibrate; therefore, serum creatinine concentration changes more slowly compared with urea. Endogenous creatinine clearance may be used as a measure of glomerular filtration rate [111]. There are a number of nonrenal causes for elevations in serum concentrations of urea and creatinine. Urea concentrations can vary with hydration status, diet, gastrointestinal hemorrhage, or protein catabolism. Serum creatinine concentration is influenced by muscle mass, but is relatively independent of dietary influences and protein catabolism compared with urea. The urea to creatinine ratio, while not definitive, can be of value in the differential diagnosis of azotemia. Thus, in renal azotemia, urea and creatinine can be anticipated to increase proportionately, while in pre-renal azotemia, urea may increase disproportionately [112]. A pronounced increase indicates liver damage and depending on the length, there are acute and chronic impairments [113]. Clinically, the detection of urea is due to its ability to react at high temperature in an acid medium with ferric chloride as the oxidizing agent and thiosemicarbazide used as indicator. The intensity of the red color developed and measured spectrophotometrically at 520 nm is proportional to the amount of urea of the medium [113].

23.4.3 Malondialdehyde

MDA is a highly reactive three-carbon dialdehyde produced as a byproduct of polyunsaturated fatty acid peroxidation and arachidonic acid metabolism [114]. MDA is also a reactive species occurs naturally and is a one of a biological marker for oxidative stress, mostly existing in the enol form. MDA come from polyunsaturated lipids after degradation by reactive oxygen species (ROS) [115]. It has been reported to occur in animal tissues, especially under conditions of antioxidant deficiency. Recent studies have confirmed the presence of significant quantities of MDA in foods of animal origin where it apparently arises mainly from the oxidation of arachidonic acid in cell membranes. It is present in blood platelets and in serum [116]. Among the several byproducts of lipid peroxidation processes, MDA is one of the most frequently used biomarkers providing an indication of the overall lipid peroxidation level [117]. Lipid peroxidation plays a role in the pathogenesis of many types of tissue injuries and especially in the tissue damage induced by several toxic substances [118]. As one of the most known secondary products of lipid peroxidation, MDA can be used as a marker of cell membrane injury. Increased levels of lipid peroxidation products have been associated with a variety of chronic diseases in both humans and model systems [119]. A possible relationship between lipid peroxidation levels in the kidney and renal toxicity have been established [120]. MDA forms a DNA adduct after reacting with deoxyadenosine and deoxyguanosine in DNA. The primary one is the mutagenic heterocyclic compound pyrimido (1,2-a).purin-10(3H)-one (M_1G) [121]. The condensation of MDA and guanidine group of arginine residues yields 2-aminopyrimidines. The laboratory detection and quantification of MDA is based on its ability to condense with two equivalents of thiobarbituric acid to give a fluorescent red derivative that can be assayed by spectrophotometry [122]. Alternatively, 1-methyl-2-phenylindole with a higher selectivity can be used [122]. Clinically, an increased level of MDA is observed in the corneas of patients suffering from keratoconus and bulloskeratopathy. It also can be found in tissue sections of joints from patients with osteoarthritis [123,124].

23.4.4   Glutathione

Glutathione (GSH) is a tripeptide derived from glycine, glutamate, and cysteine (γ-glutamyl-cysteinyl-glycine). The γ-glutamyl part means that the amino group of the cysteine is attached to the side chain carboxyl group of the glutamic acid rather than to the α-carboxyl [125]. GSH, present in plants, animals, and some bacteria, often at high levels, can be thought of as a redox buffer. The γ-carboxyl group of glutamate is activated by ATP to form an acyl phosphate intermediate, which is then attacked by the α-amino group of cysteine. A second condensation reaction follows, with the α-carboxyl group of cysteine activated to an acyl phosphate to permit reaction with glycine. The oxidized form of glutathione (GSSG), produced in the course of its redox activities, contains two glutathione molecules linked by a disulfide bond [126]. GSH is the prevalent low-molecular-weight thiol in mammalian cells. It is formed in a two-step enzymatic process including, first, the formation of gamma-glutamylcysteine from glutamate and cysteine, by the activity of the gamma-glutamylcysteine synthetase; and second, the formation of GSH by the activity of GSH synthetase which uses gamma-glutamylcysteine and glycine as substrates.

While its synthesis and metabolism occur intracellularly, its catabolism occurs extracellularly through a series of enzymatic and plasma membrane transport steps. GSH provides reducing capacity for several reactions and plays an important role in detoxification of hydrogen peroxide, other peroxides, and free radicals [127]. Once oxidized, it can be reduced back by glutathione reductase, using NADPH as an electron donor [128]. It displays remarkable metabolic and regulatory versatility. GSH/GSSG is the most important oxydo-reduction couple in animal cells and plays crucial roles in antioxidant defense, nutrient metabolism and the regulation of pathways essential for whole body homeostasis. Glutathione deficiency contributes to oxidative stress, and, therefore, may play a key role in aging and the pathogenesis of many diseases [129]. The ratio of reduced glutathione to oxidized glutathione within cells is often used as a measure of cellular toxicity [130]. A decrease in blood-reduced glutathione (GSH) has been reported in patients affected by deficiencies of the enzymes involved in the synthesis of glutathione [127]. The toxicological importance of GSH should be considered in serum ALT and AST activities in studies on both humans and rats. Unlike MDA, glutathione deficiency contributes to oxidative stress, however an increased GSSG-to-GSH ratio is considered indicative of oxidative stress. Cellular levels of antioxidants respond to levels of oxygen and oxyradicals; which enable cells to defend against increased free radical production. If produced in excess, or not removed effectively, free radicals result in cellular damage. Therefore determination of plasma glutathione may be valuable in the evaluation of the ability of xenobiotics as inducers or inhibitors of detoxification systems. Several forms of toxic liver damage have been claimed to involve free radical mechanisms [131].

23.4.5 Triglycerides

Triglycerides (TGs, also called neutral fats, triacylglycerols, or triacylglycerides) are a common, simple type of lipid consisting of three long-chain fatty acids esterified to glycerol [126]. There are exogenous (chylomicrons) and endogenous (pre-β-lipoproteins) triglycerides. Exogenous triglycerides originate from food, while endogenous triglycerides are formed in the liver. Two main biosynthetic pathways are known, the sn-glycerol-3-phosphate pathway, which predominates in the liver and adipose tissue, and a monoacylglycerol pathway in the intestines [132]. Hence, TGs constitute the main source of energy for the body, apart from being the main and most reliable energy reserves of the human body, triacylglycerols take part in metabolic processes that determine the rate of fatty acid oxidation, the plasma levels of free fatty acids, the biosynthesis of other lipid molecules and the metabolic fate of lipoproteins [133].

Many cell types and organs have the ability to synthesize triacylglycerols, but in animals the liver, intestines, and adipose tissue are most active with most of the body stores in the last. Within the cell, triacylglycerols are stored as cytoplasmic lipid droplets (adiposomes) enclosed by a monolayer of phospholipids and hydrophobic proteins [126]. Only small amounts are found in the blood. The determination of serum triglyceride concentration is used to assess the possible presence of hypertriglyceridemia (increased blood and serum levels of triglycerides). An elevated level of serum triglycerides increases blood viscosity and precipitates platelet aggregation, which in turn results in diminished vascular flow. In addition, increased triglyceride concentrations in the blood coincide with decreased levels of HDL cholesterol [134]. Increased in the serum concentration of triglycerides, may be a risk factor for coronary artery disease (CAD) because triglycerides are used to estimate concentrations of LDL cholesterol, which definitely has been shown to be a risk factor for CAD [135]. In toxicity studies, there may be several associated perturbations of lipoprotein metabolism. In acute hepatic injury, levels of hepatic enzymes may be reduced, for example, triglyceride lipase, thus elevating plasma triglycerides with a concomitant decrease in cholesterol [43].

23.4.6   Cholesterol

Cholesterol is a steroidal lipid built from four linked hydrocarbon rings. It is absent from prokaryotes, but is ubiquitous in all animal tissues, where much of it is located in the membranes, although it is not evenly distributed [136]. Cholesterol has vital structural roles in membranes and in lipid metabolism in general. It is a biosynthetic precursor of bile acids, vitamin D and steroid hormones (glucocorticoids, oestrogens, progesterones, androgens, and aldosterone) [137].

It is generally believed that the main function of cholesterol is to modulate the fluidity of membranes by interacting with their complex lipid components, specifically the phospholipids. In addition, it contributes to the development and working of the central nervous system, and it has major functions in signal transduction and sperm development [138]. Cholesterol travels through the blood attached to a protein. This cholesterol-protein package is called a lipoprotein. Lipoprotein analysis (lipoprotein profile or lipid profile) measures blood levels of total cholesterol, LDL cholesterol, HDL cholesterol, and triglycerides.

Cholesterol levels are maintained through tightly regulated and complex mechanisms. It is well known that insufficient or excessive cellular cholesterol results in a wide range of pathologies. An excess of plasma cholesterol leads to its accumulation in the artery wall causing atherosclerosis. Cholesterol and triglycerides blood tests measure the total amount of fatty substances in the blood. High cholesterol, also known as hypercholesterolemia, is a major risk factor for heart disease and stroke. The concentration of cholesterol in serum depends on a number of factors. Estrogen hormones enhance the synthesis of cholesterol; however, they play an even greater role in its clearance by lowering the concentration of serum cholesterol [139]. Sometimes relatively small changes of plasma cholesterol (and triglycerides) are found in toxicological studies. Impaired hepatic protein metabolism affects lipoprotein synthesis resulting in hypolipoproteinaemia and hypotriglyceridaemia [43].

23.4.7   Bile Acids

The bile acids are the end products of cholesterol metabolism in animals, the main functions of which are to act as powerful detergents or emulsifying agents in the intestines to aid the digestion and absorption of fatty acids, monoacylglycerols, and other fatty products and to prevent the precipitation of cholesterol in bile [140]. They are produced in the liver and are stored in the gallbladder. Gallbladder contraction with feeding releases bile acids into the intestine. Bile acids undergo enterohepatic circulation, that is, they are absorbed in the intestine and taken up by hepatocytes for re-excretion into bile. Total bile acids are implicated in various signal transduction pathways and are elevated with liver injury and functional change [141]. Although there are a number of different biosynthetic routes to bile acids from cholesterol, there are four main steps, and the liver is the only organ concerned in the production of the "primary" bile acids. In fact, there are at least 16 enzymes that catalyze up to 17 reactions to convert insoluble cholesterol into a highly soluble conjugated bile salt [140]. Bile acids functionally contribute to the catabolism and elimination of cholesterol; are the primary determinant of bile flow; regulate pancreatic secretions; and release of gastrointestinal peptides, and contribute to the digestion and absorption of fat (and indirectly fat-soluble vitamins) in the small intestine. Total bile acids are also implicated in various signal transduction pathways and are elevated with liver injury and functional change [141]. Measurement of bile acid concentrations is, therefore, a good indicator of hepatobiliary function [27].

23.4.8 Glucose

Glucose (also known as dextrose) is a carbohydrate compound consisting of six carbon atoms and an aldehyde group and they are referred to as aldohexose. The glucose structure can exist in an open-chain (acyclic) and ring (cyclic) form. It is a carbohydrate and is the most important simple sugar (monosaccharide) in animal metabolism. Glucose is one of the main products of photosynthesis and is involved in cellular respiration to produce ATP and NADH. Glucose is one of the primary molecules that cells use as an energy source and a metabolic intermediate. Glucose metabolism is critical to normal physiological functioning; glucose also acts as a source of starting material for many biosynthetic reactions. The cellular metabolism of glucose involves glycolysis, gluconeogenesis, glycogenolysis, glycogenesis, pentose-phosphate pathway, and citric acid cycle [142].

Maintenance of a normal plasma glucose concentration requires precise matching of glucose utilization and endogenous glucose production or dietary glucose delivery. Glucose is derived from three sources: the intestinal absorption that follows the digestion of dietary carbohydrates, glycogenolysis, and gluconeogenesis. Glucose is transported into cells through multiple metabolic pathways: it may be stored as glycogen; it may undergo glycolysis to pyruvate or ethanol. Finally, it may be released into the circulation by the liver and kidneys, the sole organs containing glucose-6-phosphatase, the enzyme necessary for the release of glucose into the circulation [143]. The primary aim of the determination of blood glucose concentration is to assess the carbohydrate metabolism.

23.4.9 Total Protein

Proteins are large biological molecules composed of amino acids linked with peptide bonds. Amino acids are nitrogenous compounds which consists of an acidic carboxyl (-COOH) and basic amino (-NH2) groups attached to a carbon atom. It also contains a carbon radical (R) group for side chains that differ from each amino acid. Proteins are constitutional material of the human body and are the most abundant intracellular macromolecules. According to their composition, proteins are classified into simple and complex proteins. Simple proteins are composed exclusively of amino acids, whereas complex proteins contain so-called nonprotein or prosthetic group in addition to their protein structure. Proteins have been named according to their prosthetic groups, for example, glucoproteins, lipoproteins, hemoproteins, etc. Proteins are very different but specific for particular organs and tissues. Many plasma proteins including ALB and most of the globulins are synthesized in the liver, while the immunoglobulins are synthesized in the plasma cells and B lymphocytes of the spleen, bone marrow and lymph nodes [43]. Plasma proteins represent a heterogeneous group with ALB constituting the major portion. ALB serves as a regulator of osmotic equilibrium. Globulins are also important plasma proteins and they are primarily associated with antibodies. Protein in the plasma is made up mainly of ALB and globulin. ALB is the most abundant blood plasma protein and is produced in the liver and forms a large proportion of all plasma protein. The human version is human serum ALB, and it normally constitutes about 50% of human plasma protein [144]. The globulin in turn is made up of α1, α2, β, and γ globulins. Certain globulins bind with hemoglobin. Other globulins transport metals, such as iron, in the blood and help fight infection. Serum globulins can be separated into several subgroups by serum protein electrophoresis. Acute phase proteins are associated with the acute inflammatory response and are useful markers for acute and chronic active inflammation.

Total protein also called serum total protein or plasma total protein is a biochemical test for measuring the total amount of protein in blood plasma or serum. ALB–Globulin (A/G) ratio is the ratio of ALB present in serum in relation to the amount of globulin. The ratio can be interpreted only in light of the total protein concentration. Total protein, ALB, and globulin estimations are useful in the assessment of general bodily condition, nutritional status, and the response to the toxicity of xenobiotics. Xenobiotics can bind to the various plasma protein fractions and the degrees to which this binding occurs can have a marked effect on the pharmacological action of a drug [145].

23.4.10 Troponin

Troponins are the protein filament components of the contractile cardiac and skeletal muscles, but which are not present in smooth muscle [43]. The contractile apparatus of skeletal and myocardial striated muscle is composed of myosin-containing thick filaments, surrounded by an octagonal array of actin-containing thin filaments. The thin filament includes an actin helix bordered by tropomyosin strands that periodically contain a three-member "troponin complex." The sliding filament model of muscle contraction proposes that an ATP-dependent actin and myosin interactions is "triggered" by release of calcium ions during electrical depolarization. Released calcium ions bind to the troponin complex (consisting of troponin I, troponin T, and troponin C) to change their conformation and result is contraction [146]. Contractile proteins (myosin and actin) and regulatory proteins (tropomyosin and troponin complex) are structural parts of the sarcomere. Contractions takes place when the heads of myosin molecules, which form cross-bridges of thick-filament, bind to actin, followed by shift in orientation of cross-bridge that pulls thin filament toward center of sarcomere. Activation requires calcium binding to troponin complex, reversing inhibition of interaction between myosin and actin. In the cycle of chemical reactions underlying contraction, hydrolysis of ATP produces the cross-bridge motion. Relaxation occurs when calcium becomes dissociated from troponin. Functionally, troponin T serves to bind the troponin complex to the tropomyosin strand of the actin thin filament. Troponin I functions to inhibit the activity of actinomycin ATPase. Troponin C serves to bind four calcium ions and regulates contraction [147]. The structures (amino acids) of troponins T and I found in cardiac muscle differ from the structures of troponins T and I found in skeletal muscle, and these differences can be used to distinguish the two types (cardiac vs. skeletal).

Troponins are an important tool for the assessment of acute coronary syndromes. An increased level of the cardiac protein isoform of troponin circulating in the blood has shown to be a biomarker of heart disorders, the most important of which is myocardial infarction [148]. Raised troponin levels indicate cardiac muscle cell death as the enzyme is released into the blood upon injury to the heart. Normal levels of troponin should be zero and even small levels of detectable troponin indicate cardiac muscle cell death. Serial determinations of cardiac troponins are also useful for assessing reperfusion following thrombolysis [149].

23.5 Electrolytes of Toxicological Relevance

Serum electrolyte concentrations are among the most commonly used biochemical parameters by toxicologist for assessment of drug safety. Sodium, potassium, and chloride are among the most commonly monitored electrolytes in toxicity studies. Magnesium, calcium, and phosphate are also monitored. This section present the physiological basis of the major electrolytes recorded during the assessment of the toxicity of some African medicinal plants.

23.5.1 Sodium

Sodium (Na^+) is the most abundant cation in the body and its concentration is closely related to osmotic homeostasis, maintenance of body fluid volumes and neuromuscular excitability. Plasma sodium is freely filtered *via* the renal glomeruli, but about 70% is reabsorbed in the proximal tubules and 25% in the loop of Henle together with chloride ions and water. In the distal convoluted tubules, aldosterone modulates sodium reabsorption and sodium ions are exchanged for potassium and hydrogen ions [150]. While sodium is essential for maintaining the appropriate transmembrane electric potential for action potential and neuromuscular functioning, the principal role of sodium is to regulate serum osmolality as well as fluid balance. Serum osmolality is an estimate of the water-solute ratio in the vascular fluid. It is useful in determining volume status, especially the intravascular volume [132]. Changes in body water and plasma volume can directly or indirectly affect the serum sodium concentration. The serum sodium is determined by the total body sodium plus potassium and the total body water. Sodium is the primary extracellular osmole, so the serum sodium concentration usually reflects the osmolality of the extracellular fluid [151]. Sodium content is the key factor in controlling the volume of the extracellular fluid. Gain of sodium leads to seeking and retaining water with expansion of the extracellular fluid volume. Loss of sodium, at least initially, leads to loss of water and contraction in the extracellular fluid volume. Thus, sodium metabolism governs the volume of the extracellular fluid [152].

A wide variety of xenobiotics influence sodium metabolism, leading to retention or depletion. The body attempts to maintain a constant extracellular fluid volume, as major changes in extracellular fluid volume can have profound effects on the cell. The kidney plays a critical role in maintenance of extracellular fluid volume, *via* sodium and water retention. Regulation of body water is accomplished by monitoring of plasma osmolality (determined primarily by sodium concentration) and blood volume. This is achieved by osmoreceptors and baroreceptors. They are attributable to disturbance of the water metabolism. Hyponatremia is almost always a condition of water excess while hypernatremia is due water deficiency. Physiological normonatremia (normal plasma osmolality) is maintained by an integrated system involving regulated water intake *via* thirst and control of water excretion via antidiuretic hormone secretion [153].

23.5.2 Potassium

Potassium (K^+) is the major intracellular osmole, and water equilibrates across cell membranes such that the intra- and extracellular osmolality will balance. It is also important for maintaining resting membrane potential of cells. Potassium is also known as cellular electrolyte. Total body potassium is approximately 55 mEq/kg body weight. Of this amount, 98% is in the intracellular compartment (primarily in the muscle, skin, subcutaneous tissue, and red blood cells) and 2% is in the extracellular compartment [154]. The concentration of potassium in the intracellular space is about 40-fold of the concentration in the extracellular space. The intracellular potassium gradient is maintained by an ATP-dependent active extrusion of sodium, which is balanced by the calcium pumping of potassium and hydrogen ions into the cells [155]. The most important physiological role of potassium is in the regulation of muscle and nerve excitability, especially on muscle and nervous tissue excitability. During periods of potassium imbalance, the cardiovascular system is of principal concern. Cardiac muscle cells depend on their ability to change their electrical potentials, with accompanying potassium flux when exposed to the proper stimulus, to result in muscle contraction and nerve conduction [156]. Potassium also may play important roles in the control of intracellular volume (similar to the ability of sodium in controlling extracellular volume), protein synthesis, enzymatic reactions, and carbohydrate metabolism [157]. Between 60% and

75% of total body potassium is found within muscle cells, with the remainder in bone. Only 5% of potassium is located in the extracellular fluid, therefore potassium concentration in blood is not always an indication of total body potassium levels. Plasma (extracellular fluid) K^+ concentration is tightly regulated; fairly small changes can have marked effects on organ function, explaining its importance in toxicity studies.

23.5.3 Chloride

Chloride (Cl^-) is an inorganic anionic halogen that is distributed exclusively within the extracellular fluid compartment (ECF), which comprises the blood/plasma (or serum) compartment and the interstitial fluid compartment. It is the major anion associated with sodium in the ECF [158]. Chloride is the major extracellular anion and it is important in the maintenance of electroneutrality and osmolality together with sodium. After filtration through the renal glomeruli, chloride is passively reabsorbed in the proximal convoluted tubules, actively reabsorbed in the loop of Henle by a "chloride pump,", and it is also reabsorbed with sodium in the distal tubules. This ability of the kidneys to vary daily chloride excretion keeps total body chloride values relatively constant and maintains serum chloride concentrations within a narrow range despite marked daily variations in chloride intake. The presence of specific clinical disorders can affect the ability of the kidneys to maintain chloride balance. The result is hyperchloremia (elevated serum chloride concentrations) or hypochloremia (reduced serum chloride concentrations). Changes in chloride should always be interpreted with changes in free water, which alters Na^+ and Cl^- concentrations proportionally. Serum chloride values are used as confirmatory tests to identify fluid balance and acid-base abnormalities [159]. Like sodium, a change in the serum chloride concentration does not necessarily reflect a change in total body content. Rather, it indicates an alteration in fluid status and/or acid-base balance.

23.5.4 Calcium

Calcium (Ca^{2+}) is the most abundant positively charged ion (cation) in the body and is a constituent of the principal mineral of the skeleton. It plays a pivotal role in the physiology and biochemistry of organisms and the cell. Calcium plays an important role in signal transduction pathways, where it acts as a second messenger, in neurotransmitter release from neurons, contraction of all muscle cell types, coagulation, cell growth, membrane transport mechanisms, and fertilization [160]. This cation also has a role in cardiac action potentials and pacemaker activity, and the contraction of cardiac, skeletal and smooth muscle, with implications for myocardial infarction and drug therapies. Many enzymes require calcium ions as a cofactor, those of the blood-clotting cascade being notable examples. Extracellular calcium is also important for maintaining the potential difference across excitable cell membranes, as well as proper bone formation [161]. Total serum calcium comprises three major forms: ionized calcium (about 50% of total), protein bound (about 40% of total), and calcium complexed with anions such as bicarbonate, citrate, lactate, and phosphate (about 10% of total). Most of the protein-bound calcium is bound to ALB. The ionized, or free, calcium is the metabolically active form of calcium. Approximately 40−50% of the plasma calcium is free or ionized, with the remaining plasma fraction bound to plasma proteins, mainly ALB. Acidosis increases plasma ionized calcium concentrations, whereas alkalosis causes a decrease due to the effects of pH in the ECF or on protein binding. Calcium is absorbed in the intestine with phosphate under the action of vitamin D. Corticosteroids inhibit absorption of calcium. It is stored in the body in bone and excreted through the kidneys. Renal excretion is influenced by parathyroid hormone, vitamin D, and calcitonin [162]. When evaluating calcium, it is important to relate total calcium to the quantity of ALB in the serum and the acid-base status of the animal. The total calcium concentration can increase in hyperalbuminaemia and decrease in hypoalbuminaemia. Acid-base changes alter the ratio of ionized to protein-bound calcium. Acidosis increases the ionized calcium fraction, whereas alkalosis increases the protein-bound fraction. Therefore total calcium, ALB and bicarbonate levels are important in evaluating calcium concentrations and related diseases.

23.5.5 Magnesium

Magnesium is the fourth most important cation in the body and the second most important intracellular cation. It occurs typically as the Mg^{2+} ion. Magnesium is an essential mineral nutrient in biological system and is present in every cell types in every organism. Magnesium affects many cellular functions. In addition to energy production and maintaining electrolyte balance, magnesium is essential for normal neuromuscular function as well as calcium and potassium transport [163,164]. Magnesium is essential for many enzyme reactions; it activates approximately 300 enzyme systems, including many involved in energy metabolism. It is essential for the production and functioning of ATP, which is fully functional only when chelated to magnesium. It acts as a cofactor for phosphorylation of ATPs from ADPs and is also vital for binding macromolecules to organelles (e.g., messenger ribonucleic acid (mRNA) to ribosomes) [157,164]. Other processes dependent on magnesium include the production of DNA, RNA, and protein synthesis. Magnesium is an essential regulator of calcium access into the cell and of the actions of calcium within the cell [165]. The human body contains 24 g magnesium of which 60% is present in bone. The remaining 40% is nearly equally distributed between skeletal muscle and other tissues like heart and liver. Approximately 1% of total body magnesium is extracellular and its concentration is equal to the magnesium concentration in the vascular compartment. The serum magnesium can be categorized into three fractions: protein-bound, ionized (the biologically active form) and complexed with anions such as phosphate, bicarbonate, and citrate [164]. Magnesium homeostasis is determined largely by the balance between intestinal absorption and renal excretion. Overall, very little is known about the factors that control magnesium homeostasis; indeed magnesium is referred to as the "forgotten" element. The kidneys play a pivotal role in controlling serum Mg^{2+} levels by modulating tubular reabsorption. Seventy percent to eighty percent of Mg is filtered through the glomerulus [166]. It is less commonly measured in toxicological studies, but there is a growing interest in its measurement [166].

23.5.6 Phosphate

Phosphorus is an abundant element that is widespread in its distribution. Phosphate (PO_4^{3-}) is a major intracellular anion in mammals with several functions [167]. Phosphate is widely distributed in the body throughout the plasma, extracellular fluid, cell membrane structures, intracellular fluid, collagen, and bone. Bone contains 85% of the phosphate in the body. About 90% of plasma phosphate is filtered at the glomeruli, and the majority is actively reabsorbed at the proximal tubule. The majority of intracellular PO_4^{3-} ion is either bound or exists as inorganic phosphate esters, phospholipids in cell membranes, or phosphorylated intermediate molecules. Cytosolic free PO_4^{3-} ion concentration is quite low, whereas mitochondrial PO_4^{3-} represents a large proportion of total cellular PO_4^{3-}, mainly in the form of calcium phosphate salts [168]. Total body phosphate is found mostly in bone (80−85%), with smaller amounts in muscle and the ECF (<1%). Serum phosphate consists of both organic and inorganic phosphate. Organic phosphate is present in phospholipids, phosphate esters, phosphoproteins, nucleic acids, etc. Inorganic phosphate is measured in assays, most of which is present as mono- and dihydrophosphate. Between 10% and 25% of phosphate is protein bound, and the rest is free or complexed to cations such as calcium, sodium and magnesium [169]. In biological systems, phosphate functions to store and release energy via high energy bonds in ATP and is integral to the structure of proteins, lipids, and bone. Phosphate is important for intracellular metabolism of proteins, lipids, and carbohydrates and it is a major component in phospholipid membranes, RNAs, nicotinamide diphosphate (an enzyme cofactor), cyclic adenine and guanine nucleotides (second messengers), and phosphoproteins. Factors favoring cellular uptake include glucose, fructose, alkalosis, insulin, β-adrenergic stimulation, and anabolism. Phosphate occurs in either organic or inorganic forms. Most of the intracellular phosphate is organic. Plasma contains lipid phosphates, organic ester phosphates, and inorganic phosphates, including divalent (HPO_4^{2-}) and monovalent ($H_2PO_4^{-}$) phosphate [157]. Many of the factors that influence serum calcium concentrations also affect serum phosphate, either directly or indirectly. Laboratory values for calcium and phosphate should, therefore, be interpreted together. The kidney remains the most significant organ regulating phosphate homeostasis. Therefore, significant variation of phosphate is considered as a result of kidney dysfunction [167].

23.5.7 Iron

Iron is a trace element that is essential for life, being required for important cell processes such as DNA synthesis, energy production, and defense. Many different structural and enzymatic proteins contain iron, which is essential to their function [170]. Iron is taken by food where it is found in a divalent (Fe^{2+}) or trivalent

(Fe^{3+}) form. In the intestinal lumen, it is reduced to a divalent form, as it can cross the cellular membrane in this form. In the intestinal epithelial cells, iron is transformed back to the trivalent form, and is transferred to plasma where it binds to the protein named transferrin (siderophylline). The transferrin bound iron ($7-28$ μmol/L) is normally transferred to the organs, where it is accepted by the protein named apoferritin, and iron is then stored in the form of ferritin and hemosiderin. A major part of iron stores are found in the liver, and some in the bone marrow and spleen. About $66-70\%$ of body iron is found in hemoglobin, 4% in myoglobin, and 30% in the storage form (e.g., hemosiderin and ferritin). A small portion of iron is found in the composition of some body compounds (e.g., flavoprotein, catalase) [171]. Iron is measured in serum or plasma, where it is most commonly used as a marker of iron status (deficiency or excess) and inflammation [172].

23.6 Effect of African Medicinal Plants on Biochemical Indices

A perusal of Table 23.2 shows that, in nine African countries, over 70 medicinal plants belonging to 43 families have been screened biochemically for their potential toxic effects. The types of toxicity studied include acute, subacute, chronic and subchronic toxicity as well as hepatotoxicity with mice and rats as experimental models. Two biochemical indices namely AST and ALT are the most commonly evaluated enzymes. AST and ALT have been assessed in all the toxicological study performed in African medicinal plants as reported in the present Chapter.

Nonenzymatic parameters mostly assessed are TP, BUN, TBL, and CRE. Others parameters studied include ALP, GGT, LDH, SDH,CPK, TP, ALB, GLO, TG, CHO, HDL, LDL, TBL, CBL, UBL, CRE, BUN, MDA, and GSH. The most evaluated electrolytes and minerals include Na^+, Cl^-, K^+, Ca^{2+}, and PO_4^{3-}.

Asteraceae, Fabaceae, and Euphorbiaceae were found to be the most represented families with seven and five plants species screened in each family respectively. Others families include: Apocynaceae, Lamiaceae, Meliaceae, Acanthaceae, Caryophyllaceae, Mimosaceae, Agavaceae, Anacardiaceae, Annonaceae, Asclepiadaceae, Aspodelaceae, Bignoniaseae, Capparaceae, Ceasalpiniaceae, Cecropiaceae, Combretaceae, Ebenaceae, Gentianaceae, Guttiferae, Hydnoraceae, Labiaceae, Lauraceae, Leeaceae, Malvaceae, Menispermaceae, Molluginaceae, Myrtaceae, Nyctaginaceae, Olacaceae, Papilionaceae, Ranunculaceae, Rosaceae, Rubiaceae, Rubiaceae, Rutaceae, Rutaceae, Solanaceae, and Zingiberaceae.

23.6.1 Asteraceae

The biochemical changes induced by seven plants of the Asteraceae family harvested in Africa are documented in Table 23.2. The studied plants include *Ageratum conyzoides, Artemisia annua, Aspilia africana, Bidens pilosa, Chromolaena odorata, Felicia muricata, Vernonia amygdalina*. The overall toxicological profile reveals that, plants species from this family are relatively safe. Some species such as *Artemisia annua* induced increase in AST and ALT levels. Members of the Asteraceae especially those from *Artemisia* genus produce a wide array of toxic compounds, one of the most famous is *Artemesia absinthium*, which produces thujone.

However, two species (*Ageratum conyzoides* and *Vernonia amygdalina*) appears to be safe, with no induced biochemical changes in expermental animals. Besides, *Vernonia amygdalina* further had hepatoprotective properties. Extracts of *Vernonia amygdalina* have been used traditionally as a tonic, and in the treatment of sexually transmitted diseases, feverish condition, cough, constipation, and hypertension. Nutritionally, *Vernonia* species grow in many parts of the world as a food vegetable and as a culinary herb and it contains phytochemical principles, which include bitter sesquiterpene lactones, vernolepin, vernodalin, vernomygdin, and steroid glucosides from which its biological properties were derived [173]. In regard to its extensive use in traditional medicine, in addition to its wide spectrum range of the biological activities, *Vernonia amygdalina* can be considered as a lead herbal drug candidate with good efficacy, fewer side effects and reduced toxicity and therefore deserve to be selected for clinical trials.

23.6.2 Fabaceae

Toxicity studies of plants from this family were performed in Nigeria and Cameroon. These included *Bauhinia monandra, Berlina grandiflora, Cylicodiscus gabunensis, Erythrina senegalensis, Faidherbia albida, Glycine max, Piptadeniastum africana*. The overall toxicological profile shows that, some plants species from Fabaceae family are potentially toxic. Significant biochemical changes were observed in animal studies. A part from Bauhinia monandra and Erythrina senegalensis, all other plant species induced increased in ALT, AST, and ALP levels. However, the levels of TBL and CBL were not affected. Fabaceae is the third largest family among the angiosperms after Orchidaceae and Asteraceae, consisting of more than 700 genera and about 20,000 species of trees, shrubs, vines, and herbs and is worldwide in distribution [174]. It is the second largest family of medicinal plants, containing over 490 medicinal plant species, most of which is being used as traditional medicine [175]. Plants of this family are found throughout the world, growing in many different environments and climates. A number are important agricultural and food plants such as beans, peas, and soybeans. Overall, the plants of the Fabaceae family range from being barely edible to barely poisonous. Some species do contain toxic alkaloids, especially in the seed coats. As deduced from the biochemical analysis depicted in Table 23.2, caution should be taken when using of some plants of this family for medicinal purposes.

23.6.3 Euphorbiaceae

The biochemical changes recorded as the results of animal's gavage with extracts of five plants of the family Euphorbiaceae, harvested in four African countries namely Cameroon, Ghana, Nigeria and South Africa were documented (Table 23.2).

The tested plants include *Croton membranaceus, Drypetes gossweileri, Hymenocardia acida, Jatropha curcas*, and *Phyllanthus muellerianus*. Apart from *Phyllanthus muellerianus* which was relatively toxic, others were not toxic. Most of the species of Euphorbiaceae are known to be toxic and poisonous plants due to their milky latex, having strong skin irritant activity, and chronic exposure can result carcinogenic effect. The toxic constituents of Euphorbiaceae species are specific diterpenes, called phorboids. These compounds (tigliane, ingenane and daphnane derivatives) possess extreme pro-inflammatory and tumor-promoting effects due to the activation of protein kinase C enzyme [176].

Interestingly, the present work reveals that many species in this family use in African traditional medicine appears to be safe. This observation was previously made by Hohmann and Molnár [176], who divided Euphorbiaceae diterpenes into three groups: (i) one group of diterpenes, such as most of phorbol and ingenol esters which can be considered exclusively as toxins without any possible medicinal use; (ii) the second group of diterpenes comprising compounds, which display toxicity, but which in adequate dose, have therapeutic perspective (e.g., the resiniferatoxin with capsaicine-like effect); and (iii) the third group of compounds such as diterpenes of nonphorboid type with macrocyclic or polycyclic structures which do not have toxic effect or this property is markedly reduced. These latter compounds may be promising lead compounds for natural product based drug developments.

One interesting plant species found to be nontoxic in this work is *Jatropha curcas*. Ethnomedicinal uses of this plant have been reported from many countries in Africa, Asia, South America, and the Middle East for almost 100 different types of ailments. The phytochemical studies have shown the presence of many secondary metabolites including diterpenoids, sesquiterpenoids, alkaloids, flavonoids, phenols, lignans, coumarins, and cyclic peptides. Crude extracts and isolated compounds from *Jatropha curcas* showed a wide range of pharmacological activities, such as anti-inflammatory, antioxidant, antimicrobial, antiviral, anticancer, antidiabetic, anticoagulant, hepatoprotective, analgesic, and abortifacient effects. It has been a source of medicine for decades in many cultures [177]. *Jatropha curcas* is a valuable source of medicinally important compounds; all these data provide plausible support for its entry into clinical trials for a phytomedicinal drug development.

23.6.4   Apocynaceae

The overall toxicological profile of the three plant species (*Carissa edulis, Tabernaemontana crassa, Rauvolfia vomitoria*) studied in this family is safe. The biochemical parameters assessed included AST, ALT, ALP, TP, TBL, CBL, CRE, GLU, BUN, GSH, MDA, Na^+, K^+, Cl^-, and Ca^{2+}. Although almost all the analyzed biochemical parameters varied, levels remained within acceptable limits. Nonetheless, it was demonstrated that ethanol etract of the stem bark of *Tabernaemontana crassa* was fairly nontoxic, but also that it should be taken with caution in therapeutic use, as higher dose were able to induce organs damage [10]. Such situation also appears as a result of a prolonged treatment.

23.6.5 Other Families

Several African medicinal plants belonging to many other families as depicted in Table 23.2 were also screened for their potential effects on biochemical indices of toxicological relevance. Among these, *Turraeanthus africanus* (Meliacaeae), *Spathodea campanulata* (Bignoniaseae), *Glinus lotoides* (Molluginaceae), *Fagara macrophylla* (Rutaceae), *Aloe ferox* Mill. (Aspodelaceae), *Ajuga iva* (Labiaceae) displayed low toxicities. No significant change was observed with the investigated biochemical indices. For other plants about 50 other species belonging to 36 families as shown in Table 23.2 exerted varied degrees of toxicity. It should be noted that, beside the study of single plants, herbal formulation made up with a mixture of many plants has also been screened for toxicity; as for most of the individual plant studied, slight harmful effect was observed.

23.7 Conclusion

This chapter attempted to provide information retrieved from literature on the biochemical parameters assessed in the study of the toxicological profile of African medicinal plants. The general feeling is that the assessment of biochemical parameters have been for a great importance in the establishment of the safety of many African plants species. The toxicological profile of about 70 medicinal plants belonging to 43 families have been documented on the basis of their effect on electrolytes, enzymes, and nonenzymatic markers levels.

It is well known that plants commonly used in traditional medicine are assumed to be safe because of their long history in the treatment of diseases according to knowledge accumulated over centuries, though scientifically, no xenobiotic can be classified as safe. The present work confirms that many plants used in traditional medicine are not completely safe, most are potentially toxic especially when used at high doses or over a long term. However, few number of African medicinal plants species were found to be potentially nontoxic. Among these safe plants, those with significant biological activities such as *Vernonia amygdalina* and *Jatropha curcas* are promising drug candidates and are recommended for clinical trials.

Despite the considerable analysis of biochemical parameters in the establishment of African medicinal plants profile, it is worth noting that number of important parameters have not yet been assessed in Africa. These include GDH, OCT which liver-specific enzyme, GST, PON, NAG, 5'NT, ODC, arginase, bile acids, and troponin. Therefore, more work needs to be done in particular on the previous mentioned under-assessed parameters. This could offer some of the most promising possibilities for increasing the sensitivity of currently available parameters and further aid in the detection of even minor toxicity. It would also be pretentious to think that the study of biochemical indices ensures completely the harmlessness of a plant. A very illustrative example in this chapter is that of a poisonous plant, *Datura stramonium* (see Chapter 22), known to have many side effects, but that biochemical analyzes as summarized in this chapter suggests a low toxicity.

Table 23.2 Effect of African Medicinal Plants on Biochemical Parameters

Plant Name (Family)	Doses (g/kg bw) and Studied Parts	Toxicity Assessed (Animals)	Area of Collection	Observed Effects
Acanthus montanus (Nees) T. Anderson (Acanthaceae)	0.5–8 g/kg bw; aqueous extract of the whole plant.	Acute (Wistar rat)	Cameroon	No toxic effects at therapeutic doses. AST[a], ALT[a], TP[a], CRE[b], CHO[b] [178].
	0.125–1 g/kg bw; aqueous extract of the whole plant.	Subacute (Wistar rat)	Cameroon	Nephrotoxic and hypercreatinemic: AST[a], ALT[b], GLU[a], CHO[a], serum protein[a], renal protein[a], hepatic protein[b] [179].
Ageratum conyzoides L. (Asteraceae)	5 g/kg bw (acute), 0.5–1 g/kg bw (subacute); hydroalcohol leaf extract	Acute and subchronic (Wistar rat)	Togo	Safe: AST[a], ALT[a], ALP[a], TP[a], CRE[a], BUN[a] [180].
Ajuga iva (L.) Schreber (Labiaceae)	1.5–5.5 g/kg bw (acute), 0.1–0.6 g/kg bw (chronic); aqueous extract of whole plant.	Acute and chronic (IOPS OFA mice and Wistar rat)	Morroco	Safe: AST[a], ALT[a], CHO[a], CRE[a] [181].
Aloe ferox Mill. (Aspodelaceae)	0.05–0.2; aqueous leaves extract.	7-days study (Rat)	South Africa	Safe: AST[a], ALT[a], ALP[a], GGT[a], TP[a], ALB[a], TBL[a], BUN[a], CRE[a], UA[a], K[+a], Ca[2+a] [182].
Anacardium occidentale Linn (Anacardiaceae)	6–26 g/kg bw (acute); 6–14 g/kg bw (chronic); hexane leaves extract.	Acute and subchronic (Mice)	Cameroon	Toxic at higher doses: AST[c], ALT[c], BUN[b], TP[c], CRE[b] [183].
Annona senegalensis Pers. (Annonaceae)	0.01–5 g/kg bw (acute), 0.05–0.4 g/kg bw (subacute); MeOH–CH$_2$Cl$_2$ (1:1) extract of bark roots.	Acute and subacute (Albino rat)	Nigeria	Caution in use at higher doses but safe at therapeutic doses: AST[a], ALT[a] [184].

Plant (Family)	Dose/Extract	Study	Country	Findings
Artemisia annua L. (Asteraceae)	0.1–0.3 g/kg bw; ethanol leaves extract	11 days study (Rat)	Nigeria	Relatively toxic: AST[b], ALT[b], CRE[a], TBL[c], CBL[c], TP[c], ALB[a], GLU[a], HDL[b], LDL[a], CHO[a], TG[a], Na[+a], Cl[−c], K[+c], HCO$_3$[−c] [185].
Aspilia africana (Pers.) C.D. Adams (Asteraceae)	2–16 g/kg bw (acute) 0.5 and 1 g/kg bw (subacute); aqueous leaves extract.	Acute and subacute (Swiss albino mice)	Cameroon	Low toxicity: AST[b], ALT[b], ALP[a], GSH[a], TP[a], CRE[a] [186].
Azadirachta indica (A. Juss) (Meliaceae)	0.05–0.3 g/kg bw; ethanol stem bark extract.	21-days study (Wistar rat)	Nigeria	Not completely safe: AST[c], ALT[a], ALP[b], GGT[c], CHO[a], HDL[b], LDL[b], TG[c], TP[a], ALB[a], GLO[b], TBL[b], CBL[b], Na[+c], K[+a], Ca^{2+b} [187].
Bauhinia monandra (Fabaceae)	1 and 2 g/kg bw; ethanol extract of leaves.	Hepatotoxicity (Rat)	Nigeria	Hepatoprotective effect: AST[c], ALT[c], TBL[a], CBL[a], TP[a], ALB[a] [188].
Berlina grandiflora. Hutch & Dalz (Fabaceae)	0.2–2 (acute), 0.125–0.5 (subacute); methanol extract of stem bark	Acute and subacute (Wistar rat)	Nigeria	Exert varied toxicological effects: AST[b], ALT[b], ALP[c], CPK[c], GGT[b], TP[a], ALB[a], TBL[a], BUN[a], TG[b], CHO[b], UA[a], Ca^{2+a} [189].
Bidens pilosa (Asteraceae)	0.6–10 g/kg bw (acute), 0.3–1.3 g/kg bw (28-day repeated dose); water extract of leaves	Acute and subacute (Albino rat)	Cameroon	Low toxicity: CHO[b], GLU[b], CRE[a], urine urea[a] [190].
Boerhavia diffusa (Nyctaginaceae)	0.5–2 g/kg bw (subchronic); aqueous leaves extract.	Acute and subchronic (Albino rat)	Nigeria	Not toxic: AST[a], ALT[a], ALP[a], TBL[a] [191].
Buchholzia coriacea (Capparaceae)	0.125–0.5 g/kg bw; methanol seeds extract.	Subacute (Rat)	Nigeria	Low toxicity at high doses: AST[a], ALT[a], TP[a], ALB[a], GLO[a], GLU[a], CHO[a], BUN[a], CRE[a], Na[+a], K[+a] [192].

(Continued)

Table 23.2 (Continued)

Plant Name (Family)	Doses (g/kg bw) and Studied Parts	Toxicity Assessed (Animals)	Area of Collection	Observed Effects
Canthium mannii Hiern (Rubiaceae)	2–16 g/kg bw (acute), 0.3–1.2 g/kg bw (subacute); ethanol stem barks extract.	Acute and subacute (Mice: *Mus musculus*)	Cameroon	Safe: AST[a], ALT[a] [193].
Carissa edulis (Apocynaceae)	0.01–2.9 g/kg bw and 5 g/kg bw (acute), 0.25–1 (subacute); standardized extract.	Acute and subacute (Rat)	Nigeria	No severe toxic effects: AST[b], ALT[b], ALP[b], TBL[b], CBL[b], CRE[a], Na[+a], K[+a], Cl[−a], Ca[2+ a] [194].
Centaurium erythraea (L.) Rafn. (Gentianaceae)	1–5 orally and 2–14 g/kg bw intraperitoneally (acute), 0.1–1.2 g/kg bw (subchronic); aqueous lyophylyzed extract of whole plant.	Acute and subchronic (IOPS OFA strain mice and Wistar rat).	Morocco	Wide margin of safety: AST[a], ALT[a], GLU[c], CHO[a], TG[c], CRE[a], BUN[a], TBL[a], TP[a] [195].
Chromolaena odorata (L.) King and Robinson (Asteraceae)	1–20 g/kg bw (acute) and 0.05–0.5 g/kg bw (subchronic); hydroethanol leaves extract.	Acute and subchronic (Swiss albino mice and Wistar rat).	Nigeria	Long-term use of high dose could have deleterious effect on the heart. ALT[c], AST[b], GLU[c], CHO, TG, HDL[b], LDL[c], CRE[b], TP[a] [196].
Chromolaena odorata (continued, Cameroon row)	4–20 g/kg bw (acute), 0.5 and 1 (subacute); aqueous extracts of leaves	Acute and subacute (Mice and rat)	Cameroon	Low toxicity: AST[b], ALT[b], ALP[b], GSH[a], TP[a], CRE[a] [197].
Corrigiola telephiifolia Pourr. (Caryophyllaceae)	5–15 g/kg bw (acute), 0.05–2 g/kg bw (40 day study); aqueous ethanol roots extract	Acute, 40 days study (Swiss albino mice and Wistar rat).	Morocco	Safe at the doses used ethnomedicinally: AST[a], ALT[a], ALP[b], GGT[b], TBL[a], TG[a], CHO[a], UA[a], CRE[b], BUN[a], GLU[a], TP[c], Na[+c], Cl[−c], K[+a], Ca[2+a], Mg[2+a], PO$_4^{3−b}$, HCO$_3^{−a}$, Fe[a] [198].

Species (Family)	Dose/Extract	Study	Country	Effects
Coula edulis Bail. (Olacaceae)	2–28 g/kg bw (acute), 0.025–0.2 g/kg bw (subacute); MeOH–CH$_2$Cl$_2$ (1:1) stem bark extract.	Acute and subacute (Swiss albino mice and Wistar rat).	Cameroon	Hepatotoxic and nephrotoxic at high doses. Serum: AST[b], ALT[b], TP[b], GLU[a], CHO[c], CRE[b]. Liver: AST[a], ALT[a], TP[b] [199].
Croton membranaceus Müll.Arg. (Euphorbiaceae)	1.5–3 g/kg bw; ethanol roots extract.	Acute (S−D rat)	Ghana	No general acute toxicity: AST[a], ALT[a], ALP[a], CBL[a], GGT[b], LDH[a], CPK[a], BUN[a], total CRE[c], brain/muscle CRE[c] [12].
Cylicodiscus gabunensis (Fabaceae)	4–16 g/kg bw (acute), 0.75–6 g/kg bw (subacute); ethyl acetate extract of stem bark	Acute and subacute (Albino Wistar rat)	Cameroon	Varied toxicological effects. Serum: AST[b], ALT[b], ALP[a], TP[a], CHO[b], GLU[b], BUN[a], CRE[a]. Liver: TP[a], MDA[c], GSH[a]. Kidney: urea[c], CRE[c] [200].
Datura stramonium (Solanaceae)	0.05–0.2 g/kg bw; ethanol extract of the leaves	5 weeks study (Rat)	Nigeria	Complete safety not established: AST[a], ALT[a], TBL[a], CRE[b], Na^{+a}, Cl^{-a}, K^{+a}, HCO$_3$$^{-a}$ [201].
Diospyros canaliculata (De Wildeman) (Ebenaceae).	0.5–2 g/kg bw; methanol extract of stem bark.	4-week repeated oral dose (Wistar albino rat, *Rattus norvegicus*)	Cameroon	Slightly toxic: AST[b], ALT[a], ALP[b], TBL[a], CBL[a], GLU[a], CHO[a], TP[b], CRE[a], BUN[b], MDA[a], GSH[a] [9].
Drypetes gossweileri (Euphorbiaceae)	4–12 g/kg bw (acute), 0.5 and 1 g/kg bw (subacute); MeOH–CH$_2$Cl$_2$ (1:1) extract of stem bark.	Acute and subacute (Albino Wistar rat)	Cameroon	Safe: AST[a], ALT[a], CRE[a], CHO[a], CBL[a], TBL[a], TP[a] [202].

(Continued)

Table 23.2 (Continued)

Plant Name (Family)	Doses (g/kg bw) and Studied Parts	Toxicity Assessed (Animals)	Area of Collection	Observed Effects
Erythrina senegalensis DC (Fabaceae)	1.25–12.5 g/kg bw (acute), 0.3–1.2 g/kg bw (subchronic); aqueous stem bark extract.	Acute and subchronic (Mice and rat)	Cameroon	Wide margin of safety: AST^c, ALT^c, ALP^c, GLU^a, BUN^c, TP^c, ALB^a, TBL^a, CBL^a, CHO^c, LDL, HDL^a, TG^c, CRE^c, UA^c, Na^{+a}, Cl^{-c}, K^{+a}, Ca^{2+c}, Mg^{2+a}, PO_4^{3-c} [203].
Fagara macrophylla (Rutaceae)	0.6–10 g/kg bw (acute), 0.3–1.3 g/kg bw (28-day repeated dose); water extract of stem bark and root bark.	Acute and subacute (Albino rat)	Cameroon	Safe; CHO^a GLU^b, CRE^a, urine urea^a [190].
Faidherbia albida (DEL) A. chev. (Fabaceae)	0.01–1 g/kg bw then 1.6–5 g/kg bw (acute), 0.25–0.5 g/kg bw (subacute); ethanol extract of the stem bark.	Acute and subacute (Rat)	Nigeria	Relatively safe: AST^b, ALT^b, ALP^b, TBL^a, CBL^a [204].
Felicia muricata Thunb. (Asteraceae)	0.05–2 g/kg bw; aqueous extract of leaves	14-days study (Wistar rat)	South Africa	Selective toxicity: AST^a, ALT^c, ALP, GGT^a, TP^a GLO^a, ALB^a, TBL^a, CBL^c TG^b, CHO^a HDL^a, LDL, CRE^a, BUN^a, Na^{+a}, K^{+a}, Cl^{-a}, Ca^{2+a}, PO_4^{3-a} [205].
Glinus lotoides L. (Molluginaceae)	1 and 5 g/kg bw (single dose), 0.25–1 g/kg bw (28 days repeated study); MeOH seeds extract	Acute and subacute (Albino rat)	Ethiopia	Safe. Male: AST^a, ALT^a, GGT^a, BUN^a, CRE^a, GLU^a, TP^a ALB^a, CHO^a, Ca^{2+a}. Female: AST^c, ALT^a, GGT^a, BUN^a, CRE^a, GLU^a, TP^a, ALB^a, CHO^b, Ca^{2+a} [206].

Plant (Family)	Dose; extract	Study type	Country	Findings
Glycine max (L.) Merr. (Fabaceae)	0.25–0.5 g/kg bw; oil emulsion	Subchronic (Rat)	Nigeria	Caution when used for a long period: Plasma: AST[b], ALT[b], LDH[a], TP[a], CHO[a], TG[b]. Liver: AST[a], ALT[a], LDH[b], TP[c], GSH[a] [207].
Gongronema latifolium (Asclepiadaceae)	0.2–0.4 g/kg bw; aqueous extract of leaves.	Hepatotoxicity (Wistar albino rat)	Nigeria	Offers protection against drug toxicity: AST[b], ALT[b], ALP[b], TP[b].
Herniaria glabra (Caryophyllaceae)	2.5–14.5 g/kg bw; (acute) 0 (control); 1, 2 and 4 (subacute); aqueous extract of the whole plant.	Acute and subchronic (IOPS OFA mice and Wistar rat).	Morocco	No significant toxicity (except at high doses), wide margin of safety: AST[b], ALT[b], GLU[c], CRE[b] [11].
Hydnora johannis Becca. (Hydnoraceae)	0.05–0.4 g/kg bw; ethanol roots extract.	14-days study (Rat)	Sudan	Toxic: AST[b], ALT[a], ALP[b], CHO[a], BUN[a] [208].
Hymenocardia acida (Tul.) (Euphorbiaceae)	0.1–0.8 g/kg bw; aqueous ethanolic extract of stem bark.	28 days repeated dose administration (Rat)	Nigeria	Relatively safe: AST[a], ALT[b], ALP[a], BUN[a], CRE[a], TP[a], GLU[a], ALB[a], CHO[a], TG[a], CRE[a], GLU[a], TG[a], TP[a], TBL[a], ALB[a], GLO[a], CHO[a] [209].
Jateorhiza macrantha (Hook.f) Exell & Mendonça (Menispermaceae)	2.5 and 5 g/kg bw (acute), 0.15–0.6 g/kg bw (subacute); aqueous leaves extract	Acute and subacute (Albino mice *Mus musculus* Swiss)	Cameroon	Low toxicity: AST[b], ALT[b], CRE[a], CHO[b], HDL[a], LDL[b], TG[b] [210].
Jatropha curcas (Linn) (Euphorbiaceae)	0.5–2 g/kg bw; methanol leaves extract.	21-days study (Rat)	South Africa	Not toxic: AST[a], ALT[a], ALP[a], GGT[a], ALB[a], TBL[a], TP[a], BUN[a], GLU[a], CRE[a], Na[+a], Cl[−a], K[+a], Ca[2+a] [211].
Khaya senegalensis (Desv) A. Juss (Meliaceae)	0.01–0.04 g/kg bw; aqueous stem barks extract.	Hepatotoxicity and subchronic (Albino rat)	Nigeria	Toxic potential: AST[b], ALT[b], ALP[b], TBL and TP[c] [212].

(Continued)

Table 23.2 (Continued)

Plant Name (Family)	Doses (g/kg bw) and Studied Parts	Toxicity Assessed (Animals)	Area of Collection	Observed Effects
Leea guineensis (Leeaceae)	0.5–1 g/kg bw; aqueous ethanol extract of leaves.	Subacute (Albino rat)	Cameroon	Relatively safe: Blood: AST[a], ALT[a], ALP[a], TP[a], CRE[a]. Liver: AST[b], ALT[b], ALP[b], TP[a] [213].
Leonotis leonurus (L.) R.Br. (Lamiaceae)	0.2–3.2 g/kg bw (acute); 0.4–1.6 g/kg bw (subchronic); 0.2 and 0.4 g/kg bw (chronic); aqueous shoots extract.	Acute, subacute and chronic (Rat)	South Africa	Exercise caution is needed in the use: AST[c], ALT[c], ALP[a], GGT[c], GLU[c], CRE[c], BUN[c], TBL[c], CBL[a], UBL[a], ALB[c], TP[c], GLO[c], Na^{+a}, Cl^{-a}, K^{+a}, PO$_4^{3-a}$ [214].
Mammea africana Sabine (Guttiferae)	0.019–0.3 g/kg bw; MeOH–CH$_2$Cl$_2$ (1:1) extract of stem bark.	Acute and subacute (Rat)	Cameroon	Improved the metabolic alterations: AST[a], ALT[a], CRE[a], TP[a], CHO[a] [215].
Morinda lucida Benth. (Rubiaceae)	2–20 g/kg bw (acute), 0.1 and 1 g/kg bw (subacute); aqueous extract of stem bark.	Acute and subacute (Mice and Wistar albino rat)	Cameroon	Well tolerated at low doses but toxic at high doses: AST[b], ALT[b], CHO[a], TG[b], GLU[a], CRE[b], BUN[a] [216].
Murraya koenigii (L.) Spreng. (Rutaceae)	0.2–2 g/kg bw (acute), 0.25–0.45 g/kg bw (subchronic); methanol leaves extract	Acute and subchronic (Swiss albino rat)	Nigeria	Moderately toxic: AST[c], ALT[c], ALP[c], ALB[c], GLU[c], BUN[c], TP[c], GLO[c], BL[c], CHO[c] [217].
Musanga cecropioides (Cecropiaceae)	3 g/kg bw (acute), 0.75 g/kg bw (subacute); aqueous stem bark extract.	Acute and subacute (Rat)	Nigeria	Relatively safe: AST[a], ALT[a], ALP[a], TBL[a], CBL[a], TP[a], ALB[a], CHO[a], BUN[a], CRE[b], Na^{+a}, K^{+a}, HCO$_3^{-a}$ [218].

Plant (species/family)	Dose	Study type	Country	Findings
Nigella sativa L. (Ranunculaceae)	0.01–0.05 mL/kg bw (acute), 2 mL/kg bw (chronic); fixed oil from seeds.	Acute and chronic (IOPS OFA mice and Wistar kyoto rat)	Morroco	Wide margin of safety: AST[a], ALT[a], ALP[a], TBL[a], TG[c], CHO[c], HDL[c], UA[a], CRE[a], GLU[c] [219].
Ocimum gratissimum Linn. (Lamiaceae)	0.66–2.62 and 1.07–3.73 mL/kg of bw (acute), 0.08–0.213 mL/kg of bw (subchronic); essential oil from leaves.	Acute and subchronic (Swiss albino mice)	Nigeria	Toxic: AST[c], ALT[c], TP[c] [13].
Ocimum suave Wild (Lamiaceae)	2–8 g/kg bw (acute), 0.25–1 g/kg bw (subacute and teratogenic assay); leaves aqueous extract	Acute, subchronic and teratogenic (Rat)	Cameroon	Nontoxic: AST[a], ALT[a], TP[a], CRE[a], BUN[a] [220].
Persea americana Mill (Lauraceae)	2–10 (acute), 2.5 (subacute); aqueous seedsextract.	Acute and subacute (Rat)	Nigeria	Safe at low dose: AST[a], ALT[a], CRE[a], TP[b], ALB[a] [221].
Phyllanthus muellerianus (O. Ktze) Exel. (Euphorbiaceae)	4–20 g/kg bw; methanol extract of stem barks.	Acute (Wistar albino rat)	Cameroon	Relatively toxic: Serum: AST[b], ALT[b], CRE[a], BUN[a], Liver: AST[c], ALT[c], MD[b], GSH[c] [222].
Piptadeniastum africana (Hook. f.) Bren. (Fabaceae)	4–20 g/kg bw; methanol extract of leaves.	Acute (Wistar albino rat)	Cameroon	Relatively toxic. Serum: AST[b], ALT[b], CRE[a], BUN[a], Liver: AST[c], ALT[c], MD[b], GSH[c] [222].
Prunus Africana (Hook f.) (Rosaceae)	0.01–2 g/kg bw; chloroform extract of stem Bark	8 weeks study (Rat)	Kenya	Over toxicity in multiple doses of 3.3 g/kg bw: AST[b], ALT[b], ALP[b], LDH[b], CPK[b], BUN[b] [223].
Pteleopsis hylodendron Mildbr. (Combretaceae)	2–8 g/kg bw (acute), 0.028–0.68 g/kg bw (subacute); methanol stem bark extract.	Acute and subacute (Swiss albino mice and Wistar rat).	Cameroon	Hepatotoxic and nephrotoxic: AST, ALT[b], TP[b], CRE[b], TG[a], CHO[b], HDL[b], LDL[b] [224].

(Continued)

Table 23.2 (Continued)

Plant Name (Family)	Doses (g/kg bw) and Studied Parts	Toxicity Assessed (Animals)	Area of Collection	Observed Effects
Pterocarpus soyauxii Taub (Papilionaceae)	2.5–12.5 g/kg bw (acute) 0.15–0.6 g/kg bw (subchronic); aqueous stem bark extract.	Acute and subchronic (BALB/c mice and Wistar rat)	Cameroon	Very low toxicity in oral acute and no toxicity in oral subchronic. Serum: AST[a], ALT[a], CRE[a], TP[a], GSH[a]. Liver: TP[a], GSH[a], MDA[a], catalase[b]. Kidney: TP[a], GSH[a], MDA[a], catalase[a] [225].
Rauvolfia vomitoria (Apocynaceae)	0.6–10 g/kg bw (acute), 0.3–1.3 g/kg bw (28-day repeated dose); water extract of leaves	Acute and subacute (Albino rat)	Cameroon	Relatively safe: CHO[b], CRE[a], urine urea[a] [190].
Sanseviera liberica (Agavaceae)	0.5–2 g/kg bw orally and 0.1–1.6 g/kg bw intraperitoneally (acute), 0.08–2 (and 52-days subchronic; aqueous roots extract.	Acute and 52-days subchronic oral toxicity (Albino mice)	Nigeria	Relatively safe: AST[a], ALT[a], ALP[b], UA[b] [226].
Senna alata (L.) Roxb. (Ceasalpiniaceae)	0.04–1 g/kg bw (acute), 0.5–1 g/kg bw (subacute); aqueous ethanol extract of leaves.	Acute and subacute (Wistar albino rat and Swiss albino mice)	Cameroon	Relatively safe: Blood: AST[a], ALT[a], ALP[a], TP[a], CRE[a], CHO[a]. Liver: AST[b], ALT[b], ALP[b], TP[a], GSH[b] [227].
Sida rhombifolia Linn. (Malvaceae)	4–16 g/kg bw; aqueous-methanol extract	Acute (Albino Wistar rat)	Cameroon	Toxic effect: ALT[a], AST[b], ALP[b], CRE[b], TP[b], GSH[b] [228].
Spathodea campanulata P. Beauv (Bignoniaseae)	1–5 g/kg bw (acute0, 0.75–3 g/kg bw (subchronic); aqueous ethanol leaves extract.	Acute and subchronic (Dawley rat)	Nigeria	Safe: AST[a], ALT[a], ALP[a] [229].

Plant (Family)	Dose; extract	Model (animal)	Country	Findings
Syzigium aromaticum (L.) Merr. & Perry (Myrtaceae)	0.1–0.52 g/kg bw intraperitoneally and 0.5–5 orally (acute), 0.3 and 0.7 g/kg bw (subacute); aqueous buds extract.	Acute and subchronic (Swiss albino mice and Wistar rat)	Nigeria	Long-term use could be hazardous: AST[c], ALT[c], ALP[c], TP[c], ALB[a], CHO[a], CBL[a], BUN[b], CRE[a], Na[+c], Cl[−c], K[+c], HCO3[−c] [230].
Tabernaemontana crassa (Apocynaceae)	2–8 (acute), 0.1–1 (subacute); hydroethanol extract of stem bark.	Acute and subacute (Albino Wistar rat)	Cameroon	Fairly toxic: Blood: AST[a], ALT[b], ALP[a], TP[a], TBL[a], CBL[a], CRE[a], GLU[a], BUN[a]. Liver: AST[a], ALT[b], ALP[b], TP[a], GSH[a], MDA[b] [10].
Turraeanthus africanus (Meliaceae)	5–30 orally 3–12 intraperitoneally (acute), 1.5–6 (subacute); aqueous stem barks extract.	Acute and subacute (Mice)	Cameroon	Not toxic orally, parenteral use should be avoided. AST[a], ALT[a], GGT[a], TG[c], CHO[c], HDL[c], LDL[c], TP[b], CRE[a] tissue creatinine[b] [231].
Vernonia amygdalina Del (Asteraceae)	0.5–2.5 mL/kg bw; aqueous extract of leaves	Hepatotoxicity (Rat)	Nigeria	Potent anti-hepatotoxic action: AST[a], ALT[a], ALP[a] [232].
Zingiber officinale (Zingiberaceae)	0.01–0.0312 mL/kg of bw; ethanol extract of rhizomes.	Hepatotoxicity(Rat)	Nigeria	Protective effect in hepatotoxicity: AST[b], ALT[b], ALP[b], LDH[b], SDH[b], GDH[b], CBL[b] UA[b] [233].
	2–7 (acute), 0.6–1.8 (subacute); essential oil	Acute and subacute (Swiss albino mice and Wistar rat)	Cameroon	Should be consumed at a dose less than 0.6 g/kg bw: AST[b], ALT[b], TP[b], CRE[b] [234].
Mixture: Alstonia congensis Engler (Apocynaceae) and Xylopia aethiopica (Dunal) A. Rich (Annonaceae)	1–20 (acute), 0.05–0.5 (subacute); herbal formulation prepared with Alstonia congensis bark and Xylopia aethiopica fruits in equal proportion.	Acute and subacute (Wistar albino rat and Swiss albino mice)	Nigeria	Subacute study cause kidney problems on a long-term use: AST[a], ALT[b], GLU[c], TP[a], TG[c], CHO[a], HDL[b], LDL[c], CRE[b] [235].

(Continued)

Table 23.2 (Continued)

Plant Name (Family)	Doses (g/kg bw) and Studied Parts	Toxicity Assessed (Animals)	Area of Collection	Observed Effects
Mixture: *Maytenus senegalensis* (Celastraceae), *Annona senegalensis* (Annonaceae), *Kigelia africana* (Bignoniaceae) and *Lannea welwitschii* Anacardiaceae).	0.1–0.5 g/kg bw; herbal formulation prepared with bark of *Maytenus senegalensis*, root of *Annona senegalensis*, fruit of *Kigelia africana* and bark of *Lannea welwitschii*.	Subchronic (Wistar albino rat)	Ghana	No overt organ-specific toxicity: Blood: AST[a], ALT[a], ALP[a], CPK[a], BUN[a], CRE[a], ALB[a]. Urine: nitrite[a], proteins[a], GLU[a], TBL[a], ketones[a], uribilinogen[a] [236].
Mixture: *Aloe buettneri* (Liliaceae), *Dicliptera verticillata* Acanthaceae), *Hibiscus macranthus* (Malvaceae) and *Justicia insularis* (Acanthaceae	2–32 g/kg bw (acute), 0.0125–0.1 g/kg bw (subacute); aqueous extract of leaves mixture.	Acute and subacute (Wistar albino rat and Swiss albino mice)	Cameroon	Might be harmful in the long used. AST[a], ALT[c], CRE[a] [237].

ALB, albumin; ALP, alkaline phosphatase; ALT, alanine aminotransferase; AST, aspartate aminotransferase; BUN, blood urea; CBL, conjugated bilirubin (direct bilirubin); ChE, cholinesterase; CHO, cholesterol; CPK, creatine phosphokinase; CRE, creatinine; GDH, glutamate dehydrogenase; GGT, γ-glutamyltransferase; GLO, globulin; GLU, glucose; GSH, glutathione; HDL, high density lipoprotein; IDH, isocitrate dehydrogenase; LDH, lactate dehydrogenase; LDL, low density lipoprotein; LOAEL, lowest observed adverse effect levels; MDA, malondialdehyde; NOAEL, no-observed-adverse-effect level; TBL, total bilirubin; TG, triglyceride; TP, total protein; UA, uric acid; UBL, unconjugated bilirubin (indirect bilirubin).

The toxicity of extracts of plants is obviously associated with the dose. If the dosage can be accurately determined, a dose that would have therapeutic value without damage to the organism could be used even with toxic plant species. An important parameter to investigate is therefore the therapeutic index. If a plant extract with relatively low activity but a very low toxcicity, it may still become a useful product.

Furthermore, the extractant used is extremely important because a plant used for centuries as an aqueous extract may be toxic if an organic extract is used. It is clear that widely different compounds are extracted by different extractants [238].

In investigating the toxicity of plant extracts, much work has been done on cellular toxicity (see Chapter 8) and some work on genotoxicicty [239]. This contribution shows the importance of looking at biochemical parameters in determining toxicity and to obtain information on the possible site of toxicity.

Like all pharmaceuticals, the safety of medicinal plants in acute and chronic animal models should be determined. It is only after these studies that their use could be recommended and this includes the possible side effects, and their notices must also be revised after several years as a result of clinical observations. Also, a considerable effort has to be done in African countries, to systematically screen the toxicity of medicinal plants. If African traditional medicine hopes to have a reputation comparable to that of Western herbal medicine, it is therefore imperative that all efforts be combined between governments, herbalists and traditional healers, and researchers to find plants that are both active and safe.

References

[1] Iwu MM. Handbook of African medicinal plants. Boca Raton, FL: CRC Press; 1993.
[2] World Health Organisation. Traditional medicine, fact sheet no 134. Geneva: WHO; 2003.
[3] Weingand K, Brown G, Hall R, Davies D, Gossett K, Neptun D, et al. Harmonization of animal clinical pathology testing in toxicity and safety studies. Fundam Appl Toxicol 1996;29:198−201.
[4] Stewart MJ, Steenkamp V, Zuckerman M. The toxicology of African herbal remedies. Ther Drug Monit 1998;20:510−6.
[5] Marshall WJ. Illustrated textbook of clinical chemistry. 2nd ed. London: Gower Medical Publishing; 1992.
[6] Fennell CW, Lindsey KL, McGaw LJ, Sparg SG, Stafford GI, Elgorashi EE, et al. Assessing African medicinal plants for efficacy and safety: pharmacological screening and toxicology. J Ethnopharmacol 2004;94:205−17.
[7] Bodenstein JW. Toxicity of traditional herbal remedies. SAMJ 1977;2:790.
[8] Kothari SC, Shivarudraiah P, Babu Venkataramaiah S, Gavara S, Soni MG. Subchronic toxicity and mutagenicity/genotoxicity studies of irvingia gabonensis extract (IGOB131). Food Chem Toxicol 2012;50:1468−79.
[9] Dzoyem JP, Nkegoum B, Kuete V. A 4-week repeated oral dose toxicity study of the methanol extract from *Diospyros canaliculata* in rats. Comp Clin Pathol 2013;22:75−81.

[10] Kuete V, Manfouo RN, Beng VP. Toxicological evaluation of the hydroethanol extract of
      Tabernaemontana crassa (Aspocynaceae) stem bark. J Ethnopharmacol 2010;130:470–6.
[11] Rhiouani H, El-Hilaly J, Israili ZH, Lyoussi B. Acute and sub-chronic toxicity of an
      aqueous extract of the leaves of *Herniaria glabra* in rodents. J Ethnopharmacol
      2008;118:378–86.
[12] Asare GA, Sittie A, Bugyei K, Gyan BA, Adjei S, Addo P, et al. Acute toxicity studies
      of *Croton membranaceus* root extract. J Ethnopharmacol 2011;134:938–43.
[13] Orafidiya LO, Agbani EO, Iwalewa EO, Adelusola KA, Oyedapo OO. Studies on the
      acute and sub-chronic toxicity of the essential oil of *Ocimum gratissimum* L. leaf.
      Phytomedicine 2004;11:71–6.
[14] Woodman DD. Assessment of hepathotoxicity. In: Evans GO, editor. Animal clinical
      chemistry: a primer for toxicologists. London: Taylor & Francis; 1996. p. 66–82.
[15] Waner T, Nyska A. the toxicological significance of decreased activities of blood ala-
      nine and aspartate-aminotransferase. Vet Res Commun 1991;15:73–8.
[16] Vroon DH, Israili Z. Aminotransferases. In: Walker HK, Hall WD, Hurst JW, editors.
      Clinical methods: the history, physical, and laboratory examinations. 3rd ed. Boston,
      MA: Butterworth-Heinemann; 1990. p. 492–3.
[17] Jadhao SB, Yang R, Lin Q, Hu H, Anania FA, Shuldiner AR, et al. Murine alanine
      aminotransferase: CDNA cloning, functional expression, and differential gene regula-
      tion in mouse fatty liver. Hepatology 2004;39:1297–302.
[18] Yang RZ, Blaileanu G, Hansen BC, Shuldiner AR, Gong DW. cDNA cloning, genomic
      structure, chromosomal mapping, and functional expression of a novel human alanine
      aminotransferase. Genomics 2002;79:445–50.
[19] Lindblom P, Rafter I, Copley C, Andersson U, Hedberg JJ, Berg A, et al. Isoforms of
      alanine aminotransferases in human tissues and serum: differential tissue expression
      using novel antibodies. Arch Biochem Biophys 2007;466:66–77.
[20] Sakagishi Y. Alanine aminotransferase (ALT). *Nippon rinsho.* Jpn J Clin Med
      1995;53:1146–50.
[21] Yang R, Park S, Reagan WJ, Goldstein R, Zhong S, Lawton M, et al. Alanine amino-
      transferase isoenzymes: Molecular cloning and quantitative analysis of tissue expres-
      sion in rats and serum elevation in liver toxicity. Hepatology 2009;49:598–607.
[22] Hayashi H, Wada H, Yoshimura T, Esaki N, Soda K. Recent topics in pyridoxal 5′-
      phosphate enzyme studies. Annu Rev Biochem 1990;59:87–110.
[23] Muriana FJG, Alvarezossorio MC, Relimpio AM. Purification and characterization of
      aspartate-aminotransferase from the halophile archaebacterium *Haloferax mediterranei.*
      Biochem J 1991;278:149–54.
[24] Okamoto A, Kato R, Masui R, Yamagishi A, Oshima T, Kuramitsu S. An aspartate
      aminotransferase from an extremely thermophilic bacterium, *Thermus thermophilus*
      HB8. J Biochem 1996;119:135–44.
[25] Rej R. Aminotransferases in disease. Clin Lab Med 1989;9:667–87.
[26] Herlong HF. Approach to the patient with abnormal liver-enzymes. Hosp Pract
      1994;29:32–8.
[27] Singh A, Bhat TK, Sharma OP. Clinical biochemistry of hepatotoxicity. J Clinic
      Toxicol 2011;S4:001.
[28] Pratt DS. Liver chemistry and function tests. In: Feldman M, Friedman LS, Brandt LJ,
      editors. Sleisenger and Fordtran's gastrointestinal and liver disease. Philadelphia, PA:
      Saunders Elsevier; 2010.
[29] Rei R. Measurement of aminotransferase: Part I. asparate aminotransferase. CRC Crit
      Rev Clin Lab Sci 1984;21:99–186.

[30] Huang X, Choi Y, Im H, Yarimaga O, Yoon E, Kim H. Aspartate aminotransferase (AST/GOT) and alanine aminotransferase (ALT/GPT) detection techniques. Sensors 2006;6:756−82.

[31] Berg JM, Tymoczko JL, Stryerand L. Biochemistry. 5th ed. New York, NY: W.H. Freeman and Company; 2002.

[32] Baker PJ, Waugh ML, Wang XG, Stillman TJ, Turnbull AP, Engel PC, et al. Determinants of substrate specificity in the superfamily of amino acid dehydrogenases. Biochemistry 1997;36:16109−15.

[33] Watanabe M, Yumi O, Itoh Y, Yasuda K, Kamachi K, Ratcliffe RG. Deamination role of inducible glutamate dehydrogenase isoenzyme 7 in *Brassica napus* leaf protoplasts. Phytochemistry 2011;72:587−93.

[34] Duncan PA, White BA, Mackie RI. Purification and properties of nadp-dependent glutamate-dehydrogenase from *Ruminococcus flavefaciens* FD-1. Appl Environ Microbiol 1992;58:4032−7.

[35] Maulik P, Ghosh S. NADPH/NADH-dependent cold-labile glutamate-dehydrogenase in *Azospirillum brasilense*: purification and properties. Eur J Biochem 1986;155:595−602.

[36] Coulton JW, Kapoor M. Studies on kinetics and regulation of glutamate dehydrogenase of *Salmonella typhimurium*. Can J Microbiol 1973;19:439−50.

[37] Botton B, Msatef Y. Purification and properties of NADP-dependent glutamate-dehydrogenase from *Sphaerostilbe repens*. Physiol Plantarum 1983;59:438−44.

[38] O'Brien PJ, Slaughter MR, Polley SR, Kramer K. Advantages of glutamate dehydrogenase as a blood biomarker of acute hepatic injury in rats. Lab Anim 2002;36:313−21.

[39] El-Kabbani O, Darmanin C, Chung RPT. Sorbitol dehydrogenase: structure, function and ligand design. Curr Med Chem 2004;11:465−76.

[40] Khayrollah AA, Altamer YY, Taka M, Skursky L. Serum alcohol-dehydrogenase activity in liver-diseases. Ann Clin Biochem 1982;19:35−42.

[41] Dooley JF, Turnquist LJ, Racich L. Kinetic determination of serum sorbitol dehydrogenase-activity with a centrifugal analyzer. Clin Chem 1979;25:2026−9.

[42] Ozer J, Ratner M, Shaw M, Bailey W, Schomaker S. The current state of serum biomarkers of hepatotoxicity. Toxicology 2008;245:194−205.

[43] Evans GG. Animal clinical chemistry: a primer for toxicologists. London: Taylor & Francis; 1996.

[44] Ramaiah SK. A toxicologist guide to the diagnostic interpretation of hepatic biochemical parameters. Food Chem Toxicol 2007;45:1551−7.

[45] Wright TM, Vandenberg AM. Risperidone-and quetiapine-induced cholestasis. Ann Pharmacother 2007;41:1518−23.

[46] Goldberg DM. Structural, functional, and clinical aspects of gamma-glutamyltransferase. CRC Crit Rev Clin Lab Sci 1980;12:1−58.

[47] Tate SS, Meister A. Gamma-glutamyl-transferase transpeptidase from kidney. Meth Enzymol 1985;113:400−19.

[48] Albert Z, Szewczuk A, Orlowski M, Orlowska J. Histochemical and biochemical investigations of gamma-glutamyl transpeptidase in the tissues of man and laboratory rodents. Acta Histochem 1964;18:78−89.

[49] Sheehan M, Haythorn P. Predictive values of various liver-function tests with respect to the diagnosis of liver-disease. Clin Biochem 1979;12:262−3.

[50] Leonard TB, Neptun DA, Popp JA. Serum gamma glutamyl transferase as a specific indicator of bile-duct lesions in the rat-liver. Am J Pathol 1984;116:262−9.

[51] Cabaniss CD. Creatine kinase. In: Walker HK, Hall WD, Hurst JW, editors. Clinical methods: the history, physical, and laboratory examinations. 3rd ed. Boston, MA: Butterworths; 1990. p. 161−3.

[52] Goldblat. H. Effect of high salt intake on blood pressure of rabbits. Lab Invest 1969;21:126−8.

[53] Braun S. Isoformen der creatinkinase-isoenzyme. Klinische Labor 1992;38:549−54.

[54] Mcbride JH, Rodgerson DO, Hilborne LH. Human, rabbit, bovine, and porcine creatine-kinase isoenzymes are glycoproteins. J Clin Lab Anal 1990;4:196−8.

[55] Wallimann T, Wyss M, Brdiczka D, Nicolay K, Eppenberger HM. Intracellular compartmentation, structure and function of creatine-kinase isoenzymes in tissues with high and fluctuating energy demands—the phosphocreatine circuit for cellular-energy homeostasis. Biochem J 1992;281:21−40.

[56] Dawson DM, Kaplan NO, Goodfriend TL. Lactic dehydrogenases—functions of 2 types—rates of synthesis of 2 major forms can be correlated with metabolic differentiation. Science 1964;143:929−33.

[57] Li SSL. Lactate-dehydrogenase isoenzyme-A (muscle), isoenzyme-B (heart) and isoenzyme-C (testis) of mammals and the genes-coding for these enzymes. Biochem Soc Trans 1989;17:304−7.

[58] Milne EM, Doxey DL. Lactate-dehydrogenase and its isoenzymes in the tissues and sera of clinically normal dogs. Res Vet Sci 1987;43:222−4.

[59] Evans GO. Clinical pathology testing recommendations for nonclinical toxicity and safety studies. Toxicol Pathol 1993;21:513−4.

[60] Carakostas MC. What is serum ornithine decarboxylase. Clin Chem 1988;34:2606−7.

[61] Hayes JD, Pulford DJ. The glutathione S-transferase supergene family: Regulation of GST and the contribution of the isoenzymes to cancer chemoprotection and drug resistance. Crit Rev Biochem Mol Biol 1995;30:445−600.

[62] Sherratt PJ, Hayes JD. Glutathione S-transferases. In: Cotas I, editor. Enzyme systems that metabolise drugs and other xenobiotics. New York, NY: John Wiley & Sons; 2001. p. 319−52.

[63] Hayes JD, Strange RC. Glutathione S-transferase polymorphisms and their biological consequences. Pharmacology 2000;61:154−66.

[64] Mannervik B. The isoenzymes of glutathione transferase. Adv Enzymol Relat Areas Mol Biol 1985;57:357−417.

[65] Board PG, Baker RT, Chelvanayagam G, Jermiin LS. Zeta, a novel class of glutathione transferases in a range of species from plants to humans. Biochem J 1997; 328:929−35.

[66] Hayes JD, McLellan LI. Glutathione and glutathione-dependent enzymes represent a co-ordinately regulated defence against oxidative stress. Free Radic Res 1999;31:273−300.

[67] Josephy PD. Genetic variations in human glutathione transferase enzymes: significance for pharmacology and toxicology. Hum Genomics Proteomics 2010;2010:876940.

[68] Minarik P, Tomaskova N, Kollarova M, Antalik M. Malate dehydrogenases: structure and function. Gen Physiol Biophys 2002;21:257−65.

[69] Musrati RA, Kollarova M, Mernik N, Mikulasova D. Malate dehydrogenase: distribution, function and properties. Gen Physiol Biophys 1998;17:193−210.

[70] Zelewski M, Swierczynski J. Malic enzyme in human liver: intracellular-distribution, purification and properties of cytosolic isozyme. Eur J Biochem 1991;201:339−45.

[71] Bergmeyer H, Gawehn K. \. In: 2nd ed. Bergmeyer H, editor. Methods of enzymatic analysis, vol. 2. New York, NY: Academic Press; 1974. p. 613−8.

[72] La Du BN. Human serum paraoxonase/arylesterase. In: Kalow W, editor. Pharmacogenetics of drug metabolism. New York, NY: Pergamon Press; 1992. p. 51–91.

[73] Bergmeier C, Siekmeier R, Gross W. Distribution spectrum of paraoxonase activity in HDL fractions. Clin Chem 2004;50:2309–15.

[74] Li HL, Liu DP, Liang CC. Paraoxonase gene polymorphisms, oxidative stress, and diseases. J Mol Med-JMM 2003;81:766–79.

[75] Ng CJ, Wadleigh DJ, Gangopadhyay A, Hama S, Grijalva VR, Navab M, et al. Paraoxonase-2 is a ubiquitously expressed protein with antioxidant properties and is capable of preventing cell-mediated oxidative modification of low density lipoprotein. J Biol Chem 2001;276:44444–9.

[76] Reddy ST, Wadleigh DJ, Grijalva V, Ng C, Hama S, Gangopadhyay A, et al. Human paraoxonase-3 is an HDL-associated enzyme with biological activity similar to paraoxonase-1 protein but is not regulated by oxidized lipids. Arterioscler Thromb Vasc Biol 2001;21:542–7.

[77] Mackness MI, Walker CH. Multiple forms of sheep serum A-esterase activity associated with the high-density lipoprotein. Biochem J 1988;250:539–45.

[78] Durrington PN, Mackness B, Mackness MI. Paraoxonase and atherosclerosis. Arterioscler Thromb Vasc Biol 2001;21:473–80.

[79] Meneses-Lorente G, Guest PC, Lawrence J, Muniappa N, Knowles MR, Skynner HA, et al. A proteomic investigation of drug-induced steatosis in rat liver. Chem Res Toxicol 2004;17:605–12.

[80] Furuhata N, Shiba K, Nara N. N-acetyl-beta-D-glucosaminidase). *Nippon rinsho.* Jpn J Clin Med. 1995;53:1267–76.

[81] Tucker SM, Pierce RJ, Price RG. Characterization of human N-acetyl-beta-D-glucosaminidase isoenzymes as an indicator of tissue-damage in disease. Clinica Chimica Acta 1980;102:29–40.

[82] Price RG. The role of nag (N-acetyl-beta-D-glucosaminidase) in the diagnosis of kidney-disease including the monitoring of nephrotoxicity. Clin Nephrol 1992;38: S14–S9.

[83] Skalova S. The diagnostic role of urinary N-acetyl-beta-D-glucosaminidase (NAG) activity in the detection of renal tubular impairment. Acta Medica (Hradec Kralove) 2005;48:75–80.

[84] Kavukcu S, Soylu A, Turkmen M. The clinical value of urinary N-acetyl-beta-D-glucosaminidase levels in childhood age group. Acta Med Okayama 2002;56:7–11.

[85] Rosi F, Agostinho AB, Carlucci F, Zanoni L, Porcelli B, Marinello E, et al. Behaviour of human lymphocytic isoenzymes of 5'-nucleotidase. Life Sci 1998;62:2257–66.

[86] Sunderman FW. The clinical biochemistry of 5'-nucleotidase. Ann Clin Lab Sci 1990;20:123–39.

[87] Zimmermann H. 5'-nucleotidase: molecular-structure and functional-aspects. Biochem J 1992;285:345–65.

[88] Asada Y, Tanizawa K, Nakamura K, Moriguchi M, Soda K. Stereochemistry of ornithine decarboxylase reaction. J Biochem 1984;95:277–82.

[89] Pegg AE. Recent advances in the biochemistry of polyamines in eukaryotes. Biochem J 1986;234:249–62.

[90] Wu GY, Morris SM. Arginine metabolism: nitric oxide and beyond. Biochem J 1998;336:1–17.

[91] Jenkinson CP, Grody WW, Cederbaum SD. Comparative properties of arginases. Comp Biochem Physiol B Biochem Mol Biol 1996;114:107–32.

[92] Morris Jr. SM. Regulation of enzymes of the urea cycle and arginine metabolism. Annu Rev Nutr 2002;22:87−105.

[93] Di Costanzo L, Moulin M, Haertlein M, Meilleur F, Christianson DW. Expression, purification, assay, and crystal structure of perdeuterated human arginase I. Arch Biochem Biophys 2007;465:82−9.

[94] Ashamiss F, Wierzbicki Z, Chrzanowska A, Scibior D, Pacholczyk M, Kosieradzki M, et al. Clinical significance of arginase after liver transplantation. Ann Transplant 2004;9:58−60.

[95] Massoulie J, Pezzementi L, Bon S, Krejci E, Vallette FM. Molecular and cellular biology of cholinesterases. Prog Neurobiol 1993;41:31−91.

[96] Wang R, Tang XC. Neuroprotective effects of huperzine A-A natural cholinesterase inhibitor for the treatment of Alzheimer's disease. Neurosignals 2005;14:71−82.

[97] Miller RD. Anesthesia. 6th ed. Philadelphia, PA: Elsevier, Churchill, Livingstone; 2005.

[98] Chatterjea MN, Shinde R. Textbook of medical biochemistry. 6th ed. India: Jaypee; 2005.

[99] Berk PD, Jones EA, Howe RB, Berlin NI. Disorders of bilirubin metabolism. In: Bondy PK, Rosenberg LE, editors. Metabolic control and disease. 8th ed. Philadelphia, PA: W.B. Saunders; 1979. p. 1009−88.

[100] Jones EA, Shrager R, Bloomer JR, Berk PD, Howe RB, Berlin NI. Quantitation of hepatic bilirubin synthesis in man. In: Berk PD, Berlin NI, editors. Bile pigments: chemistry and physiology. Washington, DC: US Government Printing Office; 1977. p. 189−205.

[101] Dufour DR, Lott JA, Nolte FS, Gretch DR, Koff RS, Seeff LB. Diagnosis and monitoring of hepatic injury. II. Recommendations for use of laboratory tests in screening, diagnosis, and monitoring. Clin Chem 2000;46:2050−68.

[102] Stempfel R, Zetterstrom R. Concentration of bilirubin in cerebrospinal fluid in hemolytic disease of the newborn. Pediatrics 1955;16:184−95.

[103] Thapa BR, Walia A. Liver function tests and their interpretation. Indian J Pediatr 2007;74:663−71.

[104] Narayanan S, Appleton HD. Creatinine: a review. Clin Chem 1980;26:1119−26.

[105] Taylor EH. Clinical chemistry. New York, NY: John Wiley and Sons; 1989.

[106] Evans GO. Post-prandial changes in canine plasma creatinine. J Small Anim Pract 1987;28:311−5.

[107] Taylor AJ, Vadgama P. Analytical reviews in clinical biochemistry: the estimation of urea. Ann Clin Biochem 1992;29:245−64.

[108] Teitz NW. Clinical guide to laboratory tests. 3rd ed. Philadelphia, PA: W.B. Saunders; 1995.

[109] Walser M. Urea metabolism in chronic renal-failure. J Clin Invest 1974;53:1385−92.

[110] Dossetor JB. Creatininemia versus uremia: relative significance of blood urea nitrogen and serum creatinine concentrations in azotemia. Ann Intern Med 1966;65:1287−99.

[111] Cockcroft DW, Gault MH. Prediction of creatinine clearance from serum creatinine. Nephron 1976;16:31−41.

[112] Haschek WM, Rousseaux CG, Wallig MA. Clinical pathology. In: Haschek WM, Rousseaux CG, Wallig MA, editors. Fundamentals of toxicologic pathology. 2nd ed. London NW1, UK: Academic Press; 2009. p. 43−65.

[113] Cheesbrough M. Medical laboratory manual for tropical countries. Oxford, Cambridge: Butterworth-Heinemann Ltd; 1991.

[114] Hartman PE. Review: putative mutagens and carcinogens in foods.4. Malonaldehyde. Environ Mutagen 1983;5:603−7.

[115] Stancliffe RA, Thorpe T, Zemel MB. Dairy attentuates oxidative and inflammatory stress in metabolic syndrome. Am J Clin Nutr 2011;94:422−30.
[116] International Agency for Research on Cancer. IARC monographs on the evaluation of the carcinogenic risk of chemicals to humans, vol. 36, allyl compounds, aldehydes, epoxides and peroxides. Lyon: IARC; 1985.
[117] Nielsen F, Mikkelsen BB, Nielsen JB, Andersen HR, Grandjean P. Plasma malondialdehyde as biomarker for oxidative stress: reference interval and effects of life-style factors. Clin Chem 1997;43:1209−14.
[118] Kappus H, Sies H. Toxic drug effects associated with oxygen-metabolism: redox cycling and lipid-peroxidation. Experientia 1981;37:1233−41.
[119] Lovric J, Mesic M, Macan M, Koprivanac M, Kelava M, Bradamante V. Measurement of malondialdehyde (MDA) level in rat plasma after simvastatin treatment using two different analytical methods. Period Biol 2008;110:63−7.
[120] Del Rio D, Stewart AJ, Pellegrini N. A review of recent studies on malondialdehyde as toxic molecule and biological marker of oxidative stress. Nutr Metab Cardiovasc Dis 2005;15:316−28.
[121] Marnett LJ. Lipid peroxidation: DNA damage by malondialdehyde. Mutat Res Fundam Mol Mech Mutag 1999;424:83−95.
[122] Nair V, O'Neil CL, Wang PG.Malondialdehyde. e-EROS encyclopedia of reagents for organic synthesis; 2008. http://dx.doi.org/doi:10.1002/047084289X.rm013.pub2.
[123] Buddi R, Lin B, Atilano SR, Zorapapel NC, Kenney MC, Brown DJ. Evidence of oxidative stress in human corneal diseases. J Histochem Cytochem 2002;50:341−51.
[124] Tiku ML, Narla H, Jain M, Yalamanchili P. Glucosamine prevents in vitro collagen degradation in chondrocytes by inhibiting advanced lipoxidation reactions and protein oxidation. Arthritis Res Ther 2007;9:R76.
[125] Brown B. Basic concepts in biochemistry: a student's survival guide by Hiram F Gilbert. New York, NY: McGraw-hill; 1992pp 298, Biochemical Education. 20, 186.
[126] Nelson DL, Cox MM. Lehninger. Principles of biochemistry. 3th ed. New York, NY: Worth Publishing; 2000.
[127] Meister A, Larsson A. Glutathione synthetase deficiency and other disorders of the g-glutamyl cycle. In: Scriver CR, Beaudet AL, Sly WS, Valle D, editors. the Metabolic bases of inherited disease. 6th ed. New York, NY: McGraw-Hill; 1989. p. 855−68.
[128] Couto N, Malys N, Gaskell SJ, Barber J. Partition and turnover of glutathione reductase from Saccharomyces cerevisiae: a proteomic approach. J Proteome Res 2013;12:2885−94.
[129] Wu GY, Fang YZ, Yang S, Lupton JR, Turner ND. Glutathione metabolism and its implications for health. J Nutr 2004;134:489−92.
[130] Pastore A, Piemonte F, Locatelli M, Lo Russo A, Gaeta LM, Tozzi G, et al. Determination of blood total, reduced, and oxidized glutathione in pediatric subjects. Clin Chem 2001;47:1467−9.
[131] Dianzani MU. The role of free-radicals in liver-damage. Proc Nutr Soc 1987;46:43−52.
[132] Guyton AC, Hall JE. Textbook of medical physiology. 10th ed. W.B. Saunders; 2000.
[133] Karantonis HC, Nomikos T, Demopoulos CA. Triacylglycerol metabolism. Curr Drug Targets 2009;10:302−19.
[134] Semenkovich CF. Disorders of lipid metabolism. In: Goldman L, Ausiello D, editors. Cecil Medicine. 24th ed. Philadelphia, PA: Saunders Elsevier; 2011.
[135] Klotzsch SG, Mcnamara JR. Triglyceride measurements: a review of methods and interferences. Clin Chem 1990;36:1605−13.

[136] Grundy SM. Cholesterol-metabolism in man. West J Med 1978;128:13–25.

[137] Lewis B. The metabolism of cholesterol. Postgrad Med J 1959;35:208–15.

[138] Fernandez C, Lobo MDVT, Gomez-Coronado D, Lasuncion MA. Cholesterol is essential for mitosis progression and its deficiency induces polyploid cell formation. Exp Cell Res 2004;300:109–20.

[139] Durrington P. Dyslipidaemia. Lancet 2003;362:717–31.

[140] Agellon LB. Metabolism and function of bile acids in biochemistry of lipids, lipoproteins and membranesIn: Vance DE, Vance J, editors. 5th ed. Amsterdam: Elsevier; 2008. p. 423–40.

[141] Zollner G, Marschall H, Wagner M, Trauner M. Role of nuclear receptors in the adaptive response to bile acids and cholestasis: Pathogenetic and therapeutic considerations. Mol Pharm 2006;3:231–51.

[142] Stryer, L. (1995) Biochemistry 4th ed., E.H. Freeman and Company, New York, NY.

[143] Giugliano D, Ceriello A, Esposito K. Glucose metabolism and hyperglycemia. Am J Clin Nutr 2008;87:217S–22S.

[144] Farrugia A. Albumin usage in clinical medicine: tradition or therapeutic? Transfus Med Rev 2010;24:53–63.

[145] Lindup WE, Orme MCLE. Clinical-pharmacology: plasma-protein binding of drugs. Br Med J 1981;282:212–4.

[146] Katus HA, Scheffold T, Remppis A, Zehlein J. Proteins of the troponin complex. Lab Med 1992;23:311–7.

[147] Christenson RH, Newby LK, Ohman EM. Cardiac markers in the assessment of acute coronary syndromes. Md Med J 1997;(Suppl.)18–24.

[148] Malouf NN, Mcmahon D, Oakeley AE, Anderson PAW. A cardiac troponin-T epitope conserved across phyla. J Biol Chem 1992;267:9269–74.

[149] Saggin L, Gorza L, Ausoni S, Schiaffino S. Cardiac troponin-T in developing, regenerating and denervated rat skeletal-muscle. Development 1990;110:547–54.

[150] Sterns RH, Spital A, Clark EC. Disorders of water balance. In: Kokko JP, Tannen RL, editors. Fluids and Electrolytes. 3rd ed. Philadelphia, PA: W.B. Saunders; 1996. p. 63–109.

[151] Thurman JM, Berl T. Disorders of water metabolism. In: Mount DB, Sayegh MH, Singh AK, editors. Core concepts in the disorders of fluid, electrolytes and acid-base balance, 29. New York, NY: Springer; 2013. p. 29–48.

[152] Ackerman GL. Serum sodium. In: Walker HK, Hall WD, Hurst JW, editors. Clinical methods: the history, physical, and laboratory examinations. 3rd ed. Boston, MA: Butterworths; 1990. p. 879–83.

[153] Palm C, Reimann D, Gross P. Hyponatremia—with comments on hypernatremia. Ther Umsch 2000;57:400–7.

[154] Rastegar A. Serum potassium. In: Walker HK, Hall WD, Hurst JW, editors. Clinical methods: the history, physical, and laboratory examinations. 3rd ed. Boston, MA: Butterworths; 1990. p. 884–7.

[155] Wingo CS, Cain BD. The renal H-K-atpase: physiological significance and role in potassium homeostasis. Annu Rev Physiol 1993;55:323–47. Available from: http://dx.doi.org/doi:10.1146/annurev.physiol.55.1.323.

[156] Oh MS, Carroll HJ. Electrolyte and acid–base disorders. In: Chernow B, editor. The pharmacologic approach to the critically Ill Patient. 3rd ed. Baltimore, MD: Williams & Wilkins; 1994. p. 957–68.

[157] Lau A, Chan LA. Electrolytes, other minerals and trace elements. In: Lee M, editor. Basic skills in interpreting laboratory data. 4th ed. Bethesda, MD: Harvey Whitney Books; 2009. p. 119–50.

[158] Morrison G. Serum chloride. In: Walker HK, Hall WD, Hurst JW, editors. Clinical methods: the history, physical, and laboratory examinations. 3rd ed. Boston, MA: Butterworths; 1990. p. 890—4.

[159] Koch SM, Taylor RW. Chloride-ion in intensive-care medicine. Crit Care Med 1992;20:227—40.

[160] Deluca HF. The vitamin-D story: a collaborative effort of basic science and clinical medicine. FASEB J 1988;2:224—36.

[161] Brini M, Calì T, Ottolini D, Carafoli E. Intracellular calcium homeostasis and signaling. In: Banci L, editor. Metallomics and the cell. metal ions in life sciences, vol. 12. Springer; 2013. p. 119—68.

[162] Boron WF, Boulpaep EL. The parathyroid glands and vitamin D. Medical physiology: a cellular and molecular approach. Philadelphia, PA: Saunders/Elsevier; 2003.

[163] Rude RK. Physiology of magnesium-metabolism and the important role of magnesium in potassium-deficiency. Am J Cardiol 1989;63:G31—4.

[164] Saris NEL, Mervaala E, Karppanen H, Khawaja JA, Lewenstam A. Magnesium: an update on physiological, clinical and analytical aspects. Clin Chim Acta 2000;294:1—26.

[165] Cowan JA. The biological chemistry of magnesium. New York, NY: VCH; 1995.

[166] Fawcett WJ. Magnesium: physiology and pharmacology. Br J Anaesth 1999;83:973.

[167] Bansal V. Serum inorganic phosphorus. In: Walker HK, Hall WD, Hurst JW, editors. Clinical methods: the history, physical, and laboratory examinations. 3rd ed. Boston, MA: Butterworths; 1990. p. 895—9.

[168] Tenenhouse HS, Portale AA. Phosphate homeostasis. In: Feldman D, Pike JW, Glorieu FH, editors. Vitamin D. 2nd ed. San Diego, CA: Elsevier; 2005. p. 453—76.

[169] Levine B, Kleenman C. Clinical disorders of fluid and electrolyte metabolism. New York, NY: McGraw Hill; 1994.

[170] Dlouhy AC, Outten CE. The iron metallome in eukaryotic organisms. In: Banci L, editor. Metallomics and the cell. metal ions in life sciences, vol 12. Springer; 2013. p. 242—63.

[171] Andrews NC. Medical progress: disorders of iron metabolism. N Engl J Med 1999;341:1986—95.

[172] Aisen P, Wessling-Resnick M, Leibold EA. Iron metabolism. Curr Opin Chem Biol 1999;3:200—6.

[173] Iwalokun BA, Efedede BU, Alabi-Sofunde JA, Oduala T, Magbagbeola OA, Akinwande AI. Hepatoprotective and antioxidant activities of *Vernonia amygdalina* on acetaminophen-induced hepatic damage in mice. J Med Food 2006;9:524—30.

[174] Stevens PF. Angiosperm phylogeny website. version 12, July 2012, 2001.

[175] Gao T, Yao H, Song J, Liu C, Zhu Y, Ma X, et al. Identification of medicinal plants in the family Fabaceae using a potential DNA barcode ITS2. J Ethnopharmacol 2010;130:116—21.

[176] Hohmann J, Molnar J. (Euphorbiaceae diterpenes: plant toxins or promising molecules for the therapy?). Acta Pharm Hung 2004;74:149—57.

[177] Abdelgadir HA, Van Staden J. Ethnobotany, ethnopharmacology and toxicity of *Jatropha curcas* L. (Euphorbiaceae): a review. S Afr J Bot 2013;88:204—18.

[178] Nana P, Asongalem EA, Foyet HS, Dimo T, Kamtchouing P. Acute toxicolgical studies of *Acanthus montanus* (Nees) T. Anderson (Acanthaceae) in Wistar rats. Pharmacologyonline 2007;1:339—48.

[179] Djami TAT, Asongalem EA, Nana P, Choumessi A, Kamtchouing P, Asonganyi T. Subacute toxicity study of the aqueous extract from *Acanthus montanus*. eJBio 2011;7:11—5.

[180] Diallo A, Eklu-Gadegkeku K, Agbonon A, Aklikokou K, Creppy EE, Gbeassor M. Acute and sub-chronic (28-day) oral toxicity studies of hydroalcohol leaf extract of *Ageratum conyzoides* L (Asteraceae). Trop J Pharm Res 2010;9:463−7.

[181] El Hilaly J, Israili ZH, Lyoussi B. Acute and chronic toxicological studies of *Ajuga iva* in experimental animals. J Ethnopharmacol 2004;91:43−50.

[182] Wintola OA, Sunmonu TO, Afolayan AJ. Toxicological evaluation of aqueous extract of *Aloe ferox* Mill. in loperamide-induced constipated rats. Hum Exp Toxicol 2011;30:425−31.

[183] Leonard T, Dzeufiet PDD, Dimo T, Asongalem EA, Sokeng SN, Flejou JF, et al. Acute and subchronic toxicity of *Anacardium occidentale* Linn (Anacardiaceae) leaves hexane extract in mice. Afr J Tradit Complement Altern Med 2007;4:140−7.

[184] Okoye TC, Akah PA, Ezike AC, Okoye MO, Onyeto CA, Ndukwu F, et al. Evaluation of the acute and sub acute toxicity of *Annona senegalensis* root bark extracts. Asian Pac J Trop Med 2012;5:277−82.

[185] Abolaji AO, Eteng MU, Ebong PE, Brisibe EA, Dar A, Kabir N, et al. A safety assessment of the antimalarial herb *Artemisia annua* during pregnancy in Wistar rats. Phytother Res 2013;27:647−54.

[186] Taziebou LC, Etoa F -, Nkegoum B, Pieme CA, Dzeufiet DPD. Acute and subacute toxicity of *Aspilia africana* leaves. Afr J Tradit Complement Altern Med 2007;4:127−34.

[187] Ashafa AOT, Orekoya LO, Yakubu MT. Toxicity profile of ethanolic extract of *Azadirachta indica* stem bark in male Wistar rats. Asian Pac J Trop Biomed 2012;2:811−7.

[188] Onyije FM, Avwioro OG. Effect of ethanolic extract of *Bauhinia monandra* leaf on the liver of alloxan induced diabetic rats. J Phys Pharm Adv 2012;2:59−63.

[189] Aniagu SO, Nwinyi FC, Olanubi B, Akumka DD, Ajoku GA, Izebe KS, et al. Is *Berlina grandiflora* (Leguminosae) toxic in rats?. Phytomedicine 2004;11:352−60.

[190] Ngogang J, Nkongmeneck BA, Essam LFBB, Oyono JLE, Tsabang N, Zapfack L, et al. Evaluation of acute and sub acute toxicity of four medicinal plants extracts used in Cameroon. Toxicol Lett 2008;180:S185−6.

[191] Orisakwe OE, Afonne OJ, Chude MA, Obi E, Dioka CE. Sub-chronic toxicity studies of the aqueous extract of *Boerhavia diffusa* leaves. J Health Sci 2003;49:444−7.

[192] Nweze NE, Anene BM, Asuzu IU, Ezema WS. Subacute toxicity study of the methanolic seed extract of *Buchholzia coriacea* (Capparaceae) in rats. Comp Clin Pathol 2012;21:967−74.

[193] Pone JW, Mbida M, Bilong CFB. Acute and sub-acute toxicity of ethanolic extract of *Canthium mannii* Hiern stem bark on *Mus musculus*. Indian J Exp Biol 2011;49:146−50.

[194] Ya'u J, Chindo BA, Yaro AH, Okhale SE, Anuka JA, Hussaini IM. Safety assessment of the standardized extract of *Carissa edulis* root bark in rats. J Ethnopharmacol 2013;147:653−61.

[195] Tahraoui A, Israili ZH, Lyoussi B. Acute and sub-chronic toxicity of a lyophilised aqueous extract of *Centaurium erythraea in rodents*. J Ethnopharmacol 2010;132:48−55.

[196] Ogbonnia SO, Mbaka GO, Anyika EN, Osegbo OM, Igbokwe NH. Evaluation of acute toxicity in mice and subchronic toxicity of hydroethanolic extract of *Chromolaena odorata* (L.) King and Robinson (Fam. Asteraceae) in rats. Agric Biol J N Am 2010;1:859−65.

[197] Taziebou LC, Shang DJ, Pieme CA, Essia NJJ, Etoa FX, Nkegoum B. Toxicological investigation of aqueous extracts of *Chromolaena odorata* leaves on rats and mice. Res J Biotechnol 2008;3.

[198] Lakmichi H, Bakhtaoui FZ, Gadhi CA, Ezoubeiri A, El Jahiri Y, El Mansouri A, et al. Toxicity profile of the aqueous ethanol root extract of *Corrigiola telephiifolia* Pourr. (Caryophyllaceae) in rodents. Evid Based Complement Alternat Med 2011;2011:317090.

[199] Tamokou JDD, Kuiate JR, Gatsing D, Efouet APN, Njouendou AJ. Antidermatophytic and toxicological evaluations of dichloromethane-methanol extract, fractions and compounds isolated from *Coula edulis*. Iran J Med Sci 2011;36:111−21.

[200] Kouitcheu Mabeku LB, Penlap Beng V, Kouam J, Essame O, Etoa FX. Toxicological evaluation of ethyl acetate extract of *Cylicodiscus gabunensis* stem bark (Mimosaceae). J Ethnopharmacol 2007;111:598−606.

[201] Gidado A, Zainab AA, Hadiza MU, Serah DP, Anas HY, Milala MA. Toxicity studies of ethanol extract of the leaves of *Datura stramonium* in rats. Afr J Biotechnol 2007;6:1012−5.

[202] Gouana V, Tsouh PVF, Donfack VFD, Boyom FF, Zollo PHA. Acute and subacute toxicity of the MeOH/methylene chloride bark extractfrom *Drypetes gossweileri* (Euphorbiaceae) in Wistar rat. Pharmacologyonline 2010;3:240−62.

[203] Atsarno AD, Nguelefack TB, Datte JY, Kamanyi A. Acute and subchronic oral toxicity assessment of the aqueous extract from the stem bark of *Erythrina senegalensis* DC (Fabaceae) in rodents. J Ethnopharmacol 2011;134:697−702.

[204] Oluwakanyinsola SA, Adeniyi TY, Akingbasote JA, Florence OE. Acute and subacute toxicity study of ethanolic extract of the stem bark of *Faidherbia albida* (DEL) A. Chev (Mimosoidae) in rats. Afr J Biotechnol 2010;9:1218−24.

[205] Ashafa AOT, Yakubu MT, Grierson DS, Afolayan AJ. Toxicological evaluation of the aqueous extract of *Felicia muricata* Thunb. leaves in Wistar rats. Afr J Biotechnol 2009;8:949−54.

[206] Demma J, Gebre-Mariam T, Asres K, Ergetie W, Engidawork E. Toxicological study on *Glinus lotoides*: a traditionally used taenicidal herb in Ethiopia. J Ethnopharmacol 2007;111:451−7.

[207] Akanmu MA, Orafidiya FA, Adekunle SA, Obuotor EM, Osasan SA. Subchronic toxicity and behavioural effects of *Glycine max* (L.) oil emulsion in male rats. Int J Phytomedicine 2011;3:227−39.

[208] Yagi S, Yagi AI, Gadir EHA, Henry M, Chapleur Y, Laurain-Mattar D. Toxicity of *Hydnora johannis* Becca. dried roots and ethanol extract in rats. J Ethnopharmacol 2011;137:796−801.

[209] Abu AH, Uchendu CN. Safety assessment of aqueous ethanolic extract of *Hymenocardia acida* stem bark in Wistar rats. Arch Appl Sci Res 2010;2:56−68.

[210] Oumarou BA, Tchuemdem LM, Djomeni PDD, Bilanda DC, Tom ENL, Ndzana MTB, et al. Mineral constituents and toxicological profile of *Jateorhiza macrantha* (Menispermaceae) aqueous extract. J Ethnopharmacol 2013;149:117−22.

[211] Igbinosa OO, Oviasogie EF, Igbinosa EO, Igene O, Igbinosa IH, Idemudia OG. Effects of biochemical alteration in animal model after short-term exposure of *Jatropha curcas* (Linn) leaf extract. Sci World J 2013;798096.

[212] Abubakar MG, Lawal A, Usman MR. Hepatotoxicity studies of sub-chronic administration of aqueous stem bark of *Khaya senegalensis* in albino rats. Bayero J Pure Appl Sci 2010;3:26−8.

[213] Pieme CA, Penlap VN, Nkegoum B, Taziebou CL, Ngogang J. *In vivo* antioxidant and potential antitumor activity extract of *Leea guineensis* Royen ex. L. (Leeaceae) on carcinomatous cells. Pharmacologyonline 2008;1:538−47.

[214] Maphosa V, Masika PJ, Adedapo AA. Safety evaluation of the aqueous extract of *Leonotis leonurus* Shoots in rats. Hum Exp Toxicol 2008;27:837−43.

[215] Tchamadeu M -, Dzeufiet PDD, Nouga CCK, Azebaze AGB, Allard J, Girolami J -, et al. Hypoglycaemic effects of *Mammea africana* (Guttiferae) in diabetic rats. J Ethnopharmacol 2010;127:368−72.

[216] Agbor GA, Tarkang PA, Fogha JVZ, Biyiti LF, Tamze V, Messi HM, et al. Acute and subacute toxicity studies of aqueous extract of *Morinda lucida* stem bark. J Pharmacol Toxicol 2012;7:158−65.

[217] Adebajo AC, Ayoola OF, Lwalewa EO, Akindahunsi AA, Omisore NOA, Adewunmi CO, et al. Anti-trichomonal, biochemical and toxicological activities of methanolic extract and some carbazole alkaloids isolated from the leaves of *Murraya koenigii* growing in Nigeria. Phytomedicine 2006;13:246−54.

[218] Adeneye AA, Ajagbonna OP, Adeleke TI, Bello SO. Preliminary toxicity and phyto-chemical studies of the stem bark aqueous extract of *Musanga cecropioides* in rats. J Ethnopharmacol 2006;105:374−9.

[219] Zaoui A, Cherrah Y, Mahassini N, Alaoui K, Amarouch H, Hassar M. Acute and chronic toxicity of *Nigella sativa* fixed oil. Phytomedicine 2002;9:69−74.

[220] Tan PV, Mezui C, Enow-Orock G, Njikam N, Dimo T, Bitolog P. Teratogenic effects, acute and sub chronic toxicity of the leaf aqueous extract of *Ocimum suave* wild (Lamiaceae) in rats. J Ethnopharmacol 2008;115:232−7.

[221] Ozolua RI, Anaka ON, Okpo SO, Idogun SE. Acute and sub-acute toxicological assessment of the aqueous seed extract of *Persea americana* Mill (Lauraceae) in rats. Afr J Tradit Complement Altern Med 2009;6:573−8.

[222] Assob JCN, Kamga HLF, Nsagha DS, Njunda AL, Nde PF, Asongalem EA, et al. Antimicrobial and toxicological activities of five medicinal plant species from Cameroon traditional medicine. BMC Complement Altern Med 2011;11.

[223] Gathumbi PK, Mwangi JW, Mugera GM, Njiro SM. Toxicity of chloroform extract of *Prunus africana* stem bark in rats: gross and histological lesions. Phytother Res 2002;16:244−7.

[224] Nana HM, Ngane RAN, Kuiate JR, Mogtomo LMK, Tamokou JD, Ndifor F, et al. Acute and sub-acute toxicity of the methanolic extract of *Pteleopsis hylodendron* stem bark. J Ethnopharmacol 2011;137:70−6.

[225] Tchamadeu MC, Dzeufiet PDD, Nana P, Nouga CCK, Tsofack FN, Allard J, et al. Acute and sub-chronic oral toxicity studies of an aqueous stem bark extract of *Pterocarpus soyauxii* Taub (Papilionaceae) in rodents. J Ethnopharmacol 2011;133:329−35.

[226] Amida MB, Yemitan OK, Adeyemi OO. Toxicological assessment of the aqueous root extract of *Sanseviera liberica* Gerome and Labroy (Agavaceae). J Ethnopharmacol 2007;113:171−5.

[227] Pieme CA, Penlap VN, Nkegoum H, Taziebou CL, Tekwu EM, Etoa FX, et al. Evaluation of acute and subacute toxicities of aqueous ethanolic extract of leaves of *Senna alata* (L.) Roxb (Ceasalpiniaceae). Afr J Biotechnol 2006;5:283−9.

[228] Assam JPA, Dzoyem JP, Pieme CA, Penlap VB. *In vitro* antibacterial activity and acute toxicity studies of aqueous-methanol extract of *Sida rhombifolia* Linn. (Malvaceae). BMC Complement Altern Med 2010;10.

[229] Ilodigwe EE, Akah PA, Nworu CS. Evaluation of the acute and subchronic toxicities of ethanol leaf extract of *Spathodea campanulata*. Int J Appl Res Nat Prod 2010;3:17−21.

[230] Agbaje EO, Adeneye AA, Daramola AO. Biochemical and toxicological studies of aqueous extract of *Syzigium aromaticum* (L.) Merr. & Perry (Myrtaceae) in rodents. Afr J Tradit Complement Altern Med 2009;6:241−54.

[231] Lembe DM, Sonfack A, Gouado I, Dimo T, Dongmo A, Demasse MFA, et al. Evaluations of toxicity of *Turraeanthus africanus* (Meliaceae) in mice. Andrologia 2009;41:341−7.

[232] Arhoghro EM, Ekpo KE, Anosike EO, Ibeh GO. Effect of aqueous extract of bitter leaf (*Vernonia amygdalina* Del) on carbon tetrachloride (CCl4) induced liver damage in albino Wistar rats. Eur J Sci Res 2009;26:122−30.

[233] Yemitan OK, Izegbu MC. Protective effects of *Zingiber officinale* (Zingiberaceae) against carbon tetrachloride and acetaminophen-induced hepatotoxicity in rats. Phytother Res 2006;20:997−1002.

[234] Biapa NPC, Kuiate JR, Wankeu M, Ntiokam D. Acute and subacute toxicity studies of *Zinginber officinalis* Roscoe essential oil on mice (swiss) and rats (Wistar). Afr J Pharm Sci Pharm 2010;1:39−49.

[235] Ogbonnia S, Adekunle AA, Bosa MK, Enwuru VN. Evaluation of acute and subacute toxicity of *Alstonia congensis* Engler (Apocynaceae) bark and *Xylopia aethiopica* (Dunal) A. rich (Annonaceae) fruits mixtures used in the treatment of diabetes. Afr J Biotechnol 2008;7:701−5.

[236] Nyarkao AK, Okine LKN, Wedzi RK, Addo PA, Ofosuhene M. Subchronic toxicity studies of the antidiabetic herbal preparation ADD-199 in the rat: absence of organ toxicity and modulation of cytochrome P450. J Ethnopharmacol 2005;97:319−25.

[237] Pone KB, Telefo PB, Tchouanguep FM. Acute and subacute toxicity of aqueous extract of leaves mixture of *Aloe buettneri* (Liliaceae), *Dicliptera verticillata* (Acanthaeae), *Hibiscus macranthus* (Malvaceae) and *Justicia insularis* (Acanthaceae) on swiss mice and albinos Wistar female rats. Indian J Forensic Med Toxicol 2013;7:257−87.

[238] Eloff JN, Famakin JO, Katerere DRP. *Combretum woodii* (Combretaceae) leaf extracts have high activity against Gram-negative and Gram-positive bacteria. Afr J Biotechnol 2005;4:1161−6.

[239] Makhafola TJ, McGaw LJ, Eloff JN. *In vitro* cytotoxicity and genotoxicity of five *Ochna* species (Ochnaceae) with excellent antibacterial activity. S Afr J Bot 2014;91:9−13.

Printed and bound by CPI Group (UK) Ltd, Croydon, CR0 4YY

08/05/2025

01864995-0001